Advances in Biochemistry in Health and Disease

Series Editor: Naranjan S. Dhalla
Winnipeg, Manitoba, Canada

Volume 1: S. K. Cheema (ed), *Biochemistry of Atherosclerosis*

Volume 2: S. W. Schaffer and M-Saadeh Suleiman (eds), *Mitochondria: The Dynamic Organelle*

Volume 3: A. K. Srivastava and M. B. Anand-Srivastava (eds), *Signal Transduction in the Cardiovascular System in Health and Disease*

Volume 4: B. Ostadal and N. S. Dhalla (eds), *Cardiac Adaptations-Molecular Mechanisms*

Volume 5: B. I. Jugdutt and N. S. Dhalla (eds), *Cardiac Remodeling-Molecular Mechanisms*

Volume 6: Jawahar L. Mehta and N. S. Dhalla (eds), *Biochemical Basis and Therapeutic Implications of Angiogenesis*

For further volumes:
http://www.springer.com/series/7064

Jawahar L. Mehta • Naranjan S. Dhalla
Editors

Biochemical Basis and Therapeutic Implications of Angiogenesis

Springer

Editors
Jawahar L. Mehta
Division of Cardiovascular Medicine
University of Arkansas for Medical Sciences and the Central Arkansas Veterans Healthcare System
Little Rock, AR, USA

Naranjan S. Dhalla
Institute of Cardiovascular Sciences
St. Boniface Hospital Research,
University of Manitoba
Winnipeg, MB, Canada

ISBN 978-1-4614-5856-2 ISBN 978-1-4614-5857-9 (eBook)
DOI 10.1007/978-1-4614-5857-9
Springer New York Heidelberg Dordrecht London

Library of Congress Control Number: 2012956286

Printed on acid-free paper

Springer is part of Springer Science+Business Media (www.springer.com)

This book is dedicated to
Dr. Wolfgang Schaper for his pioneering
work in the field of angiogenesis,
arteriogenesis, and collateral circulation.
He was bestowed upon the highest honour
(Medal of Merit) by the International
Academy of Cardiovascular Sciences
for his outstanding contributions in the
area of ischemic heart disease.
Dr. Schaper was Director of the
Max Planck Institute for Heart and Lung
Research for a long period and
is currently serving as Division Head
of the Arteriogenesis Research at the
same Institute in Bad Nauheim,
Germany.

Preface

Angiogenesis plays a key role in human physiology. For example, in utero angiogenesis is the backbone of vascular development as is hematopoiesis. Angiogenesis is a critical feature of wound healing and growth of uterine intima during the menstrual cycle. It has also become evident that angiogenesis is relevant in the genesis of diabetic retinopathy and growth of several tumors. Absent or suboptimal angiogenesis may relate to the syndrome of myocardial ischemia.

In the last several years, a number of investigators have described various signals and pathways leading to the evolution and persistence of angiogenesis. These include a number of growth factors, receptors, redox state, and pro-inflammatory state. Based on microarray technology, a host of known pathways leading to angiogenesis have been confirmed and novel pathways postulated.

As our understanding of the role of various triggers and inhibitors of angiogenesis has expanded, several novel therapies for a myriad of disease states have been proposed. For example, angiogenesis inhibitors have been approved and are used to arrest the growth of some tumors. On the other hand, a number of trials have been conducted to test the value of pro-angiogenic factors in myocardial ischemia. The results of these studies are not consistent and not as revealing as once thought.

We thought it timely to get several world class experts on different aspects of angiogenesis to present their work in one book. Some chapters relate to the definition of angiogenesis (vs. arteriogenesis) and some on the mechanisms of angiogenesis, while others relate to the role of angiogenesis in cancers and ischemic heart disease. Futuristic strategies to manipulate angiogenesis using evolving information from microarray technology and use of microRNAs are also discussed in some chapters.

While the book does not cover each and every aspect of angiogenesis, it covers most relevant issues in the biology of this interesting phenomenon that plays a critical role in physiology and pathology. We have organized 26 chapters, prepared by highly respected scientists in the field of angiogenesis, in the form of three sections, namely (a) Molecular Mechanisms of Angiogenesis, (b) Pathophysiological Aspect of Angiogenesis, and (c) Pharmacology and Therapy of Angiogenesis. Eleven chapters in the first part of this book mainly deal with the role of growth factors, neuropeptides and signal transduction mechanisms as well as cellular regulation by mast cells, integrins, and stem cells. The second part of this book has nine chapters dealing with the role of angiogenesis in different diseases such as atherosclerosis, diabetes, stroke, ischemic heart disease, and cancer. The third section of this book has six chapters which are primarily devoted to the therapeutic implications of growth factors, endothelial progenitor cells, microRNA, and angiogenesis inhibitors from natural sources.

We are grateful to Dr. Vijayan Elimban and Ms. Eva Little of Winnipeg for their help in editing the manuscripts. Cordial thanks are also due to Ms. Melanie Tucker and Ms. Portia Formento, Springer, USA, for their continuous advice and understanding during the editorial process. Lastly, we thank the authors for contributing the best of their work for inclusion in this book. We hope the readers will find this compilation of work from several laboratories useful in understanding the pathobiology of angiogenesis and designing new therapies. Dr. Mehta would like to thank his wife, Paulette, and children, Asha and Jason for their eternal support, and his colleagues for inspiration and all their help over the years.

Little Rock, AR, USA Jawahar L. Mehta
Winnipeg, MB, Canada Naranjan S. Dhalla

Contents

Contributors

Veeranjaneyulu Addepalli Department of Pharmacology, SPP School of Pharmacy & Technology Management, SVKM's NMIMS, Mumbai, Maharashtra, India

António Araújo Department of Medical Oncology, Portuguese Oncology Institute, Rua Dr. António Bernardino de Almeida, Porto, Portugal

Ken Arai Neuroprotection Research Laboratory, Massachusetts General Hospital, Harvard Medical School, Charlestown, MA, USA

Takayuki Asahara Department of Regenerative Medicine Science, Tokai University School of Medicine, Isehara, Japan

Junya Azuma Department of Clinical Gene Therapy, Osaka University Graduate School of Medicine, Suita, Osaka, Japan

Department of Geriatric Medicine and Nephrology, Osaka University Graduate School of Medicine, Suita, Osaka, Japan

Lina Badimon Cardiovascular Research Center, CSIC-ICCC, Hospital de la Santa Creu i Sant Pau, IIB-Sant Pau, Barcelona, Spain

CIBEROBN-Pathophysiology of Obesity and Nutrition, Barcelona, Spain

Cardiovascular Research Chair, UAB, Barcelona, Spain

Elena Cavarretta Department of Medical-Surgical Sciences and Biotechnologies, University of Rome "Sapienza", Latina, Italy

Gianluigi Condorelli Humanitas Clinical and Research Center and Institute of Genetics and Biomedical Research, National Research Council of Italy, Rozzano (Mi), Italy

Department of Medicine, University of California San Diego, La Jolla, CA, USA

David J. Bruce Oxford University Hospitals NHS Trust, John Radcliffe Hospital, Oxford, UK

Brenda L. Coomber Department of Biomedical Sciences, Ontario Veterinary College, University of Guelph, Guelph, ON, Canada

Enrico Crivellato Department of Experimental and Clinical Medicine, Section of Anatomy, University of Udine Medical School, Udine, Italy

Hiranmoy Das Cardiovascular Stem Cell Research Laboratory, The Dorothy M. Davis Heart and Lung Research Institute, Wexner Medical Center at The Ohio State University, Columbus, OH, USA

Abhishek Deshmukh Division of Cardiovascular Medicine, University of Arkansas for Medical Sciences and the Central Arkansas Veterans Healthcare System, Little Rock, AR, USA

Naranjan S. Dhalla Institute of Cardiovascular Sciences St. Boniface Hospital Research, University of Manitoba, Winnipeg, MB, Canada

University of Manitoba, Winnipeg, MB, Canada

Zufeng Ding Division of Cardiovascular Medicine, University of Arkansas for Medical Sciences and the Central Arkansas Veterans Healthcare System, Little Rock, AR, USA

Todd A. Duhamel Institute of Cardiovascular Sciences, St. Boniface General Hospital Research Centre, Winnipeg, MB, Canada

Health, Leisure and Human Performance Research Institute, University of Manitoba, Winnipeg, MB, Canada

Faculty of Kinesiology and Recreation Management, Winnipeg, MB, Canada

Yasuyuki Fujita Division of Vascular Regeneration, Unit of Regenerative Medicine, Institute of Biomedical Research and Innovation, Kobe, Japan

Giacomo Frati Department of Medical-Surgical Sciences and Biotechnologies, University of Rome "Sapienza", Latina, Italy

Department of Angiocardioneurology, IRCCS Neuromed, Pozzilli, Isernia, Italy

Dipti Gatne Department of Pharmaceutical Chemistry, SPP School of Pharmacy & Technology Management, SVKM's NMIMS, Mumbai, Maharashtra, India

Karsten Grote Department of Cardiology and Angiology, Hannover Medical School, Hannover, Germany

Kazuhide Hayakawa Neuroprotection Research Laboratory, Massachusetts General Hospital, Harvard Medical School, Charlestown, MA, USA

Hideki Hayakawa Molecular Medical Research Project, Tokyo Metropolitan Institute of Medical Science, Setagaya-Ku, Tokyo, Japan

Venceslau Hespanhol Department of Medicine, University of Porto, Porto, Portugal

Service of Pneumology, Hospital São João, Porto, Portugal

Kazuma Iekushi Department of Clinical Gene Therapy, Osaka University Graduate School of Medicine, Suita, Osaka, Japan

Department of Geriatric Medicine and Nephrology, Osaka University Graduate School of Medicine, Suita, Osaka, Japan

Masaaki Iwabayashi Department of Clinical Gene Therapy, Osaka University Graduate School of Medicine, Suita, Osaka, Japan

Mai Jarad Department of Biomedical Sciences, Ontario Veterinary College, University of Guelph, Guelph, ON, Canada

Atsuhiko Kawamoto Division of Vascular Regeneration, Unit of Regenerative Medicine, Institute of Biomedical Research and Innovation, Kobe, Japan

Dustin Scott Kehler Institute of Cardiovascular Sciences, St. Boniface General Hospital Research Centre, Winnipeg, MB, Canada

Health, Leisure and Human Performance Research Institute, University of Manitoba, Winnipeg, MB, Canada

Magomed Khaidakov Division of Cardiovascular Medicine, University of Arkansas for Medical Sciences and the Central Arkansas Veterans Healthcare System, Little Rock, AR, USA

Raj Kishore Feinberg Cardiovascular Research Institute, Feinberg School of Medicine, Northwestern University, Chicago, IL, USA

Yuen Ting Lam Heart Research Institute, Sydney, Australia

Laura Lecce Heart Research Institute, Sydney, Australia

Eng H. Lo Neuroprotection Research Laboratory, Massachusetts General Hospital, Harvard Medical School, Charlestown, MA, USA

Jingwei Lu Cardiovascular Stem Cell Research Laboratory, The Dorothy M. Davis Heart and Lung Research Institute, Wexner Medical Center at The Ohio State University, Columbus, OH, USA

Michael Luis Department of Medical Oncology, Portuguese Oncology Institute, Rua Dr. António Bernardino de Almeida, Porto, Portugal

Reinhold J. Medina Centre for Vision and Vascular Science, School of Medicine, Dentistry, and Biomedical Science, Queen's University Belfast, Royal Victoria Hospital, Belfast, Northern Ireland, UK

Jawahar L. Mehta Division of Cardiovascular Medicine, University of Arkansas for Medical Sciences and the Central Arkansas Veterans Healthcare System, Little Rock, AR, USA

Ramon Andrade de Mello Department of Medical Oncology, Portuguese Oncology Institute, Rua Dr. António Bernardino de Almeida, Porto, Portugal

Department of Medicine, University of Porto, Porto, Portugal

Nobukazu Miyamoto Neuroprotection Research Laboratory, Massachusetts General Hospital, Harvard Medical School, Charlestown, MA, USA

Shinya Mizuno Division of Virology, Department of Microbiology and Immunology, Osaka University Graduate School of Medicine, Suita, Japan

Ryuichi Morishita Department of Clinical Gene Therapy, Osaka University Graduate School of Medicine, Suita, Osaka, Japan

Christoph Nabzdyk Department of Surgery, Tufts Medical Center, Boston, MA, USA

Toshikazu Nakamura Division for Regenerative Drug Discovery, Center for Advanced Science and Innovation, Osaka University, Suita, Japan

Deepti Navaratna Neuroprotection Research Laboratory, Massachusetts General Hospital, Harvard Medical School, Charlestown, MA, USA

Martin K. C. Ng Heart Research Institute, Sydney, Australia

Department of Cardiology, Royal Prince Alfred Hospital, Sydney, NSW, Australia

Department of Medicine, University of Sydney, Sydney, Australia

Blanca Oñate Cardiovascular Research Center, CSIC-ICCC, Hospital de la Santa Creu i Sant Pau, IIB-Sant Pau, Barcelona, Spain

CIBEROBN-Pathophysiology of Obesity and Nutrition, Barcelona, Spain

Hiroyuki Ohnishi Division for Regenerative Drug Discovery, Center for Advanced Science and Innovation, Osaka University, Suita, Japan

College of Pharmacy, Kinjo Gakuin University, Nagoya, Japan

Kiyomasa Oka Division for Regenerative Drug Discovery, Center for Advanced Science and Innovation, Osaka University, Suita, Japan

Sadip Pant Division of Cardiovascular Medicine, University of Arkansas for Medical Sciences and the Central Arkansas Veterans Healthcare System, Little Rock, AR, USA

Loc-Duyen D. Pham Neuroprotection Research Laboratory, Massachusetts General Hospital, Harvard Medical School, Charlestown, MA, USA

Vincent J. Pompili Cardiovascular Stem Cell Research Laboratory, The Dorothy M. Davis Heart and Lung Research Institute, Wexner Medical Center at The Ohio State University, Columbus, OH, USA

Leena Pradhan-Nabzdyk Department of Surgery, Division of Vascular and Endovascular Surgery, Beth Israel Deaconess Medical Center, Harvard Medical School, Boston, MA, USA

Gerald W. Prager Clinical Division of Oncology, Department of Medicine I, Medical University of Vienna, Comprehensive Cancer Center Vienna, Vienna, Austria

Hiromi Rakugi Department of Geriatric Medicine and Nephrology, Osaka University Graduate School of Medicine, Suita, Osaka, Japan

Rui Manuel Reis Life and Health Sciences Research Institute (ICVS), School of Health Sciences, University of Minho, Braga, Portugal

Molecular Oncology Research Center, Barretos Cancer Hospital, Barretos, SP, Brazil

Domenico Ribatti Department of Basic Medical Sciences, Neurosciences, and Sensory Organs, Section of Human Anatomy and Histology, University of Bari Medical School, Bari, Italy

Amy Richard Department of Biomedical Sciences, Ontario Veterinary College, University of Guelph, Guelph, ON, Canada

Fumihiro Sanada Department of Clinical Gene Therapy, Osaka University Graduate School of Medicine, Suita, Osaka, Japan

Harald Schütt Department of Cardiology and Angiology, Hannover Medical School, Hannover, Germany

Wolfgang Schaper Division of Arteriogenesis Research, The Max Plank Institute for Heart Research and Lung Research, Bad Nauheim, Germany

Bernhard Schieffer Department of Cardiology and Angiology, Hannover Medical School, Hannover, Germany

Ji Hae Seo Neuroprotection Research Laboratory, Massachusetts General Hospital, Harvard Medical School, Charlestown, MA, USA

Futoshi Shibasaki Molecular Medical Research Project, Tokyo Metropolitan Institute of Medical Science, Setagaya-Ku, Tokyo, Japan

Sandeep Singla Division of Cardiovascular Medicine, University of Arkansas for Medical Sciences and the Central Arkansas Veterans Healthcare System, Little Rock, AR, USA

Alan W. Stitt Centre for Vision and Vascular Science, School of Medicine, Dentistry, and Biomedical Science, Queen's University Belfast, Royal Victoria Hospital, Belfast, Northern Ireland, UK

Ghazaleh Tabatabai Department of Neurology, University Hospital Zurich, Zurich, Switzerland

Peng H. Tan Department of Immunology, Division of Medicine, Imperial College London, Hammersmith Hospital, London, UK

Yoshiaki Taniyama Department of Clinical Gene Therapy, Osaka University Graduate School of Medicine, Suita, Osaka, Japan

Department of Geriatric Medicine and Nephrology, Osaka University Graduate School of Medicine, Suita, Osaka, Japan

Melissa A. Thal Feinberg Cardiovascular Research Institute, Feinberg School of Medicine, Northwestern University, Chicago, IL, USA

Gemma Vilahur Cardiovascular Research Center, CSIC-ICCC, Hospital de la Santa Creu i Sant Pau, IIB-Sant Pau, Barcelona, Spain

CIBEROBN-Pathophysiology of Obesity and Nutrition, Barcelona, Spain

Alicia Viloria-Petit Department of Biomedical Sciences, Ontario Veterinary College, University of Guelph, Guelph, ON, Canada

Xianwei Wang Division of Cardiovascular Medicine, University of Arkansas for Medical Sciences and the Central Arkansas Veterans Healthcare System, Little Rock, AR, USA

Michael Weller Department of Neurology, University Hospital Zurich, Zurich, Switzerland

Christoph C. Zielinski Clinical Division of Oncology, Department of Medicine I, Medical University of Vienna, Comprehensive Cancer Center Vienna, Vienna, Austria

Sonja Zours Department of Biomedical Sciences, Ontario Veterinary College, University of Guelph, Guelph, ON, Canada

Part I
Molecular Mechanisms of Angiogenesis

Chapter 1
Endothelial Growth Factor Receptors in Angiogenesis

David J. Bruce and Peng H. Tan

Abstract It is hard to underestimate the role of endothelial growth factor receptors in the generation of new blood vessels. This axis is involved in vascular development in embryos and angiogenesis in adults. As the signaling of these tyrosine kinase receptors has been elucidated, we have gained an appreciation of the complex interactions with other receptors, co-receptors, and downstream pathways.

Its involvement in pathology makes it a particularly tempting therapeutic target with its manipulation offering several theoretical benefits. The most intensely studied is the role of anti-VEGFR drugs in cancer chemotherapy. Initial trials were disappointing but a decade ago the first drug targeting the vascular endothelial growth factor (VEGF) axis was approved, providing a vital proof of concept. Therapies specifically targeting the receptor are in early development for prevention of neovascular diseases of the eye. Conversely, promotion of revascularization following vascular occlusion is another possible application being studied.

While these therapies show promise, the manipulation of VEGF receptors themselves remains a relatively small niche in the therapeutic armory. A deeper understanding of the receptor, its co-receptors, and the downstream web of signaling is required to complete the pieces of the puzzle and unlock the potential of this receptor pathway.

Keywords Angiogenesis • VEGFR • Neovascularization • Tumorigenesis • Growth factor • Signaling interactions • Tyrosine kinase inhibitor • Receptor

D.J. Bruce (✉)
Oxford University Hospitals NHS Trust, John Radcliffe Hospital, Oxford OX3 9DU, UK
e-mail: davebruce@doctors.org.uk

P.H. Tan
Department of Immunology, Division of Medicine, Imperial College London, Hammersmith Hospital, London, UK

J.L. Mehta and N.S. Dhalla (eds.), *Biochemical Basis and Therapeutic Implications of Angiogenesis*, Advances in Biochemistry in Health and Disease 6,
DOI 10.1007/978-1-4614-5857-9_1,

1.1 Introduction: The Signaling Axes Behind Angiogenesis

Angiogenesis begins with the remodeling of the initial lattice of vascular endothelial cell precursors, which results in differential growth of the vessels to form branches and sprout new vessels. This requires coordinated communication within the preliminary lattice of homogenous vessels. Intercellular signaling continues to play a part in the recruitment of supporting cells including smooth muscle, pericytes, and fibroblasts [1] and in the breakdown and deposition of the extracellular matrix [2]. This development and remodeling of the vascular network is controlled by the interaction of angiogenic growth factors with their receptors and the subsequent downstream signaling networks [3] (Table 1.1).

One of the most important and closely studied of these is the vascular endothelial growth factor (VEGF) pathway which plays a central role. This signaling cascade influences numerous cellular events involved in angiogenesis including endothelial cell proliferation and migration, remodeling of the extracellular matrix, increased vascular permeability, and survival of new blood vessels [4]. In addition, the angiopoietin-Tie pathway has been identified as a second vascular tyrosine kinase system that is essential during vasculogenesis and adult vascular homeostasis [5].

This chapter will aim to cover some of the evidence for the role of the endothelial growth factor receptors in angiogenesis and the interactions of these receptors with complementary signaling pathways. The therapeutic potential of VEGF receptor manipulation will be discussed, with particular emphasis on antitumor therapies where most work has been focused.

1.2 The VEGF Axis

The VEGF family of signaling molecules was originally identified as a potent mediator of vascular permeability [6] but is now known to stimulate vasculogenesis in embryos and angiogenesis in adults [7]. The downstream pathways form a complex network involving cross talk with other signaling axes [8]. The end result of these pathways is an increase in vascular permeability and the stimulation of cell survival, proliferation, and migration, ultimately leading to angiogenesis or lymphangiogenesis [3,7,9,10].

Table 1.1 Growth factors, receptors, and co-receptors involved in angiogenesis [66,91]

Receptor	Role
VEGFR	Receptor for VEGF, VEGF, PlGF
Tie-1, Tie-2	Receptor for angiopoietin (Ang)
FGFR	Receptor for fibroblast growth factor (FGF)
PDGFR	Receptor for platelet-derived growth factor (PDGF)
NRP	Co-receptor for VEGFR
HSPG	Co-receptor for VEGFR

Table 1.2 Isoforms of human VEGF-A [13,25]

Isoform	Receptors and co-receptors	
VEGF 121	VEGFR1, VEGFR2	Predominant in humans. Secreted into systemic circulation
VEGF 145	VEGFR1, VEGFR2, HSPG	
VEGF 165	VEGFR1, VEGFR2, NRP, HSPG	Predominant in humans
VEGF 183	VEGFR1, HSPG	
VEGF 189	VEGFR1, HSPG	Bound to ECM proteoglycans
VEGF 206	VEGFR1, HSPG	Bound to ECM proteoglycans

1.2.1 VEGF Structure and Function

VEGF is part of the cysteine knot growth factor superfamily [11] and is found in mammals as a family of structurally homologous secreted glycoproteins: VEGF-A, VEGF-B, VEGF-C, VEGF-D, and placental growth factor (PlGF). VEGF is usually secreted as a dimeric glycoprotein [7].

VEGF-A is the best characterized of this family. The *VEGF-A* gene is found on chromosome 6 (locus 6p21.1) [12], and alternative exon splicing produces six isoforms (Table 1.2). All of them contain binding domains for the receptors VEGFR-1 and VEGFR-2 [13], but it is the splicing of exons encoding different C-terminal domains that gives each isoform their own unique biochemical properties [12]. Proteolytic processing and other posttranslational modifications act to further refine VEGF activity [14].

The different isoforms of VEGF have different solubility and bioavailability. This is due to the ability of VEGF to tether to the extracellular matrix to varying degrees, allowing it to act in a paracrine fashion [15]. The effect of the localization of VEGF is reflected in embryos expressing a VEGF splice variant lacking heparin-binding and ECM interaction domains. The disruption of VEGF-A signaling, due to failure to generate concentration gradients, leads to endothelial cells failing to form additional branches and impaired filopodia function [16]. Matrix metalloproteinases (MMP) cleave VEGF-A releasing the receptor binding domain. Blocking this activity can arrest the angiogenic switch associated with carcinogenesis [17].

1.2.1.1 The VEGF Receptors

VEGF binds to a family of receptor tyrosine kinases, known as the VEGF receptors (VEGFRs). There are three characterized VEGFRs (Table 1.3) which are structurally similar, with an extracellular ligand binding domain, a transmembrane helix, and a cytoplasmic region containing a kinase domain (Fig. 1.1) [18]. When the ligand has bound the extracellular domain, the intracellular domain autophosphorylates, and the receptors dimerize. VEGFRs activate various downstream signaling

Table 1.3 The VEGFR receptors and their ligands [25]

Receptor	Gene	Ligands	Localization
VEGFR1	*FLT-1*	VEGF-A, VEGF-B, PlGF	Vascular endothelium
VEGFR2	*KDR*	VEGF-A, VEGF-C, VEGF-D, VEGF-E	Vascular and lymphatic endothelium
VEGFR-3	*FLT-4*	VEGF-C, VEGF-D	Lymphatic endothelium

pathways (Fig. 1.2) [7]. The VEGFRs are expressed by endothelial cells with each member of the family showing distinct expression patterns, as revealed by in situ hybridization and northern blot [19].

VEGFR-1

VEGFR-1 binds VEGF-A and is the only known cell surface receptor for VEGF-B and PlGF [20]. While VEGF-A binds VEGFR-1 with higher affinity than VEGFR-2, the tyrosine kinase activity of VEGFR-1 is weaker than VEGFR-2. It has been suggested that VEGFR-1 acts either as a "decoy" receptor to reduce the bioavailability of VEGF-A by forming inert complexes or as a dominant antagonist of the axis [21]. Additionally, PlGF may act by displacing VEGF-A from VEGFR-1, thus increasing VEGF-A bioavailability [22]. Both in vitro fluorescence studies and in vivo experiments showed that cells which lack VEGFR-1 had decreased sprout formation and reduced migration. However, a soluble isoform of VEGFR-1 (sVEGFR-1) rescued angiogenesis indicating that VEGFR-1 may have a positive regulatory role, possibly related to its effect on the localization of VEGF [10]. This effect has been noted in $FLT^{-/-}$ mutant vessels, where both membrane-bound and soluble forms of VEGFR-1 rescued aberrant endothelial proliferation but only soluble VEGFR-1 rescued vessel branching. Hence, it may be that heterogeneous sVEGFR-1 expression underlies the spatial variation of VEGFR-2 signaling which are required for correct branching development [23].

Heterodimers form between VEGFR-1 and VEGFR-2, which may augment certain signaling pathways. In cell lines expressing both receptors, signaling pathways leading to PLCγ activation were enhanced, and these cells showed more efficient migration towards VEGF-A compared to cells expressing VEGFR-1 alone [24].

VEGFR-2

VEGFR-2 is vital to angiogenesis. In murine models, knockout of VEGFR-2 leads to aberrant vascular development and embryonic lethality [8]. VEGFR-2 binds VEGFR-A, VEGF-C, VEGF-D, and VEGF-E. It is found on both vascular and

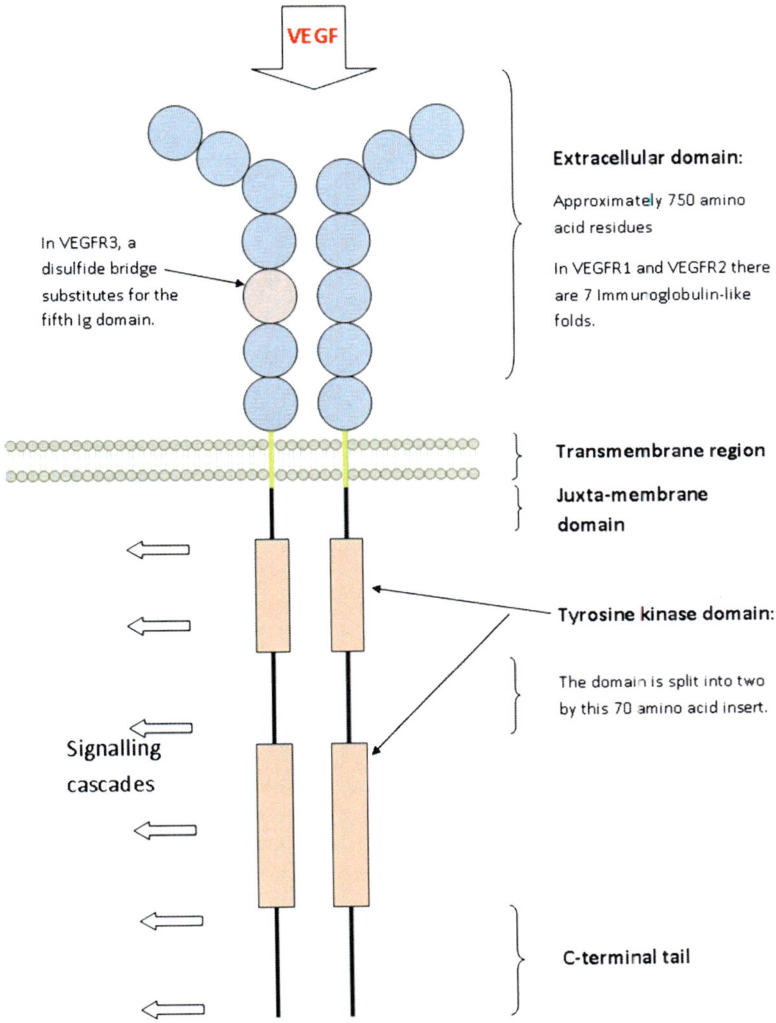

Fig. 1.1 The structure of VEGF receptor (VEGFR). The VEGFR has an extracellular domain that consists of seven immunoglobulin (Ig)-like domains, except in VEGFR3 where the fifth domain is replaced by a disulfide bridge. The intracellular domain consists of a tyrosine kinase domain, interrupted by a 70 amino acid insert, a juxta-membrane region, and a C-terminal tail. The binding of signaling molecules to phosphorylation sites present on the intracellular domain initiates cell signaling (see Fig. 1.2). Studies of the crystal structure of VEGFR1 show that the second Ig domain is the ligand binding site. The third Ig domain determines ligand binding specificity in VEGFR2 [9]

lymphatic endothelium [25]. VEGFR-2 has at least four autophosphorylation sites. The tyrosine residues Tyr1054 and Tyr1059 are required for maximal kinase activity [26], and Tyr1175 is known to be crucial for PLCγ/PKC/MAPK signaling which

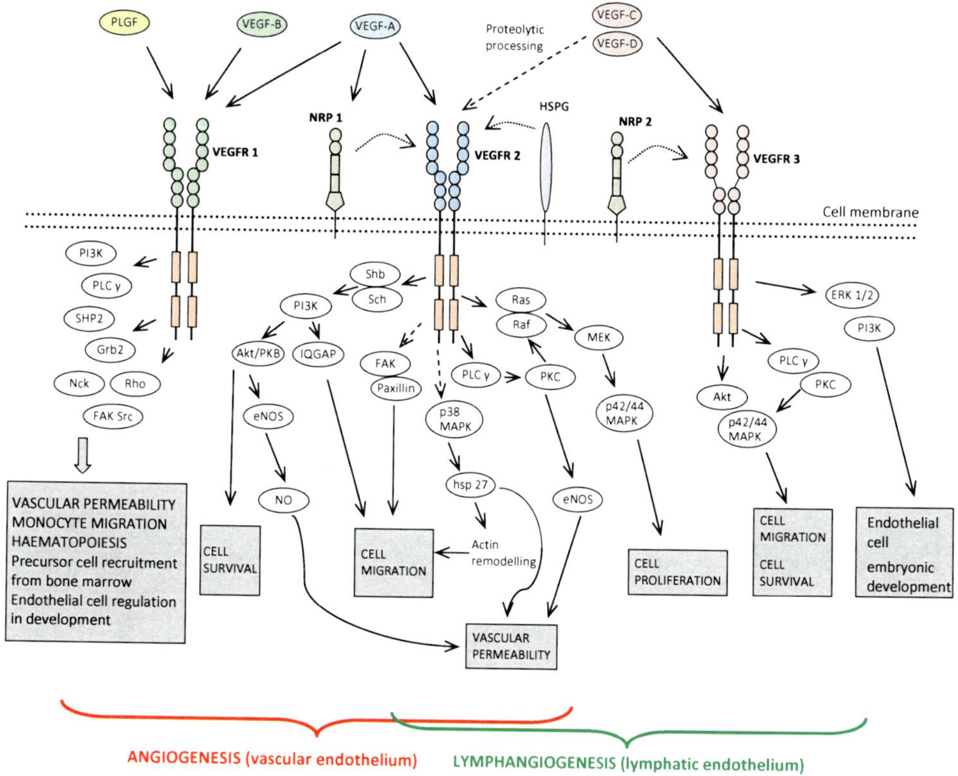

Fig. 1.2 VEGFR signaling pathways. VEGF ligands bind to particular VEGFR homodimers and heterodimers. Co-receptors may modify this interaction, such as the neuropilins and heparan sulfate proteoglycans. VEGFR1 is found on vascular endothelium, VEGFR3 is found on lymphatic endothelium, and VEGFR2 is found on both. These signaling pathways stimulate cell survival, proliferation, and migration and increase vascular permeability, ultimately leading to angiogenesis or lymphangiogenesis. Adapted from [9,20]

is necessary for endothelial cell proliferation [27]. Tyr1175 is also bound by the adaptor protein Shb, and inhibition of Shb reduces PI3K activation and subsequent VEGF-induced cell migration [28].

A number of pathways are known to be necessary for VEGFR-2-mediated cell migration. Two complementary pathways mediate VEGFR-2 activation of FAK; one involves the Src kinase while the other involves RhoA and ROCK. This in turn initiates paxillin and vinculin recruitment [29].

Studies on human umbilical vein endothelial (HUVEC) cells indicate that VEGFR-2 is involved in the regulation of the actin cytoskeleton via Ras activation. Inhibition of this signaling pathway inhibits branching morphogenesis. The Ras pathway mediates ERK activation which is required for VEGF-mediated cell proliferation [30]. ERK signaling is also involved in promoting cell survival and proliferation but is inhibited by p38MAPK suggesting cross talk between these pathways influences the cellular response to angiogenic stimuli [31]. VEGF-mediated activation of cdc42 and p38MAPK regulates actin polymerization and stress fiber reorganization leading to endothelial cell migration via heat shock protein 27 activation [31].

In HUVEC cells, VEGFR-2 co-localizes with the scaffolding protein IQGAP1 to the leading edge of migrating cells. IQGAP1 interacts with the cell cytoskeleton to manipulate cell motility and morphogenesis [32].

In summary then, VEGFR-2 plays a vital role in endothelial cell survival, proliferation, and migration and stimulates vascular permeability and invasion. These processes are crucial for angiogenesis [25].

VEGFR-3

VEGFR-3 in adults is associated with the lymphatic endothelium. It is found to be expressed exclusively on lymphatic endothelial and some high endothelial venules [33].

There are two splice variants in humans and both have a high affinity for VEGF-C and VEGF-D [34]. VEGFR-3 forms homodimers or heterodimers. The binding of VEGF-C induces VEGFR-2/VEGFR-3 heterodimer formation, and the function of the regulatory tyrosine phosphorylation sites differs in the various ligand-induced dimerized complexes [35]. VEGFR-3 is responsible for inducing cell migration and prevents apoptosis via Akt and p42/p44MAPK [36]. Pathways involving PKC, ERK1/2, PI3K, PLCγ, SHP2, and the transcription factors STAT3 and STAT5 are all influenced by VEGFR-3 signaling [7].

Murine studies have shown lethal defects in vasculogenesis, severe anemia, and cardiac effusion in VEGFR-3-deficient embryos. This may be due to a direct result of reduced VEGFR-3-mediated signaling. However, it is also possible that these effects are a result of cross talk with VEGFR-2, perhaps by the reduction of VEGF-C bioavailability to VEGFR-2 due to binding of VEGF-C to VEGFR-3 [37]. It is thought that VEGFR-3 may negatively regulate VEGFR-2 signaling [38]. VEGFR plays an important role during angiogenic sprouting, where endothelial cells develop as either tip or stalk cells. VEGFR-3 has also been implicated in reinforcing Notch signaling in tip cells of angiogenic sprouts and thus influencing tip to stalk cell conversion [39].

1.3 Co-receptors

VEGFR activity may be modulated by co-receptors. Neuropilin (NRP) is a transmembrane glycoprotein involved in neuronal axon guidance but is also known to act as a co-receptor for VEGFR. However, the mechanism underlying the enhanced signaling remains uncertain. It may be that VEGFR2/NRP1 complex formation increases VEGFR affinity for VEGF, or that the intrinsic catalytic activity of the kinase domain is enhanced, or the signaling complex may be stabilized with a prolonged half-life. While there is still some disagreement in the literature, it is likely that these receptor complexes are induced by VEGF binding [7,40]. Furthermore, VEGF may directly bind NRP, and while NRPs lack intrinsic catalytic activity, they may associate with other transmembrane proteins to stimulate signal transduction [7]. This potential alternative signaling pathway may be a mechanism for resistance to anti-angiogenic therapies, as supported by the augmentation of the effects of Bevacizumab when used in combination with anti-NRP antibodies [41].

Heparan sulfate proteoglycans (HSPGs) have been identified as co-receptors to VEGFR during angiogenesis and vasculogenesis, where it acts to augment the duration and magnitude of the response and influences localization of VEGF/VEGFR complexes. In situ studies showed a direct interaction between HSPG and VEGFR, and blockade of HSPG interactions inhibited in vivo hyperpermeability. This supports the possibility of targeting HSPGs in ischemic disease [42]. Chondroitin sulfate proteoglycans bind angiogenic growth factors, and their production is increased in cells lacking HSPGs in vitro. It appears that there may be functional overlap with HSPGs during sprouting angiogenesis but the in vivo implications of this interaction remain unclear [43].

1.4 The Role of the Endothelial Growth Factor Receptors in Angiogenesis and Neovascularization

1.4.1 The Role of the Endothelial Growth Factor Receptors in Tumorigenesis

Tumor growth is restricted by the availability of a suitable vascular supply, and angiogenesis is required for a tumor to develop beyond a few millimeters in diameter. In vivo anti-VEGF antibody administration inhibits tumor growth. The VEGF-VEGFR pathway may also exert a direct effect on the proliferation and growth of tumor cells themselves [25]. However, immunohistochemical analyses of several human cancers have indicated that VEGFR-2 and VEGFR-3 are not expressed by tumor cells. Hence, animal models and in vitro experiments should be interpreted with caution [44].

1.4.1.1 VEGFR-1

VEGFR-1 is known to be upregulated in several tumors. In pancreatic cancer cells, VEGFR-1 stimulated upregulation of transcription factors associated with motility and invasion [45], and a similar effect is seen in colonic cancers [46]. In vitro multiple myeloma cells were found to express only VEGFR-1, and ablation of VEGFR-1 signaling was sufficient to inhibit cell proliferation and motility [47]. Other tumors in which VEGFR-1 is upregulated include prostate, glioblastoma, and malignant melanoma [25].

sVEGFR-1 has been identified in breast, pancreatic, lung, and ovarian cancers and leukemias [25]. sVEGFR-1 shows antitumor effects when administered in vitro and in vivo. This is likely due to its interception of VEGF-A. In a rat hepatocellular carcinoma model, tumor weight was decreased 19-fold in cells transduced with sVEGFR-1 [48]. Consequently, it may be that the soluble isoform of VEGFR-1 could be developed for cancer treatment.

1.4.1.2 VEGFR-2

A rat anti-VEGFR-2 antibody, known as DC101, inhibits the spread and growth of metastases in several models via apoptosis of vascular endothelial cells [49]. In vitro, DC101 inhibits neovascularization, while in vivo models showed inhibited growth of several cancers including breast, melanoma, and both primary and secondary lung tumors. Growth was also repressed in human xenografts of epidermoid, glioblastoma, pancreatic, and renal cancers [50]. In a murine lymphoma model, inhibition of either VEGFR-1 or VEGFR-2 led to altered vessel growth and development. However, inhibition of both was required for tumor regression [51] supporting the use of multi-targeted inhibitors in antitumor therapies.

1.4.1.3 VEGFR-3

VEGFR-3 is a key receptor in lymphangiogenesis and is strongly expressed in human tumours including lung, cervical, breast, prostate and colorectal cancers. Higher levels of both VEGF-C and VEGFR-3 correlate with increased metastases and shorter survival [52]. VEGFR-3 is upregulated on the tumor vasculature but poorly expressed in cells from a number of human tumors. Therefore, it is unlikely to affect the tumor cells directly [44]. Overexpression of VEGF-C induced hyperplasia in peri-tumor lymphatics and led to an increased lymph flow rate, an effect which was suppressed by anti-VEGFR-3 antibody administration [53].

1.4.1.4 Prognosis

VEGFR-1 and VEGF expressions were found to correlate closely to microvessel density and were important predictors of poor prognosis and clinical progression in nephroblastoma [54]. Similarly, VEGFR-1 and VEGFR-2 were upregulated in glioblastoma vascular cells, but not in low-grade blastoma [55]. In non-small cell lung cancer patients, co-expression of VEGFR-1 and VEGF was associated with a shorter survival time. Co-expression of VEGFR-1 and VEGFR-2 was an independent prognostic factor [56]. Thus, VEGF/VEGFR signaling could potentially provide useful prognostic information in the clinical setting.

1.4.1.5 Metastasis

Tumor angiogenesis is preceded by recruitment of endothelial precursor and hematopoietic cells which may act to "prepare" microenvironments for metastatic spread [57]. VEGFR-1 induces MMP9 which degrades ECM to allow remodeling. The upregulation of MMP9 in healthy lung appears to promote lung metastasis, but this effect was abolished in VEGFR-1 knockout mice [58]. However, the rate of metastases formation in murine models was not affected by VEGFR-1 blockade, which may be due to alternate signaling pathways for inducing the "pre-metastatic niche" [59]. In human gastric cancer patients, the rate of metastasis was increased in those patients expressing high levels of VEGFR-1 in bone marrow and blood [60]. Further work is needed to elucidate the precise role of VEGFR-1 in metastatic spread.

1.4.1.6 Autocrine Signaling

VEGF may act in an autocrine or paracrine manner to amplify the malignant potential of cells co-expressing VEGFR-1 and VEGFR-2 [56]. In a study of myelodysplastic patients, monocyte and myeloid precursor cells were found to co-express VEGFR-1 or VEGFR-2 in a majority of patients. It is speculated that VEGF may act in an autocrine fashion to promote leukemia cell survival [61]. Autocrine signaling has also been suggested to occur with VEGF/VEGFR-1 signaling present on breast cancer cells to increase cell invasion [62]. Autocrine signaling via the VEGF-VEGFR2-NRP1 axis has been suggested as a mechanism of resistance to anti-angiogenic therapies, associated with VEGFR2-NRP1 recycling and a pool of VEGFR2 present in the cytosolic compartment of glioblastoma multiforme (GBM) cells. Bevacizumab transiently reduces GBM tumor growth, but in vitro inhibition of VEGFR2 attenuates glioma stem-like cell viability, thus suggesting a possible role of VEGF receptor inhibition to augment current anticancer therapies [63].

1.4.1.7 Therapeutic Applications

Tumor cells are inherently heterogeneous and genetically unstable. The high mutation rate leads to the evolution of resistant cell lines. Endothelial cells are more stable with a lower mutation rate, although there is data to suggest that tumor endothelial cells may be derived from tumor stem cells. In a murine in vivo model, repeated administration and discontinuation of an angiogenesis inhibitor did not lead to drug resistance, and tumors remained dormant for longer following repeated treatment. However, there has been a lack of reproducibility of these findings [64]. Nevertheless, inhibitors of angiogenesis provide an intriguing method of restricting tumorigenesis.

Murine models of VEGFR-2 and VEGFR-3 inhibition show a reduction in metastases in lymph nodes and lung. However, simultaneous inhibition of both has a more potent effect [65]. In another experiment, gene therapy-mediated inhibition of VEGFR-1, VEGFR-3, Tie-1, and Tie-2 led to significantly decreased ovarian tumor mass [66]. Thus, targeting multiple VEGF receptors and interacting pathways, such as the angiopoietin network, may increase the efficacy of treatment.

Inhibition of VEGFR in tumors may help to normalize the vasculature and allow effective delivery of drugs and decrease resistance to radiotherapy. Blockade of VEGFR-2 creates a "normalization window" during which period combined treatment led to the most marked tumor regression in murine brain tumors [67]. However, excessive destruction of tumor vessels may hinder drug delivery and generate undue hypoxia. The inhibition of VEGFR2 leads to angiopoietin-1 upregulation, pericyte proliferation, and MMP activation and the breakdown of pathologically thick vascular basement membrane. Measurement of the levels of hypoxia could allow optimization of the schedules of anti-VEGFR and radiotherapy combination therapy [68].

Several small-molecule tyrosine kinase inhibitors (TKIs) have entered clinical development. These inhibitors were originally identified by screening large peptide libraries, but more recently the use of molecular modeling analysis using X-ray crystallographic data has provided a more targeted approach. The TKIs commonly compete for the ATP-binding site within the tyrosine kinase, thus ablating phosphorylation and consequent downstream signaling. The ATP-binding region within the kinase domain is well conserved between several receptor tyrosine kinase families; thus, several VEGFR TKIs have been found to also inhibit other receptor pathways [69].

Initial trials of inhibitors were disappointing with several drugs failing in clinical trials. However, the field has been reinvigorated following the successful licensing of the VEGF inhibitor Bevacizumab which provided the proof of concept for the field. Several inhibitors of the VEGF receptor are in development and a number of licenses have been granted for their use (Table 1.4) [69]. A TKI known as SU11248 (Sunitinib) which targets class II/V RTKs, including VEGFR, PDGFR, c-kit, and FLT3, has been licensed for gastrointestinal stromal tumors [70] and renal cell carcinoma [71] following successful Phase III trials. Further trials are underway looking at its potential use in breast, non-small cell lung, and hepatocellular cancer [69].

Table 1.4 Some antitumor VEGFR inhibitors in clinical development [69]

Drug	Trade name	Target receptors	Trial phase
AG-013736	Axitinib	VEGFR, PDGFR, c-kit	III
AMG-706	Motesanib	VEGFR, PDGFR, c-kit	III
ZD2171	Cediranib/ Recentin	VEGFR, PDGFR, c-kit	III
PTK787	Vatalanib	VEGFR	III
BMS582664	Brivanib	VEGFR, FGFR	III
SU11248	Sutent/Sunitinib	VEGFR1, VEGFR2, FLT3, PDGFR, c-kit, cRET	IV licensed for use in gastrointestinal stromal tumors and renal cell carcinoma
ZD6474	Vandetanib	VEGFR1, VEGFR2, EGFR1	III
TKI 258	Dovitinib	VEGFR1, VEGFR2, PDGFR, FGFR, c-kit	III
BAY43-9006	Sorafenib/ Nexavar	VEGFR2, VEGFR3, PDGFR, c-kit, FGFR1, B-raf	IV licensed for use in hepatocellular carcinoma and renal cell carcinoma

The multi-targeted nature of SU11248 is a key factor in its efficacy. This is demonstrated by the selective VEGFR inhibitor called SU10944 in combination with Gleevec (a PDGF receptor inhibitor) which showed similar in vitro and in vivo activity compared to SU11248. However, when only one of SU10944 or Gleevec was given, the efficacy was significantly inferior [72]. Another successful VEGFR inhibitor is BAY 43-9006 (Sorafenib) which was originally developed to inhibit Raf-1 and the RAF/MEK/ERK pathway. Again, BAY 43-9006 may owe its efficacy to the inhibition of several RTKs. BAY 43-9006 prolonged patient survival in Phase III trials in hepatocellular carcinoma and renal cell carcinoma [69], but at present its prohibitive cost has limited its use [73]. In orthotopic murine models of human non-small cell lung cancer, MEK inhibition in combination with the VEGFR inhibitor Cediranib led to superior anti-angiogenic and antitumor effects [74].

More specific targeting of VEGFRs can be achieved by the use of monoclonal antibodies. The use of monoclonal antibodies to target VEGF/VEGFR signaling has seen success in Bevacizumab, a monoclonal antibody to VEGF-A, which is licensed for use in several cancers [75]. The anti-VEGFR antibody DC101 has been shown to reduce tumor growth in murine models of colon cancer [76]. However, results of VEGFR-targeted antibodies have been less successful in humans. One of the most advanced is IMC-1121B (Ramucirumab) which entered Phase II trials for gastric cancers, breast cancer, hepatocellular carcinoma, and non-small cell lung cancer [69]. The use of small interfering RNA (siRNA) to depress VEGFR-2 expression has been shown to reduce invasion in vitro, but the use of this strategy remains a distant prospect [77].

One advantage of VEGFR inhibition is the anticipation of a relatively favorable toxicity profile compared to conventional chemotherapy. Life-threatening adverse events associated with cytotoxic agents have been rarely seen with angiogenesis inhibitors [78]. Side effects may be due to the inhibition of VEGF signaling or off-target effects, for example inhibition of other kinases. Off-target effects are more dependent on patient- or treatment-specific factors, such as comorbidities and disease stage, whereas the on-target effects tend to be seen with all TKIs [9]. While multi-targeted TKIs have shown more efficacy in the treatment of tumors, this is associated with an increase in off-target adverse events [79]. Therefore, a careful balance must be met between improving efficacy and minimizing toxicity.

1.4.2 The Eye

VEGF-A is produced by retinal pigment epithelium (RPE) in humans. There is limited knowledge of the role of VEGF in the maintenance of the adult ocular vasculature, but VEGF-A may act in a paracrine fashion between the RPE and choriocapillaries. Currently, inhibition of VEGF does not appear to have adverse effects on the ocular vasculature [80]. However, intraocular angiogenesis and changes in vascular permeability underlie the development of retinal vascular disorders including retinal vein occlusion, diabetic retinopathy, and age-related macular degeneration (AMD) [81]. Disease activity reflects the balance between pro-angiogenic and anti-angiogenic factors. There is impairment of vascular autoregulation, macular edema, and aberrant development of retinal vessels [82]. Levels of VEGF-A are high in these conditions. In animal models, specific isoforms of VEGF-A are increased in certain disorders. Tissue displaying choroidal neovascularization in patients with AMD expresses both $VEGF_{121}$ and $VEGF_{165}$ isoforms of VEGF-A [80]. $VEGF_{164}$ (the murine orthologue of $VEGF_{165}$) is a potent in vivo inducer of angiogenesis and inflammation in the eye. It stimulates ICAM-1 expression on endothelial cells via VEGFR2 and chemotaxis of monocytes via VEGFR1. In vitro $VEGF_{165}$ induces activation of human VEGFR1 more efficiently than other isoforms [83].

Soluble VEGFRs are found in the vitreous in patients with vitreoretinal disease. sVEGFR-1 levels were found to increase with age, and lower levels were associated with more active proliferative diabetic retinopathy. This suggests that sVEGFR may be a useful tool in tipping the balance to hinder neovascularization [81]. However, it has also been suggested that increased levels of sVEGFR-2 in the vitreous may contribute to increased vascular permeability in macular edema. Further work is required to elucidate the role of sVEGFR in the eye and evaluate its potential for therapeutic use [82].

1.4.2.1 Therapeutic Applications

VEGF-inhibition is known to be effective for treatment of ocular diseases involving neovascularization, and anti-VEGF treatments have been licensed for use.

Table 1.5 Tyrosine kinase inhibitors (TKIs) in development for eye disease

TKI	Route	Sponsor
Pazopanib	Topical	GlaxoSmithKline, PA, USA
TG100801	Topical	TargeGen, CA, USA
TG101095	Topical	TargeGen, CA, USA
AG013958	Sub-Tenon	Allergan, CA, USA
AL39324	Intravitreal	Alcon, TX, USA
PTK787 (Vatalanib)	Oral	Novartis, CH

These are still in the early stages of development with trials in Phase II. These TKIs target VEGFR but some, such as Pazopanib, also have activity against other tyrosine kinases such as PDGFR and c-kit [80,92,93]

Pegaptanib (Macugen) is a ribonucleic aptamer which selectively binds $VEGF_{165}$ and thus attenuates VEGFR2 activation. The VISION trials validated the use of Pegaptanib in neovascular AMD with significantly reduced loss of vision compared to placebo. Subsequently, the VEGF inhibitors Ranibizumab (Lucentis) and Bevacizumab (Avastin) were developed, although only Ranibizumab is FDA approved. Both agents are derived from the same murine anti-VEGF mAb, but Ranibizumab consists of the Fab fragment and has a binding affinity 20-fold greater than Bevacizumab. The MARINA trial of Ranibizumab not only showed a reduction of vision loss but an improvement in acuity in 1 in 3 patients. Ranibizumab is superior to the previous benchmark, photodynamic therapy, and it has shown a favorable safety profile in Phase IV trials. Bevacizumab has shown similar efficacy but is not FDA approved, largely due to financial implications for the pharmaceutical industry [84].

A number of other therapies targeting the VEGF/VEGFR pathway in the eye are under investigation. VEGF trap is a fusion protein containing the VEGFR1 and VEGFR2 binding sites and thus inhibits VEGF-A and PlGF [80]. A Phase I/II study of intravitreal administration to 21 patients showed improvements in vision, and Phase III trials are currently underway [84].

Inhibitors of the VEGFRs themselves are also in development, and small-molecule TKIs have shown some promise in preclinical models of ocular neovascular disease. PTK/ZK is a TKI which acts on all the VEGFRs. A murine model of ischemia-induced retinopathy showed that intravitreal injection of this TKI significantly reduced angioproliferative retinopathy [85]. Another TKI named SU5416, which selectively inhibits VEGFR2, also inhibits neovascularization in mouse cornea [86]. Subsequently, several TKIs have entered early clinical trials (Table 1.5). Unfortunately, a Phase I/II study of the TKI AG-013958 for the treatment of choroidal neovascularization was recently terminated due to a lack of efficacy [87]. siRNA has been used to target both VEGF and VEGFR. SIRNA-027 targets the VEGFR, and while initial trials are encouraging, the overall clinical benefit is still uncertain [84].

Anti-VEGF therapy is undoubtedly beneficial in the treatment of neovascular eye disease, but there is scope for development. Other approaches to targeting this pathway have not been as successful as anti-VEGF mAb therapy, but many of these alternative drugs have shown early promise. The development of drugs which can be delivered topically or orally would be a major step forward in increasing the availability of anti-VEGF/VEGFR therapies to patients.

1.4.3 Pro-angiogenic Therapies

VEGF/VEGFR signaling is known to be involved in the maintenance of vascular function and homeostasis in the adult. VEGF stimulates endothelial production of nitric oxide and prostacyclin which act as vasodilators and inhibit platelet aggregation and smooth muscle cell proliferation within the vasculature [40]. Genetic deletion of endothelial VEGF in a murine model caused endothelial degeneration and apoptosis, which led to vascular pathology including hemorrhage, perforation, infarcts, and sudden death. Exogenous VEGF could not rescue these defects suggesting involvement of autocrine VEGF signaling [88]. Consequently, anti-angiogenic therapies targeting the VEGF/VEGFR axis have significant cardiovascular side effects which must be monitored [40].

PlGF, a member of the VEGF family, is produced during angiopoiesis in the placenta. Additionally it is produced by infarcted myocardium and its expression is also significantly higher in brain microvascular endothelial cells following oxygen and glucose deprivation in vitro. PlGF administration led to significantly higher VEGFR-2 expression and it may be that VEGFR2 plays a role in PlGF-mediated neuroprotection [89]. In vivo models have shown a beneficial effect of simvastatin, which may be due to VEGFR2 activation of Akt and nitric oxide synthase [90].

Pro-angiogenic therapies may provide an important treatment modality for those patients with vascular disease unsuitable for invasive revascularization procedures. The premise would be to induce the formation of collateral blood vessels in order to revascularize ischemic tissue. At present most research is focused on the use of angiogenic growth factors such as VEGF-A. Preclinical animal models have provided the evidence for the development of these therapies, but so far there has been limited success in translating this success in clinical trials [40].

1.5 Conclusion

VEGF-mediated signaling influences a range of cellular events underpinning blood vessel growth and has a firmly established role in vasculogenesis and angiogenesis. While this axis has been intensely dissected, many of the subtleties are yet to be understood. The activity of the VEGF receptor is modulated by co-receptors such as NRP [41] and HSPG [42] and downstream there is a web of interacting signaling pathways, such as the angiopoietin-Tie pathway [1].

The advent of successful novel therapies targeting the VEGF/VEGFR axis gave impetus to the research in this field. Currently, clinical trials are revealing the potential of inhibiting the VEGF receptor to combat cancer [69] and neovascular eye disease [84]. Conversely, its use as a pro-angiogenic treatment following infarction has also been explored [89]. Furthermore, due to the targeted nature of VEGFR inhibitors, the side effect burden is reduced compared to standard cytotoxic chemotherapy [78].

The VEGF receptor itself provides a tempting target for novel therapeutic strategies, but despite this it has only limited clinical applications at present. The most promising drugs appear to be useful as adjuvants to chemotherapy regimens [69]. Elucidating the signaling mechanisms based around VEGF and its receptor will not only enable us to better understand the fundamental underpinnings of vascular development and regrowth; it will provide us with the tools required to manipulate this axis.

References

1. Gale NW, Yancopoulos GD (1999) Growth factors acting via endothelial cell-specific receptor tyrosine kinases: VEGFs, angiopoietins, and ephrins in vascular development. Genes Dev 13:1055–1066
2. Avraamides CJ, Garmy-Susini B, Varner JA (2008) Integrins in angiogenesis and lymphangiogenesis. Nat Rev Cancer 8:604–617
3. Cross MJ, Claesson-Welsh L (2001) FGF and VEGF function in angiogenesis: signalling pathways, biological responses and therapeutic inhibition. Trends Pharmacol Sci 22:201–207
4. Azam F, Mehta S, Harris AL (2010) Mechanisms of resistance to antiangiogenesis therapy. Eur J Cancer 46:1323–1332
5. Augustin HG, Young Koh G, Thurston G, Alitalo K (2009) Control of vascular morphogenesis and homeostasis through the angiopoietin-Tie system. Nat Rev Mol Cell Biol 10:165–177
6. Senger DR, Connolly DT, Van De Water L et al (1990) Purification and NH2-terminal amino acid sequence of guinea pig tumor-secreted vascular permeability factor. Cancer Res 50:1774–1778
7. Olsson A-K, Dimberg A, Kreuger J, Claesson-Welsh L (2006) VEGF receptor signalling—in control of vascular function. Nat Rev Mol Cell Biol 7:359–371
8. Hicklin DJ, Ellis LM (2005) Role of the vascular endothelial growth factor pathway in tumor growth and angiogenesis. J Clin Oncol 23:1011–1027
9. Ivy SP, Wick JY, Kaufman BM (2009) An overview of small-molecule inhibitors of VEGFR signaling. Nat Rev Clin Oncol 6:569–579
10. Kearney JB, Kappas NC, Ellerstrom C et al (2004) The VEGF receptor flt-1 (VEGFR-1) is a positive modulator of vascular sprout formation and branching morphogenesis. Blood 103:4527–4535
11. Muller YA, Christinger HW, Keyt BA, de Vos AM (1997) The crystal structure of vascular endothelial growth factor (VEGF) refined to 1.93 Å resolution: multiple copy flexibility and receptor binding. Structure 5:1325–1338
12. Ng Y-S, Krilleke D, Shima DT (2006) VEGF function in vascular pathogenesis. Exp Cell Res 312:527–537
13. Robinson C, Stringer S (2001) The splice variants of vascular endothelial growth factor (VEGF) and their receptors. J Cell Sci 114:853–865
14. Houck KA, Leung DW, Rowland AM et al (1992) Dual regulation of vascular endothelial growth factor bioavailability by genetic and proteolytic mechanisms. J Biol Chem 267:26031–26037

15. Park J, Keller G, Ferrara N (1993) The vascular endothelial growth factor (VEGF) isoforms: differential deposition into the subepithelial extracellular matrix and bioactivity of extracellular matrix-bound VEGF. Mol Biol Cell 4:1317–1326
16. Ruhrberg C, Gerhardt H, Golding M et al (2002) Spatially restricted patterning cues provided by heparin-binding VEGF-A control blood vessel branching morphogenesis. Genes Dev 16:2684–2698
17. Keyt BA, Berleau LT, Nguyen HV et al (1996) The carboxyl-terminal domain (111165) of vascular endothelial growth factor is critical for its mitogenic potency. J Biol Chem 271:7788–7795
18. Joukov V, Pajusola K, Kaipainen A, Chilov D et al (1996) A novel vascular endothelial growth factor, VEGF-C, is a ligand for the Flt4 (VEGFR-3) and KDR (VEGFR-2) receptor tyrosine kinases. EMBO J 15:9
19. Kaipainen A, Korhonen J, Pajusola K et al (1993) The related FLT4, FLT1, and KDR receptor tyrosine kinases show distinct expression patterns in human fetal endothelial cells. J Exp Med 178:2077–2088
20. Schwartz JD, Rowinsky EK, Youssoufian H et al (2010) Vascular endothelial growth factor receptor-1 in human cancer. Cancer 116(S4):1027–1032
21. Rahimi N, Golde TE, Meyer RD (2009) Identification of ligand-induced proteolytic cleavage and ectodomain shedding of VEGFR-1/FLT1 in leukemic cancer cells. Cancer Res 69:2607–2614
22. Park JE, Chen HH, Winer J et al (1994) Placenta growth factor. Potentiation of vascular endothelial growth factor bioactivity, in vitro and in vivo, and high affinity binding to Flt-1 but not to Flk-1/KDR. J Biol Chem 269:25646–25654
23. Kappas NC, Zeng G, Chappell JC et al (2008) The VEGF receptor Flt-1 spatially modulates Flk-1 signaling and blood vessel branching. J Cell Biol 181:847–858
24. Huang K, Andersson C, Roomans GM et al (2001) Signaling properties of VEGF receptor-1 and -2 homo- and heterodimers. Int J Biochem Cell Biol 33:315–324
25. Bruce D, Tan PH (2011) Vascular endothelial growth factor receptors and the therapeutic targeting of angiogenesis in cancer: where do we go from here? Cell Commun Adhes 18:85–103
26. Dougher M, Terman BI (1999) Autophosphorylation of KDR in the kinase domain is required for maximal VEGF-stimulated kinase activity and receptor internalization. Oncogene 18:29
27. Takahashi T, Yamaguchi S, Chida K, Shibuya M (2001) A. single autophosphorylation site on KDR/Flk-1 is essential for VEGF-A-dependent activation of PLC-[gamma] and DNA synthesis in vascular endothelial cells. EMBO J 20:2768–2778
28. Holmqvist K, Cross MJ, Rolny C et al (2004) The adaptor protein shb binds to tyrosine 1175 in vascular endothelial growth factor (VEGF) receptor-2 and regulates VEGF-dependent cellular migration. J Biol Chem 279:22267–22275
29. Le Boeuf F, Houle F, Huot J (2004) Regulation of vascular endothelial growth factor receptor 2-mediated phosphorylation of focal adhesion kinase by heat shock protein 90 and Src kinase activities. J Biol Chem 279:39175–39185
30. Meadows KN, Bryant P, Pumiglia K (2001) Vascular endothelial growth factor induction of the angiogenic phenotype requires Ras activation. J Biol Chem 276:49289–49298
31. McMullen ME, Bryant PW, Glembotski CC et al (2005) Activation of p38 has opposing effects on the proliferation and migration of endothelial cells. J Biol Chem 280:20995–21003
32. Yamaoka-Tojo M, Ushio-Fukai M, Hilenski L et al (2004) IQGAP1, a novel vascular endothelial growth factor receptor binding protein, is involved in reactive oxygen species-dependent endothelial migration and proliferation. Circ Res 95:276–283
33. Kaipainen A, Korhonen J, Mustonen T et al (1995) Expression of the fms-like tyrosine kinase 4 gene becomes restricted to lymphatic endothelium during development. Proc Natl Acad Sci USA 92:3566–3570
34. Cross MJ, Dixelius J, Matsumoto T, Claesson-Welsh L (2003) VEGF-receptor signal transduction. Trends Biochem Sci 28:488–494
35. Dixelius J, Mäkinen T, Wirzenius M et al (2003) Ligand-induced vascular endothelial growth factor receptor-3 (VEGFR-3) heterodimerization with VEGFR-2 in primary lymphatic endothelial cells regulates tyrosine phosphorylation sites. J Biol Chem 278:40973–40979

36. Makinen T, Veikkola T, Mustjoki S et al (2001) Isolated lymphatic endothelial cells transduce growth, survival and migratory signals via the VEGF-C/D receptor VEGFR-3. EMBO J 20:4762–4773
37. Hamada K, Oike Y, Takakura N et al (2000) VEGF-C signaling pathways through VEGFR-2 and VEGFR-3 in vasculoangiogenesis and hematopoiesis. Blood 96:3793–3800
38. Matsumura K, Hirashima M, Ogawa M et al (2003) Modulation of VEGFR-2-mediated endothelial-cell activity by VEGF-C/VEGFR-3. Blood 101:1367–1374
39. Tammela T, Zarkada G, Nurmi H et al (2011) VEGFR-3 controls tip to stalk conversion at vessel fusion sites by reinforcing Notch signalling. Nat Cell Biol 13:1202–1213
40. Zachary I, Morgan RD (2011) Therapeutic angiogenesis for cardiovascular disease: biological context, challenges, prospects. Heart 97:181–189
41. Staton C, Yang Z, Reed M, Brown N (2008) Bevacizumab resistance in breast cancer: are neuropilins the key? Breast Cancer Res 10(suppl 2):P75
42. Xu D, Fuster MM, Lawrence R, Esko JD (2011) Heparan sulfate regulates VEGF165 and VEGF121-mediated vascular hyperpermeability. J Biol Chem 1:9
43. Le Jan S, Hayashi M, Kasza Z et al (2012) Functional overlap between chondroitin and heparan sulfate proteoglycans during VEGF-induced sprouting angiogenesis. Arterioscler Thromb Vasc Biol 32:1255–1263
44. Smith NR, Baker D, James NH et al (2010) Vascular endothelial growth factor receptors VEGFR-2 and VEGFR-3 are localized primarily to the vasculature in human primary solid cancers. Clin Cancer Res 16:3548–3561
45. Yang AD, Camp ER, Fan F et al (2006) Vascular endothelial growth factor receptor-1 activation mediates epithelial to mesenchymal transition in human pancreatic carcinoma cells. Cancer Res 66:46–51
46. André T, Kotelevets L, Vaillant J-C et al (2000) Vegf, vegf-B, vegf-C and their receptors KDR, FLT-1 and FLT-4 during the neoplastic progression of human colonic mucosa. Int J Cancer 86:174–181
47. Vincent L, Jin DK, Karajannis MA et al (2005) Fetal stromal–dependent paracrine and intracrine vascular endothelial growth factor-a/vascular endothelial growth factor receptor-1 signaling promotes proliferation and motility of human primary myeloma cells. Cancer Res 65:3185–3192
48. Graepler F, Verbeek B, Graeter T et al (2005) Combined endostatin/sFlt-1 antiangiogenic gene therapy is highly effective in a rat model of HCC. Hepatology 41:879–886
49. Youssoufian H, Hicklin DJ, Rowinsky EK (2007) Review: monoclonal antibodies to the vascular endothelial growth factor receptor-2 in cancer therapy. Clin Cancer Res 13:5544s–5548s
50. Prewett M, Huber J, Li Y et al (1999) Antivascular endothelial growth factor receptor (fetal liver kinase 1) monoclonal antibody inhibits tumor angiogenesis and growth of several mouse and human tumors. Cancer Res 59:5209–5218
51. Lyden D, Hattori K, Dias S et al (2001) Impaired recruitment of bone-marrow–derived endothelial and hematopoietic precursor cells blocks tumor angiogenesis and growth. Nat Med 7:8
52. Su J-L, Yang P-C, Shih J-Y et al (2006) The VEGF-C/Flt-4 axis promotes invasion and metastasis of cancer cells. Cancer Cell 9:209–223
53. Hoshida T, Isaka N, Hagendoorn J et al (2006) Imaging steps of lymphatic metastasis reveals that vascular endothelial growth factor-C increases metastasis by increasing delivery of cancer cells to lymph nodes: therapeutic implications. Cancer Res 66:8065–8075
54. Ghanem MA, van Steenbrugge GJ, Sudaryo MK et al (2003) Expression and prognostic relevance of vascular endothelial growth factor (VEGF) and its receptor (FLT-1) in nephroblastoma. J Clin Pathol 56:107–113
55. Plate KH, Breier G, Weich HA et al (1994) Vascular endothelial growth factor and glioma angiogenesis: coordinate induction of VEGF receptors, distribution of VEGF protein and possible *In vivo* regulatory mechanisms. Int J Cancer 59:520–529
56. Seto T, Higashiyama M, Funai H et al (2006) Prognostic value of expression of vascular endothelial growth factor and its flt-1 and KDR receptors in stage I non-small-cell lung cancer. Lung Cancer 53:91–96
57. Kaplan RN, Rafii S, Lyden D (2006) Preparing the "soil": the premetastatic niche. Cancer Res 66:11089–11093

58. Hiratsuka S, Nakamura K, Iwai S et al (2002) MMP9 induction by vascular endothelial growth factor receptor-1 is involved in lung-specific metastasis. Cancer 2:289–300
59. Dawson MR, Duda DG, Fukumura D, Jain RK (2009) VEGFR1-activity-independent metastasis formation. Nature 461:E4
60. Mimori K, Fukagawa T, Kosaka Y et al (2008) Hematogenous metastasis in gastric cancer requires isolated tumor cells and expression of vascular endothelial growth factor receptor-1. Clin Cancer Res 14:2609–2616
61. Bellamy WT, Richter L, Sirjani D et al (2001) Vascular endothelial cell growth factor is an autocrine promoter of abnormal localized immature myeloid precursors and leukemia progenitor formation in myelodysplastic syndromes. Blood 97:1427–1434
62. Price DJ, Miralem T, Jiang S et al (2001) Role of vascular endothelial growth factor in the stimulation of cellular invasion and signaling of breast cancer cells. Cell Growth Differ 12:129–135
63. Hamerlik P, Lathia JD, Rasmussen R et al (2012) Autocrine VEGF–VEGFR2–neuropilin-1 signaling promotes glioma stem-like cell viability and tumor growth. J Exp Med 209:507–520
64. Boehm T, Folkman J, Browder T, O'Reilly MS (1997) Antiangiogenic therapy of experimental cancer does not induce acquired drug resistance. Nature 390:404–407
65. Roberts N, Kloos B, Cassella M et al (2006) Inhibition of VEGFR-3 activation with the antagonistic antibody more potently suppresses lymph node and distant metastases than inactivation of VEGFR-2. Cancer Res 66:2650–2657
66. Sallinen H, Anttila M, Grohn O et al (2011) Cotargeting of VEGFR-1 and -3 and angiopoietin receptor Tie2 reduces the growth of solid human ovarian cancer in mice. Cancer Gene Ther 18:100–109
67. Bradley DP, Tessier JJ, Lacey T et al (2009) Examining the acute effects of cediranib (RECENTIN, AZD2171) treatment in tumor models: a dynamic contrast-enhanced MRI study using gadopentate. Magn Reson Imaging 27:377–384
68. Winkler F, Kozin SV, Tong RT et al (2004) Kinetics of vascular normalization by VEGFR2 blockade governs brain tumor response to radiation: role of oxygenation, angiopoietin-1, and matrix metalloproteinases. Cancer Cell 6:553–563
69. Bruce D, Tan PH (2011) Blocking the interaction of vascular endothelial growth factor receptors with their ligands and their effector signaling as a novel therapeutic target for cancer: time for a new look? Expert Opin Investig Drugs 20:1413–1434
70. Demetri GD, van Oosterom AT, Garrett CR et al (2006) Efficacy and safety of sunitinib in patients with advanced gastrointestinal stromal tumour after failure of imatinib: a randomised controlled trial. Lancet 368:1329–1338
71. Motzer RJ, Hutson TE, Tomczak P et al (2007) Sunitinib versus interferon alfa in metastatic renal-cell carcinoma. N Engl J Med 356:115–124
72. Potapova O, Laird AD, Nannini MA et al (2006) Contribution of individual targets to the antitumor efficacy of the multitargeted receptor tyrosine kinase inhibitor SU11248. Mol Cancer Ther 5:1280–1289
73. Connock M, Round J, Bayliss S et al (2010) Sorafenib for the treatment of advanced hepatocellular carcinoma. Health Technol Assess 14(suppl 1):17–21
74. Takahashi O, Komaki R, Smith PD et al (2012) Combined MEK and VEGFR inhibition in orthotopic human lung cancer models results in enhanced inhibition of tumor angiogenesis, growth, and metastasis. Clin Cancer Res 18:1641–1654
75. Hurwitz H, Fehrenbacher L, Novotny W et al (2004) Bevacizumab plus irinotecan, fluorouracil, and leucovorin for metastatic colorectal cancer. N Engl J Med 350:2335–2342
76. Shaheen RM, Ahmad SA, Liu W et al (2001) Inhibited growth of colon cancer carcinomatosis by antibodies to vascular endothelial and epidermal growth factor receptors. Br J Cancer 85:584–589
77. Wang F-Q, Barfield E, Dutta S et al (2009) VEGFR-2 silencing by small interference RNA (siRNA) suppresses LPA-induced epithelial ovarian cancer (EOC) invasion. Gynecol Oncol 115:414–423
78. Eskens FA, Verweij J (2006) The clinical toxicity profile of vascular endothelial growth factor (VEGF) and vascular endothelial growth factor receptor (VEGFR) targeting angiogenesis inhibitors; a review. Eur J Cancer 42:3127–3139

79. Cheng H, Force T (2010) Molecular mechanisms of cardiovascular toxicity of targeted cancer therapeutics. Circ Res 106:21–34
80. Bhisitkul RB (2006) Vascular endothelial growth factor biology: clinical implications for ocular treatments. Br J Ophthalmol 90:1542–1547
81. Asato R, Kita T, Kawahara S et al (2011) Vitreous levels of soluble vascular endothelial growth factor receptor (VEGFR)-1 in eyes with vitreoretinal diseases. Br J Ophthalmol 95:1745–1748
82. Montoro-García S, Lip P-L, Chan C-C, Lip GYH (2011) Soluble vascular endothelial growth factor receptor (VEGFR)-2 in macular oedema—a mechanism for regulating angiogenesis? Br J Ophthalmol 95:757–758
83. Usui T, Ishida S, Yamashiro K et al (2004) VEGF164(165) as the pathological isoform: differential leukocyte and endothelial responses through VEGFR1 and VEGFR2. Invest Ophthalmol Vis Sci 45:368–374
84. Pieramici DJ, Rabena MD (2008) Anti-VEGF therapy: comparison of current and future agents. Eye (Lond) 22:1330–1336
85. Maier P, Unsoeld AS, Junker B et al (2005) Intravitreal injection of specific receptor tyrosine kinase inhibitor PTK787/ZK222 584 improves ischemia-induced retinopathy in mice. Graefes Arch Clin Exp Ophthalmol 243:593–600
86. Keskin U, Totan Y, Karada R et al (2012) Inhibitory effects of SU5416, a selective vascular endothelial growth factor receptor tyrosine kinase inhibitor, on experimental corneal neovascularization. Ophthalmic Res 47:13–18
87. Clinicaltrials.gov. A study of the safety and efficacy of AG-013958 in subjects with subfoveal choroidal neovascularization associated with age-related macular degeneration. Clinicaltrials.gov identifier: NCT00090532 http://clinicaltrials.gov/ct2/show/NCT00090532?term=ag+013958&rank=12011
88. Lee S, Chen TT, Barber CL et al (2007) Autocrine VEGF signaling is required for vascular homeostasis. Cell 130:691–703
89. Du H, Li P, Pan Y et al (2010) Vascular endothelial growth factor signaling implicated in neuroprotective effects of placental growth factor in an in vitro ischemic model. Brain Res 1357:1–8
90. Wu H, Jiang H, Lu D et al (2011) Induction of angiogenesis and modulation of vascular endothelial growth factor receptor-2 by simvastatin after traumatic brain injury. Neurosurgery 68:1363–1371; discussion 1371
91. Bergers G, Hanahan D (2008) Modes of resistance to anti-angiogenic therapy. Nat Rev Cancer 8:592–603
92. Barakat MR, Kaiser P (2009) VEGF inhibitors for the treatment of neovascular age-related macular degeneration. Expert Opin Investig Drugs 18:637–646
93. USNIH (2010) ClinicalTrials.gov. http://www.clinicaltrials.gov/ct2/home. Accessed 20 May 2011

Chapter 2
Role of Transforming Growth Factor Beta in Angiogenesis

Alicia Viloria-Petit, Amy Richard, Sonja Zours, Mai Jarad, and Brenda L. Coomber

Abstract Transforming growth factor beta (TGFβ) is a pleiotropic factor that plays pivotal roles in both vasculogenesis and angiogenesis and thus is indispensable for development and homeostasis of the vascular system. TGFβ drives vascular responses via its binding to a TGFβ receptor complex formed by type I and type II receptors, as well as type III co-receptors present on both endothelial and mural cells. Signaling by these receptors is context-dependent and tightly regulated, particularly on cultured endothelial cells, where TGFβ can either promote or suppress endothelial migration, proliferation, permeability, and sprouting. These, together with evidence obtained from knockout animals for different TGFβ receptor types, and genetic studies in humans linking mutations in TGFβ signaling components to cardiovascular syndromes, suggest that TGFβ is a central mediator of angiogenesis, where it may play contrasting roles depending on the stage of the process. This review presents an overview of knowledge accumulated to date on TGFβ's role in angiogenesis as well as vascular biology and vascular disease and discusses potential applications of this knowledge to the treatment of angiogenesis-dependent diseases such as cancer.

Keywords TGFβ • ALK • Endoglin • Smad • Endothelial cell • Pericyte • Angiogenesis • HHT • Fibrosis • Cancer

---

A. Viloria-Petit (✉) • A. Richard • S. Zours • M. Jarad • B.L. Coomber (✉)
Department of Biomedical Sciences, Ontario Veterinary College,
University of Guelph, Guelph, ON, Canada N1G 2W1
e-mail: aviloria@uoguelph.ca; bcoomber@uoguelph.ca

J.L. Mehta and N.S. Dhalla (eds.), *Biochemical Basis and Therapeutic Implications of Angiogenesis*, Advances in Biochemistry in Health and Disease 6,
DOI 10.1007/978-1-4614-5857-9_2,

2.1 TGFβ Molecule Family: Sources, Activation, and Regulation of Transcription

2.1.1 The TGFβ Family

The transforming growth factor beta (TGFβ) superfamily consists of 33 members, most of which are dimeric, secreted polypeptides that regulate proliferation, survival/apoptosis, migration, adhesion, invasiveness, and self-renewal properties in responsive cells [1, 2]. Depending on the cell and tissue type, modulation of these cellular properties by TGFβ superfamily members will regulate different processes ranging from gastrulation to formation of a functional vascular system during embryonic development, as well as organ morphogenesis and homeostasis at various postnatal stages [1].

The TGFβ superfamily is conserved through metazoan evolution and includes TGFβs (1–3), bone morphogenic proteins (BMPs 1–20), growth and differentiation factors (GDFs including myostatin), activins (A and B), inhibins (A and B), nodal, leftys (1 and 2), and Mullerian inhibiting substances (MIS) [2]. The TGFβ family, the focus of this review, comprises three different isoforms encoded by separate genes: TGFβ1, 2, and 3. Human TGFβ2 and TGFβ3 have a 70 % homology with TGFβ1 [3]. TGFβ1 is the predominant and most ubiquitous isoform, while the other two are expressed in a more limited spectrum of cells and tissues. The three isoforms have overlapping functions in vitro; however, mice deficient in individual isoforms show nonoverlapping phenotypes suggesting that each TGFβ isoform has distinct functions in vivo [4].

2.1.2 TGFβ Sources

TGFβ ligands are not specific to a particular cell type, as they have been shown to be synthesized by a variety of normal and malignant cells. In addition, almost every cell in the body expresses TGFβ receptors (TGFβRs) and so is capable of responding to TGFβ [5]. TGFβ was initially isolated from platelets, due to their high content of the ligand. However, bone cells, particularly osteoblasts, are the highest producers of TGFβ currently known. Strong intracellular TGFβ staining has also been reported in adrenal cortex, megakaryocytes, cardiac myocytes, chondrocytes, renal distal tubules, ovarian glandular cells, and chorionic cells of mouse placenta, among others (reviewed in [6]). In carcinomas, as well as in sites of wound healing, TGFβ is expressed by epithelial cells, associated fibroblasts and myofibroblasts, and infiltrating immune cells, such as macrophages and T lymphocytes [7].

2.1.3 Synthesis and Activation of TGFβ

All TGFβ ligands are synthesized as precursor polypeptides, containing a longer 25 kDa N-terminal pro-peptide, followed by a C-terminal, 12.5 kDa mature polypeptide.

Two of these precursors form a dimer via disulfide bonds. Pro-peptide and mature peptide are cleaved by furin-like proteases while trafficking via the exocytic pathway, but remain associated by the disulfide bonds. Once cleaved, the pro-peptide becomes the "latency associated peptide" (LAP), which acts as a chaperone during exocytosis of the complex. LAP also aids in TGFβ deposition into the extracellular matrix (ECM) and keeps TGFβ inactive within its core once the complex is secreted [1]. LAP-TGFβ is known as the small latent complex (SLC) that often exists in association with the latent TGFβ binding protein (LTBP), which, together with the SLC, forms the large latent complex (LLC) [5]. LAP's direct interaction with LTBP as well as with ECM components such as fibronectin and fibrillin, among others, mediates TGFβ's deposition into the ECM [1]. Cleavage-dependent activation of the mature C-terminal dimeric TGFβ ligand from its ECM-deposited form is mediated by a number of proteases including thrombospondin 1 (TSP-1), plasmin, cathepsin D, matrix metalloproteinase (MMP) 2 and 9, calpain, chymase, elastase, endoglycosidase F, and kallikrein. In addition, acidic environment, reactive oxygen species (ROS), heat, and sheer stress have also been shown to activate TGFβ (reviewed in [2, 5, 8]). However, in many physiological situations, integrins have been shown to be the critical players in TGFβ activation. An RGD sequence present in TGFβ's LAP mediates its binding to all αv integrins, and αvβ3, αvβ5, αvβ6, and αvβ8 have all been shown to release active TGFβ via both proteolysis-independent and proteolysis-dependent mechanisms [5].

2.1.4 Regulation of Transcription by TGFβ

Once activated TGFβ initiates signaling by inducing the activity of specific serine/threonine kinase type I and type II receptor heterotetrameric complexes. These in turn phosphorylate specific effector proteins called Smads, which translocate to the nucleus and modulate transcription of targets genes (discussed in more detail below). Nuclear Smad complexes bind to chromatin and, together with other transcription factors, regulate gene expression. A list of Smad target genes has been published elsewhere [1]. Among these, the inhibitors of differentiation (Id) family of transcription factors, vascular endothelial growth factor (VEGF), and thrombospondin-1 (TSP-1) are important modulators of angiogenesis. TGFβ also signals in a noncanonical manner to modulate the level and function of effector proteins in the absence of changes in gene transcription [9]. Misregulation of TGFβ signaling plays roles in a number of pathologies, including autoimmune and cardiovascular disorders, and cancer [10]. Cardiovascular disorders resulting from abnormal TGFβ signaling include hereditary hemorrhagic telangiectasia (HHT), cardiac remodeling/fibrosis, and pulmonary arterial hypertension, among others [11].

2.1.5 TGFβ's Role in Angiogenesis

Genetic studies in mouse and human have provided evidence for the importance of components of the TGFβ signaling pathway in vascular morphogenesis, including formation of the primitive vascular plexus, and the recruitment of pericytes/smooth muscle cells necessary for vessel wall integrity [11]. Deletion of TGFβ1 in the mouse results in embryonic lethality because of defective yolk sac vasculogenesis. Targeted deletion of ALK1, ALK5, TGβRII, and endoglin results in similar phenotypes. All of these knockout embryos die during mid-gestation due to hyper-dilated, impaired, leaky vessels [12]. These vascular abnormalities are similar to those described in patients with HHT [11]. Endothelial and smooth muscle cell-specific targeting of TGFβRII and ALK5 suggests that TGFβ signaling in both compartments is required for proper vessel development, but likely at different stages [11, 13].

In the next sections, we present a detailed overview of current knowledge on TGFβ signaling in endothelial and associated vascular cells, such as pericytes and smooth muscle cells, and the role of this signaling at the various stages of the angiogenesis process. We also provide some evidence of differential TGFβ signaling in physiologic vs. pathologic angiogenesis and discuss potential applicability in therapeutic intervention.

2.2 TGFβ Receptors and Signaling

TGFβ members signal through type I and type II serine/threonine kinase receptors. There are seven members of the type I receptor family, also known as activin receptor-like kinases (ALK) 1–7, and five members of the type II receptor family (TGFβRII, BMPRII, ActRIIA, ActRIIB, and MISRII) [14]. TGFβ also signals via accessory, type III, TGFβ receptors: endoglin and betaglycan (discussed in a later section). Reflective of its role in signaling a multitude of cell types, there are relatively few studies devoted specifically to TGFβ signaling in vascular endothelium. In the sections below, except where explicitly stated, the signaling events and outcomes described have not yet been validated in endothelial cells.

In most cell types TGFβ 1–3 isoforms signal through an ALK5-TGFβRII complex; however, endothelial cells also express and signal through an ALK1-TGFβRII complex [15]. The balance between activation of these two signaling pathways regulates endothelial cell functions such as proliferation and migration, and this balance is believed to regulate the switch of endothelial cells from quiescent mature vessels into activated angiogenic sprouts. In fact, genetic mutants of TGFβ receptors, ALK5 and endoglin, inhibit angiogenesis in vitro and result in embryonic lethality in mice due to vascular defects [12].

In the absence of ligand, type I and II receptors form homodimers with themselves, which upon ligand binding complex with each other to form a heterotetramer [16]. Formation of this tetrameric complex brings together the constitutively active type II receptor with the type I receptor, resulting in auto- and transphosphorylation at various serine residues in the receptors' GS domain. Once activated by these phosphorylation events, the TGFβ receptors become functional serine/threonine kinases and subsequently phosphorylate and activate several intracellular signaling molecules [17].

Although TGFβ receptors are classically referred to as serine/threonine kinases, upon ligand binding TGFβRII also becomes autophosphorylated at multiple tyrosine residues [18]. These phosphorylated tyrosines then act as docking sites for various Src homology 2 (SH2) and phosphotyrosine-binding (PTB) domain containing scaffolding/adaptor molecules, such as Shc, and growth factor-binding protein 2 (Grb2) [19]. These adaptor molecules function as scaffolding proteins, bringing together the TGFβ receptor's tyrosine kinase function with various protein substrates [20].

TGFβ receptors, through their action as both serine/threonine and tyrosine kinases, are able to activate several intracellular signaling cascades, including the canonical Smad signaling pathways, as well as several noncanonical signaling pathways, such as PI3K/Akt, RhoA-dependent, and JNK, p38, and Erk MAPK pathways.

2.2.1 Canonical Smad Signaling

The Smad family of proteins is composed of three classes: receptor Smads (R-Smads 1, 2, 3, 5, and 8), inhibitory Smads (I-Smads 6 and 7), and the common Smad (Co-Smad4) [14]. R-Smads are recruited to TGFβRI following receptor activation and interact indirectly via auxiliary proteins such as Smad anchor for receptor activation (SARA) [21]. Once recruited R-Smads become phosphorylated by TGFβRI in their C-terminal SSXS domain; R-Smads 1, 5, and 8 are activated by ALKs 1–3 and 6, whereas R-Smads 2–3 are activated by ALKs 4, 5, and 7 [22].

Once activated R-Smads dissociate from TGFβRI and form heterodimers with Co-Smad4 or heterotrimers containing two R-Smads and one Smad4, which translocate to the nucleus [23]. This translocation is guided by nuclear localization sequences (NLSs) in the MH1 domains of Smad3 and 4, which mediate their interaction with importin proteins β1 and α, respectively [24, 25]. In the nucleus these Smad complexes function as transcription factors and regulate transcription through their direct interaction with DNA containing Smad-binding elements (SBEs), as well as corepressors, co-activators (CBP and p300), and other transcription factors [26]. The diversity of these transcriptional complexes directs the tissue and dose-dependent regulation of transcription by the Smad proteins.

TGFβ stimulation affects the transcription of several hundred genes [27]. Targets of R-Smad and Co-Smad transcriptional regulation include proteins involved in regulating cellular proliferation, apoptosis, and the epithelial-to-mesenchymal transition (EMT). Additionally, TGFβ signaling results in expression of I-Smads 6 and 7, whose promoters contain SBEs [28]. The I-Smads then establish a negative feedback circuit on TGFβ signaling through their ability to negatively regulate signaling pathway activation on multiple levels, with Smad7 expressed in response to and antagonizing all TGFβ signaling and Smad6 expressed specifically in response to and antagonizing the Smad1, 5, and 8 signaling pathways [29].

I-Smads contain various functional domains that enable their inhibitory function. Through their MH2 domain, Smads 6 and 7 are able to compete with R-Smads for TGFβRI binding, thus inhibiting R-Smad phosphorylation and subsequent Co-Smad4 complex formation [30]. I-Smads are also capable of recruiting E3 ubiquitin ligases, Smurf1 and Smurf2, to activated TGFβRI leading to its polyubiquitination and subsequent proteasomal degradation [31]. Smad7 is also capable of interfering with TGFβ signaling at the level of receptor activation via its ability to recruit the phosphatase GADD34-PP1c to the activated receptor complex [32].

In addition to their functioning in the cytoplasm, at the level of receptor and R-Smads inhibition, I-Smads also function in the nucleus at the level of transcriptional repression. Through its MH2 domain, Smad7 is able to bind directly to DNA to prevent Smad2, 3, and 4 binding [26]. Also, once bound to DNA, I-Smads recruit histone deacetylases (HDACs) to the promoter regions of Smad target genes, leading to chromatin compaction and transcriptional inhibition [33].

2.2.2 Noncanonical Signaling Pathways

2.2.2.1 MAPK

TGFβ signaling can also lead to activation of MAPK signaling pathways, including Erk, p38, and JNK MAPK signaling. This activation is likely independent of Smad-dependent transcription, due to the rapid onset of MAPK phosphorylation (5–15 min) [34], and the ability of cells genetically deficient in Smad activation to maintain their ability to activate MAPK signaling in response to TGFβ [35]. Erk MAPK becomes activated through a receptor tyrosine kinase (RTK)/Ras/Erk pathway. Following TGFβ ligand binding to TGFβRII, type I and II receptors become phosphorylated on three tyrosine residues, Y259, Y336, and Y424, in the receptors' cytoplasmic domain [18]. These phosphorylated tyrosine residues are then bound by SH2 and PTB domain containing adaptor molecules. Grb2 is an SH2 domain-containing protein that complexes with SOS in the cytoplasm and upon RTK phosphorylation is recruited to the receptor. Once localized to the RTK the Grb2/SOS complex is able to activate membrane-localized Ras, bridging TGFβ receptor activation with the MAPK signaling pathway. In its activated, GTP-bound state, Ras is able to phosphorylate and activate the Raf-MEK-Erk MAPK cascade [19, 20].

Erk, through its functioning as a serine/threonine kinase, not only regulates intracellular mitogen signaling in the cytoplasm but also translocates to the nucleus where it regulates the activity of various transcriptional regulators [36]. Through its modulation of gene transcription, Erk mediates TGFβ-induced disassembly of adherens junctions and enhanced migration, two key events in TGFβ-induced EMT [34]. As such, Erk activation is required, however insufficient (since it must cooperate with Smad signaling), for TGFβ-induced EMT [37].

JNK and p38 MAPK signaling pathways are also activated in response to TGFβ signaling. This activation is dependent on the scaffolding protein TRAF6, which associates with activated TGFβRII through its C-terminal TRAF domain. TGFβRII-bound TRAF6 undergoes auto-polyubiquitination, leading to its association with the MAP3K, TAK1 [38]. TAK1 is required for activation of both the JNK and p38 MAPK pathways via activation of MKK4-JNK and MKK3/6-p38 cascades [39]. In fact, TAK1 is indispensable for JNK and p38 MAPK activation, and embryos deficient in TAK1 suffer from vascular defects whose phenotype is similar to ALK1 and endoglin mutants [40]. The activation of TGFβ-TAK1-JNK/p38 MAPK pathways is independent of Smad-mediated transcription; however, these signaling pathways cooperate with Smad signaling in order to regulate TGFβ-induced cellular functions such as apoptosis [41] and EMT [42]. Recent studies revealed that different isoforms of p38 MAPK are responsible for the differential effects of VEGF and TGFβ on endothelial cells. In particular, TGFβ is able to shift VEGF signaling from pro-survival to pro-apoptotic isoforms of p38. Thus in the absence of TGFβ, VEGF supports endothelial proliferation, but when TGFβ is also present, endothelial cell death can occur [43].

2.2.2.2 Rho-GTPase

TGFβ also rapidly activates RhoA signaling in a Smad-independent manner [44]. However, TGFβ signaling has also been shown to lead to a downregulation of RhoA protein in response to TGFβ activation of the Par6 polarity pathway [45]. Par6 is a scaffolding protein that complexes with TGFβRI at tight junctions. Following ligand binding to TGFβRII, it travels to the tight junction where it complexes with TGFβRI and phosphorylates Par6 at serine 345. Activated Par6 recruits the E3 ubiquitin ligase, Smurf1, to tight junctions where it ubiquitinates and targets RhoA for degradation, leading to localized RhoA downregulation at tight junctions [45]. This results in a loss of polarity and enhanced cellular motility. This localized degradation is responsible for the dissolution of tight junctions, reorganization of the actin cytoskeleton, and extension of filopodia [46], all of which are essential for EMT [45]. The potential role of the Par6 pathway in vascular biology has been recently highlighted by in vitro studies on endothelial-mesenchymal transition (EndoMT), a process induced by TGFβ (discussed in more detail in later sections), which is essential for heart valve formation in the developing embryo. It was observed that blockade of Par6 activation abrogated EndoMT in response to TGFβ2, and this was dependent on the presence of both ALK-5 and type 3 TGFβ receptor betaglycan [47]. We have recently demonstrated Par6 activation in response to TGFβ1 in bovine aortic endothelial cells, particularly at low (0.5 ng/mL and lower) TGFβ concentrations (Viloria-Petit, unpublished observations). Since low TGFβ concentrations have been previously observed to be pro-angiogenic [48], our results suggest that Par6 activation might mediate angiogenesis in response to TGFβ.

2.2.2.3 PI3K (AKT)

The Akt signaling pathway is also activated downstream of TGFβ through TGFβRI phosphorylation of PI3K, an upstream kinase of Akt. PI3K interacts with TGFβRII independent of receptor activation and upon ligand stimulation is brought in contact with TGFβRI, where it is phosphorylated [49]. Downstream activation of the Akt signaling pathway is required for TGFβ-induced EMT and does so by two proposed mechanisms: first, through its ability to mediate TGFβ-induced actin filament reorganization and enhanced cellular migration and secondly, through Akt's activation of downstream mTOR [50]. The mTOR signaling pathway regulates cellular translation levels, and Akt-mTOR activation is believed to facilitate Smad-mediated transcriptional programs.

2.3 TGFβ and Endothelial Sprouting, Proliferation, and Permeability

Extensive evidence suggests that TGFβ plays a role during the activation phase of angiogenic sprouting by promoting vascular permeability, proliferation, and migration of endothelial cells. TGFβ also mediates the reverse events that occur during the resolution phase of angiogenesis, including inhibition of endothelial cell migration and proliferation and decreased permeability, which are necessary for vessel stabilization [51].

2.3.1 Endothelial Sprouting

Endothelial sprouting involves two distinct endothelial cell phenotypes: the tip cells, which lead the newly forming vessel sprout, and the stalk cells, which proliferate and form the lumen of the new vessel [52]. These cells are initially part of a mature vessel. An increase in permeability and migratory characteristics allows these cells to delaminate from the endothelium and become involved in the newly forming vessel sprout. Proliferation must be suppressed in the tip cells and enhanced in the stalk cells to ensure their respective functions. Finally, when the new vessel is in place, there must be a reversion back to characteristics of cells in a quiescent endothelium, which includes a decrease in permeability, proliferation, and migratory capacity [52].

One of the key regulators of sprouting during angiogenesis is VEGF-A since tip cell migration largely depends on the gradient of VEGF-A, which binds to VEGF receptor 2 on endothelial cells [53]. Studies in mice and zebra fish have contributed to our understanding of the role of VEGF and Notch signaling in vessel sprouting. VEGF binding to VEGFR2 on a tip cell activates VEGFR signaling leading to increased expression of the Notch ligand Dll4, which in turn binds to

Notch1 receptor on adjacent stalk cells. Notch signaling in the latter reduces VEGR2 and VEGR3 expression, making cells insensitive to VEGF stimulation, thereby suppressing the tip cell phenotype [54].

TGFβ effects on in vitro endothelial cell sprouting are variable, including induction, repression, or no effect depending on the concentration of TGFβ, the type of endothelial cells employed, and the source of signaling activation, i.e. whether constitutively activated receptors or exogenously added ligand was used [55]. The nature of the angiogenic response to TGFβ depends on the balance of ALK1 vs. ALK5 signaling input, with ALK1 predominantly promoting sprouting and ALK5 favoring the resolution/stabilization phase of angiogenesis [55]. The inhibitory effect of an ALK1 antibody on endothelial cell sprouting in vitro and on angiogenesis in two different tumor models supports this concept [56]. However, recent studies on developmental angiogenesis in mice suggest that, rather than promoting sprouting, ALK1 signaling cooperates with Notch signaling to repress VEGF responsiveness, tip cell formation, and sprouting [57]. Whether these discrepancies represent differences in developmental vs. pathologic angiogenesis remains to be determined. Interestingly, we have recently found TGFβ to decrease endothelial VEGFR2 [58] expression via ALK5 signaling. Thus, ALK5 signaling may potentially contribute to endothelial cell insensitivity to VEGF stimulation that might be necessary for both maintenance of the stalk cell fate during sprouting and the resolution phase of angiogenesis.

Finally, EndoMT has been hypothesized to be one of the mechanisms mediating angiogenic sprouting in response to TGFβ. EndoMT is the process whereby cells from a quiescent, stable endothelium delaminate from this cell layer and take on a fibroblastoid phenotype. During this process, endothelial cells experience loss of adherens and tight junctions and their associated markers including vascular endothelial (VE)-cadherin, zonula occludens (ZO)-1, and claudin-5. The cells' transition towards a mesenchymal phenotype is associated with the gain of mesenchymal markers such as α-smooth muscle actin and fibroblast-specific protein-1, as well as motility [59, 60]. As previously mentioned, EndoMT mediates cardiac development and is also responsible for pathologic tissue fibrosis; however, its role in promoting angiogenesis is still unknown [60]. It is possible that a partial, reversible form of EndoMT facilitates angiogenic sprouting since loss of cell–cell junctions is required for endothelial cells to delaminate from the existing vessel. An EndoMT of the tip cells could similarly promote the invasive/migratory characteristics that are necessary for them to guide the vessel sprout.

2.3.2 *Endothelial Permeability*

Signaling via the ALK5 TGFβ receptor has been shown to promote and inhibit vascular permeability, depending on cell context. TGFβ can induce permeability in pulmonary endothelial cell monolayers, which is attenuated by treatment with SB-431542, an ALK5 kinase inhibitor [61]. Specifically, SB-431542 upregulates the expression of the endothelial-specific tight junction component, claudin-5 [62]. In contrast, in vivo blockade of TGFβ signaling in mouse retinal endothelial cells leads to increased permeability and decreases vessel barrier function. Both in vivo and in vitro analyses demonstrated that TGFβ signaling blockade resulted in increased endothelial permeability characterized by decreased interaction between the tight junction proteins occludin and ZO-1 [63].

As mentioned above, TGFβ-mediated EndoMT may contribute to an increase in endothelial permeability that is necessary for angiogenic sprouting. During EndoMT, there is a decrease in expression of the adherens junction protein VE-cadherin, as well as tight junction proteins ZO-1 and claudin-5. TGFβ has been shown to downregulate claudin-5 at the transcriptional level, and VE-cadherin has been observed to upregulate expression of claudin-5 [64, 65]. Thus, TGFβ-mediated downregulation of VE-cadherin during EndoMT can indirectly decrease expression of claudin-5, resulting in the loss of both adherens junctions and tight junctions with a concomitant increase in endothelial permeability.

Along with TGFβ, VEGF has also been shown to be an important mediator of endothelial permeability during angiogenesis [66]. VEGF has a demonstrated function in modulating VE-cadherin at the adherens junctions through tyrosine phosphorylation, which leads to an increase in permeability [67]. Since TGFβ induces VEGF expression in vascular endothelial cells, this relationship may provide an alternative mechanism whereby TGFβ can modulate VE-cadherin expression and therefore increase endothelial cell permeability [68]. TGFβ's ability to downregulate VEGF receptor 2 expression can also provide an additional means by which TGFβ can regulate and perhaps reverse vascular permeability during the resolution phase of angiogenesis.

2.3.3 Endothelial Proliferation and Migration

TGFβ can enhance cell proliferation at low doses and suppress proliferation at high doses. The presence of both type I receptors ALK1 and ALK5 may provide a means by which TGFβ's dual role in proliferation is regulated [69]. Activation of ALK1 has been primarily shown to stimulate proliferation and migration of endothelial cells during the activation phase of angiogenesis [69, 70]. The downstream effector of ALK1 responsible for this process is Id1, an inhibitor of differentiation that is required for proliferation and migration. When ALK1 is active, both endothelial cells and fibroblasts are induced to express Id1 [69, 71]. It is interesting to note that ALK1 in combination with ALK5 is a potential negative regulator of endothelial cell migration and proliferation, which may play a role in the resolution phase of angiogenesis [72–74]. Possible mediators of the inhibitory effects of ALK1 signaling are JNK and ERK, but the process is not fully characterized in endothelial cells [72].

In contrast to ALK1, ALK5 seems to have more defined anti-proliferative roles during both the activation and the resolution phases of angiogenesis [69]. It is believed that activated Smad2/3 proteins cooperate with nuclear corepressors to repress the transcription of c-myc and cyclin-dependent kinase (CDK) genes and with nuclear co-activators to activate transcription of p15 and p21, two major inhibitors of the cell cycle, collectively inhibiting proliferation [75, 76]. ALK5 has been specifically shown to prevent proliferation and migration in endothelial cell spheroid assays and embryonic stem cell-derived endothelial cells, whereas the ALK5 kinase inhibitor, SB-431542, has opposite effects [62, 77]. Furthermore, in vitro studies have found that ALK5-induced blood vessel maturation is mediated by the induction of plasminogen activator inhibitor (PAI)-1 in endothelial cells. PAI-1 prevents degradation of the provisional ECM that surrounds the nascent vessel, hence promoting vessel maturation during the resolution phase [69]. Thus, ALK5 likely plays roles in both inhibiting proliferation of the tip cells during activation and in both tip and stalk cells during resolution of angiogenesis. While a balance between ALK1 and ALK5 may be important to mediate the effects of TGFβ on the endothelium, their actions are not mutually exclusive, and they may serve as regulators of one another.

The variation in roles played by ALK1 and ALK5 as well as the balance between these two type I TGFβ receptors is likely dependent on cellular context, with the cross talk between them providing a mechanism whereby TGFβ can strategically regulate proliferation of the tip and stalk cells during angiogenesis. It should also be noted that VEGF has also been shown to play a role in proliferation of endothelial cells during angiogenesis which is dependent on its concentration [53]. Since VEGF is a positive regulator of proliferation and TGFβ has been shown to be an inducer of VEGF expression in endothelial cells, this interaction provides yet another regulatory mechanism for TGFβ to control proliferation [53, 68].

2.4 TGFβ Co-receptors in Angiogenesis

The human type III TGFβ co-receptors endoglin and betaglycan are type I integral membrane glycoproteins [78, 79]. Betaglycan is universally expressed on nearly all cell types and is the most highly expressed of the TGFβ superfamily receptors [80]. However, the expression of betaglycan in some cell types, specifically vascular endothelial cells with the exception of those forming the endocardium [81], appears to be weak or absent, and instead, endothelial cells predominantly express the related TGFβ co-receptor, endoglin [80]. Both endoglin and betaglycan are generally expressed on the cell surface as homodimers, with endoglin homodimers being linked by disulfide bridges; however, endoglin and betaglycan are capable of forming heteromeric complexes in microvascular endothelial cells [79, 80]. Both type III co-receptors also exist as soluble forms. Betaglycan shedding is mediated in part by membrane-type matrix metalloproteinase1 (MT1-MMP) and plasmin [82], while soluble endoglin is produced by cleavage of the membrane-bound endoglin at close proximity to the transmembrane domain by matrix metalloproteinase 14 (MMP14) [83].

Endoglin expression is potently stimulated by hypoxia, BMP9, and TGFβ via ALK1, while TNFα exerts an inhibitory effect on endoglin expression in endothelial cells [3, 84, 85]. Both betaglycan and endoglin cytoplasmic domains can be phosphorylated by serine/threonine kinases [86, 87]. ALK5 is responsible for the phosphorylation of endoglin's cytoplasmic tail, which has been shown to be necessary for the activation of TGFβ-dependent ALK1 signaling [88]. Thus, ALK5 is indirectly responsible for ALK1 activation via endoglin, which in turn is necessary for endothelial cell proliferation. The phosphorylation of endoglin has been shown to influence its subcellular localization, probably by modulating its interaction with adhesive proteins such as zyxin and zyxin-related protein 1 (ZRP-1), hence modifying the adhesive properties of endoglin-expressing cells [86].

It is not fully understood how endoglin regulates TGFβ-dependent responses. Endothelial cells that lack endoglin experience decreased proliferation due to diminished ALK1 activity and increased ALK5 activity [89]. The increase in ALK5 and subsequent TGFβ-induced growth inhibition, even at low concentrations of TGFβ, which normally promote proliferation [69], may also be due in part to decreased inhibition of ALK5 by ALK1 [89]. ALK1 has been shown to interrupt ALK5 signaling, likely acting downstream of Smad2/3 phosphorylation [70]. Thus ALK1 may be involved in a negative regulatory mechanism that is able to mediate the anti-proliferative effects of ALK5 in endothelial cells. However, endoglin association with TβRII results in alteration of its phosphorylated status, thus ensuing loss of ALK5 from the TGFβ receptor complex, possibly explaining endoglin's inhibitory effect on ALK5 signaling [90]. Furthermore, studies conducted on human umbilical vein endothelial cells demonstrate that ALK1-dependent inhibition of cell adhesion is counteracted by endoglin phosphorylation [90, 91]. These results suggest that endoglin interaction with TGFβ signaling receptors via both its extracellular and cytoplasmic domains might affect TGFβ cell responses.

2.4.1 Regulation of TGFβ Ligand Access to Co-receptors

Betaglycan binds multiple members of the TGFβ family, including TGFβ1, TGFβ2, TGFβ3, activin A, BMP2, BMP4, and BMP7 [92–94]. Betaglycan also plays a role in presenting the ligand to TβRII, leading to either enhanced or inhibited signaling [95]. Unlike betaglycan, endoglin binds TGFβ1 and TGFβ3 but not TGFβ2 [96]. Other endoglin ligands include activins and BMPs, and endoglin can also interact with activin type II receptors [97]. Therefore, functional differences and similarities found between betaglycan and endoglin could be due to differences between these two proteins' ligand binding profiles.

In the case of the type III co-receptor betaglycan, its function as a co-receptor to specific members of the TGFβ superfamily is carried out through its ectodomain, which consists of two independent ligand-binding domains. The residual carboxy-terminal half of the protein is necessary for protein anchoring to the cell membrane [94]. Comparative studies between endoglin and betaglycan intracellular responses to TGFβ signaling found a distinctive role for the extracellular domains [98]. Exchanging the extracellular domain between these two co-receptors did not alter endoglin ligand binding potential; however, in contrast to betaglycan, TβRII is essential for endoglin binding of TGFβ1, activin A, BMP2, and BMP7 [97, 98]. The soluble form of endoglin reduced binding of TGFβ1 by interfering with its interaction with TβRII, and soluble endoglin suppressed TGFβ1 signaling in endothelial cells [99].

As previously mentioned, TGFβ can signal in endothelial cells through either ALK1 or ALK5, resulting in the stimulation of endothelial cell proliferation and migration (ALK1) or inhibition of these responses (ALK5) [89]. Forced expression of endoglin led to inhibition of TGFβ/ALK5 signaling, and subsequent blockade of TGFβ induced growth inhibitory effect on endothelial cells [89, 98, 100, 101]. Moreover, endoglin can block apoptosis in response to hypoxia and TGFβ. When endoglin-expressing and endoglin-deficient endothelial cells were both exposed to TGFβ1 under hypoxic stress, the presence of endoglin was sufficient to block the synergistic pro-apoptotic effect of TGFβ1 and hypoxia [84]. Additionally, in endothelial cells the endoglin cytoplasmic tail interacts with β-arrestin2, leading to endoglin-mediated inhibitory effects on TGFβ-induced ERK activation and migration [102]. Finally, endoglin is able to inhibit cell migration through its interaction with LIM domain containing focal adhesion proteins such as zyxin, possible in a TGFβ-independent fashion [103].

During embryogenesis, inflammation, and wound healing, modifications in vascular structure occur, and endoglin expression is elevated during these modifications [104, 105]. The importance of endoglin function in maintaining normal vascular structure is underlined by the relationship between mutations in the endoglin gene and HHT, which is a disorder characterized by the formation of small dilated blood vessels and arteriovenous malformations (AVMs) in the vasculature of lung, liver, and brain [106, 107]. Studies done to elucidate the role endoglin plays in the enhancement of the TGFβ/ALK1 signaling pathway suggest that endothelial cell response to TGFβ is critically dependent on endoglin functional association with ALK1 [90]. The results from these studies agree with what is seen in cases of HHT where the predominant mutations are in either human endoglin (ENG) or ALK1 (ACVRL1) genes [108, 109].

2.5 TGFβ and Vascular Mural Cells

The structure of microvessels varies between different tissue beds, and one of the major alterations is in the nature and prevalence of mural cells. Pericytes are found in capillaries, venules, and small arterioles, while true vascular smooth muscle cells are associated with larger arterioles and the macrocirculation [110]. In addition, there are significant differences in pericyte coverage and phenotype between vascular beds [111], and the ratio of pericytes to endothelium can vary from an almost 1:2 ratio in retina [112] to less than one pericyte for every ten endothelial cells. Pericytes can also be additionally specialized for tissue-specific vascular function, becoming glomerular mesangial cells (kidney) or Ito/stellate cells (liver), for example [110].

Mural cells play significant roles in the stabilization, functionality, and phenotype determination of the microcirculation, and recruitment of these cells is an essential part of the so-called resolution stage of sprouting angiogenesis [113]. During development, platelet-derived growth factors (PDGFs) act as potent chemoattractants for mural cell precursors and are produced by endothelial cells during vasculogenesis in the embryo and during sprouting angiogenesis in adult tissue. There is evidence from in vitro studies that PDGF-B can induce TGFβ production via the MAPK/ERK pathway and angiopoietin 1 (Ang-1) production via PKC and PI3K pathways during vascular smooth muscle differentiation of 10T1/2 cells [114]. Furthermore, TGFβ can downregulate this PDGF-B induction of Ang-1, and both TGFβ and Ang-1 synergistically reduce PDGF production by vascular endothelial cells, suggesting that cross talk between endothelium and mural cell precursors is essential for maturation of the microvascular bed [114]. There is also evidence that monocyte chemoattractant protein 1 (MCP-1) is also chemoattractant for vascular smooth muscle and 10T1/2 cells [115]. MCP-1 is upregulated in ischemic regions of brain associated with endoglin positive microcirculation and in human brain microvessel endothelial cells exposed to ischemia in vitro, highlighting the potential role of TGFβ-mediated pathways in angiogenic recovery of reperfused brain after stroke [116].

Culture of 10T1/2 cells with vascular endothelial cells leads to activation of latent TGFβ (similar to what is seen with EC/smooth muscle cell coculture [117]) and subsequent TGFβ-driven 10T1/2 cell differentiation into pericyte-like cells [118]. Endothelial-mural cell precursor contact is required for this TGFβ activation, and coculture of endothelial cells with mesenchymal precursors from mutant mouse embryos demonstrates that cell coupling via gap junction protein connexin 43 is essential for this activity [119]. Coculture of endothelial cells and 10T1/2 cells enhances the survival of both cell types; ECs require active ALK5 signaling for this, while 10T1/2 cells in coculture employ other pathways for survival [63]. This TGFβ-mediated reciprocal interaction between vascular components is relevant for both vascular functionality (such as permeability/barrier function) and neural retinal cell survival in adult mice [63]. Proper pericyte/endothelial cell interactions are also essential for maintaining blood–brain barrier characteristics in cerebral vessels. This is mediated via endothelial cell Smad4 signaling, which in cooperation with Notch signaling leads to increased N-cadherin expression and stable endothelial-mural cell adhesion [120]. In mesenchymal stem cells, TGFβ induces production of the Notch ligand Jagged1 and subsequent vascular smooth muscle cell-specific gene expression via Smad3 and Rho kinase pathways [121].

TGFβ signaling via endoglin or ALK1 is able to reduce endothelial activation via TGFβ/ALK5 and therefore tends to promote vessel destabilization and proliferation/sprouting [89]. Endoglin in cooperation with αv integrin leads to TGFβ activation and also signals for subsequent reduced pericyte migration. The matricellular protein "secreted protein acidic and rich in cysteine" (SPARC) is able to interfere with TGFβ-mediated inhibition of pericyte migration via its ability to prevent endoglin from incorporating into pericyte focal contacts and associating αv integrin [122]. Interestingly, endoglin is able to associate with αv integrin independent of the formation of focal adhesions, and endoglin may interfere with pericyte focal adhesion formation or maturation, partially accounting for its ability to reduce mural cell migration upon TGFβ stimulation [122]. Rivera and Brekken propose a model whereby, as pericytes come into contact with endothelial cells, SPARC is degraded or removed from the integrin complex, leading to TGFβRII/αv integrin/TGFβ interactions and subsequent signaling [122].

Mice null for endothelial expression of the tumor suppressor LKB1 display early embryonic death associated with defective yolk sac vessel recruitment of mesenchymal precursors of vascular smooth muscle cells, similar to what is seen in endoglin knockout murine embryos [123, 124]. LKB1 null endothelial cells were defective in TGFβ production, implicating this kinase in regulation of TGFβ production, via as yet unclear mechanisms [123].

2.6 Pathological Angiogenesis

Angiogenesis plays key roles in reproduction, development, growth, and wound healing and can drive the so-called angiogenesis-dependent diseases such as diabetic retinopathy, chronic inflammation, and cancer [125, 126]. There is growing evidence that, despite underlying fundamental similarities the angiogenesis occurring under such pathological settings displays significant alterations in pathways and processes. While such differences complicate our understanding of the angiogenic process, they can also provide opportunities for therapeutic intervention specifically targeting pathological neovascularization [126–128]. In this section, we describe in more detail some examples of "pathological angiogenesis" where TGFβ plays a significant role.

2.6.1 HHT

HHT is an autosomal dominant syndrome associated with epistaxis, AVMs in multiple organs, and dilated regions of high capillary density or telangiectases [106]. There are two commonly identified genetic defects in this condition accounting for two types of HHTs; HHT1 arises due to mutation of the TGFβ type III receptor endoglin and HHT2 from mutation in the TGFβ type I receptor ALK1 [106]. Additionally, Smad4 mutations has also been identified in patients with a syndrome characterized by both juvenile polyposis and HHT [129]. Pulmonary circulation is especially affected in all HHT cases, leading to potential for life-threatening hemorrhage. Studies have found a general loss of pulmonary capillaries and gain of AVM with primarily venous identity endothelium, perhaps due to excessive endothelial proliferation. Thus, loss of TGFβ regulation of endothelial quiescence and endothelial differentiation may be an underlying molecular defect in these individuals [130]. This is highlighted by the newly identified mutation in *PTPN14* associated with HHT, especially the pulmonary manifestations [131]. *PTPN14* codes for a protein tyrosine phosphatase; its expression is modulated by both ALK1 and EphrinB2, and PTPN14 knockdown leads to increased angiogenesis in vitro due to enhanced number of tip cells [131].

2.6.2 Organ Fibrosis

Due to the known association between TGFβ signaling and fibrosis in many systems, it is perhaps not surprising that vascular manifestations of this situation arise. Both the mural cell and endothelial cell components of the microcirculation are documented targets of TFGβ-mediated fibrosis in several organs. For instance, during the development of liver cirrhosis, hepatic stellate cells or Ito cells (the vascular sinusoidal mural cells) express excessive collagen upon TGFβ signaling and in a neuropilin-1-dependent fashion [132]. In the kidney, the renal glomerulus is prone to mesangial cell proliferative glomerulonephritis. These cells are modified and specialized pericytes of the glomerular filtration capillaries, and their proliferation is driven in part by excessive TGFβ production [133]. Interestingly, the bioactive lipid mediator sphingosine-1 phosphate1 (S1P1) is able to cross activate TGFβ signaling in renal mesangial cells via Smad1, 2, and 3 [133], indicating possible transactivation of TGFβ signaling pathways.

In addition to targeting vascular mural cells to promote fibrosis, TGFβ can also induce endothelial-to-mesenchymal transition (EndoMT), a process by which endothelial cells transform into mesenchymal cells, such as fibroblasts and bone cells. This phenomenon is well documented in heart and kidney fibrosis and is mediated by the snail family of transcriptional repressors. Both canonical and non-canonical TGFβ signaling, including Smad, MEK, PI3K, p38 MAPK, c-Abl, and PKC-δ signaling, have been reported to mediate an EndoMT response to TGFβ (reviewed in [59]).

2.6.3 Cancer

Perhaps the best-studied example of pathological angiogenesis where TGFβ plays a significant role is in the neovascularization of solid tumors. TGFβ orchestrates a switch from vascular inhibition to pro-angiogenic activity, likely indirectly via stimulation of cancer cell production of pro-inflammatory and immune suppressive gene products (as reviewed by Tian et al. [113]). Proteomic comparison of angiogenesis in glioblastoma to physiological angiogenesis (endometrial tissue) found numerous TGFβ target genes overexpressed in glioma vessels compared to endometrium, in particular TGFβ-induced protein ig-h3, periostin, integrin-αv, and tenascin C [134].

We found that TGFβ was able to downregulate the expression of VEGFR2 in colorectal tumor vasculature in an Alk5/Smad2-dependent fashion [58]. VEGFR2 expression on glioma blood vessels increased with tumor progression, and the proportion of p-Smad2-positive endothelial cells was significantly higher in tumor vessels compared to normal brain vasculature [135]. There is evidence that ALK1 signaling in cancer angiogenesis may modulate cross talk between EC and pericytes, and inhibition of ALK1 may be especially effective in VEGF refractory tumors [128]. Our results support the possibility that ALK5 activation in endothelial cells may be, at least in part, responsible for development of tumor vessel refractoriness to VEGF inhibition by modulating the levels of VEGFR2 and likely VEGF dependence on endothelial cells.

Finally, endoglin, an essential modulator of TGFβ signaling in endothelial cells, has been shown to be significantly upregulated in tumor-associated endothelium and its expression correlated with poor prognosis in patients with various tumor types including breast, lung, colorectal, prostate, gastric, endometrial, hepatocellular, ovarian, cervical, and head and neck cancers, as well as glioblastoma (reviewed in [136]). Tumor growth and vascularization is reduced in *Eng*− heterozygous mice [137], and both endoglin-neutralizing antibodies [136] and soluble endoglin [138] target the tumor vasculature and inhibit tumor growth in experimental models, suggesting endoglin as another potential therapeutic target in cancer.

2.7 Conclusion

In summary, this review highlights the central actions of TGFβ on the vascular components undergoing angiogenesis, which are broad ranging and context-dependent. In particular, the pleomorphic responses to TGFβ occurring in pathological vs. physiological angiogenesis provide avenues for improved understanding and therapeutic control of these events.

References

1. Moustakas A, Heldin CH (2009) The regulation of TGFbeta signal transduction. Development 136:3699–3714
2. Gordon KJ, Blobe GC (2008) Role of transforming growth factor-beta superfamily signaling pathways in human disease. Biochim Biophys Acta 1782:197–228
3. Li C, Guo B, Ding S et al (2003) TNF alpha down-regulates CD105 expression in vascular endothelial cells: a comparative study with TGF beta 1. Anticancer Res 23:1189–1196
4. Dobaczewski M, Chen W, Frangogiannis NG (2011) Transforming growth factor (TGF)-beta signaling in cardiac remodeling. J Mol Cell Cardiol 51:600–606
5. Worthington JJ, Klementowicz JE, Travis MA (2011) TGFbeta: a sleeping giant awoken by integrins. Trends Biochem Sci 36:47–54
6. Patil AS, Sable RB, Kothari RM (2011) An update on transforming growth factor-beta (TGF-beta): sources, types, functions and clinical applicability for cartilage/bone healing. J Cell Physiol 226:3094–3103

7. Wahl SM, Wen J, Moutsopoulos N (2006) TGF-beta: a mobile purveyor of immune privilege. Immunol Rev 213:213–227
8. Beiter K, Hiendlmeyer E, Brabletz T et al (2005) Beta-catenin regulates the expression of tenascin-C in human colorectal tumors. Oncogene 24:8200–8204
9. Viloria-Petit AM, Wrana JL (2010) The TGFbeta-Par6 polarity pathway: linking the Par complex to EMT and breast cancer progression. Cell Cycle 9:623–624
10. Massague J (2008) TGFbeta in cancer. Cell 134:215–230
11. Goumans MJ, Liu Z, ten Dijke P (2009) TGF-beta signaling in vascular biology and dysfunction. Cell Res 19:116–127
12. Pardali E, Goumans MJ, ten Dijke P (2010) Signaling by members of the TGF-beta family in vascular morphogenesis and disease. Trends Cell Biol 20:556–567
13. Pardali E, ten Dijke P (2009) Transforming growth factor-beta signaling and tumor angiogenesis. Front Biosci 14:4848–4861
14. Shi Y, Massague J (2003) Mechanisms of TGF-beta signaling from cell membrane to the nucleus. Cell 113:685–700
15. Oh SP, Seki T, Goss KA, Imamura T et al (2000) Activin receptor-like kinase 1 modulates transforming growth factor-beta 1 signaling in the regulation of angiogenesis. Proc Natl Acad Sci U S A 97:2626–2631
16. Yamashita H, ten Dijke P, Franzen P et al (1994) Formation of hetero-oligomeric complexes of type I and type II receptors for transforming growth factor-beta. J Biol Chem 269:20172–20178
17. Luo K, Lodish HF (1997) Positive and negative regulation of type II TGF-beta receptor signal transduction by autophosphorylation on multiple serine residues. EMBO J 16:1970–1981
18. Lawler S, Feng XH, Chen RH et al (1997) The type II transforming growth factor-beta receptor autophosphorylates not only on serine and threonine but also on tyrosine residues. J Biol Chem 272:14850–14859
19. Ravichandran KS (2001) Signaling via Shc family adapter proteins. Oncogene 20:6322–6330
20. Galliher AJ, Schiemann WP (2007) Src phosphorylates Tyr284 in TGF-beta type II receptor and regulates TGF-beta stimulation of p38 MAPK during breast cancer cell proliferation and invasion. Cancer Res 67:3752–3758
21. Tsukazaki T, Chiang TA, Davison AF et al (1998) SARA, a FYVE domain protein that recruits Smad2 to the TGFbeta receptor. Cell 95:779–791
22. Miyazawa K, Shinozaki M, Hara T et al (2002) Two major Smad pathways in TGF-beta superfamily signalling. Genes Cells 7:1191–1204
23. Jayaraman L, Massague J (2000) Distinct oligomeric states of SMAD proteins in the transforming growth factor-beta pathway. J Biol Chem 275:40710–40717
24. Kurisaki A, Kose S, Yoneda Y et al (2001) Transforming growth factor-beta induces nuclear import of Smad3 in an importin-beta1 and Ran-dependent manner. Mol Biol Cell 12:1079–1091
25. Xiao Z, Watson N, Rodriguez C, Lodish HF (2001) Nucleocytoplasmic shuttling of Smad1 conferred by its nuclear localization and nuclear export signals. J Biol Chem 276:39404–39410
26. Shi X, Chen F, Yu J et al (2008) Study of interaction between Smad7 and DNA by single-molecule force spectroscopy. Biochem Biophys Res Commun 377:1284–1287
27. Koinuma D, Tsutsumi S, Kamimura N et al (2009) Chromatin immunoprecipitation on microarray analysis of Smad2/3 binding sites reveals roles of ETS1 and TFAP2A in transforming growth factor beta signaling. Mol Cell Biol 29:172–186
28. Stopa M, Anhuf D, Terstegen L et al (2000) Participation of Smad2, Smad3, and Smad4 in transforming growth factor beta (TGF-beta)-induced activation of Smad7. THE TGF-beta response element of the promoter requires functional Smad binding element and E-box sequences for transcriptional regulation. J Biol Chem 275:29308–29317

29. Yan X, Liu Z, Chen Y (2009) Regulation of TGF-beta signaling by Smad7. Acta Biochim Biophys Sin (Shanghai) 41:263–272
30. Mochizuki T, Miyazaki H, Hara T et al (2004) Roles for the MH2 domain of Smad7 in the specific inhibition of transforming growth factor-beta superfamily signaling. J Biol Chem 279:31568–31574
31. Kavsak P, Rasmussen RK, Causing CG et al (2000) Smad7 binds to Smurf2 to form an E3 ubiquitin ligase that targets the TGF beta receptor for degradation. Mol Cell 6:1365–1375
32. Shi W, Sun C, He B et al (2004) GADD34-PP1c recruited by Smad7 dephosphorylates TGFbeta type I receptor. J Cell Biol 164:291–300
33. Ichijo T, Voutetakis A, Cotrim AP et al (2005) The Smad6-histone deacetylase 3 complex silences the transcriptional activity of the glucocorticoid receptor: potential clinical implications. J Biol Chem 280(51):42067–42077
34. Olsson N, Piek E, Sundstrom M et al (2001) Transforming growth factor-beta-mediated mast cell migration depends on mitogen-activated protein kinase activity. Cell Signal 13(7):483–490
35. Engel ME, McDonnell MA, Law BK, Moses HL (1999) Interdependent SMAD and JNK signaling in transforming growth factor-beta-mediated transcription. J Biol Chem 274:37413–37420
36. Zavadil J, Bitzer M, Liang D et al (2001) Genetic programs of epithelial cell plasticity directed by transforming growth factor-beta. Proc Natl Acad Sci U S A 98:6686–6691
37. Davies M, Robinson M, Smith E et al (2005) Induction of an epithelial to mesenchymal transition in human immortal and malignant keratinocytes by TGF-beta1 involves MAPK, Smad and AP-1 signalling pathways. J Biol Chem 95:918–931
38. Wang C, Deng L, Hong M et al (2001) TAK1 is a ubiquitin-dependent kinase of MKK and IKK. Nature 412:346–351
39. Shim JH, Xiao C, Paschal AE et al (2005) TAK1, but not TAB1 or TAB2, plays an essential role in multiple signaling pathways in vivo. Genes Dev 19:2668–2681
40. Jadrich JL, O'Connor MB, Coucouvanis E (2006) The TGF beta activated kinase TAK1 regulates vascular development in vivo. Development 133(8):1529–1541
41. Liao JH, Chen JS, Chai MQ et al (2001) The involvement of p38 MAPK in transforming growth factor beta1-induced apoptosis in murine hepatocytes. Cell Res 11:89–94
42. Yamashita M, Fatyol K, Jin C et al (2008) TRAF6 mediates Smad-independent activation of JNK and p38 by TGF-beta. Mol Cell 31:918–924
43. Ferrari G, Terushkin V, Wolff MJ et al (2012) TGF-beta1 induces endothelial cell apoptosis by shifting VEGF activation of p38MAPK from the prosurvival p38beta to proapoptotic p38alpha. Mol Cancer Res 10:605–614
44. Edlund S, Landstrom M, Heldin CH, Aspenstrom P (2002) Transforming growth factor-beta-induced mobilization of actin cytoskeleton requires signaling by small GTPases Cdc42 and RhoA. Mol Biol Cell 13:902–914
45. Ozdamar B, Bose R, Barrios-Rodiles M et al (2005) Regulation of the polarity protein Par6 by TGFbeta receptors controls epithelial cell plasticity. Science 307:1603–1609
46. Wang L, Zeng H, Wang P et al (2003) Neuropilin-1-mediated vascular permeability factor/vascular endothelial growth factor-dependent endothelial cell migration. J Biol Chem 278:48848–48860
47. Townsend TA, Wrana JL, Davis GE, Barnett JV (2008) Transforming growth factor-beta-stimulated endocardial cell transformation is dependent on Par6c regulation of RhoA. J Biol Chem 283:13834–13841
48. Pepper MS, Vassalli JD, Orci L, Montesano R (1993) Biphasic effect of transforming growth factor-beta 1 on in vitro angiogenesis. Exp Cell Res 204:356–363
49. Yi JY, Shin I, Arteaga CL (2005) Type I transforming growth factor beta receptor binds to and activates phosphatidylinositol 3-kinase. J Biol Chem 280:10870–10876
50. Bakin AV, Tomlinson AK, Bhowmick NA et al (2000) Phosphatidylinositol 3-kinase function is required for transforming growth factor beta-mediated epithelial to mesenchymal transition and cell migration. J Biol Chem 275:36803–36810

51. Lebrin F, Deckers M, Bertolino P, Ten Dijke P (2005) TGF-beta receptor function in the endothelium. Cardiovasc Res 65:599–608
52. Gerhardt H (2008) VEGF and endothelial guidance in angiogenic sprouting. Organogenesis 4:241–246
53. Gerhardt H, Golding M, Fruttiger M et al (2003) VEGF guides angiogenic sprouting utilizing endothelial tip cell filopodia. J Cell Biol 161:1163–1177
54. Kume T (2012) Ligand-dependent Notch signaling in vascular formation. Adv Exp Med Biol 727:210–222
55. Holderfield MT, Hughes CC (2008) Crosstalk between vascular endothelial growth factor, notch, and transforming growth factor-beta in vascular morphogenesis. Circ Res 102:637–652
56. van Meeteren LA, Thorikay M, Bergqvist S et al (2012) An anti-human ALK1 antibody attenuates BMP9 induced ALK1 signaling and interferes with endothelial cell sprouting. J Biol Chem 287:18551–18561
57. Larrivee B, Prahst C, Gordon E et al (2012) ALK1 signaling inhibits angiogenesis by cooperating with the Notch pathway. Dev Cell 22:489–500
58. Kuczynski EA, Viloria-Petit AM, Coomber BL (2011) Colorectal carcinoma cell production of transforming growth factor beta decreases expression of endothelial cell vascular endothelial growth factor receptor 2. Cancer 117:5601–5611
59. van Meeteren LA, ten Dijke P (2012) Regulation of endothelial cell plasticity by TGF-beta. Cell Tissue Res 347:177–186
60. Yoshimatsu Y, Watabe T (2011) Roles of TGF-beta signals in endothelial-mesenchymal transition during cardiac fibrosis. Int J Inflam 2011:724080
61. Birukova AA, Adyshev D, Gorshkov B et al (2005) ALK5 and Smad4 are involved in TGF-beta1-induced pulmonary endothelial permeability. FEBS Lett 579:4031–4037
62. Watabe T, Nishihara A, Mishima K et al (2003) TGF-beta receptor kinase inhibitor enhances growth and integrity of embryonic stem cell-derived endothelial cells. J Cell Biol 163:1303–1311
63. Walshe TE, Saint-Geniez M, Maharaj AS et al (2009) TGF-beta is required for vascular barrier function, endothelial survival and homeostasis of the adult microvasculature. PLoS One 4:e5149
64. Ota T, Fujii M, Sugizaki T et al (2002) Targets of transcriptional regulation by two distinct type I receptors for transforming growth factor-beta in human umbilical vein endothelial cells. J Cell Physiol 193:299–318
65. Taddei A, Giampietro C, Conti A et al (2008) Endothelial adherens junctions control tight junctions by VE-cadherin-mediated upregulation of claudin-5. Nat Cell Biol 10:923–934
66. Hicklin DJ, Ellis LM (2005) Role of the vascular endothelial growth factor pathway in tumor growth and angiogenesis. J Clin Oncol 23:1011–1027
67. Esser S, Lampugnani MG, Corada M et al (1998) Vascular endothelial growth factor induces VE-cadherin tyrosine phosphorylation in endothelial cells. J Cell Sci 111:1853–1865
68. Ferrari G, Pintucci G, Seghezzi G et al (2006) VEGF, a prosurvival factor, acts in concert with TGF-beta1 to induce endothelial cell apoptosis. Proc Natl Acad Sci U S A 103:17260–17265
69. Goumans MJ, Valdimarsdottir G, Itoh S et al (2002) Balancing the activation state of the endothelium via two distinct TGF-beta type I receptors. EMBO J 21:1743–1753
70. Goumans MJ, Valdimarsdottir G, Itoh S et al (2003) Activin receptor-like kinase (ALK)1 is an antagonistic mediator of lateral TGFbeta/ALK5 signaling. Mol Cell 12:817–828
71. Cunha SI, Pardali E, Thorikay M et al (2010) Genetic and pharmacological targeting of activin receptor-like kinase 1 impairs tumor growth and angiogenesis. J Exp Med 207:85–100
72. David L, Mallet C, Vailhe B et al (2007) Activin receptor-like kinase 1 inhibits human microvascular endothelial cell migration: potential roles for JNK and ERK. J Cell Physiol 213:484–489

73. Lamouille S, Mallet C, Feige JJ, Bailly S (2002) Activin receptor-like kinase 1 is implicated in the maturation phase of angiogenesis. Blood 100:4495–4501
74. Mallet C, Vittet D, Feige JJ, Bailly S (2006) TGFbeta1 induces vasculogenesis and inhibits angiogenic sprouting in an embryonic stem cell differentiation model: respective contribution of ALK1 and ALK5. Stem Cells 24:2420–2427
75. Moustakas A, Pardali K, Gaal A, Heldin CH (2002) Mechanisms of TGF-beta signaling in regulation of cell growth and differentiation. Immunol Lett 82:85–91
76. Derynck R, Zhang YE (2003) Smad-dependent and Smad-independent pathways in TGF-beta family signalling. Nature 425:577–584
77. Liu Z, Kobayashi K, van Dinther M et al (2009) VEGF and inhibitors of TGFbeta type-I receptor kinase synergistically promote blood-vessel formation by inducing alpha5-integrin expression. J Cell Sci 122:3294–3302
78. Gougos A, Letarte M (1990) Primary structure of endoglin, an RGD-containing glycoprotein of human endothelial cells. J Biol Chem 265:8361–8364
79. Moren A, Ichijo H, Miyazono K (1992) Molecular cloning and characterization of the human and porcine transforming growth factor-beta type III receptors. Biochem Biophys Res Commun 189:356–362
80. Wong SH, Hamel L, Chevalier S, Philip A (2000) Endoglin expression on human microvascular endothelial cells association with betaglycan and formation of higher order complexes with TGF-beta signalling receptors. Eur J Biochem 267:5550–5560
81. Brown CB, Boyer AS, Runyan RB, Barnett JV (1999) Requirement of type III TGF-beta receptor for endocardial cell transformation in the heart. Science 283:2080–2082
82. Lamarre J, Vasudevan J, Gonias SL (1994) Plasmin cleaves betaglycan and releases a 60 kDa transforming growth factor-beta complex from the cell surface. Biochem J 302:199–205
83. Kaitu'u-Lino TJ, Palmer KR, Whitehead CL et al (2012) MMP-14 is expressed in preeclamptic placentas and mediates release of soluble endoglin. Am J Pathol 180:888–894
84. Li C, Issa R, Kumar P et al (2003) CD105 prevents apoptosis in hypoxic endothelial cells. J Cell Sci 116:2677–2685
85. Scharpfenecker M, van Dinther M, Liu Z et al (2007) BMP-9 signals via ALK1 and inhibits bFGF-induced endothelial cell proliferation and VEGF-stimulated angiogenesis. J Cell Sci 120:964–972
86. Koleva RI, Conley BA, Romero D et al (2006) Endoglin structure and function: determinants of endoglin phosphorylation by transforming growth factor-beta receptors. J Biol Chem 281:25110–25123
87. Blobe GC, Schiemann WP, Pepin MC et al (2001) Functional roles for the cytoplasmic domain of the type III transforming growth factor beta receptor in regulating transforming growth factor beta signaling. J Biol Chem 276:24627–24637
88. Ray BN, Lee NY, How T, Blobe GC (2010) ALK5 phosphorylation of the endoglin cytoplasmic domain regulates Smad1/5/8 signaling and endothelial cell migration. Carcinogenesis 31:435–441
89. Lebrin F, Goumans MJ, Jonker L et al (2004) Endoglin promotes endothelial cell proliferation and TGF-beta/ALK1 signal transduction. EMBO J 23:4018–4028
90. Blanco FJ, Santibanez JF, Guerrero-Esteo M et al (2005) Interaction and functional interplay between endoglin and ALK-1, two components of the endothelial transforming growth factor-beta receptor complex. J Cell Physiol 204:574–584
91. Warrington K, Hillarby MC, Li C et al (2005) Functional role of CD105 in TGF-beta1 signalling in murine and human endothelial cells. Anticancer Res 25(3B):1851–1864
92. Kirkbride KC, Townsend TA, Bruinsma MW et al (2008) Bone morphogenetic proteins signal through the transforming growth factor-beta type III receptor. J Biol Chem 283:7628–7637
93. Farnworth PG, Wang Y, Escalona R et al (2007) Transforming growth factor-beta blocks inhibin binding to different target cell types in a context-dependent manner through dual mechanisms involving betaglycan. Endocrinology 148:5355–5368

94. Esparza-Lopez J, Montiel JL, Vilchis-Landeros MM et al (2001) Ligand binding and functional properties of betaglycan, a co-receptor of the transforming growth factor-beta superfamily. Specialized binding regions for transforming growth factor-beta and inhibin A. J Biol Chem 276:14588–14596
95. del Re E, Babitt JL, Pirani A et al (2004) In the absence of type III receptor, the transforming growth factor (TGF)-beta type II-B receptor requires the type I receptor to bind TGF-beta2. J Biol Chem 279:22765–22772
96. Cheifetz S, Bellon T, Cales C et al (1992) Endoglin is a component of the transforming growth factor-beta receptor system in human endothelial cells. J Biol Chem 267:19027–19030
97. Barbara NP, Wrana JL, Letarte M (1999) Endoglin is an accessory protein that interacts with the signaling receptor complex of multiple members of the transforming growth factor-beta superfamily. J Biol Chem 274:584–594
98. Letamendia A, Lastres P, Almendro N et al (1998) Endoglin, a component of the TGF-beta receptor system, is a differentiation marker of human choriocarcinoma cells. Int J Cancer 76:541–546
99. Venkatesha S, Toporsian M, Lam C et al (2006) Soluble endoglin contributes to the pathogenesis of preeclampsia. Nat Med 12:642–649
100. Scherner O, Meurer SK, Tihaa L et al (2007) Endoglin differentially modulates antagonistic transforming growth factor-beta1 and BMP-7 signaling. J Biol Chem 282:13934–13943
101. Guo B, Slevin M, Li C et al (2004) CD105 inhibits transforming growth factor-beta-Smad3 signalling. Anticancer Res 24:1337–1345
102. Lee NY, Blobe GC (2007) The interaction of endoglin with beta-arrestin2 regulates transforming growth factor-beta-mediated ERK activation and migration in endothelial cells. J Biol Chem 282:21507–21517
103. Conley BA, Koleva R, Smith JD et al (2004) Endoglin controls cell migration and composition of focal adhesions: function of the cytosolic domain. J Biol Chem 279:27440–27449
104. Torsney E, Charlton R, Parums D et al (2002) Inducible expression of human endoglin during inflammation and wound healing in vivo. Inflamm Res 51:464–470
105. Jonker L, Arthur HM (2002) Endoglin expression in early development is associated with vasculogenesis and angiogenesis. Mech Dev 110:193–196
106. Abdalla SA, Letarte M (2006) Hereditary haemorrhagic telangiectasia: current views on genetics and mechanisms of disease. J Med Genet 43:97–110
107. Letteboer TG, Mager JJ, Snijder RJ et al (2006) Genotype-phenotype relationship in hereditary haemorrhagic telangiectasia. J Med Genet 43:371–377
108. McAllister KA, Grogg KM, Johnson DW et al (1994) Endoglin, a TGF-beta binding protein of endothelial cells, is the gene for hereditary haemorrhagic telangiectasia type 1. Nat Genet 8:345–351
109. Johnson DW, Berg JN, Baldwin MA et al (1996) Mutations in the activin receptor-like kinase 1 gene in hereditary haemorrhagic telangiectasia type 2. Nat Genet 13:189–195
110. Gaengel K, Genove G, Armulik A, Betsholtz C (2009) Endothelial-mural cell signaling in vascular development and angiogenesis. Arterioscler Thromb Vasc Biol 29:630–638
111. Ribatti D, Nico B, Crivellato E (2011) The role of pericytes in angiogenesis. Int J Dev Biol 55:261–268
112. Agardh CD, Agardh E, Hultberg B, Ahren B (2000) Long-standing hyperglycemia in C57BL/6J mice does not affect retinal glutathione levels or endothelial/pericyte ratio in retinal capillaries. J Diabetes Complications 14:146–153
113. Tian M, Neil JR, Schiemann WP (2011) Transforming growth factor-beta and the hallmarks of cancer. Cell Signal 23:951–962
114. Nishishita T, Lin PC (2004) Angiopoietin 1, PDGF-B, and TGF-beta gene regulation in endothelial cell and smooth muscle cell interaction. J Biol Chem 91:584–593
115. Ma J, Wang Q, Fei T et al (2007) MCP-1 mediates TGF-beta-induced angiogenesis by stimulating vascular smooth muscle cell migration. Blood 109:987–994
116. Slevin M, Krupinski J, Rovira N et al (2009) Identification of pro-angiogenic markers in blood vessels from stroked-affected brain tissue using laser-capture microdissection. BMC Genomics 10:113

117. Antonelli-Orlidge A, Saunders KB, Smith SR, D'Amore PA (1989) An activated form of transforming growth factor beta is produced by cocultures of endothelial cells and pericytes. Proc Natl Acad Sci U S A 86:4544–4548
118. Darland DC, D'Amore PA (2001) TGF beta is required for the formation of capillary-like structures in three-dimensional cocultures of 10T1/2 and endothelial cells. Angiogenesis 4:11–20
119. Hirschi KK, Burt JM, Hirschi KD, Dai C (2003) Gap junction communication mediates transforming growth factor-beta activation and endothelial-induced mural cell differentiation. Circ Res 93:429–437
120. Li F, Lan Y, Wang Y et al (2011) Endothelial Smad4 maintains cerebrovascular integrity by activating N-cadherin through cooperation with Notch. Dev Cell 20:291–302
121. Kurpinski K, Lam H, Chu J et al (2010) Transforming growth factor-beta and notch signaling mediate stem cell differentiation into smooth muscle cells. Stem Cells 28:734–742
122. Rivera LB, Brekken RA (2011) SPARC promotes pericyte recruitment via inhibition of endoglin-dependent TGF-beta1 activity. J Cell Biol 193:1305–1319
123. Londesborough A, Vaahtomeri K, Tiainen M et al (2008) LKB1 in endothelial cells is required for angiogenesis and TGFbeta-mediated vascular smooth muscle cell recruitment. Development 135:2331–2338
124. Carvalho RL, Jonker L, Goumans MJ et al (2004) Defective paracrine signalling by TGFbeta in yolk sac vasculature of endoglin mutant mice: a paradigm for hereditary haemorrhagic telangiectasia. Development 131:6237–6247
125. Folkman J (2001) Angiogenesis-dependent diseases. Semin Oncol 28(6):536–542
126. Franses JW, Edelman ER (2011) The evolution of endothelial regulatory paradigms in cancer biology and vascular repair. Cancer Res 71:7339–7344
127. Chaudhary A, Hilton MB, Seaman S et al (2012) TEM8/ANTXR1 blockade inhibits pathological angiogenesis and potentiates tumoricidal responses against multiple cancer types. Cancer Cell 21:212–226
128. Cunha SI, Pietras K (2011) ALK1 as an emerging target for antiangiogenic therapy of cancer. Blood 117:6999–7006
129. Gallione CJ, Repetto GM, Legius E et al (2004) A combined syndrome of juvenile polyposis and hereditary haemorrhagic telangiectasia associated with mutations in MADH4 (SMAD4). Lancet 363:852–859
130. Mahmoud M, Upton PD, Arthur HM (2011) Angiogenesis regulation by TGFbeta signalling: clues from an inherited vascular disease. Biochem Soc Trans 39:1659–1666
131. Benzinou M, Clermont FF, Letteboer TG et al (2012) Mouse and human strategies identify PTPN14 as a modifier of angiogenesis and hereditary haemorrhagic telangiectasia. Nat Commun 3:616
132. Cao S, Yaqoob U, Das A et al (2010) Neuropilin-1 promotes cirrhosis of the rodent and human liver by enhancing PDGF/TGF-beta signaling in hepatic stellate cells. J Clin Invest 120:2379–2394
133. Xin C, Ren S, Kleuser B et al (2004) Sphingosine 1-phosphate cross-activates the Smad signaling cascade and mimics transforming growth factor-beta-induced cell responses. J Biol Chem 279:35255–35262
134. Mustafa DA, Dekker LJ, Stingl C et al (2012) A proteome comparison between physiological angiogenesis and angiogenesis in glioblastoma. Mol Cell Proteomics 11:M111.008466
135. Kuczynski EA, Patten SG, Coomber BL (2011) VEGFR2 expression and TGF-beta signaling in initial and recurrent high-grade human glioma. Oncology 81:126–134
136. Seon BK, Haba A, Matsuno F, Takahashi N et al (2011) Endoglin-targeted cancer therapy. Curr Drug Deliv 8:135–143
137. Duwel A, Eleno N, Jerkic M et al (2007) Reduced tumor growth and angiogenesis in endoglin-haploinsufficient mice. Tumour Biol 28:1–8
138. Castonguay R, Werner ED, Matthews RG et al (2011) Soluble endoglin specifically binds bone morphogenetic proteins 9 and 10 via its orphan domain, inhibits blood vessel formation, and suppresses tumor growth. J Biol Chem 286:30034–30046

Chapter 3
Role of Cytokines in Angiogenesis: Turning It On and Off

Melissa A. Thal and Raj Kishore

Abstract The microenvironment surrounding endothelial cells is rich with secreted proteins necessary for regulating the life cycle of a blood vessel. Without these critical cytokines, initiation and growth of new blood vessels would cease and the integrity of existing blood vessels would be diminished. Cytokine-induced regulation of angiogenesis is critical for normal vessel growth and ischemic tissue repair, but if unchecked may lead to tumor progression and metastasis. In this chapter, we introduce the cytokines that play active roles during the different stages of angiogenesis and discuss their ability to coordinate both pro- and anti-angiogenic outcomes.

Keywords Cytokines • Pro-/anti-angiogenic • VEGF

3.1 Introduction

Blood vessels are a complex network of hollow tubes distributed throughout the body that exist in order to provide sufficient blood flow to all tissues. This network is so extensive that if every vessel from an adult were laid out in a straight line, it would measure over 100,000 miles. Such length is required to reach every nook and corner within the human body as a means to provide tissues with vigilant immune surveillance, sufficient oxygenation, adequate nutrient supply, and proper elimination of waste products. Angiogenesis, the growth of new blood vessels from existing vasculature, occurs during normal growth and development, wound healing, and response to a hypoxic signal. When a signal is received that tissue is hypoxic and/or in need of increased blood flow, the body must

M.A. Thal • R. Kishore, Ph.D. (✉)
Feinberg Cardiovascular Research Institute, Feinberg School of Medicine, Northwestern University, 303 East Chicago Avenue, Chicago, IL 60611, USA
e-mail: r-kishore@northwestern.edu

J.L. Mehta and N.S. Dhalla (eds.), *Biochemical Basis and Therapeutic Implications of Angiogenesis*, Advances in Biochemistry in Health and Disease 6,
DOI 10.1007/978-1-4614-5857-9_3,

respond quickly to remedy the problem. Therefore, the pro-angiogenic signals must be robust and fast acting. Conversely, the signals also need to be tightly regulated to prevent abnormal growth or tumor formation. After initiation of angiogenesis, cytokines also direct and orchestrate the formation of the budding vessel through both growth and maturation processes. Cytokine signaling is the major system in place to regulate angiogenesis to ensure the proper formation of new blood vessels. Therefore, in this chapter, the relevant pro- and anti-angiogenic factors and the stages in vessel development or maintenance in which they act are discussed.

3.2 The Sequential Steps of Angiogenesis: Vessel Branching, Maturation, and Quiescence

The majority of lifespan of endothelial cell is spent in quiescence where it is protected from insult and injury by the autocrine action of maintenance signals such as vascular endothelial growth factor (VEGF), Notch (intracellular domain), fibroblast growth factors (FGFs) and constitutive angiopoietin-1 (ANG-1)/tyrosine kinase with immunoglobulin, and epidermal growth factor homology domains (TIE-2), a receptor tyrosine kinase signaling. Quiescent endothelial cells form a smooth, single cell layer of phalanx cells in direct contact with smooth muscle cells and pericytes to prevent leakage and maintain the quiescent state. Pericytes, which are adventitial cells located within the basement membrane of capillary and post-capillary venules, ensheathe the endothelial cells to stabilize the vessel wall while they suppress endothelial cell proliferation and release cell-survival signals such as VEGF and ANG-1. Both the interaction and balance in number between endothelial cell and pericytes are tightly regulated by platelet-derived growth factor (PDGF-B), transforming growth factor-β (TGF-β), VEGF, ANG1/2, and signaling pathways such as Notch and ephrins [1, 2].

Upon stimulation by an angiogenic signal such as VEGF, VEGF-C, or ANG-2, a coordinated series of events centered on the endothelial cells proceeds to promote vessel branching. The first step is local enzymatic destruction of the basement membrane of a vessel by matrix metalloproteinases (MMPs). Pericytes become detached from the vessel wall, endothelial cells loosen their junctions, and the nascent vessel dilates. VEGF promotes an increase in the permeability of the endothelial cell layer, allowing plasma proteins to extravasate the vessel and create a provisional extracellular matrix (ECM) or scaffold for the endothelial cells to migrate onto. To build a single tube and prevent all endothelial cells from moving en masse towards the angiogenic signal, factors such as VEGF receptors and Notch ligands, Delta-like ligand 4 (DLL4), and JAGGED promote the selection of one endothelial cell to become the "tip cell." Filopodia on the tip cell allow it to help sense and guide the branching vessel towards environmental cues such as VEGF, Eph family receptor interacting proteins (ephrins), and semaphorins. Interestingly, VEGF does not promote proliferation of the tip of the cell but rather forms a gradient to guide the direction of growth. Following behind the tip cell are the stalk cells, which divide to elongate the stalk in response to Notch, Notch-regulated ankyrin repeat protein (NRARP), WNTs, placental growth factor

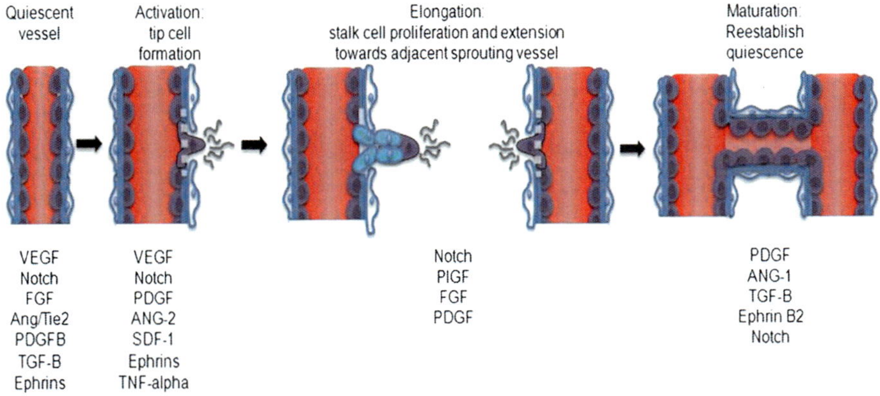

Fig. 3.1 Cytokine regulation of vessel branching. The sequential stages of angiogenesis and the relevant cytokines are listed below the corresponding stage. Blood vessels are maintained in a quiescent state, protected by a sheath of pericytes. Upon activation with the various cytokines listed, initiation of a cascade of activity including dilation of the vessel and selection of an endothelial tip cell begins. Extracellular matrix is broken down around the tip cell by matrix metalloproteinases, allowing for pericyte detachment and endothelial cell junction loosening. The tip cell is directed by a gradient of angiogenic factors towards an adjacent sprouting vessel while being pushed out by the proliferating stalk cells. The new stalk cells establish the lumen of the budding vessel, attract pericytes, and deposit basement membranes to stabilize the growing vessel. The two sprouting vessels fuse to create the perfused neovessel which matures and returns to a quiescent state with the reestablishment of junctions, deposition of basement membrane, maturation of pericytes, and production of vascular maintenance signals

(PIGF), and FGFs to establish the lumen of the branching vessel (mediated by VE-cadherin, CD34, sialomucins, VEGF, and hedgehog). Stalk cells release epidermal growth factor-like domains protein 7 (EGFL7) into the ECM to notify surrounding cells of their spatial relationship with the newly developing vessel and promote its elongation. Angiogenic signals then aid in the formation of a bridge of myeloid cells to fuse two branching vessels and allow for blood to flow. Lastly, in order for a blood vessel to mature and become fully functional, endothelial cells resume their quiescent phalanx state, and PDGF-B, ANG-1, TGF-β, ephrin-B2, and Notch signals promote the re-encapsulation of the endothelial cells by pericytes. Finally, the basement membrane and junctions are reestablished to ensure proper blood flow (Fig. 3.1).

3.3 The Players

Cytokines are regulatory proteins released by cells to promote certain biological effects within target cells. They regulate cellular functions such as viability, proliferation, motility, and cell–cell communication. Specifically, this chapter introduces and describes the different families of cytokines that both promote and inhibit angiogenesis.

3.4 VEGF

VEGFs are signaling proteins that regulate vascular development during both embryogenesis (vasculogenesis) and blood vessel generation (angiogenesis) in the adult. VEGF is the predominant family of factors responsible for vascular development, yet there are relatively few family members and they act mainly in nonredundant roles. In mammals, there are five main VEGF ligands, VEGF-A (also known as VEGF or VEGF-1), VEGF-B, VEGF-C, VEGF-D, and PlGF. VEGF-A was originally identified as vascular permeability factor [3] and has since become the most widely studied family member. Its role has been established in both angiogenesis and vasculogenesis, whereas VEGF-B plays a more central role in cardiac development [4, 5]. VEGF-C and VEGF-D both promote lymphatic vessel formation [6–8] and possibly contribute to angiogenesis as well [9]. PlGF was originally discovered as a VEGF homologue in the placenta [10]. Unlike VEGF-A, PlGF is not required for vascular development, but rather its role is to indirectly stimulate angiogenesis in pathologic conditions by activating bone marrow-derived endothelial progenitor cells, myeloid cells, and stromal cells to upregulate VEGF-A and other angiogenic factors [11, 12].

Alternative splicing of the VEGF pre-mRNA creates multiple isoforms with varying biological activities. There are five main isoforms of human VEGF-A denoted as VEGFA121, VEGFA145, VEGFA165, VEGFA189, and VEGFA206. The mouse isoforms are one amino-acid residue shorter than the corresponding human isoform. Varying binding ability and affinity to VEGFR co-receptors such as heparan sulfate proteoglycans (HSPGs) and neuropilins dictate their bioactivity.

VEGF family members interact mainly with three receptors, VEGFR-1 (Flt-1), VEGFR-2 (KDR in humans and Flk-1 in mouse), and VEGFR-3 (Flt-4). The VEGFRs are members of the receptor tyrosine kinase superfamily in the same subclass as PDGF and FGF receptors. The VEGFRs all contain an extracellular domain with seven immunoglobulin (Ig)-like folds, followed by a single transmembrane region, a juxtamembrane domain, a split tyrosine-kinase domain that is interrupted by a 70 amino-acid kinase insert, and a C-terminal tail. Based on the crystal structure of the extracellular domain of VEGFR1 alone and in complex with ligand, the Ig domain-2 constitutes the ligand-binding site on the receptor [13]. VEGF-A, VEGF-B, and PlGF bind to VEGFR-1. VEGF-A and cleaved forms of VEGF-C and VEGF-D bind VEGFR-2, whereas the uncleaved forms of VEGF-C and VEGF-D bind to VEGFR-3.

Gene-targeting studies have shown that VEGF and VEGFRs are primarily expressed in the vascular system and are critically important for vasculogenesis. Embryos deficient in VEGFR-2 die early in development at approximately E9 due to a failure to initiate vasculogenesis and hematopoiesis [14]. VEGF deficiency is similarly embryonically lethal due to severe vascular defects. The deletion of even a single allele of VEGF causes embryonic lethality at E11 with defects in formation of the dorsal aorta and development of blood cells [15, 16]. Embryos lacking VEGFR-1 are also embryonic lethal with vascular defects but result in endothelial cell overgrowth rather than inhibition [17]. VEGFR-1 exists both as a membrane-anchored signaling competent form and as a soluble secreted form (also known as fms-related tyrosine kinase 1, sFLT-1). The soluble form can trap or sequester its ligand to assist in the guidance of an emerging branch or inhibit sprouting all together. VEGFR-1 also has relatively weak tyrosine kinase activity compared to VEGFR-2 allowing it to act as a decoy for VEGF, moderating the amount of free VEGF available to activate VEGFR-2. The negative regulation of VEGFR-2 by VEGFR-1 explains why the loss of VEGFR-1 results in vessel overgrowth.

The role of VEGFR-3 in blood vessel formation is unclear. It is thought to play a more important role in lymphatic vessel development. Early in development, it is expressed on blood vessels but later becomes restricted to lymphatic vessels. Embryos lacking VEGFR-3 initiate vasculogenesis and angiogenesis, but major cardiovascular defects result in death at E9.5, well before the onset of lymphatic vessel formation.

The interaction between VEGF and its receptors induces receptor homodimerization or heterodimerization, which activates receptor kinase activity, receptor autophosphorylation, and downstream signaling. Engagement of VEGFR-2 by its ligands has been widely studied and appears to play an important role in stimulating endothelial cell migration, differentiation, proliferation, and survival.

3.5 Notch

Notch signaling is critically important in fate determination in several different lineages, but it also plays an important role in endothelial cell sprouting and directing angiogenesis. There are four mammalian Notch receptors (Notch 1, 2, 3, and 4) and five canonical Notch ligands (Jagged1, Jagged2, Delta-like ligand (Dll1, Dll3, and Dll4)). In the vasculature, Notch 1 and 4 are expressed by the endothelium while Notch 3 functions in smooth muscle cells. Engagement of ligand and receptor results in activation and proteolytic cleavage of the receptor at two sites. A disintegrin and metalloprotease family metalloprotease 10 (ADAM-10) cleaves the Notch protein just outside the membrane [18] which releases the extracellular portion of Notch still bound to the ligand [19]. The ligand plus the Notch extracellular domain is then endocytosed by the ligand-expressing cell. A second cleavage by γ-secretase releases the remaining intracellular part of the Notch protein into the Notch expressing cell. The Notch intracellular domain (NICD) then moves to the nucleus, where it can regulate gene expression by activating the transcription factor C promoter binding factor 1 (CBP).

Through genetic analysis, Notch has been implicated in regulating angiogenesis. Although Notch 1-deficient mice have a normal primary vascular plexus, they eventually display vessel degeneration and abnormal vessel remodeling and die at day E11 [20]. Double homozygous mutants of Notch 1 and Notch 4 display a more severe defect in vascular integrity compared to Notch 1 homozygous knockout mice, suggesting a role for both Notch 1 and Notch 4 in vessel development [20]. Constitutively active Notch signaling in endothelial cells similarly results in impaired vessel formation resulting in embryonic death at E10 [21]. Further, deletion of either Notch ligand, Jagged 1, or Dll4 results in embryonic death potentially due to vascular defects [22–24].

As previously discussed, VEGF is critically important for angiogenesis, but its intimate association with Notch signaling is also a major regulatory mechanism necessary for proper vessel formation. During vessel branching, tip cells typically migrate while stalk cells proliferate. Endothelial cells stimulated by VEGF-A compete for the tip cell position via Dll4/Notch signaling. VEGF activation of VEGFR-2 upregulates expression of Dll4 in tip cells [25, 26], which subsequently activates Notch in neighboring stalk cells. This results in a downregulation of VEGFR-2 and an increase in VEGFR-1 preventing the stalk cells from responding to the sprouting activity of VEGF. Therefore, the cell that produces more Dll4 than its neighbors remains the tip cell, because it can effectively suppress the same response in competing neighbors via activation of Notch signaling. This is supported in the mouse with mosaic endothelial Cre recombination of a floxed Notch 1 allele in which the majority of Notch-deficient endothelial cells adopt tip cell characteristics [27]. Furthermore, suppression of Notch signaling by γ-secretase inhibitor (GSI) treatment or genetic deletion of one Dll4 allele in the mouse dramatically augments sprouting, branching, and hyperfusion of the capillary network as a result of excessive tip cell formation [7, 27, 28].

Notch signaling is also required at low levels for the maintenance of quiescent phalanx cells. Without which, as in the chronic inhibition of Dll4 in healthy animals, vascular neoplasms form [29]. Notch signaling also plays a role in endothelial cell motility as evidenced by suppression or over-activation studies resulting in an increase or decrease in filopodial protrusion, respectively.

3.6 FGF

The FGF superfamily consists of 22 paracrine or endocrine peptide factors, which possess a homologous core domain and differentially activate several FGF tyrosine kinase receptors [30]. Presumably, due to the redundancy within the FGF superfamily, *FGF1* (acidic FGF) or *FGF2* (basic FGF, bFGF) deficiency in mice does not result in a vascular defect. In the heart, FGF-mediated signaling stimulates vessel growth through activating release of hedgehog, ANG-2, and VEGF-B. It is also important for the maintenance of quiescent endothelial cells and the integrity of the blood vessel since inhibition of FGFR causes vessel disintegration [31]. Overactive FGF signaling promotes tumor angiogenesis and can result in tumor outgrowth despite the presence of VEGF- or EGFR-inhibitor treatments [32].

3.7 Ephrin and Ephrin Receptors

Two other mechanisms that regulate angiogenic processes are (1) localization of the signal and (2) distribution of attractive vs. repulsive signals. Ephrin and Eph receptors are prime examples. Ephrin B2 is expressed on arteries while EphB4 is present on veins during early developmental stages. The reciprocal distribution helps define the border between the two types of vessels and prevents them from entwining [33–35].

Unlike typical receptor tyrosine kinases, Eph receptors/ephrins are activated by both dimerization and multimerization of the ligand, and the strength of the signal and biological effect is dependent on the degree of multimerization [36]. Also unique to Eph receptor/ephrin signaling is the ability to proceed bidirectionally, meaning that the Eph receptor can act as a ligand and the ephrin ligand can also act as a receptor. In forward signaling, Ephrin binding promotes receptor clustering and phosphorylation of tyrosine residues and results in downstream forward signaling in the receptor-bearing cell. In reverse signaling, cascades are activated in ligand-expressing cells upon clustering of ephrins on the cell surface and receptor engagement. Since ephrins are almost exclusively found membrane bound, cell-to-cell contact is required for both signaling directions.

Eph/Ephrin signaling directly and indirectly affects several stages of angiogenesis. For example, EphB2 and EphB4 cooperate with stromal cell-derived factor 1 (SDF-1) to promote endothelial cell chemotaxis and directional guidance of tip cells [37, 38]. In order to facilitate maturation of a newly formed vessel, EphB4/ephrinB2 mediates monocyte adhesion and transmigration into the vascular endothelium [39]. During the maintenance stage of angiogenesis, expression of EphrinB2 on vascular smooth muscle cells and pericytes allows for their interaction with endothelial and mural cells to provide stability to the mature vessel [40]. EphrinB2 may also indirectly affect angiogenesis by regulating VEGFR internalization and its ability to function during angiogenesis [41, 42].

3.8 TGF-β

TGF-β is a cytokine that functions during vasculogenesis, but its specific role remains unclear. There are several members of the TGF-β family including the TGFβs (TGFβ1, TGFβ2, TGFβ3), bone morphogenetic proteins, activins, and inhibins [43] that bind to two types of receptors (type I and type II). Endothelial cells express TGFβRII and the type I receptors activin receptor-like kinase (Alk1 and Alk5). The co-receptor endoglin [3] is also expressed on endothelial cells and is capable of altering TGFβ signaling effects [44, 45]. Mutations in the genes that encode ENG or (ALK-1) are linked to a disorder called human hereditary hemorrhagic telangiectasia, which results in vascular malformations [46]. Similarly in mice, loss of TGF-β receptors ALK-1, ALK-5, TGFR2, or ENG results in arteriovenous malformations suggesting a role for TGF-β signaling in vasculogenesis. However, due to conflicting data and the sheer complexity of the system, the molecular characterization of this pathway has yet to be elucidated. In vitro studies initially suggested that TGF-β signaling inhibited endothelial cell proliferation and migration [47, 48]. However, additional studies have shown the opposite effect [49–51]. The discrepancy may simply be due to a dose response where low doses of TGF-β activate angiogenic properties while high doses are inhibitory [52]. Results derived from in vivo experimentation are similarly context-dependent and TGF-β can be both pro- and anti-angiogenic [53]. The target cell of TGF-β signaling is also in question since TGF-β regulates smooth muscle cell differentiation while ENG or ALK deficiency manifests in mural cell developmental defects [54]. Therefore, cellular context, local concentration of ligand, receptors, co-receptors, antagonists, and their interactions play crucial roles in the apparently contradictory behavior of TGF-β signaling during angiogenesis.

3.9 Platelet-Derived Growth Factor

Platelet-derived growth factor (PDGF) is a major facilitator in stabilizing endothelial cells and promoting the maturation of a vessel. The PDGF family consists of four different isoforms (PDGFA, PDGFB, PDGFC, and PDGFD). Homodimers, or a PDGF-AB heterodimer, bind and activate their corresponding tyrosine kinase receptor, PDGFRα or PDGFRβ. Proliferating endothelial cells secrete PDGFB which binds PDGFRβ on local undifferentiated mesenchymal cells or pericytes and recruits them to blood vessels. Mice deficient in either PDGFB or PDGFRβ have a significant decrease in the number of smooth muscle cells and pericytes associated with vessels, leading to edema and embryonic death [55]. During activation and elongation of branching vessels, PDGF is expressed highly in the tip cells where secretion establishes a gradient of PDGF allowing for efficient pericyte recruitment [56].

3.10 Angiopoietins and Tie Receptors

ANG/TIE signaling plays a binary role in endothelial cell biology by allowing cells to remain in quiescence yet still able to respond to angiogenic stimuli. This dual role is achieved by antagonizing angiopoietin molecules competing for TIE2 receptor binding. Tie (tyrosine kinase with immunoglobulin and epidermal growth factor homology domains) is a novel class of receptor tyrosine kinases that predominantly are expressed by vascular endothelial cells. There are two members, Tie1 (also known as Tie) and Tie2 (also known as Tek) [57–62]. Both play a significant role in vascular development based on extensive malformations of the vasculature and embryonic death in genetic deletions of either gene. Tie1 is an orphan receptor with no known ligand and may act as a negative regulator of Tie2, but its precise role is still unknown [63]. There are three known human ligands for Tie2: angiopoietin 1, 2, and 4 (ANG1, ANG2, ANG4). ANG1 binds and induces receptor phosphorylation of TIE2 while ANG2 acts as a competitive ANG1 antagonist and prevents TIE2 receptor activation [60, 64]. ANG4 has not been well studied but is postulated to behave similarly as ANG1. ANG1 is expressed by mural cells, which promotes TIE-2 clustering at cell–cell junctions to maintain endothelial cell quiescence and facilitates the intimate connection between the two cell types to maintain integrity of the blood vessel [61, 65]. This interaction is destabilized by ANG2, which is released by sprouting endothelial cells. By antagonizing the ANG1/TIE2 signal, ANG2 enhances mural cell detachment, vascular permeability, and endothelial cell sprouting [63].

ANG/TIE2 signaling on tumor growth is also a conflicting duality of pro- and anti-angiogenesis that appears to be context-dependent. ANG-1 promotes endothelial cell survival and vessel maturation, both of which will support tumor growth. However, ANG-1 also maintains vessel integrity keeping healthy vessels tumor-free while also inhibiting tumor extravasation. ANG-2, on the other hand, would be a more suitable antitumor target because it promotes angiogenesis as evidenced by their recruitment of pro-angiogenic TIE-2 expressing monocytes (TEMs) [66] and inhibition of ANG-2 results in vessel regression and normalization [67]. Blocking agents of either TIE-2 or ANG-2 are in early phase clinical trials.

3.11 Chemokines and G Protein-Coupled Receptors

During the early stages in blood vessel development, organization is dependent on pro-angiogenic factors present in the surrounding tissue. Chemokines attract pro-angiogenic immune cells and endothelial progenitor cells to the site of sprouting and growing blood vessels. Immune cells secrete pro-angiogenic factors to act on the surrounding endothelium to help regulate various aspects of angiogenesis including mediating fusion between endothelial tip cells [68] and maintenance of quiescent endothelial cells. Activation of endothelial cells by VEGF, bFGF, or tumor necrosis factor-alpha (TNF-α) promotes the upregulation of a well-studied chemokine, SDF-1α (CXCL12), and its receptors on the surface of endothelial cells. Upon engagement and activation of its receptors, CXCR4 [69] and CXCR7 [70], VEGF production and secretion is enhanced. This resultant positive-feedback loop in which VEGF induces CXCR4 and SDF-1α expression while engagement of ligand and receptor enhances VEGF production amplifies the pro-angiogenic signal in the target tissue.

3.12 Inhibitors of Angiogenesis

Cytokines are not only pro-angiogenic; there are also several proteins required for negatively regulating angiogenesis in order to prevent vascular chaos and tumor growth. Using several of the same mechanisms as pro-angiogenic cytokines, inhibitory factors can be used in both an autocrine and a paracrine fashion to shut down the synthesis of new blood vessels. The balance between pro- and anti-angiogenic factors has to be tightly regulated to prevent uncontrolled tumor growth but still allow for proper wound healing and other normal, necessary blood vessel development. Therefore, the identification and characterization of anti-angiogenic factors has proven to be significant.

Endogenous inhibitors of angiogenesis such as α-interferon and platelet factor-4 were first identified for their ability to inhibit endothelial cell chemotaxis and proliferation, respectively [71]. The most notable anti-angiogenic factor, thrombospondin-1 (TSP-1), was first discovered when a spontaneously transformed cell line was found to produce significantly lower levels of this inhibitor than the same cells prior to the transformation causing mutation. TSP-1 is a matricellular protein, meaning it associates with the ECM to regulate cell interactions within the environment, but it does not contribute structurally to the ECM. It is a large, modular glycoprotein capable of interacting with different cell receptors, soluble cytokines, ECM, and proteases simultaneously, which explains how they can be pleiotropic. TSP-1 directly inhibits angiogenesis by interacting with endothelial cell-specific receptors (CD36, CD47, integrins, HSPG, and low-density lipoprotein receptor-related protein (LRP)) to affect cell viability [72]. TSP-1 has also been shown to decrease the mobilization of circulating endothelial cells and their progenitors to angiogenic sites [73]. Indirectly, TSP-1 affects the bioavailability of different pro-angiogenic mediators such as FGF-2, VEGF, hepatocyte growth factor (HGF), and PDGF by binding and sequestering them from their targets. The TSP-1 signal can override the proliferative and chemotactic signals of both VEGF and bFGF suggesting that when sufficient levels of TSP-1 are present, the positive signals are negated, which either maintains or turns the angiogenic switch to the off position [74, 75].

3.13 Cytokines in Tumor Angiogenesis

It has been proposed that transformation of a normal cell into a cancer cell requires the acquisition of at least six traits or "hallmarks" [76], one of which is the ability to obtain and sustain sufficient blood flow through angiogenesis. At first, tumors are supported by nearby blood vessels but, once they reach a certain size, new blood vessels are required to sustain growth of the tumor. Although physiological angiogenesis is tightly regulated and orchestrated, dysregulation of cytokine and other pro-angiogenic factor expression causes tumor angiogenesis to be irregular and often results in poor quality and even leaky vessels. One of the many ways tumors escape the regulatory mechanisms meant to protect the body from such outgrowth is through overexpression of pro-angiogenic growth factors and cytokines in response to hypoxia. As tumor cells proliferate, the local oxygen supply gets rapidly depleted which can create a hypoxic microenvironment within the tumor. This hypoxic microenvironment then permits the activation of the transcription factor hypoxia-inducible factor (HIF) leading to transcription of its target genes, namely, VEGF and other pro-angiogenic factors [77]. By hijacking this endogenous method of inducing pro-angiogenic signals, tumors can grow unchecked (Table 3.1).

Table 3.1 Overview of pro- and anti-angiogenic cytokines during the different stages of neovascularization

	Activation	Elongation	Maintenance
Pro-angiogenic	VEGF, VEGF-C	Notch	VEGF
	Notch, DLL4, JAGGED	Wnts	Notch
	PDGF	PIGF	FGF
	ANG-2	FGH	Ang/Tie
	Ephrins	EGFL7	PDGFB
	SDF-1	PDGF	TGF-β
	TNF-alpha		Ephrins
Anti-angiogenic	TSP-1	Alpha interferon	
	TGF-β (at high doses)	TSP-1	
	ANG-1		

The idea of targeting angiogenesis to combat growth of solid tumors has been around for several decades. Because cytokines play a critical role in promoting angiogenesis, they are an obvious choice for targeting anti-angiogenic therapy. Bevacizumab is a humanized monoclonal antibody against VEGF-A which occludes receptor binding and thereby prevents downstream signal activation. It was the first FDA-approved drug developed solely for anti-angiogenesis use in humans. The anticancer activity has had mixed results, but the addition of bevacizumab to the traditional chemotherapy regimen for colorectal cancer significantly increased the progression-free survival (PFS) and median overall survival (OS) [78]. Several other compounds are currently in preclinical development with the intent of targeting regulatory and signaling molecules that govern angiogenesis including VEGF, PlGF, FGF, and epidermal growth factor [79].

Despite the initial excitement over anti-angiogenic approaches to cancer, several hurdles remain. It has yet to be shown whether anti-angiogenic drugs can eliminate or even shrink tumors or if instead they prevent further growth and spread. To address this potential deficiency, traditional chemotherapy agents are still in use in conjunction with anti-angiogenic drugs. It has also been shown than most tumors develop resistance to anti-angiogenic therapy over time. This very well could be contributed to by the redundant nature of angiogenic signaling pathways. Blocking a single pathway might cause the tumors to seek out other angiogenic mechanisms and recruit other pro-angiogenic signaling cascades. Therefore, further studies are needed to determine if targeting multiple pathways will overcome or delay this acquired resistance making the drugs more effective long term.

3.14 Conclusion

Angiogenesis is a tightly orchestrated dance dependent on spatial awareness, timing, and intensity of signal. It is therefore understandable why so many different families of proteins and signaling pathways are involved in its regulation. The redundancy in signaling pathways ensures that even if one route to regulation fails, others will compensate for the desired outcome. Our current understanding of the individual cytokines is limited with only a cursory, at best, understanding of how they overlap and interact with other cytokines and signaling pathways. Therefore, it is of obvious importance to research not only the basic mechanisms that govern vessel formation, development, and maintenance but also the cross talk and attractive/repulsive nature of interacting pathways. Ultimately, further research is needed to elucidate the mechanisms regulating vascular development so that what is learned can be applied to improving treatment of human disease.

Acknowledgments Work described in this manuscript was in part supported by National Institute of Health grants HL091983, HL105597, HL095874, HL053354 and HL108795 (R.K.) and NRSA F32 postdoctoral award HL107093 (M.T.).

References

1. Hughes CC (2008) Endothelial-stromal interactions in angiogenesis. Curr Opin Hematol 15:204–209
2. von Tell D, Armulik A, Betsholtz C (2006) Pericytes and vascular stability. Exp Cell Res 312:623–629
3. Senger DR, Connolly DT, Van de Water L et al (1990) Purification and NH2-terminal amino acid sequence of guinea pig tumor-secreted vascular permeability factor. Cancer Res 50:1774–1778
4. Aase K, von Euler G, Li X et al (2001) Vascular endothelial growth factor-B-deficient mice display an atrial conduction defect. Circulation 104:358–364
5. Bellomo D, Headrick JP, Silins GU et al (2000) Mice lacking the vascular endothelial growth factor-B gene (Vegfb) have smaller hearts, dysfunctional coronary vasculature, and impaired recovery from cardiac ischemia. Circ Res 86:E29–E35
6. Karkkainen MJ, Haiko P, Sainio K et al (2004) Vascular endothelial growth factor C is required for sprouting of the first lymphatic vessels from embryonic veins. Nat Immunol 5:74–80
7. Tammela T, Alitalo K (2010) Lymphangiogenesis: molecular mechanisms and future promise. Cell 140(4):460–476
8. Tammela T, Enholm B, Alitalo K, Paavonen K (2005) The biology of vascular endothelial growth factors. Cardiovasc Res 65:550–563
9. Cao Y, Linden P, Farnebo J et al (1998) Vascular endothelial growth factor C induces angiogenesis in vivo. Proc Natl Acad Sci U S A 95:14389–14394
10. Maglione D, Guerriero V, Viglietto G et al (1991) Isolation of a human placenta cDNA coding for a protein related to the vascular permeability factor. Proc Natl Acad Sci U S A 88:9267–9271
11. Carmeliet P, Moons L, Luttun A et al (2001) Synergism between vascular endothelial growth factor and placental growth factor contributes to angiogenesis and plasma extravasation in pathological conditions. Nat Med 7:575–583
12. Fischer C, Mazzone M, Jonckx B, Carmeliet P (2008) FLT1 and its ligands VEGFB and PlGF: drug targets for anti-angiogenic therapy? Nat Rev Cancer 8:942–956
13. Christinger HW, Fuh G, de Vos AM, Wiesmann C (2004) The crystal structure of placental growth factor in complex with domain 2 of vascular endothelial growth factor receptor-1. J Biol Chem 279:10382–10388
14. Shalaby F, Rossant J, Yamaguchi TP et al (1995) Failure of blood-island formation and vasculogenesis in Flk-1-deficient mice. Nature 376:62–66
15. Carmeliet P, Ferreira V, Breier G et al (1996) Abnormal blood vessel development and lethality in embryos lacking a single VEGF allele. Nature 380:435–439

16. Ferrara N, Bunting S (1996) Vascular endothelial growth factor, a specific regulator of angiogenesis. Curr Opin Nephrol Hypertens 5:35–44
17. Fong GH, Rossant J, Gertsenstein M, Breitman ML (1995) Role of the Flt-1 receptor tyrosine kinase in regulating the assembly of vascular endothelium. Nature 376:66–70
18. Brou C, Logeat F, Gupta N et al (2000) A novel proteolytic cleavage involved in Notch signaling: the role of the disintegrin-metalloprotease TACE. Mol Cell 5:207–216
19. Jakobsson L, Bentley K, Gerhardt H (2009) VEGFRs and Notch: a dynamic collaboration in vascular patterning. Biochem Soc Trans 37:1233–1236
20. Krebs LT, Xue Y, Norton CR et al (2000) Notch signaling is essential for vascular morphogenesis in mice. Genes Dev 14:1343–1352
21. Uyttendaele H, Ho J, Rossant J, Kitajewski J (2001) Vascular patterning defects associated with expression of activated Notch4 in embryonic endothelium. Proc Natl Acad Sci U S A 98:5643–5648
22. Hrabe de Angelis M, McIntyre J II, Gossler A (1997) Maintenance of somite borders in mice requires the Delta homologue DII1. Nature 386:717–721
23. Krebs LT, Shutter JR, Tanigaki K et al (2004) Haploinsufficient lethality and formation of arteriovenous malformations in Notch pathway mutants. Genes Dev 18:2469–2473
24. Xue Y, Gao X, Lindsell CE et al (1999) Embryonic lethality and vascular defects in mice lacking the Notch ligand Jagged1. Hum Mol Genet 8:723–730
25. Benedito R, Roca C, Sorensen I et al (2009) The notch ligands Dll4 and Jagged1 have opposing effects on angiogenesis. Cell 137:1124–1135
26. Liu ZJ, Shirakawa T, Li Y et al (2003) Regulation of Notch1 and Dll4 by vascular endothelial growth factor in arterial endothelial cells: implications for modulating arteriogenesis and angiogenesis. Mol Cell Biol 23:14–25
27. Hellstrom M, Phng LK, Gerhardt H (2007) VEGF and Notch signaling: the yin and yang of angiogenic sprouting. Cell Adh Migr 1:133–136
28. Suchting S, Freitas C, le Noble F et al (2007) The Notch ligand Delta-like 4 negatively regulates endothelial tip cell formation and vessel branching. Proc Natl Acad Sci U S A 104:3225–3230
29. Yan M, Callahan CA, Beyer JC et al (2010) Chronic DLL4 blockade induces vascular neoplasms. Nature 463:E6–E7
30. Beenken A, Mohammadi M (2009) The FGF family: biology, pathophysiology and therapy. Nat Rev Drug Discov 8:235–253
31. Murakami M, Nguyen LT, Zhuang ZW et al (2008) The FGF system has a key role in regulating vascular integrity. J Clin Invest 118:3355–3366
32. Bergers G, Hanahan D (2008) Modes of resistance to anti-angiogenic therapy. Nat Rev Cancer 8:592–603
33. Brantley-Sieders DM, Chen J (2004) Eph receptor tyrosine kinases in angiogenesis: from development to disease. Angiogenesis 7:17–28
34. Kim YH, Hu H, Guevara-Gallardo S et al (2008) Artery and vein size is balanced by Notch and ephrin B2/EphB4 during angiogenesis. Development 135(22):3755–3764
35. Wang HU, Chen ZF, Anderson DJ (1998) Molecular distinction and angiogenic interaction between embryonic arteries and veins revealed by ephrin-B2 and its receptor Eph-B4. Cell 93:741–753
36. Surawska H, Ma PC, Salgia R (2004) The role of ephrins and Eph receptors in cancer. Cytokine Growth Factor Rev 15:419–433
37. Salvucci O, de la Luz Sierra M, Martina JA et al (2006) EphB2 and EphB4 receptors forward signaling promotes SDF-1-induced endothelial cell chemotaxis and branching remodeling. Blood 108:2914–2922
38. Adams RH, Eichmann A (2010) Axon guidance molecules in vascular patterning. Cold Spring Harb Perspect Biol 2:a001875
39. Pfaff D, Heroult M, Riedel M et al (2008) Involvement of endothelial ephrin-B2 in adhesion and transmigration of EphB-receptor-expressing monocytes. J Cell Sci 121:3842–3850

40. Foo SS, Turner CJ, Adams S et al (2006) Ephrin-B2 controls cell motility and adhesion during blood-vessel-wall assembly. Cell 124:161–173
41. Sawamiphak S, Seidel S, Essmann CL et al (2010) Ephrin-B2 regulates VEGFR2 function in developmental and tumour angiogenesis. Nature 465:487–491
42. Wang Y, Nakayama M, Pitulescu ME et al (2010) Ephrin-B2 controls VEGF-induced angiogenesis and lymphangiogenesis. Nature 465:483–486
43. Rossant J, Howard L (2002) Signaling pathways in vascular development. Annu Rev Cell Dev Biol 18:541–573
44. Barbara NP, Wrana JL, Letarte M (1999) Endoglin is an accessory protein that interacts with the signaling receptor complex of multiple members of the transforming growth factor-beta superfamily. J Biol Chem 274:584–594
45. Letamendia A, Lastres P, Botella LM et al (1998) Role of endoglin in cellular responses to transforming growth factor-beta. A comparative study with betaglycan. J Biol Chem 273:33011–33019
46. Azuma H (2000) Genetic and molecular pathogenesis of hereditary hemorrhagic telangiectasia. J Med Invest 47:81–90
47. Baird A, Durkin T (1986) Inhibition of endothelial cell proliferation by type beta-transforming growth factor: interactions with acidic and basic fibroblast growth factors. Biochem Biophys Res Commun 138:476–482
48. Frater-Schroder M, Muller G, Birchmeier W, Bohlen P (1986) Transforming growth factor-beta inhibits endothelial cell proliferation. Biochem Biophys Res Commun 137:295–302
49. Iruela-Arispe ML, Sage EH (1993) Endothelial cells exhibiting angiogenesis in vitro proliferate in response to TGF-beta 1. J Cell Biochem 52:414–430
50. RayChaudhury A, D'Amore PA (1991) Endothelial cell regulation by transforming growth factor-beta. J Cell Biochem 47:224–229
51. Sutton AB, Canfield AE, Schor SL et al (1991) The response of endothelial cells to TGF beta-1 is dependent upon cell shape, proliferative state and the nature of the substratum. J Cell Sci 99:777–787
52. Hofer E, Schweighofer B (2007) Signal transduction induced in endothelial cells by growth factor receptors involved in angiogenesis. Thromb Haemost 97:355–363
53. Roberts AB, Sporn MB, Assoian RK et al (1986) Transforming growth factor type beta: rapid induction of fibrosis and angiogenesis in vivo and stimulation of collagen formation in vitro. Proc Natl Acad Sci U S A 83:4167–4171
54. Pardali E, Goumans MJ, ten Dijke P (2010) Signaling by members of the TGF-beta family in vascular morphogenesis and disease. Trends Cell Biol 20:556–567
55. Hellstrom M, Gerhardt H, Kalen M et al (2001) Lack of pericytes leads to endothelial hyperplasia and abnormal vascular morphogenesis. J Cell Biol 153:543–553
56. Gerhardt H, Golding M, Fruttiger M et al (2003) VEGF guides angiogenic sprouting utilizing endothelial tip cell filopodia. J Cell Biol 161:1163–1177
57. Dumont DJ, Gradwohl G, Fong GH et al (1994) Dominant-negative and targeted null mutations in the endothelial receptor tyrosine kinase, tek, reveal a critical role in vasculogenesis of the embryo. Genes Dev 8:1897–1909
58. Partanen J, Armstrong E, Makela TP et al (1992) A novel endothelial cell surface receptor tyrosine kinase with extracellular epidermal growth factor homology domains. Mol Cell Biol 12:1698–1707
59. Iwama A, Hamaguchi I, Hashiyama M et al (1993) Molecular cloning and characterization of mouse TIE and TEK receptor tyrosine kinase genes and their expression in hematopoietic stem cells. Biochem Biophys Res Commun 195:301–309
60. Maisonpierre PC, Goldfarb M, Yancopoulos GD, Gao G (1993) Distinct rat genes with related profiles of expression define a TIE receptor tyrosine kinase family. Oncogene 8:1631–1637
61. Sato TN, Qin Y, Kozak CA, Audus KL (1993) Tie-1 and tie-2 define another class of putative receptor tyrosine kinase genes expressed in early embryonic vascular system. Proc Natl Acad Sci U S A 90:9355–9358

62. Schnurch H, Risau W (1993) Expression of tie-2, a member of a novel family of receptor tyrosine kinases, in the endothelial cell lineage. Development 119:957–968
63. Augustin HG, Koh GY, Thurston G, Alitalo K (2009) Control of vascular morphogenesis and homeostasis through the angiopoietin-Tie system. Nat Rev Mol Cell Biol 10:165–177
64. Valenzuela DM, Griffiths JA, Rojas J et al (1999) Angiopoietins 3 and 4: diverging gene counterparts in mice and humans. Proc Natl Acad Sci U S A 96:1904–1909
65. Suri C, Jones PF, Patan S et al (1996) Requisite role of angiopoietin-1, a ligand for the TIE2 receptor, during embryonic angiogenesis. Cell 87:1171–1180
66. De Palma M, Venneri MA, Galli R et al (2005) Tie2 identifies a hematopoietic lineage of proangiogenic monocytes required for tumor vessel formation and a mesenchymal population of pericyte progenitors. Cancer Cell 8:211–226
67. Falcon BL, Hashizume H, Koumoutsakos P et al (2009) Contrasting actions of selective inhibitors of angiopoietin-1 and angiopoietin-2 on the normalization of tumor blood vessels. Am J Pathol 175:2159–2170
68. Fantin A, Vieira JM, Gestri G et al (2010) Tissue macrophages act as cellular chaperones for vascular anastomosis downstream of VEGF-mediated endothelial tip cell induction. Blood 116:829–840
69. Salcedo R, Wasserman K, Young HA et al (1999) Vascular endothelial growth factor and basic fibroblast growth factor induce expression of CXCR4 on human endothelial cells: in vivo neovascularization induced by stromal-derived factor-1alpha. Am J Pathol 154:1125–1135
70. Burns JM, Summers BC, Wang Y et al (2006) A novel chemokine receptor for SDF-1 and I-TAC involved in cell survival, cell adhesion, and tumor development. J Exp Med 203:2201–2213
71. Brouty-Boye D, Zetter BR (1980) Inhibition of cell motility by interferon. Science 208(4443):516–518
72. Kazerounian S, Yee KO, Lawler J (2008) Thrombospondins in cancer. Cell Mol Life Sci 65:700–712
73. Shaked Y, Bertolini F, Man S et al (2005) Genetic heterogeneity of the vasculogenic phenotype parallels angiogenesis; implications for cellular surrogate marker analysis of antiangiogenesis. Cancer Cell 7:101–111
74. Rastinejad F, Polverini PJ, Bouck NP (1989) Regulation of the activity of a new inhibitor of angiogenesis by a cancer suppressor gene. Cell 56:345–355
75. Good DJ, Polverini PJ, Rastinejad F et al (1990) A tumor suppressor-dependent inhibitor of angiogenesis is immunologically and functionally indistinguishable from a fragment of thrombospondin. Proc Natl Acad Sci U S A 87:6624–6628
76. Hanahan D, Weinberg RA (2000) The hallmarks of cancer. Cell 100:57–70
77. Pugh CW, Ratcliffe PJ (2003) Regulation of angiogenesis by hypoxia: role of the HIF system. Nat Med 9:677–684
78. Hurwitz H, Fehrenbacher L, Novotny W et al (2004) Bevacizumab plus irinotecan, fluorouracil, and leucovorin for metastatic colorectal cancer. N Engl J Med 350:2335–2342
79. Cook KM, Fig. WD (2010) Angiogenesis inhibitors: current strategies and future prospects. CA Cancer J Clin 60:222–243

Chapter 4
Neuropeptides and Angiogenesis

Leena Pradhan-Nabzdyk and Christoph Nabzdyk

Abstract Neuropeptides are one of the most conserved proteins across different species and are ubiquitously expressed in different organs. In the peripheral nervous system, neuropeptides are secreted by the sensory and autonomic nerves and participate in a wide range of functions including immune surveillance, cardiovascular homeostasis, regulation of endocrine function, cytokine and growth factor release, and importantly angiogenesis. Neuropeptides including neuropeptide Y, substance P, calcitonin gene-related peptide, vasoactive intestinal peptide, and somatostatin (SS) are some of the neuropeptides that have been investigated regarding their role in modulating the vascular system and angiogenesis. All of these neuropeptides are pro-angiogenic except SS, which has anti-angiogenic properties. This chapter aims to present up-to-date evidence on the various mechanisms of action of the aforementioned neuropeptides and their clinical implications.

Keywords Neuropeptides • Wound healing • Cancer • Ischemia

L. Pradhan-Nabzdyk, Ph.D. (✉)
Department of Surgery, Division of Vascular and Endovascular Surgery,
Beth Israel Deaconess Medical Center, Harvard Medical School,
330 Brookline Avenue, Dana Building 805, Boston, MA 02215, USA
e-mail: lpradhan@bidmc.harvard.edu

C. Nabzdyk
Department of Surgery, Tufts Medical Center, Boston, MA USA

J.L. Mehta and N.S. Dhalla (eds.), *Biochemical Basis and Therapeutic Implications of Angiogenesis*, Advances in Biochemistry in Health and Disease 6,
DOI 10.1007/978-1-4614-5857-9_4,

4.1 Introduction

Neuropeptides are some of the most conserved and abundant peptides in vertebrates and function as neurotransmitters or neuromodulators or both within the central nervous system (CNS) and peripheral nervous system (PNS). Neuropeptides released from the PNS participate in major physiological functions ranging from cardiovascular homeostasis and gastrointestinal motility to immune cell trafficking. Neuropeptides including neuropeptide Y (NPY), substance P (SP), calcitonin gene-related peptide (CGRP), vasoactive intestinal peptide (VIP), and somatostatin (SS) have emerged as important regulators that affect the vascular system by modulating the vascular tone, angiogenesis, and vascular remodeling [1–10]. Most of these neuropeptides are pro-angiogenic, with the exception of SS, which has anti-angiogenic properties. Numerous studies have investigated the role of neuropeptides in angiogenesis, but there has been limited success in the development of neuropeptide-based therapies targeting angiogenesis. While some of the neuropeptides or neuropeptide-based therapeutics is in the pipeline for treatment of other diseases, so far, SS analogues are the only neuropeptide-based therapies used for modulating/inhibiting angiogenesis in cancer. This suggests that the ubiquitous and complicated neuropeptide signaling system in the vasculature needs a more in-depth understanding. Additionally, delivery of these peptides and their receptor agonists/antagonists to the target organ also poses various challenges.

This chapter aims to provide the most up-to-date information on the widely studied angiogenic neuropeptides, NPY, SP, CGRP, VIP, and SS. The reader has to note that this is not an exhaustive review but rather a concise summary that includes those neuropeptides that have received the most attention for their role in modulating angiogenesis.

4.2 Neuropeptide Y

NPY is a 36-amino acid peptide and is one of the most abundantly and ubiquitously distributed neurotransmitters in the CNS and PNS [11]. In the CNS, it is mainly expressed in the hypothalamus, [12] and in the PNS, it is found in sympathetic nerves, where it is stored either alone in small vesicles or in combination with catecholamines in larger vesicles [13, 14]. Moreover, NPY is also produced by vascular smooth muscle cells, endothelial cells (ECs), vas deferens cells, and pancreatic acinar cells [12, 15].

NPY signals through six subtypes of G protein-coupled receptors (GPCRs) including NPY1R, NPY2R, NPY3R, NPY4R, NPY5R, and NPY6R [16]. NPY regulates vascular tone by inducing vasoconstriction [17], stimulates vascular smooth muscle cell growth and hypertrophy of ventricular cardiomyocytes [18, 19], and is also involved in immune cross talk [20]. Signaling through NPY1R,

NPY2R, and NPY5R is implicated in diabetes, heart failure, hypertension, peripheral arterial disease, and feeding disorders [21]. Furthermore, these receptors are involved in angiogenesis [22], calcium homeostasis [21], renin-angiotensin-aldosterone system (RAAS), and protein kinase C (PKC) activation of diabetes and heart failure [23].

4.3 NPY and Angiogenesis

In recent years, the contribution of NPY-associated angiogenesis to cardiovascular disease, wound healing, and cancer has received a lot of attention [1, 2, 24–26]. The first report that described the sympathetic co-transmitter NPY, as a pro-angiogenic molecule, was presented by Zukowska et al. and was based on work in human umbilical vein endothelial cells (HUVECs). The authors reported that not only the receptors NPY1R and NPY2R but also NPY itself and the enzyme, dipeptidyl peptidase (DPPIV) that converts NPY_{36} to the angiogenic NPY_{3-36}, are expressed in these cells. Of the NPY receptors, NPY1R, NPY2R, and NPY5R are involved in angiogenesis. So far the only known mechanisms by which NPY leads to angiogenesis are release of vascular endothelial growth factor (VEGF), platelet-derived growth factor (PDGF), and NO [26, 27].

NPY1R antagonism is shown to inhibit angioplasty-induced atherosclerotic-like vascular remodeling, without affecting ischemic revascularization; NPY2R activation is shown to stimulate ischemic angiogenesis. NPY5R enhances the effect of both NPY1R and NPY2R [6, 25]. NPY1R is normally implicated in pathologies of the cardiovascular system including cardiac hypertrophy and vasoconstriction, whereas NPY2R and NPY5R are known to ameliorate ischemia and impaired wound healing. In a recent swine study of chronic myocardial ischemia, treatment with exogenous NPY_{3-36} increased capillary and arteriole formation along with upregulation of NPY1R, NPY2R, NPY5R, VEGF, endothelial nitric oxide synthase (eNOS), phospho-eNOS (p-eNOS) on Ser1177, and PDGF and a downregulation of anti-angiogenic factors endostatin and angiostatin in the NPY-treated group [26]. This suggests that NPY signaling occurs in a feed-forward pro-angiogenic manner where it enhances the entire NPY signaling system by upregulating its own receptors.

NPY has been demonstrated to have a role in the angiogenesis phase of wound healing in both cutaneous and tendon healing by acting through NPY2R and NPY5R [1, 28, 29]. In fact, in NPY2R knockout mice, angiogenesis and thereby wound healing are disrupted [8].

In cancer, NPY and its receptors are expressed in both the tumor cells and the tumor vasculature. In human neuroblastoma tissues, NPY is predominant in the tumor cells whereas NPY2R is expressed in both tumor and ECs, making NPY2R a promising target for neuroblastoma therapy [30]. Similarly, prostate tumors and neural crest-derived tumors release NPY that can facilitate tumor vascularization [31, 32]. These studies demonstrate an important role for NPY in the treatment of wound healing, ischemic revascularization, and cancer.

4.4 Substance P

SP is an 11-amino acid neurotransmitter and neuromodulator widely distributed in the CNS and PNS [13, 33]. SP is released from C-fiber sensory nerves, in primary sensory neurons and neurons intrinsic to the gastrointestinal, respiratory, and genitourinary tracts [34, 35]. SP binds to three G protein-coupled receptors, NK1R, NK2R, and NK3R, out of which NK1R is its high-affinity receptor that is present on a variety of cell types including immune cells, keratinocytes, ECs, neurons, and glial cells [36]. SP is involved in several physiological processes, including maintenance of cardiovascular tone, smooth muscle activity, vomiting reflex, defensive behavior, and stimulation of salivary secretion [9, 37]. SP has been shown to be mitogenic towards smooth muscle cells, fibroblasts, and ECs [1, 35, 38–40]. SP has shown to affect different pathways in different cell types including activation of phospholipase C producing a net rise in intracellular (Ca^{2+}) and cAMP via AC [41, 42]; induction of cytokines, tumor necrosis factor-alpha (TNF-α), interleukin (IL)-6, IL-8, IL-1β, and IL-2, from T lymphocytes, macrophages, and neutrophils [43]; activation of protein kinase B (Akt) pathway in dopaminergic neurons [44]; and induction of transforming growth factor-beta (TGF-β) from fibroblasts [45, 46].

4.5 SP and Angiogenesis

The vasodilatory effect of SP was evident long before nitric oxide (NO) was recognized as the endothelium-derived relaxation factor (EDRF) [47]. Only later, it was demonstrated that SP-induced vasodilation was indeed mediated by NO via the NK1R [48, 49]. The first evidence indicating the role of SP in EC proliferation and neovascularization was demonstrated by Ziche and colleagues in vitro in HUVECs and in vivo in an avascular cornea rabbit model [40]. Similar to NPY, SP-mediated angiogenesis via NK1R contributes to wound healing, tumor angiogenesis, and ischemic angiogenesis [1, 50–53]. SP plays a major role in cutaneous wound healing by participating in the inflammatory and angiogenic phases [35]. In vitro studies indicate that SP treatment leads to proliferation and tube formation of dermal microvascular ECs in normal and hyperglycemic conditions [1, 54, 55]. Moreover, inhibition of the enzyme neprilysin that breaks down SP also enhances angiogenesis [56]. In cutaneous wound healing, NO release is the only known mechanism through which SP has been shown to lead to angiogenesis. In addition to cutaneous wound healing, SP has also shown to aid in tendon healing after injury by augmenting angiogenesis [51, 57].

SP and NK1R are implicated in different cancers including brain tumors, melanomas, and breast cancer [58–60]. NK1R receptors are upregulated in brain tumors, and therefore NK1R antagonists that cause tumor cell apoptosis and also inhibit tumor vascularization hold great promise in treating these tumors [58, 61]. Aprepitant, the NK1R antagonist approved for the treatment of chemotherapy-induced nausea and vomiting, has shown some promising results in inhibiting cell growth of different melanoma cell lines [62]. However, at this time aprepitant has not been approved for any cancer treatment. The genes that encode SP, preprotachykinin 1, and NK1R and NK2R have shown to be upregulated in breast cancer, and NK1R antagonists have shown to inhibit breast cancer cell growth in in vivo models [60, 63]. Currently there are no studies targeting breast tumor angiogenesis with NK1R antagonists.

In a recent mouse hind limb ischemia study, SP has shown to mobilize NK1R-expressing progenitor cells from bone marrow and promote reparative angiogenesis [50]. In the same study, authors showed that patients with acute myocardial infarction had high circulating levels of SP and NK1R-positive cells that co-express progenitor cell antigens, which are abundant in infarcted hearts, but interestingly not in hearts that developed an infarct after transplantation [50].

These studies demonstrate that in addition to its role as a sensory neuropeptide, SP and its receptors also play an important role in wound healing, ischemic revascularization, and cancer.

4.6 Calcitonin Gene-Related Peptide

CGRP is a 37-amino acid peptide generated in the CNS and PNS from the alternate splicing of calcitonin gene mRNA in a tissue-specific manner [64]. In the PNS, CGRP is expressed in the nerves innervating the skin, gut, pancreas, heart, and the vasculature. It is co-localized with SP in the sensory nervous system and, similar to SP, is a potent vasodilator [65].

CGRP receptor is a heterodimer of calcitonin receptor-like receptor (CRLR), a GPCR, which is linked to receptor activity modifying protein (RAMP)1, RAMP2, or RAMP3 [66]. CRLR-RAMP1 is specific for CGRP while CRLR-RAMP2 and RAMP3 can also bind adrenomedullin, a related peptide [66]. CGRP is known to be involved in vasodilation, nociception, glucose uptake, and the stimulation of glycolysis in skeletal muscles [67]. Binding of CGRP to its receptors leads to an increase of cAMP and activation of protein kinase A (PKA), phospholipase C beta (PLCβ1), mitogen-activated protein kinase (MAPK), and production of NO [68].

4.7 CGRP and Angiogenesis

Similar to SP, CGRP was found to be a potent vasodilator that acted through an endothelial-dependent mechanism which later was identified to be NO [69, 70]. The first connection between CGRP and angiogenesis was identified in a rat ischemic skin flap partial denervation study where angiogenesis preceded reinnervation with CGRP positive nerves [71]. In a subsequent study, the same group showed that in fact CGRP at least partially contributed towards angiogenesis [72]. Similar to SP, CGRP-associated angiogenesis is involved in wound healing, ischemic revascularization and tumor vascularization. In a rat knee-joint model intra-articular CGRP injection increased endothelial cell proliferation while inhibition of CRLR-RAMP1 attenuated EC proliferation in capsaicin-induced knee-joint synovitis [73]. In a model of rat hind limb ischemia, CGRP levels were elevated in the ischemic tissue, and over-expression of CGRP enhanced blood flow recovery and increased capillary density in ischemic hind limbs most likely via activation of AMP-activated protein kinase (AMPK) [74]. In vitro, CGRP induced p-eNOS in HUVECs at Ser1177 and Ser633 in a time-dependent manner, and these effects were abolished by AMPK inhibition [74]. In addition to AMPK and eNOS, CGRP has also shown to enhance angiogenesis in cutaneous and gastric mucosal healing via VEGF [75, 76]. In CGRP knockout mice, tumor growth and tumor-associated angiogenesis of implanted Lewis lung carcinoma (LLC) cells along with downregulation of VEGF expression in tumor stroma were significantly reduced compared with those in wild-type mice [77].

CGRP is very similar to SP in its angiogenic profile and can serve as an important target for wound healing and cancer.

4.8 Vasoactive Intestinal Peptide

VIP is a 28-amino acid peptide that is expressed all throughout the CNS including the hypothalamus, PNS, intestines, pancreas, urogenital tract, thyroid, and adrenal glands [78].

VIP affects a wide range of biological activities including vasodilation and smooth muscle relaxation, stimulation of pepsinogen secretion by the chief cells of the gut, secretion of water and electrolytes into the intestines, enhancement of glycogen metabolism in the cerebral cortex, regulation of embryonic growth, promotion of neuronal survival, and modulation of the immune system as well as mammalian circadian rhythm [78]. VIP exerts its biological effects through its receptors, vasoactive intestinal polypeptide receptor 1 (VPAC1) and VPAC2 which are GPCRs. Receptor activation leads to modulating the activity of phospholipase D (PLD) and also increased cAMP and intracellular calcium levels [78].

4.9 VIP and Angiogenesis

Similar to SP and CGRP, VIP was identified as a vasodilator in a canine model where VIP was injected in the glandular artery of the submandibular gland [79]. Later it was demonstrated that VIP-induced vasodilation is also mediated via l-arginine-NO pathway [80]. VIP-associated angiogenesis is implicated in tumor vascularization with some studies demonstrating its protective role in ischemic injury. Although

VIP is shown to aid in wound healing, most of this effect is attributed to its role in bronchial epithelial cell proliferation [81] and keratinocyte proliferation [82, 83] but not through angiogenesis. In addition to increasing EC proliferation, VIP is also known to increase EC migration [84]. In a model of acute cerebral ischemia, a single dose of intracerebroventricular injection of VIP at the beginning of reperfusion led to increased number of ECs and microvessels at the boundary of the ischemic lesion. VIP further significantly increased VEGF levels in the ischemic hemisphere as well as VEGF receptor flt-1 and flk-1 immunoreactivity in Ecs [85, 86].

In the human small lung cancer cell line H446 VIP administration increased c-fos and VEGF mRNA expression that was reversed by c-fos antisense oligodeoxynucleotide [87]. In a mouse xenograft model in which human prostate cancer cells were transplanted, VIP administration resulted in increased tumor growth along with increased VEGF expression [88]. In another study, the same group showed that VIP induces VEGF mRNA expression via c-fos, which in turn is induced by calcium signaling in human prostate LNCaP cells [89]. In a colon carcinoma model, systemic VIP treatment reduced angiogenesis within tumor masses by cAMP-dependent mechanism [90].

Similar to NPY, SP, and CGRP, VIP has been shown to play a role in ischemic vascularization and in cancer angiogenesis; however, its role in the angiogenic phase of wound healing has not yet been investigated. Given the angiogenic profile of VIP, it would not be too far-fetched to surmise that VIP could promote wound healing.

4.10 Somatostatin

Somatostatin (SS) which was discovered in 1974 by Brazeau et al. [91] is a major inhibitory neuropeptide produced in the hypothalamus and the arcuate nucleus but also secreted by peripheral nerves in the gut [92, 93]. In fact, the gastrointestinal tract contains about 70 % of the total SS [94]. SS exists in three forms, the 14 amino acid, SS-14 (clinical analogues are made towards SS-14); the 28 amino acid, SS-28; and the 25 amino acid, SS-25 [95]. SS-28 and SS-25 are in fact SS-14 with a 14-amino acid and an 11-amino acid extension at the N-terminus, respectively [96]. SS-14 and SS-28 have similar physiological activities but bind to the receptors with different potencies, while not much is known about SS-25 [95].

SS exerts its effects via five different receptor subtypes (sst1–sst5) that are high-affinity membrane-bound GPCRs and are expressed throughout the CNS and on the pancreas, gut, pituitary, kidneys, thyroid, lung, and endothelial and immune cells [92, 93]. SS affects various tissues and is known to inhibit endocrine and exocrine secretions, intestinal motility, modulate neurotransmission, motor and cognitive functions, vascular contractility, and cell proliferation [92]. All five receptors are known to play a role in the anti-proliferative effect of SS. Binding of SS to sst2 and sst3 leads to apoptosis via p53-independent pathways. On the other hand, binding of SS to sst1, sst2, sst4, and sst5 leads to cytostasis via different pathways including increase of $p21^{Cip1/Waf1}$, induction of $p27^{Kip1}$, hypophosphorylation of retinoblastoma (Rb) protein, inhibition of cyclin E/cdk2, inhibition of MAPK cascade, and inhibition of guanylyl cyclase [93]. Additionally, SS also inhibits cellular proliferation by inhibiting secretion of numerous growth factors such as growth hormone, epidermal growth factor (EGF), basic fibroblast growth factor (bFGF), insulin-like growth factors I and II (IGF-I and IGF-II), insulin-like growth factor binding protein, and PDGF [93].

4.11 Somatostatin and Angiogenesis

In contrast to all the aforementioned neuropeptides, one of the earliest studies investigating the role of SS on vasculature suggested that SS decreases blood flow in the gut [96]. Later on it was suggested that this was a direct effect of SS on vascular smooth muscle cells [97]. It is now known that the vasomodulatory effects of SS are confined to the splanchnic circulation and that SS decreases gut motility most likely by causing the release of vasoconstrictors such as endothelin-1 [98].

SS as an anti-angiogenic peptide has a therapeutic potential in the treatment of diseases such as cancer, proliferative retinopathy, and endometriosis where the goal is to inhibit angiogenesis. Using a chick embryo-chorioallantoic membrane (CAM) assay, Woltering and colleagues conducted one of the first studies indicating a role for SS as an anti-angiogenic neuropeptide, suggesting a role for SS in tumor angiogenesis [99].

sst2 is the most commonly involved receptor eliciting the anti-angiogenic effects of SS along with sst1 and sst5. SS is shown to have anti-angiogenic effect by suppressing the expression of pro-angiogenic factors such as VEGF and VEGFR2 [100], IGF-R1, and angiopoietin 2 (Ang 2) and its receptor Tie-2 [101] or by increasing the expression of anti-angiogenic factors such as thrombospondin-1 (TSP-1) [102].

The antitumor effect of SS could be direct by leading to tumor cell apoptosis or indirect by suppressing angiogenesis and inhibiting growth factors [103]. sst2 is commonly involved in cancers. Over-expression of sst2 in human pancreatic (capan-2) and lung cancer cells (A549) transplanted in mice inhibited the growth of both sst2-positive and sst2-negative xenografts by affecting the cellular levels of signaling molecules in apoptotic pathways, MAPK pathway, and angiogenesis [104]. SS analogue, octreotide, has been therapeutically used in cancer treatment, especially neuroendocrine tumors [105–107]. Octreotide has also shown beneficial effects in a mouse hepatocellular carcinoma model where it inhibited the incidence of second primary tumors, decreased lung metastasis, and prolonged the life span by decreasing intratumoral angiogenesis [108]. In a gastric cancer trial, compared to patients that received placebo, patients that received SS had significant decrease in serum VEGF level, which was partially dependent on the synthesis and degradation of the protein but not the transcription of mRNA

[109]. SS treatment has also demonstrated beneficial effects in prostate cancer, pancreatic adenocarcinoma models, and an ovarian cancer study [110–113].

In addition to cancer treatment, SS analogues are used for the treatment of refractory bleeding in gastrointestinal angiodysplasias [111, 114] and have also shown beneficial effects in models of endometriosis where receptors sst1, sst2, and sst5 are highly expressed [115]. SS is also used in treatment of proliferative retinopathy [116, 117] where another SS receptor agonist, non-peptide imidazolidine-2,4-dione (NISA), similar to octreotide, inhibited growth factor-induced EC proliferation, migration, and tube formation [118].

Unlike the other neuropeptides discussed, SS is the only one that is proven to be anti-angiogenic and the only one that has been successfully used as an anti-angiogenic agent in cancers. In addition, these data demonstrate that SS-based therapy can also be promising in the treatment of other diseases such as proliferative retinopathy and endometriosis.

4.12 Conclusions

Neuropeptides have a tremendous potential to be developed as therapies for treating myriads of diseases in which modulation of angiogenesis is essential. Thus in ischemia and impaired wound healing, therapies should aim at enhancing NPY, SP, CGRP, and VIP signaling and to inhibit SS signaling system (Fig. 4.1). Conversely in cancer, the goal ought to be to block the NPY, SP, CGRP, and VIP signaling system and to enhance SS signaling (Fig. 4.1). At time there is lack of in-depth understanding regarding the signaling mechanisms of some neuropeptides therefore, the common and overlapping pathways through which some of these neuropeptides signal should be further explored for their potential therapeutic use. Another area that needs investigation is delivery of neuropeptides, their analogues or receptor antagonists at the disease site. For example, one of the challenges in treating chronic

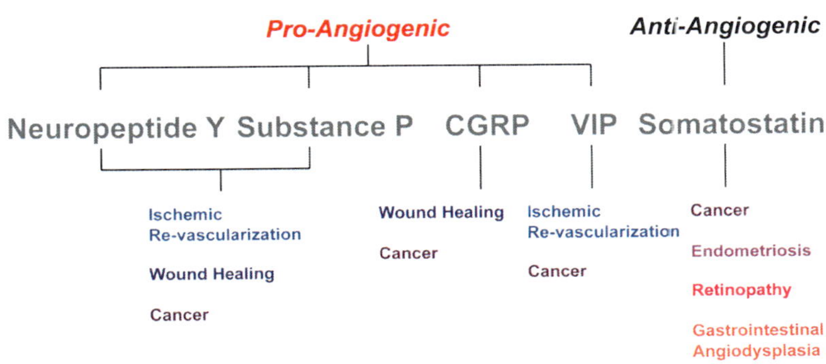

Fig. 4.1 Summary of the role played by neuropeptides in different angiogenesis-based functions and pathologies

wounds such as diabetic foot ulcers is that the wound microenvironment is very hostile with high expression of proteases, and therefore direct application of neuropeptides to the wound site is not feasible. Another example is in treatment of ischemic revascularization where a systemic delivery of neuropeptide and/or analogue is necessary. One way to overcome this problem is by inhibiting proteases that could cause endogenous neuropeptide degradation instead of systemic injection of neuropeptides that is not feasible. One such protease, neprilysin, that breaks down substance P has already received some attention. However, this too could pose a problem because systemic increase of neuropeptides has a high potential for side effects as these neuropeptides participate in several different physiological processes.

Overall, neuropeptides play an important role in angiogenesis and their ubiquitous nature emphasizes that they can affect angiogenesis in almost all organs. Thus further in-depth exploration is necessary to develop neuropeptide-based angiogenesis therapies and develop strategies for their efficacious delivery.

References

1. Hulagu S, Senturk O, Erdem A et al (2002) Effects of losartan, somatostatin and losartan plus somatostatin on portal hemodynamics and renal functions in cirrhosis. Hepatogastroenterology 49:783–787
2. Ruohonen ST, Abe K, Kero M et al (2009) Sympathetic nervous system-targeted neuropeptide Y overexpression in mice enhances neointimal formation in response to vascular injury. Peptides 30:715–720
3. Kuo LE, Abe K, Zukowska Z (2007) Stress, NPY and vascular remodeling: implications for stress-related diseases. Peptides 28:435–440
4. Abe K, Tilan JU, Zukowska Z (2007) NPY and NPY receptors in vascular remodeling. Curr Top Med Chem 7:1704–1709
5. Zukowska Z, Grant DS, Lee EW (2003) Neuropeptide Y: a novel mechanism for ischemic angiogenesis. Trends Cardiovasc Med 13:86–92
6. Lee EW, Grant DS, Movafagh S et al (2003) Impaired angiogenesis in neuropeptide Y (NPY)-Y2 receptor knockout mice. Peptides 24:99–106
7. Ejaz A, LoGerfo FW, Pradhan L (2011) Diabetic neuropathy and heart failure: role of neuropeptides. Expert Rev Mol Med 13:e26
8. Jain M, LoGerfo FW, Guthrie P et al (2011) Effect of hyperglycemia and neuropeptides on interleukin-8 expression and angiogenesis in dermal microvascular endothelial cells. J Vasc Surg 53:1654–1660 e2
9. Pradhan L, Cai X, Wu S et al (2011) Gene expression of pro-inflammatory cytokines and neuropeptides in diabetic wound healing. J Surg Res 167:336–342
10. Zhang S, Liu Y, Guo S et al (2010) Vasoactive intestinal polypeptide relaxes isolated rat pulmonary artery rings through two distinct mechanisms. J Physiol Sci 60:389–397
11. Blomqvist AG, Herzog H (1997) Y-receptor subtypes—how many more? Trends Neurosci 20:294–298
12. Adrian TE, Allen JM, Bloom SR et al (1983) Neuropeptide Y distribution in human brain. Nature 306:584–586
13. Fried G, Terenius L, Hokfelt T et al (1985) Evidence for differential localization of noradrenaline and neuropeptide Y in neuronal storage vesicles isolated from rat vas deferens. J Neurosci 5:450–458

14. Ekblad E, Edvinsson L, Wahlestedt C et al (1984) Neuropeptide Y co-exists and co-operates with noradrenaline in perivascular nerve fibers. Regul Pept 8:225–235
15. Zukowska Z, Pons J, Lee EW et al (2003) Neuropeptide Y: a new mediator linking sympathetic nerves, blood vessels and immune system? Can J Physiol Pharmacol 81:89–94
16. Brothers SP, Wahlestedt C (2010) Therapeutic potential of neuropeptide Y (NPY) receptor ligands. EMBO Mol Med 2:429–439
17. Franco-Cereceda A, Lundberg JM, Dahlof C (1985) Neuropeptide Y and sympathetic control of heart contractility and coronary vascular tone. Acta Physiol Scand 124:361–369
18. Erlinge D, Brunkwall J, Edvinsson L (1994) Neuropeptide Y stimulates proliferation of human vascular smooth muscle cells: cooperation with noradrenaline and ATP. Regul Pept 50:259–265
19. Zukowska-Grojec Z, Pruszczyk P, Colton C et al (1993) Mitogenic effect of neuropeptide Y in rat vascular smooth muscle cells. Peptides 14:263–268
20. Bedoui S, Kawamura N, Straub RH et al (2003) Relevance of neuropeptide Y for the neuroimmune crosstalk. J Neuroimmunol 134:1–11
21. Pedrazzini T, Pralong F, Grouzmann E (2003) Neuropeptide Y: the universal soldier. Cell Mol Life Sci 60:350–377
22. Zukowska-Grojec Z, Karwatowska-Prokopczuk E, Rose W et al (1998) Neuropeptide Y: a novel angiogenic factor from the sympathetic nerves and endothelium. Circ Res 83:187–195
23. Kohno D, Gao HZ, Muroya S et al (2003) Ghrelin directly interacts with neuropeptide-Y-containing neurons in the rat arcuate nucleus: Ca2+ signaling via protein kinase A and N-type channel-dependent mechanisms and cross-talk with leptin and orexin. Diabetes 52:948–956
24. Ejaz A, LoGerfo FW, Khabbaz K et al (2011) Expression of neuropeptide Y, substance P, and their receptors in the right atrium of diabetic patients. Clin Transl Sci 4:346–450
25. Robich MP, Matyal R, Chu LM et al (2010) Effects of neuropeptide Y on collateral development in a swine model of chronic myocardial ischemia. J Mol Cell Cardiol 49:1022–1030
26. Zatelli MC, Minoia M, Martini C et al (2010) Everolimus as a new potential antiproliferative agent in aggressive human bronchial carcinoids. Endocr Relat Cancer 17:719–729
27. Lee EW, Michalkiewicz M, Kitlinska J et al (2003) Neuropeptide Y induces ischemic angiogenesis and restores function of ischemic skeletal muscles. J Clin Invest 111:1853–1862
28. Movafagh S, Hobson JP, Spiegel S et al (2006) Neuropeptide Y induces migration, proliferation, and tube formation of endothelial cells bimodally via Y1, Y2, and Y5 receptors. FASEB J 20:1924–1926
29. Ackermann PW, Ahmed M, Kreicbergs A (2002) Early nerve regeneration after Achilles tendon rupture—a prerequisite for healing? A study in the rat. J Orthop Res 20:849–856
30. Ekstrand AJ, Cao R, Bjorndahl M et al (2003) Deletion of neuropeptide Y (NPY) 2 receptor in mice results in blockage of NPY-induced angiogenesis and delayed wound healing. Proc Natl Acad Sci U S A 100:6033–6038
31. Lu C, Everhart L, Tilan J et al (2010) Neuropeptide Y and its Y2 receptor: potential targets in neuroblastoma therapy. Oncogene 29:5630–5642
32. Lenkinski RE, Bloch BN, Liu F et al (2008) An illustration of the potential for mapping MRI/MRS parameters with genetic over-expression profiles in human prostate cancer. MAGMA 21:411–421
33. Kitlinska J, Abe K, Kuo L et al (2005) Differential effects of neuropeptide Y on the growth and vascularization of neural crest-derived tumors. Cancer Res 65:1719–1728
34. Hokfelt T, Kellerth JO, Nilsson G et al (1975) Experimental immunohistochemical studies on the localization and distribution of substance P in cat primary sensory neurons. Brain Res 100:235–252
35. Maggi CA (2000) The troubled story of tachykinins and neurokinins. Trends Pharmacol Sci 21:173–175
36. Pradhan L, Nabzdyk C, Andersen ND et al (2009) Inflammation and neuropeptides: the connection in diabetic wound healing. Expert Rev Mol Med 11:e2

37. Maggi CA (1995) The mammalian tachykinin receptors. Gen Pharmacol 26:911–944
38. Nieber K, Oehme P (1982) [Substance P—a neuropeptide transmitter]. Z Gesamte Inn Med 37:577–582
39. Nilsson J, von Euler AM, Dalsgaard CJ (1985) Stimulation of connective tissue cell growth by substance P and substance K. Nature 315:61–63
40. Ziche M, Morbidelli L, Pacini M et al (1990) Substance P stimulates neovascularization in vivo and proliferation of cultured endothelial cells. Microvasc Res 40:264–278
41. Rameshwar P, Poddar A, Zhu G et al (1997) Receptor induction regulates the synergistic effects of substance P with IL-1 and platelet-derived growth factor on the proliferation of bone marrow fibroblasts. J Immunol 158:3417–3424
42. Khawaja AM, Rogers DF (1996) Tachykinins: receptor to effector. Int J Biochem Cell Biol 28:721–738
43. Harrison S, Geppetti P (2001) Substance p. Int J Biochem Cell Biol 33:555–576
44. Delgado AV, McManus AT, Chambers JP (2003) Production of tumor necrosis factor-alpha, interleukin 1-beta, interleukin 2, and interleukin 6 by rat leukocyte subpopulations after exposure to substance P. Neuropeptides 37:355–361
45. Chu JM, Chen LW, Chan YS et al (2011) Neuroprotective effects of neurokinin receptor one in dopaminergic neurons are mediated through Akt/PKB cell signaling pathway. Neuropharmacology 61:1389–1398
46. Bulut K, Felderbauer P, Deters S et al (2008) Sensory neuropeptides and epithelial cell restitution: the relevance of SP- and CGRP-stimulated mast cells. Int J Colorectal Dis 23:535–541
47. Felderbauer P, Bulut K, Hoeck K et al (2007) Substance P induces intestinal wound healing via fibroblasts—evidence for a TGF-beta-dependent effect. Int J Colorectal Dis 22:1475–1480
48. Couture R, Gaudreau P, St-Pierre S et al (1980) The dog common carotid artery: a sensitive bioassay for studying vasodilator effects of substance P and of kinins. Can J Physiol Pharmacol 58:1234–1244
49. Persson MG, Hedqvist P, Gustafsson LE (1991) Nerve-induced tachykinin-mediated vasodilation in skeletal muscle is dependent on nitric oxide formation. Eur J Pharmacol 205:295–301
50. Gustafsson LE, Wiklund CU, Wiklund NP et al (1990) Modulation of autonomic neuroeffector transmission by nitric oxide in guinea pig ileum. Biochem Biophys Res Commun 173:106–110
51. Amadesi S, Reni C, Katare R et al (2012) Role for substance P-based nociceptive signaling in progenitor cell activation and angiogenesis during ischemia in mice and in human subjects. Circulation 125(14):1774–1786
52. Andersson G, Backman LJ, Scott A et al (2011) Substance P accelerates hypercellularity and angiogenesis in tendon tissue and enhances paratendinitis in response to Achilles tendon overuse in a tendinopathy model. Br J Sports Med 45:1017–1022
53. Munoz M, Covenas R (2010) Neurokinin-1 receptor: a new promising target in the treatment of cancer. Discov Med 10:305–313
54. Munoz M, Rosso M, Covenas R (2011) The NK-1 receptor: a new target in cancer therapy. Curr Drug Targets 12:909–921
55. Wiedermann CJ, Auer B, Sitte B et al (1996) Induction of endothelial cell differentiation into capillary-like structures by substance P. Eur J Pharmacol 298:335–338
56. Kohara H, Tajima S, Yamamoto M et al (2010) Angiogenesis induced by controlled release of neuropeptide substance P. Biomaterials 31:8617–8625
57. Scott JR, Muangman P, Gibran NS (2007) Making sense of hypertrophic scar: a role for nerves. Wound Repair Regen 15(suppl 1):S27–S31
58. Burssens P, Steyaert A, Forsyth R et al (2005) Exogenously administered substance P and neutral endopeptidase inhibitors stimulate fibroblast proliferation, angiogenesis and collagen organization during Achilles tendon healing. Foot Ankle Int 26:832–839

59. Khare VK, Albino AP, Reed JA (1998) The neuropeptide/mast cell secretagogue substance P is expressed in cutaneous melanocytic lesions. J Cutan Pathol 25:2–10
60. Singh D, Joshi DD, Hameed M et al (2000) Increased expression of preprotachykinin-I and neurokinin receptors in human breast cancer cells: implications for bone marrow metastasis. Proc Natl Acad Sci U S A 97:388–393
61. Harford-Wright E, Lewis KM, Vink R (2011) Towards drug discovery for brain tumours: interaction of kinins and tumours at the blood brain barrier interface. Recent Pat CNS Drug Discov 6:31–40
62. Munoz M, Covenas R (2011) NK-1 receptor antagonists: a new paradigm in pharmacological therapy. Curr Med Chem 18:1820–1831
63. Munoz M, Rosso M, Robles-Frias MJ et al (2010) The NK-1 receptor is expressed in human melanoma and is involved in the antitumor action of the NK-1 receptor antagonist aprepitant on melanoma cell lines. Lab Invest 90:1259–1269
64. Mayordomo C, Garcia-Recio S, Ametller E et al (2012) Targeting of substance P induces cancer cell death and decreases the steady state of EGFR and Her2. J Cell Physiol 227:1358–1366
65. Adeghate E, Ponery A (2003) Pancreatic peptides, neuropeptides and neurotransmitters in diabetes mellitus: a review. Int J Diabetes Metab 11:1–6
66. Bergdahl A, Valdemarsson S, Nilsson T et al (1999) Dilatory responses to acetylcholine, calcitonin gene-related peptide and substance P in the congestive heart failure rat. Acta Physiol Scand 165:15–23
67. McLatchie LM, Fraser NJ, Main MJ et al (1998) RAMPs regulate the transport and ligand specificity of the calcitonin-receptor-like receptor. Nature 393:333–339
68. van Rossum D, Hanisch UK, Quirion R (1997) Neuroanatomical localization, pharmacological characterization and functions of CGRP, related peptides and their receptors. Neurosci Biobehav Rev 21:649–678
69. Walker CS, Conner AC, Poyner DR et al (2010) Regulation of signal transduction by calcitonin gene-related peptide receptors. Trends Pharmacol Sci 31:476–483
70. Brain SD, Williams TJ, Tippins JR et al (1985) Calcitonin gene-related peptide is a potent vasodilator. Nature 313:54–56
71. Gardiner SM, Compton AM, Kemp PA et al (1991) Haemodynamic effects of human alpha-calcitonin gene-related peptide following administration of endothelin-1 or NG-nitro-L-arginine methyl ester in conscious rats. Br J Pharmacol 103:1256–1262
72. Manek S, Terenghi G, Shurey C et al (1993) Neovascularisation precedes neural changes in the rat groin skin flap following denervation: an immunohistochemical study. Br J Plast Surg 46:48–55
73. Manek S, Terenghi G, Shurey C et al (1994) Angiogenesis and reinnervation in skin flaps: the effects of ischaemia examined in an animal model. Int J Exp Pathol 75:243–255
74. Mapp PI, McWilliams DF, Turley MJ et al (2012) A role for the sensory neuropeptide calcitonin gene-related peptide in endothelial cell proliferation in vivo. Br J Pharmacol 166:1261–1271
75. Zheng S, Li W, Xu M et al (2010) Calcitonin gene-related peptide promotes angiogenesis via AMP-activated protein kinase. Am J Physiol Cell Physiol 299:C1485–C1492
76. Toda M, Suzuki T, Hosono K et al (2008) Roles of calcitonin gene-related peptide in facilitation of wound healing and angiogenesis. Biomed Pharmacother 62:352–359
77. Ohno T, Hattori Y, Komine R et al (2008) Roles of calcitonin gene-related peptide in maintenance of gastric mucosal integrity and in enhancement of ulcer healing and angiogenesis. Gastroenterology 134:215–225
78. Toda M, Suzuki T, Hosono K et al (2008) Neuronal system-dependent facilitation of tumor angiogenesis and tumor growth by calcitonin gene-related peptide. Proc Natl Acad Sci U S A 105:13550–13555
79. White CM, Ji S, Cai H et al (2010) Therapeutic potential of vasoactive intestinal peptide and its receptors in neurological disorders. CNS Neurol Disord Drug Targets 9:661–666

80. Gaw AJ, Aberdeen J, Humphrey PP et al (1991) Relaxation of sheep cerebral arteries by vasoactive intestinal polypeptide and neurogenic stimulation: inhibition by L-NG-monomethyl arginine in endothelium-denuded vessels. Br J Pharmacol 102:567–572
81. Guan CX, Cui YR, Sun GY et al (2009) Role of CREB in vasoactive intestinal peptide-mediated wound healing in human bronchial epithelial cells. Regul Pept 153:64–69
82. Wollina U, Huschenbeck J, Knoll B et al (1997) Vasoactive intestinal peptide supports induced migration of human keratinocytes and their colonization of an artificial polyurethane matrix. Regul Pept 70:29–36
83. Wollina U, Knopf B (1993) Vasoactive-intestinal-peptide (vip) modulates early events of migration in human keratinocytes. Int J Oncol 2:229–232
84. Marion-Audibert AM, Nejjari M, Pourreyron C et al (2000) [Effects of endocrine peptides on proliferation, migration and differentiation of human endothelial cells]. Gastroenterol Clin Biol 24:644–648
85. Yang J, Zong CH, Zhao CH et al (2009) [Vasoactive intestinal peptide enhances angiogenesis after focal cerebral ischemia]. Nan Fang Yi Ke Da Xue Xue Bao 29:619–622
86. Yang J, Zong CH, Zhao ZH et al (2009) Vasoactive intestinal peptide in rats with focal cerebral ischemia enhances angiogenesis. Neuroscience 161:413–421
87. Zhao Z, Cheng Q, Li X et al (2006) [c-fos antisense oligodeoxynucleotide reduces VIP-induced upregulation of VEGF expression in small cell lung cancer cells]. Zhongguo Fei Ai Za Zhi 9:312–315
88. Collado B, Carmena MJ, Clemente C et al (2007) Vasoactive intestinal peptide enhances growth and angiogenesis of human experimental prostate cancer in a xenograft model. Peptides 28:1896–1901
89. Collado B, Sanchez MG, Diaz-Laviada I et al (2005) Vasoactive intestinal peptide (VIP) induces c-fos expression in LNCaP prostate cancer cells through a mechanism that involves Ca2+ signalling. Implications in angiogenesis and neuroendocrine differentiation. Biochim Biophys Acta 1744:224–233
90. Ogasawara M, Murata J, Kamitani Y et al (1999) Inhibition by vasoactive intestinal polypeptide (VIP) of angiogenesis induced by murine colon 26-L5 carcinoma cells metastasized in liver. Clin Exp Metastasis 17:283–291
91. Brazeau P, Vale W, Burgus R et al (1974) Isolation of somatostatin (a somatotropin release inhibiting factor) of ovine hypothalamic origin. Can J Biochem 52:1067–1072
92. Ferjoux G, Bousquet C, Cordelier P et al (2000) Signal transduction of somatostatin receptors negatively controlling cell proliferation. J Physiol Paris 94:205–210
93. Dasgupta P (2004) Somatostatin analogues: multiple roles in cellular proliferation, neoplasia, and angiogenesis. Pharmacol Ther 102:61–85
94. Shulkes A (1994) Somatostatin: physiology and clinical applications. Baillieres Clin Endocrinol Metab 8:215–236
95. McIntosh CH, Bakich V, Kwok YN et al (1986) A comparison of the inhibitory effects of somatostatin-14, -25, and -28 on motility of the guinea pig ileum. Can J Physiol Pharmacol 64:303–306
96. Koch H (1976) [Gastrointestinal hormones and blood circulation in the gastric mucosa]. Z Gastroenterol 14(suppl):105–109
97. Tyden G, Samnegard H, Thulin L et al (1979) Circulatory effects of somatostatin in anesthetized man. Acta Chir Scand 145:443–446
98. Huang HC, Lee FY, Chan CC et al (2002) Effects of somatostatin and octreotide on portal-systemic collaterals in portal hypertensive rats. J Hepatol 36:163–168
99. Woltering EA, Barrie R, O'Dorisio TM et al (1991) Somatostatin analogues inhibit angiogenesis in the chick chorioallantoic membrane. J Surg Res 50:245–251
100. Bocci G, Culler MD, Fioravanti A et al (2007) In vitro antiangiogenic activity of selective somatostatin subtype-1 receptor agonists. Eur J Clin Invest 37:700–708
101. Dal Monte M, Cammalleri M, Martini D et al (2007) Antiangiogenic role of somatostatin receptor 2 in a model of hypoxia-induced neovascularization in the retina: results from transgenic mice. Invest Ophthalmol Vis Sci 48:3480–3489

102. Laklai H, Laval S, Dumartin L et al (2009) Thrombospondin-1 is a critical effector of oncosuppressive activity of sst2 somatostatin receptor on pancreatic cancer. Proc Natl Acad Sci U S A 106:17769–17774
103. Hasskarl J, Kaufmann M, Schmid HA (2011) Somatostatin receptors in non-neuroendocrine malignancies: the potential role of somatostatin analogs in solid tumors. Future Oncol 7:895–913
104. Zhou T, Xiao X, Xu B et al (2009) Overexpression of SSTR2 inhibited the growth of SSTR2-positive tumors via multiple signaling pathways. Acta Oncol 48:401–410
105. Faivre S, Sablin MP, Dreyer C et al (2010) Novel anticancer agents in clinical trials for well-differentiated neuroendocrine tumors. Endocrinol Metab Clin North Am 39:811–826
106. Valentino J, Evers BM (2011) Recent advances in the diagnosis and treatment of gastrointestinal carcinoids. Adv Surg 45:285–300
107. Florio T (2008) Molecular mechanisms of the antiproliferative activity of somatostatin receptors (SSTRs) in neuroendocrine tumors. Front Biosci 13:822–840
108. Jia WD, Xu GL, Wang W et al (2009) A somatostatin analogue, octreotide, inhibits the occurrence of second primary tumors and lung metastasis after resection of hepatocellular carcinoma in mice. Tohoku J Exp Med 218:155–160
109. Zhao B, Yang P, Yang J et al (2011) A randomized trial of somatostatin to regulate the VEGFs/VEGFRs in patients with gastric cancer. Hepatogastroenterology 58:1425–1430
110. Sun LC, Luo J, Mackey LV et al (2007) A conjugate of camptothecin and a somatostatin analog against prostate cancer cell invasion via a possible signaling pathway involving PI3K/Akt, alphaVbeta3/alphaVbeta5 and MMP-2/-9. Cancer Lett 246:157–166
111. Kumar M, Liu ZR, Thapa L et al (2004) Anti-angiogenic effects of somatostatin receptor subtype 2 on human pancreatic cancer xenografts. Carcinogenesis 25:2075–2081
112. Hall GH, Turnbull LW, Bedford K et al (2005) Neuropilin-1 and VEGF correlate with somatostatin expression and microvessel density in ovarian tumours. Int J Oncol 27:1283–1288
113. Carrere N, Vernejoul F, Souque A et al (2005) Characterization of the bystander effect of somatostatin receptor sst2 after in vivo gene transfer into human pancreatic cancer cells. Hum Gene Ther 16:1175–1193
114. Molina-Infante J, Perez-Gallardo B (2011) Somatostatin analogues for bleeding gastrointestinal angiodysplasias: when should thalidomide be prescribed? Dig Dis Sci 56:266–267
115. Fasciani A, Quilici P, Biscaldi E et al (2010) Overexpression and functional relevance of somatostatin receptor-1, -2, and -5 in endometrium and endometriotic lesions. J Clin Endocrinol Metab 95:5315–5319
116. Prokosch V, Fink J, Heiduschka P et al (2011) VEGF, Ang-2 and SRIF associated abnormal postnatal development of the retinal capillary network in the Royal College of Surgeons rat. Exp Eye Res 92:128–137
117. Simo R, Carrasco E, Garcia-Ramirez M et al (2006) Angiogenic and antiangiogenic factors in proliferative diabetic retinopathy. Curr Diabetes Rev 2:71–98
118. Palii SS, Afzal A, Shaw LC et al (2008) Nonpeptide somatostatin receptor agonists specifically target ocular neovascularization via the somatostatin type 2 receptor. Invest Ophthalmol Vis Sci 49:5094–5102

Chapter 5
Role of Integrins in Angiogenesis

Ghazaleh Tabatabai and Michael Weller

Abstract Angiogenesis critically depends on environmental factors. In particular, cellular adhesion and migration events play a critical role in the formation of new blood vessels from preexisting cells in multiple pathological conditions. Integrins are a large family of cell surface receptors that transfer signals from the extracellular microenvironment into the intracellular compartment of endothelial cells or tumor cells. In this chapter, we review the role of integrins in inducing and maintaining angiogenesis by regulating the survival, proliferation, and migration of endothelial cells as well as of tumor cells. Furthermore, we summarize some pharmacological approaches for modulating integrin signaling in tumor angiogenesis.

Keywords Integrins • Angiogenesis • Cancer • Glioma • Integrin inhibition

5.1 Introduction

The growth of new vessels from preexisting vessels depends on the migration and invasion of endothelial cells through the extracellular matrix (ECM). During this process, cellular adhesion to the ECM plays a crucial role. Endothelial cells are connected to the ECM via adhesion molecules that are critical for their survival, growth, and migration [1].

Integrins constitute a large family of cell surface transmembrane molecules and are the main cell surface receptors mediating adhesion to the ECM. They are composed of α and β subunits. Eighteen α subunits, 10 β subunits, and 24 different heterodimeric integrin molecules have been identified. Alpha and β subunits can form heterodimers in multiple ways, for example, αv can associate with many different β-subunits. Other

G. Tabatabai, M.D., Ph.D. (✉) • M. Weller, M.D.
Department of Neurology, University Hospital Zurich, Frauenklinikstrasse 26, 8091 Zurich, Switzerland
e-mail: Ghazaleh.tabatabai@usz.ch

J.L. Mehta and N.S. Dhalla (eds.), *Biochemical Basis and Therapeutic Implications of Angiogenesis*, Advances in Biochemistry in Health and Disease 6,
DOI 10.1007/978-1-4614-5857-9_5, © Springer Science+Business Media New York 2013

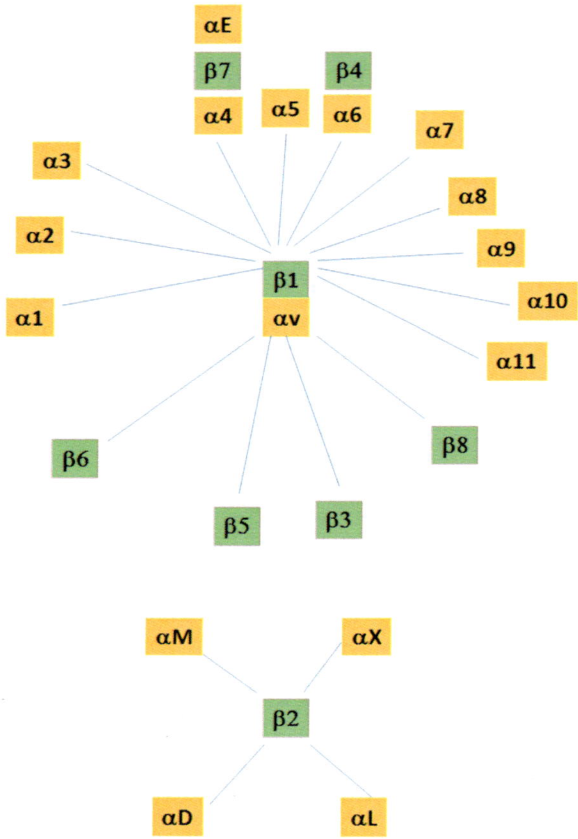

Fig. 5.1 The integrin family is a diverse family of heterodimeric proteins each containing an α and a β subunit (Modified from Tabatabai et al. [50])

α-subunits, however, only form heterodimers with one specific β-subunit partner (Fig. 5.1). Of note, the α-subunit seems to be important for determining the ligand-binding properties of the integrin heterodimer. Each heterodimer is capable of binding to a subset of ligands. In turn, a single ECM ligand can bind to several different integrins. In addition to ECM ligands, integrins mediate binding to matrix metalloproteinases (MMPs) and cell surface immunoglobulin-type receptors such as vascular adhesion molecule (VCAM). Taken together, integrin-induced endothelial cell migration is a key regulator of physiological and pathological angiogenesis. In tumors, integrins enable the cross talk between tumor cells and the surrounding stromal components [2, 3].

5.2 Bidirectional Signaling by Integrins Regulates Cellular Fate

Many integrins are not constitutively activated, i.e., the ligand binding site of the integrins is closed, and the affinity for ligand binding is low. Activation occurs by binding of extracellular ligands, i.e., laminins or vitronectins. Engagement and clustering of integrin receptors lead to the formation of focal contact sites where cells adhere to ECM. These contact sites are called focal adhesions. They function as a nexus for attracting elements of the cytoskeleton, signaling molecules, and adaptor proteins. The transmembrane helices and short cytoplasmic tails of the integrin α-subunit and β-subunit are important for initiating and coordinating cellular responses. Integrin cytoplasmic tails are between 15 and 78 amino acids in length for α-subunits and between 46 and 68 amino acids in length for β-subunits. These cytoplasmic tails of integrins do not contain enzymatic activity. Therefore, interaction with cytoplasmic adaptor proteins is crucial for mediating intracellular signaling [3]. These adaptor proteins can associate with the cytoplasmic tails of integrins and form a bridge between the cell surface and the cytoplasm.

Classical focal adhesions contain signaling complexes composed of growth factor receptors, focal adhesion kinase, integrin-linked kinase, Src, phosphoinositide 3-kinase (PI3K), and actin-associated cytoskeletal proteins [4]. Often, the activation of the Rho family of GTPases occurs by integrins upon the formation of focal adhesions. This activation induces immediate cellular migration events including formation of lamellipodia at the invading front of the cell as well as releasing of ECM adhesion contacts at other parts of the cell to enable directed migration and invasion. Not only migration and invasion events are the results of integrin-induced signaling that occur in these focal adhesions. The induction of the Ras GTPase family, for example, mediates signals to the PI3K/Akt and the Ras/mitogen-activated protein (MAP) kinase signaling pathway leading to the activation of several transcription factors, including NF-kB, HoxD3, or Id proteins [5, 6], that are key players in cell cycle regulation and cell survival. These examples illustrate that integrin-mediated signaling from the ECM to the intracellular compartment, the so-called outside-in signaling, regulates migration, invasion, cytoskeletal organization, cell survival, and cell cycle progression. These are critical events during angiogenesis. In reverse direction, integrin-mediated binding can be modulated from the intracellular to the extracellular compartment, the so-called inside-out signaling. Certain changes from the intracellular domain of the integrin heterodimers can regulate the binding affinity of the extracellular integrin receptor to the ECM components [7]. This bidirectional signaling leads to the formation of a complex cross talk network regulating the activation status of integrins. Thus, depending on the activation status, even opposite effects might occur, for example, either enhancement of cell survival or induction of apoptosis, indicating that integrin signaling is among the key regulators for determining cell fate [8, 9].

5.3 Integrin Binding to Proteolytic Protein Fragments

Migration and invasion of endothelial cells is facilitated by MMP-mediated degradation of ECM. However, these proteolysis events also lead to the formation of proteolyzed protein fragments including tumstatin [10], endostatin [11], and PEX [12] that have antiangiogenic activity by antagonizing integrin signaling in endothelial cells.

5.4 Cross talk of Integrins with Other Signaling Pathways

Since angiogenesis is a highly regulated process involving several signaling pathways, a coordinated cross talk between the key signaling pathways is a necessity. This holds also true for integrin signaling in angiogenesis. Cross talk with several other signaling pathways exists, for example, with Notch, VEGF, or TGF-β signaling.

5.4.1 Notch Signaling

Notch signaling is an evolutionary highly conserved pathway that plays crucial roles, for example, for cellular fates during embryogenesis. A role of Notch signaling in angiogenesis has recently been established including vascular development, vessel patterning, and vascular maturation [13]. Notch signaling leads to the recruitment of vascular smooth muscle cells to newly formed vessels. This leads to perivascular coverage and stabilization of the newly formed vessels. In this process of vascular maturation, vascular smooth muscle cells interact with the Notch ligand Jagged 1 on endothelial cells. This interaction leads to upregulation and activation of $\alpha v\beta 3$. Thus, $\alpha v\beta 3$ acts downstream upon Notch activation and allows vascular smooth muscle cells to adhere to endothelial basement membrane [14].

5.4.2 VEGF Signaling

The VEGF family has several members and mediates multiple functions including vascular permeability, angiogenesis, lymphangiogenesis, and tumorigenesis. In angiogenesis, VEGF-A is the predominant VEGF family member. Due to several gene splicing events, diverse mature VEGF-A isoforms exist. The best-studied isoform so far is VEGF-A165. It binds to VEGF-R2 for mediating downstream signaling. The small GTPase Rap1 activates and promotes VEGF-A signaling in endothelial cells. This activation occurs partly via $\alpha v\beta 3$ activation [15].

Recent evidence suggests that VEGF-A can directly bind to integrins, specifically to integrin $\alpha 9\beta 1$ or $\alpha v\beta 3$. The direct interaction of VEGF-A with $\alpha 9\beta 1$ occurs via a three-amino acid sequence, EYP, that is encoded by the exon 3 of VEGF-A. Upon binding of VEGF-A to $\alpha 9\beta 1$, endothelial cell migration is induced [16]. Direct interaction of VEGF-A with $\alpha v\beta 3$ contributes to the adhesion, migration, and survival of human umbilical arterial endothelial cells [17, 18].

5.4.3 *TGF-β Signaling*

TGF-β is a key molecule controlling a variety of cellular processes, including proliferation, differentiation, apoptosis, and migration. The TGF-β isoforms and the corresponding receptors are expressed by many different cell types. TGF-β is released as an inactive cytokine. The latent complex must be activated. The regulation of this activation step provides a site-specific control of TGF-β function and might explain the diversity of TGF-β effects depending on the composition of the respective microenvironment. For this activation step, integrins seem to play an important role. Six integrins can bind latent TGF-β including αvβ1, αvβ3, αvβ5, αvβ6, αvβ8, and α8β1. The binding of integrins is mediated by an RGD motif that is present in the latency-associated peptide region of the latent complex. Of note, the latent forms of TGF-β 1 and TGF-β 3, but not TGF-β 2, contain the RGD motif. TGF-β 1 activity is highly dependent on the activation step by integrins: Transgenic mice were generated carrying a single point mutation in the RGD integrin binding motif of latent TGF-β 1 that changed RGD to RGE. TGF-β $1^{RGE/RGE}$ mice expressed latent TGF-β 1 in a form that cannot bind integrins. TGF-β $1^{RGE/RGE}$ mice developed defects identical with those seen in mice that are TGF-β 1-deficient, i.e., vasculogenesis defects during embryonic development and multiorgan inflammation at the age of 2–3 weeks postnatal. These data strongly indicated that integrin-mediated activation of latent TGF-β 1 is absolutely required for TGF-β 1 functions in vivo [19]. Inhibition of β8 integrin in glioma cells leads to reduced activation of latent TGF-β [20]. Moreover, treatment of glioma cells with the integrin inhibitor cilengitide (see below) results in detachment and decreased TGF-$β_1$ and TGF-$β_2$ mRNA and protein expression, reduced phosphorylation of Smad2, and reduced TGF-β-mediated reporter gene activity [21].

5.5 Integrin Inhibition in Tumor Angiogenesis

Angiogenesis is necessary for tumor growth, dissemination, and metastasis. Integrins are key regulatory proteins for tumor angiogenesis. In view of current clinical applications of integrin inhibitors, we will focus on the role of integrins and integrin inhibition in brain tumor models.

In malignant gliomas, the integrins αvβ3 and αvβ5 are predominantly expressed. They are detected on angiogenic tumor endothelial cells and on tumor cells [22, 23]. Interestingly, αvβ3 is selectively expressed in gliomas and is absent in normal brain as demonstrated by positron emission tomography (PET) imaging studies using the tracer [18F]Galacto-RGD. The positive PET signals have been confirmed by biopsy and histological analysis of the tumor tissue [24]. These findings support αvβ3 as a rational candidate for specific therapeutic targeting in glioblastomas. Currently, the standard therapeutic strategies for patients with glioblastoma include surgical resection as feasible or biopsy and radiotherapy plus concomitant and adjuvant temozolomide chemotherapy [25, 26]. Even with this multimodality approach, the median survival of these patients does not exceed 15 months. Therefore, new therapeutic targets are needed to improve the prognosis of patients with glioblastoma. In this regard, integrins have become attractive candidate molecules for therapeutic intervention and have been explored in preclinical and clinical studies.

Preclinical studies have suggested important biological functions of integrins in malignant gliomas via tumor cell migration, invasion and adhesion, and angiogenesis [27–29]. The correlation of expression levels and activities of MMP-2 and -9 of Bcl-2 family members and of αvβ3 integrin in 12 human glioma cell lines to glioma cell migration and invasion revealed that αvβ3 integrins alone did not predict a migratory or invasive phenotype. A neutralizing αvβ3 integrin antibody, however, inhibited migration and invasion selectively in cell lines with αvβ3 integrin expression. This indicates that αvβ3 integrin has a key role in migration and invasion of malignant glioma cells.

Most glioma cell lines detach from the cell culture dish, but do not die when exposed to integrin αvβ3/5 antagonists, either antibodies or the RGD-mimetic peptide cilengitide. These observations suggest that the potential anti-glioma activity of integrin antagonists (see below) are unlikely to be related to direct cytolytic activity against glioma cells. However, others have reported that cilengitide leads to detachment of glioma cells from the ECM with subsequent apoptosis [28].

Treatment of animals with combined treatment of experimental U87MG gliomas with the RGD peptide mimetic S247 and fractionated radiation therapy has been more effective than either treatment alone, suggesting antiangiogenic activity with reduced microvessel density resulting in improved survival. Tumor cell proliferation was also significantly reduced after S247 or irradiation of U87 tumors as assessed by Ki67 staining of histological sections. Importantly, the combination of S247 and radiotherapy resulted in better results. CD31 staining showed reduced vessels after combined treatment as compared to S247 treatment or irradiation alone. Further, histological sections from S247-treated and irradiated xenografts showed pronounced reduction of Akt phosphorylation and increased TUNEL immunostaining indicating endothelial cell apoptosis. This suggests that αvβ3 antagonism might confer sensitization towards radiotherapy. It remains to be clarified whether the combination of S247 and radiation therapy results in synergistic or additive effects.

Concurrent treatment of orthotopic U251 gliomas with cilengitide and radiotherapy in vivo increased the rate of apoptotic cell death and the survival of animals compared with either treatment modality alone. Interestingly, a critical parameter for a successful combination of integrin antagonism and radiotherapy seems to be the timeline of treatment modalities. For example, single dose of cilengitide led to synergistic effects when administered 4–8 h before radiation, but not when administered 2 h before or after radiation [30]. Combined treatment with cilengitide and radiation therapy significantly increased the rate of apoptotic and autophagic cells. It remains unclear, however, whether the observed treatment-induced apoptosis in vivo in this study was occurring primarily in the

U251 glioma cells or in the endothelial cell compartment and whether the induction of apoptosis may be considered a valid surrogate marker for prolonged survival of the animals. In vitro, the treatment of glioma cell lines with cilengitide induces detachment in a concentration-dependent manner without significantly affecting proliferation and survival [28]. The effect of integrin inhibition prior to radiation might indicate that cilengitide induces changes in the vascular architecture. Of note, tumor vessels that are formed by highly proliferating angiogenic endothelial cells are often functionally inefficient leading to reduced perfusion, hypoxia, and resistance to irradiation. Thus inhibition of tumor angiogenesis by $\alpha v\beta 3$ antagonism may restore normal perfusion and reduce tumor hypoxia by normalizing the vasculature. A recent MRI-based study by Muldoon and colleagues [31], however, demonstrated that inhibition of αv integrins actually increased vascular permeability questioning the value of vascular normalization. Another explanation might be the interference with hypoxia-regulated pathways. Indeed, hypoxia increases $\alpha v\beta 3$ and $\alpha v\beta 5$ expression. Depletion of $\alpha v\beta 3$ and $\alpha v\beta 5$ integrins by siRNA decreased the transcriptional activity of hypoxia-inducible factor-1 α (HIF-1α) and reduced hypoxia [32]. Thus targeting $\alpha v\beta 3$ and $\alpha v\beta 5$ might sensitize tissues to radiation by reduction of hypoxia. Taken together, the "correct" scheduling of radiation therapy and cilengitide might reduce hypoxia by induction of vascular changes and enhances the beneficial effects of radiation. This, however, remains to be proven by imaging studies such as refined MRI modalities or vascular-specific PET imaging in humans. Indeed, a clinical trial using MRI and PET imaging studies during cilengitide treatment has just been launched in patients with glioblastoma to address the question of vascular changes in patients. The above-mentioned observations, nonetheless, demonstrate important role of integrins for glioma angiogenesis. Integrin inhibition is, therefore, considered a promising target for the treatment of malignant glioma because both endothelial and tumor cells express this target. The following paragraphs will briefly review clinical experience with integrin inhibitors against malignant glioma.

5.6 Integrin Inhibitors in Clinical Development

5.6.1 Cilengitide

Cilengitide (Merck Serono, Darmstadt, Germany) is a synthetic Arg–Gly–Asp (RGD) pentapeptide with a molecular weight of 588.7 mass units. In general, maximal plasma concentrations are reached within 1 h after the injection and the half-life is approximately 3–5 h. Cilengitide binds to the RGD ligand-binding motif (ligand binding site) on the integrin receptors $\alpha v\beta 3$ and $\alpha v\beta 5$ and has no effect on adhesions mediated by $\alpha 1\beta 1$, $\alpha 2\beta 1$, and $\alpha 5\beta 1$ integrins [33]. Cilengitide reduces vascular endothelial growth factor (VEGF)-induced angiogenesis in chorioallantoic membranes [34] and the proliferation of human umbilical vein endothelial cells [35].

Phase I studies with cilengitide demonstrated that the drug is well tolerated when given intravenously twice a week. No dose limiting toxicity was observed up to a dose of 2,400 mg/m^2. Cilengitide was also well tolerated in adult and pediatric patients with recurrent glioma or other brain tumors. Tumor responses were observed both at lower and higher doses. Enzyme-inducing anticonvulsants did not interfere with the pharmacokinetics of cilengitide. In a study by Gilbert and colleagues [36], cilengitide was given for 2 weeks at two doses before surgical intervention. Drug concentrations were significantly higher in the tumor than in the corresponding plasma. Of note, cerebrospinal fluid concentrations were approximately in the range of 1% of the plasma concentrations and were reached within 3 h after injection. This study supports the concept that cilengitide penetrates blood–brain barrier and is enriched in the tumor tissue of patients with glioblastoma [36].

In a randomized phase II trial enrolling 81 patients with glioblastoma, utility of a lower and a higher dose of cilengitide (500 mg and 2,000 mg, flat dosing) was explored. Objective responses were observed in 5% (500 mg) and 13% (2,000 mg) of the patients. Progression-free survival at 6 months was 10 and 15%, and overall survival was 6.5 vs. 9.9 months in the low-dose and high-dose groups, respectively [37]. Four-year survival rate was 2.4% in patients treated with 500 mg and 10% in patients treated with 2,000 mg [38].

A European phase II trial in 52 newly diagnosed patients with glioblastoma evaluated the addition of cilengitide (500 mg) to standard radiation and chemotherapy with temozolomide. Treatment was well tolerated. Progression-free survival at 6 months was 69%, median survival 16.1 months, with a 2-year survival rate of 35%. Compared with historical controls, the 23 patients with a methylated O^6-methylguanine–DNA methyltransferase (*MGMT*) promoter appeared to benefit most from the addition of cilengitide. Overall survival at 15 months was 75% in patients with a methylated and 47% in patients with an unmethylated *MGMT* promoter [39]. The relation between *MGMT* methylation status and benefit from cilengitide is, however, controversial. Glioma cell response in in vitro studies was unaffected by cilengitide alone or cilengitide in combination with temozolomide after modulation of *MGMT* expression levels, i.e., by ectopic expression of *MGMT* in *MGMT*-negative or by shRNA-mediated *MGMT* silencing in *MGMT*-positive glioma cells [28]. It seems that vascular normalization and improved tumor perfusion might be a key mechanism of action of cilengitide in this setting. Enhanced perfusion allows better "delivery" of temozolomide chemotherapy to which tumor cells with a methylated *MGMT* promoter are particularly sensitive [40].

Recently, Nabors and colleagues [41] reported on a safety run-in and randomized phase II study of cilengitide and standard radio and temozolomide chemotherapy in newly diagnosed glioblastomas comparing the addition of 500 mg vs. 2,000 mg cilengitide. The primary endpoint, median overall survival was compared with historical controls. The combination of cilengitide with radiation and temozolomide chemotherapy was well tolerated. The median overall survival was 19.7 months for all patients, 17.4 months for those who received 500 mg of cilengitide, and 20.8 months for those who received 2,000 mg cilengitide. For future trials, the authors suggested use of 2,000 mg cilengitide. However, these

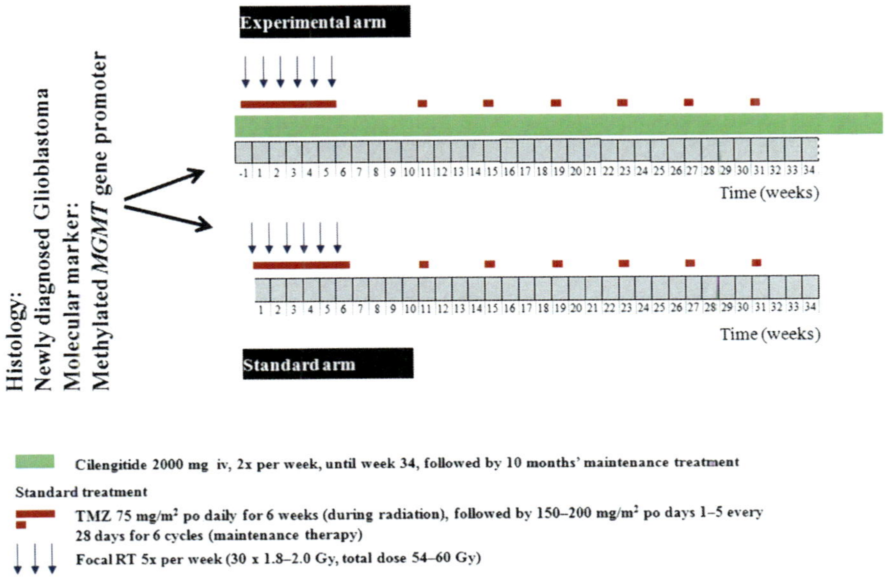

Fig. 5.2 Design of the phase III trial CENTRIC trial (NCT00689221). *MGMT*, Methylguanine methyltransferase; *TMZ*, temozolomide, *iv*, intravenous; *po*, per os; *RT*, radiation therapy (Modified from Tabatabai et al. [50])

data from uncontrolled phase II clinical trials need to be interpreted with caution as there was no control arm without the new agent. Larger randomized trials are required to confirm the efficacy of these novel treatment approaches.

Based on the benefit observed in particular in patients with a methylated *MGMT* promoter in the tumor, a pivotal phase III trial was conducted in patients with glioblastoma with methylated *MGMT* promoter. Patients were screened upfront for *MGMT* promoter methylation by methylation-specific PCR. Eligible patients were randomized either to standard therapy consisting of temozolomide and radiation or to 2,000 mg cilengitide twice weekly in addition to standard therapy (Fig. 5.2). Maintenance cilengitide therapy was continued for up to 18 months. This trial has completed its accrual goal in January 2011, and results are expected in the middle of 2013.

In parallel, a randomized phase II study in patients with newly diagnosed glioblastomas with an unmethylated *MGMT* promoter is ongoing. The CORE trial evaluates the safety, feasibility, and efficacy of intensified daily cilengitide 2,000 mg with radiation and temozolomide. Thereafter, all patients will receive 6 cycles of standard 5 days out of 28 days temozolomide plus cilengitide dosed at 2,000 mg twice weekly. The treatment will be continued until progression or toxicity. Patients will be randomized into the following 3 arms: first arm, cilengitide 2,000 mg from Monday through Friday with radiation and concomitant temozolomide chemotherapy, followed by cilengitide twice weekly with six adjuvant temozolomide cycles followed by cilengitide maintenance; second arm, cilengitide 2,000 mg twice weekly with standard radiation and temozolomide chemotherapy and with six adjuvant cycles of temozolomide followed by cilengitide maintenance; third arm, standard radiation and temozolomide chemotherapy with 6 cycles of temozolomide.

5.6.2 ATN-161

ATN-161 (Tactic Pharma, Evanston, IL) is an $\alpha5\beta1$ integrin antagonist and decreased the phosphorylation of mitogen-activated protein kinase [42, 43]. Treatment with ATN-161 in a breast cancer model induced a significant dose-dependent decrease in tumor volume and metastasis to bone and soft tissues. Histological analysis revealed reduced microvessel density and cell proliferation. In a phase I study, ATN-161 was administered to patients with advanced solid tumors excluding brain tumors. Treatment was well tolerated at all dose levels from 0.1 to 16 mg/kg [44]. The clinical development, however, seems not to be further continued.

5.6.3 DI17E6

DI17E6 (Merck Serono, Darmstadt, Germany) is a pan anti-αV antibody. In a preclinical melanoma study, DI17E6 was covalently coupled to doxorubicin-loaded human serum albumin nanoparticles. These tailored nanoparticles specifically targeted $\alpha v\beta3$-positive melanoma cells and displayed significantly higher cytotoxic activity than the free drug alone [45]. Phase I evaluation has recently been completed, and it will be further investigated in combination with cetuximab in colorectal cancer.

5.7 Conclusions

Integrins play a crucial role in physiological and pathological angiogenesis. Endothelial cell invasion events critically depend on integrin-mediated signaling. Bidirectional integrin signaling regulates cell proliferation, cell cycle progression, migration, invasion of tissues, and cell survival. Integrin antagonists occur naturally and are derived from proteolytic degradation of ECM. Integrin inhibition, particularly in combination with cytotoxic chemotherapy and/or radiation therapy, is a promising strategy for targeting angiogenesis and tumor cells in certain tumor entities. Glioblastomas are paradigmatic in this context, and the clinical development of integrin antagonists is very advanced in this special brain tumor entity.

Indeed, preclinical evidence suggests that inhibition of integrins is an attractive approach for anti-glioma therapies. It seems crucial to design treatment algorithms while paying attention to the schedule, for example, time point of injection when combined with radiation therapy or choosing the right dose. While cilengitide has been confirmed to be very safe in the clinical trials performed so far, experimental observations have recently challenged the concept of integrin antagonism. Administration of low nanomolar concentrations of RGD mimetics led to a pro-angiogenic and pro-invasive tumor phenotype in melanoma and lung cancer models. This was mainly explained by an activation of the Rab4 pathway by nanomolar RGD mimetics. This activation promotes the recycling of internalized VEGF-R2. This, in turn, inhibits the degradation of VEGF-R2 and allows relocalization of VEGF-R2 to the cell surface and thus promotes cellular responses to VEGF [46]. Yet, the plasma concentrations achieved with the doses of cilengitide used in the ongoing clinical trials exceed the potentially proangiogenic concentrations by orders of magnitude [47]. Further, the ongoing clinical trials in patients with glioblastoma applied cilengitide as an adjuvant to radiation and chemotherapy, and not as a single agent. Taking into account that RGD mimetics may have concentration-dependent biphasic effect with antagonistic activity at high concentrations and agonistic activity at low concentrations [48, 49], the risk of pro-angiogenic side effects in clinical application is probably low.

To date, cilengitide is the dominant integrin inhibition strategy in clinical trials. Clinical experience with cilengitide indicates that the safety and tolerability profile of this agent is excellent. Importantly, cilengitide use has not been associated with increased risk of intracerebral hemorrhage so far. Furthermore, toxicity does not seem not to be increased in combination with radiation and temozolomide.

These data, taken together, suggest that integrins are key mediators of angiogenesis and thus attractive targets for therapy of diseases with pathological angiogenesis. For ensuring therapeutic success, it will be crucial to carefully define which integrins to selectively target in which disease.

References

1. Heynes RO (1992) Integrins: versatility, modulation, and signaling in cell adhesion. Cell 69:11–25
2. Weis SM, Cheresh DA (2011) Tumor angiogenesis: molecular pathways and therapeutic targets. Nat Med 7:1359–1370
3. Cox D, Brennan M, Moran N (2010) Integrins as therapeutic targets: lessons and opportunities. Nat Rev Drug Discov 9:804–820
4. Sastry SK, Burridge K (2000) Focal adhesions: a nexus for intracellular signaling and cytoskeletal dynamics. Exp Cell Res 261:25–36
5. Boudreau N, Andrews C, Srebrow A et al (1997) Induction of the angiogenic phenotype by HoxD3. J Cell Biol 139:257–264
6. Lyden D, Young AZ, Zagzag D et al (1999) Id1 and Id3 are required for neurogenesis, angiogenesis and vascularization of tumor xenografts. Nature 401:670–677
7. Ginsberg MH, Du X, Plow EF (1992) Inside out integrin signaling. Curr Opin Cell Biol 4:766–771
8. Heynes R (2002) Integrins: bidirectional, allosteric signaling machines. Cell 110:673–687
9. Desgrosellier JS, Cheresh DA (2010) Integrins in cancer: biological implications and therapeutic opportunities. Nat Rev Cancer 10:9–22

10. Maeshima Y, Sudhakar A, Liverly JC et al (2002) Tumstatin, an endothelial cell-specific inhibitor of protein synthesis. Science 295:140–143
11. Sund M, Hamano Y, Sugimoto H et al (2005) Function of endogenous inhibitors of angiogenesis as endothelium-specific suppressors. Proc Natl Acad Sci 102:2934–2939
12. Pfeifer A, Kessler T, Silletti S, Cheresh DA et al (2000) Suppression of angiogenesis by lentiviral delivery of PEX, a non-catalytic fragment of matrix metalloproteinase 2. Proc Natl Acad Sci 97:12227–12232
13. Bridges E, Oon CE, Harris A (2011) Notch regulation of tumor angiogenesis. Future Oncol 7:569–588
14. Scheppke L, Murphy EA, Zarpellon A et al (2012) Notch promotes vascular maturation by inducing integrin-mediated smooth muscle cell adhesion to the endothelial basement membrane. Blood 119:2149–2158
15. Lakshmikanthan S, Sobczak M, Chun C et al (2011) Rap 1 promotes VEGFR2 activation and angiogenesis by a mechanism involving integrin avb3. Blood 118:2015–2026
16. Oommen S, Gupta S (2011) Vlahakis. Vascular endothelial growth factor (VEGF-A) induces endothelial and cancer cell migration through direct binding to integrin a9b1. J Biol Chem 286:1083–1092
17. Hutchings H, Ortega N, Plouet J (2003) Extracellular matrix-bound vascular endothelial growth factor promotes endothelial cell adhesion, migration, and survival through integrin ligation. FASEB J 17:1520–1522
18. Vlahakis NE, Young BA, Atakilit A, Sheppard D (2005) The lymphangiogenic vascular endothelial growth factor VEGF-C and –D are ligands for the integrin alpha9beta1. J Biol Chem 280:4544–4552
19. Worthington JJ, Klementowicz JE, Travis MA (2011) TGF-β: a sleeping giant awoken by integrins. Trends Biochem Sci 36:47–54
20. Tchaicha JH, Reyes SB, Shin J et al (2011) Glioblastoma angiogenesis and tumor cell invasiveness are differentially regulated by β8 integrin. Cancer Res 71:6371–6381
21. Weller W, Silginer M, Goodman SL et al (2012) Effect of the integrin inhibitor cilengitide on TGF-beta signaling. J Clin Oncol 30:(suppl; abstr 2055)
22. Schnell O, Krebs B, Wagner E et al (2008) Expression of integrin alphavbeta3 in gliomas correlates with tumor grade and is not restricted to tumor vasculature. Brain Pathol 18:378–386
23. Bello L, Francolini M, Marthyn P et al (2011) Alpha(v) beta3 and alpha(v)beta5 integrin expression in glioma periphery. Neurosurgery 49:380–389
24. Schnell O, Krebs B, Carlsen J et al (2009) Imaging of integrin alphaVbeta3 expression with malignant glioma by [18 F] Galacto-RGD positron emission tomography. Neuro Oncol 11:861–870
25. Stupp R, Mason WP, van den Bent MJ et al (2005) Radiotherapy plus concomitant and adjuvant temozolomide for glioblastoma. N Engl J Med 10:987–996
26. Stupp R, Hegi M, Mason W et al (2009) Effects of radiotherapy with concomitant and adjuvant temozolomide versus radiotherapy alone on survival in glioblastoma in a randomised phase III study: 5-yesar analysis of the EORTC-NCIC trial. Lancet Oncol 10:459–466
27. Wild-Bode C, Weller M, Wick W (2001) Molecular determinants of glioma cell migration and invasion. J Neurosurg 94:978–984
28. Maurer GD, Tritschler I, Adams B et al (2009) Cilengitide modulates attachment and viability of human glioma cells, but not sensitivity to irradiation or temozolomide *in vitro*. Neuro Oncol 11:747–756
29. Abdollahi A, Griggs DW, Zieher H et al (2005) Inhibition of alpha(V)beta3 integrin survival signaling enhances antiangiogenic and antitumor effects of radiotherapy. Clin Cancer Res 11:6270–6279
30. Mikkelsen T, Brodie C, Finniss S et al (2009) Radiation sensitization of glioblastoma by cilengitide has unanticipated schedule-dependency. Int J Cancer 124:2719–2727
31. Muldoon LL, Gahramanov S, Li X et al (2011) Dynamic magnetic resonance imaging assessment of vascular targeting agent effects in rat intracerebral tumors. Neuro Oncol 13:51–60

32. Skuli N, Monferran S, Delmas C et al (2009) Alphavbeta3/alphavbeta5 integrins-fak-rhob: a novel pathway for hypoxia regulation in glioblastoma. Cancer Res 69:3308–3316
33. Goodman SL, Holzemann G, Sulyok GA, Kessler H (2002) Nanomolar small molecule inhibitors for alphaV(beta)6, alphaV (beta)5, and alphaV(beta)3 integrins. J Med Chem 45:1045–1051
34. Friedlander M, Theesfeld CL, Sugita M et al (1996) Involvment of integrins alphaVbeta 3 and alphaVbeta5 in ocular neovascular diseases. Proc Natl Acad Sci U S A 93:9764–9769
35. Hammes HP, Brownlee M, Jonczyk A et al (1996) Subcutaneous injection of a cyclic peptide antagonist of vitronectin receptor-type integrins inhibits retinal neovascularization. Nat Med 2:529–533
36. Gilbert M, Lamborn K, Lassman A et al (2007) Tumor tissue delivery of cilengitide after intravenous administration to patients with recurrent glioblastoma. Preliminary data from NABTC protocol 03–02. Neuro Oncol 4:525
37. Reardon DA, Fink KL, Mikkelsen T et al (2008) Randomized phase II study of cilengitide, an integrin-targeting arginine-glycine-aspartic acid peptide, in recurrent glioblastoma multiforme. J Clin Oncol 26:5610–5617
38. Fink K, Mikkelsen T, Nabors LB et al. Long-term effects of cilengitide, a novel integrin inhibitor in recurrent glioblastoma: a randomized phase II a study. J Clin Oncol 28: suppl abstr, 2010.
39. Stupp R, Hegi ME, Neyns B et al (2010) Phase I/IIa study of cilengitide and temozolomide with concomitant radiotherapy followed by cilengitide and temozolomide maintenance therapy in patients with newly diagnosed glioblastoma. J Clin Oncol 28:2712–2718
40. Hegi ME, Diserens AC, Gorlia T et al (2005) MGMT gene silencing and benefit from temozolomide in glioblastoma. N Engl J Med 10:997–1003
41. Nabors LS, Mikkelsen T, Hegi ME et al; for the New Approaches to Brain Tumor Therapy (NABTT) Central Nervous System Consortium (2012) A safety run-in and randomized phase 2 study of cilengitide combined with chemoradiation for newly diagnosed glioblastomas (NABTT 0306) Cancer 118(22):5601–5607
42. Plunkett ML, Tel-Tsur Z, Bera M et al (2002) A novel anti-angiogenic/anti-metastatic peptide, ATN-161 (Ac-PHSCN-NH2), which targets multiple fully activated integrins including alpha-5 beta-1 and alpha-v beta-3, leads to increased anti-tumor activity and increased survival in multiple tumor models when combined with chemotherapy. Eur J Cancer 38(Suppl 7):79
43. Khalili P, Arakelian A, Chen G et al (2006) A non-RGD-based integrin binding peptide (ATN-161) blocks breast cancer growth and metastasis *in vivo*. Mol Cancer Ther 5:2271–2280
44. Cianfrocca ME, Kimmel KA, Gallo J et al (2006) Phase 1 trial of the antiangiogenic peptide ATN-161 (Ac-PHSCN-NH2), a beta integrin antagonist in patients with solid tumours. Br J Cancer 94:1621–1626
45. Wagner S, Rothweiler F, Anhorn MG et al (2010) Enhanced drug targeting by attachment of an anti alphaV integrin antibody to doxorubicin loaded human serum albumin nanoparticles. Biomaterials 31:2388–2398
46. Reynolds AR, Hart IR, Watson AR et al (2009) Stimulation of tumor growth and angiogenesis by low concentrations of RGD-mimetic integrin inhibitors. Nat Med 15:392–400
47. Weller M, Reardon D, Nabors B, Stupp R (2009) Will integrin inhibitors have proangiogenic effects in the clinic? Nat Med 15:726
48. Legler DF, Wiedle G, Ross FP, Imhof BA (2001) Superactivation of integrin alphaVbeta3 by low antagonist concentrations. J Cell Sci 114:1545–1553
49. Weis SM, Stupack DG, Cheresh DA (2009) Agonizing integrin antagonists? Cancer Cell 15:359–361
50. Tabatabai G, Tonn JC, Stupp R, Weller M (2011) The role of integrins in glioma biology and anti-glioma therapies. Curr Pharm Des 17:2402–2410

Chapter 6
Regulation of Angiogenesis by Hypoxia-Inducible Factors

Hideki Hayakawa and Futoshi Shibasaki

Abstract In the years since the identification of hypoxia-inducible factors (HIFs), it has been well established that HIFs regulate angiogenic genes, such as vascular endothelial growth factor. HIFs play essential roles in embryonic vascular development. Recent studies have demonstrated that HIFs can influence the adult vascular system, not only in neovascularization but also in angiogenesis in pathological conditions. In this chapter, oxygen-dependent and novel oxygen-independent regulatory systems of HIF-αs are reviewed. Hypoxia-associated factor regulates HIF-1α and HIF-2α reversely. Int6-silencing stabilizes HIF-2α and functional vessel formation. HIFs can be therapeutic targets in both anti-angiogenic and angiogenic therapy. The new findings have implications for the development of therapies for ischemic disease.

Keywords HIF-1α • HIF-2α • pVHL • Int6 • HAF • HDAC7

6.1 Regulation of Hypoxia-Inducible Factors

6.1.1 Oxygen-Dependent Regulation of Hypoxia-Inducible Factors

Hypoxia is a reduction in physiological oxygen levels. It is caused by vascular and pulmonary diseases or by the triggering of cancerous-tissue growth and leads to cellular dysfunction. There are three types of hypoxia: chronic hypoxia, acute hypoxia, and hypoxia with reperfusion. Chronic hypoxia (2–3 % oxygen tension) causes uncontrolled proliferation of cells under limited oxygen diffusion [1]. By contrast, acute hypoxia is caused by a temporary closure of irregular and aberrant tumor vasculature.

H. Hayakawa • F. Shibasaki (✉)
Molecular Medical Research Project, Tokyo Metropolitan Institute of Medical Science, 2-1-6, Kamikitazawa, Setagaya-Ku, Tokyo 156-8506, Japan
e-mail: shibasaki-ft@igakuken.or.jp

J.L. Mehta and N.S. Dhalla (eds.), *Biochemical Basis and Therapeutic Implications of Angiogenesis*, Advances in Biochemistry in Health and Disease 6,
DOI 10.1007/978-1-4614-5857-9_6,

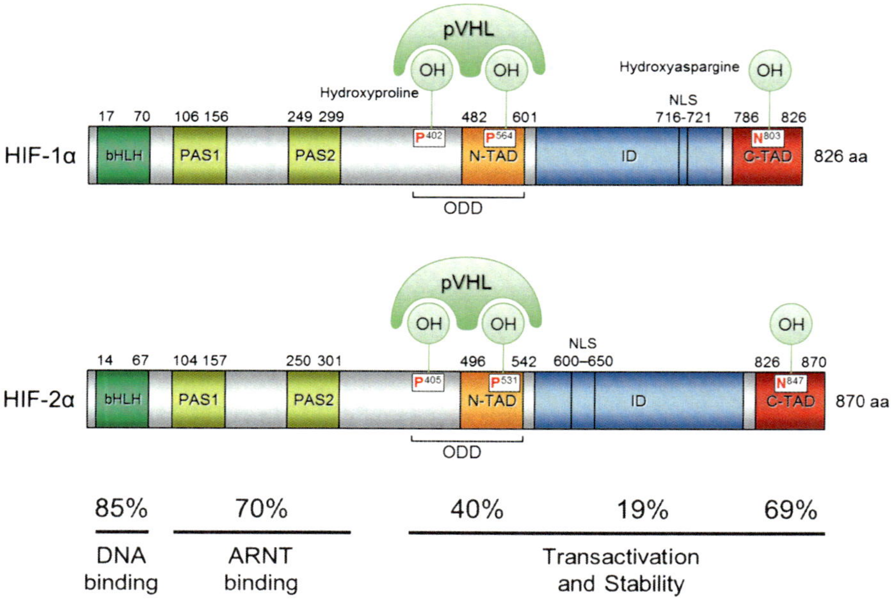

Fig. 6.1 Structures of HIF-1α and HIF-2α proteins. HIF-1α and HIF-2α exhibit a high degree of similarity in their DNA, ARNT-binding domains, and the C-TADS. The N-TAD and IDs are less conserved, except for the ~30 amino acids surrounding the two oxygen-regulated proline residues. Percent amino acid similarities between HIF-1α and HIF-2α are shown. Proline (P) and asparagine (N) residues hydroxylated under normal oxygen tension are indicated. *bHLH* basic helix-loop-helix; *PAS* Par-Arnt-SIM; *N/C-TAD* N/C-terminal transactivation domain; *ID* inhibitory domain; *ODD* oxygen-dependent degradation domain; *NLS* nuclear localization signal; *pVHL* von Hippel–Lindau protein

Reversal of blood flow (reperfusion) leads to an increase in free radical concentration, resulting in tissue damage known as "reoxygenation injury" [1, 2].

Cellular responses to low oxygen tension (hypoxia or hypoxic environment) are mainly mediated by the activation of heterodimeric transcription factors, hypoxia-inducible factors (HIFs), consisting of an oxygen-regulated subunit, mainly HIF-1α or HIF-2α (also called endothelial PAS (EPAS) protein, HIF-1α-like factor (HLF), or HIF-related factor (HRF)), and a constitutively expressed HIF-1β subunit (aryl hydrocarbon receptor nuclear translocator, ARNT). Another α-subunit, HIF-3α (or inhibitory PAS domain protein (IPAS)) lacks the transactivation domain present in HIF-1α and HIF-2α [3]. HIFs have Per-ARNT-Sim (PAS) domains of 100–120 amino acids and belong to the superfamily of basic helix-loop-helix (bHLH) transcription factors (Fig. 6.1).

HIF-1α is ubiquitously expressed, whereas HIF-2α and HIF-3α exhibit more restricted tissue distributions. HIF-2α is primarily expressed in the vasculature of the early developing embryo and subsequently in the lung, kidney interstitial cells, liver parenchyma, and neural crest cells [4–6]. HIF-3α mRNA and protein are primarily detected in the thymus, kidney, cerebellar Purkinje cells, and corneal epithelium of the eye [3, 7].

HIF-α subunits are the main sensors of hypoxia. Their expression is chiefly controlled in a posttranslational manner; under normoxic conditions, the HIF-α proteins are maintained at low levels due to their rapid degradation via the ubiquitin–proteasome pathway. This ensures quick responses to oxygen reduction within 2 min [8], compared with transcriptional regulation, which produces a delayed response due to RNA processing and subsequent cytosolic translation. In normoxia, constitutively expressed cytosolic HIF-α subunits are hydroxylated at two proline residues, Pro402 and Pro564 in human HIF-1α [9] and Pro405 and Pro531 in human HIF-2α [10], by a family of three HIF-specific prolyl hydroxylases, PHD1, PHD2, and PHD3 [11, 12]. The residues that surround these two proline residues (30 amino acids each) are highly conserved between HIF-1α, HIF-2α, and HIF-3α. PHDs exhibit some degree of specificity in HIF-α subunit recognition, with PHD2 primarily hydroxylating HIF-1α, PHD3 predominantly targeting HIF-2α, and PHD1 exhibiting a slightly higher preference for HIF-2α over HIF-1α [13].

The hydroxylated proline residues are a prerequisite for interaction with the von Hippel–Lindau (VHL) E3 ubiquitin ligase complex, which contains elongin B, elongin C, Cul2, and Rbx1 and targets the HIF-α subunits for proteasomal degradation [14] (Fig. 6.2). In addition to protein degradation signaling, HIF-1α is downregulated by another hydroxylase known as factor inhibiting HIF-1, which hydroxylates an asparagine residue in the C-TAD, preventing the interaction of HIF-1α with the transcriptional coactivator p300 [15]. When cells are exposed to hypoxia, the oxygen-requiring hydroxylation process is prevented, and HIF-α escapes proteasomal degradation, dimerizes with HIF-1β, and associates with transcriptional coactivators, such as CBP/p300 (cyclic-AMP responsive element binding (CREB) protein). The transcriptional complex subsequently recognizes the hypoxia response element (HRE; consensus sequence, A/GCGTG) in hypoxia-responsive genes. HIF-α stabilization elicits the activation of 100–200 genes, including genes involved in energy metabolism (proliferation, migration, cell cycle, apoptosis, etc.), redox homeostasis, vascular remodeling, tumorigenesis, inflammation, and other processes [16].

6.1.2 Oxygen-Independent Regulation of HIF-2α

The stability of HIF-α is also regulated by an O_2-independent pathway. Recently, using yeast two-hybrid analysis, our group identified Int6/eIF3e, a mammalian tumor suppressor, as a novel regulator of HIF-2α [17]. The *Int6/eIF3e* gene was first identified as a tumor suppressor gene during screening of mouse mammary tumor virus (MMTV) integration sites and was recognized as a target of MMTV insertional mutagenesis. The insertion in mouse Int6 creates C-terminal truncated forms of the protein (Int6-ΔC) [18], which appear to act in a dominant-negative fashion to induce breast tumor formation. Int6/eIF3e has been independently identified at least five times [19]. The name stands for the preferred chromosomal integration site (No. 6) of MMTV or the "e" subunit of eukaryotic

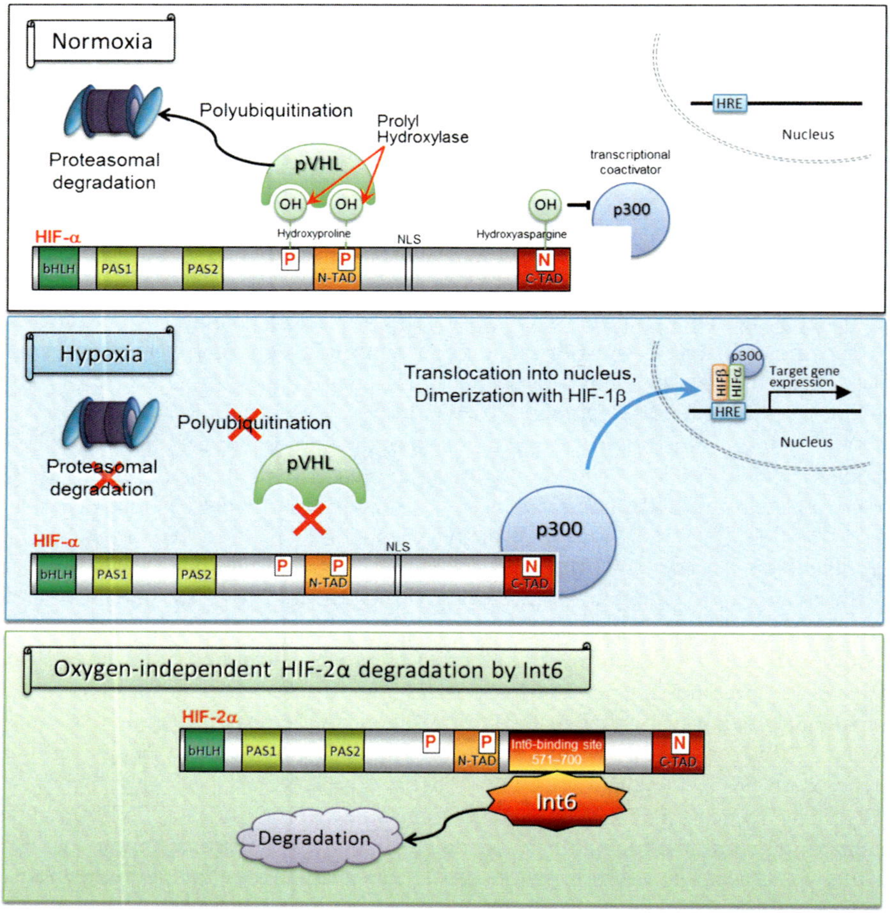

Fig. 6.2 Schematic diagram of HIF-α degradation. *Upper*: In normoxia, the proline residues (P) are hydroxylated and recognized by pVHL, which targets HIF-α for proteasomal degradation. pVHL is a part of a large complex that comprises elongin B, elongin C, CUL2, RBX1, and an ubiquitin-conjugating enzyme (E2). The asparagine residue (N) in the C-TAD of HIF-α, which is an oxygen-dependent hydroxylation-regulated binding site for p300, is inactive in its hydroxylated state. *Middle*: In hypoxia, prolyl hydroxylase cannot modify HIF-α, and the protein escapes degradation. The stabilized HIF-α is translocated to the nucleus, where it interacts with cofactors HIF-1β and p300, and transcribes hypoxia-related target genes. *Lower*: The N-terminal region of Int6 specifically binds to the C-terminal inhibitory domain of HIF-2α. Binding of Int6 induces HIF-2α instability, whereas binding of the dominant-negative mutant Int6-ΔC with deleted C-terminal PINT domain induces stable HIF-2α expression, even under normoxic conditions [17]

translation initiation factor 3 (eIF3). Studies on the role of Int6 have focused on the relationship between MMTV and breast tumor and on a subunit of eIF3 in mammals, yeasts, and plants.

Our analyses have revealed that Int6 specifically binds to HIF-2α, but not to HIF-1α and HIF-3α. Int6 binding induced HIF-2α instability in MCF-7 and HeLa cells, whereas binding of the dominant-negative mutant Int6-ΔC lacking C-terminal proteasome/Int6/Nip1/thyroid receptor-interacting protein-15 (TRIP-15) (PINT) domain [20] induced stable HIF-2α expression, even under normoxic conditions [17]. The degradation of HIF-2α following Int6 binding could be completely inhibited by the proteasome inhibitors MG132 and MG101, but Int6-induced HIF-2α degradation was still observed in 786-O cells, which lack endogenous pVHL expression. These results imply that Int6 is able to bind to HIF-2α and induce its instability via the ubiquitin–proteasome pathway, independently of pVHL. Indeed, the yeast two-hybrid binding assays demonstrated that HIF-2α has different binding sites for pVHL and Int6 and that the binding is noncompetitive [17]. The association of pVHL with HIF-2α triggers ubiquitin-dependent HIF-2α degradation in an oxygen-dependent manner. However, HIF-2α degradation by Int6 is independent of oxygen concentration. A luciferase assay performed using the HRE promoter plasmid showed that Int6-ΔC expression increased the stability and transcriptional activity of HIF-2α. Moreover, using small interfering RNA (siRNA) against Int6 [17], we found that the silencing of endogenous Int6 was sufficient to induce HIF-2α expression even under normoxic conditions. This normoxic stabilization of HIF-2α enhanced the expression of angiogenic factors, such as VEGF, angiopoietin (Angpt)-1, Angpt-2, and Angpt receptor Tie-2 (tyrosine kinase, endothelial) [17].

Our group also demonstrated the VHL-independent regulation of HIF-1α transcriptional activity by histone deacetylase (HDAC) 7, a transcriptional repressor belonging to the mammalian class II HDAC family. Using yeast two-hybrid analysis, we demonstrated that the catalytic domain of HDAC7 interacts with the inhibitory domain (ID) of HIF-1α. HDAC7 forms a complex with HIF-1α and p300 in the nucleus under hypoxic conditions, resulting in enhanced transcription of HIF-1α target genes (VEGF and Glut-1). By contrast, HDAC4 and HDAC5, the other members of the class II HDAC family, do not bind to HIF-1α. Collectively, we propose that the catalytic domain of HDAC7 may be masked by the HDAC7-HIF-1α interaction and that the binding may lead to a conformational change within the ID of HIF-1α resulting in facilitated binding of coactivators, such as CBP and p300, and increased transcriptional activity [21].

Recently, an oxygen-independent HIF-1α degradation mechanism was identified. Koh and colleagues identified hypoxia-associated factor (HAF, also known as squamous cell carcinoma antigen recognized by T cells, $SART1_{800}$) as a new E3 ligase for HIF-1α (but not HIF-2α) that mediates the proteasomal degradation of HIF-1α via a pVHL- and oxygen-independent pathway [22]. HAF binds to the oxygen-dependent degradation domain (ODD) of HIF-1α and ubiquitinates it, whereas in the case of HIF-2α, HAF binds to the region between the N-TAD and C-TAD of HIF-2α and increases HIF-2α transactivation, leading to a switch from HIF-1α- to HIF-2α-dependent transcription during tumor hypoxia [23]. Thus, the activities of the HIF-α subunits are strictly regulated by both oxygen-dependent and oxygen-independent mechanisms.

6.2 HIFs in Angiogenesis

6.2.1 Roles of HIFs in Vasculogenesis and Angiogenesis

It has been shown that HIF-1α and HIF-2α have distinct, nonredundant roles in angiogenesis (Fig. 6.3). As mentioned above, both HIF-1α and HIF-2α regulate their target genes via HREs. Many target genes are transactivated by either HIF-1α or HIF-2α, although some genes have been confirmed to be upregulated by both. However, there remains a question as to whether the roles of HIF-1α and HIF-2α overlap and whether the two proteins can compensate for each other.

Based on initial reports showing high levels of HIF-2α mRNA expression in endothelial cells (ECs) and some highly vascularized tissues, such as the heart, lung, and placenta, HIF-2α was originally named "endothelial PAS

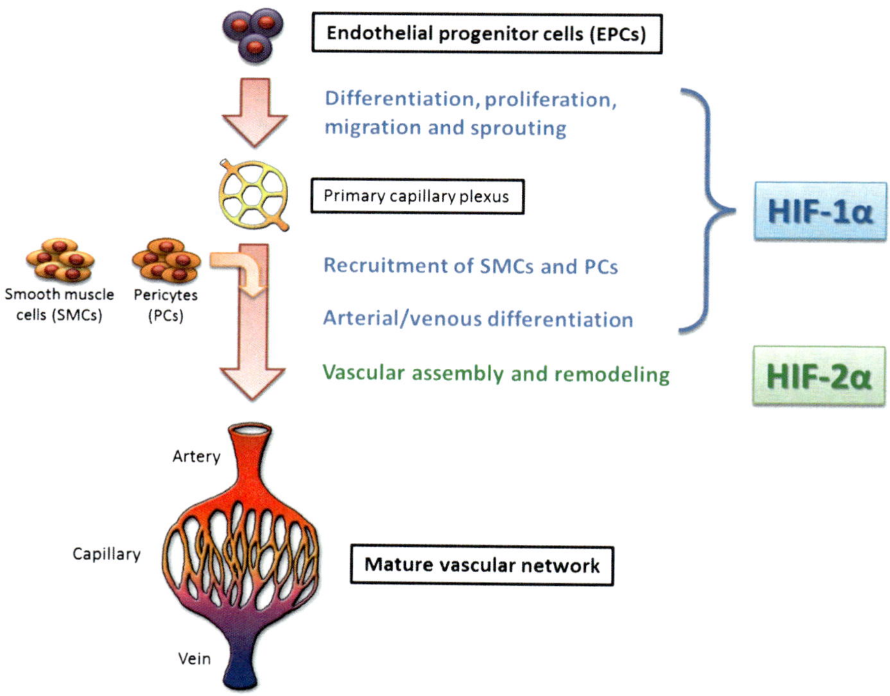

Fig. 6.3 Differential roles of HIF-1α and HIF-2α in vascular development. HIF-1α and HIF-2α have distinct, nonredundant roles in the formation of the vascular tree. While endothelial cell proliferation, migration, and vessel sprouting rely more on HIF-1α, HIF-2α plays a larger role in controlling vascular morphogenesis, integrity, and assembly. HIF-1α also promotes the recruitment of endothelial progenitor cells, which physically constitute the vessel wall. HIF-1α favors pericyte and smooth muscle cell recruitment, which cover and stabilize the vessels. HIF-1α also regulates the arterial/venous fate, promoting the formation of arteries

domain protein-1 (EPAS-1)" [5, 7]; HIF-1α is ubiquitously expressed [16]. Emerging evidence suggests that HIF-1α and HIF-2α have distinct, nonoverlapping biological functions, regulate different target genes, and often do not substitute for one another. For example, hypoxic induction of HIF-1α target genes is attenuated in *Hif-1α*-deficient ECs [24], suggesting that HIF-2α or other hypoxia-induced factors cannot sufficiently compensate for the loss of *Hif-1α*. The endothelial proliferation and migration response to hypoxia largely relies on HIF-1α [24, 25].

Genetic and gene transfer studies have shed light on the distinct roles of HIFs in vascular development. Embryonic expression of a dominant-negative HIF that inhibits both HIF-1α and HIF-2α leads to defective vascular remodeling and failed vascular sprouting in the yolk sac and embryo [26]. *Hif-1β*-deficient embryos also die in utero due to impaired angiogenesis in the yolk sac [27]. Moreover, embryos lacking *Hif-1α* display severe vascular defects [28–30]; these defects are not rescued by HIF-2α, suggesting that HIF-1α and HIF-2α control different target genes. Conversely, *Hif-1α* gene transfer stimulates revascularization and improves functional recovery of ischemic tissues in the adult [31].

Hif-2α-deficient phenotypes are variable, depending on the genetic background. In one mouse strain, loss of *Hif-2α* causes improper remodeling of nascent vessels into larger conduits; this defect is rescued by EC-specific expression of HIF-2α [32]. In other murine backgrounds, loss of *Hif-2α* causes abnormal organ development or homeostasis, without, however, any apparent vascular defects [33–35]. Furthermore, silencing of HIF-2α in ECs leads to the formation of an aberrant vascular network in tumors via reduced expression of Ephrin A1 [36]. Given the emerging role of nitric oxide (NO) in vessel normalization [37], it is worth mentioning that HIF-2α, but not HIF-1α, is able to induce expression of endothelial NO synthase (eNOS) [38]. Some of these results suggest that HIF-2α participates in microvasculature remodeling and maturation. Consistent with this hypothesis, HIF-2α elevates the expression of the junctional molecule VE-cadherin, which is not regulated by HIF-1α [39].

Hu et al. [40] demonstrated that although HIF-2α is unable to activate multiple endogenous HIF-1α-specific target genes (e.g., glycolytic enzymes), HIF-2α still binds to their promoters in vivo and activates reporter genes derived from such targets. Importantly, replacement of the N-TAD (but not the DNA-binding domain, PAS domain, or C-TAD) of HIF-2α with the analogous region of HIF-1α is sufficient to convert HIF-2α into a protein with the functional specificity of HIF-1α. They also showed, however, that some genes (VEGF, Glut-1, etc.) are transactivated by both HIF-αs [40, 41]. Furthermore, the expression level of HIF-αs and their target genes and the effect (cell migration) were markedly cell type specific [25].

Indeed, we have previously shown that HIF-2α stabilization by Int6 silencing leads to upregulation of angiogenic genes [42]. Int6 silencing by small hairpin RNA (shRNA) delays HIF-2α degradation, and the stabilized HIF-2α upregulates itself via the HRE in the HIF-2α promoter region (Fig. 6.4). This feedback loop strengthens

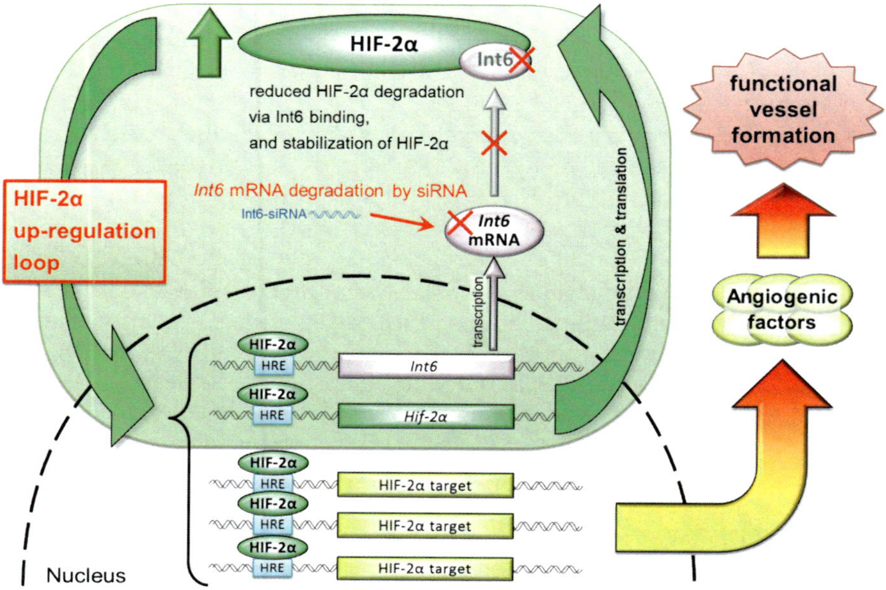

Fig. 6.4 Creation of a HIF-2α upregulation loop by siRNA-mediated Int6 silencing. Int6 degrades the HIF-2α protein via the ubiquitin/proteasome pathway. HIF-2α recognizes HREs in its own promoter and in the *Int6* promoter, thereby transcribing both genes, but Int6 binds to and degrades the newly synthesized HIF-2α through the ubiquitin/proteasome pathway even under hypoxic conditions [15]. The negative feedback regulation of HIF-2α by Int6 limits the HIF-2α protein increase and subsequent angiogenic factor expression. During Int6 silencing, newly formed HIF-2α escapes degradation and maintains transcription from its own promoter. The accumulation of HIF-2α protein leads to the expression of several angiogenic factors. Thus, Int6 silencing accelerates HIF-2α accumulation by inhibition of further de novo synthesis of Int6 [42]

HIF-2α activity and the downstream gene expression of various angiogenic factors, such as VEGF, basic fibroblast growth factor, platelet-derived growth factor, Angpts, and Tie-2, which lead to neoangiogenesis in vivo and accelerate the closure of injury in a wound healing model [42]. Importantly, the newly grown vessels following Int6 silencing are mature vessels with smooth muscle cells. This may be consistent with the observation that HIF-2α is involved in microvessel remodeling and mature vessel formation [26, 43].

In a femoral artery ligation model of hindlimb ischemia, the ischemic limb muscles in *Hif-1*$\alpha^{+/-}$ mice exhibit impaired HIF-1α induction, defective activation of angiogenic factors, and decreased reperfusion [44]. Conversely, adenoviral delivery of constitutively active HIF-1α to the site of ligation leads to enhanced reperfusion. Thus, HIF-1α expression is essential and sufficient to promote reperfusion in ischemic skeletal muscles.

6.2.2 HIFs and Tumor Angiogenesis

There is ample evidence that solid tumors frequently encounter hypoxic stress. Rapidly, proliferating cancer cells may outgrow their vascular network, limiting O_2 diffusion within the tumor. Hypoxic stress can also be caused by perfusion defects resulting from abnormal tumor blood vessel structure and function. Not surprisingly, therefore, solid tumors often exhibit high levels of HIF-α accumulation [45]. It should be noted that HIF-α expression in cancer cells is also increased via hypoxia-independent mechanisms. Various genetic alterations, such as VHL mutation in renal cell carcinoma, mutations in the Wnt/β-catenin signaling pathway in colon carcinoma, and other oncogenic events, have been reported to elicit HIF-α stabilization [15]. Collectively, these findings indicate that HIF-α expression and the downstream activation of hypoxic stress response are widespread in many cancers.

Studies from several laboratories have revealed that HIF-regulated gene responses play key roles in various aspects of cancer development, including proliferation (MYC), angiogenesis (VEGF, PDGF), apoptosis/autophagy (NDRG2, BNIP3), metabolism (PDK1, LDHA), DNA damage response (GADD45A), microRNAs (MIR210), extracellular matrix remodeling (LOX, MMP1), and cell migration and invasion (CXCR4, SDF1(CXCL12)) [15, 45, 46]. The importance of HIF activity in cancer is demonstrated by the fact that increased HIF-α expression correlates with poor clinical prognosis in many cancer types [47].

HIF exerts similar effects on ECs in both tumor and nonmalignant tissues to mediate angiogenesis. However, unlike "normal" blood vessels, tumor-associated vasculature is leaky, tortuous, and noncontiguous [48]. Tumor-associated endothelium interacts with tumor cells as well as nonmalignant stromal cells, such as fibroblasts and infiltrating bone marrow-derived cells. These cell types differ widely in their responses to hypoxic stress and may therefore contribute differently to tumor angiogenesis. For example, HIF activity in glioblastoma promotes tumor angiogenesis, as HIF-1α inhibition in glioblastoma cells reduces vascular remodeling and normalizes tumor vasculature [49]. Paradoxically, HIF-1α depletion in these cells also increases perivascular invasion due to the direct effect of decreased VEGF levels on glioblastoma cell migration [49]. The development of tumor vasculature also appears to specifically require myeloid-derived VEGF. Deletion of the HIF target gene *Vegf* in myeloid cells increases murine mammary tumor growth, tumor oxygenation, and tumor sensitivity to chemotherapy, most likely due to "normalization" of tumor vessels [50]. By contrast, haploinsufficiency of the HIF regulator PHD2 in nonmalignant tissues allows the "normalization" of xenograft tumor vasculature, improves oxygenation, and reduces metastasis [51]. However, the dependence of these effects on HIF stabilization remains uncertain. These studies illustrate that the tumor vasculature responds to distinct and perhaps opposing HIF activities in different cell types. Therefore, selective manipulation of the hypoxic stress response in distinct tumor sub-compartments may be more effective as an antitumor strategy than systemic HIF inhibition.

6.3 HIF-Targeted Therapy

6.3.1 Anti-angiogenic Therapy: Combination of Radiotherapy and HIF Inhibition

Reduced oxygen tension far below physiological levels is a characteristic of solid tumors [1, 52, 53]. Polarographic needle electrode analysis and immunohistochemical analysis have demonstrated that solid tumors contain severely hypoxic regions, in which pO_2 values are less than 10 mmHg [54, 55]. HIF-1α is transiently induced in tumors following radiotherapy in response to the reoxygenation stress after irradiation [56]. HIF-1α transactivates various genes that confer malignant properties on tumors, such as apoptosis resistance, enhanced tumor growth, invasion, and metastasis [16]. In addition, HIF-1α activates pro-angiogenic cytokines, such as VEGF and PDGF, which confer radiation resistance on ECs and increase the proliferation and regrowth of tumor blood vessels [56–59]. Neovascularization is a particularly important contribution of HIF-1α to the survival and regrowth of tumors after irradiation [59]. Harada et al. [60] have demonstrated that the combination of radiotherapy and reduction of HIF-1α activity efficiently reduces tumor cells via hypoxia and irradiation, leading to long-term suppression of tumor growth and angiogenesis.

6.3.2 Angiogenic Therapy for Ischemic Disease

Knowledge of HIF biology is being rapidly translated into clinical applications and, in particular, drug discovery. The most advanced HIF pathway-targeted pharmaceuticals in terms of clinical development, to date, are the PHD inhibitors. These compounds, FG-2216 and FG-4592, are being evaluated for treatment of anemia and are currently in clinical trials (http://clinicaltrials.gov/). In addition to PHD inhibition, several strategies to promote HIF-1α activity and angiogenesis are in development for use in ischemic disease. In hindlimb ischemia models, adenoviral delivery of constitutively active HIF-1α has shown promise when administered alone or in combination with bone marrow-derived angiogenic cells [44, 61]. HIF-1α adenoviral therapy has also shown benefit in limb ischemia models in aged and diabetic mice [44, 62]. These findings are significant given that diabetics and the elderly are the two major patient populations afflicted by atherosclerosis and associated ischemic diseases [63].

HIF-2α stabilization by Int6 knockdown promotes functional vessel formation. Int6 silencing by siRNA injection not only accelerated wound healing in diabetic mice [42] but also led to recovery of peripheral circulation in a rat ischemic hindlimb model, resulting in restoration of hindlimb function [64]. These results indicate that HIF-2α stabilization may also be of therapeutic use in ischemic diseases, such as myocardial infarction and cerebral infarction.

6.4 Conclusions

Given that oxygen is an essential nutrient that serves as a key substrate in cellular metabolism and bioenergetics, hypoxia and HIFs are strongly associated with angiogenesis. Although the oxygen-dependent and oxygen-independent regulatory mechanisms of HIF-α subunits and their roles in normal and cancerous tissues have been increasingly clarified, a further understanding of how HIF-α subunits are uniquely regulated and how they can be selectively modulated will provide essential information to translate our knowledge of HIF pathway-mediated vessel formation into clinical settings.

References

1. Harris AL (2002) Hypoxia—a key regulatory factor in tumour growth. Nat Rev Cancer 2:38–47
2. Brown JM, Giaccia AJ (1998) The unique physiology of solid tumors: opportunities (and problems) for cancer therapy. Cancer Res 58:1408–1416
3. Makino Y, Cao R, Svensson K et al (2001) Inhibitory PAS domain protein is a negative regulator of hypoxia-inducible gene expression. Nature 414:550–554
4. Jain S, Maltepe E, Lu MM, Simon C, Bradfield CA (1998) Expression of ARNT, ARNT2, HIF1α, HIF2α and Ah receptor mRNAs in the developing mouse. Mech Dev 73:117–123
5. Tian H, McKnight SL, Russell DW (1997) Endothelial PAS domain protein 1 (EPAS1), a transcription factor selectively expressed in endothelial cells. Genes Dev 11:72–82
6. Gu YZ, Moran SM, Hogenesch JB, Wartman L, Bradfield CA (1998) Molecular characterization and chromosomal localization of a third α-class hypoxia inducible factor subunit, HIF3α. Gene Expr 7:205–213
7. Wiesener MS, Jürgensen JS, Rosenberger C et al (2002) Widespread hypoxia-inducible expression of HIF-2α in distinct cell populations of different organs. FASEB J 17:271–273
8. Jewell UR, Kvietikova I, Scheid A, Bauer C, Wenger RH, Gassmann M (2001) Induction of HIF-1α in response to hypoxia is instantaneous. FASEB J 15:1312–1314
9. Hon WC, Wilson MI, Harlos K et al (2002) Structural basis for the recognition of hydroxyproline in HIF-1α by pVHL. Nature 417:975–978
10. Kondo K, Kim WY, Lechpammer M, Kaelin WG Jr (2003) Inhibition of HIF2α is sufficient to suppress pVHL-defective tumor growth. PLoS Biol 1:439–444
11. Epstein AC, Gleadle JM, McNeill LA et al (2001) *C. elegans* EGL-9 and mammalian homologs define a family of dioxygenases that regulate HIF by prolyl hydroxylation. Cell 107:43–54
12. Bruick RK, McKnight SL (2001) A conserved family of prolyl-4-hydroxylases that modify HIF. Science 294:1337–1340
13. Appelhoff RJ, Tian YM, Raval RR et al (2004) Differential function of the prolyl hydroxylases PHD1, PHD2, and PHD3 in the regulation of hypoxia-inducible factor. J Biol Chem 279:38458–38465
14. Kaelin WG Jr (2002) Molecular basis of the VHL hereditary cancer syndrome. Nat Rev Cancer 2:673–682
15. Kaelin WG Jr, Ratcliffe PJ (2008) Oxygen sensing by metazoans: the central role of the HIF hydroxylase pathway. Mol Cell 30:393–402
16. Semenza GL (2003) Targeting HIF-1 for cancer therapy. Nat Rev Cancer 3:721–732
17. Chen L, Uchida K, Endler A, Shibasaki F (2007) Mammalian tumor suppressor Int6 specifically targets hypoxia inducible factor 2α for degradation by hypoxia- and pVHL-independent regulation. J Biol Chem 282:12707–12716

18. Marchetti A, Buttitta F, Miyazaki S, Gallahan D, Smith GH, Callahan R (1995) *Int-6*, a highly conserved, widely expressed gene, is mutated by mouse mammary tumor virus in mammary preneoplasia. J Virol 69:1932–1938
19. Kim T, Hofmann K, von Arnim AG, Chamovitz DA (2001) PCI complexes: pretty complex interactions in diverse signaling pathways. Trends Plant Sci 6:379–386
20. von Armin AG, Chamovitz DA (2003) Protein Homeostasis: a degrading role for Int6/eIF3e. Curr Biol 13:R323–R325
21. Kato H, Tamamizu-Kato S, Shibasaki F (2004) Histone deacetylase 7 associates with hypoxia-inducible factor 1α and increases transcriptional activity. J Biol Chem 279:41966–41974
22. Koh MY, Darnay BG, Powis G (2008) Hypoxia-associated factor, a novel E3-ubiquitin ligase, binds and ubiquitinates hypoxia-inducible factor 1α, leading to its oxygen-independent degradation. Mol Cell Biol 28:7081–7095
23. Koh MY, Lemos R Jr, Liu X, Powis G (2011) The hypoxia-associated factor switches cells from HIF-1α- to HIF-2α-dependent signaling promoting stem cell characteristics, aggressive tumor growth and invasion. Cancer Res 71:4015–4027
24. Tang N, Wang L, Esko J et al (2004) Loss of HIF-1α in endothelial cells disrupts a hypoxia-driven VEGF autocrine loop necessary for tumorigenesis. Cancer Cell 6:485–495
25. Sowter HM, Raval RR, Moore JW et al (2003) Predominant role of hypoxia-inducible transcription factor (Hif)-1α versus Hif-2α in regulation of the transcriptional response to hypoxia. Cancer Res 63:6130–6134
26. Licht AH, Muller-Holtkamp F, Flamme I, Breier G (2006) Inhibition of hypoxia-inducible factor activity in endothelial cells disrupts embryonic cardiovascular development. Blood 107:584–590
27. Maltepe E, Schmidt JV, Baunoch D, Bradfield CA, Simon MC (1997) Abnormal angiogenesis and responses to glucose and oxygen deprivation in mice lacking the protein ARNT. Nature 386:403–407
28. Carmeliet P, Dor Y, Herbert JM et al (1998) Role of HIF-1α in hypoxia-mediated apoptosis, cell proliferation and tumour angiogenesis. Nature 394:485–490
29. Iyer NV, Kotch LE, Agani F et al (1998) Cellular and developmental control of O_2 homeostasis by hypoxia-inducible factor 1α. Genes Dev 12:149–162
30. Ryan HE, Lo J, Johnson RS (1998) HIF-1α is required for solid tumor formation and embryonic vascularization. EMBO J 17:3005–3015
31. Vincent KA, Shyu KG, Luo Y et al (2000) Angiogenesis is induced in a rabbit model of hindlimb ischemia by naked DNA Encoding an HIF-1α/VP16 hybrid transcription factor. Circulation 102:2255–2261
32. Duan LJ, Zhang-Benoit Y, Fong GH (2005) Endothelium-intrinsic requirement for Hif-2α during vascular development. Circulation 111:2227–2232
33. Compernolle V, Brusselmans K, Acker T et al (2002) Loss of HIF-2α and inhibition of VEGF impair fetal lung maturation, whereas treatment with VEGF prevents fatal respiratory distress in premature mice. Nat Med 8:702–710
34. Scortegagna M, Ding K, Oktay Y et al (2003) Multiple organ pathology, metabolic abnormalities and impaired homeostasis of reactive oxygen species in *Epas1-/-* mice. Nat Genet 35:331–340
35. Tian H, Hammer RE, Matsumoto AM et al (1998) The hypoxia-responsive transcription factor EPAS1 is essential for catecholamine homeostasis and protection against heart failure during embryonic development. Genes Dev 12:3320–3324
36. Yamashita T, Ohneda K, Nagano M et al (2008) Hypoxia-inducible transcription factor-2α in endothelial cells regulates tumor neovascularization through activation of ephrin A1. J Biol Chem 283:18926–18936
37. Kashiwagi S, Tsukada K, Xu L et al (2008) Perivascular nitric oxide gradients normalize tumor vasculature. Nat Med 14:255–257
38. Coulet F, Nadaud S, Agrapart M, Soubrier F (2003) Identification of hypoxia-response element in the human endothelial nitric-oxide synthase gene promoter. J Biol Chem 278:46230–46240

39. Le Bras A, Lionneton F, Mattot V, Lelievre E, Caetano B, Spruyt N, Soncin F (2007) HIF-2α specifically activates the *VE-cadherin* promoter independently of hypoxia and in synergy with Ets-1 through two essential ETS-binding sites. Oncogene 26:7480–7489
40. Hu CJ, Sataur A, Wang L, Chen H, Simon MC (2007) The N-terminal transactivation domain confers target gene specificity of hypoxia-inducible factors HIF-1α and HIF-2α. Mol Biol Cell 18:4528–4542
41. Hu CJ, Wang LY, Chodosh LA et al (2003) Differential roles of hypoxia-inducible factor 1α (HIF-1α) and HIF-2α in hypoxic gene regulation. Mol Cell Biol 23:9361–9374
42. Chen L, Endler A, Uchida K et al (2010) *Int6*/eIF3e silencing promotes functional blood vessel outgrowth and enhances wound healing by upregulating hypoxia-induced factor 2α expression. Circulation 122:910–919
43. Peng J, Zhang L, Drysdale L, Fong GH (2000) The transcription factor EPAS-1/hypoxia-inducible factor 2α plays an important role in vascular remodeling. Proc Natl Acad Sci U S A 97:8386–8391
44. Bosch-Marce M, Okuyama H, Wesley JB et al (2007) Effects of aging and hypoxia-inducible factor-1 activity on angiogenic cell mobilization and recovery of perfusion after limb ischemia. Circ Res 101:1310–1318
45. Bertout JA, Patel SA, Simon MC (2008) The impact of O_2 availability on human cancer. Nat Rev Cancer 8:967–975
46. Huang X, Ding L, Bennewith KL et al (2009) Hypoxia-inducible mir-210 regulates normoxic gene expression involved in tumor initiation. Mol Cell 35:856–867
47. Semenza GL (2007) Evaluation of HIF-1 inhibitors as anticancer agents. Drug Discov Today 12:853–859
48. Jain RK (2005) Normalization of tumor vasculature: an emerging concept in antiangiogenic therapy. Science 307:58–62
49. Du R, Lu KV, Petritsch C et al (2008) HIF1α induces the recruitment of bone marrow-derived vascular modulatory cells to regulate tumor angiogenesis and invasion. Cancer Cell 13:206–220
50. Stockmann C, Doedens A, Weidemann A et al (2008) Deletion of vascular endothelial growth factor in myeloid cells accelerates tumorigenesis. Nature 456:814–818
51. Mazzone M, Dettori D, Leite de Oliveira R et al (2009) Heterozygous deficiency of *PHD2* restores tumor oxygenation and inhibits metastasis via endothelial normalization. Cell 136:839–851
52. Vaupel P, Kallinowski F, Okunieff P (1989) Blood flow, oxygen and nutrient supply, and metabolic microenvironment of human tumors: a review. Cancer Res 49:6449–6465
53. Brown JM, Wilson WR (2004) Exploiting tumour hypoxia in cancer treatment. Nat Rev Cancer 4:437–447
54. Höckel M, Knoop C, Schlenger K et al (1993) Intratumoral pO_2 predicts survival in advanced cancer of the uterine cervix. Radiother Oncol 26:45–50
55. Raleigh JA, Calkins-Adams DP, Rinker LH et al (1998) Hypoxia and vascular endothelial growth factor expression in human squamous cell carcinomas using pimonidazole as a hypoxia marker. Cancer Res 58:3765–3768
56. Moeller BJ, Cao Y, Li CY, Dewhirst MW (2004) Radiation activates HIF-1 to regulate vascular radiosensitivity in tumors: role of reoxygenation, free radicals, and stress granules. Cancer Cell 5:429–441
57. Gorski DH, Beckett MA, Jaskowiak NT et al (1999) Blockage of the vascular endothelial growth factor stress response increases the antitumor effects of ionizing radiation. Cancer Res 59:3374–3378
58. Geng L, Donnelly E, McMahon G et al (2001) Inhibition of vascular endothelial growth factor receptor signaling leads to reversal of tumor resistance to radiotherapy. Cancer Res 61:2413–2419
59. Abdollahi A, Lipson KE, Han X et al (2003) SU5416 and SU6668 attenuate the angiogenic effects of radiation-induced tumor cell growth factor production and amplify the direct anti-endothelial action of radiation *in vitro*. Cancer Res 63:3755–3763

60. Harada H, Kizaka-Kondoh S, Li G et al (2007) Significance of HIF-1-active cells in angiogenesis and radioresistance. Oncogene 26:7508–7516
61. Rey S, Lee K, Wang CJ et al (2009) Synergistic effect of HIF-1α gene therapy and HIF-1-activated bone marrow-derived angiogenic cells in a mouse model of limb ischemia. Proc Natl Acad Sci U S A 106:20399–20404
62. Sarkar K, Fox-Talbot K, Steenbergen C, Bosch-Marce M, Semenza GL (2009) Adenoviral transfer of HIF-1α enhances vascular responses to critical limb ischemia in diabetic mice. Proc Natl Acad Sci U S A 106:18769–18774
63. Beckman JA, Creager MA, Libby P (2002) Diabetes and atherosclerosis: epidemiology, pathophysiology, and management. JAMA 287:2570–2581
64. Okamoto N, Tanaka A, Jung K et al (2011) Silencing of *int6* gene restores function of the ischaemic hindlimb in a rat model of peripheral arterial disease. Cardiovasc Res 92:209–217

Chapter 7
Role of Mast Cells in Angiogenesis

Enrico Crivellato and Domenico Ribatti

Abstract Angiogenesis is a process that leads to the development of a new vascular network starting from pre-existing capillaries and venules. Mast cells (MCs) are tissue resident cells which actively participate in the angiogenic scenario, playing an important role in normal angiogenesis, angiogenesis that accompanies inflammation and tumour-associated angiogenesis. This chapter will focus on the specific role of MCs in the different types of angiogenic processes.

Keywords Angiogenesis • Inflammation • Mast cells • Tumour growth

7.1 Introduction

Angiogenesis is a complex and highly orchestrated process leading to the formation of new blood vessels from pre-existing capillaries and venules. This process occurs in different conditions, such as embryo development and postnatal tissue growth, inflammation and cancer. Both structural cells and inflammatory cells in the different tissues are involved in the mechanisms of endothelial cell activation, proliferation and migration, through the production and release of a large spectrum of pro-angiogenic mediators. These create the specific microenvironment that favours an increased rate of tissue vascularization. Among inflammatory cells, increasing attention has recently been devoted to mast cells (MCs).

E. Crivellato
Department of Experimental and Clinical Medicine, Section of Anatomy, University of Udine Medical School, Udine, Italy

D. Ribatti (✉)
Department of Basic Medical Sciences, Neurosciences, and Sensory Organs, Section of Human Anatomy and Histology, University of Bari Medical School, Piazza G. Cesare, 11, Policlinico, 70124 Bari, Italy
e-mail: domenico.ribatti@uniba.it

J.L. Mehta and N.S. Dhalla (eds.), *Biochemical Basis and Therapeutic Implications of Angiogenesis*, Advances in Biochemistry in Health and Disease 6,
DOI 10.1007/978-1-4614-5857-9_7, © Springer Science+Business Media New York 2013

MCs are bone marrow-derived, tissue-homing secretory cells, firstly identified by Paul Ehrlich in 1878 [1, 2]. These cells have unique features when viewed by light microscopy. They appear as ubiquitous, round or oval, mononuclear elements, whose cytoplasm is filled with numerous metachromatic granules. MCs are located in particular in the connective tissues belonging to lying surfaces of the body, such as the skin and the gastrointestinal and respiratory tracts that are common portals for pathogen, allergen and toxin entry [3].

MCs have been long recognized as initiators of IgE-dependent allergic diseases, but recently our perception of these cells has basically changed mostly on the ground of studies in rodents. MCs are currently conceived as tissue "sentinels" or "effectors" operating in the context of both innate and adaptive immunity to infections [4, 5]. They also participate to inflammatory autoimmune diseases and may either facilitate or retard tumour growth depending on the type of cancer [6]. Besides defensive functions, MCs are capable to express functional properties which can be reconciled with the concept of "tissue homeostasis". In particular, they contribute to organize angiogenic responses and protect tissue damage by arranging tissue repair processes after injury [7]. In general terms, MCs responses may have beneficial or detrimental effects to the host. These cells may either favour or worsen such conditions like coronary spasm, cardiomyopathy, atherosclerosis, myocardial ischemia and aortic aneurysms.

There are emerging data assigning MCs an important role in orchestrating angiogenesis, both in normal and pathological conditions. Angiogenesis associated with physiological events, such as embryonic development, ovulation and wound healing, has been found to be accompanied by tissue accumulation of MCs. Besides, increased number of MCs has been reported in angiogenesis associated with chronic inflammatory diseases, like asthma, rheumatoid arthritis and nasal polyposis. Lastly, MC accumulation has been documented in different histological types and clinical stages of tumour masses.

In this chapter, we will present and discuss the most recent findings on the contribution of MCs to the development and progression of normal and inflammation-associated angiogenesis as well as angiogenesis related to tumour growth.

7.2 The Process of Angiogenesis

Angiogenesis is a sequential progression of vessel formation from pre-existing capillaries and venules, leading to the development of a new vascular network [8]. Under physiological conditions, angiogenesis depends on the balance of the positive and negative angiogenic mediators within the vascular microenvironment and requires the functional involvement of a number of molecules, including angiogenic factors, extracellular matrix proteins, adhesion receptors and proteolytic enzymes [8]. Thus, vascular development in a given tissue is controlled in a remarkably complex way by many microenvironmental factors.

This process is fundamental for organ formation. During the later stages of embryogenesis, the vascularization of many tissues occurs by angiogenesis. The mechanism of vessel production consists in a multistep and highly orchestrated process, which is under control of different genetic and epigenetic mechanisms [9]. This multistep process includes different phases: basement membrane degradation, endothelial cell migration and invasion of the surrounding extracellular matrix, endothelial cell proliferation and capillary lumen formation. From a morphological, purely descriptive point of view, at least two types of angiogenic processes are to be distinguished: (a) the so-called sprouting angiogenesis, which is characterized by the proliferation and migration of endothelial cells into avascular sites, and (b) the "non-sprouting" angiogenesis or intussusceptive microvascular growth, which occurs by splitting of the existing vasculature by transluminal pillars or transendothelial bridges [8, 10].

Endothelial cell tubes formed by sprouting angiogenesis are soon stabilized by pericytes and mural cells recruited from the surrounding interstitium. Platelet-derived growth factor (PDGF) and transforming growth factor (TGF)-beta signals are involved in the recruitment of pericytes. In larger vessels, arterioles and venules, the vascular wall is made up of endothelial cells and smooth muscle cells, which are recruited mainly through the Tie-2 and angiopoietin-1 (Ang-1) receptor-ligand pair, although neuropilins and Notch pathways are also involved in mural cell formation.

Angiogenesis associated with physiological, inflammatory and tumour events has been found to be accompanied by tissue accumulation of MCs.

7.3 Mast Cell Biology

MCs are haematopoietic cells which acquire their definite phenotype after entering peripheral tissues. These cells originate from progenitor cells in the bone marrow, which move through the circulation and become mature MCs after homing to destination tissue under the influence of local chemotactic and microenvironment factors [11, 12]. These include the stem cell factor (SCF) secreted by fibroblasts, stromal cells and endothelial cells; cell adhesion receptors, mainly cell surface receptors of the integrin family; chemokine receptors CXCR2, CCR3, CXCR4 and CCR5 [13]; and Notch receptor-mediated signalling; interleukins (ILs) and cytokines, such as IL-3, IL-4, IL-9, TGF-beta and nerve growth factor (NGF).

MCs reside in almost all of the major organs and tissues of the body. Normally, they localize in proximity to surfaces that interface the external environment: skin, intestinal and respiratory mucosae. MCs also populate the connective tissues, particularly in association with structures such as blood vessels, lymphatic vessels and nerves, in a position which make them key elements in processes like wound healing, tissue regeneration and remodelling after injury, fibrosis and angiogenesis [14–16]. MCs also reside in proximity to vulnerable spaces such as the peritoneum and the joint cavities. MCs are not found in avascular tissues such as mineralized bone, cartilage and the cornea.

7.4 Mast Cell Angiogenic Factors

MCs produce a large spectrum of pro-angiogenic factors. Human, rat and mouse MCs release preformed fibroblast growth factor (FGF)-2 from their secretory granules [17]. Human cord blood-derived MCs release vascular endothelial growth factor (VEGF) upon stimulation through FcεRI and KIT receptors. Both FGF-2 and VEGF have also been identified by immunohistochemistry in mature MCs in human tissues [18, 19]. It has recently been shown that human MCs are a potent source of VEGF in the absence of degranulation through activation of the EP(2) receptor by prostaglandin E2 (PGE2) [20]. Recent studies have highlighted the contribution of MCs in angiogenesis during allergic inflammation. For example, following IgE-dependent activation, MCs released several pro-angiogenic mediators stored in their granules, such as VEGF [21] and FGF-2 [22], that promote angiogenesis even in the early phase of allergic inflammation. Together with eosinophils, MCs can also migrate in vivo [23] and in vitro [24] in response to VEGF. It has also been demonstrated that rat peritoneal MCs contain angiogenic factors stored in their secretory granules. Granulated MCs and their granules, but not degranulated MCs, are able indeed to stimulate an intense angiogenic reaction in the chick embryo chorioallantoic membrane (CAM) assay. This angiogenic activity is partly inhibited by anti-FGF-2 and -VEGF antibodies, suggesting that these cytokines are involved in the angiogenic reaction [25]. Similarly it has been demonstrated, using the rat-mesenteric window angiogenic assay, that intraperitoneal injection of compound 48/80—a potent MC secretagogue—causes a vigorous angiogenic response [26]. The same treatment in mice also causes angiogenesis.

MCs store large amounts of preformed active serine proteases, such as tryptase and chymase, in their secretory granules. A role in the angiogenic process for these proteolytic enzymes has been established. Tryptase, in particular, stimulates the proliferation of human vascular endothelial cells, promotes vascular tube formation in culture and also degrades connective tissue matrix to provide space for neovascular growth. Tryptase also acts indirectly in the angiogenic scenario by activating latent matrix metalloproteases (MMPs) and plasminogen activator (PA), which in turn degrade the extracellular connective tissue with consequent release of VEGF or FGF-2 from their matrix-bound state [27]. MC-derived chymase degrades extracellular matrix components and therefore matrix-bound VEGF may be potentially released.

Other MC-specific mediators with angiogenic properties include histamine and heparin [28]. Both molecules have been shown to stimulate proliferation of endothelial cells and to induce the formation of new blood vessels in the CAM assay [29, 30]. Histamine, the major preformed mediator stimulates new vessel formation by acting trough both H1 and H2 receptors. Heparin, the main glycosaminoglycan constituent of MC granules, may act directly on blood vessels or indirectly by inducing release of FGF-2 from the extracellular storage site. In addition, other cytokines produced by MCs, such as IL-8 [31] tumour necrosis factor (TNF)-alpha, TGF-beta, NGF and urokinase-type PA, have been implicated in normal and tumour-associated angiogenesis [32, 33]. Lastly, MCs also contain preformed MMPs, such as MMP-2 and MMP-9, and tissue inhibitors of MMPs (TIMPs), which enable MCs to directly modulate extracellular matrix degradation. This, in turn, allows for tissue release of extracellular matrix-bound angiogenic factors.

Interestingly, PGE2 dose-dependently induces primary MCs to release the pro-angiogenic chemokine monocyte chemoattractant protein (MCP)-1 [34]. Release of MCP-1 is complete by 2h after PGE2 exposure and is not accompanied by degranulation or increased MCP-1 gene expression. By immunoelectron microscopy, MCP-1 is detected within MCs at a cytoplasmic location distinct from the secretory granules. Dexamethasone and cyclosporine A inhibit PGE2-induced MCP-1 secretion by approximately 60 %. Agonists of PGE2 receptor subtypes reveal that the EP1 and EP3 receptors can independently mediate MCP-1 release from MCs. These observations identify PGE2-induced MCP-1 release from MCs as a pathway underlying inflammation-associated angiogenesis.

Thus, MCs can contribute to tissue angiogenesis either directly, by releasing pro-angiogenic molecules, or indirectly, by stimulating other structural and inflammatory cells—such as epithelial cells fibroblasts, endothelial cells, macrophages, eosinophils, lymphocytes and neutrophils—to secrete angiogenic factors and cytokines as well as extracellular matrix-degrading proteases.

7.5 The Process of Physiological Angiogenesis

The cardiovascular system is the first functional organ system to develop in the vertebrate embryo. Blood vessel formation during embryonic development is achieved by two partly consecutive processes, called vasculogenesis and angiogenesis. Blood vessels first appear as the result of vasculogenesis [35]. The earliest blood vessels in the yolk sac and in the embryo originate from mesodermal cells that are specified into angioblasts, most likely in response to FGF-2 and VEGF signals. Angioblasts begin to differentiate into endothelial cells and assemble into tubes, principally as the result of VEGF signals from surrounding tissues and the expression of cell-matrix adhesion molecules. Vasculogenesis leads to the formation of the first major intra-embryonic blood vessels which are established before the onset of heart beat. Development of the vascular network of certain endodermal organs—including liver, lung, pancreas, stomach, intestine and spleen—occurs by vasculogenesis. Remarkably, blood vessels also act as signalling centres to help specification of organs like the liver, pancreas and bone during embryogenesis [36]. Disruption of blood vessel architecture by altering the expression of vasculogenic molecules often results in a parallel disruption of specific tissue types as consequence of loss of endothelial cell signalling. Vasculogenesis occurs also in postnatal life, as "postnatal vasculogenesis", which is de novo vessel formation by in situ incorporation, differentiation, migration and/or proliferation of bone marrow-derived endothelial precursor cells [37].

The majority of vessels of the developing embryo are formed through angiogenesis. A complex series of molecules orchestrates the process of angiogenesis. VEGF-A signalling appears a major inducer, but Sprouty (Spry) family proteins, class 3 semaphorins and the axon-guidance netrins play additional roles in stimulating or modulating the angiogenic process. In general terms, the specific pattern of vasculature which can be recognized in each organ is established by a complex interaction of vasculogenic and angiogenic steps. In organs like the kidney, the liver or the heart, differentiation pathways of organ tissues are closely accompanied by vasculogenic and angiogenic processes which are specific of a given structure [38].

7.6 Mast Cells and Physiological Angiogenesis

In adult life, angiogenesis associated with physiological events—such as ovulation, endometrial vascularization in menstrual cycle and pregnancy, as well as wound healing—has been found to be accompanied by tissue accumulation of MCs.

MC-derived IL-8 has recently been implicated in the ovarian mechanisms of follicle growth and ovulation [39]. These processes progress in parallel with the formation of new blood vessels. MCs have also been found in mammalian uterus. The number of MCs in the uterus fluctuates during menstrual cycle in man and oestrus cycle in rat and mouse, indicating a hormonal influence on their recruitment. In human uterus the majority of MCs are located close to smooth muscle cells, which might indicate a potential role for MCs in tissue remodelling during the menstrual cycle [40]. A major factor governing MC activity in the uterus might be histamine. It is suggested that histamine is important for normal ovulation, blastocyst implantation, placental blood flow regulation, lactation and contractile activity of uterus. Histamine may also play a role in pathological processes such as pre-eclampsia or preterm delivery. Participation of MCs and histamine in blastocyst implantation is very controversial because MC-deficient W/W^v mice show normal implantation. Recently, it has been proposed that estradiol and progesterone modulate the migration of MCs from the periphery to the uterus and their degranulation, which may prepare the uterus for implantation [41].

Inflammation-associated angiogenesis also occurs during pathophysiological reactions, like wound healing and scar formation. The process of extracellular matrix remodelling that accompanies this kind of tissue responses is strictly dependent upon angiogenic events. MCs are thought to regulate tissue remodelling during wound healing. Tryptase and chymase released by MCs stimulate collagen production and angiogenesis. In addition, MC-deficient W/W^v mice show an impaired skin wound closure [15].

7.7 The Process of Inflammatory Angiogenesis

There is increasing evidence to support the view that angiogenesis is an integral component of a diverse range of chronic inflammatory and autoimmune diseases, including atherosclerosis, rheumatoid arthritis, diabetic retinopathy, psoriasis, airway inflammation, peptic ulcers and Alzheimer's disease. Indeed, angiogenesis is intrinsic to chronic inflammation and is associated with structural changes, including activation and proliferation of endothelial cells, capillary and venule remodelling, all of which result in expansion of the tissue microvascular bed. Chronic inflammation in the airways, for instance, is associated with dramatic architectural changes in the walls of the airways and in the vasculature they contain. Therefore, it seems that an imbalance in favour of pro-angiogenic factors leads to the abnormal growth of new blood vessels in asthma. Inflammatory diseases, such as rheumatoid arthritis and psoriasis, are characterized by proliferating tissue containing an abundance of inflammatory cells and newly formed blood vessels. During prolonged inflammatory reactions, many structural and resident cells, such as fibroblasts, epithelial cells, smooth muscle cells, MCs and/or infiltrating cells, such as monocytes/macrophages, neutrophils, lymphocytes and eosinophils, synthesize and secrete pro-angiogenic factors that promote neovascularization. The anatomic expansion of the microvascular bed combined with its increased functional activation can therefore foster further recruitment of inflammatory cells, and angiogenesis and inflammation become chronically co-dependent processes. In addition, many of the mediators that are fundamental players in angiogenesis are also inflammatory molecules.

7.8 Mast Cells and Inflammatory Angiogenesis

Increasing evidence suggests that MCs are involved in the pathogenesis of chronic inflammatory diseases and, in particular, in rheumatoid arthritis. MCs reside in connective tissues and in synovial tissue of joints, where they produce an array of pro-inflammatory mediators, tissue destructive proteases and cytokines, most prominently TNF-alpha, which is one of the key cytokines in the pathogenesis of rheumatoid arthritis [42]. Inhibition of the most important growth factor receptor of human MCs, the KIT receptor, by the selective tyrosine kinase inhibitor imatinib mesylate, induces apoptosis of synovial tissue MCs. As MCs are long-lived cells, induction of their apoptosis could be a feasible approach to inhibit their functions. MCs may also participate to the development of secondary or amyloid A amyloidosis, as the partial degradation of the serum amyloid A protein by MCs leads to the generation of a highly amyloidogenic N-terminal fragment of serum amyloid A. MCs may contribute to the pathogenesis of connective tissue diseases, scleroderma, vasculitic syndromes and systemic lupus erythematosus, although the data available are limited.

MCs appear to be actively involved in post infarction inflammation. Myocardial infarction is associated with an acute inflammatory response, leading to replacement of injured cardiomyocytes with granulation tissue [43]. Myocardial necrosis is associated with complement activation and free radical generation, triggering a cytokine cascade and chemokine upregulation. IL-8 and C5a are released in the ischemic myocardium and may have a crucial role in neutrophil recruitment. MCs are actively involved in post infarction inflammation by releasing histamine and TNF-alpha, triggering a cytokine cascade. During the proliferative phase of healing, MCs accumulate in the infarct and may regulate angiogenesis and fibrous tissue deposition by releasing growth factors, angiogenic mediators and proteases. In the healing infarct, MCs are associated with other cell types that are important for granulation tissue formation. Inflammatory mediators may induce recruitment of blood-derived primitive stem cells in the healing infarct, which may differentiate into endothelial cells and even lead to limited myocardial regeneration.

MCs have been implicated in the neovascularization associated with skin diseases, like rosacea and atopic dermatitis. Angiogenesis seems to play an important role in the pathogenesis of rosacea, especially of the most severe clinical forms. MCs seem to participate in evolution to disease chronicity by contributing to inflammation, angiogenesis and tissue fibrosis [44]. Atopic dermatitis skin lesions are characterized by inflammatory changes and epithelial hyperplasia requiring angiogenesis. MCs may participate to this process via bidirectional secretion of tissue-damaging enzymes and pro-angiogenic factors. It has been shown that MCs are abundantly localized in the papillary dermis and migrate through the basal lamina into the epidermis of atopic dermatitis lesions [45]. Approximately 80% were chymase positive. A high number of MCs express KIT receptor. Most papillary and epidermal MCs localize close to endothelial cells. Vascular expression of endoglin (CD105) demonstrates neoangiogenic processes. MC stimulation leads to the expression of pro-angiogenic factors and tissue-damaging factors such MMPs. These data suggest that in atopic dermatitis, MCs close to papillary vessels and within the epidermis may be implicated in stimulation of neoangiogenesis.

MCs have been implicated in the neovascularization associated with asthma. Increased vascularity in the airways has been recognized not only in patients with severe asthma but also in those with mild disease [46, 47]. The major structural and functional changes of the airway microcirculation include the proliferation of new vessels, increased vascular area of the medium and small airways, increased blood flow and microvascular permeability, and oedema formation in the airway wall [48]. Indeed, persistent chronic allergic inflammation can result in remodelling of the affected tissues, and these structural changes are often associated with activation of the angiogenic process. MCs infiltrate the bronchial submucosa and epithelium [49].

MCs appear to be implicated in subacute thyroiditis, in particular in thyroid tissue repair and regeneration. Along with thyrocytes, endothelial cells and fibroblasts, MCs release tissue factors such as VEGF, FGF-2, platelet derived growth factor B (PDGF-B) and epidermal growth factor (EGF), which act as angiogenic factors and contribute to follicle regeneration through their growth-promoting effect on thyrocytes [50]. Tryptase, histamine and MMPs from MCs play important roles in extracellular matrix remodelling.

7.9 The Process of Tumour Angiogenesis

The importance of a potential functional link between chronic inflammation and cancer has long been recognized, and the question as to the possible contribution of MCs to tumour development has progressively been emerged. It was Rudolf Virchow in 1863, which critically recognized the presence of inflammatory cells infiltrating neoplastic tissues and first established a causative connection between the "lymphoreticular infiltrate" at sites of chronic inflammation and the development of cancer. More than a century later, Dvorak noted that wound healing and tumour stroma formation shared many important properties [51]. They both were rich of newly formed vessels and fibrin matrix. He likened tumours to "wounds that do not heal", because tumour cells secreted a vascular permeability factor, now referred to as VEGF, which could lead to persistent extravasation of fibrin and fibronectin and continuous generation of extracellular matrix.

Neoplastic cells produce and secrete in the tumour microenvironment cytokines such as VEGF, FGF-2, IL-8, placental growth factor (PlGF), TGF-beta, PDGF and Angs, which exert potent angiogenic activity, as well as pro-inflammatory factors and mediators for inflammatory cells. As a consequence, tumour cells are surrounded by an infiltrate of inflammatory cells, namely, lymphocytes, neutrophils, macrophages and MCs [52], which communicate via a complex network of intercellular signalling pathways, mediated by surface adhesion molecules, cytokines and their receptors. Results point to the importance of a crosstalk between several host cells for promoting angiogenic effects in tumour areas. In tumour areas, angiogenesis creates a new vascular supply that conveys oxygen and nutrients to the rapidly proliferating tissue and removes the by-products of cellular metabolism. Moreover, the special structure of tumour blood vessels, with their often incomplete endothelial envelope and their increased permeability, favours escape of neoplastic cells and generation of metastases.

The characteristics of blood vessels running through or surrounding tumour masses have attracted attention of investigators. Tumour blood vessels display many structural and functional abnormalities. They are irregular in size, shape and branching pattern, lack the normal vessel hierarchy and do not display the recognizable features of arterioles, capillaries and venules [53]. Like normal blood vessels, they consist of endothelial cells, mural cells and their enveloping basement membrane, but each component of the vessel wall presents abnormalities. Tumour-associated endothelial cells proliferate 50–200 times faster than normal endothelial cells. They present a defective organization of the endothelium, which shows discontinuities or gaps that allow haemorrhage, and facilitate permeability of macromolecules and the traffic of tumour cells into the bloodstream. The basement membrane that envelops endothelial cells and pericytes of tumour vessels may have extra layers that have no apparent association with the cells. Pericytes of tumour vessels are loosely associated with endothelial cells, have abnormal shape, paradoxically extend cytoplasmic processes away from the vessel wall, and have extra layers of loosely fitting basement membrane. Thus, tumour vasculature is typically aberrant, disordered and leaky.

7.10 Mast Cells and Tumour Angiogenesis

Experimental evidence indicates MCs as key host cells in the tumour infiltrate, with important consequence on tumour cell fate [6]. They accumulate and are activated at sites of tumour growth in response to numerous chemoattractants, like small tumour-derived peptides such as the purine nucleoside adenosine and adrenomedullin [54], factors like FGF-2, VEGF, PDGF, RANTES and MCP-1 [55]. A major point linking MCs to cancer is the well-recognized capacity of these cells to synthesize and release potent angiogenic compounds. The role played by MCs in tumour development is complex and controversial. On the one hand, they express detrimental effects on tumour growth by producing molecules killing tumour cells and inducing an inflammatory reaction. On the other hand, they favour tumour progression by promoting expansion of its vascular supply, degradation of the tumour extracellular matrix and immunosuppression [56]. Thus, some evidence suggests that MCs can promote tumorigenesis and tumour progression. By contrast, some clinical conditions and experimental models suggest that MCs may have functions that favour the host. In conclusion, involvement of MCs in tumour onset and tumour progress is multifaceted and far from being settled.

MCs play a role in tumour angiogenesis. We have already seen that MCs synthesize and release a vast array of pro-inflammatory and pro-angiogenic molecules that favour new vessel formation either directly or via local recruitment of activated inflammatory cells. MC-deficient W/W^v mice exhibit indeed a decreased rate of tumour angiogenesis [57]. Molecules like heparin could facilitate tumour vascularization not only by a direct pro-angiogenic effect but also through its anti-clotting effect [56]. In addition, MC-derived MMPs can degrade the interstitial tumour stroma and hence release matrix-bound angiogenic factors. An increased number of MCs has indeed been reported in angiogenesis associated with vascular neoplasms, like haemangioma and haemangioblastoma, as well as a number of solid and haematopoietic tumours. At present, it is not known if tumour-associated MCs are involved in "sprouting" angiogenesis or in "non-sprouting" angiogenesis, or in both. In general, MC density correlates with angiogenesis and poor tumour outcome. Association between MCs and new vessel formation has been reported in breast cancer, colorectal cancer and uterine cervix cancer. Tryptase-positive MCs increase in number, and vascularization increases in a linear fashion from dysplasia to invasive cancer of the uterine cervix [58]. An association of VEGF and MCs with angiogenesis has been demonstrated in laryngeal carcinoma [59] and in small lung carcinoma, where most intratumoural MCs express VEGF [60–62] MC accumulation has also been noted repeatedly around melanomas, especially invasive melanoma [63]. MC accumulation was correlated with increased neovascularization, MC expression of VEGF [64] and FGF-2 [65], tumour aggressiveness and poor prognosis. Indeed, a prognostic significance has been attributed to MC and microvascular density not only in melanoma [63] but also in squamous cell cancer of the oesophagus [66]. Angiogenesis has been shown to correlate with tryptase-positive MC count in human endometrial cancer. Both parameters were found to increase in agreement with tumour progression [67].

MC density, new vessel rate and clinical prognosis have also been found to correlate in haematological tumours. In benign lymphadenopathies and B cell non-Hodgkin's lymphomas, angiogenesis correlates with total and tryptase-positive MC counts, and both increase in step with the increase with Working Formulation malignancy grades [68]. In non-Hodgkin's lymphomas, a correlation has been found between vessel count and the number of MCs and VEGF-expressing cells [69]. In the bone marrow of patients with inactive and active multiple myeloma as well as those with monoclonal gammopathies of undetermined significance, angiogenesis highly correlates with MC counts [70]. A similar pattern of correlation between bone marrow microvessels' count, total and tryptase-positive MC density and tumour progression has been found in patients with myelodysplastic syndrome [71] and B cell chronic lymphocytic leukaemia [72]. In the early stages of B cell chronic lymphocytic leukaemia, the density of tryptase-positive MCs in the bone marrow has been shown to predict the outcome of the disease [73].

As MCs have the potency to express either favourable or detrimental effects on tumour cell growth, the hypothesis has been proposed that such dual role may depend on the way MCs release their bioactive molecules from secretory granules. Frank exocytosis would export secretory cytokines mainly involved in promoting tumour cell apoptosis whilst piecemeal degranulation, a particulate and possibly selective way of MC secretion, would allow for release of mediators and growth factors principally responsible of angiogenesis, immunosuppression and extracellular matrix disruption [56]. Interestingly, lymph node and bone marrow MCs in B cell non-Hodgkin's lymphomas and multiple myeloma show ultrastructural features of slow and particulate secretion as it occurs in piecemeal degranulation [74, 75].

7.11 Conclusions

Increasing evidence suggests that MCs play an important role in tissue angiogenesis through release of various preformed and newly formed angiogenic factors. MC involvement in blood vessel formation has been suggested in physiological angiogenesis as well as inflammatory and tumour angiogenesis (Fig. 7.1). Many aspects of MC recruitment and activation remain to be elucidated, in particular under what circumstances MCs represents a crucial source of angiogenic factors. These aspects are important for building new therapeutic interventions aimed at downregulating MC-related angiogenic potential, in particular in tumour-associated angiogenesis.

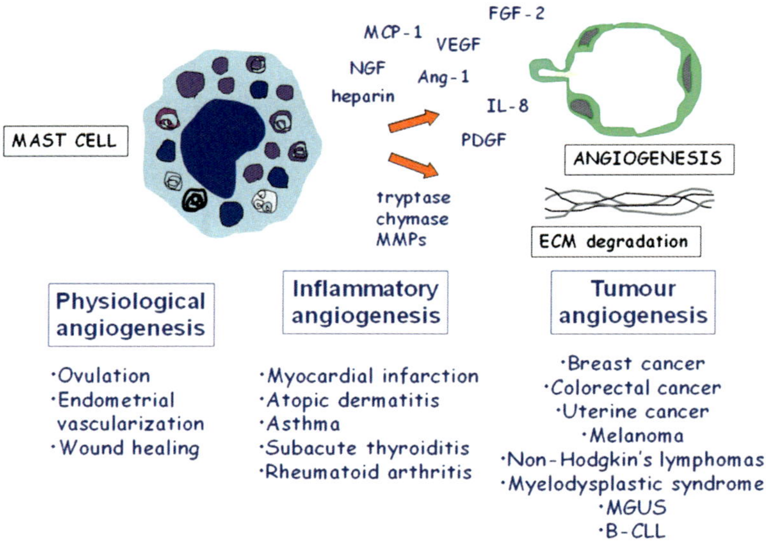

Fig. 7.1 Different roles of mast cells in physiological, inflammatory and tumour angiogenesis. *Ang-1* angiopoietin-1; *B-CLL* B cell chronic lymphocytic leukaemia; *ECM* extracellular matrix; *FGF-2* fibroblast growth factor-2; *IL-8* interleukin-8; *MCP-1* monocyte chemotactic protein-1; *MGUS* monoclonal gammopathy of undetermined significance; *NGF* nerve growth factor; *PDGF* platelet-derived growth factor

Acknowledgements This work is supported by European Union Seventh Framework Programme (FPT7/2007-2013) under grant agreement no. 278570 to DR.

References

1. Ehrlich P (1878) Beiträge zur Theorie und Praxis der histologiscen Färbung. Thesis, Leipzig University, Leipzig
2. Crivellato E, Nico B, Vacca A, Ribatti D (2003) Ultrastructural analysis of mast cell recovery after secretion by piecemeal degranulation in B-cell non-Hodgkin's lymphoma. Leuk Lymphoma 44:517–521
3. Crivellato E, Beltrami CA, Mallardi F, Ribatti D (2004) The mast cell: an active participant or an innocent bystander? Histol Histopathol 19:259–270
4. Galli SJ, Maurer M, Lantz CS (1999) Mast cells as sentinels of innate immunity. Curr Opin Immunol 11:53–59
5. Galli SJ, Nakae S, Tsai M (2005) Mast cells in the development of adaptive immune responses. Nat Immunol 6:135–142
6. Ribatti D, Crivellato E (2012) Mast cells, angiogenesis and tumour growth. Biochim Biophys Acta 1822:2–8
7. Saito H (2005) Role of mast cell proteases in tissue remodeling. Chem Immunol Allergy 87:80–84

8. Risau W (1997) Mechanisms of angiogenesis. Nature 386:671–674
9. Ribatti D (2006) Genetic and epigenetic mechanisms in the early development of the vascular system. J Anat 208:139–152
10. Burri P, Djonov D (2002) Intussusceptive angiogenesis. The alternative to capillary sprouting. Mol Aspects Med 23:1–27
11. Gurish MF, Austen KF (2001) The diverse role of mast cells. J Exp Med 194:1–5
12. Kitamura Y, Ito A (2005) Mast cell-committed progenitors. Proc Natl Acad Sci U S A 102:11129–11130
13. Ochi H, Hirani WM, Yuan Q et al (1999) T helper cell type 2 cytokine-mediated comitogenic responses and CCR3 expression during differentiation of human mast cells in vitro. J Exp Med 190:267–280
14. Gonzalez S, Moran M, Kochevar IE (1999) Chronic photodamage in skin of mast cell-deficient mice. Photochem Photobiol 70:248–253
15. Weller K, Foitzik K, Paus R et al (2006) Mast cells are required for normal healing of skin wounds in mice. FASEB J 20:2366–2368
16. Grimbaldeston MA, Nakae S, Kalesnikoff K et al (2007) Mast cell-derived interleukin 10 limits skin pathology in contact dermatitis and chronic irradiation with ultraviolet B. Nat Immunol 38:1095–1104
17. Qu Z, Lieber JM, Powers MR et al (1995) Mast cells are a major source of basic fibroblast growth factor in chronic inflammation and cutaneous hemangioma. Am J Pathol 147:564–573
18. Qu Z, Kayton RJ, Ahmadi P et al (1998) Ultrastructural immunolocalization of basic fibroblast growth factor in mast cell secretory granules: morphological evidence for bFGF release through degranulation. J Histochem Cytochem 46:1119–1128
19. Grutzkau A, Kruger-Krasagakes S, Baumesteir H et al (1998) Synthesis, storage, and release of vascular endothelial growth factor/vascular permeability factor (VEGF/VPF) by human mast cells: implications for the biological significance of VEGF 206. Mol Biol Cell 9:875–884
20. Abdel-Majid RM, Marshall JS (2004) Prostaglandin E2 induces degranulation-independent production of vascular endothelial growth factor by human mast cells. J Immunol 172:1227–1236
21. Boesiger J, Tsai M, Maurer M et al (1998) Mast cells can secrete vascular permeability factor/ vascular endothelial cell growth factor and exhibit enhanced release after immunoglobulin E-dependent upregulation of Fcε receptor I expression. J Exp Med 188:1135–1145
22. Kanbe N, Kurosawa M, Nagata H et al (2000) Production of fibrogenic cytokines by cord blood-derived cultured human mast cells. J Allergy Clin Immunol 106:S85–S90
23. Detmar M, Brown LF, Schon MP et al (1998) Increased microvascular density and enhanced leukocyte rolling and adhesion in the skin of VEGF transgenic mice. J Invest Dermatol 11:1–6
24. Gruber BL, Marchese MJ, Kaw R (1995) Angiogenic factors stimulate mast cell migration. Blood 86:2488–2493
25. Ribatti D, Crivellato E, Candussio L et al (2001) Mast cells and their secretory granules are angiogenic in the chick embryo chorioallantoic membrane. Clin Exp Allergy 31:602–608
26. Norrby K, Jakobsson A, Sörbo J (1986) Mast-cell-mediated angiogenesis: a novel experimental model using the rat mesentery. Virchows Arch B Cell Pathol Incl Mol Pathol 52:195–206
27. Blair RJ, Meng H, Marchese MJ et al (1997) Human mast cells stimulate vascular tube formation. Tryptase is a novel, potent angiogenic factor. J Clin Invest 99:2691–2700
28. Azizkhan RG, Azizkhan JC, Zetter BR, Folkman J (1980) Mast cell heparin stimulates migration of capillary endothelial cells in vitro. J Exp Med 152:931–944
29. Ribatti D, Roncali L, Nico B, Bertossi M (1987) Effects of exogenous heparin on the vasculogenesis of the chorioallantoic membrane. Acta Anat (Basel) 130:257–263
30. Sörbo J, Jakobsson A, Norrby K (1994) Mast cell histamine is angiogenic through receptors for histamine 1 and histamine 2. Int J Exp Pathol 75:43–50

31. Möller A, Lippert U, Lessmann D et al (1993) Human mast cells produce IL-8. J Immunol 151:3261–3266
32. Walsh LJ, Trinchieri G, Waldorf HA, Whiraker D, Murphy GF (2001) Human dermal mast cells contain and release tumor necrosis factor alpha, which induces endothelial leukocyte adhesion molecule 1. Proc Natl Acad Sci U S A 88:4220–4224
33. Nilsson G, Forsberg-Nilsson K, Xiang Z et al (1997) Human mast cells express functional TrkA and are a source of nerve growth factor. Eur J Immunol 27:2295–2301
34. Nakayama T, Mutsuga N, Yao L, Tosato G (2006) Prostaglandin E2 promotes degranulation-independent release of MCP-1 from mast cells. J Leukoc Biol 79:95–104
35. Risau W, Flamme I (1995) Vasculogenesis. Annu Rev Cell Dev Biol 11:73–91
36. Crivellato E, Nico B, Ribatti D (2007) Contribution of endothelial cells to organogenesis: a modern reappraisal of an old Aristotelian concept. J Anat 211:415–427
37. Ribatti D, Vacca A, Nico B, Roncali L, Dammacco F (2001) Postnatal vasculogenesis. Mech Dev 100:157–163
38. Ribatti D, Nico B, Crivellato E (2009) Morphological and molecular aspects of physiological vascular morphogenesis. Angiogenesis 12:101–111
39. Szukiewicz D, Pyzlak M, Klimkiewicz J, Szewczyk G, Maslinska D (2007) Mast cell-derived interleukin-8 may be involved in the ovarian mechanisms of follicle growth and ovulation. Inflamm Res 56:S35–S36
40. Szelag A, Merwid-Lad A, Troche M (2002) Histamine receptors in the female reproductive system. Part I. Role of the mast cells and histamine in female reproductive system. Ginekol Pol 73:627–635
41. Jensen F, Woudwyk M, Teles A et al (2010) Estradiol and progesterone regulate the migration of mast cells from the periphery to the uterus and induce their maturation and degranulation. PLoS One 5:e14409
42. Eklund KK (2007) Mast cells in the pathogenesis of rheumatic diseases and as potential targets for anti-rheumatic therapy. Immunol Rev 217:38–52
43. Ren G, Dewald O, Frangogiannis NG (2003) Inflammatory mechanisms in myocardial infarction. Curr Drug Targets Inflamm Allergy 2:242–256
44. Aroni K, Tsagroni E, Kavantzas N, Patsouris E, Ioannidis E (2008) A study of the pathogenesis of rosacea: how angiogenesis and mast cells may participate in a complex multifactorial process. Arch Dermatol Res 300:125–131
45. Groneberg DA, Bester C, Grützkau A et al (2005) Mast cells and vasculature in atopic dermatitis. Potential stimulus of neoangiogenesis. Allergy 60:90–97
46. Li X, Wilson JM (1997) Increased vascularity of the bronchial mucosa in mild asthma. Am J Respir Crit Care Med 156:229–233
47. Orsida BE, Li X, Hickey B et al (1999) Vascularity in asthmatic airways: relation to inhaled steroid dose. Thorax 54:289–295
48. Wilson J (2000) The bronchial microcirculation in asthma. Clin Exp Allergy 30:51–53
49. Holgate ST (2002) Airway inflammation and remodeling in asthma: current concepts. Mol Biotechnol 22:179–189
50. Hiromatsu Y, Toda S (2003) Mast cells and angiogenesis. Microsc Res Tech 60:64–69
51. Dvorak HF (1986) Tumors: wounds that do not heal. Similarities between tumor stroma generation and wound healing. N Engl J Med 315:1650–1659
52. Balkwill F, Mantovani A (2001) Inflammation and cancer: back to Virchow? Lancet 357:539–545
53. Ribatti D, Nico B, Crivellato E, Vacca A (2007) The structure of the vascular network in tumors. Cancer Lett 248:18–23
54. Kessler DA, Langer RS, Pless NA, Folkman J (1976) Mast cells and tumor angiogenesis. Int J Cancer 18:703–709
55. Conti P, Pang X, Boucher W et al (1997) Impact of Rantes and MCP-1 chemokines on in vivo basohilic mast cell recruitment in rat skin injection model and their role in modifying the protein and mRNA levels for histidine decarboxylase. Blood 89:4120–4127

56. Theoharides T, Conti P (2004) Mast cells: the Jekyll and Hyde of tumor growth. Trends Immunol 25:235–241
57. Starkey JR, Crowle PK, Taubenberger S (1998) Mast cell-deficient W/W^v mice exhibit a decreased rate of tumor angiogenesis. Int J Cancer 42:48–52
58. Benitez-Bribiesca L, Wong A, Utrera D, Castellanos E (2001) The role of mast cell tryptase in neoangiogenesis of premalignant and malignant lesions of the uterine cervix. J Histochem Cytochem 49:1061–1062
59. Sawatsubashi M, Yamada T, Fukushima N et al (2000) Association of vascular endothelial growth factor and mast cells with angiogenesis in laryngeal squamous cell carcinoma. Virchows Arch 436:243–248
60. Imada D, Shijubo N, Kojima H, Abe S (2000) Mast cells correlate with angiogenesis and poor outcome in stage I lung adenocarcinoma. Eur Respir J 15:1087–1093
61. Takanami I, Takeuchi K, Narume M (2000) Mast cell density is associated with angiogenesis and poor prognosis in pulmonary adenocarcinoma. Cancer 88:2686–2692
62. Tomita M, Matsuzaki Y, Onitsuka T (2000) Effect of mast cell on tumor angiogenesis in lung cancer. Ann Thorac Surg 69:1686–1690
63. Ribatti D, Ennas MG, Vacca A et al (2003) Tumor vascularity and tryptase positive-mast cells correlate with a poor prognosis in melanoma. Eur J Clin Invest 33:420–425
64. Toth T, Toth-Jakatics R, Jimi S, Takebayashi S, Kawamoto N (2000) Cutaneous malignant melanoma: correlation between neovascularization and peritumor accumulation of mast cells overexpressing vascular endothelial growth factor. Hum Pathol 31:955–960
65. Ribatti D, Vacca A, Ria R et al (2003) Neovascularization, expression of fibroblast growth factor-2, and mast cell with tryptase activity increase simultaneously with pathological progression in human malignant melanoma. Eur J Cancer 39:666–675
66. Elpek GO, Gelen T, Aksoy NH et al (2001) The prognostic relevance of angiogenesis and mast cells in squamous cell carcinoma of the oesophagus. J Clin Pathol 54:940–944
67. Ribatti D, Finato N, Crivellato E et al (2005) Neovascularization and mast cells with tryptase activity increase simultaneously with pathologic progression in human endometrial cancer. Am J Obstet Gynecol 193:1961–1965
68. Ribatti D, Vacca A, Marzullo A et al (2000) Angiogenesis and mast cell density with tryptase activity increase simultaneously with pathological progression in B-cell non-Hodgkin's lymphomas. Int J Cancer 82:171–175
69. Fukushima N, Satoh T, Sano M, Tokunaga O (2001) Angiogenesis and mast cells in non-Hodgkin's lymphoma: strong correlation in angioimmunoblastic T-cell lymphoma. Leuk Lymphoma 42:709–720
70. Ribatti D, Vacca A, Nico B et al (1999) Bone marrow angiogenesis and mast cell density increase simultaneously with progression of human multiple myeloma. Br J Cancer 79:451–455
71. Ribatti D, Polimero G, Vacca A et al (2002) Correlation of bone marrow angiogenesis and mast cells with tryptase activity in myelodysplastic syndromes. Leukemia 16:1680–1684
72. Ribatti D, Molica S, Vacca A et al (2003) Tryptase-positive mast cells correlate positively with bone marrow angiogenesis in B-cell chronic lymphocytic leukemia. Leukemia 17:1428–1430
73. Molica S, Vacca A, Crivellato E, Cuneo A, Ribatti D (2003) Tryptase-positive mast cells predict clinical outcome of patients with early B-cell chronic lymphocytic leukemia. Eur J Haematol 71:137–139
74. Crivellato E, Nico B, Vacca A, Dammacco F, Ribatti D (2002) Mast cell heterogeneity in B-cell non-Hodgkin's lymphomas: an ultrastructural study. Leuk Lymphoma 43:2201–2205
75. Crivellato E, Beltrami C, Mallardi F, Ribatti D (2003) Paul Ehrlich's doctoral thesis: a milestone in the study of mast cells. Br J Haematol 123:19–21

Chapter 8
Vascular Stem Cells in Regulation of Angiogenesis

Jingwei Lu, Vincent J. Pompili, and Hiranmoy Das

Abstract Angiogenesis is the process by which new vessels are generated from the preexisting blood vessels, which is the major contributor of postnatal neovascularization process. Disruption or dysregulation of angiogenesis is involved in various pathological conditions, such as ischemia and tumor progression. Stimulation of angiogenesis was proposed to be able to restore the blood flow and contribute to the tissue recovery in ischemia, while inhibition of angiogenesis can impede tumor progression. The importance of angiogenesis has generated tremendous interest in studying the mechanisms and to find out major contributors of the process. The current stem cell research has significantly improved our understanding of angiogenesis and its possible therapeutic application. Hypoxia is the most important driving force of angiogenesis, while other factors, such as chemokines and cytokines, haptotaxis, and mechanotaxis, are also important in regulating neovascularization process. In this chapter, we will focus on the progenitor cells that contribute to the angiogenesis and the underlining mechanisms involved in this process.

Keywords Angiogenesis • Stem cells • Hypoxia • Cytokines • Hypotaxis • Mechanotaxis • Signaling molecules • MicroRNA

J. Lu • V.J. Pompili • H. Das (✉)
Cardiovascular Stem Cell Research Laboratory, The Dorothy M. Davis Heart and Lung Research Institute, Wexner Medical Center at The Ohio State University, 460 West 12th Avenue, BRT 394, Columbus, OH 43210, USA
e-mail: hiranmoy.das@osumc.edu

J.L. Mehta and N.S. Dhalla (eds.), *Biochemical Basis and Therapeutic Implications of Angiogenesis*, Advances in Biochemistry in Health and Disease 6,
DOI 10.1007/978-1-4614-5857-9_8, © Springer Science+Business Media New York 2013

8.1 Introduction

Galen, the second-century physician, speculated that the vascular system served to carry blood and provide nutrition to the human body [1]. It is now well established that the vascular system provides the main network of channels for nutrients (such as amino acids, electrolytes, oxygen, and hormones) to all the body tissues. Disturbances in the vascular system, mainly blocking the blood supply to the tissues, cause a variety of circulatory diseases, from peripheral artery disease to peripheral venous disease, and include among them vascular diseases like aneurysms, renal artery stenoses, and Buerger's disease [2]. Disruptions of angiogenesis play a critical role in the pathological progression of various ischemic diseases, such as stroke, ischemic heart disease, and the multiple peripheral vascular disease syndromes, resulting in a shortage of blood supply and which eventually induce apoptosis and necrosis of cells and the tissues of the vascular system. Angiogenesis, however, plays an important role in the regeneration of such ischemic tissues. In a seemingly contradictory role to that in the ischemic diseases, angiogenesis contributes to damage caused by the progressive growth of malignant tumors. Targeting tumor growth by targeting tumor angiogenesis, as in using various drugs to reduce blood supply to the tumor, is one of the major therapeutic considerations for effective control. Rapid proliferation of tumor cells, with lack of blood supply and lack of oxygen, triggers upregulation of vascular endothelial growth factor (VEGF) secretion, which promotes the angiogenesis process. The importance of angiogenesis in pathological conditions has generated interest in studying the mechanisms and signaling pathways for angiogenesis. Various stem cells were proposed to be important for initiation of the angiogenesis process. Mesenchymal stem cells (MSCs) and hematopoietic stem cells (HSCs), which were shown to repair ischemic tissues, have great ability to promote angiogenesis via neovascularization and thereby to reduce the amount of ischemic tissue damage [3, 4]. Other cell types such as smooth muscle stem cells and vascular pericytes were also shown to be beneficial for the process of angiogenesis. In this chapter, we will focus on the role of various stem cells on the angiogenesis process and on the molecular mechanisms that promote these stem cells to form new blood vessels.

8.2 Angiogenesis and Stem Cells

The wall of blood vessels contain endothelial cells, mural cells, and extracellular matrix (ECM). The inner lining of blood vessel is the endothelium, which is a thin layer of endothelial cells. Mural cells are specified as determined by the location of the vessel; they could be pericytes, smooth muscle cells, and fibroblasts. The mural cells are embedded in the extracellular matrices [5]. The various types of cells forming blood vessels could be derived from multiple stem/progenitor cells. Circulating endothelial progenitor cells (EPCs) and HSCs could differentiate into endothelial

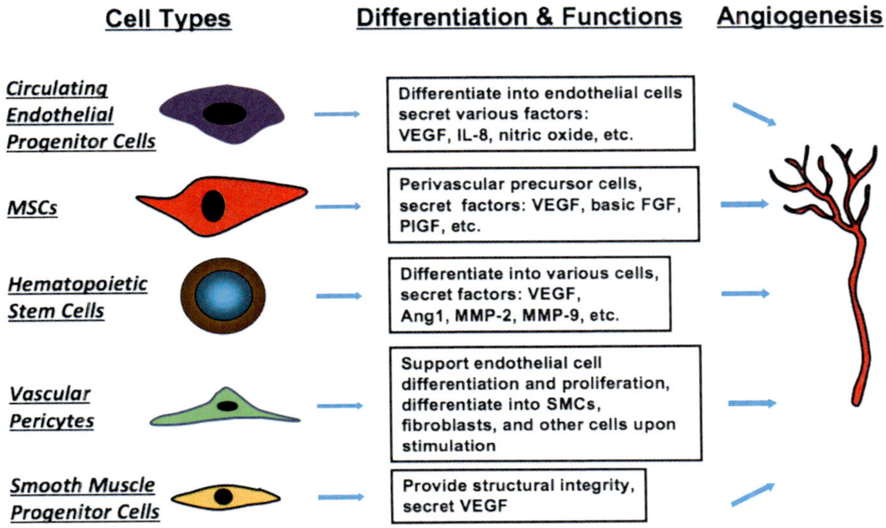

Fig. 8.1 Contribution of various stem/progenitor cells and their secretory molecules in the angiogenesis process. *MSCs* mesenchymal stem cells; *SMCs* smooth muscle cells; *VEGF* vascular endothelial growth factor; *IL-8* interleukin-8; *FGF* fibroblast growth factor; *PlGF* placental growth factor; *Ang1* angiopoietin 1; *MMP* matrix metalloproteinases

cells thus directly contributing to the angiogenesis process. MSCs, though they may not be able to directly differentiate into endothelial cells, can secrete factors, such as VEGF, and promote the neovascularization process. Other progenitor cells, such as vascular pericytes and smooth muscle progenitor cells, can also contribute to angiogenesis (Fig. 8.1).

8.2.1 Circulating Endothelial Progenitor Cells

The first study on putative EPCs was based on isolation of CD34+ mononuclear blood cells. The isolated cells were adhered to plastic and differentiated into endothelial cells upon culture [6]. Since the discovery of EPCs, various markers have been proposed to identify EPCs, such as CD34, CD133, expression of both CD133 and vascular endothelial growth factor receptor (VEGFR) 2, and expression of monocyte/macrophage-related molecule CD14 with minimal CD34 molecule [7]. The functional role of circulating EPCs has been actively investigated during the past few years. It was shown that higher level of VEGF may induce a rapid mobilization of HSCs and bone marrow-derived circulating endothelial precursor cells, which contribute to postnatal angiogenesis and hematopoiesis [8]. However, further study has shown that bone marrow-derived cells do not significantly contribute to tumor- or cytokine-induced angiogenesis rather tumor- or VEGF-induced angiogenesis is involved [9]. Based on their proliferation properties, two different categories of EPCs were identified in the peripheral blood, early EPCs and late EPCs. Early EPCs secrete more angiogenic cytokines, such as VEGF and interleukin (IL)-8 than do late EPCs; however, late EPCs produce more nitric oxide and incorporate more readily into human umbilical vein endothelial cell monolayers and form capillary tubes as compared to early EPCs [10].

8.2.2 Hematopoietic Stem Cells

HSCs and EPCs develop in close proximity to each other within the embryo. HSCs share the same ancestor with EPCs, called the hemangioblast. The existence of hemangioblasts was supported by various experimental observations, but its role during development is still controversial. Even though evidence has shown that single cell-derived colonies could produce both hematopoietic and endothelial cells in vitro, only a small portion of hematopoietic and endothelial cells were derived from hemangioblasts during development, which indicated that hemangioblasts might not be as significant as originally expected [11]. However, these studies illustrated the relationship of hematopoietic and endothelial lineage and indicated the possibility that HSCs might facilitate the angiogenesis during embryonic development and postnatal development. Indeed in acute myeloid leukemia (AML)-1-deficient embryos, which lack definitive hematopoiesis, defective angiogenesis in the head and in the pericardium was observed. The disruption in angiogenesis of para-aortic splanchnopleural (P-Sp) explant culture was rescued by addition of HSCs [12]. The recruitment of myeloid cells was found to be associated with formation of new blood vessel during pathological angiogenesis, and depletion of circulating myeloid cells significantly reduced the density of microvessels in a bioengineered human vascular implant [13]. The functional role of HSCs during angiogenesis may come from expression of proangiogenic factors such as VEGF and Ang1 and remodeling factors such as matrix metalloproteinase (MMP)-2 and MMP-9, which promote angiogenesis and guide the migration of endothelial cells [12]. It was found that hematopoietic cytokines SDF-1, induced by soluble Kit ligand, thrombopoietin, erythropoietin, and granulocyte-macrophage colony-stimulating factor (GM-CSF) released from platelets, enhanced neovascularization through mobilization of chemokine receptor (CXCR)-4+ VEGFR1+ hemangiocytes [14]. The important role of hematopoietic cells in angiogenesis has received great attention and proposed to be important target for anti-angiogenesis therapy following radiotherapy during treatment of tumor progression [15].

8.2.3 Mesenchymal Stem Cells

MSCs are present in many organs and function to maintain and regenerate connective tissues and replace damaged tissues following injury or inflammation. MSCs could efficiently stabilize nascent blood vessels in vivo acting as perivascular precursor cells, although differentiation of MSCs into endothelial cells was not detectable [16]. Co-implant human primary endothelial cells with human bone marrow MSCs showed enhanced formation of a network of functional, mature blood vessels accessed by in vivo whole body bioluminescence imaging in immunodeficient mice [17]. Transplantation of MSCs was shown to be able to decrease fibrosis and myocardial scarring and improve myocardial regeneration in infarct-damaged hearts, through paracrine effects, via secretion of VEGF, basic fibroblast growth factor (bFGF), and placental growth factor (PlGF), even though MSC differentiation into ECs was not clearly demonstrated [18].

8.2.4 *Smooth Muscle Progenitor Cells*

Smooth muscle cells in the vascular system provide the structural integrity of the vessel wall. Recent study has shown that smooth muscle progenitor cells may have a potential role in angiogenesis. In a murine stroke model, it was shown that co-injection of smooth muscle progenitor cells with EPCs gave better results than administration of EPCs alone for vascular remodeling, cell proliferation, and neuroblast migration [19]. Perturbation in the signaling of transforming growth factor (TGF)-β, which is a multifunctional cytokine and plays an important role in carcinogenesis, was reported to affect endothelial and smooth muscle cell function and to contribute to tumor angiogenesis and tumor progression [20]. Smooth muscle cells can also contribute to angiogenesis by secreting mitogens, such as VEGF upon response to the hypoxia [21].

8.2.5 *Vascular Pericytes*

Pericytes are located surrounding the endothelial cells of the capillaries. Clonally isolated cells expressing pericyte markers were shown to be myogenic in culture in vivo [22]. It was proposed that pericytes derived from MSCs retain nascent stem cell properties, were recruited to the nascent microvascular wall during development and postnatal growth, and remained in a growth-arrested state until triggered to resume proliferation and differentiation later [23].

8.3 External Factors Regulate Angiogenesis Process

There are three distinct mechanisms, which promote cell migration during angiogenesis, chemotaxis, haptotaxis, and mechanotaxis. Chemotaxis directs cell migration toward a gradient of soluble chemoattractants, such as VEGF and bFGF. Haptotaxis attracts cells toward a gradient of immobilized ligands such as integrins

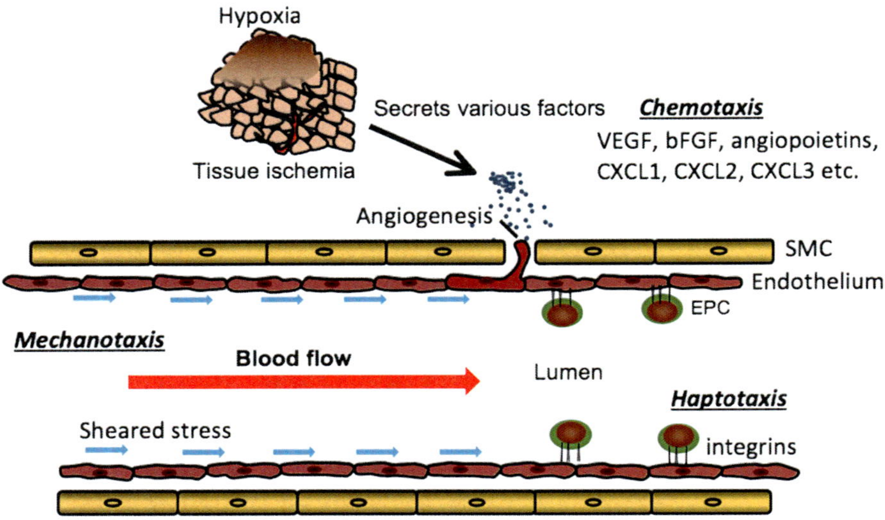

Fig. 8.2 Factors regulating angiogenesis process. Hypoxia, chemokines and cytokines, hypotaxis, and mechanotaxis are the major factors induce and regulate angiogenesis process. *EPCs* endothelial progenitor cells; *SMC* smooth muscle cells; *VEGF* vascular endothelial growth factor; *bFGF* basic fibroblast growth factor; *CXCL* chemokine (CXC-motif) ligand; *MMP* matrix metalloproteinases

binding to ECM components. Mechanotaxis promotes cell migration by mechanical forces, such as fluid shear stress [24]. Other factors including hypoxia will also be discussed here (Fig. 8.2).

8.3.1 Chemotaxis: Cytokines, Chemokines, and Growth Factors

Various cytokines and soluble proteins, such as VEGF, bFGF, angiopoietins, FGF-2, hepatocyte growth factor (HGF), platelet-derived growth factor (PDGF), epidermal growth factor (EGF), TGF-β, interleukins, and tumor necrosis factor (TNF)-α, promote the migration of endothelial cells during angiogenesis. VEGF is a major factor that regulates angiogenesis. Various factors can induce the production of VEGF, and hypoxia was reported to be one of them. Hypoxia is able to enhance the production of VEGF and its receptors [25]. Production of reactive oxygen species (ROS), for example, hydrogen peroxide (H_2O_2), also upregulates the gene expression of VEGF in endothelial cells [26]. VEGF was also found to be expressed by almost all solid tumor as an angiogenic mitogen and so is now targeted for anti-angiogenesis therapy for tumor metastasis [27]. VEGF and its family members stimulate cellular responses by binding to the tyrosine kinase receptors called VEGFRs. VEGFR1 (Flt-1) is required for the recruitment of hematopoietic precursors and migration of monocytes and macrophages [28].

VEGFR1-deficient mice die in utero between 8.5 and 9.5 days post-coitum due to early defects in the development of hematopoietic and endothelial cells [29]. The functional role of VEGFR2 (KDR/Flk-1) has been linked with proliferation, migration survival, and increased permeability, all of which contributes to the angiogenesis process [30].

VEGF plays critical roles in endothelial differentiation, in acquisition of arterial endothelial cell identity, and in the vascular patterning of vertebrate embryos. VEGF ligands and receptors such as VEGF-A, the prototype of VEGF ligand, VEGFR1, VEGFR2, and VEGFR3 regulate vasculogenesis and angiogenesis during various stages of growth [31]. By studying a series of nerve-specific Cre lines, it was shown that peripheral nerve-derived VEGF promotes arterial differentiation through the $VEGF^{164}$-NRP1 positive-feedback loop [32]. It was further demonstrated that VEGF acted downstream of sonic hedgehog (Shh) and upstream of Notch pathway in the differentiation of endothelial cells to arterial fate [33].

Other factors also play important roles in promoting angiogenesis including bFGF, angiopoietins, HGF, PDGF, EGF, TGF-β, TNF-α, etc. Slow release of bFGF (using gelatin hydrogels) can promote new blood vessel formation compared with a control group in a murine limb ischemia model [34]. Angiopoietin was required for endothelial development from progenitors circulating in human cord blood. More specifically, endogenous angiopoietin-1 regulates initial endothelial cell commitment, while angiopoietin-2 improves expansion of the endothelial cell progeny [35]. Angiopoietin-1 and angiopoiein-2 may also play important role in regulating recruitment of mural cells during angiogenesis [36]. It was shown that overexpression of HGF in smooth muscle cells can be beneficial in EPC differentiation, proliferation, and migration [37]. Further study has shown that HGF stimulates migration and tube formation of human umbilical vein endothelial cells in a Nox2-dependent manner [38]. However, transplantation of bone-derived MSCs showed no significant differences in promoting angiogenesis with or without HGF, which indicated that further study is needed to investigate the interplay between HGF and MSCs [39].

Chemokines are a family of small chemotactic cytokines and are classified by the presence of four cysteine residues in conserved locations. Members of the chemokine family are divided into four groups CC chemokines, CXC chemokines, C chemokines, and CX_3C chemokines. Many chemokines were proven to be angiogenic such as CXCL1, CXCL2, and CXCL3. These chemokines activate endothelial cells upon binding with their receptors. It was reported that functional differences among endothelial cells is dependent on the level of expression of CXC chemokine receptors [40]. It was also proposed that CXC chemokine IL-8; growth-related oncogenes alpha, beta, and gamma; granulocyte chemotactic protein 2; and epithelial neutrophil-activating protein-78 mediate angiogenesis in the absence of preceding inflammation partially through interaction with CXC chemokine receptor 2 (CXCR2) [41]. CXCR2 is a member of the G-protein-coupled receptor family and is expressed in endothelial cells. CXCR2 knockout mice exhibited defective neutrophil recruitment, an altered temporal pattern of monocyte recruitment, significant delay in epithelialization, and decreased neovascularization in wound-healing processes [42]. It was shown that upon binding to IL-8, CXCR2 activates the Rac pathway, which leads to cell retraction and formation of gaps between neighboring cells. Translocation of Rac into the plasma membrane eventually results in endothelial activation [43]. These experiments suggest that CXCR2 plays an important role in the recruitment of cells and promoting angiogenesis. Other than CXCR2, VEGF- and bFGF-activated angiogeneses were also partially mediated through CXCR4. Stimulation of human umbilical vein endothelial cells with VEGF or bFGF was shown to be able to induce upregulation of CXCR4. It was further shown that chemokine SDF-1α, which specifically bind CXCR4, is a potent chemoattractant for endothelial cells and participates in angiogenesis stimulated by VEGF and bFGF [44].

8.3.2 Haptotaxis

Haptotaxis is the directional motility of cells by the ligands typically presented in the ECM. Exposure of ECM and binding to integrin help homing and recruitment of the immune cells during the angiogenesis process. These ECM and integrin molecules are also critical for homing of transplanted HSCs to the bone marrow and the recruitment of inflammatory cells to the sites of inflammation [45]. It was shown that hematopoietic progenitor cells of β_2 integrin-deficient mice are less capable of homing to the ischemic site and that improving neovascularization and preactivation of the β_2 integrins expressed on EPCs augmented the EPC-induced neovascularization [46]. Antagonists of integrin $\alpha_4\beta_1$ were shown to be able to block the adhesion of monocytes to endothelium and prevented monocyte stimulation during angiogenesis [47]. It was further shown that administration of α_4 integrin antibody resulted in increased numbers of circulating EPCs in vivo and systemic administration of anti-α_4 integrin antibody increased recruitment and the incorporation of bone marrow EPCs in newly formed vasculature of hind-limb ischemia and myocardial infarction models [48]. Integrin-dependent homing of progenitor cells can be enhanced by various factors. It was reported that high-mobility group box 1 (HMGB1) activated EPC migration in a RAGE (HMGB1 receptor expressed on EPCs)-dependent manner and was inhibited by β_1 and β_2 integrin inhibition. HMGB1 could rapidly increase the affinity of integrin and induce polarization of integrin, which might be related to the corresponding enhanced adhesion capability of EPCs [49]. Pharmacologic activation of Epac1, a nucleotide-exchange protein for Rap1, could increase Rap1 activity and stimulate the adhesion of various human progenitor cells. EPCs, CD34+ hematopoietic progenitor cells, and MSCs are activated through increased β_2 and β_1 integrin-dependent adhesion and activated progenitor cells home to the ischemic muscles in an increased amount as a result, neovascularization occurs [50].

8.3.3 Mechanotaxis

Mechanotaxis is the directed movement of cells by mechanical cues, such as fluidic shear stress and stiffness of substrate. Endothelial cells which make up the inner lining of blood vessels are constantly under fluid-mediated shear stress in vivo, and it was shown that this mechanical stress-mediated signaling contributes to each step of endothelial migration, cell-ECM adhesion, and cell–cell adhesion processes [51]. Shear stresses were reported able to induce changes in the shape of endothelial cells and partial disassembly of adherent junctions [52]. It was shown that endothelial cells, cultured on type I collagen-coated coverslip and wounded later, enhanced wound healing under higher shear stress [53]. The endothelial cell alignments induced by fluid shear stress were proposed to act through the p38/mitogen-activated protein (MAP) kinase-activated protein kinase 2 (MAPKAP kinase 2)/heat shock protein (HSP) 25/27 pathway due to its critical role in actin dynamics. It was shown that by inhibiting p38 signaling, endothelial elongation and alignment were blocked in the direction of flow, elicited by shear stress [54]. Other mechanisms involving G protein have also been studied. It was shown that shear stress-induced cytoskeletal reorientation was abolished in cells overexpressing dominant negative Rac 1. This indicated that the Rac GTPase might play a role in regulating endothelial cytoskeleton by shear stress [55]. The endothelial cell reorientation in response to shear stress was further studied and was proposed to follow a two-step process involving Rho-induced depolarization, followed by Rho-/Rac-mediated polarization and migration in the direction of flow [56].

8.3.4 Hypoxia

Hypoxia plays a critical role in neovascularization, both in embryonic development and in postnatal development. During embryonic development, the vascular system is stimulated by an inadequate supply of oxygen, which is caused by rapid expansion of embryonic tissues. In adult tissues, the blood vessels do not undergo significant growth, and the oxygen concentrations remain relatively constant between 30 and 50 mm of Hg. In pathological conditions, however, as in ischemia, hypoxia is created by the lack of blood, which is the main carrier of oxygen, and reduction of the oxygen level triggers angiogenesis. Important molecules involved in the hypoxia response include prolyl hydroxylase domain-containing proteins (PHDs) and hypoxia-inducible factors (HIFs). PHDs play an important role in oxygen sensing by inhibiting HIFs expression and by promoting HIFs degradation. HIF is a key transcription factor governing a large set of gene expressions for hypoxia adaptation, for example, the inhibition of PHD suppressed lipopolysaccharide-induced TNF-α expression. Reducing oxygen will lead to poor hydroxylation activities by PHDs and thus lead to accumulation of HIF-α. Hundreds of proteins were regulated by HIFs in response to hypoxia. It was shown that hypoxia, by regulating HIF, stimulates the production of various angiogenic cytokines such as VEGF and angiopoietin-1 and promotes proliferation of embryonic hemangioblasts [57]. Hypoxia can also promote recruitment of bone marrow-derived vascular modulatory cells through HIF-1α, which enhances the synthesis and secretion of endothelial molecules on vascular progenitor cells, such as CD31, VEGFR2, and endothelial NO synthase (eNOS) [58]. Even though hypoxia has been demonstrated to be useful in maintaining undifferentiated stem cells, researchers have found that hypoxia can also stimulate differentiation of stem cells in certain condition [59]. Hypoxia may stimulate adipose stromal cells (ASCs) into endothelial-like cells. It was shown that secretion of VEGF correlates inversely with oxygen concentration, and ASCs assumed an endothelial phenotype characterized by their ability to form tubes when seeded with differentiated endothelial cells on Matrigel assays [60]. ASCs were reported to be able to express endothelial markers when cultured with VEGF and differentiated in response to local cues into endothelial cells, which contributed to neoangiogenesis in a hind-limb ischemic model [61]. HIF-α, in response to hypoxia, regulates a variety of genes such as uPAR, collagen prolyl 4-hydroxylases, matrix metalloproteinases, and tissue inhibitors of matrix metalloproteinases, which were proposed to facilitate endothelial transition from a stable growth-arrested state to a plastic proliferative phenotype [62].

8.4 Signaling Molecules Involved in Angiogenesis

Several complex signaling pathways are involved in angiogenesis. However, two major signaling pathways play critical roles in angiogenesis, the Notch-signaling pathway and the hedgehog-signaling pathway, and these will be discussed here. We shall also discuss miRNAs, which are involved in the angiogenesis process (Fig. 8.3).

8.4.1 Notch and Delta Signaling

Notch-signaling pathway is highly conserved with four different Notch receptors, NOTCH1, NOTCH2, NOTCH3, and NOTCH4, and five ligands from the jagged (Jagged-1 and Jagged-2) and Delta (Delta-like 1, Delta-like 3, and Delta-like 4) families plus modifier proteins from the Fringe family (lunatic, manic, and radical fringe) [63]. Notch proteins play critical role throughout embryonic development, such as cell survival, self-renewal for stem cells, and lineage determination for developing cells. Upon ligand activation, the cytoplasmic domain of Notch is proteolytically released, translocates into the nucleus, activates CSL [CBF1, Su (H), Lag-2], and converts them to transcriptional activators. The Notch/CSL-dependent signaling directly targets HERP families of transcriptional repressors, which are involved in multiple aspects of vascular development including muscle

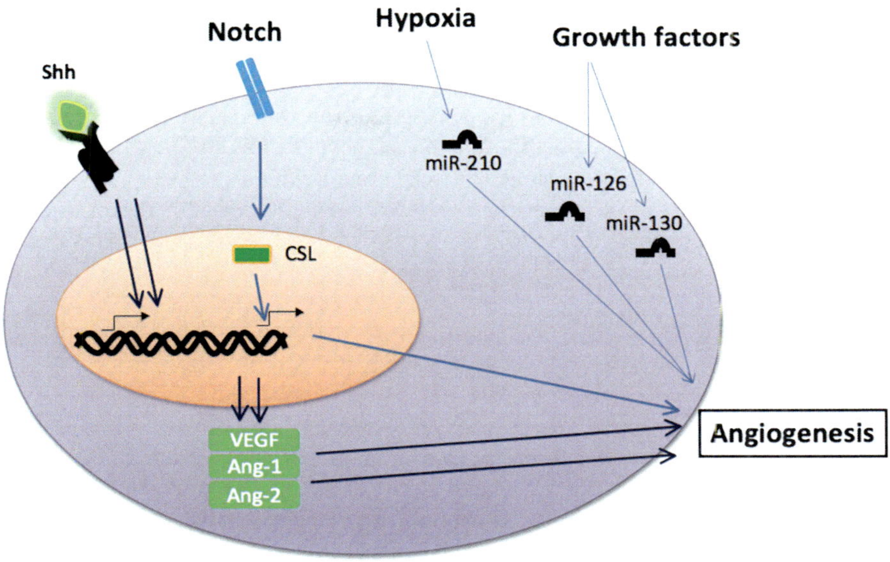

Fig. 8.3 Signaling molecules involved in cellular angiogenesis. Various cellular signaling molecules are involved in the angiogenesis process includes notch pathway, hedgehog pathway, hypoxia, and growth factors. MicroRNAs are also participating in the regulation of angiogenesis process. *Shh* sonic hedgehog; *CSL* combination of three proteins CBF1, Su (H), and Lag-2; *miR* microRNA; *VEGF* vascular endothelial growth factor; *Ang* angiopoietin

differentiation, angiogenic processes, arterial-venous cell fate determination, and vascular morphogenesis in mice [64]. The Delta-Notch-signaling pathway also targets members of the Hey family, the loss of which led to global lack of vascular remodeling and massive hemorrhage [65]. It was also shown that the differentiation-associated growth arrest in endothelial cells activated by Notch pathway was mediated by mitogen-activated protein kinase (MAPK) and phosphatidylinositol 3-kinase (PI3K)/Akt pathway [66].

8.4.2 *Hedgehog Signaling*

Hedgehogs interact with heparin on the cell surface through N-terminal basic domains. The molecular weight of this class is around 19 kDa. The role of hedgehog signaling in angiogenesis was brought to attention a decade ago. It was shown that Shh, a hedgehog homolog in mammals, can induce expression of two families of angiogenic cytokines, including all three VEGF-1 isoforms and angiopoietins-1 and -2 in interstitial mesenchymal cells. Shh was able to induce robust angiogenesis and augment blood flow recovery and limb salvage in an induced hind-limb ischemia model of aged mice [67]. By studying murine brain capillary endothelial cells (IBE cells) and human umbilical endothelial cells, it was shown that Shh-induced capillary morphogenesis through stimulating PI3-kinase activity [68]. During development, it was demonstrated that hedgehog proteins participate in the embryonic endothelial and fibroblast cell migration and play a role in the angiogenesis process [69]. In a diabetic wound-healing murine model, gene therapy of Shh together with bone marrow transplantation resulted in accelerated wound recovery partially by enhanced recruitment of bone marrow-derived progenitor cells and promoting production of angiogenic cytokines [70].

8.4.3 MicroRNA

In recent years it was found that microRNAs play an important role in regulating endothelial differentiation and in promoting angiogenesis. By studying zebra fish embryos, it was found that mechano-sensitive zinc finger transcription factor klf2 activates the VEGF-signaling pathway by inducing expression of endothelial-specific microRNA mir-126 [71]. Dicer is key enzyme, which contributes to the maturation of microRNA. Specific silencing of Dicer using siRNA has led to altered expression of key regulators of angiogenesis such as TEK/Tie-2, KDR/VEGFR2, Tie-1, endothelial nitric oxide synthase, and IL-8 in endothelial cells [72]. Furthermore, reduction of endothelial microRNAs by inactivation of Dicer reduces postnatal angiogenic response to exogenous VEGF, tumors, limb ischemia, and wound-healing models [73]. These findings indicate that microRNAs play important roles in regulating endothelial cells during the angiogenesis process. Multiple microRNAs have been found to influence the angiogenesis process including microRNA-17, 92, 23, 27, 24, 130a, 181a, and 210. Till recently, few microRNAs have been identified to regulated endothelial differentiation, and microRNA-mediated control of endothelial differentiation remains to be explored [74].

8.5 Conclusions and Future Directions

Major efforts were given in studying the mechanisms of angiogenesis in various pathological conditions. These efforts will significantly improve our understanding of therapeutic angiogenesis. Various regulating factors including microRNAs were found to be important during angiogenesis. Numerous treatments are under development targeting appropriate regulatory factors of angiogenesis in the context of pathological condition of the disease. Results are now available from many clinical trials using various stem cells for the treatment of ischemia [3, 75, 76]. It was shown that HSCs and MSCs were indeed able to improve the vascularization process in ischemic tissues and to improve clinical outcomes in both animal model and in clinical use [77]. Nevertheless, the future role of stem cell treatment compared to current pharmacologic treatment remains undetermined.

Moreover, the best timing for the possible administration of stem cells is still unknown. As we learn more about the molecular mechanisms of angiogenesis, we are likely to find an effective window for future stem cell therapy to improve the outlook for the recovery of ischemic tissues.

Acknowledgements This work was supported in part by National Institutes of Health grants, K01 AR054114 (NIAMS), SBIR R44 HL092706-01 (NHLBI), R21 CA143787 (NCI), Pelotonia idea award and the Ohio State University start-up fund for stem cell research. The funders had no role in study design, data collection and analysis, decision to publish, or preparation of the manuscript.

References

1. Patan S (2000) Vasculogenesis and angiogenesis as mechanisms of vascular network formation, growth and remodeling. J Neurooncol 50:1–15
2. Halperin JL (2002) Evaluation of patients with peripheral vascular disease. Thromb Res 106:V303–V311
3. Lee JS, Hong JM, Moon GJ et al (2010) A long-term follow-up study of intravenous autologous mesenchymal stem cell transplantation in patients with ischemic stroke. Stem Cells 28:1099–1106
4. Kwon SM, Lee YK, Yokoyama A et al (2011) Differential activity of bone marrow hematopoietic stem cell subpopulations for EPC development and ischemic neovascularization. J Mol Cell Cardiol 51:308–317
5. Jain RK (2003) Molecular regulation of vessel maturation. Nat Med 9:685–693
6. Asahara T, Murohara T, Sullivan A et al (1997) Isolation of putative progenitor endothelial cells for angiogenesis. Science 275:964–967
7. Kovacic JC, Moore J, Herbert A et al (2008) Endothelial progenitor cells, angioblasts, and angiogenesis—old terms reconsidered from a current perspective. Trends Cardiovasc Med 18:45–51
8. Hattori K, Dias S, Heissig B et al (2001) Vascular endothelial growth factor and angiopoietin-1 stimulate postnatal hematopoiesis by recruitment of vasculogenic and hematopoietic stem cells. J Exp Med 193:1005–1014
9. Rajantie I, Ilmonen M, Alminaite A et al (2004) Adult bone marrow-derived cells recruited during angiogenesis comprise precursors for periendothelial vascular mural cells. Blood 104:2084–2086
10. Hur J, Yoon CH, Kim HS et al (2004) Characterization of two types of endothelial progenitor cells and their different contributions to neovasculogenesis. Arterioscler Thromb Vasc Biol 24:288–293
11. Vogeli KM, Jin SW, Martin GR, Stainier DY (2006) A common progenitor for haematopoietic and endothelial lineages in the zebrafish gastrula. Nature 443:337–339
12. Takakura N, Watanabe T, Suenobu S et al (2000) A role for hematopoietic stem cells in promoting angiogenesis. Cell 102:199–209
13. Melero-Martin JM, De Obaldia ME, Allen P et al (2010) Host myeloid cells are necessary for creating bioengineered human vascular networks in vivo. Tissue Eng Part A 16:2457–2466
14. Jin DK, Shido K, Kopp HG et al (2006) Cytokine-mediated deployment of SDF-1 induces revascularization through recruitment of CXCR4(+) hemangiocytes. Nat Med 12:557–567
15. Melero-Martin JM, Dudley AC (2011) Concise review: vascular stem cells and tumor angiogenesis. Stem Cells 29:163–168
16. Au P, Tam J, Fukumura D, Jain RK (2008) Bone marrow-derived mesenchymal stem cells facilitate engineering of long-lasting functional vasculature. Blood 111:4551–4558
17. Sanz L, Santos-Valle P, Alonso-Camino V et al (2008) Long-term in vivo imaging of human angiogenesis: critical role of bone marrow-derived mesenchymal stem cells for the generation of durable blood vessels. Microvasc Res 75:308–314

18. Kim SW, Kim H, Yoon YS (2011) Advances in bone marrow-derived cell therapy: CD31-expressing cells as next generation cardiovascular cell therapy. Regen Med 6:335–349
19. Nih LR, Deroide N, Lere-Dean C et al (2012) Neuroblast survival depends on mature vascular network formation after mouse stroke: role of endothelial and smooth muscle progenitor cell co-administration. Eur J Neurosci 35:1208–1217
20. Pardali E, ten Dijke P (2009) Transforming growth factor-beta signaling and tumor angiogenesis. Front Biosci 14:4848–4861
21. Okuda Y, Tsurumaru K, Suzuki S et al (1998) Hypoxia and endothelin-1 induce VEGF production in human vascular smooth muscle cells. Life Sci 63:477–484
22. Crisan M, Yap S, Casteilla L et al (2008) A perivascular origin for mesenchymal stem cells in multiple human organs. Cell Stem Cell 3:301–313
23. Bianco P, Robey PG, Simmons PJ (2008) Mesenchymal stem cells: revisiting history, concepts, and assays. Cell Stem Cell 2:313–319
24. Lamalice L, Le Boeuf F, Huot J (2007) Endothelial cell migration during angiogenesis. Circ Res 100:782–794
25. Brogi E, Schatteman G, Wu T et al (1996) Hypoxia-induced paracrine regulation of vascular endothelial growth factor receptor expression. J Clin Invest 97:469–476
26. Chua CC, Hamdy RC, Chua BH (1998) Upregulation of vascular endothelial growth factor by H2O2 in rat heart endothelial cells. Free Radic Biol Med 25:891–897
27. Barleon B, Siemeister G, Martiny-Baron G et al (1997) Vascular endothelial growth factor up-regulates its receptor fms-like tyrosine kinase 1 (FLT-1) and a soluble variant of FLT-1 in human vascular endothelial cells. Cancer Res 57:5421–5425
28. Olsson AK, Dimberg A, Kreuger J, Claesson-Welsh L (2006) VEGF receptor signalling—in control of vascular function. Nat Rev Mol Cell Biol 7:359–371
29. Shalaby F, Rossant J, Yamaguchi TP et al (1995) Failure of blood-island formation and vasculogenesis in Flk-1-deficient mice. Nature 376:62–66
30. Holmes K, Roberts OL, Thomas AM, Cross MJ (2007) Vascular endothelial growth factor receptor-2: structure, function, intracellular signalling and therapeutic inhibition. Cell Signal 19:2003–2012
31. Shibuya M, Claesson-Welsh L (2006) Signal transduction by VEGF receptors in regulation of angiogenesis and lymphangiogenesis. Exp Cell Res 312:549–560
32. Mukouyama YS, Gerber HP, Ferrara N et al (2005) Peripheral nerve-derived VEGF promotes arterial differentiation via neuropilin 1-mediated positive feedback. Development 132:941–952
33. Lawson ND, Vogel AM, Weinstein BM (2002) Sonic hedgehog and vascular endothelial growth factor act upstream of the Notch pathway during arterial endothelial differentiation. Dev Cell 3:127–136
34. Matsui M, Tabata Y (2012) Enhanced angiogenesis by multiple release of platelet-rich plasma contents and basic fibroblast growth factor from gelatin hydrogels. Acta Biomater 8:1792–1801
35. Hildbrand P, Cirulli V, Prinsen RC et al (2004) The role of angiopoietins in the development of endothelial cells from cord blood CD34+ progenitors. Blood 104:2010–2019
36. Iurlaro M, Scatena M, Zhu WH et al (2003) Rat aorta-derived mural precursor cells express the Tie2 receptor and respond directly to stimulation by angiopoietins. J Cell Sci 116:3635–3643
37. Zhu G, Huang L, Song M et al (2010) Over-expression of hepatocyte growth factor in smooth muscle cells regulates endothelial progenitor cells differentiation, migration and proliferation. Int J Cardiol 138:70–80
38. Schroder K, Schutz S, Schloffel I et al (2011) Hepatocyte growth factor induces a proangiogenic phenotype and mobilizes endothelial progenitor cells by activating Nox2. Antioxid Redox Signal 15:915–923
39. Yang ZJ, Ma DC, Wang W et al (2006) Experimental study of bone marrow-derived mesenchymal stem cells combined with hepatocyte growth factor transplantation via noninfarct-relative artery in acute myocardial infarction. Gene Ther 13:1564–1568

40. Salcedo R, Resau JH, Halverson D et al (2000) Differential expression and responsiveness of chemokine receptors (CXCR1-3) by human microvascular endothelial cells and umbilical vein endothelial cells. FASEB J 14:2055–2064
41. Addison CL, Daniel TO, Burdick MD et al (2000) The CXC chemokine receptor 2, CXCR2, is the putative receptor for ELR+ CXC chemokine-induced angiogenic activity. J Immunol 165:5269–5277
42. Devalaraja RM, Nanney LB, Du J et al (2000) Delayed wound healing in CXCR2 knockout mice. J Invest Dermatol 115:234–244
43. Schraufstatter IU, Chung J, Burger M (2001) IL-8 activates endothelial cell CXCR1 and CXCR2 through Rho and Rac signaling pathways. Am J Physiol Lung Cell Mol Physiol 280:L1094–L1103
44. Salcedo R, Wasserman K, Young HA et al (1999) Vascular endothelial growth factor and basic fibroblast growth factor induce expression of CXCR4 on human endothelial cells: in vivo neovascularization induced by stromal-derived factor-1alpha. Am J Pathol 154:1125–1135
45. Real C, Caiado F, Dias S (2008) Endothelial progenitors in vascular repair and angiogenesis: how many are needed and what to do? Cardiovasc Hematol Disord Drug Targets 8:185–193
46. Chavakis E, Aicher A, Heeschen C et al (2005) Role of beta2-integrins for homing and neovascularization capacity of endothelial progenitor cells. J Exp Med 201:63–72
47. Jin H, Su J, Garmy-Susini B et al (2006) Integrin alpha4beta1 promotes monocyte trafficking and angiogenesis in tumors. Cancer Res 66:2146–2152
48. Qin G, Ii M, Silver M et al (2006) Functional disruption of alpha4 integrin mobilizes bone marrow-derived endothelial progenitors and augments ischemic neovascularization. J Exp Med 203:153–163
49. Chavakis E, Hain A, Vinci M et al (2007) High-mobility group box 1 activates integrin-dependent homing of endothelial progenitor cells. Circ Res 100:204–212
50. Carmona G, Chavakis E, Koehl U et al (2008) Activation of Epac stimulates integrin-dependent homing of progenitor cells. Blood 111:2640–2646
51. Li S, Huang NF, Hsu S (2005) Mechanotransduction in endothelial cell migration. J Cell Biochem 96:1110–1126
52. Noria S, Cowan DB, Gotlieb AI, Langille BL (1999) Transient and steady-state effects of shear stress on endothelial cell adherens junctions. Circ Res 85:504–514
53. Albuquerque ML, Waters CM, Savla U et al (2000) Shear stress enhances human endothelial cell wound closure in vitro. Am J Physiol Heart Circ Physiol 279:H293–H302
54. Azuma N, Akasaka N, Kito H et al (2001) Role of p38 MAP kinase in endothelial cell alignment induced by fluid shear stress. Am J Physiol Heart Circ Physiol 280:H189–H197
55. Birukov KG, Birukova AA, Dudek SM et al (2002) Shear stress-mediated cytoskeletal remodeling and cortactin translocation in pulmonary endothelial cells. Am J Respir Cell Mol Biol 26:453–464
56. Wojciak-Stothard B, Ridley AJ (2003) Shear stress-induced endothelial cell polarization is mediated by Rho and Rac but not Cdc42 or PI 3-kinases. J Cell Biol 161:429–439
57. Ramirez-Bergeron DL, Runge A, Dahl KDC et al (2004) Hypoxia affects mesoderm and enhances hemangioblast specification during early development. Development 131:4623–4634
58. Du R, Lu KV, Petritsch C et al (2008) HIF1 alpha induces the recruitment of bone marrow-derived vascular modulatory cells to regulate tumor angiogenesis and invasion. Cancer Cell 13:206–220
59. Abdollahi H, Harris LJ, Zhang P et al (2011) The role of hypoxia in stem cell differentiation and therapeutics. J Surg Res 165:112–117
60. Thangarajah H, Vial IN, Chang E et al (2009) IFATS collection: adipose stromal cells adopt a proangiogenic phenotype under the influence of hypoxia. Stem Cells 27:266–274
61. Cao Y, Sun Z, Liao L et al (2005) Human adipose tissue-derived stem cells differentiate into endothelial cells in vitro and improve postnatal neovascularization in vivo. Biochem Biophys Res Commun 332:370–379

62. Weidemann A, Johnson RS (2008) Biology of HIF-1alpha. Cell Death Differ 15:621–627
63. Radtke F, Wilson A, Mancini SJ, MacDonald HR (2004) Notch regulation of lymphocyte development and function. Nat Immunol 5:247–253
64. Iso T, Kedes L, Hamamori Y (2003) HES and HERP families: multiple effectors of the Notch signaling pathway. J Cell Physiol 194:237–255
65. Fischer A, Schumacher N, Maier M et al (2004) The Notch target genes Hey1 and Hey2 are required for embryonic vascular development. Genes Dev 18:901–911
66. Liu ZJ, Xiao M, Balint K et al (2006) Inhibition of endothelial cell proliferation by Notch1 signaling is mediated by repressing MAPK and PI3K/Akt pathways and requires MAML1. FASEB J 20:1009–1011
67. Pola R, Ling LE, Silver M et al (2001) The morphogen Sonic hedgehog is an indirect angiogenic agent upregulating two families of angiogenic growth factors. Nat Med 7:706–711
68. Kanda S, Mochizuki Y, Suematsu T et al (2003) Sonic hedgehog induces capillary morphogenesis by endothelial cells through phosphoinositide 3-kinase. J Biol Chem 278:8244–8249
69. Hochman E, Castiel A, Jacob-Hirsch J et al (2006) Molecular pathways regulating pro-migratory effects of Hedgehog signaling. J Biol Chem 281:33860–33870
70. Asai J, Takenaka H, Kusano KF et al (2006) Topical sonic hedgehog gene therapy accelerates wound healing in diabetes by enhancing endothelial progenitor cell-mediated microvascular remodeling. Circulation 113:2413–2424
71. Nicoli S, Standley C, Walker P et al (2010) MicroRNA-mediated integration of haemodynamics and Vegf signalling during angiogenesis. Nature 464:1196–1200
72. Suarez Y, Fernandez-Hernando C, Pober JS, Sessa WC (2007) Dicer dependent microRNAs regulate gene expression and functions in human endothelial cells. Circ Res 100:1164–1173
73. Suarez Y, Fernandez-Hernando C, Yu J et al (2008) Dicer-dependent endothelial microRNAs are necessary for postnatal angiogenesis. Proc Natl Acad Sci U S A 105:14082–14087
74. Howard L, Kane NM, Milligan G, Baker AH (2011) MicroRNAs regulating cell pluripotency and vascular differentiation. Vascul Pharmacol 55:69–78
75. Tongers J, Roncalli JG, Losordo DW (2010) Role of endothelial progenitor cells during ischemia-induced vasculogenesis and collateral formation. Microvasc Res 79:200–206
76. Shumiya T, Shibata R, Shimizu Y et al (2010) Evidence for the therapeutic potential of ex vivo expanded human endothelial progenitor cells using autologous serum. Circ J 74:1006–1013
77. Wollert KC, Drexler H (2010) Cell therapy for the treatment of coronary heart disease: a critical appraisal. Nat Rev Cardiol 7:204–215

Chapter 9
Toll-Like Receptor-Linked Signal Transduction in Angiogenesis

Karsten Grote, Harald Schütt, and Bernhard Schieffer

Abstract Mammalian Toll-like receptors (TLRs) represent pattern recognition receptors of the immune system and are related to the Toll protein of *Drosophila*. Pathogen-associated molecular patterns (PAMPs) of microbial and viral origin bind to TLRs and initiate the innate and adaptive immune response. However, TLRs are not solely found on cells of the immune system but also on nonmyeloid cells in various tissues, e.g., on vascular cells. In addition to PAMPs, there is increasing evidence that TLRs also recognize endogenous ligands. Recent studies demonstrate the contribution of distinct TLRs in different inflammatory disorders such as cardiovascular diseases, rheumatoid arthritis, systemic lupus erythematosus, and cancer. Many of these disorders are characterized by enhanced angiogenesis which is mainly trigged by inflammation. However, this inflammation-induced angiogenesis is not only important for pathogen defense during acute infection or chronic inflammatory disorders but as well involved in regenerative processes during wound healing and tissue repair. There is cumulative evidence that TLR activation by exogenous as well as endogenous ligands especially contributes to angiogenic process in this scenario. The present chapter will summarize the current understanding of TLR-linked signal transduction in angiogenesis during inflammatory processes with future prospects for pro- or antiangiogenic therapy.

Keywords Toll-like receptors • Angiogenesis • Pathogen-associated pattern • Leucine-rich repeats • Damage-associated molecular patterns • Inflammation

K. Grote (✉) • H. Schütt • B. Schieffer
Department of Cardiology and Angiology, Hannover Medical School,
Carl-Neuberg-Strasse 1, 30625 Hannover, Germany
e-mail: grote.karsten@mh-hannover.de

J.L. Mehta and N.S. Dhalla (eds.), *Biochemical Basis and Therapeutic Implications of Angiogenesis*, Advances in Biochemistry in Health and Disease 6,
DOI 10.1007/978-1-4614-5857-9_9,

9.1 Introduction: Toll! Everything Started in *Drosophila*

A group of maternal effect genes are necessary for the embryo patterning of the fruit fly *Drosophila melanogaster* including the Toll gene. Lack of function experiments revealed that the Toll gene product provides the source for a morphogen gradient in the dorsal–ventral axis of the *Drosophila* embryo [1, 2]. Mutants in this gene were originally identified in 1985 by the group of Christiane Nüsslein-Volhard at the Max-Planck-Institute in Tübingen/Germany. The name of the gene derives from her exclamation "Das ist ja toll!," which translates as "That's amazing!" during microscopic observation of the drosophila mutants. Three years later, the Toll gene of *Drosophila* was cloned in the lab of Kathryn Anderson, the first author of the initial studies [3]. In 1992, Christiane Nüsslein-Volhard was awarded with the Nobel Prize for her groundbreaking research. Later on, Toll was found to play an important role in the fly's immune response by the group of Jules Hoffmann [4, 5]. In total, nine Toll receptors are encoded in the *Drosophila* genome, including the Toll pathway receptor Toll. The induction of the Toll pathway by fungi or by gram-positive bacteria leads to the activation of antimicrobial peptides. After proteolytically cleavage, binding of the extracellular ligand Spaetzle to the Toll receptor controls the expression of the antifungal peptide gene drosomycin. Mutations in the Toll signaling pathway dramatically reduce survival after fungal infection demonstrating the importance of this pathway for immune response.

9.2 Toll-Like Receptors in Mammalians

The identification of the *Drosophila* Toll pathway and the subsequent characterization of Toll-like receptor (TLR) function have reshaped the current understanding of the immune system. Mammalian homologues of the *Drosophila* Toll protein have been discovered 10 years later in the mid-1990s of last century which were consequently named TLRs [6, 7]. A scientific highlight in mammalian TLR discovery was the identification of TLR4 as the functional receptor for bacterial lipopolysaccharide (LPS) in mice carrying a mutation in the TLR4 gene by Bruce Beutler [8]. Based on this important discovery, Bruce Beutler was awarded with the Nobel Prize in 2011 which he shared with Jules Hoffmann for his findings in *Drosophila* and Ralph Steinmann who discovered dendritic cells. The discoveries of Hoffmann and Beutler triggered an explosion of research in innate immunity. Around a dozen different TLRs have now been identified in humans and mice comprising an entire receptor protein family [9]. All of them have initially been described as guardians of the innate immunity recognizing invading pathogens in the front line on the plasma membrane or after phagocytosis and processing on endosomal membranes, respectively. TLRs represent cognate pattern recognition receptors (PRRs) of the innate immunity and recognizing a high diversity of molecules common in pathogens of bacterial and viral origin referred to as pathogen-associated molecular patterns (PAMPs) (Fig. 9.1).

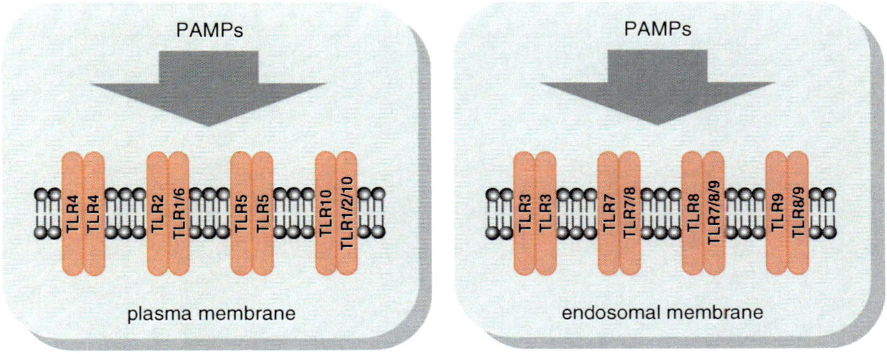

Fig. 9.1 Toll-like receptors (TLRs) form homo- and heterodimers

The specificity of TLRs for their ligands was mainly investigated in mice with functional mutations carrying an increased risk of infection. TLR ligation induces the activation of inflammatory pathways such as the mitogen-activated protein kinase (MAPK) cascade or nuclear factor-κB (NF-κB) and finally leads to the expression of cytokines and co-stimulatory molecules [10]. Thus, TLRs activate a potent immunostimulatory response and the signal that is transmitted from TLRs must therefore be tightly controlled. Structurally, all TLRs are type I integral membrane proteins consisting of an ectodomain comprised of leucine-rich repeats (LRRs) and a cytoplasmic domain containing a Toll/interleukin-1 receptor homology domain (TIR), which is required for signaling. TLRs occur as dimers; different receptor assemblies as mono- or heterodimers are known [11–13]. TLR2 builds heterodimers; in this regard TLR2/1 dimers sense bacterial triacylated lipopeptides, and TLR2/6 dimers sense bacterial diacylated lipopeptides. The LPS receptor TLR4 and TLR9, the receptor for unmethylated CpG-motifs in bacterial and viral DNA, homodimerize. Homodimerization is also presumed to be the case for TLR3 which senses synthetic and double-stranded RNA of viral origin (dsRNA) as well as for TLR5 which detects flagellin from bacteria. TLR7 and TLR8 recognize synthetic imidazoquinolines components and single-stranded RNA (ssRNA) and TLR8 has been shown to dimerize with TLR7 and TLR9. The ligand for TLR10 is currently unknown. However, it has been demonstrated that TLR10 can heterodimerize with TLR1 or TLR2 [14]. However, the number of putative TLR interaction partners and identified PAMPs that bind to TLRs is already large and is still growing (Fig. 9.1, Table 9.1) [15]. More recently, TLR11, 12 and 13 have been identified but there is not much known about their specific ligands so far.

A fundamental basis of TLR signaling is dependent upon the recruitment and association of adaptor molecules that contain the structurally conserved TIR domain. Signaling by TLRs involves five so far identified adaptor proteins known as myeloid differentiation primary response gene 88 (MyD88), MyD88-adaptor-like (MAL, also known as TIRAP), TIR-domain-containing adaptor protein inducing interferon-β (TRIF; also known as TICAM1), TRIF-related adaptor molecule (TRAM; also

Table 9.1 TLRs and their PAMPs

TLR	PAMP	Pathogen
TLR1	Triacylated lipopeptides	Bacteria
TLR2	Diacylated lipopeptides	Bacteria
	Triacylated lipopeptides	Bacteria
	Peptidoglycan	Bacteria
	Lipoteichoic acid	Gram-positive bacteria
TLR3	Double-stranded RNA	Viruses
TLR4	Lipopolysaccharide	Gram-negative bacteria
	Heat shock proteins	Bacteria
	Viral proteins	Viruses
TLR5	Flagellin	Bacteria
TLR6	Diacylated lipopeptides	Bacteria
TLR7	Single-stranded RNA	Viruses
TLR8	Single-stranded RNA	Viruses
TLR9	Unmethylated CpG DNA	Bacteria
TLR10	Unknown	

known as TICAM2), and sterile α- and armadillo-motif-containing protein (SARM). These adaptor molecules provide the necessary framework to recruit and activate downstream kinases and transcription factors that regulate the host inflammatory response. The canonical TIR pathway is dependent on MyD88, the immediate adapter molecule that is common to all TLRs, except TLR3. An alternative MyD88-independent pathway is controlled by TRIF, the only TLR3 adaptor, whereas TLR4 binds both MyD88 and TRIF. The remaining three adaptor proteins serve as coadaptors (MAL, TRAM) or even as a negative regulator (SARM). MAL and TRAM are just used by few TLRs. MAL recruits MyD88 to TLR2 and TLR4, whereas TRAM recruits TRIF to TLR4 [13]. After ligand binding to the specific TLR and assembly of the adaptor proteins, the activated membrane receptor complex induces the interleukin-1 receptor-associated kinase (IRAK) and tumor necrosis factor receptor-associated factor (TRAF) family members. The IRAK family—with their four members: IRAK1, IRAK2, IRAK4, and IRAKM—plays a pivotal role in mediating almost all TLR-mediated functions. All IRAK family members contain an amino-terminal death domain and a serine/threonine kinase domain. IRAK4 is known to be essential for TLR-mediated cellular responses. After TLR ligation, IRAK4 phosphorylates IRAK1 [16]. IRAK activation results in the recruitment/activation of TRAF family members such as TRAF3 and TRAF6, along with other E2 ubiquitin protein ligases which activate a complex containing transforming growth factor-β-activated kinase 1 (TAK1), TAK1-binding protein 1 (TAB1), TAB2, and TAB3 [17]. MyD88-dependent TAK1 activation induces the NF-κB pathway and MAPK members such as the extracellular signal-regulated kinase (ERK)1/2, p38, and stress-activated protein kinase/c-Jun N-terminal kinase (SAPK/JNK) to initiate the expression of inflammatory cytokines [9–11]. The TLR3 pathway is MyD88-indendent but TRIF-dependent that activates TRAF6 and NF-κB, resulting in the expression of inflammatory cytokines [18]. But TLR3 engagement also induces the expression of type I interferons (IFNs) via interferon regulatory transcription factor (IRF) 3 [19]. TLR7 and TLR9

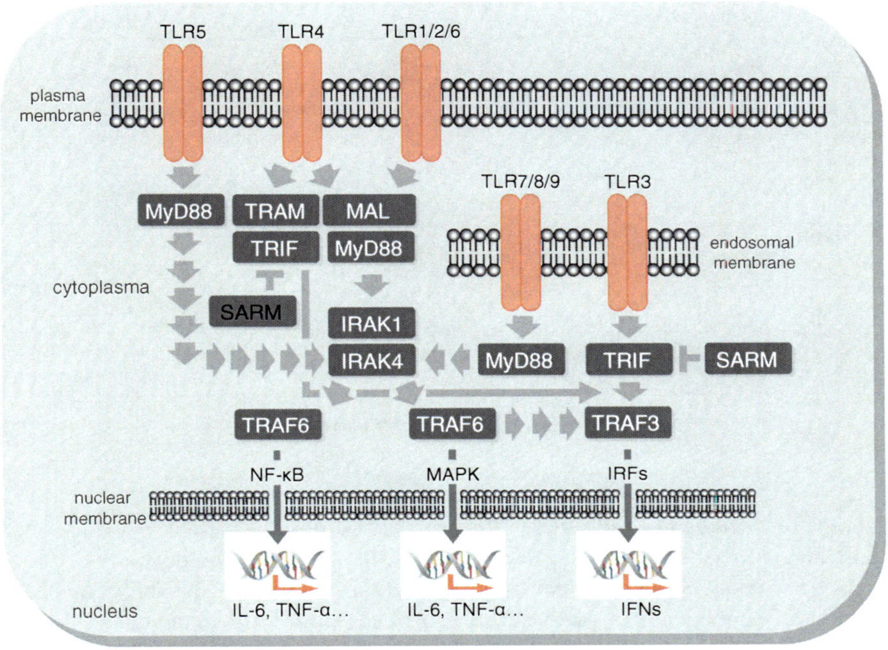

Fig. 9.2 TLR adaptor molecules and signaling pathways

engagement induces the secretion of inflammatory cytokines through the activation of NF-κB via MyD88. However, TLR7 and TLR9 can also induce the expression of type I IFNs through the activation of IRF7 [20]. Taken together, PAMP ligation to TLRs leads to the expression of effector molecules which finally organize the body's immune responds to pathogens (Fig. 9.2).

There is accumulating evidence from recent research that TLRs have distinct different functions beyond simple pathogen recognition. In a more complex immunologic view, an important role in dendritic cell maturation and T cell activation established TLRs as a link between innate and adaptive immunity [21]. Furthermore, the detection of several TLR members in multiple tissues and cell types—besides cells of the immune system—led to a more wide-ranging view on TLRs. Especially inflammatory disorders such as ischemic coronary artery disease [22] and liver disease [23] but also autoimmune diseases [24] are critically influenced by TLRs. Moreover, an involvement of TLRs in allograft acceptance/rejection during transplantation [25] or contact allergy to nickel [26] has been shown. Of interest, an interaction of TLRs with endogenous ligands released during tissue damage and fibrosis or from apoptotic cells has recently been discovered and seems to regulate many sterile inflammatory processes [27]. In this regard the term danger or damage-associated molecular patterns (DAMPs) has been introduced. These ligands include proteins and peptides, polysaccharides and proteoglycans, nucleic acids, and phospholipids, which are cellular components or extracellular matrix (ECM)

Table 9.2 Endogenous TLR ligands

TLR	Endogenous Ligand
TLR2	hyaluoran biglycan heat shock proteins high-mobility-group-protein B1
TLR3	RNA
TLR4	fibronectin fibrinogen hyaluoran biglycan heparin sulfate heat shock proteins high-mobility-group-protein B1 oxidized low density lipoprotein
TLR9	mitochondrial DNA high-mobility-group-protein B1

degradation products (Table 9.2). Recent studies provided clear evidence that endogenous ligand-mediated TLR signaling is involved in pathological conditions such as tissue injury, autoimmune diseases, and tumorigenesis. The ability of TLRs to recognize endogenous ligands appears to be essential for their function in regulating noninfectious inflammation. Furthermore, a novel role for TLRs in wound healing [28, 29] and liver regeneration [30] also in response to endogenous ligands [27] has been reported, suggesting even a regenerative aspect in TLR biology.

9.3 Angiogenesis: General Remarks

Physiological tissue function depends on adequate supply of nutrients and oxygen through blood vessels. Consequently, the cardiovascular system is the first organ system that develops during embryogenesis. Blood vessels in the embryo form the hemangioblast by differentiation of common mesodermal progenitor cells. The hemangioblast forms aggregates that evolve into hematopoietic precursor cells and angioblasts which further assemble the primary capillary plexus as differentiated endothelial cells. The formation of this primitive network on the basis of progenitor cells is called vasculogenesis. On the contrary, angiogenesis describes the enlargement of capillaries which sprout or become divided by pillars of periendothelial cells (intussusception) or by transendothelial cell bridges followed by remodeling and maturation processes that transform the primary capillary plexus into a complex network of functional vessels [31]. Further covering and stabilization of vessels by smooth muscle cells as well as the enlargement of preexisting collateral arterioles is summarized as arteriogenesis. Although in the adult most vessels arise through angiogenesis, vasculogenesis may also be involved to some extent. Therefore, both processes are summarized in the hypernym neovascularization, which is involved in organ growth and wound healing but also contributes to pathological processes in malignant and inflammatory disorders [32–34].

Angiogenesis through sprouting and subsequent remodeling of capillaries into larger vessels has been extensively studied, and several essential steps have been described. Vascular endothelial growth factor (VEGF) and its receptors (VEGFR) have been identified as central regulators of both vasculogenesis and angiogenesis [35]. Until now, five VEGF ligands have been identified which occur in different spliced and processed variants and all of them represent secreted dimeric glycoproteins of ~40 kDa. In addition to VEGF A–D, placenta growth factor (PLGF) belongs to the VEGF family as well. These ligands bind to the three receptor tyrosine kinases VEGFR1–3 with an overlapping pattern and co-receptors such as heparan sulfate, proteoglycans, and neuropilins. Different VEGFRs have distinct different functions; VEGFR1 is involved in the recruitment of hematopoietic progenitor cells and migration of monocytic cells whereas VEGFR2 and 3 are essential for the function of endothelial cells, especially during angiogenesis. Initially, VEGF increases vascular permeability [36], thereby permitting extravasation of plasma proteins that establish a preliminary scaffold for migrating endothelial cells. For the emigration of endothelial cells from their resident site, interendothelial cell contacts and periendothelial cell support have to be dissolved, leading to destabilization of the mature vessel. Angiopoietin (Ang) 2, an inhibitor for tyrosine kinase with Ig and epidermal growth factor (EGF) homology domains (Tie) 2 signaling [37, 38], may be involved in detaching smooth muscle cells and breaking up the ECM. Especially during angiogenesis, the interaction of the Ang-Tie system with the VEGF system becomes apparent. Capillaries sprout and subsequently grow alongside a VEGF gradient. Endothelial cells at the leading edge of the migration front, so-called tip cells, exhibit numerous filopodia and express members of the VEGFR family. Subjacent endothelial cells could be subdivided in highly proliferative and differentiating stalk cells and resting phalanx cells which both express components of the Ang-Tie system [38]. Furthermore, proteinases of the plasminogen activator, matrix metalloproteinases (MMPs), and chymase families influence angiogenesis by degrading ECM and by liberating growth factors, e.g., VEGF, basic fibroblast growth factor (bFGF), and insulin-like growth factor (IGF)-1, sequestered within the ECM. When the path has been cleared, endothelial cells can proliferate and migrate to remote sites [32].

Angiogenic sprouting is controlled by a tightly regulated balance of activators and inhibitors. In addition to VEGF, Tie2 phosphorylation by Ang1 is chemotactic for endothelial cells and stabilizes VEGF-initiated endothelial networks by stimulating the interaction between endothelial cells and periendothelial cells [38]. Members of the FGF and platelet-derived growth factor (PDGF) family support angiogenesis presumably by recruitment of mesenchymal or inflammatory cells. Another key component of sprouting angiogenesis by regulating tip cell vs. stalk cell communication is the highly conserved Delta/Notch signaling pathway. Mammalians possess four different notch receptors, referred to as Notch1–4. Notch receptors are single-pass transmembrane receptors and capable of binding the membrane-bound ligands Delta-like (Dll) 1–4 and Jagged. Notch signaling in the stalk cells induces a quiescent and non-sprouting phenotype in endothelial cells whereas adjacent tip cells express Dll4, therefore promoting sprouting activity [39]. In addition, molecules are involved which mediate cell–cell or cell–matrix interactions, e.g., $\alpha_v\beta_3$ which localizes MMP-2 at the endothelial cell surface and promotes endothelial cell spreading. Moreover, a continuously number of molecules are discovered which are proangiogenic upon exogenous administration, including erythropoietin, leptin, hepatocyte growth factor (HGF), EGF, IGF-1, tissue factor (TF), and several other cytokines, chemokines, and growth factors [32, 40]. Even hematopoietic growth factors such as granulocyte colony-stimulating factor (G-CSF) and granulocyte-macrophage colony-stimulating factor (GM-CSF) have been shown to exhibit proangiogenic potential [41]. On the contrary, angiogenesis inhibitors suppress endothelial cell proliferation and migration, e.g., angiostatin, endostatin, antithrombin III, IFN-β, leukemia inhibitory factor (LIF), and platelet factor 4 [32, 42]. Thus, various pro- and antiangiogenic factors cooperate to regulate the angiogenic process.

Finally, vessel maturation finalizes the angiogenic process. Proliferating endothelial cells initially assemble as solid cords which acquire additional lumen formation. Lumen formation is accomplished by thinning of endothelial cells or fusion of preexisting vessels, mediated by VEGF, Ang1, and integrins such as $\alpha_v\beta_3$ or α_5 and controlled by the inhibitory effects of thrombospondin-1. Additional important steps after lumen establishment involve the differentiation of endothelial cells according to the environmental demands, maturation into a functional three-dimensional endothelial network, and the protection of quiescent endothelial cells against apoptosis [32, 40]. Periendothelial cells are essential for vascular maturation and completion of angiogenesis. Nascent vessels are stabilized by pericytes in case of capillaries. In case of arteries, arterioles, veins, and venules, smooth muscle cell recruitment and growth mediated by VEGF or PDGF are crucial for vessel stabilization. They thereby provide hemostatic control and protect the new endothelium-lined vessel against rupture and regression [32, 42].

Blood vessel formation in the adult includes vasculogenesis, angiogenesis, and arteriogenesis. Impaired neovascularization represents a therapeutic target in several pathologies associated with insufficient blood supply, e.g., acute myocardial infarction or chronic peripheral artery disease. Formation of new vessels and remodeling of the preexisting vasculature are essential for a successful therapy. Therefore, different treatment strategies involving administration of growth factors, cytokines, or progenitor cells are considered [43]. Important for the understanding of angiogenic mechanisms in these pathologies is the knowledge of variations from physiologic angiogenesis. In contrast to the physiological processes, pathologic angiogenesis is often promoted by inflammation. Monocytes, platelets, mast cells, and other leukocytes are recruited to sites of inflammation or wound healing, partly by proangiogenic factors such as VEGF [32, 34]. Moreover, development of solid tumors strictly depends on a growing capillary network—termed as tumor angiogenesis—ensuring sufficient supply with oxygen and nutrients. Accordingly, antiangiogenic concepts aim at the inhibition of tumor angiogenesis and thereby tumor nutrient supply [44]. In this regard, the first antitumor therapy with a VEGF-neutralizing monoclonal antibody for the treatment of metastasizing bowel cancer was approved by the US Food and Drug Administration in 2004.

9.4 Inflammation-Induced Angiogenesis

Disorders associated with perpetuated angiogenesis are considered to be angiogenic inflammatory diseases. Inflammation plays not only a key role in pathogen defense during infection; it also plays a key role in repair mechanisms, e.g., wound healing and subsequent tissue regeneration. Physiological wound healing requires the integration of complex cellular and molecular events. The repair process is tightly controlled involving different cell types during the phases of initial inflammation as well as the successive cell migration, cell proliferation, and angiogenesis. Several angiogenic mediators, including growth factors, cytokines, MMPs, matrix macromolecules, cell adhesion receptors, chemokines, and chemokine receptors, have been implicated in the process of capillary formation [45]. Of note, cytokines and growth factors released at the site of injury are essential for the repair process [46]. In this regard, angiogenesis, the reestablishment of a capillary network by endothelial cells, is mainly initiated and maintained by the major proangiogenic factor VEGF. Besides endothelial cells, the angiogenic process involves also other cell types including inflammatory cells which represent a major source of growth factors and critically contribute to angiogenesis [47]. Platelets, mast cells, primarily monocytes/macrophages, neutrophils, and other leukocytes are recruited to sites of wound healing, partly by the action of the proangiogenic factors such as VEGF. All these cells in turn release proangiogenic factors such as VEGF, bFGF, TGF-β, PDGF, tumor necrosis factor (TNF)-α, insulin-like growth factor (IGF)-1, monocyte chemotactic protein (MCP)-1, interleukin (IL)-6, IL-8, and many more. All these factors finally attract endothelial cells, smooth muscle cells, pericytes, and fibroblasts to accomplish vessel growth in order to restore sufficient blood supply [34]. Newly formed blood vessels again enhance inflammatory cell recruitment setting up a stimulating forward loop. In this regard, inflammation often promotes angiogenesis establishing the term inflammation-induced angiogenesis.

In inflamed tissues a regulatory network is involved in the control of angiogenesis. Accumulating evidence suggests an association between angiogenesis and inflammation in pathological situations. Therefore, angiogenesis and inflammation seem to be intimately involved in many chronic inflammatory disorders with distinct etiopathogenic origin, including rheumatoid arthritis, diabetes, cancer, and many more. For example, there is considerable evidence of an interrelationship between the mechanisms of angiogenesis and chronic inflammation in inflammatory bowel disease (IBD). The increased expression of endothelial junction adhesion molecules found in IBD patients indicates the presence of active angiogenesis. Evidence that angiogenesis is involved in IBD was also obtained from animal models of colitis, most notably from studies of angiogenesis inhibition. Moreover, serum levels of VEGF correlate with disease activity in human IBD [48]. This concept has been further supported by the finding that several previously established noninflammatory disorders, such as obesity, display both inflammation and angiogenesis in an exacerbated manner [49]. In addition, the interplay between recruited inflammatory cells and local endothelial cells and fibroblasts at sites of chronic inflammation, together with the fact that inflammation and angiogenesis can actually be triggered by the same molecular events, further strengthen this association. Angiogenesis might be targeted by several specific approaches that could be therapeutically used to control inflammatory diseases.

9.5 Toll-Like Receptors in Inflammation-Induced Angiogenesis

It is experimentally well established that angiogenesis and inflammation represent two prominent processes involved in normal physiologic responses and pathological states. Emerging evidence also suggests that TLRs have an important role in maintaining tissue homeostasis by regulating the inflammatory and tissue repair responses to injury. Infectious disorders result in inflammation which in turn promotes angiogenesis mainly by the action of growth factors released by different leucocytes. Even though the association of inflammation and angiogenesis has been established for a while the knowledge about the role of TLRs in this context is still limited [50]. However, a significant number of publications demonstrate that several TLR agonists are able to induce the expression and secretion of angiogenic factors from different cell types in vitro. The majority of these studies remain rather descriptive in this context and are very much focused on LPS and VEGF. Up to know, only few data document a direct involvement of TLRs in angiogenesis, both in physiological and in pathophysiological settings.

9.6 Toll-Like Receptors in Infection-Induced Angiogenesis

Accumulating evidence points to a direct contribution of TLRs to the angiogenic process following bacterial infections, also referred to as infection-induced angiogenesis. In this regard, PAMPs from various bacterial species—superabound in an infection setting—are known to act via different TLRs. A possible influence of TLRs on angiogenic processes was first discovered in the context of adenosine and its A_{2A} receptor (A_{2A}R). The nucleoside adenosine was found to stimulate angiogenesis through upregulation of VEGF, thereby participating in tissue protection following ischemic events. In 2002 Leibovich et al. described a synergistic interaction of A_{2A}R agonists with LPS through the TLR4 pathway. This interaction resulted in a strong upregulation of VEGF and downregulation of TNF-α in macrophages [51] and could also be demonstrated for TLR2, 7, and 9 [52], representing an angiogenic switch. This synergy observed in vitro seems to play an important role in vivo, too. Given the fact that MyD88-deficient mice showed markedly slower wound healing and reduced generation of new capillaries in response to an $A_{2A}R$ agonist [29]. In terms of TLR4, it is very likely that LPS induces adenosine which in turn promotes angiogenesis through $A_{2A}R$ by the upregulation of VEGF expression in macrophages [53].

Independent of the $A_{2A}R$ system, Pollet et al. showed that the TLR4 ligand LPS directly stimulates endothelial sprouting in vitro via a TRAF6-, NFκB-, and JNK-dependent mechanism. However, the responsible angiogenic growth factor remained elusive in this context [54]. Furthermore, a so far unidentified TLR ligand seems to be involved in the formation of angiogenic lesions resulting from infection with the facultative intracellular bacterium *Bartonella henselae*. This bacterial infection leads to enhanced MCP-1 production in endothelial cells which in turn induces chemotaxis of monocytes in order to initiate angiogenesis by VEGF production. Interestingly, MCP-1 production was independent of LPS/TLR4 but dependent on NFκB [55]. A serious problem of severe ocular infection is pathological corneal neovascularization which could finally lead to visual disorders. In this regard, it has been shown that VEGF and TLR4 expression are upregulated in response to LPS and that VEGF expression is TLR4-dependent [56].

But angiogenesis also contributes to the regeneration process during liver fibrosis which is associated with increased endotoxin levels in the gut and portal circulation. Jagavelu et al. recently demonstrated a key role for the TLR4/MyD88 axis during VEGF production and the subsequent angiogenic process in liver endothelial cells following LPS stimulation [57]. Likewise, mycoplasma infections could be accompanied by enhanced angiogenesis and microvascular remodeling which are features of the chronic inflammation as elicited by *Mycoplasma pulmonis* infections of the respiratory tract [58]. In this regard, we recently investigated the highly angiogenic properties of the specific TLR2/6 agonist macrophage-activating lipopeptide of 2 kDa (MALP-2), a diacylated lipopeptide which occurs in *Mycoplasma* species and gram-positive bacteria. Interestingly, this process seems to be independent of VEGF. We discovered a TLR2/6-dependent induction of the MAPK cascade and NFκB and strong secretion of GM-CSF in particular from endothelial cells and to a lesser degree from monocytes. Accordingly, MALP-2-induced angiogenesis in vitro and in vivo could be suppressed by inhibition of GM-CSF [59]. Similarly, human bone marrow mesenchymal stem cells (MSCs) secreted growth factors in response to a TLR2/6-dependent stimulation by MALP-2. This process in turn promoted proangiogenic properties of endothelial cells such as migration, proliferation, and tube formation in vitro in a paracrine manner. Finally, MSCs isolated from the bone marrow of sheep and cocultivated with MALP-2 ex vivo significantly enhanced capillary density of skeletal muscle after autogenic implantation of these MSCs (unpublished data). This renders MALP-2 potentially eligible for therapeutic angiogenesis or cell therapy.

In addition to an acute infection upon injury, there are different acute or chronic inflammatory disorders which are also associated with bacterial infection independent of an initial injury. Arthritis is characterized by inflammatory cell infiltration into the concerned joint. Progression of the disease includes self-perpetuating destruction of articular cartilage and extensive angiogenesis in the synovial membrane. Especially TLR2 ligands of gram-positive bacteria such as peptidoglycan (PGN) seem to be responsible for this angiogenic phenotype characterized by the induction of VEGF in chondrocytes [60] and accordingly VEGF and IL-8 in fibroblasts [61]. In light of immune defense, infection-induced angiogenesis might represent a general mechanism to restore blood flow in order to recruit immune cells for pathogen clearance and tissue regeneration with implication for future angiogenic therapy.

9.7 Toll-Like Receptors in Tumor Angiogenesis

The development of cancer has been associated with microbial infection, injury, inflammation, angiogenesis, and tissue repair. The role of TLRs in tumor angiogenesis is quite diverse just as cancer itself. Tumor inflammation could promote tumor angiogenesis, immunosuppression, and finally tumor growth. However, the mechanism controlling inflammatory cell recruitment to the tumor is not well understood. Cyclooxygenase (COX)-2 is known to play a crucial role in *Helicobacter pylori*-associated gastric cancer. In this regard, Chang et al. demonstrated that *H. pylori* acts through TLR2 and TLR9 to activate the MAPK cascade leading to COX-2-dependent prostaglandin E_2 (PGE_2) release and thereby contributing to cancer cell invasion and angiogenesis [62]. On the other hand, stimulation of TLRs with particular agonists can also cause antitumor activity, interfering with cancer proliferation and angiogenesis by mechanism still incompletely understood. For instance, the immunomodulatory TLR9 agonist IMO inhibited microvessel formation and tumor growth [63]. Likewise, TLR3 agonists not only affect tumor microenvironment by suppressing angiogenesis but also directly induce tumor cell apoptosis and inhibit tumor cell migration [64]. Interestingly, siRNAs may produce therapeutic effects in a target-independent manner through the stimulation of the TLR3/interferon pathway and suppression of angiogenesis. Injection of siRNAs against different targets led to a comparable reduction in liver tumors and to an inhibition of tumor vasculature remodeling. In addition, polyI:C treatment reduced liver tumors and decreased hepatic arterial blood flow, indicating that TLR3 may mediate antiangiogenic and antitumor properties [65].

In all likelihood, there are two different possibilities for TLR agonists to limit tumor growth. First, by altering the tumor microenvironment and inhibiting angiogenesis and second, by clearing tumor cells due to enhanced activity of natural killer and tumor-reactive T cells. In this regard, the TLR7 agonist imidazoquinoline and the TLR9 agonist unmethylated CpG oligonucleotides were shown to exhibit strong local activity against leukemia, and respective phase I trials are currently in progress at different centers [66]. We recently identified proangiogenic properties for the TLR2/6 ligand MAL-2 [59]. Interestingly, there are also antitumor activities reported for MALP-2 [67–69]. However, whether MALP-2 affects tumor angiogenesis is currently unknown. TLR4 expression in the tumor microenvironment was found to be associated with adenocarcinoma in human samples and in the murine model. Adenocarcinoma patients with higher TLR4 expression in stromal compartment had a significantly increased risk in disease progression. These data suggest that high TLR4 expression in the tumor microenvironment represents a possible marker for disease progression in colon cancer [70]. So far, there are different polymorphisms in several TLR gene clusters known which may shift balance between pro- and anti-inflammatory cytokines, modulating the risk of infection, chronic inflammation, and cancer. This may offers the possibility for improved diagnostics in patients. Future studies in large populations should shed light on the significance of TLR polymorphisms for cancer prevention [71].

9.8 Endogenous Toll-Like Receptor Ligands in Angiogenesis

Sustained proinflammatory responses in diseases such as rheumatoid arthritis, atherosclerosis, diabetic retinopathy, and cancer are often associated with increased angiogenesis that contributes to tissue disruption and disease progression. In recent years, there was accumulating evidence that also endogenous ligands which are released during ECM breakdown or by apoptotic cells could bind to different TLRs (Table 2). In this context, the high-mobility group B1 (HMGB1) which is released by necrotic cells has been recognized to signal through the receptor for advanced glycation end products (RAGE) and via TLR2 and TRL4. Activation of these receptors resulted in the activation of NFκB and the upregulation of angiogenic factors like VEGF in both hematopoietic and endothelial cells [72]. HMGB1 released at wound sites initiates TLR4-dependent responses that contribute to angiogenesis [73]. Recent data by van Beijnum et al. identified HMGB1 even as an important modulator of tumor angiogenesis [74]. Thus, targeting the HMGB1 signaling cascade may constitute a novel therapeutic approach to angiogenesis-related diseases. Following this line, inflammation-induced oxidative stress and angiogenesis is emerging as an important mechanism underlying numerous processes from tissue regeneration and remodeling to cancer progression. Interestingly, West et al. recently reported that end products of lipid oxidation such as ω-(2-carboxyethyl)pyrrole (CEP) are generated and accumulate during inflammation, wound healing, and in tumors. CEP is specifically recognized by TLR2 but not TLR4 or scavenger receptors in endothelial cells, leading to a MyD88-dependent angiogenic response that is independent of VEGF [75]. In this regard, stress-sensing by TLR2 seems to be a major driver of angiogenesis [76]. Apparently, also endogenous ligands, which accumulate during inflammatory tissue disruption and enhanced oxidative stress conditions, are capable of promoting angiogenesis via a TLR-dependent pathway. Thus, TLRs are activated not only in response to tissue-invading pathogens but also pathogen-independent. In both cases TLRs have important functions in the recruitment of immune cells in order to initiate a regenerative program: in the first case mainly to eliminate invading pathogens and in the second case to clear the affected tissue from apoptotic cells and cellular debris. Obviously, angiogenic processes are involved in both scenarios (Fig. 9.3).

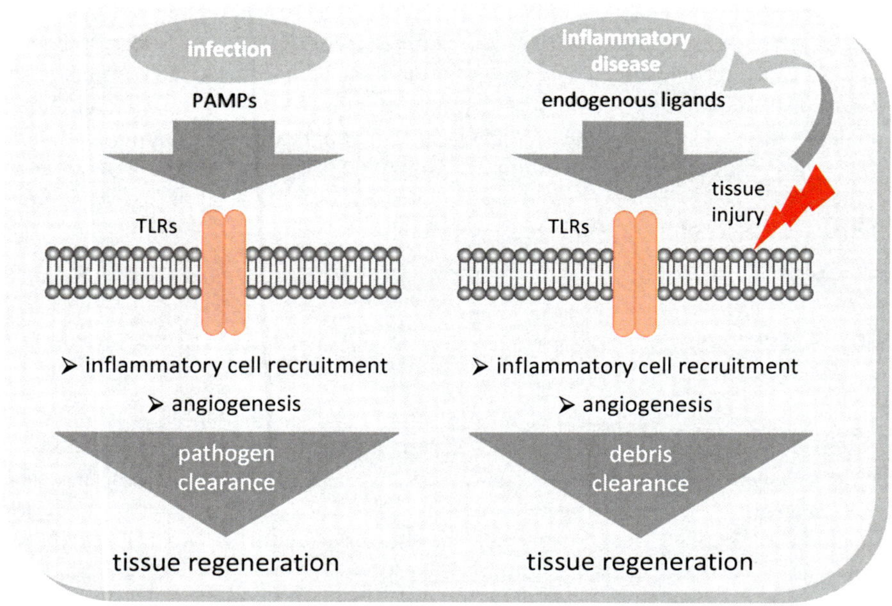

Fig. 9.3 Tissue regeneration through TLRs

9.9 Summary and Therapeutic Perspectives

Accumulating evidence points to a crucial role of TLRs in angiogenesis. However, the mode of action of TLRs in this context is quite diverse. TLR activation consistently promotes angiogenesis in various inflammatory settings in response to both exogenous and endogenous ligands. In regard to an acute local infectious scenario, the angiogenic process seems to be important for sufficient blood supply and the recruitment of immune competent cells for pathogen clearance and subsequent tissue regeneration. In contrast, chronic local infection or prolonged pathogen-independent inflammation leads to excessive angiogenesis with eventually pathological consequences. The role of TLRs in tumor angiogenesis is even more contradictory since pro- and antiangiogenic properties are reported.

In the future, modulation of TLR signaling could provide the basis for the development of novel therapeutic approaches in diverse settings. Stimulation of TLRs with specific ligands could be used for future therapeutic angiogenesis. However, beneficial effects of therapeutic angiogenesis may be negatively impacted by side effects of pharmacological substances such as statins or nonpharmacologic hormones such as erythropoietin. Moreover, certain requirements for this therapeutic process are warranted. First, as simple as it may sound but no harm should be induced especially tumor induction or tumor growth should be avoided. Second, in order to promote a sustained recovery, endogenous mechanisms of angiogenesis should be induced rather using an excessive administration of exogenous factors which may also act as antigens or inducing tolerance when applied over a long period of time. Finally, organ-specific requirements for recovery should be considered, e.g., for cerebral reconstitution angiogenesis, neurogenesis, synaptogenesis, and neuronal and synaptic plasticity should be induced in parallel [77].

Thus, therapeutic modulation of TLR signaling is a very attractive and novel but also sophisticated therapeutic approach to promote angiogenesis. In order to induce long-term organ repair and restoration after ischemic events, for example, detrimental TLR signaling should be inhibited and in parallel beneficial TLR signaling should be induced. From this point of view, inhibitory strategies targeting TLR signaling seem to be plausible in chronic and persistent infectious situations such as rheumatoid arthritis. Small molecules or siRNA against specific TLRs or their downstream targets may provide novel tools to combat local inflammation via inhibition of angiogenesis. Especially advanced tissue penetration properties of those engineered molecules render them applicable and superior for the use in tissues which are inaccessible for antibiotics. Likewise, inhibitory strategies targeting TLRs could be used to inhibit pathological tumor angiogenesis in order to limit tumor growth. In particular, modulation of TLR3, TLR7, and TLR9 activity seems to be a potential future therapeutic target [63–66]. However, great caution is required since pro- and antiangiogenic properties with subsequent pro- or antitumorigenic properties of different agonists recognized by the same TLR are reported.

The potentially most promising future therapeutic approach is the application of specific TLR agonists in damaged ischemic or hypoxic tissues in order to promote angiogenesis and subsequent tissue regeneration, especially when the tissue damage is not initiated or accompanied by severe infection, e.g., in peripheral arterial occlusive disease. In such settings, a single application of TLR agonists mimics an infectious scenario without prolonged local pathogen presences. Such an initial therapeutic boost of the immune system with specific TLR agonist aims to launch a defined regenerative program including enhanced angiogenesis. Of note, the application of single proangiogenic growth factors has already been tested in clinical trials. However, in the case of VEGF monotherapy, large-scale trials have not yet yielded consistent beneficial results [78, 79]. This may be related to recent observations that several other potent proangiogenic factors act in concert with VEGF for proper vessel formation and maturation [38, 39, 42]. In this regard, stimulation of specific TLRs (e.g., TLR2/6) may provide an opportunity to induce a specific pattern of proangiogenic growth factors for sufficient vessel growth and tissue regeneration. Thus, we raised our hope on biologicals such as the lipopeptide and TLR2/6 agonist MALP-2. Recent results from our group indicated that the proangiogenic properties of MALP-2 critically depended on the induction of the growth factor GM-CSF in endothelial cells and monocytes [59]. Additional experiments in a vascular endothelial denudation model in mice revealed promising effects of MALP-2 on endothelial regeneration after vascular injury (unpublished data). Those experimental data are the basis for studies in larger experimental animals and future applications using MALP-2 in patients, who suffer from peripheral vascular damage or occlusion in diabetes or post percutaneous vascular interventions or even following stroke. Nevertheless, the question remains how to apply a substance like MALP-2 since local delivery is preferred in order to avoid side effects and promote endogenous proangiogenic restoration effects downstream of the site of application. Therefore, we are in the process of testing coating procedures on traditional devices such as drug-eluting stents or coated balloons widely used in interventional cardiovascular medicine. However more innovative devices/treatment approaches such as nanofibers, polymer biodegradable stents soaked with MALP-2, and endovascular MALP-2 patches placed in the occluded vessel or as seal on the balloon-disrupted vascular segment are in the focus of our interest.

In summary, modulation of TLR activity may offer the possibility for different future therapeutic concepts. Inhibition of TLRs is maybe favorable in settings of prolonged infection/inflammation to rescue the inflamed tissue or to inhibit pathological tumor angiogenesis to limit tumor growth. The contrary concept, TLR stimulation, offers a promising option to promote therapeutic angiogenesis for tissue regeneration.

Acknowledgments The work of KG and BS is supported by grants from the German Research Foundation (DFG) KFO 136 and SFB 566/b9 and from the Federal Ministry of Education and Research (BMBF) 01GU0711.

References

1. Anderson KV, Jürgens G, Nüsslein-Volhard C (1985) Establishment of dorsal-ventral polarity in the Drosophila embryo: genetic studies on the role of the Toll gene product. Cell 42:779–789
2. Anderson KV, Bokla L, Nüsslein-Volhard C (1985) Establishment of dorsal-ventral polarity in the Drosophila embryo: the induction of polarity by the Toll gene product. Cell 42:791–798
3. Hashimoto C, Hudson KL, Anderson KV (1988) The Toll gene of Drosophila, required for dorsal-ventral embryonic polarity, appears to encode a transmembrane protein. Cell 52:269–279
4. Lemaitre B, Nicolas E, Michaut L et al (1996) The dorsoventral regulatory gene cassette spätzle/Toll/cactus controls the potent antifungal response in Drosophila adults. Cell 86:973–983
5. Valanne S, Wang JH, Rämet M (2011) The Drosophila Toll signaling pathway. J Immunol 186:649–656
6. Taguchi T, Mitcham JL, Dower SK et al (1996) Chromosomal localization of TIL, a gene encoding a protein related to the Drosophila transmembrane receptor Toll, to human chromosome 4p14. Genomics 32:486–488
7. Medzhitov R, Preston-Hurlburt P, Janeway CA Jr (1997) A human homologue of the Drosophila Toll protein signals activation of adaptive immunity. Nature 388:394–397
8. Poltorak A, He X, Smirnova I et al (1998) Defective LPS signaling in C3H/HeJ and C57BL/10ScCr mice: mutations in Tlr4 gene. Science 282:2085–2088
9. Takeda K, Kaisho T, Akira S (2003) Toll-like receptors. Annu Rev Immunol 21:335–376
10. Oda K, Kitano H (2006) A comprehensive map of the toll-like receptor signaling network. Mol Syst Biol 2:2006.0015

11. Brikos C, O'Neill LA (2008) Signalling of toll-like receptors. In: Bauer S, Hartmann G (eds) Toll-like receptors (TLRs) and innate immunity, vol 183, Handbook of experimental pharmacology. Springer, Heidelberg, pp 21–50
12. Brown J, Wang H, Hajishengallis GN, Martin M (2011) TLR-signaling networks: an integration of adaptor molecules, kinases, and cross-talk. J Dent Res 90:417–427
13. O'Neill LA, Bowie AG (2007) The family of five: TIR-domain-containing adaptors in Toll-like receptor signalling. Nat Rev Immunol 7:353–364
14. Hasan U, Chaffois C, Gaillard C, Saulnier V, Merck E, Tancredi S, Guiet C, Brière F, Vlach J, Lebecque S, Trinchieri G, Bates EE (2005) Human TLR10 is a functional receptor, expressed by B cells and plasmacytoid dendritic cells, which activates gene transcription through MyD88. J Immunol 174:2942–2950
15. Lee CC, Avalos AM, Ploegh HL (2012) Accessory molecules for Toll-like receptors and their function. Nat Rev Immunol 12:168–179
16. Li S, Strelow A, Fontana EJ, Wesche H (2002) IRAK-4: a novel member of the IRAK family with the properties of an IRAK-kinase. Proc Natl Acad Sci U S A 99:5567–5572
17. Chen ZJ (2005) Ubiquitin signalling in the NF-kappaB pathway. Nat Cell Biol 7:758–765
18. Sato S, Sugiyama M, Yamamoto M et al (2003) Toll/IL-1 receptor domain-containing adaptor inducing IFN-beta (TRIF) associates with TNF receptor-associated factor 6 and TANK-binding kinase 1, and activates two distinct transcription factors, NF-kappa B and IFN-regulatory factor-3, in the Toll-like receptor signaling. J Immunol 171:4304–4310
19. Oshiumi H, Matsumoto M, Funami K et al (2003) TICAM-1, an adaptor molecule that participates in Toll-like receptor 3-mediated interferon-beta induction. Nat Immunol 4:161–167
20. Honda K, Yanai H, Mizutani T et al (2004) Role of a transductional-transcriptional processor complex involving MyD88 and IRF-7 in Toll-like receptor signaling. Proc Natl Acad Sci U S A A101:15416–15421
21. Kaisho T, Akira S (2003) Regulation of dendritic cell function through toll-like receptors. Curr Mol Med 3:759–771
22. Satoh M, Ishikawa Y, Minami Y et al (2008) Role of Toll like receptor signaling pathway in ischemic coronary artery disease. Front Biosci 13:6708–6715
23. Seki E, Brenner DA (2008) Toll-like receptors and adaptor molecules in liver disease: update. Hepatology 48:322–335
24. Marshak-Rothstein A (2006) Toll-like receptors in systemic autoimmune disease. Nat Rev Immunol 6:823–835
25. Obhrai J, Goldstein DR (2006) The role of toll-like receptors in solid organ transplantation. Transplantation 81:497–502
26. Schmidt M, Raghavan B, Muller V et al (2010) Crucial role for human Toll-like receptor 4 in the development of contact allergy to nickel. Nat Immunol 11:814–819
27. Yu L, Wang L, Chen S (2010) Endogenous toll-like receptor ligands and their biological significance. J Cell Mol Med 14:2592–2603
28. Deiters U, Barsig J, Tawil B, Muhlradt PF (2004) The macrophage-activating lipopeptide-2 accelerates wound healing in diabetic mice. Exp Dermatol 13:731–739
29. Macedo L, Pinhal-Enfield G, Alshits V et al (2007) Wound healing is impaired in MyD88-deficient mice: a role for MyD88 in the regulation of wound healing by adenosine A2A receptors. Am J Pathol 171:1774–1788
30. Seki E, Tsutsui H, Iimuro Y et al (2005) Contribution of Toll-like receptor/myeloid differentiation factor 88 signaling to murine liver regeneration. Hepatology 41:443–450
31. Ribatti D (2010) The seminal work of Werner Risau in the study of the development of the vascular system. Int J Dev Biol 54:567–572
32. Carmeliet P (2000) Mechanisms of angiogenesis and arteriogenesis. Nat Med 6:389–395
33. Risau W (1997) Mechanisms of angiogenesis. Nature 386:671–674
34. Carmeliet P (2005) Angiogenesis in life, disease and medicine. Nature 438:932–936
35. Olsson AK, Dimberg A, Kreuger-J, Claesson-Welsh L (2006) VEGF receptor signalling—in control of vascular function. Nat Rev Mol Cell Biol 7:359–371

36. Senger DR, Galli SJ, Dvorak AM et al (1983) Tumor cells secrete a vascular permeability factor that promotes accumulation of ascites fluid. Science 219:983–985
37. Partanen J, Armstrong E, Makela TP et al (1992) A novel endothelial cell surface receptor tyrosine kinase with extracellular epidermal growth factor homology domains. Mol Cell Biol 12:1698–1707
38. Augustin HG, Koh GY, Thurston G, Alitalo K (2009) Control of vascular morphogenesis and homeostasis through the angiopoietin-Tie system. Nat Rev Mol Cell Biol 10:165–177
39. Gridley T (2010) Notch signaling in the vasculature. Curr Top Dev Biol 92:277–309
40. Karamysheva AF (2008) Mechanisms of angiogenesis. Biochemistry (Mosc) 73:751–762
41. Bussolino F, Ziche M, Wang JM et al (1991) In vitro and in vivo activation of endothelial cells by colony-stimulating factors. J Clin Invest 87:986–995
42. Distler JH, Hirth A, Kurowska-Stolarska M et al (2003) Angiogenic and angiostatic factors in the molecular control of angiogenesis. Q J Nucl Med 47:149–161
43. Vandervelde S, van Luyn MJ, Tio RA, Harmsen MC (2005) Signaling factors in stem cell-mediated repair of infarcted myocardium. J Mol Cell Cardiol 39:363–376
44. Cao Y (2009) Tumor angiogenesis and molecular targets for therapy. Front Biosci 14:3962–3973
45. Yamaguchi Y, Yoshikawa K (2001) Cutaneous wound healing: an update. J Dermatol 28:521–534
46. Gharaee-Kermani M, Phan SH (2001) Role of cytokines and cytokine therapy in wound healing and fibrotic diseases. Curr Pharm Des 7:1083–1103
47. Frantz S, Vincent KA, Feron O, Kelly RA (2005) Innate immunity and angiogenesis. Circ Res 96:15–26
48. Koutroubakis IE, Tsiolakidou G, Karmiris K, Kouroumalis EA (2006) Role of angiogenesis in inflammatory bowel disease. Inflamm Bowel Dis 12:515–523
49. Costa C, Incio J, Soares R (2007) Angiogenesis and chronic inflammation: cause or consequence? Angiogenesis 10:149–166
50. Grote K, Schuett H, Schieffer B (2011) Toll-like receptors in angiogenesis. ScientificWorldJournal 11:981–991
51. Leibovich SJ, Chen JF, Pinhal-Enfield G et al (2002) Synergistic up-regulation of vascular endothelial growth factor expression in murine macrophages by adenosine A(2A) receptor agonists and endotoxin. Am J Pathol 160:2231–2244
52. Pinhal-Enfield G, Ramanathan M, Hasko G et al (2003) An angiogenic switch in macrophages involving synergy between Toll-like receptors 2, 4, 7, and 9 and adenosine A(2A) receptors. Am J Pathol 163:711–721
53. Hara Y, Kuroda N, Inoue K, Sato T (2009) Up-regulation of vascular endothelial growth factor expression by adenosine through adenosine A2 receptors in the rat tongue treated with endotoxin. Arch Oral Biol 54:932–942
54. Pollet I, Opina CJ, Zimmerman C et al (2003) Bacterial lipopolysaccharide directly induces angiogenesis through TRAF6-mediated activation of NF-kappaB and c-Jun N-terminal kinase. Blood 102:1740–1742
55. McCord AM, Burgess AW, Whaley MJ, Anderson BE (2005) Interaction of Bartonella henselae with endothelial cells promotes monocyte/macrophage chemoattractant protein 1 gene expression and protein production and triggers monocyte migration. Infect Immun 73:5735–5742
56. Rodriguez-Martinez S, Cancino-Diaz ME, Miguel PS, Cancino-Diaz JC (2006) Lipopolysaccharide from Escherichia coli induces the expression of vascular endothelial growth factor via toll-like receptor 4 in human limbal fibroblasts. Exp Eye Res 83:1373–1377
57. Jagavelu K, Routray C, Shergill U et al (2010) Endothelial cell toll-like receptor 4 regulates fibrosis-associated angiogenesis in the liver. Hepatology 52:590–601
58. McDonald DM (2001) Angiogenesis and remodeling of airway vasculature in chronic inflammation. Am J Respir Crit Care Med 164:S39–S45

59. Grote K, Schuett H, Salguero G et al (2010) Toll-like receptor 2/6 stimulation promotes angiogenesis via GM-CSF as a potential strategy for immune defense and tissue regeneration. Blood 115:2543–2552
60. Varoga D, Paulsen F, Mentlein R et al (2006) TLR-2-mediated induction of vascular endothelial growth factor (VEGF) in cartilage in septic joint disease. J Pathol 210:315–324
61. Cho ML, Ju JH, Kim HR et al (2007) Toll-like receptor 2 ligand mediates the upregulation of angiogenic factor, vascular endothelial growth factor and interleukin-8/CXCL8 in human rheumatoid synovial fibroblasts. Immunol Lett 108:121–128
62. Chang YJ, Wu MS, Lin JT, Chen CC (2005) Helicobacter pylori-induced invasion and angiogenesis of gastric cells is mediated by cyclooxygenase-2 induction through TLR2/TLR9 and promoter regulation. J Immunol 175:8242–8252
63. Damiano V, Caputo R, Bianco R et al (2006) Novel toll-like receptor 9 agonist induces epidermal growth factor receptor (EGFR) inhibition and synergistic antitumor activity with EGFR inhibitors. Clin Cancer Res 12:577–583
64. Guo Z, Chen L, Zhu Y et al (2012) Double-stranded RNA-induced TLR3 activation inhibits angiogenesis and triggers apoptosis of human hepatocellular carcinoma cells. Oncol Rep 27:396–402
65. Bergé M, Bonnin P, Sulpice E et al (2010) Small interfering RNAs induce target-independent inhibition of tumor growth and vasculature remodeling in a mouse model of hepatocellular carcinoma. Am J Pathol 177:3192–3201
66. Spaner DE, Masellis A (2007) Toll-like receptor agonists in the treatment of chronic lymphocytic leukemia. Leukemia 21:53–60
67. Shingu K, Kruschinski C, Lührmann A et al (2003) Intratracheal macrophage-activating lipopeptide-2 reduces metastasis in the rat lung. Am J Respir Cell Mol Biol 28:316–321
68. Schneider C, Schmidt T, Ziske C et al (2004) Tumour suppression induced by the macrophage activating lipopeptide MALP-2 in an ultrasound guided pancreatic carcinoma mouse model. Gut 53:355–361
69. Schmidt J, Welsch T, Jäger D et al (2007) Intratumoural injection of the toll-like receptor-2/6 agonist 'macrophage-activating lipopeptide-2' in patients with pancreatic carcinoma: a phase I/II trial. Br J Cancer 97:598–604
70. Cammarota R, Bertolini V, Pennesi G et al (2010) The tumor microenvironment of colorectal cancer: stromal TLR4 expression as a potential prognostic marker. J Transl Med 8:112
71. Kutikhin AG (2011) Association of polymorphisms in TLR genes and in genes of the Toll-like receptor signaling pathway with cancer risk. Hum Immunol 72:1095–1116
72. van Beijnum JR, Buurman WA, Griffioen AW (2008) Convergence and amplification of toll-like receptor (TLR) and receptor for advanced glycation end products (RAGE) signaling pathways via high mobility group B1 (HMGB1). Angiogenesis 11:91–99
73. Lin Q, Yang XP, Fang D et al (2011) High-mobility group box-1 mediates toll-like receptor 4-dependent angiogenesis. Arterioscler Thromb Vasc Biol 31:1024–1032
74. van Beijnum JR, Nowak-Sliwinska P, van den Boezem E et al (2012) Tumor angiogenesis is enforced by autocrine regulation of high-mobility group box 1. Oncogene 32:363–374
75. West XZ, Malinin NL, Merkulova AA et al (2010) Oxidative stress induces angiogenesis by activating TLR2 with novel endogenous ligands. Nature 467:972–976
76. Wang XY, Sarkar D, Fisher PB (2011) Stress-sensing toll-like receptor as a driver of angiogenesis. Pigment Cell Melanoma Res 24:7–9
77. Ergul A, Alhusban A, Fagan SC (2012) Angiogenesis: a harmonized target for recovery after stroke. Stroke 43:2270–2274
78. Freedman SB, Vale P, Kalka C, Kearney M et al (2002) Plasma vascular endothelial growth factor (VEGF) levels after intramuscular and intramyocardial gene transfer of VEGF-1 plasmid DNA. Hum Gene Ther 13:1595–1603
79. Henry TD, Annex BH, McKendall GR et al (2003) The VIVA trial: vascular endothelial growth factor in ischemia for vascular angiogenesis. Circulation 107:1359–1365

Chapter 10
Role of Sex Steroids in Angiogenesis

Laura Lecce, Yuen Ting Lam, and Martin K.C. Ng

Abstract Sex steroids such as estrogen and testosterone are key mediators of angiogenesis. They are implicated in both physiological and pathological angiogenesis such as during the menstrual cycle, wound healing, and cancer growth and progression. Sex steroids regulate many aspects of angiogenesis through both classic genomic signaling pathways which regulate gene expression and also rapid action nongenomic pathways. In this capacity, sex steroids are able to modulate endothelial and progenitor cell functions such as proliferation, migration, and attachment, which are all essential components involved in neovascularization. Since sex steroids are known to augment angiogenesis which is vital to tumor progression and growth, common treatment of hormone-responsive tumors is through sex steroid receptor antagonists. Due to the involvement of sex steroids in necessary physiological functions as well as the potential to promote pathological angiogenesis, it is fundamental that the mechanisms behind sex steroid-mediated neovascularization are understood.

Keywords Angiogenesis • Androgens • Estrogen • Testosterone • Endothelial cells • Growth factors

L. Lecce • Y.T. Lam
Heart Research Institute, Sydney, Australia

M.K.C. Ng, M.B.B.S., Ph.D. (✉)
Heart Research Institute, Sydney, Australia

Department of Cardiology, Royal Prince Alfred Hospital, Missenden Road, Camperdown, Sydney, NSW 2050, Australia

Department of Medicine, University of Sydney, Sydney, Australia
e-mail: mkcng@med.usyd.edu.au

J.L. Mehta and N.S. Dhalla (eds.), *Biochemical Basis and Therapeutic Implications of Angiogenesis*, Advances in Biochemistry in Health and Disease 6,
DOI 10.1007/978-1-4614-5857-9_10,

Abbreviations

AR	Androgen receptor
ARE	Androgen response element
bFGFR	Basic fibroblast growth factor receptor
BSA	Bovine serum albumen
CDK	Cyclin-dependent kinases
CFU	Colony-forming unit
cGMP	Cyclic guanosine monophosphate
DHT	Dihydrotestosterone
EDC	Estrogen-dendrimer conjugates
EGF	Epidermal growth factor
eNOS	Endothelial nitric oxide synthase
EPCs	Endothelial progenitor cells
ER	Estrogen receptor
ERE	Estrogen response element
ERK	Extracellular signal-regulated kinases
FGF-2	Fibroblast growth factor-2
Flt1	VEGF receptor-1
Grb2	Growth factor receptor-bound protein 2
HIF-1	Hypoxia-inducible factor-1
HRE	Hormone-responsive elements
IL-1	Interleukin-1
IL-1Ra	IL-1 receptor antagonist
KDR/Flk-1	VEGF receptor-2
MAPK	Mitogen-activated protein kinases
mTOR	Mammalian target of rapamycin
NF-κB	Nuclear factor-κB
PAF	Platelet-activating factor
PDGF	Platelet-derived growth factor
RhoA	Ras homolog gene family member A
ROCK	Rho-associated protein kinase
SDF-1	Stromal-derived factor-1
SHC-1	SHC-transforming protein 1
Sos	Son of sevenless
TERT	Telomerase reverse transcriptase
TNF-α	Tumor necrosis factor-α
VEGF	Vascular endothelial growth factor

10.1 Introduction

There is accumulating evidence that sex steroids regulate key events in angiogenesis and vasculogenesis. In particular, there is considerable data for a direct effect of estrogens in the modulation of both physiological angiogenesis such as the recurrent hormone-regulated neovascularization of the female reproductive tract. Estrogens have also been implicated in pathological angiogenesis such as tumor neovascularization in the context of breast cancer. While far less studied than estrogen, emerging data also demonstrates a role for androgens on the regulation of angiogenesis. This chapter will discuss the role of sex steroids in angiogenesis and the mechanisms by which they function.

10.2 Sex Steroids and Angiogenesis: Basic Mechanisms

10.2.1 The Classic Sex Steroid Signaling Pathway

The classic sex steroid signaling pathway involves genomic signaling via steroid hormone binding to specific intracellular receptors which regulate gene expression and protein synthesis as illustrated in Fig. 10.1. Ligand-bound receptors become active through dimerization before translocation to the nucleus. Once in the nucleus, the hormone receptors interact with their specific hormone-responsive elements (HRE) present on DNA. This interaction causes the induction or repression of target genes through gene transcription which ultimately regulate many cellular functions. Sex steroids also act via rapid signaling pathways (nongenomic signaling) which involve sex steroid binding to specific membrane-associated receptors. This induces the modulation and subsequent initiation of signaling pathways which involve other membrane associated or cytoplasmic proteins.

10.2.2 Estrogen and Angiogenesis

Estrogen directly modulates the growth and survival of blood vessels in vivo. Estrogen administration in ovariectomized rats improves reendothelialization following endothelial injury [1]. In oophorectomized rabbits, the withdrawal of endogenous estrogen production causes vascular degeneration in the bladder, which is restored with exogenous estrogen administration [2]. Estrogen administration also enhances the recovery of blood perfusion after ischemic injury by increasing the capacity of capillary regeneration within the tissue [3]. Estrogen-enhanced blood vessel growth is accompanied by an increase in the production of hypoxia-inducible factor-1 (HIF-1) and vascular endothelial growth factor (VEGF) expression which are key inducers of angiogenesis. Estrogen-mediated neovascularization is accomplished by complex molecular pathways that regulate specific endothelial and progenitor cell functions which collectively enable new blood vessel growth.

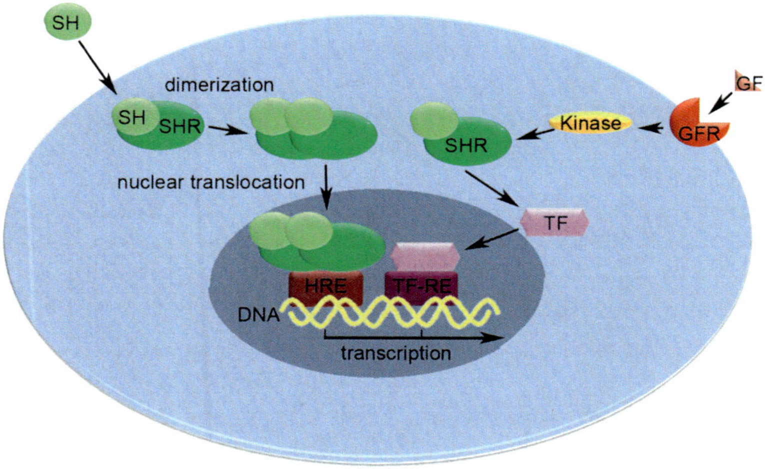

Fig. 10.1 Genomic actions of sex hormones. Sex hormones (SH) bind to sex hormone receptors (SHR) which translocate to the nucleus following dimer formation. The SHR binds to a hormone response element (HRE) within the promoter region of specific genes to initiate transcription. Transcription can also be initiated through growth factor (GF) binding to growth factor receptors (GFR) which activate specific kinases. These active kinases then lead to SHR activation of other transcription factors (TF) which bind to transcription factor response elements (TF-RE) causing gene transcription

10.2.2.1 Estrogens and Endothelial Cells

Human vascular endothelial cells express estrogen receptors (ERs) which enable endothelial cells to respond to estrogen [4, 5]. In this way, ERs function as transcription factors that initiate and regulate the production of many proangiogenic proteins [6]. Of the two known ERs, ERα is most often implicated in angiogenesis and plays a larger role in endothelial progenitor stimulation than ERβ [7]. In the absence of both receptors, the vascular protective mechanisms of estrogen are ineffective [8]. Estrogen regulates many facets of endothelial functions such as migration, attachment, and proliferation through specific factors listed in Table 10.1.

Through receptor binding, estrogen initiates an increase in the transcription and translation of many proangiogenic growth factors and accompanying receptors. These growth factors collectively enhance the angiogenic capacity of endothelial cells by promoting cell proliferation, migration, and attachment to extracellular matrix [9]. Estrogen also regulates many of the pathways that stimulate each of the events necessary for angiogenic stimulation and these pathways are summarized in Table 10.1.

Table 10.1 The role of estrogen in endothelial function

	Proliferation/cell survival	Migration/adhesion
Estrogen	*Increases*	*Increases*
	Cyclins	RhoA/ROCK/RhoGEF1
	Cyclin-dependent kinases	Notch1/Jagged1
	MAPK/ERK1, 2	VEGF/VEGFRs
	bFGF	Focal adhesions
	VEGF/VEGFRs	Focal adhesion kinase
	PI3K/Akt	LIM kinase
	HIF1α	Cofilin
	eNOS	eNOS
	Decreases	
	p27	

10.2.2.2 Estrogen and Endothelial Cell Migration and Attachment

Estrogen promotes endothelial cell migration and capillary formation by stimulating pathways that regulate changes in adhesion molecules and the actin cytoskeleton. Estrogen upregulates the mRNA and protein expression of integrins β1, α5, and α6 which allow endothelial cells to attach to extracellular matrix such as laminin and fibronectin [10]. Estrogen enhances integrin-mediated signaling via stimulation of focal adhesion kinase phosphorylation which occurs during estrogen-mediated cell attachment.

Estrogen also upregulates the expression and activity of Ras homolog gene family member A (RhoA), and RhoA-regulatory proteins such as RhoGEF1 [11]. RhoA is a small GTPase protein which regulates the actin cytoskeleton by activating Rho-associated protein kinase (ROCK), LIM kinase, and cofilin. These proteins allow endothelial cells to form stress fibers which function in cell spreading during migration and attachment. Increased endothelial tube formation and capillary stabilization by estrogen is also regulated through Notch1, Jagged1 pathway which ensure effective communication between adjacent cells [12].

10.2.2.3 Estrogen and Endothelial Cell Proliferation

Estrogen promotes increased cell proliferation in culture by causing a downregulation of genes that inhibit the cell cycle, such as p27, which specifically inhibits cyclin-dependent kinases (CDK) [11]. Estrogen-stimulated cell proliferation is further enhanced by increasing the genes that promote the cell cycle such as Cyclin-D1, Cyclin-A2, Cyclin-B1, CDK1, CDK2, and CDK4. Estrogen activation of the mitogen-activated protein kinases (MAPK), extracellular signal-regulated kinases

1 and 2 (ERK1/2), and phosphorylation of basic fibroblast growth factor receptor (bFGFR) promotes cell proliferation [13]. VEGF is also increased through the binding of estrogen to ERs, which activate the VEGF gene promoter via the estrogen response element (ERE) [6]. VEGF stimulates endothelial cell proliferation and migration mainly by binding VEGFR2 which activates the MAPK/ERK pathway and also regulates focal adhesion proteins and kinases.

10.2.3 Estrogen and Nongenomic Pathways

In addition to genomic signaling, increasing evidence suggests that steroid hormones such as estrogen exert a variety of rapid actions by via nongenomic signaling [14, 15]. By rapid signaling involving membrane-associated estrogen receptors, estrogen stimulates vasodilation as well as stimulates endothelial cell growth, migration, and survival [16].

Through the use of membrane-impermeable estrogen conjugates, estrogen-BSA (bovine serum albumen), or estrogen-dendrimer conjugates (EDC), estrogen has been shown to bind membrane-associated receptors, most likely a form of ERα, which initiate fast acting intracellular pathways. Via this mechanism, estrogen stimulates the PI3-kinase/Akt pathway which includes Akt phosphorylation and PI3-kinase signaling events. This is achieved through the binding of ERs with Gαi, c-Src (cellular-src) tyrosine kinase, and P85 (a regulatory subunit of PI3-kinase). Through this pathway endothelial nitric oxide synthase (eNOS) is also activated which subsequently leads to rapid nitric oxide release from endothelial cells as shown in Fig. 10.2 [17, 18]. Via the same ER membrane-binding mechanism, MAPK pathways are also rapidly activated, which corresponds to an increase in cyclic guanosine monophosphate (cGMP) [19]. Both pathways stimulate the activation of eNOS resulting in the production and release of nitric oxide. Nitric oxide is known to be vascular protective by controlling vascular tone and vasodilation but also stimulates angiogenesis through cell growth, proliferation, and migration of endothelial cells.

Interestingly, membrane-impermeable EDC does not stimulate the growth of Ishikawa cells (a uterine endometrial carcinoma cell line) or MCF-7 breast cancer cells, which are normally stimulated by membrane-permeable estrogen [20]. This implicates membrane-impermeable estrogen signaling in vascular protection, which may not stimulate estrogen-responsive cancers, making selective activation of membrane-associated estrogen receptors a potentially attractive therapeutic target.

10.2.4 Estrogen and Endothelial Progenitor Cells

Endothelial progenitor cells (EPCs) are directly involved in vascular repair by facilitating reendothelialization and angiogenesis [21–23]. In women, there is a correlation between an increased plasma concentration of estrogen and increased levels of

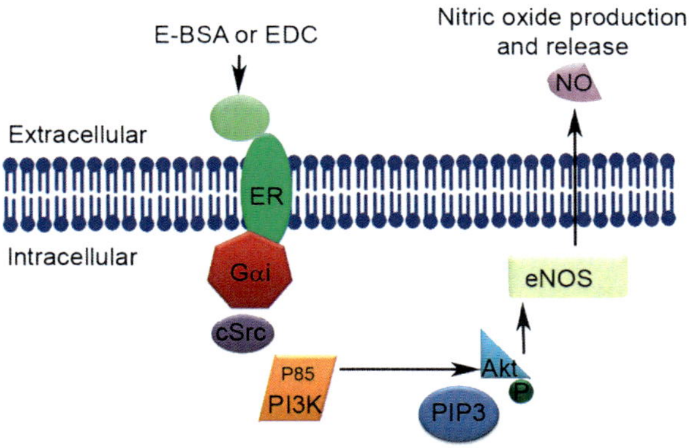

Fig. 10.2 Nongenomic estrogen signaling. Estrogen binding to membrane-associated estrogen receptors (ER) activates the PI-3 kinase pathway though Gαi, Src tyrosine kinase, and P85 subunit binding. This leads to the phosphorylation of Akt and activation of PIP3 which increases nitric oxide (NO) production through endothelial nitric oxide synthase (eNOS)

circulating EPCs [24]. Male EPCs only contain ERα, whereas female EPCs express both receptors, even so, both male and female EPCs are responsive to estrogen [25].

Similar to its effects in differentiated endothelial cells, estrogen directly stimulates the key angiogenic activities of EPCs including migration, proliferation, and tubulogenesis as outlined in Fig. 10.3 [26, 27]. The effects are mediated by both ERα and ERβ with a stronger contribution by ERα, at least in part, due to its higher expression in EPCs. Estrogen also prolongs the lifespan of EPCs in culture through a number of mechanisms. Estrogen decreases the rate of cell apoptosis by decreasing caspase-8 activity. The senescence of EPCs is also delayed by an estrogen-mediated increase in telomerase activity, which increases telomerase reverse transcriptase (TERT) and consequently increases telomere length [28]. This is all achieved through estrogen binding to ERs and the consequent activation of the PI3-kinase/Akt pathway and Akt phosphorylation [29]. This pathway is also involved in producing estrogen-enhanced cell proliferation and migration of EPCs, which are necessary processes in facilitating EPC-mediated angiogenesis.

Estrogen also stimulates EPC mobilization from bone marrow and EPC homing to sites of vascular injury or ischemia. In this manner, estrogen enhances EPC-mediated reendothelialization in murine models of carotid artery denudation model [24, 30] and enhances recovery after murine myocardial infarction by augmentation of EPC incorporation into sites of ischemia-mediated neovascularization [31]. Estrogen has been found to enhance myocardial expression of stromal-derived factor-1 (SDF-1), a critical regulator of EPC homing to sites of ischemia, following rat myocardial infarction [32]. Consistent with estrogen effects on intrinsic EPC angiogenic activities, the estrogen contribution to EPC mobilization and homing is mediated by both ERα and ERβ, with a greater role for ERα [26]. Estrogen-mediated augmentation of EPC mobilization and homing after arterial injury or in response to ischemia is eNOS-dependent and has been noted to be absent in eNOS null mice [30].

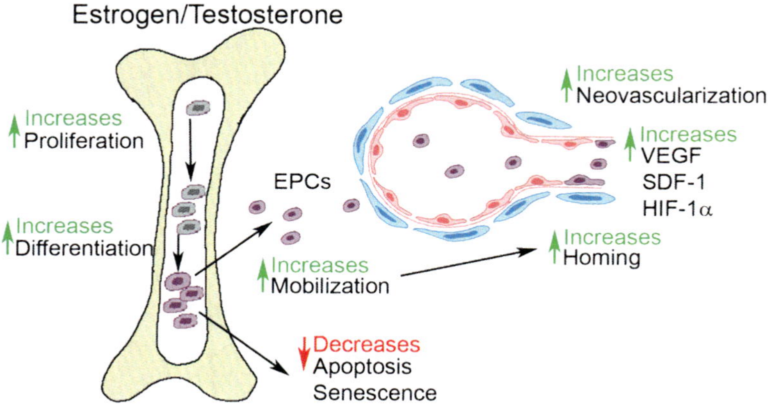

Fig. 10.3 The role of sex steroids in endothelial progenitor cell (EPC) functions. EPCs are stimulated by sex steroids which lead to an increase in EPC proliferation, differentiation, mobilization, and homing, resulting in increased neovascularization. This is further enhanced through hormone-mediated upregulation of HIF-1α and subsequent proangiogenic factors such as VEGF, SDF-1. Sex steroids also decrease apoptosis and senescence of EPCs

10.2.5 Androgens and Angiogenesis

By comparison to estrogens, the role of androgens in angiogenesis has been much less studied. Dihydrotestosterone (DHT), a potent natural androgen that is non-aromatizable to estrogen, stimulates key angiogenic processes in male endothelial cells including cell migration, proliferation, and tubulogenesis in an androgen receptor (AR)-dependent manner [33, 34]. Androgen effects on cell migration and tubulogenesis are more striking than those on proliferation. Interestingly, the effects of androgens on angiogenesis appear to be sex specific, in that DHT exposure has no effect on cell migration, proliferation, or tubulogenesis by female-donor cells [34, 35]. This sex specificity in androgen action is due, in large part, to a higher level of androgen receptor expression in male-donor endothelial cells [36]. Consistent with this, overexpression of the AR in female-donor endothelial cells confers a degree of sensitivity to the proangiogenic effects of androgens in vitro.

DHT exposure increases the mRNA expression of the AR, which is found intracellularly. Like ERs, once ARs are ligand bound they translocate to the nucleus to initiate transcription via an androgen response element (ARE) [37]. Using this mechanism, androgens increase cell proliferation through an upregulation of cyclins A and D1, which promote the cell cycle by activating CDK [33]. DHT also increases mRNA and protein secretion of VEGF, in addition to an increase in VEGFR1 (Flt-1) and VEGFR2 (KDR/Flk1) mRNA expression [33, 38]. VEGFR2 is mainly expressed on the surface of endothelial cells, which when bound to VEGF increases DHT-mediated tubulogenesis through the PI3-kinase signaling pathway. Androgens also have the capacity for rapid membrane signaling by binding a membrane-associated AR which, like estrogen, initiates a complex with Gαi, and this interaction is increased by DHT [17].

In keeping with cellular data, endogenous androgens regulate angiogenesis in vivo in a sex-specific fashion. Male murine castration results in markedly attenuated vascularization of matrigel plugs, a finding that is reversed by androgen replacement with DHT [38]. This corresponds with decreased expression of angiogenic mediators such as HIF1α, SDF-1, and KDR mRNA in ischemic tissue of castrated mice, which are increased with DHT administration. By contrast DHT has no effect on the reduced vascularization of matrigel plugs of ovariectomized female mice. Rats fitted with a human skin graft and treated with topical application of testosterone also showed improved angiogenesis with increased vascular density in the dermis of the skin graft compared with placebo-treated rats [39].

10.2.6 Androgens and Endothelial Progenitor Cells

The exact mechanisms behind androgen-mediated angiogenesis are still largely unknown; however, emerging data suggests that androgens promote EPC-mediated angiogenesis as seen in Fig. 10.3. EPCs respond to androgen treatment by translocation of the AR to the nucleus [40]. In vivo, DHT treatment after ischemia increases the number of EPCs and CFU progenitors in the bone marrow, corresponding with increased blood flow recovery and capillary density at the site of ischemia [34].

10.2.7 Sex Specificity of Steroid Actions

There are striking sex differences regarding the incidence of cardiovascular disease, atherogenesis, angiogenesis, and cardiovascular adaptation and repair in response to ischemia and infarcts [41–44]. These differences have lead researchers to explore the impact of sex steroids on the regulation and function of the cardiovascular system. In both male and female mice gonadectomy severely impairs neovascularization in vivo, findings which indicate an important role for endogenous sex steroids in regulation of angiogenic processes [34]. Further, gonadectomized male and female mice treated with exogenous DHT or estrogen, respectively, improve angiogenesis following hind limb ischemic injury. Interestingly, ovariectomized female mice do not benefit from exogenous DHT treatment following ischemia [38]. In contrast to this, male mice receiving estrogen treatment recover faster from hind limb ischemic injury than placebo-treated mice through progenitor cell recruitment [45].

Gender-specific differences in sex hormone-mediated angiogenesis are due at least in part to differences in sex steroid receptor expression in vascular tissues. Both male and female mice exhibit high binding affinity to estrogen in aortic tissue, and surprisingly male mice contain twice as many binding sites for estrogen than females, which is independent of the fact that females have higher circulating estrogen levels [46]. In male mice, ERα mRNA is expressed at low levels in vascular endothelial and smooth muscle cells following aortic denudation injury, whereas ERβ mRNA is greatly increased on endothelial cells by this procedure [47]. These studies demonstrate that male vasculature has the ability to respond to estrogen stimulus. In contrast to this, female-donor human endothelial cells express much lower androgen receptor levels compared with male cells which have a 2–5-fold higher androgen receptor expression level than female cells [34, 36]. The same is true for male-donor monocyte-derived macrophages which have been noted to express significantly higher levels of androgen receptors than those from female donors [48]. In macrophages these differences in androgen receptor expression translate to marked sex differences in transcriptional responses to androgen exposure, with male-donor macrophages exhibiting upregulation of genes in multiple processes including angiogenesis while genomic responses in female-donor cells to androgens are much less striking [49]. These studies collectively demonstrate the sex specificity of sex steroid function and cellular responses with particular relevance for sex-specific regulation of angiogenic processes by androgens.

10.3 Sex Steroids and Angiogenesis: Role in Health and Disease

10.3.1 Estrogen and Menstruation

Blood vessel growth and regression is a major part of the ovarian and menstrual cycles which are regulated by the female hormones estrogen and progesterone [50–52]. An increase in estrogen production also accompanies the development of the placenta during pregnancy, establishing an extensive vascular network [53]. These processes are mediated by complex signaling and the production of numerous growth factors and angiogenesis-promoting proteins as demonstrated in Fig. 10.4.

Estrogen regulates angiogenesis in the endometrium by influencing the proliferation and stimulation of many endometrial cells including uterine endothelial and epithelial cells, as well as stromal cells, and smooth muscle cells associated with the vascular wall [54]. These uterine cells express estrogen receptors which fluctuate in expression levels throughout the menstrual cycle and thereby allowing vascular growth and regression [55]. Estrogen stimulation of isolated endometrial vascular cells with both ERs enhance the expression of VEGF and VEGFR2, leading to enhanced cell proliferation [56]. Human endometrial endothelial cells are more sensitive to VEGF stimulation than human dermal, coronary, and umbilical endothelial

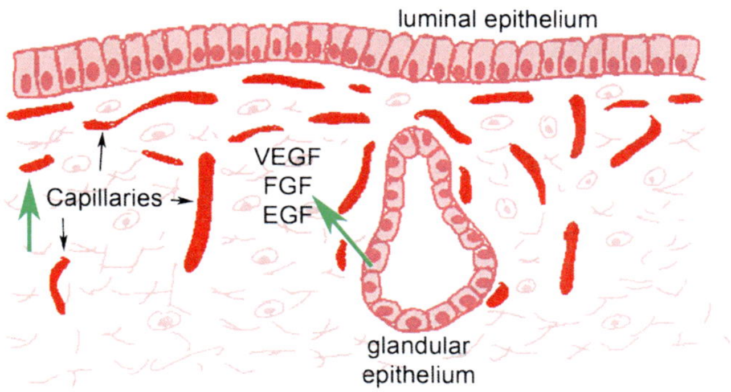

Fig. 10.4 Estrogen in endometrial angiogenesis. During specific phases of the menstrual cycle, estrogen stimulates the growth of uterine blood vessels and capillaries through glandular epithelial cell secretion of growth factors such as VEGF, FGF, and EGF

cells, which results in faster endometrial endothelial cell proliferation and angiogenic capacity [57]. This may be attributed to that fact that uterine endometrial cells having higher mRNA expression levels of the VEGF receptors, KDR, and Flt-1 [58]. VEGF also increases endometrial endothelial cell tubulogenesis and transcriptome changes which are associated with angiogenesis. The production and secretion of VEGF in the endometrium are largely attributed to the glandular epithelial cells, which produce increased amounts of VEGF with increased estrogen stimulation [59]. VEGF then exerts its proangiogenic effects by binding the VEGF receptors present on endometrial endothelial cells, causing increased proliferation and blood vessel growth in the endometrium. VEGF production is also augmented by the production of HIF-1α by luminal epithelial cells in response to estrogen stimulation and binding with ERα, which relies on estrogen activation the PI3-kinase/Akt pathway [60, 61].

Along with VEGF, fibroblast growth factor-2 (FGF-2), epidermal growth factor (EGF), and platelet-derived growth factor (PDGF) are also produced in the normal human endometrium and also influence endometrial angiogenesis [62]. Their receptors VEGFR1/2, FGFR2 and EGFR, and PDGFR are all expressed within and close to endometrial blood vessels with the strongest expression coinciding with the start of the secretory phase, which is when subepithelial capillary density increases.

10.3.2 Sex Steroids and Tumor Angiogenesis

Sex steroids promote angiogenesis through cell proliferation, migration, and capillary formation, which are all key factors in tumor growth, progression, and metastasis. It is therefore paramount that the role of sex steroids in the establishment and outcome of cancer is well understood. The roles of estrogen and testosterone in tumorigenesis

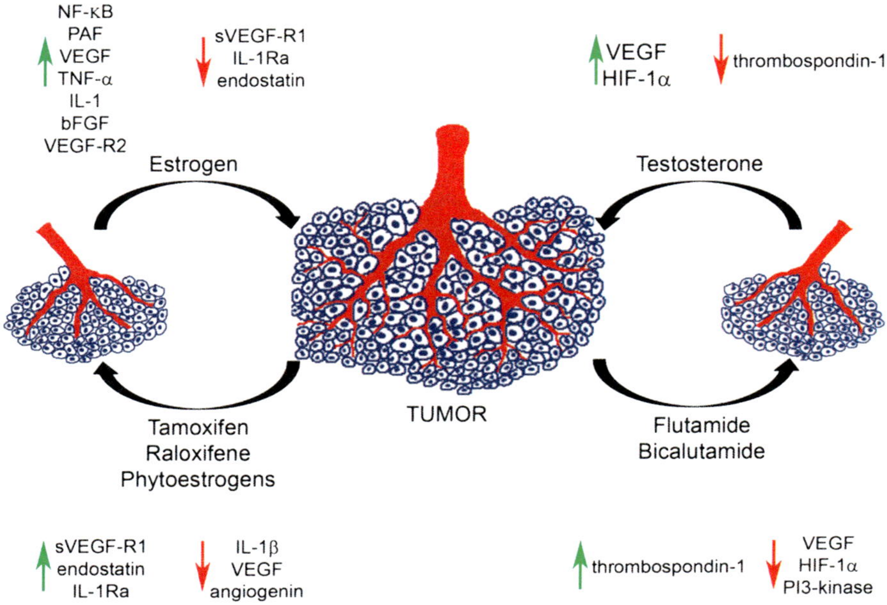

Fig. 10.5 The involvement of sex steroids in tumor growth and angiogenesis. Estrogen and testosterone both stimulate the growth of hormone-specific tumors such as breast and prostate, respectively, by increasing tumor angiogenesis. Tumor angiogenesis is augmented through hormone-initiated upregulation of proangiogenic mediators and growth factors, in addition to the downregulation of anti-angiogenic factors. Treatment of hormone-responsive tumors with antiestrogens or anti-androgens suppress tumor growth by a reduction of tumor angiogenesis

have been mostly studied in the context of breast and prostate cancer, respectively. However, they are may also be implicated in the progression of tumors in other endocrine organs such as in the uterus, ovaries [63], pituitary, adrenal glands, thyroid, and parathyroid [64]. Figure 10.5 details the association of sex steroids in tumor growth and angiogenesis and also includes common treatments that target hormone-mediated angiogenesis which restricts tumor growth and metastasis.

10.3.2.1 Estrogen and Tumor Angiogenesis

The level of tumor angiogenesis has been suggested as an indicator of relapse risk and overall survival of node-negative and node-positive breast cancers [65, 66], demonstrating that higher vessel density indicates a significantly increased risk of relapse and death. Patients treated with tamoxifen, an antiestrogen, have the best prognosis when tumors are ER-positive and exhibit low vascularization [67]. The risk of relapse also correlates with an increased secretion of VEGF from tumor cells which induces microvessel proliferation [66]. Ovariectomized Fischer 344 rats which develop spontaneous pituitary tumors when exposed to estrogen show an induction of neovascularization corresponding with growth and enlargement of tumor blood vessels [68]. This vessel growth is stimulated by an elevated production of VEGF by both endothelial and non-endothelial tumor cells. The production of VEGF is accompanied by elevated levels of VEGFR-2 on tumor endothelial cells which respond by increased tumor angiogenesis. Similarly, in MCF-7 breast cancer cells, treatment with estrogen also increases VEGF secretion, which leads to an increased expression of VEGFR-2 on HUVECs cultured with breast cancer conditioned medium [69]. In addition to the promotion of angiogenic factors, estrogen also functions to reduce negative regulators of VEGF-mediated angiogenesis such as soluble VEGFR-1 production [69, 70]. Estrogen treatment of cultured breast cancer cells downregulates the expression of VEGFR-1 in ER-positive but not ER-negative breast cancer cells. This effect is blocked with the pretreatment of cells with an ER antagonist. In matrigel plugs containing breast cancer cells given to mice with an estrogen or placebo implant, decreased VEGFR-1 expression in estrogen-treated mice was accompanied by a significant increase in angiogenesis [70]. Estrogen receptors are also implicated in controlling tumor growth and, more specifically, tumor angiogenesis. ERα expressing tumor cells increase proliferation and tumor growth when exposed to estrogen [71]. However, when tumor cells are forced to express ERβ, this is associated with reduction in tumor volume in estrogen-treated mice which is associated with reduced microvessel density and decreased expression of angiogenic factors such as VEGF and PDGF-β.

In addition to increased VEGF production, estrogen enhances expression of many other proangiogenic factors including tumor necrosis factor-α (TNF-α), interleukin-1 (IL-1), and bFGF [72]. The expression of these factors is mediated through the activation of nuclear factor (NF)-κB, via platelet-activating factor (PAF), which leads to increased angiogenesis. Importantly, long-term estrogen deprivation, such as in postmenopausal women, causes breast tumors to develop a heightened sensitivity to estrogen [73]. These cells adapt through an upregulation of ERα at the membrane, which activates growth factor pathways via SHC-transforming protein-1 (SHC1), growth factor receptor-bound protein-2 (Grb-2), and son of sevenless homolog-1 (Sos-1). Estrogen deprivation also enhances the upregulation of many proangiogenic pathways and cell proliferation growth factors including MAP-kinase, PI3-kinase, and mammalian target of rapamycin (mTOR).

Tamoxifen and raloxifene are the most commonly used ER-antagonists to treat hormone-responsive breast cancers, despite having agonistic effects on other target tissues such as the endometrium [74]. Antiestrogens, including tamoxifen, nafoxidine, and clomiphene, inhibit angiogenesis in the chorioallantoic membrane in chick embryos [75], demonstrating that antiestrogens are effective inhibitors of angiogenesis and may be beneficial in inhibiting tumor growth. The prevention of angiogenesis is a known therapeutic target to block tumor growth, and tamoxifen has been shown to be effective in inhibiting and reducing angiogenesis [76].

Tamoxifen directly reduces endothelial cell proliferation in culture and when given orally to rats with a matrigel plug containing VEGF, impairs endothelial cell proliferation and migration in vivo [77]. Tamoxifen also reduces the effects of estrogen on tumor angiogenesis by decreasing extracellular VEGF released by tumor cells which results in decreased tumor vessel area and overall tumor vasculature [78]. Exposure to tamoxifen also inhibits the estrogen-mediated reduction of sVEGFR-1, which functions as a negative regulator of VEGF stimulated angiogenesis [69]. Tamoxifen reduces other angiogenesis-promoting factors such as angiogenin which is secreted by breast cancer cells in response to estrogen stimulation and found to be higher in breast tumors than surrounding normal tissue [79]. In addition, tamoxifen treated tumor-bearing mice also showed decreased levels of angiogenin production with a corresponding decrease in angiogenesis [79]. In women, higher estradiol levels correspond with higher breast extracellular VEGF and angiogenin levels and lower levels of endostatin. With tamoxifen treatment, VEGF and angiogenin levels were decreased and endostatin increased, demonstrating tamoxifen is successful in promoting an anti-angiogenic state [80].

Tamoxifen, along with phytoestrogens (dietary estrogens) such as flaxseed and enterolactone, is able to decrease tumor microvessel density with a decrease in pro-angiogenic factor IL-1β [81]. Additionally, these treatments increased IL-1 receptor antagonist (IL-1Ra) levels which reduces endothelial cell proliferation. In contrast, estrogen reduces IL-1Ra in culture, thereby promoting endothelial proliferation [81]. The phytoestrogens, flaxseed, enterolactone, and enterodiol are also able to counteract estrogen-induced tumor growth and angiogenesis in ovariectomized mice injected with human breast cancer cells [82]. This was achieved through a phytoestrogen-mediated decrease in estradiol-induced VEGF secretion from cancer cells and counteracting estrogen-enhanced VEGFR-2 expression in HUVECs.

Although most research involving tamoxifen use is limited to ER-positive tumors which respond to estrogen stimulation, tamoxifen also exhibits anti-angiogenic effects in ER-negative fibrosarcoma tumors [83]. It was found to decrease tumor vessel formation in ovariectomized rats, promote less vascular sprouting from rat aortic rings, and decrease vascular length and vessel area in corneal pocket assays. In contrast estrogen treatment exhibited the opposite effect of increasing angiogenesis in each of these models. The most striking finding being that the anti-angiogenic effects of tamoxifen may be partly ER-independent.

Recently, a study investigating the effects of estrogen on ERα and ERβ-negative breast cancers found that estrogen was able to promote tumor growth and angiogenesis without directly influencing the ER-negative tumor cells [84]. These tumors in mice treated with estrogen exhibited increased matrix deposition, stromal condensation, and angiogenesis through an estrogen-mediated recruitment of hematopoietic and smooth muscle lineages such as macrophages to the tumor site. Estrogen treatment resulted in an upregulation of ERs on bone marrow-derived cells, which decreased from the marrow in WT and ERβ-knockout mice but not ERα-knockout mice. This suggests that ERα on bone marrow cells is necessary for cells to mobilize from the bone marrow to the tumor and participate in neovascularization. Further evidence for this is that estrogen-treated mice with ERα-knockout bone marrow failed to grow tumors. This study does, however, implicate an indirect effect of estrogen in promoting estrogen-independent breast cancers by enhancing tumor angiogenesis.

10.3.2.2 Testosterone and Tumor Angiogenesis

The pathogenesis of prostate cancer is highly dependent on the presence of androgens. This is highlighted by the fact that men who undergo castration before puberty, or men with 5α-reductase deficiency, have never been reported to develop prostate cancer [85]. However, among men with newly diagnosed prostate cancer, low serum testosterone levels are associated with a higher Gleason score, higher androgen receptor density, and higher microvessel density within the tumor which is associated with a higher metastatic potential [86]. This may be due to low serum testosterone levels corresponding with a shorter CAG repeat length within the first exon of the AR gene. A shorter CAG repeat length results in greater transactivation function of the AR and corresponds to a higher risk and earlier onset of prostate cancer [87–90]. In addition, despite the age-related decline in testosterone production in men, total DHT concentrations do not change with age and may be the main contributor to prostate cancer progression [91, 92].

Androgen ablation is the main therapy for prostate cancer as initial tumor growth is androgen-dependent [93]. Androgens directly stimulate proliferation and inhibit apoptosis of prostate cancer cells but also largely regulate prostate tumor growth by promoting angiogenesis. In humans, low-grade carcinomas with lower Gleason scores strongly express thrombospondin-1 and have few blood vessels, whereas high-grade carcinomas show no thrombospondin-1 staining and a higher number of blood vessels [94]. Patients which undergo androgen-ablation therapy show an increase in thrombospondin-1 expression and a decrease in tumor microvessel density. Similarly in rats, androgen withdrawal via castration increases thrombospondin-1 synthesis and decreases vascularization of the normal prostate, which is reversed with androgen replacement [94]. In castrated rats, prostatic endothelial cell proliferation rate, weight of total blood vessels and endothelial cell numbers decrease in addition to a decrease in epithelial cells and total organ weight [95]. Following testosterone treatment, endothelial cell proliferation rate, number, and vessel weight normalize, followed later by an increase in epithelial and total organ weight. This shows that androgen-stimulated angiogenesis within the prostate precedes the overall increased growth of the organ. A study which implanted mice with androgen-dependent Shionogi tumors showed that androgen withdrawal following castration lead to blood vessel regression from the tumor periphery within 24 h, followed by tumor regression after 1–2 days [96]. This corresponded to TUNEL-positive apoptotic endothelial cells on day 1 after castration followed by additional TUNEL-positive neoplastic cells on day 2. Blood vessel regression caused by castration was accompanied by decreased vessel diameter and less tortuous vessels, along with decreased vascular permeability and leukocyte rolling. However, erythrocyte velocity was not altered by androgen withdrawal suggesting that the death of tumor cells is due to lack of androgens rather than reduced tissue perfusion.

Much of the effects of androgens on prostate angiogenesis are attributed to the ability of androgens to regulate VEGF production. Androgens upregulate VEGF mRNA and VEGF secretions from normal prostatic cell lines as well as prostatic tumor cell lines [97, 98]. In rats given testosterone injections, VEGF activity in the prostate increased with an increase in ventral lobe weight [98]. Castrated rats show a reduction in VEGF mRNA and protein levels in the normal prostate which is increased upon exogenous androgen replacement [99]. Androgen ablation in rats also causes a reduction in both human and rat androgen-sensitive tumor growth rate with a reduction in VEGF levels [97]. Similarly in mice with androgen-sensitive LNCaP tumors, castration stopped subsequent tumor growth which continued in tumors implanted into sham and control mice [100]. This was accompanied by a reduction in baseline VEGF levels and decreased microvessel density, followed by a decrease in overall tumor size. The reduction in VEGF levels following androgen deprivation is accomplished through a destabilization of VEGF mRNA transcripts [100]. The reduced expression of VEGF with androgen withdrawal was seen in androgen-responsive LNCaP prostate cancer cells, however is not observed in hormone-independent PC3 cells. The mechanism behind androgen-mediated upregulation of VEGF is likely through the activation of HIF-1α which occurs in LNCaP but not PC3 cells [101]. Androgens increase HIF-1α protein through a translation-dependent pathway which activates the subsequent production of VEGF. Nonsteroidal anti-androgens such as flutamide and bicalutamide inhibit the ability of DHT to upregulate HIF-1α, as does inhibiting the PI3-kinase pathway. Similarly, in humans there is correlation between HIF levels, VEGF expression, and AR expression in prostate cancers [102].

Even though early stage prostate cancers begin through androgen driven growth and survival, inevitably over time the cancers become androgen insensitive and grow independent of androgen availability [93]. The change between androgen-dependent to androgen-independent tumors is demonstrated in mice given androgen-sensitive Shionogi tumors and subsequently castrated [96]. These mice initially show tumor and vessel regression with decreased VEGF transcription, but a subsequent regrowth of tumors after 2 weeks with a new upregulation VEGF expression. In a similar study in mice using a CWR22Rv1, human androgen-responsive but androgen-independent tumor model. Castration in mice with this tumor also showed initial decreased VEGF expression, decreased cell proliferation, and a decrease in microvessel density [103]. Following this, recurrent cancers were found to have a significantly higher expression of ARs. This indicates that androgen-independent tumors continue to maintain a functional androgen signaling pathway which may be activated through other growth factors and cytokines which may continue to regulate tumor growth and angiogenesis following androgen withdrawal. Considering these findings, mice with androgen-dependent LNCaP tumors treated with both androgen ablation concurrently with a VEGF-receptor tyrosine kinase inhibitor showed the greatest decrease in tumor regrowth and volume with an increase in tumor necrosis compared to androgen withdrawal alone [104]. This demonstrates that a combined therapeutic treatment of prostate cancer using androgen withdrawal or anti-androgens combined with anti-angiogenic mediators may provide the best outcome.

10.4 Conclusion

It is evident that sex steroids are fundamentally involved in the regulation of key angiogenic processes. This chapter has detailed the known mechanisms by which sex steroids mediate basic endothelial and progenitor cell functions that are necessary in both physiological and pathological neovascularization. Given the complex nature of sex steroid-mediated angiogenesis, further research is necessary in understanding the pathways which contribute to angiogenesis in both health and disease. This will lead to improved treatment of hormone-responsive cancers, as well as the potential to utilize sex steroids in promoting angiogenesis following injury such as during wound healing.

References

1. Krasinski K, Spyridopoulos I, Asahara T et al (1997) Estradiol accelerates functional endothelial recovery after arterial injury. Circulation 95:1768–1772
2. Lin AD-Y, Mannikarottu A, Kogan BA et al (2006) Estrogen induces angiogenesis of the female rabbit bladder. J Endocrinol 190:241–246
3. Kyriakides ZS, Petinakis P, Kaklamanis L et al (2001) Intramuscular administration of estrogen may promote angiogenesis and perfusion in a rabbit model of chronic limb ischemia. Cardiovasc Res 49:626–633
4. Kim-Schulze S, McGowan KA, Hubchak SC et al (1996) Expression of an estrogen receptor by human coronary artery and umbilical vein endothelial cells. Circulation 94:1402–1407
5. Venkov CD, Rankin AB, Vaughan DE (1996) Identification of authentic estrogen receptor in cultured endothelial cells. A potential mechanism for steroid hormone regulation of endothelial function. Circulation 94:727–733
6. Mueller MD, Vigne JL, Minchenko A et al (2000) Regulation of vascular endothelial growth factor (VEGF) gene transcription by estrogen receptors alpha and beta. Proc Natl Acad Sci USA 97:10972–10977
7. Brouchet L, Krust A, Dupont S et al (2001) Estradiol accelerates reendothelialization in mouse carotid artery through estrogen receptor-alpha but not estrogen receptor-beta. Circulation 103:423–428
8. Karas RH, Schulten H, Pare G et al (2001) Effects of estrogen on the vascular injury response in estrogen receptor alpha, beta (double) knockout mice. Circ Res 89:534–539
9. Morales DE, McGowan KA, Grant DS et al (1995) Estrogen promotes angiogenic activity in human umbilical vein endothelial cells in vitro and in a murine model. Circulation 91:755–763
10. Cid MC, Esparza J, Schnaper HW et al (1999) Estradiol enhances endothelial cell interactions with extracellular matrix proteins via an increase in integrin expression and function. Angiogenesis 3:271–280
11. Oviedo PJ, Sobrino A, Laguna-Fernandez A et al (2011) Estradiol induces endothelial cell migration and proliferation through estrogen receptor-enhanced RhoA/ROCK pathway. Mol Cell Endocrinol 335:96–103
12. Soares R, Balogh G, Guo S et al (2004) Evidence for the notch signaling pathway on the role of estrogen in angiogenesis. Mol Endocrinol 18:2333–2343
13. Kim-Schulze S, Lowe WL Jr, Schnaper HW (1998) Estrogen stimulates delayed mitogen-activated protein kinase activity in human endothelial cells via an autocrine loop that involves basic fibroblast growth factor. Circulation 98:413–421

14. Chow RWY, Handelsman DJ, Ng MKC (2010) Minireview: rapid actions of sex steroids in the endothelium. Endocrinology 151:2411–2422
15. Simoncini T, Genazzani AR (2003) Non-genomic actions of sex steroid hormones. Eur J Endocrinol 148:281–292
16. Simoncini T, Mannella P, Genazzani AR (2006) Rapid estrogen actions in the cardiovascular system. Ann N Y Acad Sci 1089:424–430
17. Kumar P, Wu Q, Chambliss KL et al (2007) Direct interactions with G alpha i and G beta-gamma mediate nongenomic signaling by estrogen receptor alpha. Mol Endocrinol 21:1370–1380
18. Haynes MP, Li L, Sinha D et al (2003) Src kinase mediates phosphatidylinositol 3-kinase/Akt-dependent rapid endothelial nitric-oxide synthase activation by estrogen. J Biol Chem 278:2118–2123
19. Russell KS, Haynes MP, Sinha D, Clerisme E, Bender JR (2000) Human vascular endothelial cells contain membrane binding sites for estradiol, which mediate rapid intracellular signaling. Proc Natl Acad Sci U S A 97:5930–5935
20. Chambliss KL, Wu Q, Oltmann S et al (2010) Non-nuclear estrogen receptor alpha signaling promotes cardiovascular protection but not uterine or breast cancer growth in mice. J Clin Invest 120:2319–2330
21. Urbich C, Dimmeler S (2004) Endothelial progenitor cells: characterization and role in vascular biology. Circ Res 95:343–353
22. Shantsila E, Watson T, Tse H-F, Lip GYH (2008) New insights on endothelial progenitor cell subpopulations and their angiogenic properties. J Am Coll Cardiol 51:669–671
23. Eguchi M, Masuda H, Asahara T (2007) Endothelial progenitor cells for postnatal vasculogenesis. Clin Exp Nephrol 11:18–25
24. Strehlow K, Werner N, Berweiler J et al (2003) Estrogen increases bone marrow-derived endothelial progenitor cell production and diminishes neointima formation. Circulation 107:3059–3065
25. Foresta C, Zuccarello D, Biagioli A et al (2007) Oestrogen stimulates endothelial progenitor cells via oestrogen receptor-alpha. Clin Endocrinol 67:520–525
26. Hamada H, Kim MK, Iwakura A et al (2006) Estrogen receptors alpha and beta mediate contribution of bone marrow-derived endothelial progenitor cells to functional recovery after myocardial infarction. Circulation 114:2261–2270
27. Baruscotti I, Barchiesi F, Jackson EK et al (2010) Estradiol stimulates capillary formation by human endothelial progenitor cells: role of estrogen receptor-{alpha}/{beta}, heme oxygenase 1, and tyrosine kinase. Hypertension 56:397–404
28. Imanishi T, Hano T, Nishio I (2005) Estrogen reduces endothelial progenitor cell senescence through augmentation of telomerase activity. J Hypertens 23:1699–1706
29. Zhao X, Huang L, Yin Y et al (2008) Estrogen induces endothelial progenitor cells proliferation and migration by estrogen receptors and PI3K-dependent pathways. Microvasc Res 75:45–52
30. Iwakura A, Luedemann C, Shastry S et al (2003) Estrogen-mediated, endothelial nitric oxide synthase-dependent mobilization of bone marrow-derived endothelial progenitor cells contributes to reendothelialization after arterial injury. Circulation 108:3115–3121
31. Iwakura A, Shastry S, Luedemann C et al (2006) Estradiol enhances recovery after myocardial infarction by augmenting incorporation of bone marrow-derived endothelial progenitor cells into sites of ischemia-induced neovascularization via endothelial nitric oxide synthase-mediated activation of matrix metalloproteinase-9. Circulation 113:1605–1614
32. Chen Y, Jin X, Zeng Z et al (2009) Estrogen-replacement therapy promotes angiogenesis after acute myocardial infarction by enhancing SDF-1 and estrogen receptor expression. Microvasc Res 77:71–77
33. Cai J, Hong Y, Weng C et al (2011) Androgen stimulates endothelial cell proliferation via an androgen receptor/VEGF/cyclin A-mediated mechanism. Am J Physiol Heart Circ Physiol 300:H1210–H1221

34. Sieveking DP, Lim P, Chow RWY et al (2010) A sex-specific role for androgens in angiogenesis. J Exp Med 207:345–352
35. Liu J, Wu S, Wei H et al (2002) Effects of sex hormones and their balance on the proliferation of rat vascular endothelial cells. Horm Res 58:16–20
36. Death AK, McGrath KCY, Sader MA et al (2004) Dihydrotestosterone promotes vascular cell adhesion molecule-1 expression in male human endothelial cells via a nuclear factor-kappaB-dependent pathway. Endocrinology 145:1889–1897
37. Zhou X (2010) Roles of androgen receptor in male and female reproduction: lessons from global and cell-specific androgen receptor knockout (ARKO) mice. J Androl 31:235–243
38. Sieveking DP, Chow RWY, Ng MKC (2010) Androgens, angiogenesis and cardiovascular regeneration. Curr Opin Endocrinol Diabetes Obes 17:277–283
39. Stern JM, Chen J, Peters SB et al (2004) Testosterone treatment of human foreskin in a novel transplant model. Urology 63:999–1003
40. Foresta C, Zuccarello D, De Toni L et al (2008) Androgens stimulate endothelial progenitor cells through an androgen receptor-mediated pathway. Clin Endocrinol 68:284–289
41. Wu FCW, von Eckardstein A (2003) Androgens and coronary artery disease. Endocr Rev 24:183–217
42. Liu PY, Death AK, Handelsman DJ (2003) Androgens and cardiovascular disease. Endocr Rev 24:313–340
43. Mendelsohn ME, Karas RH (2005) Molecular and cellular basis of cardiovascular gender differences. Science 308:1583–1587
44. Rubinow KB, Amory JK, Page ST (2011) Androgens exert sexually dimorphic effects on angiogenesis: novel insight into the relationship between androgens and cardiovascular disease. Asian J Androl 13:626–627
45. Ruifrok W-PT, de Boer RA, Iwakura A et al (2009) Estradiol-induced, endothelial progenitor cell-mediated neovascularization in male mice with hind-limb ischemia. Vasc Med 14:29–36
46. Rubanyi GM, Freay AD, Kauser K et al (1997) Vascular estrogen receptors and endothelium-derived nitric oxide production in the mouse aorta. Gender difference and effect of estrogen receptor gene disruption. J Clin Invest 99:2429–2437
47. Lindner V, Kim SK, Karas RH et al (1998) Increased expression of estrogen receptor-beta mRNA in male blood vessels after vascular injury. Circ Res 83:224–229
48. McCrohon JA, Death AK, Nakhla S et al (2000) Androgen receptor expression is greater in macrophages from male than from female donors. A sex difference with implications for atherogenesis. Circulation 101:224–226
49. Ng MKC, Nakhla S, Baoutina A et al (2003) Dehydroepiandrosterone, an adrenal androgen, increases human foam cell formation: a potentially pro-atherogenic effect. J Am Coll Cardiol 42:1967–1974
50. Smith SK (2001) Regulation of angiogenesis in the endometrium. Trends Endocrinol Metab 12:147–151
51. Girling JE, Rogers PAW (2005) Recent advances in endometrial angiogenesis research. Angiogenesis 8:89–99
52. Rogers PAW, Donoghue JF, Walter LM, Girling JE (2009) Endometrial angiogenesis, vascular maturation, and lymphangiogenesis. Reprod Sci 16:147–151
53. Albrecht ED, Pepe GJ (2010) Estrogen regulation of placental angiogenesis and fetal ovarian development during primate pregnancy. Int J Dev Biol 54:397–408
54. Heryanto B, Rogers PAW (2002) Regulation of endometrial endothelial cell proliferation by oestrogen and progesterone in the ovariectomized mouse. Reproduction 123:107–113
55. Lecce G, Meduri G, Ancelin M, Bergeron C, Perrot-Applanat M (2001) Presence of estrogen receptor beta in the human endometrium through the cycle: expression in glandular, stromal, and vascular cells. J Clin Endocrinol Metab 86:1379–1386
56. Kayisli UA, Luk J, Guzeloglu-Kayisli O et al (2004) Regulation of angiogenic activity of human endometrial endothelial cells in culture by ovarian steroids. J Clin Endocrinol Metab 89:5794–5802

57. Shifren JL, Tseng JF, Zaloudek CJ et al (1996) Ovarian steroid regulation of vascular endothelial growth factor in the human endometrium: implications for angiogenesis during the menstrual cycle and in the pathogenesis of endometriosis. J Clin Endocrinol Metab 81:3112–3118
58. Iruela-Arispe ML, Rodriguez-Manzaneque JC, Abu-Jawdeh G (1999) Endometrial endothelial cells express estrogen and progesterone receptors and exhibit a tissue specific response to angiogenic growth factors. Microcirculation 6:127–140
59. Albrecht ED, Babischkin JS, Lidor Y et al (2003) Effect of estrogen on angiogenesis in co-cultures of human endometrial cells and microvascular endothelial cells. Hum Reprod 18:2039–2047
60. Kazi AA, Koos RD (2007) Estrogen-induced activation of hypoxia-inducible factor-1alpha, vascular endothelial growth factor expression, and edema in the uterus are mediated by the phosphatidylinositol 3-kinase/Akt pathway. Endocrinology 148:2363–2374
61. Kazi AA, Molitoris KH, Koos RD (2009) Estrogen rapidly activates the PI3K/AKT pathway and hypoxia-inducible factor 1 and induces vascular endothelial growth factor A expression in luminal epithelial cells of the rat uterus. Biol Reprod 81:378–387
62. Moller B, Rasmussen C, Lindblom B, Olovsson M (2001) Expression of the angiogenic growth factors VEGF, FGF-2, EGF and their receptors in normal human endometrium during the menstrual cycle. Mol Hum Reprod 7:65–72
63. Spillman MA, Manning NG, Dye WW et al (2010) Tissue-specific pathways for estrogen regulation of ovarian cancer growth and metastasis. Cancer Res 70:8927–8936
64. Turner HE, Harris AL, Melmed S, Wass JAH (2003) Angiogenesis in endocrine tumors. Endocr Rev 24:600–632
65. Fox SB, Leek RD, Smith K et al (1994) Tumor angiogenesis in node-negative breast carcinomas—relationship with epidermal growth factor receptor, estrogen receptor, and survival. Breast Cancer Res Treat 29:109–116
66. Toi M, Inada K, Suzuki H, Tominaga T (1995) Tumor angiogenesis in breast cancer: its importance as a prognostic indicator and the association with vascular endothelial growth factor expression. Breast Cancer Res Treat 36:193–204
67. Gasparini G, Fox SB, Verderio P et al (1996) Determination of angiogenesis adds information to estrogen receptor status in predicting the efficacy of adjuvant tamoxifen in node-positive breast cancer patients. Clin Cancer Res 2:1191–1198
68. Banerjee SK, Sarkar DK, Weston AP, De A, Campbell DR (1997) Over expression of vascular endothelial growth factor and its receptor during the development of estrogen-induced rat pituitary tumors may mediate estrogen-initiated tumor angiogenesis. Carcinogenesis 18:1155–1161
69. Garvin S, Nilsson UW, Dabrosin C (2005) Effects of oestradiol and tamoxifen on VEGF, soluble VEGFR-1, and VEGFR-2 in breast cancer and endothelial cells. Br J Cancer 93:1005–1010
70. Elkin M, Orgel A, Kleinman HK (2004) An angiogenic switch in breast cancer involves estrogen and soluble vascular endothelial growth factor receptor 1. J Natl Cancer Inst 96:875–878
71. Hartman J, Lindberg K, Morani A et al (2006) Estrogen receptor beta inhibits angiogenesis and growth of T47D breast cancer xenografts. Cancer Res 66:11207–11213
72. Seo KH, Lee H-S, Jung B et al (2004) Estrogen enhances angiogenesis through a pathway involving platelet-activating factor-mediated nuclear factor-kappaB activation. Cancer Res 64:6482–6488
73. Santen RJ, Song RX, Masamura S et al (2008) Adaptation to estradiol deprivation causes up-regulation of growth factor pathways and hypersensitivity to estradiol in breast cancer cells. Adv Exp Med Biol 630:19–34
74. Gottardis MM, Robinson SP, Satyaswaroop PG, Jordan VC (1988) Contrasting actions of tamoxifen on endometrial and breast tumor growth in the athymic mouse. Cancer Res 48:812–815
75. Gagliardi A, Collins DC (1993) Inhibition of angiogenesis by antiestrogens. Cancer Res 53:533–535

76. Tosetti F, Ferrari N, De Flora S, Albini A (2002) Angioprevention': angiogenesis is a common and key target for cancer chemopreventive agents. FASEB J 16:2–14
77. McNamara DA, Harmey J, Wang JH et al (2001) Tamoxifen inhibits endothelial cell proliferation and attenuates VEGF-mediated angiogenesis and migration in vivo. Eur J Surg Oncol 27:714–718
78. Garvin S, Dabrosin C (2003) Tamoxifen inhibits secretion of vascular endothelial growth factor in breast cancer in vivo. Cancer Res 63:8742–8748
79. Nilsson UW, Abrahamsson A, Dabrosin C (2010) Angiogenin regulation by estradiol in breast tissue: tamoxifen inhibits angiogenin nuclear translocation and antiangiogenin therapy reduces breast cancer growth in vivo. Clin Cancer Res 16:3659–3669
80. Aberg UWN, Saarinen N, Abrahamsson A et al (2011) Tamoxifen and flaxseed alter angiogenesis regulators in normal human breast tissue in vivo. PLoS One 6:e25720
81. Lindahl G, Saarinen N, Abrahamsson A, Dabrosin C (2011) Tamoxifen, flaxseed, and the lignan enterolactone increase stroma- and cancer cell-derived IL-1Ra and decrease tumor angiogenesis in estrogen-dependent breast cancer. Cancer Res 71:51–60
82. Bergman Jungestrom M, Thompson LU, Dabrosin C (2007) Flaxseed and its lignans inhibit estradiol-induced growth, angiogenesis, and secretion of vascular endothelial growth factor in human breast cancer xenografts in vivo. Clin Cancer Res 13:1061–1067
83. Blackwell KL, Haroon ZA, Shan S et al (2000) Tamoxifen inhibits angiogenesis in estrogen receptor-negative animal models. Clin Cancer Res 6:4359–4364
84. Iyer V, Klebba I, McCready J et al (2012) Estrogen promotes ER-negative tumor growth and angiogenesis through mobilization of bone marrow-derived monocytes. Cancer Res 72:2705–2713
85. Zhu Y-S, Imperato-McGinley JL (2009) 5alpha-reductase isozymes and androgen actions in the prostate. Ann N Y Acad Sci 1155:43–56
86. Schatzl G, Madersbacher S, Haitel A et al (2003) Associations of serum testosterone with microvessel density, androgen receptor density and androgen receptor gene polymorphism in prostate cancer. J Urol 169:1312–1315
87. Chamberlain NL, Driver ED, Miesfeld RL (1994) The length and location of CAG trinucleotide repeats in the androgen receptor N-terminal domain affect transactivation function. Nucleic Acids Res 22:3181–3186
88. Beilin J, Ball EM, Favaloro JM, Zajac JD (2000) Effect of the androgen receptor CAG repeat polymorphism on transcriptional activity: specificity in prostate and non-prostate cell lines. J Mol Endocrinol 25:85–96
89. Krithivas K, Yurgalevitch SM, Mohr BA et al (1999) Evidence that the CAG repeat in the androgen receptor gene is associated with the age-related decline in serum androgen levels in men. J Endocrinol 162:137–142
90. Giovannucci E, Platz EA, Stampfer MJ et al (1999) The CAG repeat within the androgen receptor gene and benign prostatic hyperplasia. Urology 53:121–125
91. Feldman HA, Longcope C, Derby CA et al (2002) Age trends in the level of serum testosterone and other hormones in middle-aged men: longitudinal results from the Massachusetts male aging study. J Clin Endocrinol Metab 87:589–598
92. Belanger A, Candas B, Dupont A et al (1994) Changes in serum concentrations of conjugated and unconjugated steroids in 40- to 80-year-old men. J Clin Endocrinol Metab 79: 1086–1090
93. Feldman BJ, Feldman D (2001) The development of androgen-independent prostate cancer. Nat Rev Cancer 1:34–45
94. Colombel M, Filleur S, Fournier P et al (2005) Androgens repress the expression of the angiogenesis inhibitor thrombospondin-1 in normal and neoplastic prostate. Cancer Res 65:300–308
95. Franck-Lissbrant I, Haggstrom S, Damber JE, Bergh A (1998) Testosterone stimulates angiogenesis and vascular regrowth in the ventral prostate in castrated adult rats. Endocrinology 139:451–456

96. Jain RK, Safabakhsh N, Sckell A et al (1998) Endothelial cell death, angiogenesis, and microvascular function after castration in an androgen-dependent tumor: role of vascular endothelial growth factor. Proc Natl Acad Sci U S A 95:10820–10825
97. Joseph IB, Nelson JB, Denmeade SR, Isaacs JT (1997) Androgens regulate vascular endothelial growth factor content in normal and malignant prostatic tissue. Clin Cancer Res 3:2507–2511
98. Sordello S, Bertrand N, Plouet J (1998) Vascular endothelial growth factor is up-regulated in vitro and in vivo by androgens. Biochem Biophys Res Commun 251:287–290
99. Haggstrom S, Lissbrant IF, Bergh A, Damber JE (1999) Testosterone induces vascular endothelial growth factor synthesis in the ventral prostate in castrated rats. J Urol 161:1620–1625
100. Stewart RJ, Panigrahy D, Flynn E, Folkman J (2001) Vascular endothelial growth factor expression and tumor angiogenesis are regulated by androgens in hormone responsive human prostate carcinoma: evidence for androgen dependent destabilization of vascular endothelial growth factor transcripts. J Urol 165:688–693
101. Mabjeesh NJ, Willard MT, Frederickson CE, Zhong H, Simons JW (2003) Androgens stimulate hypoxia-inducible factor 1 activation via autocrine loop of tyrosine kinase receptor/phosphatidylinositol 3′-kinase/protein kinase B in prostate cancer cells. Clin Cancer Res 9:2416–2425
102. Boddy JL, Fox SB, Han C et al (2005) The androgen receptor is significantly associated with vascular endothelial growth factor and hypoxia sensing via hypoxia-inducible factors HIF-1a, HIF-2a, and the prolyl hydroxylases in human prostate cancer. Clin Cancer Res 11:7658–7663
103. Cheng L, Zhang S, Sweeney CJ et al (2004) Androgen withdrawal inhibits tumor growth and is associated with decrease in angiogenesis and VEGF expression in androgen-independent CWR22Rv1 human prostate cancer model. Anticancer Res 24:2135–2140
104. Nicholson B, Gulding K, Conaway M, Wedge SR, Theodorescu D (2004) Combination antiangiogenic and androgen deprivation therapy for prostate cancer: a promising therapeutic approach. Clin Cancer Res 10:8728–8734

Chapter 11
Biochemical Mechanisms of Exercise-Induced Angiogenesis

Dustin Scott Kehler, Naranjan S. Dhalla, and Todd A. Duhamel

Abstract Angiogenesis refers to the growth of new capillaries from a pre-existing capillary bed which can occur during normal physiological and pathological conditions by sprouting and non-sprouting processes, which are activated by different stimuli. Various studies have demonstrated that exercise increases the expression of several growth factors for both sprouting and non-sprouting angiogenesis, including vascular endothelial growth factor and other cytokines in skeletal and cardiac muscle, which are associated with an increase in the number of capillaries in the heart and skeletal muscle. Exercise is known to stimulate the release of several pro- and anti-angiogenic proteins and transcription factors and it appears that hypoxia and/or ischemia play a major role in the growth and expansion of new capillaries and has also been suggested that mechanical forces, such as shear stress or muscle overload, stimulate exercise-induced angiogenesis. More importantly, an in-depth understanding of the factors that influence exercise-induced angiogenesis may contribute to the development of potential therapeutic strategies for the treatment of different diseases including hypertension and ischemic heart disease.

Keywords Angiogenesis • Physical activity • Exercise training • Vascular endothelial growth factor • Matrix metalloproteinase • Angiopoietins • Endostatin • Thrombospondin • Tissue inhibitors of matrix metalloproteinases • Endothelial progenitor cells • Chronic disease

---

D.S. Kehler • T.A. Duhamel, Ph.D. (✉)
Institute of Cardiovascular Sciences, St. Boniface General Hospital
Research Centre, 351 Tache Avenue, Winnipeg, MB, Canada R2H 2A6

Health, Leisure and Human Performance Research Institute, University of Manitoba, Winnipeg, MB, Canada University of Manitoba, Winnipeg, MB, Canada
e-mail: tduhamel@sbrc.ca

N.S. Dhalla
Institute of Cardiovascular Sciences St. Boniface Hospital Research, University of Manitoba, 351 Tache Avenue, Winnipeg, Manitoba, Canada, R2H 2A6

University of Manitoba, Winnipeg, MB, Canada

J.L. Mehta and N.S. Dhalla (eds.), *Biochemical Basis and Therapeutic Implications of Angiogenesis*, Advances in Biochemistry in Health and Disease 6,
DOI 10.1007/978-1-4614-5857-9_11,

11.1 Introduction

Blood vessels and capillaries play a vital role in supplying oxygen and nutrients to metabolic tissues as well as play a functional role in the endocrine and immune systems [1]. Much of the current understanding of the anatomy and physiology of capillaries stems from the early work of Krogh [2–5], who examined the distribution and number of capillaries in organs and tissues, the structure of the capillary wall, and the exchange of substances through the capillary wall [5]. His work has also eluded to the fact that blood vessels have the capacity to grow (i.e., arteriogenesis) as well as to form new blood vessels from a pre-existing capillary bed, which is known as angiogenesis. The term angiogenesis was first used to describe the formation of new blood vessels during placental growth [6] and was first observed by implanting a transparent chamber into the ear of a rabbit [7]. Angiogenesis is important for normal physiological growth, such as in the female reproductive tract [8], wound healing, and muscle remodeling [9]. Deviations from normal vessel growth and maintenance can lead to several chronic diseases and angiogenesis plays an important role in the healing process after conditions such as stroke [10], myocardial infarction [11], ulcers [12], and neurodegeneration [13]. On the other hand, abnormal blood vessel growth or remodeling is associated with cancer [14], inflammatory disorders [15], pulmonary hypertension [16], and in eye diseases [17].

Physical activity has been shown to be a powerful stimulus of cardiac and skeletal muscle adaptation and physiological remodeling. For example, exercise training plays a major role in the remodeling process to increase capillary density of muscle [18] and mitochondrial enzyme activity after a single bout of exercise [19] as well as after exercise training [20]. These changes may occur because skeletal and cardiac muscle are metabolically active tissues that can increase their metabolic requirements during exercise 30–50-fold compared to basal conditions [21]. Therefore, adequate circulation to supply oxygen and nutrients to these tissues and to remove "metabolic waste" from these tissues is essential for their optimal performance and health. Since its discovery, angiogenesis has been studied extensively and several investigators have attempted to elucidate the mechanisms to explain how exercise stimulates angiogenesis. Hudlicka et al. [22] have provided data on some of the possible stimuli for exercise-induced angiogenesis, including increases in blood flow (hyperemia), shear stress, and muscle stretch. Other suggested stimuli include hypoxia and metabolic disturbance [23] (Fig. 11.1).

Angiogenesis is a highly regulated process and is controlled by several pro- and anti-angiogenic factors which can be turned on when needed and completely shut down [22, 24]. Over 30 years ago, Dvorak et al. [25] discovered a new molecule involved in tumor growth they termed *vascular permeability factor*, which was later renamed as *vascular endothelial growth factor* (VEGF) [26]. VEGF is thought to be the key mediator in the angiogenic process and there is a growing body of evidence indicating that exercise increases the tran-

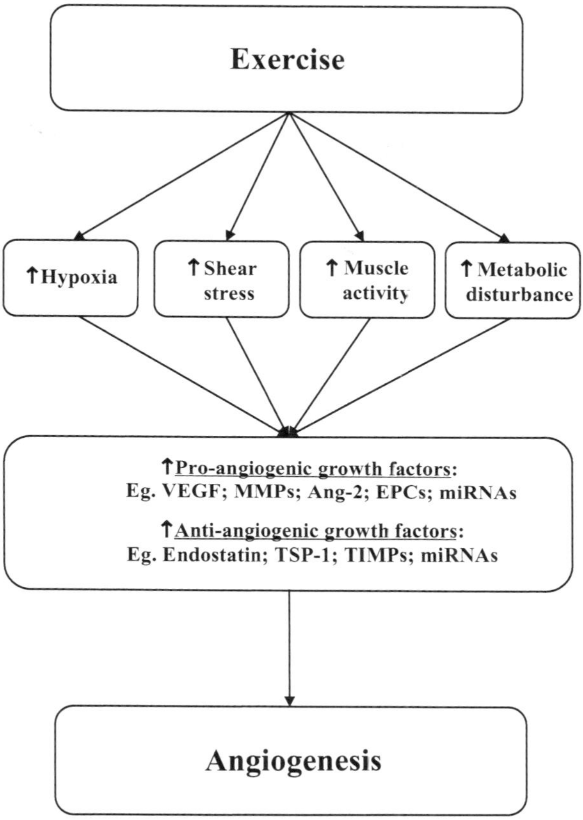

Fig. 11.1 Simplified pathway of exercise angiogenesis and the suggested mechanisms that exercise affects, including hypoxia, shear stress, muscle activity/stretch, and metabolic disturbance. *VEGF* vascular endothelial growth factor; *MMPs* matrix metalloproteinases; *Ang* angiopoietin; *EPCs* endothelial progenitor cells; *miRNAs* microRNAs; *TSP-1* thrombospondin-1; *TIMPs* tissue inhibitors of matrix metalloproteinases

scription and proteins of VEGF in cardiac and skeletal muscle [27–34]. Two other pro-angiogenic factors that increase with exercise are *matrix metalloproteinases* (MMPs) and *angiopoietins* (Ang-1, Ang-2). The three most investigated anti-angiogenic factors that are upregulated with exercise include endostatin, *thrombospondin-1* (TSP-1), and *tissue inhibitors of matrix metalloproteinases* (TIMPs). Accordingly, evidence for the role of acute and chronic exercise training for upregulating these factors will be reviewed in this article. We will discuss a few other pro- and anti-angiogenic factors to determine if these proteins change in response to acute exercise or chronic exercise training. Since *endothelial progenitor cells* (EPCs) as well as microRNAs (miRNAs) also increase with acute exercise and chronic exercise training, their role in angiogenesis will be reviewed.

11.2 Sprouting and Non-sprouting Angiogenesis Pathways

Animal studies have revealed at least two forms of angiogenesis to occur, at least in skeletal muscle: sprouting angiogenesis and non-sprouting angiogenesis [35]. Irrespective of the cause of angiogenesis, it is assumed that there is a common sequence of events; the activation of endothelial cells followed by the degradation of the basement membrane and the extracellular matrix allowing endothelial cells to migrate to sites where capillaries are needed. Sprouting angiogenesis forms new capillaries via stalk cells and tip cells, whereas non-sprouting angiogenesis occurs when capillaries split into two via intussusception. Non-sprouting angiogenesis is suggested to be a more efficient form of angiogenesis as it permits a more rapid expansion of capillaries and does not require an initial proliferation of endothelial cells [36]. Several reviews have described the details for both sprouting [1, 9, 21, 22, 24, 37–44] and non-sprouting angiogenesis [21, 35, 45–49].

Sprouting angiogenesis is considered to be the major type of angiogenic growth induced by exercise [21] and may be induced by stimuli such as hypoxia, metabolic disturbance, and muscle stretch [35]. It involves the activation of normally quiescent endothelial cells [50], which branch out from an existing capillary and extend through a surrounding matrix to form a cord-like structure. Initiation of sprouting angiogenesis requires the proteolytic degradation of the basement membrane [51], formed by inactive endothelial cells and pericytes. Normally quiescent blood vessels are built with angiogenic sensors, and following the angiogenic signal, pericytes detach from the vessel wall [52] and are removed from the basement membrane [53]. Endothelial cells begin to loosen their junctions from the vessel and the vessel begins to dilate [54]. Permeability of the endothelial cells' layer forces plasma proteins to enter the extracellular matrix and stored angiogenic molecules begin to remodel the extracellular matrix [54]. An endothelial "tip cell" leads the migration of endothelial cells and neighboring stalk cells have filopodia to guide their elongation [55]. In order for the abluminal sprouting capillary to become functional, it must fuse with a neighboring blood vessel and become mature [54]. Endothelial cells return to their inactive state. A myeloid bridge aids the elongation of the stalk as well as the fusion with another blood vessel and allows the initiation of blood flow [54]. Pericytes cover the newly formed capillary and protease inhibitors deposit a basement membrane. Junctions are re-established to ensure optimal blood flow. Regression occurs in the absence of angiogenic stimuli [56, 57].

New capillaries can also form by non-sprouting angiogenesis by intussusception and this was first observed in the rat lung [58]. Opposing capillary walls protrude towards the vessel lumen, and once contact has been established between the capillary walls, the inter-endothelial cell junctions are reorganized and the endothelial bilayer of the capillary becomes centrally perforated [58]. An interstitial pillar-like structure is formed, which is invaded by pericytes and myofibroblasts and these lay down collagen fibrils [59]. The final phase of intussusception is when the pillars increase in girth without any other change in structure [58]. When comparing sprouting vs. non-sprouting angiogenesis, it would appear that non-sprouting angiogenesis is a more efficient way of expanding the capillary network as there is much less endothelial cell proliferation and capillary maturation involved [60].

11.3 Factors That Stimulate Angiogenesis in Response to Exercise

Angiogenesis is a complex, multi-step process and is thought to involve multiple signaling pathways which are coordinated by pro- and anti-angiogenic mediators to generate effective angiogenesis. The following are various pro-angiogenic factors which are considered to be the key mediators in the angiogenic process specifically relevant to exercise-induced angiogenesis.

11.3.1 Vascular Endothelial Growth Factor in Acute Exercise and Chronic Exercise Training

VEGF has been shown to play a crucial role in exercise-induced angiogenesis [61–63]. VEGF is a 35–45 kDa peptide growth factor and includes a number of isoforms (VEGF-A to VEGF-D), with the most relevant to exercise-induced angiogenesis being VEGF-A (commonly known as just VEGF). Sources of VEGF include skeletal muscle fibers, endothelial cells, fibroblasts, macrophages, pericytes, as well as mast cells and smooth muscle cells [33, 63]. VEGF plays a major role in endothelial cell stimulation, survival and differentiation, vascular smooth muscle proliferation and migration, and induces capillary permeability and dilation of arterioles [21]. An increase in VEGF transcription is thought to be regulated by several factors. One of these factors includes *hypoxia-inducible factor-*α (HIF-1α) during hypoxia [64]. VEGF contains a hypoxic-response element [65] and binds to HIF-1α during hypoxic states. This results in an increase in VEGF transcription when there is reduced oxygen availability and it would seem that HIF-1α would also regulate VEGF protein during exercise. For example, in response to moderate to heavy exercise (knee extension exercise at 75–100% maximum work rate), intracellular PO_2 decreases from 30 mm Hg at rest [66] to approximately 3–4 mmHg during exercise in human quadriceps muscle [67]. However, the exercise-induced increase in VEGF is more complex than by activation of HIF-1α alone. It is pointed out that VEGF expression was still observed in HIF-1α knockout mice following exercise training [68]. Moreover, there is a reduced VEGF transcription response with restricted blood flow; however, there was still a similar activation in HIF-1α between the restricted and non-restricted exercising skeletal muscle [69]. Therefore, VEGF must be regulated by some factors other than HIF-1α.

More recently, metabolic activity has been highlighted as playing a role in VEGF regulation [70]. For example, *AMP-activated protein kinase* (AMPK) has been suggested to play a role in exercise-induced angiogenesis. AMPK is involved in the regulation of the stress response as well as metabolic homeostasis and is associated with increased glucose uptake in muscle [71]. It also leads to the promotion of angiogenesis by increasing the expression of VEGF in skeletal muscle cell culture (i.e., C2C12 cell line) through activation of the AMPK-p38 MAPK-dependant pathway [72]. Skeletal muscle cell culture has also indicated that adenosine enhances mRNA expression of VEGF and this may be mediated by the A_{2B} receptor [73]. Several studies in animal models have revealed that *peroxisome-proliferator activated receptor* γ *co-activator-1*α (PGC-1α) through the activation of *estrogen-related receptor* α (ERR-α) serves as a principal regulator of increased expression of VEGF through exercise [72–74].

There is a growing body of evidence that VEGF and its receptors VEGFR1, VEGFR2, and *neuropilin 1* (NRP-1), which is also a receptor of VEGF, increase in response to a single bout of exercise [27–34]. Moreover, an acute bout of exercise in untrained subjects releases higher amounts of VEGF mRNA and protein response compared with trained subjects [75]. Several studies have shown that an acute bout of single-leg knee extension exercise lasting for 45 min at approximately 25% maximum workload increases mRNA expression of VEGF by at least twofold and VEGF protein expression ~1.5-fold [27–29]. Splice variants $VEGF_{121}$ and $VEGF_{165}$ increase initially after an acute bout of exercise by 3- and 3.5-fold, respectively, after 45 min of one-leg knee extension exercise at 25% of maximum workload, followed by a 3.5-fold increase in $VEGF_{189}$ [33]. The relative change in expression after exercise may be represented by endothelial activation with $VEGF_{121}$ and $VEGF_{165}$ as they are diffusible [76], whereas $VEGF_{189}$ contains a domain encoded by exon 6 and is bound to cell-surface heparin sulfate proteoglycans which represent a later chemoattractant and differentiation role in the VEGF system [76].

Prolonged moderate intensity exercise training is known to upregulate VEGF mRNA expression and protein content. For example, 60 min of cycling at 50% of maximal aerobic capacity (i.e., VO_{2max}) increases VEGF mRNA expression by approximately 4.5-fold [30], whereas a 60 min bout of cycling at approximately 60% of VO_{2max} increased the concentration of interstitial VEGF by >6-fold compared to resting levels [77]. These data are also supported by Rullman et al. [78] who demonstrated a 1.5-fold increase in VEGF protein content in subjects who exercised for 65 min at ~65 VO_{2max}. The type and intensity of acute endurance exercise may also influence the VEGF response to exercise. In fact, high intensity sprint interval exercise has recently been investigated and appears to be at least as effective as prolonged moderate intensity exercise training for inducing a VEGF response [79]. Specifically, a greater increase in VEGF protein was observed when subjects participated in four 30 s "all out" exercise bouts which were each followed by 5 min of passive rest when compared to subjects who completed 1 h of cycling at 50% of their peak power output. With the high intensity protocol, there was a 90 pg/mL increase in VEGF 4 h after exercise, whereas there was only a 35 pg/mL increase after moderate intensity exercise. Notably, this physiological difference in VEGF content occurred even though the absolute amount of time spent exercising was 97% lower in the high intensity sprint interval group compared to the low intensity group (i.e., 2 min vs. 60 min) [79].

There is also some evidence to suggest that acute resistance exercise induces a VEGF response. In this regard, three sets of ten repetitions at 60–80% of one-repetition maximum (1RM) induce a threefold increase in skeletal muscle VEGF mRNA and 1.5-fold increase in VEGF protein 4 h post-exercise compared to rest [80]. However, this response may be dependent on the type of resistance training (i.e., isotonic, isometric, or isokinetic). Specifically, three repetitions of either 40% of 1RM or maximal isokinetic knee extension did not increase serum VEGF [81]. High volume resistance training of 3×30 repetitions in combination with a 30% muscle blood flow restriction also induces a VEGF response [82]. Notably, the VEGF response is attenuated 30 min after exercise when muscle blood flow is normal following exercise training (i.e., a 1.5-fold increase), as compared to the same exercise condition when muscle blood flow is restricted (i.e., a 2.5-fold increase). This observation suggests that intermittent hypoxia or ischemia may be a key factor in regulating exercise-induced angiogenesis.

There appears to be an attenuated VEGF response to an acute bout of exercise following chronic exercise training. For example, the arterial and venous plasma levels of VEGF in response to an acute bout of exercise decreased by 12% and 15%, respectively, after seven exercise training sessions over a 10-day period consisting of 45-min of knee extension exercise at 60–70% maximum work rate, as compared to baseline [29]. However, basal VEGF mRNA and protein expression increased twofold after the 7-day training period. The VEGF response after acute exercise appears to be further attenuated after a longer training period. Particularly, 8 weeks of single-leg knee extension exercise three times per week for 60 min at 50% maximum work rate showed an attenuated VEGF response compared to the untrained muscle. This could be explained by the 18% increase in capillary to fiber ratio of the quadriceps and therefore a further angiogenic response may not be required for the given training stimulus [75].

Receptors VEGFR1, VEGFR2, and NRP-1 are known to increase after a single bout of exercise [30, 33, 34, 83], but respond differently after chronic exercise training, which is characterized by a selective increase in NRP-1 and VEGFR2 [34, 83]. Intermittent hypoxia or ischemia may also influence the regulation of VEGF receptors. For example, resting VEGF protein and VEGFR2 protein content increased threefold under restricted blood flow conditions and remained unchanged under the non-restricted blood flow conditions in response to 5 weeks of knee extension exercise training for 45 min per day, four times per week under 20% restricted blood flow on one leg and non-restricted blood flow to the other leg [34]. This is an interesting finding because VEGFR2 is essential for a majority of the angiogenic actions of VEGF [76, 84]. Moreover, subjects who were "high responders" (i.e., an improvement of +0.71 ± 0.1 L/min of O_2 improvement) after 6 weeks of cycling exercise at 75% of their peak VO_2 for 45 min four times per week had an eightfold increase in NRP-1and a threefold increase in VEGFR-2, while "non-responders" (i.e., +0.17 ± 0.1 L/min of O_2 improvement) had no change [83].

11.3.2 Matrix Metalloproteinases in Acute Exercise and Chronic Exercise Training

It is now well known that MMPs play an important role in the formation, remodeling, and degradation of the extracellular matrix [85]. They are members of the protease family and 25 enzymes of the MMP family have been identified and are expressed in most tissues [86–88], including skeletal muscle. Specifically, MMP-2 and MMP-9 have been shown to be upregulated in response to exercise [77, 78, 89–93]. They are both collagenases which degrade collagen type IV, which is the most prevalent protein in skeletal muscle basal lamina [89]. It has been suggested that MMP-2 plays a role in the cleavage of the basal lamina and extracellular matrix during angiogenesis, making a path for the migration of endothelial sprouts and the formation of new capillaries [88, 94]. The promoter region of MMP-9 contains a variety of response elements sensitive to a variety of growth factors and cytokines [95] and may also be a regulator of growth factor bioavailability via proteolytic release. There is also evidence indicating that MMP-1, -3, and -14 also increase in response to chronic exercise training [78, 89, 93].

Evidence indicates that the MMP system is activated during an acute bout of exercise as well as following exercise in skeletal muscle [77, 78]. The same exercise protocol used by Hoier et al. [77] discussed in the VEGF section increased MMP-9 mRNA by three and fourfold at 1 and 3 h post-exercise, respectively. However, there were no changes in MMP-2 mRNA. Furthermore, 65 min of cycle ergometer exercise, consisting of 20 min at 50% of VO_{2max}, 40 min at 65% of VO_{2max}, and 5 min at the subject's maximal tolerable work rate elevated vastus lateralis MMP-9 protein content by ~2.5-fold during exercise and MMP-9 mRNA was elevated ~2.5-fold 2 h after the exercise bout [78]. These data indicate that MMP activity occurs before any measurable changes in MMP-9 transcription, suggesting a post-transcriptional mechanism. As observed with VEGF, the intensity of exercise may influence the MMP response to an acute bout of exercise. Moreover, it has been hypothesized that shear stress is increased during body vibration (as reviewed by Mester et al. [96]) and has been hypothesized that the utilization of vibration training in conjunction with hypoxia and cycling exercise will induce an angiogenic response [92]. Specifically, subjects performed 90 min of cycling; 10 min warm up at 50% VO_{2max} followed by ten intervals of 3 min at a workload of 85% VO_{2max} and 5 min recovery at 60% VO_{2max}. This protocol was carried under four different conditions: normal conditions without vibration; normal conditions with vibration; hypoxic conditions without vibration; and hypoxic conditions with vibration. With the exception of normal conditions with vibration, circulating MMP-2 was increased by approximately 10% immediately after an acute bout of exercise. On the other hand, circulating MMP-9 was shown to increase by approximately twofold in all training conditions 4 h post-exercise, therefore indicating a different activating mechanism for these proteins.

MMP-2 and MMP-9 mRNA and protein levels are known to be regulated by chronic exercise training. For example, 10 days of single-leg exercise training (which consisted of 45 min at the subjects' maximum tolerable workload four times per week) increased MMP-2 mRNA and protein content by 3.5- and 5.5-fold, respectively [89]. However, partial leg ischemia (i.e., 20%) did not exert an additional effect above that of exercise training alone [89]. Even so, basal MMP-2 levels remained elevated after 5 weeks of training in both exercise training protocols. Rullman et al. [89] also found a ~3–7-fold increase in MMP-9 in the normal and restricted blood flow conditions, respectively, after an acute bout of exercise, but the group changes were not different. Moreover, basal MMP-9 mRNA after 10 days of training increased ~5- and ~13-fold after 5 weeks of training in both groups, with no change in protein. Interestingly, basal MMP-14, which activates MMP-2 [94], mRNA, was increased ~5-fold after 10 days of training in both groups and was maintained after 5 weeks.

Chronic exercise training consisting of aerobic and sprint training in conjunction with either calisthenic or resistance training has been shown to increase circulating MMPs [93]. Calisthenic exercises consist of body weight exercises such as push-ups and pull-ups and squats and lunges. Both the calisthenic and resistance exercise groups exercised 5 days per week for 8 weeks. Exercises were performed until exhaustion in the calisthenic group; whereas the resistance exercise group performed three sets of eight repetitions at 80% of the subject's 1RM followed by 30 s of rest between sets. Immediately after an acute resistance exercise test, MMP-1 concentration increased ~1.2-fold and MMP-2 remained unchanged in both groups compared to baseline. Interestingly, MMP-3 concentration decreased ~1.5-fold and MMP-9 concentration increased ~4-fold in response to an acute resistance exercise test after exercise training in the calisthenic exercise group, but no changes were observed in the resistance training group. However, the resistance training group increased basal MMP-1 ~1.2-fold and MMP-2 increased twofold after chronic exercise training, whereas the calisthenic exercise-trained group demonstrated a ~1.4-fold decrease in basal MMP-2, a 1.4-fold decrease in basal MMP-3, and a 1.3-fold increase in MMP-9. These data suggest that the mode of exercise elicits a specific training adaptation of MMPs following chronic exercise training.

11.3.3 Angiopoietins in Acute Exercise and Chronic Exercise Training

The angiopoietin family (Ang-1 and Ang-2) has been suggested to play a role in the angiogenic process. Angiopoietins are vascular endothelial-specific factors that are activated via RTK pathways and modulate vessel development and remodeling [97]. Ang-1 is expressed in many tissues including the myocardium and perivascular cells and plays a role in the maturation of new blood vessels and vascular stabilization [97]. Ang-2 promotes vascular remodeling by either facilitating growth factor actions or vascular regression when no growth signal is present [98] and is thought to lead to vessel degradation in the absence of VEGF. Ang-1 and Ang-2 compete for the tyrosine kinase receptor Tie2, where Ang-1 acts as an agonist for Tie2 [99] and Ang-2 acts as an antagonist [98]. An increase in the Ang-2/Ang-1 ratio is suggested to have a permissive effect on angiogenesis [100].

Evidence indicates that an acute aerobic exercise bout induces a higher Ang-2/Ang-1 ratio in skeletal muscle [77]. The protocol utilized by Hoier et al. [77] found that Ang-2 mRNA was increased ~2-fold 3 h post-exercise while Ang-1 remained unchanged. As a result, the Ang-2/Ang-1 ratio was increased by ~3.5-fold. Moreover, Tie-2 mRNA was increased ~2-fold 3 h post-exercise. However, these data contradict previous findings by Gustafsson et al. [34], who reported no changes in Ang-1 or Ang-2 mRNA following exercise. These different observations could be explained by differences in exercise mode and intensity utilized in each of the two studies (i.e., cycling at 60% VO_{2max} utilized by Hoier et al. [77] vs. one-legged knee extension at the highest tolerable workload utilized by Gustafsson et al. [34]).

Endurance-trained athletes may have a different angiopoietin response to an acute bout of exercise. Moreover, the response to a lower-intensity, longer duration (e.g., marathon running) vs. a high-intensity, short bout (e.g., 1,500 m sprint) may induce different Ang-1 and Ang-2 responses. Specifically, circulating Ang-1 increased ~2.2-fold immediately after a marathon and remained unchanged after a 1,500 m field test [101]. However, immediately after the marathon and 1,500 m field test, Ang-2 increased 2- and 1.3-fold, respectively. Even so, the Ang-2/Ang-1 ratio remained unchanged after both events. Receptor Tie2 expression may also be altered by resistance exercise. The protocol utilized by Gavin et al. [80] found that vastus lateralis Tie2 mRNA increased twofold 4 h after the acute resistance exercise test. However, Ang-1, Ang-2, and the Ang-2/Ang-1 ratio remained unchanged.

Chronic exercise training appears to alter the response of Ang-1 and Ang-2 to an acute exercise bout and elicits a change in resting levels. Forty-five minutes of cycling four times per week for 6 weeks at 75% VO_{2max} increased vastus lateralis Ang-1 mRNA approximately 2.5-fold in subjects identified as "high responders" by Timmons et al. [83], while "low responders" had no change. In contrast, Ang-2 remained unchanged in both groups and therefore indicating a reversal in the Ang-2/Ang-1 ratio after an acute exercise bout prior to exercise training. However, Hoier et al. [77] demonstrated that Ang-2 mRNA after 4 weeks of aerobic training increased ~1.8-fold 3 h after an acute bout of exercise compared to rest. These data suggest that angiopoietins Ang-1 and Ang-2 tend to change in a time-dependent manner in response to an acute bout of exercise after chronic exercise training. It should also be pointed out that single-leg knee extension under non-restricted and restricted blood flow conditions for 45 min four times per week for 5 weeks did not alter vastus lateralis Ang-1, Ang-2 mRNA, or the Ang-2/Ang-1 ratio, while basal vastus lateralis Ang-1 mRNA was reduced 10 days after training and returned to baseline after 5 weeks [34]. Moreover, basal Ang-2 mRNA and protein increased ~1.5-fold after 10 days of training. Although Ang-2 mRNA remained elevated, Ang-2 protein returned to baseline levels after 5 weeks of training. Even so, basal Ang-2/Ang-1 ratio was elevated ~1.5-fold after 10 days of training and remained elevated after 5 weeks of training.

11.3.4 Other Factors That Stimulate Angiogenesis

Endothelial cells and mural cells produce and activate *transforming growth factor*-β (TGF-β) which plays a critical role in the maturation of blood vessels by covering the vessel with mural cells (see review by Jain [40]). Mice lacking the TGF-β gene die from defects of the vasculature in utero, suggesting an important role in angiogenesis prior to birth [102]. One hour of one-leg kicking exercise at 67% maximum workload increased TGF-β in the connective tissue of the vastus lateralis approximately twofold [103]. A maximal graded exercise test has also shown to increase serum TGF-β 2.7-fold immediately after exercise and 1.9-fold 2 h after exercise compared to baseline [104]. To our knowledge, TGF-β in response to chronic exercise training has not been investigated. Although *basic fibroblast growth factor* (bFGF or FGF-2), a 18–23 kDa cationic mitogen, has been shown to upregulate VEGF [105] as well as NO production [106], some studies have failed to demonstrate its upregulation in response to acute or chronic exercise training in humans [32, 79, 107].

11.4 Factors That Inhibit Angiogenesis in Response to Exercise

In addition to several pro-angiogenic molecules, angiogenesis is also regulated by several key anti-angiogenic molecules; however, their response to acute exercise and chronic exercise has been investigated in less depth than the pro-angiogenic factors. Some of these anti-angiogenic factors are described below.

11.4.1 Endostatin in Acute and Chronic Exercise Training

Endostatin, a 20-kDa fragment released from collagen XVIII C-terminal fragment, blocks angiogenesis by preventing the release of VEGF and VEGF receptor signaling [108, 109] as well as MMP-2 activity [110]. It can thus be seen to restrict endothelial cell proliferation and migration. Endostatin also plays a role in relaxation of the vessel walls through *endothelial nitric oxide synthase* (eNOS) phosphorylation and prostacyclin release [111, 112]. A 65 min bout of cycling at 65% VO_{2max} utilized by Rullman et al. [78] increased plasma arterial endostatin ~1.7-fold 17 min into the exercise protocol, which was higher than plasma venous endostatin, which increased ~1.6-fold but then returned to resting values at 57 min into exercise and remained at basal levels 120 min post-exercise. Interestingly, plasma venous endostatin continued to increase at 57 min by ~2-fold and was significantly higher compared to plasma arterial endostatin, suggesting an uptake from the circulation into the muscle. Cycling under hypoxic conditions in conjunction with vibration training, utilized by Suhr et al. [92], increased circulating endostatin by 1.1–1.25-fold immediately after exercise, with the greatest increase found under normal conditions and the lowest increase under hypoxic conditions with vibration. Circulating endostatin remained elevated 30 and 60 min post-exercise only in the normal conditions protocol. A maximal exercise test may induce a greater endostatin response. Specifically, plasma endostatin concentrations increased at 30, 120, and 360 min after exercise by 43%, 73%, and 33%, respectively, compared to rest [107]. These data indicate that endostatin increases to a greater extent following heavy intensity exercise, as compared to moderate intensity exercise. On the other hand, chronic exercise training may reduce resting plasma endostatin, which would thereby promote angiogenesis. Participants who either ran for 60 min or cycled for 90 min three times per week at moderate intensity for 6 months reduced basal circulating endostatin levels by 14% in both groups [113]. Furthermore, participants in both exercise groups significantly reduced basal endostatin levels when compared to a sedentary control group. Nonetheless, further research is needed to more clearly demonstrate the influence of chronic exercise training on the regulation of endostatin mRNA and protein content in humans.

11.4.2 Thrombospondin-1 in Acute and Chronic Exercise Training

Another endogenous inhibitor of angiogenesis is thrombospondin-1 (TSP-1), which is a large 450 kDa matrix glycoprotein produced by a number of cells, including platelet α-granules, smooth muscle cells, endothelial cells, fibroblasts, neutrophils, and macrophages and located in the extracellular compartment of several organs and tissues, including skeletal muscle, bone, skin, lung, and connective tissue [114, 115]. Its primary anti-angiogenic role is to inhibit cell migration and to induce apoptosis in endothelial cells [115–117]. Few data exist on the response of TSP-1 to acute exercise in humans. A 60 min bout of treadmill running at 20 m/min at 10° incline increased rat gastrocnemius TSP-1 mRNA 3.5-fold immediately after exercise and was further increased 6.2-fold 1 h after exercise, compared to resting levels [118]. More recently, 60 min of cycling at 60% VO_{2max} was shown to increase human vastus lateralis TSP-1 mRNA ~2.75-fold at 1 h after exercise and remained elevated for up to 3 h after exercise [77]. Exercise-trained rats, under both normoxic and hypoxic conditions for 60 min a day, 5 days per week for 8 weeks at 18 m/min and 10° incline, were shown to upregulate gastrocnemius TSP-1 mRNA 3.4 and 3.3-fold, respectively [118]. However, this response was not altered by hypoxia. Humans who completed 4 weeks of cycling for 60 min at 60% VO_{2max} also showed a ~2.5-fold increase in TSP-1 mRNA in the vastus lateralis when sampled 1 h after an acute bout of exercise; however, this acute effect was not different from the pre-training response, suggesting that TSP-1 may not be affected by chronic exercise training [77].

11.4.3 Tissue Inhibitors of Matrix Metalloproteinases in Acute Exercise and Chronic Exercise Training

TIMPs are polypeptides containing 184–194 amino acids and are shown to regulate MMP activity by suppressing extracellular matrix turnover and activation of proteins on the cell surface [119]. TIMPs also have other anti-angiogenic properties without inhibiting MMPs through TIMP-2 integrin-mediated binding to endothelial cells (see review by Stetler-Stevenson et al. [120]). Human studies indicate that TIMPs are not upregulated in the circulation [121, 122] or skeletal muscle [77, 78, 89] in response to acute exercise. This is in contrast to the evidence from an animal study indicating that rat soleus TIMP-1 and -2 mRNA increased twofold 6 h after a bout of eccentric exercise consisting of 17 m/min for 130 min at 13.5° decline, which was sustained for 2 days after running and soleus TIMP-1 protein increased 1.5-fold [123]. The lack of research examining the influence of exercise on the regulation of TIMPs suggests that this area may be investigated more fully in the future. Chronic exercise training appears to increase TIMP concentration in human skeletal muscle. Rullman et al. [89] found that 10 days of single-leg knee extension training increased vastus lateralis TIMP-1 mRNA ~4- and 6-fold 24 h after an acute bout of exercise in the non-restricted and restricted blood flow conditions, respectively, compared to pre-training levels. Moreover, Hoier et al. [77] found that 4 weeks of cycle ergometer training increased vastus lateralis TIMP-1 mRNA ~2-fold 1 and 3 h after an acute bout of exercise. These data indicate that chronic exercise training increases the expression of TIMP-1 mRNA after an acute bout of exercise compared to pre-training states, suggesting that the inhibition of extracellular remodeling is greater after training when there is sufficient capillary growth, which was demonstrated by Hoier et al. [77].

11.4.4 Other Factors That Inhibit Angiogenesis

Angiostatin was the first discovered anti-angiogenic factor found to block tumor neovascularization and growth of metastases [124]. Angiostatin is a 38-kDa proteolytic fragment derived from degradation of MMP. It is plausible that angiostatin is upregulated by exercise because MMP is known to increase in response to acute exercise and chronic exercise training, but this possibility remains to be investigated. Another factor, namely *vasohibin-1* (VASH-1), is secreted by endothelial cells and contributes to vessel stabilization and maturation [125] and is more highly expressed in less vascularized muscle compared to more vascularized muscle [126]. Rats that completed a maximal running treadmill test that lasted 60–90 min increased VASH-1 by ~1.5-fold immediately after exercise in the hindlimb but returned to baseline 4 h after exercise [126]. In contrast, myocardial VASH-1 did not increase immediately after exercise but increased 4 h after the acute exercise bout. The influence that chronic exercise training has on VASH-1 remains to be fully elucidated because this study indicated that 5 days of training for 60 min at 25–30 m/min at 12° incline did not increase VASH-1 levels in rat hindlimb or heart, as compared to baseline.

11.5 Endothelial Progenitor Cells in Acute Exercise and Chronic Exercise Training

Bone marrow contains EPCs which have the ability to multiply and differentiate into endothelial cells (see review by Ribatti [127]) and were first discovered in adult peripheral blood [128]. EPCs have also been shown to reside in interstitial tissue surrounding distressed tissues and organs in need of vascular regeneration where the EPCs produce a variety of pro-angiogenic cytokines and growth factors, including VEGF, *hepatic growth factor* (HGF), Ang-1, *stroma-derived growth factor* (SDF)-1α, *insulin-like growth factor* (IGF)-1, and eNOS [129]. They have also been suggested to play an important role in the regulation of angiogenesis through the differentiation and maturation to endothelial cells and by coordination of paracrine signaling (see review by Carmeliet and Jain [37]). Moreover, endothelial progenitors may play a role in exercise-induced angiogenesis; however, the mechanism of their migration from bone marrow is not fully understood.

Several markers of EPCs have been identified to increase after an acute bout of exercise including $CD34^+VEGFR\text{-}2^+$ cells, $CD34^+CD133^+$ cells, $CD133^+$VE-cadherin$^+$ cells, and monocyte/macrophage-derived angiogenic cells (CACs) in healthy subjects [101, 130–132] and appear to be associated with intensity of exercise. Specifically, immediately after a marathon run circulating $CD34^+VEGFR\text{-}2^+$, $CD133^+$VE-cadherin$^+$, and CAC cell count was found to increase ~1.8-, 3.5-, and 4-fold, respectively, compared to rest, and returned to baseline 20 h after the race [101]. After completion of a 1,500 m field test, $CD34^+VEGFR\text{-}2^+$, $CD133^+$VE-cadherin$^+$, and CAC increased ~2.25-, 5-, and 4-fold, respectively, compared to rest. Moreover, group differences were observed, where $CD34^+VEGFR\text{-}2^+$ was ~1.8-fold higher in the 1,500 m group compared to the marathon group and CAC cell count was ~2.3-fold higher in the marathon group compared to the 1,500 m group [101]. After a maximal graded exercise test on a cycle ergometer, $CD34^+VEGFR\text{-}2^+$ increased by ~1.7-fold, compared to rest [131]. Thirty minutes of running at anaerobic threshold and 80% of anaerobic threshold increased circulating endothelial progenitors by 235% and 263%, respectively [133]. However, no changes in circulating EPCs were observed after 10 min of running at 80% of anaerobic threshold. Therefore, these data suggest that EPCs may be mobilized differently from the bone marrow in response to different intensities of exercise. Moreover, duration of exercise may play an important role in the release of endothelial progenitors from the bone marrow. To our knowledge, there is no data that exists to determine if chronic exercise training increases circulating EPCs in healthy subjects. However, several studies have demonstrated that endothelial progenitors increase after chronic exercise training in patients with cardiovascular disease [132, 134–136].

11.6 miRNAs in Acute Exercise and Chronic Exercise Training

miRNAs are highly conserved 20–30 nucleotide RNAs that regulate the translation of proteins and enhance messenger RNA (mRNA) degradation (see review by Bartel [137]). miRNAs are formed by two major enzymes; Dicer and Drosha. Drosha regulates nuclear processing of the primary miRNAs into precursors, or pre miRNAs [138], and Dicer cleaves pre-miRNA into the mature miRNA [139]. A number of miRNAs have been identified within cardiac and skeletal muscle which may be involved in cardiogenesis during embryonic development [140] and skeletal muscle myogenesis [141] and remodeling [142]. Several miRNAs including miR-20a [143], miR-210 [144], miR-221/miR-222 [145], and miR-328 [146] have also been identified which have anti-angiogenic effects. Specifically, miR-20a expression is increased by VEGF but has an inhibitory effect on endothelial cell migration and overall capillary formation [147]. Increased miR-328 expression is found to reduce cell adhesion, aggregation, and migration and also regulates the formation of the capillary structure [146]. miR-210 upregulation under normoxic conditions increases endothelial cell tubulogenesis and migration, but under hypoxic conditions, miR-210 decreases capillary formation [144]. It has also been shown that supplementation of miR-210 to the myocardium to the pre-infarct zone induces angiogenesis in a mouse model [148]. miR-221/miR-222 mediate the angiogenic activity of *stem-cell factor* (SCF) [145]. These miRNAs have only more recently been investigated in response to an acute bout of exercise and chronic exercise training. The response of miRNAS to an acute bout of exercise has been investigated in competitive rowers [149]. Specifically, it has been shown that circulating miR-221 and miR-222 are increased ~3.6- and 2.5-fold immediately after a maximal, incremental cycling exercise test, respectively. Further investigations may be warranted to determine if acute exercise has similar effects on miRNA regulation in non-athlete populations.

Baggish et al. [149] investigated the effects of chronic exercise training in rowing athletes. After 90 days of training for an average 13.1 h per week, miR-221 and miR-222 increased ~5.8- and 2.4-fold compared to baseline resting levels, respectively. After the chronic exercise training period, miR-20a was 3.1-fold higher after an acute bout of exercise compared to baseline resting levels. An additional increase in circulating miR-222 was found (fourfold increase) after an acute bout of exercise after training but not in miR-221. These data indicate a different pattern in regulation between several miRNAs before and after chronic exercise training and further investigations are needed among different populations.

11.7 Exercise-Induced Angiogenesis in Chronic Disease

From the foregoing discussion, it is evident that chronic exercise training in healthy individuals may promote angiogenesis by affecting the expression of several angiogenic factors (Table 11.1). The role of chronic exercise training in the primary and secondary prevention of different diseases has also been investigated extensively. Interestingly, regular exercise may reduce the risk of developing over 25 chronic conditions [150] including coronary heart disease, stroke, hypertension, breast and colon cancer, type 2 diabetes, and osteoporosis. It is likely that research examining the biological mechanisms that regulate exercise-induced angiogenesis may guide the identification of novel therapeutic targets for the treatment of chronic diseases. For example, patients with cardiovascular disease who exercise train to a level

Table 11.1 Angiogenic factors that increase with chronic exercise training

Authors	Training conditions: Special conditions	Training conditions: Frequency	Exercise protocol: Intensity	Exercise protocol: Time	Exercise protocol: Type	Factors
Gustafsson et al. [34]	Normal Restricted blood flow	4×/week; 5 weeks	Highest tolerable WL	45 min	Single-leg knee extension	↑Ang-2; ↑Tie-1 ↑VEGF; ↑VEGFR-1;
Richardson et al. [75]	Normal	3×/week; 8 weeks	50% max WR	60 min	Single-leg knee extension	↑VEGF
Hoier et al. [77]	Normal	3×/week; 4 weeks	68% VO_{2max}	60 min	Cycle ergometer	↑MMP-9; ↑TIMP-1; ↑Ang-2; ↑Tie-2
Timmons et al. [83]	Normal	4×/week; 6 weeks	75% VO_{2max}	45 min	Cycle ergometer	High responders: ↑VEGFR-1; ↑VEGFR-2; ↑NRP-1↑Tie-1 Low responders: none
Rullman et al. [89]	Normal Restricted blood flow	4×/week; 5 weeks	Highest tolerable WL	45 min	Single-leg knee extension	↑MMP-2; ↑MMP-9; ↑MMP-14; ↑TIMP-1 ↑MMP-2; ↑MMP-9; ↑MMP-14; ↑TIMP-1
Urso et al. [93]	Calisthenics RT	5×/week; 8 weeks	To fatigue 80% 1RM	Not indicated	Calisthenics and AT RT and AT	↑MMP-9 None
Brixius et al. [113]	Running Cycling	3×/week; 6 month	Pulse 2–4 mmol/L lactate	60 min 90 min	Running Cycling	↓Endostatin ↓Endostatin
Baggish et al. [149]	Elite rowing athlete training	90 days	Not indicated	5.9 h/week	Rowing	↑miR-146a; ↑miR-222; ↑miR-21; ↑miR-221; ↑miR-20a

VO_2max maximum oxygen consumption; *VEGF* vascular endothelial growth factor; *VEGFR* vascular endothelial growth factor receptor; *WL* workload; *Ang* angiopoietin; *Tie* angiopoietin receptor; *WR* work rate; *NRP* neuropilin; *eNOS* endothelial nitric oxide synthase; *MMP* matrix metalloproteinase; *TIMP* tissue inhibitors of matrix metalloproteinases; *RM* repetition maximum; *miR* microR

sufficient to stimulate angiogenesis and the formation of collateral vessels in ischemic areas of the heart and skeletal muscle are known to decrease the severity of ischemic events [151–153]. The potential of chronic exercise training to upregulate and normalize circulating pro-angiogenic growth factors and EPCs in patients with cardiovascular disease has been documented previously [132, 135, 154–162]; it also has the potential to improve collateral flow in the heart [163] and increase capillary density in skeletal muscle [160]. Various studies examining the role of chronic exercise training and its ability to upregulate several pro- and anti-angiogenic growth factors as well as EPCs are listed in Table 11.2. Similar health benefits may also be attained by peripheral arterial disease patients, which is a population characterized with ischemia of the limbs that often leads to tissue necrosis requiring amputation [164]. Notably, chronic exercise training consisting of 50 min of walking exercise at an intensity below that which induces claudication symptoms twice a week for 6 months significantly elevated EPCs and VEGF in patients diagnosed with peripheral artery disease [135]. Moreover, exercise-induced angiogenesis may have an important therapeutic role in heart failure. It may be noted that patients with heart failure often experience dyspnea and exercise intolerance, which is not only due to insufficient cardiac output and impaired lung function but also due to peripheral pathophysiological maladaptations such as endothelial dysfunction [165], abnormal skeletal morphology and metabolism [166] and ventilatory control [167]. Chronic exercise training consisting of 55 min of upright cycling three times a week for 8 weeks at an intensity that elevated blood lactate to 2–2.5 mmol/L markedly increased circulating EPCs and pro-angiogenic factors, including VEGF amongst a cohort of heart failure patients. This beneficial effect of exercise training washed out 8 weeks after the heart failure patients stopped exercise training [136].

One goal of cancer therapy is to prevent the blood supply of a tumor using anti-angiogenic drugs [168]. Interestingly, even though chronic exercise training induces a positive angiogenic response in the heart and skeletal muscle, it may have an anti-angiogenic effect on tumors. To our knowledge there are currently no human models for investigating the effects of chronic exercise training and tumor blood vessel supply. Furthermore, the potential mechanisms to explain how exercise training exerts an anti-angiogenic effect on tumors still remain to be elucidated. Nonetheless, the effects of chronic exercise training have been examined using a Dalton's lymphoma mouse model [169]. Data from that study [169] indicate that blood supply to the peritoneal cavity was significantly reduced in mice that exercised 30 min per day three times per week at approximately 80% VO_{2max} [169]. Moreover, mice that exercised 60 min a day had a further decrease in blood supply to the peritoneal cavity compared to the 30 min group. One of the potential mechanisms for the reduced blood supply was that VEGF was significantly reduced in the exercise-trained mice vs. the sedentary controls. However, this may be specific to the type of cancer, as it was found that 44 days of voluntary wheel running did not reduce blood supply or tumor growth in a mouse model of human breast cancer [170]. Further studies should continue to examine the influence that chronic exercise training exerts on tumor growth and angiogenesis.

Table 11.2 Chronic exercise training studies in different disease populations

		Exercise protocol				
Disease	References	Frequency	Intensity	Time	Type	Factors
Coronary artery disease	Laufs et al. [133]	4 weeks	60–80% VO_{2max}	25 min	Cycle ergometer; walking	↑EPC
	Steiner et al. [134]	5×/week; 12 weeks	Not indicated	30–60 min	Cycle ergometer; running	↑EPC; ↑FMD; ↑NO_x
	Lee et al. [161]	3×/week; 12 weeks	55–70% VO_{2max}	30 min	Cycle ergometer	↑MBF; ↓VEGF; ↓SDF-1
Peripheral arterial disease	Schlager et al. [135]	2×/week; 6 months	At symptoms of claudication	50 min	Walking	↑EPC; ↓ADMA
	Duscha et al. [160]	3×/week; 12–16 weeks	Not indicated	30–40 min	Walking	↑CD
	Sandri et al. [162]	5 days/week; 4 weeks	At symptoms of claudication	Not indicated	Walking	↑VEGF; ↑EPC
Hypertension	Hansen et al. [159]	3×/week; 16 weeks	Not indicated	60 min	Endurance and strength exercise	↑C/F ratio; ↑VEGF
Heart failure	Sarto et al. [136]	3×/week; 8 weeks	60% HRR	55 min	Cycle ergometer	↑EPC; ↑VEGF/SDF-1 ratio
	Gustafsson et al. [156]	3×/week; 8 weeks	50% peak work rate	~40 min	Single-leg knee extension	↑Citrate synthase; ↑VEGF
	Gatta et al. [158]	6×/week; 3 weeks	85% max HR	~30 min	Calisthenics; cycle ergometer	↑EPC; ↓MMP-1; ↓TIMP-1; ↑MMP-2/TIMP-1 ratio; ↑MMP-9/TIMP-1 ratio

EPCs endothelial progenitor cells; *FMD* flow-mediated dilation; NO_x nitrites/nitrites; *VEGF* vascular endothelial growth factor; *SDF-1* stroma-derived factor 1; *MBF* myocardial blood flow; *ADMA* asymmetric dimethylarginine; *CD* capillary density; *C/F* capillary-to-fiber; *HRR* heart-rate reserve; *MMP* matrix metalloproteinase; *TIMP-1* tissue inhibitors of matrix metalloproteinases-1

11.8 Conclusions

Angiogenesis is a process regulated by pro- and anti-angiogenic factors and plays an important role for normal growth and physiological function. Several disease states are also characterized by deficient or dysregulated angiogenesis. In general, the literature indicates acute and chronic exercise training stimulates angiogenesis by upregulating pro- and anti-angiogenic factors that ultimately increase capillary density in both heart and skeletal muscle. Chronic exercise training may also have an important therapeutic role for regulating angiogenesis healthy as well as clinical populations.

Acknowledgments TAD is supported by research funding from Manitoba Health and Research Council (MHRC) and the Canadian Institutes for Health Research (CIHR). DSK is supported by a MHRC Graduate Studentship. The infrastructural support during this study was provided by the St. Boniface Hospital Research Foundation.

References

1. Carmeliet P (2003) Angiogenesis in health and disease. Nat Med 9:653–660
2. Krogh A (1919) The number and distribution of capillaries in muscles with calculations of the oxygen pressure head necessary for supplying the tissue. J Physiol 52:409–415
3. Krogh A (1919) The rate of diffusion of gases through animal tissues, with some remarks on the coefficient of invasion. J Physiol 52:391–408
4. Krogh A (1919) The supply of oxygen to the tissues and the regulation of the capillary circulation. J Physiol 52:457–474
5. Krogh A (1922) The anatomy and physiology of capillaries. Yale University Press, New Haven, CT
6. Hertig A (1935) Angiogenesis in the early human chorion and in the primary placenta of the macaque monkey. Contrib Embryol 25:37–81
7. Sandison JC (1932) Contraction of blood vessels and observations on the circulation in the transparent chamber in the rabbit's ear. Anat Rec 24:105–127
8. Findlay JK (1986) Angiogenesis in reproductive tissues. J Endocrinol 111:357–366
9. Hudlicka O, Wright AJ, Ziada AM (1986) Angiogenesis in the heart and skeletal muscle. Can J Cardiol 2:120–123
10. Kawamata T, Speliotes EK, Finklestein SP (1997) The role of polypeptide growth factors in recovery from stroke. Adv Neurol 73:377–382
11. Kumar S, West D, Shahabuddin S et al (1983) Angiogenesis factor from human myocardial infarcts. Lancet 13:364–368
12. Burgos H, Herd A, Bennett JP (1989) Placental angiogenic and growth factors in the treatment of chronic varicose ulcers: preliminary communication. J R Soc Med 82:598–599
13. Baird A (1994) Fibroblast growth factors: activities and significance of non-neurotrophin neurotrophic growth factors. Curr Opin Neurobiol 4:78–86
14. Folkman J, Merler E, Abernathy C et al (1971) Isolation of a tumor factor responsible for angiogenesis. J Exp Med 1:275–288
15. Folkman J (1995) Angiogenesis in cancer, vascular, rheumatoid and other disease. Nat Med 1:27–31
16. Christou H, Yoshida A, Arthur V et al (1998) Increased vascular endothelial growth factor production in the lungs of rats with hypoxia-induced pulmonary hypertension. Am J Respir Cell Mol Biol 18:768–776

17. Patz A (1978) Current concepts in ophthalmology. Retinal vascular diseases. N Engl J Med 298:1451–1454
18. Laughlin MH, Bowles DK, Duncker DJ (2012) The coronary circulation in exercise training. Am J Physiol Heart Circ Physiol 302:H10–H23
19. Savard R, Smith LJ, Palmer JE et al (1988) Site specific effects of acute exercise on muscle and adipose tissue metabolism in sedentary female rats. Physiol Behav 43:65–71
20. Dohm GL, Huston RL, Askew EW et al (1973) Effects of exercise, training, and diet on muscle citric acid cycle enzyme activity. Can J Biochem 51:849–854
21. Egginton S (2009) Invited review: activity-induced angiogenesis. Pflugers Arch 457:963–977
22. Hudlicka O, Brown M, Egginton S (1992) Angiogenesis in skeletal and cardiac muscle. Physiol Rev 72:369–417
23. Sundberg CJ, Kaijser L (1992) Effects of graded restriction of perfusion on circulation and metabolism in the working leg; quantification of a human ischaemia-model. Acta Physiol Scand 146:1–9
24. Milkiewicz M, Ispanovic E, Doyle JL et al (2006) Regulators of angiogenesis and strategies for their therapeutic manipulation. Int J Biochem Cell Biol 38:333–357
25. Dvorak HF, Orenstein NS, Carvalho AC et al (1979) Induction of a fibrin-gel investment: an early event in line 10 hepatocarcinoma growth mediated by tumor-secreted products. J Immunol 122:166–174
26. Dvorak HF, Brown LF, Detmar M et al (1995) Vascular permeability factor/vascular endothelial growth factor, microvascular hyperpermeability, and angiogenesis. Am J Pathol 146:1029–1039
27. Gustafsson T, Puntschart A, Kaijser L et al (1999) Exercise-induced expression of angiogenesis-related transcription and growth factors in human skeletal muscle. Am J Physiol 276:H679–H685
28. Richardson RS, Wagner H, Mudaliar SR et al (1999) Human VEGF gene expression in skeletal muscle: effect of acute normoxic and hypoxic exercise. Am J Physiol 277:H2247–H2252
29. Gustafsson T, Knutsson A, Puntschart A et al (2002) Increased expression of vascular endothelial growth factor in human skeletal muscle in response to short-term one-legged exercise training. Pflugers Arch 444:752–759
30. Gavin TP, Robinson CB, Yeager RC et al (2004) Angiogenic growth factor response to acute systemic exercise in human skeletal muscle. J Appl Physiol 96:19–24
31. Jensen L, Pilegaard H, Neufer PD et al (2004) Effect of acute exercise and exercise training on VEGF splice variants in human skeletal muscle. Am J Physiol Regul Integr Comp Physiol 287:R397–R402
32. Jensen L, Bangsbo J, Hellsten Y (2004) Effect of high intensity training on capillarization and presence of angiogenic factors in human skeletal muscle. J Physiol 557:571–582
33. Gustafsson T, Ameln H, Fischer H et al (2005) VEGF-A splice variants and related receptor expression in human skeletal muscle following submaximal exercise. J Appl Physiol 98:2137–2146
34. Gustafsson T, Rundqvist H, Norrbom J et al (2007) The influence of physical training on the angiopoietin and VEGF-A systems in human skeletal muscle. J Appl Physiol 103:1012–1020
35. Egginton S, Zhou AL, Brown MD et al (2001) Unorthodox angiogenesis in skeletal muscle. Cardiovasc Res 49:634–646
36. Gustafsson T (2011) Vascular remodelling in human skeletal muscle. Biochem Soc Trans 39:1628–1632
37. Carmeliet P, Jain RK (2011) Molecular mechanisms and clinical applications of angiogenesis. Nature 473:298–307
38. Folkman J, Klagsbrun M (1987) Angiogenic factors. Science 235:442–447
39. Folkman J (1985) Toward an understanding of angiogenesis: search and discovery. Perspect Biol Med 29:10–36
40. Jain RK (2003) Molecular regulation of vessel maturation. Nat Med 9:685–693

41. Prior BM, Yang HT, Terjung RL (2004) What makes vessels grow with exercise training? J Appl Physiol 97:1119–1128
42. Qutub AA, Mac Gabhann F, Karagiannis ED et al (2009) Multiscale models of angiogenesis. IEEE Eng Med Biol Mag 28:14–31
43. Yancopoulos GD, Davis S, Gale NW et al (2000) Vascular-specific growth factors and blood vessel formation. Nature 14:242–248
44. Egginton S (2011) Physiological factors influencing capillary growth. Acta Physiol (Oxf) 202:225–239
45. Djonov V, Baum O, Burri PH (2003) Vascular remodeling by intussusceptive angiogenesis. Cell Tissue Res 314:107–117
46. Gianni-Barrera R, Trani M, Reginato S et al (2011) To sprout or to split? VEGF, Notch and vascular morphogenesis. Biochem Soc Trans 39:1644–1648
47. Styp-Rekowska B, Hlushchuk R, Pries AR et al (2011) Intussusceptive angiogenesis: pillars against the blood flow. Acta Physiol (Oxf) 202:213–223
48. Makanya AN, Hlushchuk R, Djonov VG (2009) Intussusceptive angiogenesis and its role in vascular morphogenesis, patterning, and remodeling. Angiogenesis 12:113–123
49. Burri PH, Djonov V (2002) Intussusceptive angiogenesis—the alternative to capillary sprouting. Mol Aspects Med 23:S1–S27
50. Hobson B, Denekamp J (1984) Endothelial proliferation in tumours and normal tissues: continuous labelling studies. Br J Cancer 49:405–413
51. Haas TL, Milkiewicz M, Davis SJ et al (2000) Matrix metalloproteinase activity is required for activity-induced angiogenesis in rat skeletal muscle. Am J Physiol Heart Circ Physiol 279:H1540–H1547
52. Hansen-Smith FM, Hudlicka O, Egginton S (1996) In vivo angiogenesis in adult rat skeletal muscle: early changes in capillary network architecture and ultrastructure. Cell Tissue Res 286:123–136
53. Paku S, Paweletz N (1991) First steps of tumor-related angiogenesis. Lab Invest 65:334–346
54. Ausprunk DH, Folkman J (1977) Migration and proliferation of endothelial cells in preformed and newly formed blood vessels during tumor angiogenesis. Microvasc Res 14:53–65
55. Gerhardt H, Golding M, Fruttiger M et al (2003) VEGF guides angiogenic sprouting utilizing endothelial tip cell filopodia. J Cell Biol 161:1163–1177
56. Benjamin LE, Golijanin D, Itin A et al (1999) Selective ablation of immature blood vessels in established human tumors follows vascular endothelial growth factor withdrawal. J Clin Invest 103:159–165
57. Dor Y, Djonov V, Abramovitch R et al (2002) Conditional switching of VEGF provides new insights into adult neovascularization and pro-angiogenic therapy. EMBO J 21:1939–1947
58. Caduff JH, Fischer LC, Burri PH (1986) Scanning electron microscope study of the developing microvasculature in the postnatal rat lung. Anat Rec 216:154–164
59. Patan S, Haenni B, Burri PH (1996) Implementation of intussusceptive microvascular growth in the chicken chorioallantoic membrane (CAM): 1. pillar formation by folding of the capillary wall. Microvasc Res 51:80–98
60. Kauffman SL, Burri PH, Weibel ER (1974) The postnatal growth of the rat lung. II. Autoradiography. Anat Rec 180:63–76
61. Lloyd PG, Prior BM, Li H et al (2005) VEGF receptor antagonism blocks arteriogenesis, but only partially inhibits angiogenesis, in skeletal muscle of exercise-trained rats. Am J Physiol Heart Circ Physiol 288:H759–H768
62. Williams JL, Cartland D, Rudge JS et al (2006) VEGF trap abolishes shear stress- and overload-dependent angiogenesis in skeletal muscle. Microcirculation 13:499–509
63. Olfert IM, Howlett RA, Wagner PD et al (2010) Myocyte vascular endothelial growth factor is required for exercise-induced skeletal muscle angiogenesis. Am J Physiol Regul Integr Comp Physiol 299:R1059–R1067

64. Forsythe JA, Jiang BH, Iyer NV et al (1996) Activation of vascular endothelial growth factor gene transcription by hypoxia-inducible factor 1. Mol Cell Biol 16:4604–4613
65. Shweiki D, Itin A, Soffer D et al (1992) Vascular endothelial growth factor induced by hypoxia may mediate hypoxia-initiated angiogenesis. Nature 29:843–845
66. Richardson RS, Duteil S, Wary C et al (2006) Human skeletal muscle intracellular oxygenation: the impact of ambient oxygen availability. J Physiol 571:415–424
67. Richardson RS, Noyszewski EA, Kendrick KF et al (1995) Myoglobin O2 desaturation during exercise. Evidence of limited O2 transport. J Clin Invest 96:1916–1926
68. Mason SD, Howlett RA, Kim MJ et al (2004) Loss of skeletal muscle HIF-1alpha results in altered exercise endurance. PLoS Biol 2:e288
69. Ameln H, Gustafsson T, Sundberg CJ et al (2005) Physiological activation of hypoxia inducible factor-1 in human skeletal muscle. FASEB J 19:1009–1011
70. Fraisl P, Mazzone M, Schmidt T et al (2009) Regulation of angiogenesis by oxygen and metabolism. Dev Cell 16:167–179
71. Hardie DG (2003) Minireview: the AMP-activated protein kinase cascade: the key sensor of cellular energy status. Endocrinology 144:5179–5183
72. Ouchi N, Shibata R, Walsh K (2005) AMP-activated protein kinase signaling stimulates VEGF expression and angiogenesis in skeletal muscle. Circ Res 29:838–846
73. Jensen L, Schjerling P, Hellsten Y (2004) Regulation of VEGF and bFGF mRNA expression and other proliferative compounds in skeletal muscle cells. Angiogenesis 7:255–267
74. Arany Z, Foo SY, Ma Y et al (2008) HIF-independent regulation of VEGF and angiogenesis by the transcriptional coactivator PGC-1alpha. Nature 451:1008–1012
75. Richardson RS, Wagner H, Mudaliar SR, Saucedo E, Henry R, Wagner PD (2000) Exercise adaptation attenuates VEGF gene expression in human skeletal muscle. Am J Physiol Heart Circ Physiol 279:H772–H778
76. Neufeld G, Cohen T, Gengrinovitch S et al (1999) Vascular endothelial growth factor (VEGF) and its receptors. FASEB J 13:9–22
77. Hoier B, Nordsborg N, Andersen S et al (2012) Pro- and anti-angiogenic factors in human skeletal muscle in response to acute exercise and training. J Physiol 590:595–606
78. Rullman E, Rundqvist H, Wagsater D et al (2007) A single bout of exercise activates matrix metalloproteinase in human skeletal muscle. J Appl Physiol 102:2346–2351
79. Wahl P, Zinner C, Achtzehn S et al (2011) Effects of acid–base balance and high or low intensity exercise on VEGF and bFGF. Eur J Appl Physiol 111:1405–1413
80. Gavin TP, Drew JL, Kubik CJ et al (2007) Acute resistance exercise increases skeletal muscle angiogenic growth factor expression. Acta Physiol (Oxf) 191:139–146
81. Rojas Vega S, Knicker A, Hollmann W et al (2010) Effect of resistance exercise on serum levels of growth factors in humans. Horm Metab Res 42:982–986
82. Takano H, Morita T, Iida H et al (2005) Hemodynamic and hormonal responses to a short-term low-intensity resistance exercise with the reduction of muscle blood flow. Eur J Appl Physiol 95:65–73
83. Timmons JA, Jansson E, Fischer H et al (2005) Modulation of extracellular matrix genes reflects the magnitude of physiological adaptation to aerobic exercise training in humans. BMC Biol 3:19
84. Olsson AK, Dimberg A, Kreuger J et al (2006) VEGF receptor signalling—in control of vascular function. Nat Rev Mol Cell Biol 7:359–371
85. Chen X, Li Y (2009) Role of matrix metalloproteinases in skeletal muscle: migration, differentiation, regeneration and fibrosis. Cell Adh Migr 3:337–341
86. Baum O, Ganster M, Baumgartner I et al (2007) Basement membrane remodeling in skeletal muscles of patients with limb ischemia involves regulation of matrix metalloproteinases and tissue inhibitor of matrix metalloproteinases. J Vasc Res 44:202–213
87. Carmeli E, Moas M, Lennon S et al (2005) High intensity exercise increases expression of matrix metalloproteinases in fast skeletal muscle fibres. Exp Physiol 90:613–619
88. Deryugina EI, Quigley JP (2006) Matrix metalloproteinases and tumor metastasis. Cancer Metastasis Rev 25:9–34

89. Rullman E, Norrbom J, Stromberg A et al (2009) Endurance exercise activates matrix metalloproteinases in human skeletal muscle. J Appl Physiol 106:804–812
90. Deus AP, Bassi D, Simoes RP et al (2012) MMP(−2) expression in skeletal muscle after strength training. Int J Sports Med 33:137–141
91. Heinemeier KM, Olesen JL, Haddad F et al (2007) Expression of collagen and related growth factors in rat tendon and skeletal muscle in response to specific contraction types. J Physiol 582:1303–1316
92. Suhr F, Brixius K, de Marees M et al (2007) Effects of short-term vibration and hypoxia during high-intensity cycling exercise on circulating levels of angiogenic regulators in humans. J Appl Physiol 103:474–483
93. Urso ML, Pierce JR, Alemany JA et al (2009) Effects of exercise training on the matrix metalloprotease response to acute exercise. Eur J Appl Physiol 106:655–663
94. Haas TL, Davis SJ, Madri JA (1998) Three-dimensional type I collagen lattices induce coordinate expression of matrix metalloproteinases MT1-MMP and MMP-2 in microvascular endothelial cells. J Biol Chem 273:3604–3610
95. Yan C, Boyd DD (2007) Regulation of matrix metalloproteinase gene expression. J Cell Physiol 211:19–26
96. Mester J, Kleinoder H, Yue Z (2006) Vibration training: benefits and risks. J Biomech 39:1056–1065
97. Davis S, Aldrich TH, Jones PF et al (1996) Isolation of angiopoietin-1, a ligand for the TIE2 receptor, by secretion-trap expression cloning. Cell 87:1161–1169
98. Maisonpierre PC, Suri C, Jones PF et al (1997) Angiopoietin-2, a natural antagonist for Tie2 that disrupts in vivo angiogenesis. Science 277:55–60
99. Holash J, Maisonpierre PC, Compton D et al (1999) Vessel cooption, regression, and growth in tumors mediated by angiopoietins and VEGF. Science 284:1994–1998
100. Asahara T, Chen D, Takahashi T et al (1998) Tie2 receptor ligands, angiopoietin-1 and angiopoietin-2, modulate VEGF-induced postnatal neovascularization. Circ Res 83:233–240
101. Bonsignore MR, Morici G, Riccioni R et al (2010) Hemopoietic and angiogenetic progenitors in healthy athletes: different responses to endurance and maximal exercise. J Appl Physiol 109:60–67
102. Dickson MC, Martin JS, Cousins FM et al (1995) Defective haematopoiesis and vasculogenesis in transforming growth factor-beta 1 knock out mice. Development 121:1845–1854
103. Heinemeier KM, Bjerrum SS, Schjerling P et al (2011) Expression of extracellular matrix components and related growth factors in human tendon and muscle after acute exercise. Scand J Med Sci Sports 23:1–12
104. Czarkowska-Paczek B, Bartlomiejczyk I, Przybylski J (2006) The serum levels of growth factors: PDGF, TGF-beta and VEGF are increased after strenuous physical exercise. J Physiol Pharmacol 57:189–197
105. Stavri GT, Zachary IC, Baskerville PA et al (1995) Basic fibroblast growth factor upregulates the expression of vascular endothelial growth factor in vascular smooth muscle cells. Synergistic interaction with hypoxia. Circulation 92:11–14
106. Babaei S, Teichert-Kuliszewska K, Monge JC et al (1998) Role of nitric oxide in the angiogenic response in vitro to basic fibroblast growth factor. Circ Res 82:1007–1015
107. Gu JW, Gadonski G, Wang J et al (2004) Exercise increases endostatin in circulation of healthy volunteers. BMC Physiol 4:2
108. Hajitou A, Grignet C, Devy L et al (2002) The antitumoral effect of endostatin and angiostatin is associated with a down-regulation of vascular endothelial growth factor expression in tumor cells. FASEB J 16:1802–1804
109. Kim YM, Hwang S, Kim YM et al (2002) Endostatin blocks vascular endothelial growth factor-mediated signaling via direct interaction with KDR/Flk-1. J Biol Chem 277:27872–27879
110. Lee SJ, Jang JW, Kim YM et al (2002) Endostatin binds to the catalytic domain of matrix metalloproteinase-2. FEBS Lett 519:147–152
111. Li C, Harris MB, Venema VJ et al (2005) Endostatin induces acute endothelial nitric oxide and prostacyclin release. Biochem Biophys Res Commun 329:873–878

112. Wenzel D, Schmidt A, Reimann K et al (2006) Endostatin, the proteolytic fragment of collagen XVIII, induces vasorelaxation. Circ Res 98:1203–1211
113. Brixius K, Schoenberger S, Ladage D et al (2008) Long-term endurance exercise decreases antiangiogenic endostatin signalling in overweight men aged 50–60 years. Br J Sports Med 42:126–129
114. Bornstein P (1992) Thrombospondins: structure and regulation of expression. FASEB J 6:3290–3299
115. Iruela-Arispe ML, Luque A, Lee N (2004) Thrombospondin modules and angiogenesis. Int J Biochem Cell Biol 36:1070–1078
116. Bonnefoy A, Moura R, Hoylaerts MF (2008) The evolving role of thrombospondin-1 in hemostasis and vascular biology. Cell Mol Life Sci 65(5):713–727
117. Kazerounian S, Yee KO, Lawler J (2008) Thrombospondins in cancer. Cell Mol Life Sci 65:700–712
118. Olfert IM, Breen EC, Gavin TP, Wagner PD (2006) Temporal thrombospondin-1 mRNA response in skeletal muscle exposed to acute and chronic exercise. Growth Factors 24:253–259
119. Murphy G, Houbrechts A, Cockett MI et al (1991) The N-terminal domain of tissue inhibitor of metalloproteinases retains metalloproteinase inhibitory activity. Biochemistry 30:8097–8102
120. Stetler-Stevenson WG, Seo DW (2005) TIMP-2: an endogenous inhibitor of angiogenesis. Trends Mol Med 11:97–103
121. Frederiksen CB, Lomholt AF, Lottenburger T et al (2008) Assessment of the biological variation of plasma tissue inhibitor of metalloproteinases-1. Int J Biol Markers 23:42–47
122. Koskinen SO, Heinemeier KM, Olesen JL et al (2004) Physical exercise can influence local levels of matrix metalloproteinases and their inhibitors in tendon-related connective tissue. J Appl Physiol 96:861–864
123. Koskinen SO, Wang W, Ahtikoski AM et al (2001) Acute exercise induced changes in rat skeletal muscle mRNAs and proteins regulating type IV collagen content. Am J Physiol Regul Integr Comp Physiol 280:R1292–R1300
124. O'Reilly MS, Holmgren L, Shing Y et al (1994) Angiostatin: a novel angiogenesis inhibitor that mediates the suppression of metastases by a Lewis lung carcinoma. Cell 79:315–328
125. Sonoda H, Ohta H, Watanabe K et al (2006) Multiple processing forms and their biological activities of a novel angiogenesis inhibitor vasohibin. Biochem Biophys Res Commun 342:640–646
126. Kishlyansky M, Vojnovic J, Roudier E et al (2010) Striated muscle angio-adaptation requires changes in Vasohibin-1 expression pattern. Biochem Biophys Res Commun 399:359–364
127. Ribatti D (2007) The discovery of endothelial progenitor cells. An historical review. Leuk Res 31:439–444
128. Asahara T, Murohara T, Sullivan A et al (1997) Isolation of putative progenitor endothelial cells for angiogenesis. Science 275:964–967
129. Ii M, Nishimura H, Iwakura A et al (2005) Endothelial progenitor cells are rapidly recruited to myocardium and mediate protective effect of ischemic preconditioning via "imported" nitric oxide synthase activity. Circulation 111:1114–1120
130. Rehman J, Li J, Parvathaneni L, Karlsson G et al (2004) Exercise acutely increases circulating endothelial progenitor cells and monocyte-/macrophage-derived angiogenic cells. J Am Coll Cardiol 43:2314–2318
131. Van Craenenbroeck EM, Vrints CJ, Haine SE et al (2008) A maximal exercise bout increases the number of circulating CD34+/KDR+endothelial progenitor cells in healthy subjects. Relation with lipid profile. J Appl Physiol 104:1006–1013
132. Laufs U, Werner N, Link A et al (2004) Physical training increases endothelial progenitor cells, inhibits neointima formation, and enhances angiogenesis. Circulation 109:220–226
133. Laufs U, Urhausen A, Werner N et al (2005) Running exercise of different duration and intensity: effect on endothelial progenitor cells in healthy subjects. Eur J Cardiovasc Prev Rehabil 12:407–414

134. Steiner S, Niessner A, Ziegler S et al (2005) Endurance training increases the number of endothelial progenitor cells in patients with cardiovascular risk and coronary artery disease. Atherosclerosis 181:305–310
135. Schlager O, Giurgea A, Schuhfried O et al (2011) Exercise training increases endothelial progenitor cells and decreases asymmetric dimethylarginine in peripheral arterial disease: a randomized controlled trial. Atherosclerosis 217:240–248
136. Sarto P, Balducci E, Balconi G et al (2007) Effects of exercise training on endothelial progenitor cells in patients with chronic heart failure. J Card Fail 13:701–708
137. Bartel DP (2004) MicroRNAs: genomics, biogenesis, mechanism, and function. Cell 116:281–297
138. Lee Y, Ahn C, Han J et al (2003) The nuclear RNase III Drosha initiates microRNA processing. Nature 425:415–419
139. Lund E, Guttinger S, Calado A et al (2004) Nuclear export of microRNA precursors. Science 303:95–98
140. Zhao Y, Samal E, Srivastava D (2005) Serum response factor regulates a muscle-specific microRNA that targets Hand2 during cardiogenesis. Nature 436:214–220
141. Chen JF, Mandel EM, Thomson JM et al (2006) The role of microRNA-1 and microRNA-133 in skeletal muscle proliferation and differentiation. Nat Genet 38:228–233
142. Keller P, Vollaard NB, Gustafsson T et al (2011) A transcriptional map of the impact of endurance exercise training on skeletal muscle phenotype. J Appl Physiol 110:46–59
143. Dews M, Homayouni A, Yu D et al (2006) Augmentation of tumor angiogenesis by a Myc-activated microRNA cluster. Nat Genet 38:1060–1065
144. Fasanaro P, D'Alessandra Y, Di Stefano V et al (2008) MicroRNA-210 modulates endothelial cell response to hypoxia and inhibits the receptor tyrosine kinase ligand Ephrin-A3. J Biol Chem 283:15878–15883
145. Poliseno L, Tuccoli A, Mariani L et al (2006) MicroRNAs modulate the angiogenic properties of HUVECs. Blood 108:3068–3071
146. Wang CH, Lee DY, Deng Z et al (2008) MicroRNA miR-328 regulates zonation morphogenesis by targeting CD44 expression. PLoS One 3:e2420
147. Pin AL, Houle F, Guillonneau M et al (2012) miR-20a represses endothelial cell migration by targeting MKK3 and inhibiting p38 MAP kinase activation in response to VEGF. Angiogenesis 14:1–14
148. Hu S, Huang M, Li Z et al (2010) MicroRNA-210 as a novel therapy for treatment of ischemic heart disease. Circulation 122:S124–S131
149. Baggish AL, Hale A, Weiner RB et al (2011) Dynamic regulation of circulating microRNA during acute exhaustive exercise and sustained aerobic exercise training. J Physiol 589:3983–3994
150. Warburton DE, Katzmarzyk PT, Rhodes RE et al (2007) Evidence-informed physical activity guidelines for Canadian adults. Can J Public Health 98:S16–S68
151. Elsman P, van't Hof AW, de Boer MJ et al (2004) Role of collateral circulation in the acute phase of ST-segment-elevation myocardial infarction treated with primary coronary intervention. Eur Heart J 25:854–858
152. Habib GB, Heibig J, Forman SA et al (1991) Influence of coronary collateral vessels on myocardial infarct size in humans. Results of phase I thrombolysis in myocardial infarction (TIMI) trial. The TIMI Investigators. Circulation 83:739–746
153. Sabia PJ, Powers ER, Ragosta M et al (1992) An association between collateral blood flow and myocardial viability in patients with recent myocardial infarction. N Engl J Med 327:1825–1831
154. Beck EB, Erbs S, Mobius-Winkler S et al (2012) Exercise training restores the endothelial response to vascular growth factors in patients with stable coronary artery disease. Eur J Prev Cardiol 19:412–418
155. Onkelinx S, Cornelissen V, Defoor J et al (2011) The CAREGENE study: genetic variants of the endothelium and aerobic power in patients with coronary artery disease. Acta Cardiol 66:407–414

156. Gustafsson T, Bodin K, Sylven C et al (2001) Increased expression of VEGF following exercise training in patients with heart failure. Eur J Clin Invest 31:362–366
157. Testa M, Ennezat PV, Vikstrom KL et al (2000) Modulation of vascular endothelial gene expression by physical training in patients with chronic heart failure. Ital Heart J 1:426–430
158. Gatta L, Armani A, Iellamo F et al (2012) Effects of a short-term exercise training on serum factors involved in ventricular remodelling in chronic heart failure patients. Int J Cardiol 155:409–413
159. Hansen AH, Nielsen JJ, Saltin B et al (2010) Exercise training normalizes skeletal muscle vascular endothelial growth factor levels in patients with essential hypertension. J Hypertens 28:1176–1185
160. Duscha BD, Robbins JL, Jones WS et al (2011) Angiogenesis in skeletal muscle precede improvements in peak oxygen uptake in peripheral artery disease patients. Arterioscler Thromb Vasc Biol 31:2742–2748
161. Lee BC, Hsu HC, Tseng WY et al (2009) Effect of cardiac rehabilitation on angiogenic cytokines in postinfarction patients. Heart 95:1012–1018
162. Sandri M, Adams V, Gielen S et al (2005) Effects of exercise and ischemia on mobilization and functional activation of blood-derived progenitor cells in patients with ischemic syndromes: results of 3 randomized studies. Circulation 111:3391–3399
163. Zbinden R, Zbinden S, Meier P et al (2007) Coronary collateral flow in response to endurance exercise training. Eur J Cardiovasc Prev Rehabil 14:250–257
164. Troidl K, Schaper W (2012) Arteriogenesis versus angiogenesis in peripheral artery disease. Diabetes Metab Res Rev 28:27–29
165. Ferrari R, Bachetti T, Agnoletti L et al (1998) Endothelial function and dysfunction in heart failure. Eur Heart J 19:G41–G47
166. Drexler H, Riede U, Munzel T et al (1992) Alterations of skeletal muscle in chronic heart failure. Circulation 85:1751–1759
167. Piepoli M, Ponikowski P, Clark AL et al (1999) A neural link to explain the "muscle hypothesis" of exercise intolerance in chronic heart failure. Am Heart J 137:1050–1056
168. Folkman J (1971) Tumor angiogenesis: therapeutic implications. N Engl J Med 285:1182–1186
169. Verma VK, Singh V, Singh MP et al (2009) Effect of physical exercise on tumor growth regulating factors of tumor microenvironment: implications in exercise-dependent tumor growth retardation. Immunopharmacol Immunotoxicol 31:274–282
170. Jones LW, Viglianti BL, Tashjian JA et al (2010) Effect of aerobic exercise on tumor physiology in an animal model of human breast cancer. J Appl Physiol 108:343–348

Part II
Pathophysiological Aspects of Angiogenesis

Chapter 12
Angiogenesis in Atherosclerosis: An Overview

Sadip Pant, Abhishek Deshmukh, and Jawahar L. Mehta

Abstract Angiogenesis, the process of formation of new blood vessels, involves a complex interplay of various growth signals and cellular milieu. It plays an important role in many physiological and pathological processes. The study of angiogenesis has gained momentum from two different perspectives. On the one hand, angiogenesis is essential for physiologic processes such as menstrual bleeding, wound healing, and embryonic development; on the other hand, histopathologic studies suggest that angiogenesis has an important role in the growth of the atherosclerotic plaque and several tumors. From a clinical perspective, drugs that inhibit angiogenesis seem to increase the incidence of thromboembolic events. Recent data suggests that endothelial-derived growth factor (VEGF) constitutes only a small proportion of total body VEGF and does not contribute significantly to the overall angiogenic response. However, such autocrine VEGF signaling is required for endothelial cell survival and maintenance of vascular homeostasis. In parallel, the discovery of VEGF has guided us to therapeutic angiogenesis as a possible treatment for ischemic heart and peripheral disease. Results of numerous preclinical studies have provided evidence that angiogenic growth factors can promote collateral channel development which can reduce ischemia. Nonetheless, clinical trials on therapeutic angiogenesis have not been very impressive as expected. Bench and bedside research continues to bring insight into the mechanisms of tissue ischemia and tumor growth. Further understanding of different facets of angiogenesis may help in the development of novel and specific therapies.

Keywords Neovascularization • Atherosclerosis • VEGF • FGF • Arteriogenesis • Therapeutic angiogenesis

---

S. Pant • A. Deshmukh • J.L. Mehta (✉)
Division of Cardiovascular Medicine, University of Arkansas for Medical Sciences and the Central Arkansas Veterans Healthcare System, 4301 West Markham, Little Rock, AR 72205, USA
e-mail: mehtajl@uams.edu

J.L. Mehta and N.S. Dhalla (eds.), *Biochemical Basis and Therapeutic Implications of Angiogenesis*, Advances in Biochemistry in Health and Disease 6,
DOI 10.1007/978-1-4614-5857-9_12,

12.1 Introduction

During embryogenesis, angioblasts differentiate into endothelial cells leading to the formation of primary capillary plexus; this process is called vasculogenesis [1]. Vasculogenesis begins very early after the initiation of gastrulation in the mammalian embryo with the formation of angioblast precursors in the mesenchyma. Angiogenesis, unlike vasculogenesis, represents growth of new blood vessels from the preexisting channels. During embryonic development, angiogenesis results in growth and remodeling of the primitive network laid down by vasculogenesis to form an intricate vascular system. Distinct from angiogenesis, arteriogenesis is a process that refers to maturation of blood vessels via increase in the luminal diameter. The end result of these three processes (Fig. 12.1) is the formation of new blood vessels. However, their trigger, cellular mechanisms, and the ability to restore blood flow are quite different. By and large, the most efficient adaptive response for endurance against ischemia is arteriogenesis since the ability of these channels to contribute to blood flow per unit of time or tissue mass is relatively large compared to angiogenesis [2].

Angiogenesis plays an important role in many physiological processes in adults (Fig. 12.2). This includes ovulation, wound healing, tissue repair after injury, exercise-induced skeletal and cardiac muscle hypertrophy, and reconstruction of endometrium for implantation and placentation. Physiological angiogenesis is highly regulated at genetic and molecular level. Dysregulation of this process, also known as pathological angiogenesis, results in consequences ranging from mild disease state, such as dysfunctional uterine bleeding, to grave entity, such as cancer. Exaggerated angiogenesis is the key in the pathogenesis of many diseases like tumor growth and metastasis, proliferative retinopathy, vascular malformations, psoriasis, rheumatoid arthritis, and other chronic inflammatory diseases. On the

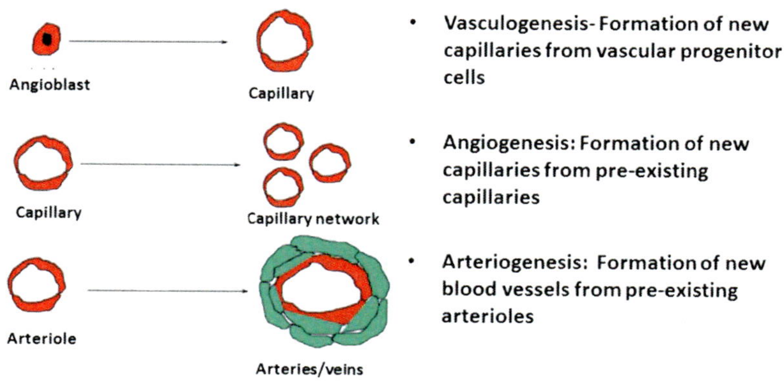

Fig. 12.1 Three processes of neovascularization in human beings: vasculogenesis, angiogenesis, and arteriogenesis

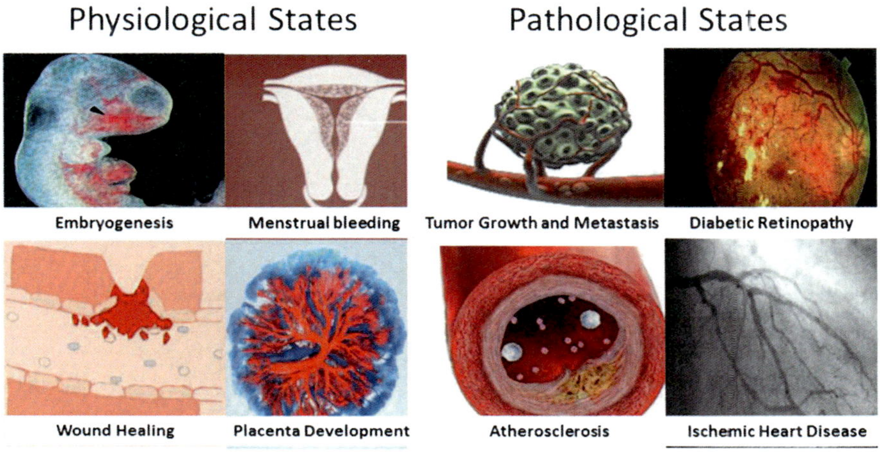

Fig. 12.2 Role of angiogenesis in health and diseases

other extreme, insufficient vessel growth results in disease processes like ischemic heart disease, cerebrovascular disease, peripheral arterial disease, delayed wound healing, and scleroderma. Hence, understanding mechanisms and regulation of angiogenesis is necessary in understanding various disease processes and their therapies. Most of these phenomena are separately described in this book. Here, we provide a general overview of angiogenesis.

12.2 Mechanism of New Vessel Formation

Angiogenesis is a cascade rather than a linear process involving a complex interplay of various growth signals (Fig. 12.3). Formation of new blood vessels depends on three major factors: angiogenesis switch, expression of receptors. and second messenger systems in the target cells and the matrix environment. The process begins when the delicate balance between angiogenesis and anti-angiogenesis factors shifts towards angiogenesis. This is generally characterized by upregulation of growth factors and/or their receptors in the endothelial cells.

Endothelial cells are one of the most quiescent and genetically stable cells of the body with a very prolonged turnover time of about 100 days. When these quiescent cells detect pro-angiogenic signal/s, their cell–cell junctional contacts develop gaps, proteases are activated, basement membrane is degraded, and the cells acquire motile behavior; this initiates new blood vessel sprouting. Vascular endothelial growth factor A (VEGF-A) which is secreted by tissues in response to inflammation, hypoxia, or ischemia is a major pro-angiogenic factor [3]. Platelet-derived growth factor (PDGF), fibroblast growth factor (FGF), and other pro-angiogenic factors act synergistically with VEGF.

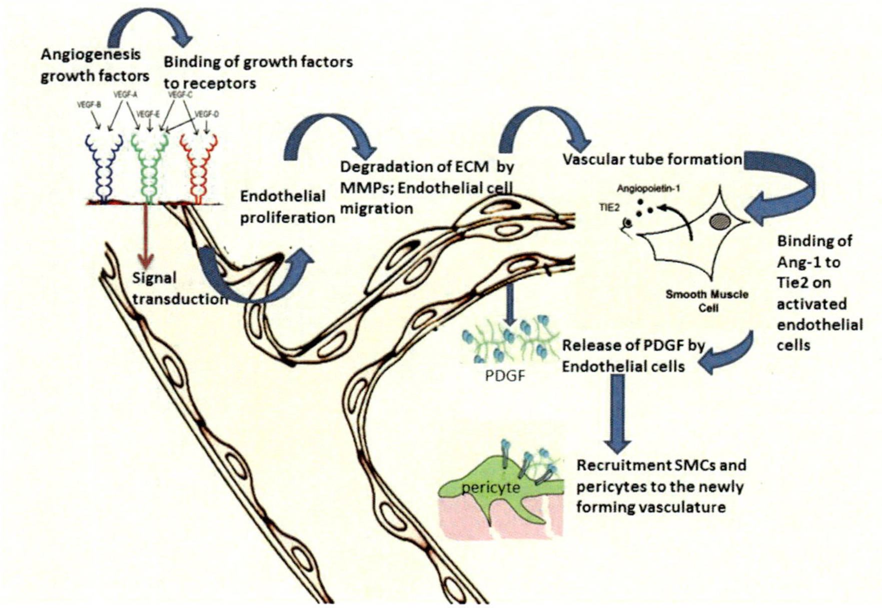

Fig. 12.3 Summary of mechanisms of angiogenesis. *ECM* extracellular matrix; *MMPs* matrix metalloproteinases; *VEGF* vascular endothelial growth factor; *PDGF* platelet-derived growth factor; *Ang-1* angiopoietin-1; *Tie-2* tyrosine kinase with immunoglobulin-like and EGF-like domains 2

VEGF binds to three structurally related receptor tyrosine kinases (RTKs) named VEGFR-1, VEGFR-2, and VEGFR-3. VEGFRs have an extracellular ligand-binding portion consisting of seven immunoglobulin-like domains, a transmembrane region, a juxtamembrane domain, an intracellular split tyrosine kinase domain, and a C-terminal tail [4]. VEGFR-3 is a regulator of lymphoendothelial function. VEGF-A, the principle regulator of angiogenesis, binds to VEGFR-1 and VEGFR-2. VEGFR-1 is expressed in endothelial and non-endothelial cells. It acts as a dummy receptor sequestering VEGF from binding to VEGFR-2 and is important during vasculogenesis in embryo. In addition, it plays a role in inflammation-induced angiogenesis by recruiting inflammatory cells followed by deposition of angiogenic growth factors. VEGFR-2 is expressed only on endothelial cells and regulates endothelial survival proliferation, migration, and the formation of the vascular tubes [4]. Integrins mediate cell matrix adhesion by binding to extracellular components. VEGF induces complex formation between integrin $\alpha V\beta 3$ and VEGFR-2 which is required for angiogenesis.

Proteinases, like matrix metalloproteinases (MMPs), chymase, and heparanase, degrade extracellular matrix and liberate growth factors from matrix. Proliferating solid cords of endothelial cells reach at distant sites once the matrix has been disrupted. PDGF is a chemoattractant for smooth muscle cells. Once recruited angiopoietin 1 and transforming growth factor $\beta 1$ (TGF$\beta 1$) stabilize the newly formed vessels by increasing interactions between endothelial cells and peri-endothelial

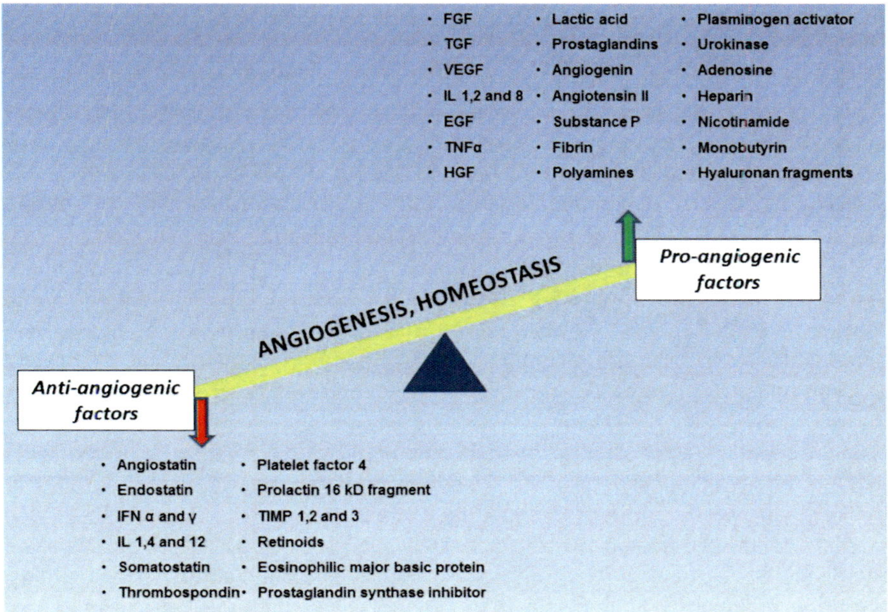

Fig. 12.4 Angiogenesis is regulated by balance in pro-angiogenic factors and anti-angiogenic factors. Increase in pro-angiogenic factors stimulates angiogenesis, while increase in anti-angiogenic factors leads to arrest of angiogenesis

vascular smooth muscle cells and pericytes [5]. These newly assembled endothelial cells develop lumen and anastomose with other buds or capillaries. They persist as long as they are required or differentiate into mature venules and arterioles as per metabolic requirements of local tissue. A complex interaction of endothelial cells, perivascular smooth muscle cells and pericytes is essential for angiogenesis. Angiogenesis is regulated not only by activating signals but also by inhibitory signals, such as thrombospondin-1, interferon-α, platelet factor-4, and angiostatin (Fig. 12.4).

The process of plaque angiogenesis in atherosclerosis is regulated by multiple signaling such as hypoxia, release of reactive oxygen species (ROS), and local inflammation [6]. Hypoxia within the thickened atherosclerotic plaque is a major driving force for growth of new blood vessels in and around the plaque. NADPH oxidase-mediated ROS production plays an important role in angiogenesis which is an adaptive response to hypoxia. Hypoxia leads to an increase in hypoxia inducible factor (HIF)-1 and -2 which in turn upregulate the expression of VEGF-A, PDGF, and other growth factors [7]. ROS also upregulate VEGF expression at both the mRNA and protein levels [8]. NADPH oxidases (Nox1, Nox2, and Nox4) are the major source of ROS in the cardiovascular system. Nox1 in smooth muscle cells and Nox4 in endothelial cells are upregulated by hypoxia. Hypoxia-induced upregulation of Nox1 increases HIF-1 and Nox4 stabilizes HIF-2α. Additionally, Nox1 upregulates

VEGF and increases the production of MMPs. Nox4 increases the expression of VEGF and PDGF in endothelial cells. PDGF in turn upregulates Nox1 in smooth muscle cells leading to recruitment of smooth muscle cells and pericytes around newly formed blood vessels. Nox2 in monocytes and endothelial cells in angiogenic tips also plays an important role in angiogenesis [7]. Nox-induced angiogenesis may play a tissue protective role in chronic ischemia; however, time for angiogenic response in the setting of acute ischemia is limited, and acute burst of ROS during the early stages of reperfusion may be detrimental to the tissues.

Lectin-like oxidized low-density lipoprotein receptor-1 (LOX-1), a receptor for oxidized low-density lipoprotein (ox-LDL), is mainly expressed in endothelial cells. Endothelial cells when exposed to ox-LDL show expression of LOX-1 and upregulation of adhesion molecules, proinflammatory proteins, tissue factor, and tissue remodeling proteins like metalloproteinases and collagens. Many of the MMPs are stimulators of angiogenesis [9]. Small concentrations of ox-LDL (<5 μg/mL) promote capillary tube formation by transcription of LOX-1, which activates NADPH oxidase/mitogen-activated protein kinases/NF-κB pathway (Fig. 12.3) [10]. Ox-LDL has been reported to markedly increase the expression of VEGF and activate a peroxisome proliferator-activated receptor (PPAR)-gamma; both are attenuated by anti-LOX-1 antibody [11]. Hu et al. demonstrated that capillary sprouting induced by angiotensin II is also related to the elevated expression of VEGF, and anti-LOX-1 antibody markedly inhibits angiotensin II-induced VEGF expression. Moreover, angiotensin II-induced capillary sprouting from aortic rings from LOX-1 null mice is minimal compared to the sprouting from aortic rings of wild-type mice [12]. These authors suggested that angiotensin II, which is overexpressed in atherosclerotic region, plays a major role in angiogenesis in the rapidly growing plaque by LOX-1 upregulation. These observations collectively suggest that LOX-1 may turn out to be an important regulator of angiogenesis in atherosclerosis. Hypoxia, a primary driver of angiogenesis in atherogenesis, is also found to induce angiogenesis through upregulation of LOX-1 and the LOX-1-mediated $p47^{phox}$ component of NADPH oxidase transcription [13]. The factors that regulate angiogenesis are shown in a simple form in Fig. 12.3.

12.3 Angiogenesis in Atherosclerosis: A Historical Perspective

Association of neovascularization and atherosclerosis was first noted by Koester [14] and others [15] over a century ago. Wilens and Plair [16] found that vasa vasorum are the source of neovascularization in the media of the arterial wall in 1965. Moreno et al. [17] recently emphasized a correlation between neovascularization in arterial wall and atherosclerosis progression in coronary arteries in human autopsy studies. They found that microvessel density was higher in lipid-rich and ruptured plaques as well as in lesions with intraplaque hemorrhage and a thin fibrous cap. This was further substantiated by Virmani and colleagues [18] who noticed the vessel density to be increased twofold in plaques categorized as vulnerable and fourfold in those that ruptured when compared with stable plaques having significant luminal narrowing. Further, they observed that increased microvesssel density is associated with intraplaque hemorrhage, a characteristic of unstable lesions. Of note, Barger and Beeuwkes [19] had suggested in 1990 that the atherosclerotic plaque with neovascularization is fragile, prone to rupture, and therefore leads to plaque destabilization and clinical manifestation of the disease. More convincing evidence for the concept that angiogenesis is not just a bystander rather a key player in atherosclerosis progression came from mouse models of atherosclerosis. Moulton et al. [20, 21] found that angiogenesis inhibitors endostatin, TNP-470, and angiostatin reduced plaque area in $ApoE^{-/-}$mice significantly. In 2001, Celletti et al. [22] reported that intraperitoneal administration of recombinant human VEGF protein enhanced atherosclerotic plaque progression in $ApoE^{-/-}ApoB100^{-/-}$ mice. Likewise, Khurana et al. [23] using two different models showed that pro-angiogenic molecules cannot initiate but definitely promote the growth of the intimal lesions. Additional evidence that VEGF has a proinflammatory and pro-angiogenic cytokine function in plaque development was provided by Zhao et al. [24].

12.4 Plaque Neovascularization and Its Consequences

Recent studies suggest that plaque angiogenesis makes the plaque vulnerable by promoting plaque inflammation, destabilization, progression, and rupture. Histological characteristic of plaque angiogenesis reveals that these microvessels often lack pericytes and smooth muscle cells and have poorly formed endothelial cell junctions. Therefore, the neovascular network in coronary atherosclerotic plaques is more fragile and prone to rupture. Rupture of plaque capillaries can trigger intraplaque hemorrhage, leading to plaque destabilization and its consequences. Close observation of such sprouting vessels in the atherosclerotic plaque reveals their clustering in the shoulder regions and to some extent at the plaque base (the intima–media border) and in the fibrous cap of the atherosclerotic plaque. These are areas that are more prone to fissuring and rupture. Such sprouting vessels supply oxygen and nutrients to the core of the atherosclerotic plaques, thereby increasing its size [25]. Finally, these microvessels are generally surrounded by macrophages (foam cells), mast cells, T cells, red blood cells, albumin, and/or lipoproteins. This suggests that these vessels actually bring in more inflammatory cells to the atherosclerotic plaque and thus play a significant role in the perpetuation of atherosclerosis.

Similar concept of plaque neovascularization leading to plaque instability holds true for atherosclerotic disease elsewhere. For example, recent studies of carotid artery plaque emphasize the significance of intraplaque hemorrhage in causation of symptomatic cerebrovascular ischemia [26]. Relative to the origin of the hemorrhage, pathologic studies suggest that angiogenesis is significantly increased in carotid artery plaques from symptomatic patients when compared with matched individuals with asymptomatic plaques. It has been shown that plaque angiogenesis and angiogenetic factors, such as VEGF, are critical in the progression of atherosclerotic carotid plaque and intraplaque hemorrhage [27].

Diabetic retinopathy is yet another prototype disease which is characterized by the occurrence of pathological angiogenesis on the retina. There is enhanced secretion of VEGF in the vitreous fluid of the eye as well as enhanced expression of VEGFR-1 and VEGFR-2 receptors in diabetic retinopathy [28, 29]. This leads to the formation of new capillaries, which are leaky, have immature extracellular matrix, and have greater tendency to rupture leading to retinal hemorrhage and, subsequently, blindness. Evidence from a variety of angiogenesis-related diseases supports the fact that neovascularization leads to the formation of frail and leaky blood vessels.

12.5 The Angiogenesis Paradox

Angiogenesis research has gained momentum from different perspectives. On the one hand, data from histopathology of atherosclerotic plaques suggests that angiogenesis has an important role in the pathogenesis of atherosclerotic disease and its clinical sequelae. On the other hand, drugs that inhibit angiogenesis paradoxically increase the incidence of thromboembolic events. In normal tissue, there is a check and balance system between pro-angiogenic factors like VEGF, bFGF, and IL-8 and anti-angiogenic molecules like thrombospondin-1 and -2 and angiostatin. If this balance tips in favor of pro-angiogenic molecules, abnormal angiogenesis ensues contributing to plaque destabilization and rupture. Hence, at first glance, anti-angiogenic agents might be of benefit to prevent neovascularization and normalize the leaky vasculature that promotes plaque destabilization and hemorrhage. A variety of agents have been developed in the last few years that can inhibit the process of angiogenesis at various stages. Action of some of these agents vis-a-vis angiogenesis is represented in Fig. 12.3. Our understanding of these agents and their effects on cardiovascular system mainly comes from their use in oncology. Some of these agents have been approved for halting tumor progression for over a decade now, and bevacizumab is a prototype example of these agents. Bevacizumab binds to VEGF and blocks its activity. Pooled analysis of five randomized controlled trials with a total of 1,745 patients with various metastatic carcinomas treated with either conventional chemotherapy or chemotherapy with bevacizumab demonstrated significantly more thromboembolic events in the latter group (hazard ratio = 2.0, $p = 0.031$). Most thromboembolic events were coronary or cerebrovascular. Further, upon multivariate analysis, a history of atherosclerosis was found to be an important risk factor for such adverse events. Similarly increased cardiovascular risk with bevacizumab has been replicated in other studies as well [30, 31]. Of note, thalidomide and lenalidomide have also been associated with thrombotic complications ranging from 8 to 17% [32, 33].

Recently, there has been a growing suggestion that VEGF not only plays a significant role in human disease as mentioned earlier, but it is of utmost importance in maintaining vascular homeostasis [34]. Endothelial-derived VEGF constitutes only a small proportion of total body VEGF and does not contribute significantly to the overall angiogenic response. Nonetheless, there is compelling evidence that such autocrine VEGF signaling is required for endothelial cell survival, and removal of endothelium-derived VEGF in vivo results in systemic endothelial apoptosis [35, 36]. In murine models [35], it was shown that removal of endothelium-derived VEGF resulted in devastating systemic vascular pathologies in the form of multiple hemorrhagic and thrombotic events and sudden deaths. Interestingly, in mice with an endothelium-specific loss of VEGF, the systemic vascular catastrophe could not be compensated for by VEGF secreted from adjacent types or circulating VEGF [35].

Furthermore, a role for VEGF in adaptive cardiac growth is also being recognized. VEGF is upregulated in myocardium in pathological conditions such as myocardial ischemia and hemodynamic overload. Data from murine models of pressure-overloaded heart suggests that VEGF-derived angiogenic response is essential for coronary vascular network growth under conditions that promote compensatory cardiac hypertrophy [37]. A reduction in available VEGF contributes to the rapid progression from compensatory cardiac hypertrophy to failure in pressure-overloaded hearts. Further, treatment with VEGF receptor decoy has been shown to attenuate cardiac hypertrophy induced by pressure overload. Hence, stress-induced heart growth may depend on the status of the vascular bed in a manner that is similar to tumor growth [37]. In summary, VEGF signaling seems essential for adaptive cardiac growth, and disruption of such signaling may impair cardiac function under stress.

12.6 Collateral Arteriogenesis vis-à-vis Plaque Angiogenesis

Unlike plaque angiogenesis, collateral arteriogenesis is characterized by the enlargement of arteriolar anastomoses to collateral vessels through growth and proliferation which results in the formation of new arteries possessing fully developed tunica media. It is not a merely a passive dilation of collateral but an active proliferation and remodeling of the arteriole resulting in the formation of large vessels that can take over the role of an artery when occluded [38]. While plaque neovascularization results in the formation of the immature, pericyte-lacking capillaries, collateral arteriogenesis involves endothelial tubulogenesis, in association with pericyte recruitment and smooth muscle cell proliferation and envelopment resulting in mature vessel formation. Hence, arteriogenesis is the preferred type of neovascularization for purposes of restoring myocardial perfusion. The preexisting collateral arterioles can give rise to conductance vessels with up to 20-fold increase in their diameter. In some patients, occluded coronary arteries can be completely compensated for by collateral arteries [38].

Mechanisms of arteriogenesis involve interplay between shear stress and circulating monocytes. In response to ischemia, shear force and blood flow increase

Table 12.1 Key differences between two major processes operant in adult neovascularization: angiogenesis and collateral arteriogenesis

Characteristics	Arteriogenesis	Angiogenesis
Stimulus	Shear stress	Ischemia
Key growth factor	VEGF	FGF, PDGF, MCP-1
Substrate	Preexisting arterioles	Preexisting capillaries
Result	New arteries	Increased capillaries
Time frame	Days to weeks	Days
Max increment in blood flow	10–20 fold	1.5–1.7 fold
Ability to compensate for an occluded artery	Able	Unable

through these collaterals. This leads to the activation of normally quiescent endothelial cells in the collateral arterioles. The activated endothelium releases monocyte chemoattractant protein-1; simultaneously there is upregulation of receptors for docking the monocytes. The adherent monocytes, in turn, release a variety of growth factors and cytokines, such as monocyte chemoattractant protein-1 and granulocyte–monocyte colony-stimulating factor, transforming growth factor-β1 that creates the environment for collateral artery growth. Among these factors, PDGF and bFGF are directly mitotic for endothelial and smooth muscle cells [38]. In plaque angiogenesis, VEGF is the primary growth factor while compensatory arteriogenesis is mediated by an array of growth factors, among which FGF and PDGF appear to be of key significance [39]. The major differences between these two processes have been summarized in Table 12.1.

12.7 Specificity and Heterogeneity in the Expression of Angiogenesis

Human vasculature comprises a variety of endothelial cells forming a pattern well suited to the functional requirement of the specific organ. Although endothelial cells have a number of characteristics in common, they exhibit great variation phenotypically, genetically, and in functionality. For instance, liver sinusoids are lined by discontinuous endothelial cells that permit greater movement of materials between intercellular gaps; kidneys are lined by fenestrated endothelial cells that facilitate selective permeability required for filtration, absorption, and secretion; and central nervous system is lined by continuous thin endothelial cells that form an effective blood–brain barrier [40]. Endothelial cells also differ in their surface phenotype and protein expression [41], antigen expression [42], and response to growth factors [43]. It is a common clinical observation that when saphenous vein is grafted to coronary artery, it acquires artery-like properties and exhibits an increased tendency to develop atherosclerosis. Data suggests that endothelial cells in regions of disturbed flow acquire a proinflammatory phenotype in response to systemic insult [44, 45]. Hence, endothelial cells are not only heterogenous but also very dynamic. These differences have been attributed mainly to genetic control (nature) and environmental influences under which these cells thrive (nurture) [46, 47]. Understanding this heterogeneity in endothelial cell biology and the factors governing such heterogeneity is essential as microvascular involvement in various pathologies is strongly governed by the behavior of these cells.

Besides the differences in endothelial cell biology at different sites, there is also a significant heterogeneity in angiogenic response induced by endothelial growth factors in different sites. For example, a recent study by Petterson et al. involving adenoviral vector delivery of VEGF to heart, skin, fat, and skeletal muscle showed that there was a common initial angiogenic response in the form of pericyte-poor leaky microvessels among all these tissues followed by tissue-specific progression of angiogenesis [48]. In muscle (cardiac and skeletal), smaller caliber or disorganized tangles of daughter vessels were formed. Although similar response occurred in skin and fat with a greater intensity, some of the resultant vessels acquired muscular coat and progressed to form conduits that closely resembled medium-sized arteries and veins that persisted indefinitely. Interestingly, these microvessels were similar to the collateral vessels formed by the process of arteriogenesis in ischemic hearts [48]. These findings led the authors to postulate that the clinical benefits in ischemic limb salvage via VEGF therapy in animal models might have resulted from spillage of injected cytokine or its encoding DNA into perimuscle fat tissue [48].

Finally, data from experimental animal models suggest that heterogeneity in angiogenic response and differential sensitivity to angiogenesis inhibitors are largely governed by genetic factors [49–51]. In 2000, Rohan et al. showed in the corneal micropocket assay of various inbred mice that there exists nearly tenfold range of response angiogenesis by growth factor among different strains [52]. Further, differential sensitivity to angiogenesis inhibitors was seen between strains. Although, at this time there is paucity of information on genetic control of angiogenic heterogeneity, it is possible that a similar heterogeneity as described in inbred mouse strains by Rohan et al. [52] is operant in human beings also. This might explain the individual differences in angiogenic potential and the different rate of progression of angiogenesis-related diseases in different individuals.

12.8 Therapeutic Angiogenesis: Promises and Problems

Ischemic heart disease exerts a tremendous burden on healthcare system worldwide. It is the leading cause of morbidity and mortality. Over a million people suffer from myocardial infarction annually [53]. While many of them undergo surgical or catheter-based revascularization, many of them are not optimal candidates for any form of revascularization procedures for various reasons. With increasing longevity and baby-booming, this problem is likely to grow further. Similarly, cerebrovascular disease resulting in transient ischemic attacks and strokes leads to a major socioeconomic burden on the society [53]. In this regard, therapeutic angiogenesis has attracted interest as an alternative treatment for patients with ischemic heart disease or cerebrovascular disease. These aspects of natural or induced revascularization are being discussed elsewhere in this book. With the rapid advancement in understanding of angiogenesis and angiogenic growth factor therapy over the last decade, therapeutic angiogenesis has put forth a hope as a new treatment modality. However, as the field of therapeutic angiogenesis is budding from its instigation phase, many other issues are arising.

Therapeutic angiogenesis vis-a-vis ischemic heart disease refers to induction of new blood vessels that can effectively supply blood to the area of myocardium jeopardized by occluded native coronary arteries. Results of numerous preclinical studies have provided evidence that angiogenic growth factors can promote collateral artery development and hence reduce ischemia in animal models of peripheral and coronary circulation [54]. Yet, clinical trials on therapeutic angiogenesis have not been as impressive [55].

The first clinical trial (VIVA trial) used a combination of intracoronary and intravenous $VEGF_{165}$ protein infusions [56]. At 60 days, there were no significant differences in exercise tolerance or angina class between the treated and control groups. Similarly, FIRST (The FGF-2 Initiating Revascularization Support Trial) was a multicentered double-blinded placebo-controlled trial that randomized patients to receive a single intracoronary infusion of bFGF. Unfortunately, bFGF did not show any improvement in exercise tolerance or myocardial perfusion, although there was some symptomatic improvements on day 90 after bFGF administration, but the improvement was not observed on day 180 [57]. Although preclinical studies of VEGF and FGF gene delivery using plasmid and adenoviral vectors have shown impressive improvements in heart function and perfusion in different animal models of myocardial ischemia, results of clinical studies have not yielded impressive results [58, 59]. Wrong targets and drugs, incorrect delivery mechanisms, poor patient selection, and/or wrong metrics to measure outcomes are some of the factors that have been implicated in the failure of therapeutic angiogenesis therapy. The optimal preparation and delivery strategy for therapeutic neovascularization is the subject of ongoing clinical investigation. At current stage, there is no clinically available therapy either to induce therapeutic angiogenesis.

A major factor which might explain the failure of these trials is the use of a single growth factor. Major trials on therapeutic angiogenesis have hovered around a single growth factor. Unlike angiogenesis, the process of collateral arteriogenesis involves a cocktail of chemokines, growth factors, and proteases. It is difficult to envision that the complex process of endothelial tubulogenesis, combined with pericyte recruitment and smooth muscle cell proliferation, can be achieved with a single agent given as a single dose. Combination therapy for collateral arteriogenesis has not been studied at this time.

12.9 Conclusion

Angiogenesis is operative in human beings right from the embryonic stage and involves a complex interplay of various cells, growth factors, cytokines, and the environment in which such process takes place. While this process plays a key role in many physiological processes and in the development of cardiovascular system, inadvertent angiogenesis is also a major pathologic factor in the development of many disease states including tumors and atherosclerosis. The role of angiogenesis in atherosclerosis has recently gained much attention.

Use of therapeutic angiogenesis has been widely regarded as an attractive approach in treatment of ischemic heart disease. On the other hand, there is growing evidence that neovascularization contributes to the progression of atherosclerotic lesions and that it may play key role in intraplaque hemorrhage, plaque destabilization, and rupture. Despite the failure of early clinical trials with angiogenic agents, bench and bedside research interest in using angiogenic stimuli has continued to bring about more insight into the mechanisms of this phenomenon. Understanding the basis of only limited success in early trials of angiogenic therapy may help in the development of novel and specific therapies.

Conflict of Interest None
Disclosures None

References

1. Risau W (1997) Mechanisms of angiogenesis. Nature 386:671–674
2. Schaper W, Scholz D (2003) Factors regulating arteriogenesis. Arterioscler Thromb Vasc Biol 23:1143–1151
3. Takahashi H, Shibuya M (2005) The vascular endothelial growth factor (VEGF)/VEGF receptor system and its role under physiological and pathological conditions. Clin Sci 109:227–241
4. Koch S, Tugues S, Li X, Gualandi L, Claesson-Welsh L (2011) Signal transduction by vascular endothelial growth factor receptors. Biochem J 437:169–183
5. Carmeliet P (2000) Mechanisms of angiogenesis and arteriogenesis. Nat Med 6:389–395
6. Sluimer JC, Daemen MJ (2009) Novel concepts in atherogenesis: angiogenesis and hypoxia in atherosclerosis. J Pathol 218:7–29
7. Ago T, Kuroda J, Kamouchi M et al (2011) Pathophysiological roles of NADPH oxidase/nox family proteins in the vascular system. -Review and perspective-. Circ J 75:1791–1800
8. Jiang J, Yan M, Mehta JL, Hu C (2011) Angiogenesis is a link between atherosclerosis and tumorigenesis: role of LOX-1. Cardiovasc Drugs Ther 25:461–468
9. Li D, Mehta JL (2000) Upregulation of endothelial receptor for oxidized LDL (LOX-1) by oxidized LDL and implications in apoptosis of human coronary artery endothelial cells: evidence from use of antisense LOX-1 mRNA and chemical inhibitors. Arterioscler Thromb Vasc Biol 20:1116–1122
10. Dandapat A, Hu C, Sun L, Mehta JL (2007) Small concentrations of oxLDL induce capillary tube formation from endothelial cells via LOX-1-dependent redox-sensitive pathway. Arterioscler Thromb Vasc Biol 27:2435–2442

11. Kanata S, Akagi M, Nishimura S et al (2006) Oxidized LDL binding to LOX-1 upregulates VEGF expression in cultured bovine chondrocytes through activation of PPAR-gamma. Biochem Biophys Res Commun 348:1003–1010
12. Hu C, Dandapat A, Mehta JL (2007) Angiotensin II induces capillary formation from endothelial cells via the LOX-1 dependent redox-sensitive pathway. Hypertension 50:952–957
13. Khaidakov M, Szwedo J, Mitra S et al (2010) Antiangiogenic and antimitotic effects of aspirin in hypoxia-reoxygenation modulation of the LOX-1-NADPH oxidase axis as a potential mechanism. J Cardiovasc Pharmacol 56:635–641
14. Koester W (1876) Endareritis and arteritis. Berl Klin Wochenschr 13:454–455
15. Winternitz MC, Thomas RM, LeCompte PM (1938) The biology of arteriosclerosis. C.C. Thomas, Springfield
16. Wilens SL, Plair CM (1965) Blood cholesterol, nutrition, atherosclerosis. A necropsy study. Arch Intern Med 116:373–380
17. Moreno PR, Purushothaman KR, Fuster V et al (2004) Plaque neovascularization is increased in ruptured atherosclerotic lesions of human aorta: implications for plaque vulnerability. Circulation 110:2032–2038
18. Kolodgie FD, Virmani R, Burke AP et al (2004) Pathologic assessment of the vulnerable human coronary plaque. Heart 90:1385–1391
19. Barger AC, Beeuwkes R III (1990) Rupture of coronary vasa vasorum as a trigger of acute myocardial infarction. Am J Cardiol 66:41–43
20. Moulton KS, Heller E, Konerding MA et al (1999) Angiogenesis inhibitors endostatin or TNP-470 reduce intimal neovascularization and plaque growth in apolipoprotein E-deficient mice. Circulation 99:1726–1732
21. Moulton KS, Vakili K, Zurakowski D et al (2003) Inhibition of plaque neovascularization reduces macrophage accumulation and progression of advanced atherosclerosis. Proc Natl Acad Sci U S A 100:4736–4741
22. Celletti FL, Waugh JM, Amabile PG, Brendolan A, Hilfiker PR, Dake MD (2001) Vascular endothelial growth factor enhances atherosclerotic plaque progression. Nat Med 7:425–429
23. Khurana R, Zhuang Z, Bhardwaj S et al (2004) Angiogenesis-dependent and independent phases of intimal hyperplasia. Circulation 110:2436–2443
24. Zhao Q, Egashira K, Hiasa K et al (2004) Essential role of vascular endothelial growth factor and Flt-1 signals in neointimal formation after periadventitial injury. Arterioscler Thromb Vasc Biol 24:2284–2289
25. Barger AC, Beeuwkes R III, Lainey LL, Silverman KJ (1984) Hypothesis: vasa vasorum and neovascularization of human coronary arteries: a possible role in the pathophysiology of atherosclerosis. N Engl J Med 310:175–177
26. Mofidi R, Crotty TB, McCarthy P et al (2001) Association between plaque instability, angiogenesis and symptomatic carotid occlusive disease. Br J Surg 88:945–950
27. Hiyama T, Tanaka T, Endo S et al (2010) Angiogenesis in atherosclerotic plaque obtained from carotid endarterectomy: association between symptomatology and plaque morphology. Neurol Med Chir 50:1056–1061
28. Aiello LP, Avery RL, Arrigg PG et al (1994) Vascular endothelial growth factor in ocular fluid of patients with diabetic retinopathy and other retinal disorders. N Engl J Med 331:1480–1487
29. Kakehashi A, Inoda S, Mameuda C, Kuroki M et al (2008) Relationship among VEGF, VEGF receptor, AGEs, and macrophages in proliferative diabetic retinopathy. Diabetes Res Clin Pract 79:438–445
30. Nalluri SR, Chu D, Keresztes R, Zhu X, Wu S (2008) Risk of venous thromboembolism with the angiogenesis inhibitor bevacizumab in cancer patients: a meta-analysis. JAMA 300:2277–2285
31. Ranpura V, Hapani S, Chuang J, Wu S (2010) Risk of cardiac ischemia and arterial thromboembolic events with the angiogenesis inhibitor bevacizumab in cancer patients: a meta-analysis of randomized controlled trials. Acta Oncol 49:287–297

32. Daher IN, Yeh ET (2008) Vascular complications of selected cancer therapies. Nat Clin Pract Cardiovasc Med 5:797–805
33. Menon SP, Rajkumar SV, Lacy M, Falco P, Palumbo A (2008) Thromboembolic events with lenalidomide-based therapy for multiple myeloma. Cancer 112:1522–1528
34. Sugimoto H, Hamano Y, Charytan D et al (2003) Neutralization of circulating vascular endothelial growth factor (VEGF) by anti-VEGF antibodies and soluble VEGF receptor 1 (sFlt-1) induces proteinuria. J Biol Chem 278:12605–12608
35. Lee S, Chen TT, Barber CL et al (2007) Autocrine VEGF signaling is required for vascular homeostasis. Cell 130:691–703
36. Kamba T, Tam BY, Hashizume H et al (2006) VEGF-dependent plasticity of fenestrated capillaries in the normal adult microvasculature. Am J Physiol Heart Circ Physiol 290:H560–H576
37. Izumiya Y, Shiojima I, Sato K et al (2006) Vascular endothelial growth factor blockade promotes the transition from compensatory cardiac hypertrophy to failure in response to pressure overload. Hypertension 47:887–993
38. Van Royen N, Piek JJ, Schaper W et al (2001) Arteriogenesis: mechanisms and modulation of collateral artery development. J Nucl Cardiol 8:687–693
39. Chen CH, Walterscheid JP (2006) Plaque angiogenesis versus compensatory arteriogenesis in atherosclerosis. Circ Res 99:787–789
40. Ribatti D, Nico B, Vacca A, Roncali L, Dammacco F (2002) Endothelial cell heterogeneity and organ specificity. J Hematother Stem Cell Res 11:81–90
41. Owman C, Hardebo JE (1988) Functional heterogeneity of cerebrovascular endothelium. Brain Behav Evol 32:65–75
42. Auerbach R (1992) Endothelial cell heterogeneity: its role as a determinant of selective metastasis. In: Simionescu N, Simionescu M (eds) Endothelial cell dysfunctions. Plenum Press, New York, pp 427–437
43. Belloni PN, Carney DH, Nicolson GL (1992) Organ-derived microvessel endothelial cells exhibit differential responsiveness to thrombin and other growth factors. Microvasc Res 43:20–45
44. Chi JT, Chang HY, Haraldsen G, Jahnsen FL et al (2003) Endothelial cell diversity revealed by global expression profiling. Proc Natl Acad Sci U S A 100:10623–10628
45. Deng D, Tsalenko A, Vailaya A et al (2006) Differences in vascular bed disease susceptibility reflect differences in gene expression response to atherogenic stimuli. Circ Res 98:200–208
46. Page C, Rose M, Yacoub M, Pigott R (1991) Antigenic heterogeneity of vascular endothelium. Am J Pathol 141:677–683
47. Aird WC, Edelberg JM, Weiler-Guettler H et al (1997) Vascular bed-specific expression of an endothelial cell is programmed by the tissue microenvironment. J Cell Biol 138:1117–1124
48. Pettersson A, Nagy JA, Brown LF, Sundberg C, Morgan E et al (2000) Heterogeneity of the angiogenic response induced in different normal adult tissues by vascular permeability factor/vascular endothelial growth factor. Lab Invest 80:99–115
49. Simpson E, Linder CC, Sargent EE, Davisson MT, Mobraaten LE, Sharp JJ (1997) Genetic variation among 129 substrains and its importance for targeted mutagenesis in mice. Nat Genet 16:19–27
50. Griep AE, Krawcek J, Lee D et al (1998) Multiple genetic loci modify risk for retinoblastoma in transgenic mice. Invest Ophthalmol Vis Sci 39:2723–2732
51. Thurston G, Murphy T, Baluk P, Lindsey JR, MacDonald DM (1998) Angiogenesis in mice with chronic airway inflammation: strain-dependent differences. Am J Pathol 153:1099–1112
52. Rohan RM, Fernandez A, Udagawa T, Yuan J, D'Amato RJ (2000) Genetic heterogeneity of angiogenesis in mice. FASEB J 14:871–876
53. Roger VL, Go AS, Lloyd-Jones DM et al (2012) Heart disease and stroke statistics–2012 update: a report from the American Heart Association. Circulation 125:e2–e220
54. Ware JA, Simons M (1997) Angiogenesis in ischemic heart disease. Nat Med 3:158–164
55. Zachary I, Morgan RD (2011) Therapeutic angiogenesis for cardiovascular disease: biological context, challenges, prospects. Heart 97:181–189

56. Henry TD, Annex BH, McKendall GR et al (2003) The VIVA trial: vascular endothelial growth factor in ischemia for vascular angiogenesis. Circulation 107:1359–1365
57. Simons M, Annex BH, Laham RJ et al (2002) Pharmacological treatment of coronary artery disease with recombinant fibroblast growth factor-2: double-blind, randomized, controlled clinical trial. Circulation 105:788–793
58. Hedman M, Hartikainen J, Syvänne M et al (2003) Safety and feasibility of catheter-based local intracoronary vascular endothelial growth factor gene transfer in the prevention of post-angioplasty and in-stent restenosis and in the treatment of chronic myocardial ischemia: phase II results of the Kuopio Angiogenesis Trial (KAT). Circulation 107:2677–2683
59. Kastrup J, Jørgensen E, Rück A et al (2005) Direct intramyocardial plasmid vascular endothelial growth factor-A165 gene therapy in patients with stable severe angina pectoris. A randomized double-blind placebo-controlled study: the Euroinject One trial. J Am Coll Cardiol 45:982–988

Chapter 13
Role of Serotonin in Angiogenesis in Diabetes

Masaaki Iwabayashi, Yoshiaki Taniyama, Junya Azuma, Fumihiro Sanada, Kazuma Iekushi, Hiromi Rakugi, and Ryuichi Morishita

Abstract Serotonin (5-hydroxytryptamine, 5-HT) is a monoaminergic neurotransmitter synthesized from tryptophan, which is one of the essential amino acids, by mostly enterochromaffin cells in the gut. 5-HT receptors (15 subtypes) are distributed in tissues throughout the body and have diverse effects and play a crucial role in central and peripheral functions. The cardiovascular system is no exception; most of the 5-HT receptor subtypes have been shown to exist and be involved in cardiovascular regulation in both physiological and pathological conditions. Particularly in the vasculature, 5-HT is known to regulate vascular tonus and vascular cell proliferation by the balance of smooth muscle and endothelial 5-HT receptors. Under pathological conditions like diabetes mellitus (DM) and peripheral artery disease (PAD), serum 5-HT level is known to increase and is the hallmark of both pathological conditions. In these conditions, contractile activity of smooth muscle 5-HT2A receptor is increased; thus, 5-HT-mediated vasodilation and endothelial proliferation are inhibited, leading to diminished angiogenesis and peripheral circulation. However, blockade of 5-HT2A by clinically available

M. Iwabayashi • F. Sanada • R. Morishita (✉)
Department of Clinical Gene Therapy, Osaka University Graduate School of Medicine, 2-2 Yamada-oka, Suita, Osaka 565-0871, Japan
e-mail: morishit@cgt.med.osaka-u.ac.jp

Y. Taniyama • J. Azuma • K. Iekushi
Department of Clinical Gene Therapy, Osaka University Graduate School of Medicine, 2-2 Yamada-oka, Suita, Osaka 565-0871, Japan

Department of Geriatric Medicine and Nephrology, Osaka University Graduate School of Medicine, Suita, Osaka 565-0871, Japan

H. Rakugi
Department of Geriatric Medicine and Nephrology, Osaka University Graduate School of Medicine, Suita, Osaka 565-0871, Japan

J.L. Mehta and N.S. Dhalla (eds.), *Biochemical Basis and Therapeutic Implications of Angiogenesis*, Advances in Biochemistry in Health and Disease 6,
DOI 10.1007/978-1-4614-5857-9_13, © Springer Science+Business Media New York 2013

drugs in a diabetic state enhances angiogenesis and vasodilation. As selective 5-HT2A blocker is widely used to treat PAD patients in clinical practice, the enhancement of angiogenesis, in addition to inhibition of vasoconstriction, might provide a unique treatment combination for therapeutic angiogenesis under a diabetic condition.

Keywords Serotonin • 5-Hydroxytryptamine • Angiogenesis • Diabetes mellitus • Peripheral artery disease

13.1 Introduction

Serotonin is a monoaminergic neurotransmitter with activities that modulate central and peripheral functions [1]. Serotonin is thought not to be able to pass the blood–brain barrier [2]. Thus, there are two independent serotonin systems: one in the central nervous system and the other in the periphery. In birds and mammals, including humans, serotonin levels in the central nervous system represent only a small fraction of the total serotonin in the body. Approximately 90 % of the human body's total serotonin is located in the enterochromaffin cells in the gut, where it is used to regulate intestinal movements [3, 4]. Serotonin secreted from enterochromaffin cells eventually finds its way out of tissues into the blood. There, it is actively taken up by blood platelets, which do not synthesize 5-HT but store it. Peripherally existing serotonin operates as a peripheral hormone, affecting vascular tonus, intestinal motility, primary hemostasis, liver repair, and control of the T-cell-mediated immune system. The remainder is synthesized in serotonergic neurons of the CNS and affects food intake, sleep, anxiety, sexual behavior, and mood.

Serotonin is synthesized from tryptophan, which is one of the essential amino acids in the human diet. Serotonin levels are known to be affected by diet [5]. α-Lactalbumin is a minor constituent of milk and contains more tryptophan than most other proteins. A diet enriched in α-lactalbumin boosts mood and cognition in some instances, probably owing to increased serotonin levels [6, 7]. Long-term enhancement of the tryptophan content of a diet with α-lactalbumin is probably not a practical strategy. However, increasing the tryptophan content of the diet is a method that was used in the past and could possibly be used again in the future.

The first step in the synthesis of serotonin from tryptophan is catalyzed by the enzyme tryptophan hydroxylase (TPH), which is also the rate-limiting enzyme in its biosynthesis. TPH is known to have two isoforms, TPH1 and TPH2 [8]. TPH1 is mainly present in the pineal gland, thymus, spleen, and enterochromaffin cells of the gastrointestinal tract. TPH2 is expressed solely in neuronal cells, such as in the raphe nuclei of the brain stem. Peripheral serotonin in mice lacking TPH1 cannot be replaced with serotonin from the central nervous system synthesized by TPH2 [9].

Serotonin is thought to exert its effects through these membrane-bound receptors [10, 11]. Fifteen receptors of serotonin, all found both in the central nervous system and in many peripheral tissues, have been divided into seven distinct classes: serotonin receptors (5HTR) 1–7, largely on the basis of their structure and operational

Table 13.1 5-HT receptor distribution in cardiovascular system

Subtype	Distribution	Function
5-HT1B	Vascular smooth muscle cell	Contraction
	Endothelial cell	Relaxation
5-HT1D	Vascular smooth muscle cell	Contraction
5-HT2A	Vascular smooth muscle cell	Contraction
	Platelet	Aggregation and 5-HT release
5-HT2B	Endothelial cell	Relaxation
5-HT3	Vagus nerve	Reflex bradycardia and hypotension
	Ganglia	
5-HT4	Cardiomyocyte	Contraction
5-HT7	Vascular smooth muscle cell	Relaxation
	Cardiomyocyte	Contraction

Modified from Trends Pharmacol Sci. 2008;29(9):472–481 and Pharmacol Rev. 2012; 64(2):359–388

characteristics. With the exception of the 5-HT3 receptor, which is a ligand-gated ion channel, 5-HT receptors belong to the G-protein-coupled receptor (GPCR) superfamily. Owing to the development of selective agonists and antagonists targeting such 5-HT receptors, many, hitherto unknown functions of 5-HT in the CNS and in the periphery have come to light. These findings ushered in a new era of drug development with therapeutic usefulness in treating diseases like anxiety, schizophrenia, hypertension, migraine, and peripheral vascular diseases.

This chapter discusses the physiological and pathophysiological effects of 5-HT and the clinical implications of 5-HT modulation in the vascular system, especially in angiogenesis.

13.2 Vascular Regulatory Effect of Serotonin

The cardiovascular system was probably the first area in which the effects attributed to 5-HT were observed; between 1869 and 1896 it was reported that a substance active on blood vessels and the heart appears in serum when blood is allowed to clot. For instance, Weiss [12] reported that "if serum was continuously but slowly injected into a rabbit, cat or dog until it caused death, the heart rate was increased and at first the beat was stronger but gradually failed, although it persisted until the respiration had ceased; in addition, an increase in the peristalsis of the small intestines and death due to paralysis of the respiratory and vasomotor centers was noticed" [13]. Indeed, at least seven subtypes of 5-HT receptors exist (Table 13.1) and regulate the cardiovascular system [10, 14], and selective agonists and antagonists have been described for these subtypes.

Theoretically, the effect of serotonin on vascular tension might be due to three mechanisms [15]: (1) prejunctional inhibition of noradrenaline release from vascular sympathetic nerve terminals, (2) involvement of the endothelium and endothelium-derived substances, and (3) the direct effect of serotonin on vascular smooth muscle [16]. Many studies have been performed a using variety of isolated blood vessels, and 5-HT predominantly causes direct arterial constriction [17]. $A \geq 3 \times 10^{-7}$ M concentration of 5-HT

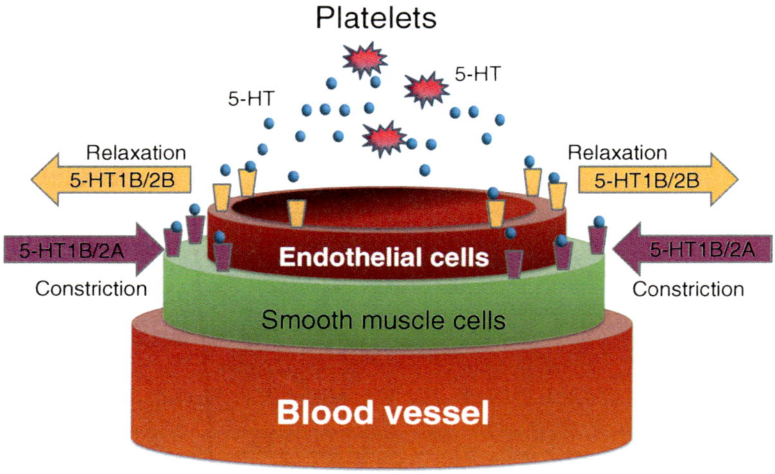

Fig. 13.1 Schematic summary of vascular regulation by 5-HT receptors. Vascular tonus is regulated by the balance between endothelial 5-HT1B/2B receptors and smooth muscle 5-HT1B/2A receptors. Modified from Trends Pharmacol Sci. 2008;29(9):472–481

contracts isolated normal human coronary arteries, these contractions being mediated by the 5-HT2A receptor [18]. The 5-HT2A receptor mainly mediates vasoconstriction, but 5-HT1B/1D receptors can also mediate constriction, as proved by the successful use of triptans (5-HT1B/1D agonists), in migraine therapy [17, 19].

In contrast, not all blood vessels constrict in response to 5-HT. In the rat jugular vein and in the pulmonary and coronary arteries of rat and greyhound, 5-HT causes relaxation by activating the 5-HT1B [20, 21] and 5-HT2B [22] receptors in endothelial cells and the 5-HT7 [23] receptor in smooth muscle cells. Haugen et al. showed that the 5-HT-induced vasorelaxation effect is absent in endothelium-denuded arteries, and no relaxant activity was seen when l-NAME (an NO inhibitor) was added [16]. Auch-Schwelk also showed that 5-HT relaxes human coronary arteries from explanted hearts by an endothelium-dependent, nitric oxide (NO)-mediated mechanism [21]. These studies show that 5-HT induces vasodilation through endothelium-dependent NOS activation.

The above findings suggest that both contractile and relaxant receptors operate in blood vessels and an intricately regulated balance of these receptors is implicated in physiological conditions (Fig. 13.1). However, in pathological conditions such as atherosclerosis, when the vasculature is impaired, this balance might be disturbed. Accordingly, intracoronary infusion of 5-HT a $\leq 10^{-5}$ M relaxed normal coronary arteries, a 10^{-4} M concentration constricted normal coronary arteries in eight subjects whereas all concentrations constricted vessels in four patients with stable angina precipitating attacks of angina. Another study showed that constrictor responses to 5-HT in patients with angina could be inhibited by ketanserin [24]. Also, there is evidence from animal studies that the sensitivity of coronary arteries to 5-HT may increase in atherosclerosis [25].

13.3 Role of Serotonin in Peripheral Artery Disease and Diabetes

13.3.1 Peripheral Artery Disease

Peripheral artery disease (PAD) is a progressive atherosclerotic disease characterized by narrowing of the peripheral arteries, most often used in reference to the arteries of the lower extremities. PAD is estimated to affect between 8 and 10 million Americans [26, 27]. It is anticipated that this prevalence will increase, not only in the United States, but worldwide along with the increasing population ages, continuing cigarette smoking, and the growing epidemic of metabolic syndrome [28]. PAD is associated with symptomatic claudication in approximately 20–40 % of patients [29]. Only a minority of patients present with classic intermittent claudication, and other symptoms include leg pain with exercise and at rest, leg weakness, and balance problems resulting in falls [30], all of which can significantly affect functional performance (i.e., the ability to perform activities of daily living, as well as exercise capacity). Impaired functional performance is associated with poor QOL [31, 32], and increased exercise endurance has been shown to positively affect QOL measures and perception of health in this patient population [33]. It is understandable that aggressive medical treatment of atherosclerotic risk factors has been shown to significantly decrease morbidity and mortality associated with PAD [34], since PAD is also known to associate with cardiovascular ischemic events (e.g., angina, myocardial infarction, and stroke) [26, 35–38]. However, it is said that PAD is underdiagnosed and undertreated, with most patients not receiving optimal therapy that has been proven to reduce mortality [39].

13.3.2 PAD in Diabetes

The risk factors for the development of PAD are the same as for other atherosclerotic diseases, including advancing age, male sex, family history, and Afro-Caribbean race. Modifiable risk factors for PAD include diabetes mellitus (DM), smoking, hypertension, and hyperlipidemia [40]. In particular, it is said that the prevalence of PAD in patients with DM is several times higher than in patients without DM [41–43]. Patients with both DM and PAD have poor lower extremity function and are prone to develop critical limb ischemia and ulceration, which may necessitate potential limb amputation [44, 45]. PAD has a more distal and generalized manifestation in Type 2 DM [46, 47]. PAD patients with DM are typically younger and have a higher body mass index (BMI) with more neuropathy and more cardiovascular comorbidity than in PAD patients without diabetes. Mortality in PAD patients is higher when accompanied by DM [48–50]. Diabetes is also considered to complicate the treatment of PAD. Patients with DM have less favorable outcomes after leg bypass surgery, a higher incidence of restenosis, more surgical complications, longer hospitalization times, and a lower rate of amputation-free survival [48, 49, 51, 52]. During angiogenesis, the presence of DM limits the extent of collateral development and the adaptive response to obstructed blood flow [53]. Type 2 DM lowers the recruitment and functional outward remodeling of preexisting collateral arterioles, exhibited clinically in the coronary circulation [54, 55] and experimentally in the lower extremities [56–60]. In a diabetic condition, it is known that substantial changes in gene expression and protein secretion occur in the local vasculature [61]. A number of factors and cytokines are known to contribute to angiogenesis and the regulation of vascular tonus. In the diabetic vasculature, the secretion of angiogenic factors, such as VEGF, FGF-2, and HGF, is decreased; on the other hand, inflammatory cytokine levels are upregulated and the expression of eNOS, a well-known vascular relaxation factor, is diminished, which leads to a decrease in peripheral circulation.

Importantly, another common factor is serotonin, an increased level of which is the hallmark of the pathology of DM [62] and PAD [63]. Under both pathological conditions, platelet activation and aggregation due to abnormal interactions between platelets and the vascular wall are known to increase [64–66]. As activated platelets in atherosclerotic lesions release large amounts of 5-HT, local and circulating 5-HT levels change markedly [67]. The released 5-HT binds to the 5-HT2A receptor in vascular smooth muscle cells (VSMC) to accelerate calcium influx and VSMC proliferation, leading to vascular occlusion. 5-HT also initiates autocrine activation and the aggregation of platelets through its receptor, 5-HT2A, on the cell surface, further augmenting vascular occlusive activity and diminishing peripheral circulation [68]; it is reported that the vascular sensitivity to 5-HT is significantly enhanced in patients with DM [69, 70], suggesting a tight link of this vicious cycle to the pathogenesis and development of PAD under diabetic conditions.

13.4 Serotonin and Angiogenesis

The critical role of the 5-HT2B receptor during heart development suggests influences on cell proliferation [71, 72]. Serotonin is classically known to stimulate growth of many types of cells including Chinese hamster lung fibroblasts [73], rat renal mesangial cells [74], rat jejunal crypt cells [75], neurons and neuroblasts [76], pancreatic carcinoid cells [77], and smooth muscle cells [78–81]. Vascular endothelial cells are no exception; Pakala et al. firstly showed that serotonin dose-dependently increased [^{3}H]thymidine incorporation into DNA and cell number in canine aortic endothelial cells [82–84]. Asada et al. reported that 5-HT plays a crucial role in tumor angiogenesis [85]. They showed that tumor size was significantly reduced in 5-HT transporter knockout (5-HTT$^{-/-}$) mice in a carcinoma cell transplantation model. Additionally, SB204741 administration significantly reduced tumor growth by suppressing angiogenesis. This is consistent with previous study reported by Tutton et al. [86]. Asada et al. further revealed that 5-HT activates ERK1/2, which is inhibited by administration of SB204741, a selective 5-HT2B receptor inhibitor, but not by RS102221, which is a specific 5-HT2C receptor inhibitor, in human umbilical vein endothelial cell (HUVEC). Another study performed by McDuffie et al. revealed that 5-HT-induced ERK1/2 phosphorylation is mediated by the 5-HT1B receptor in bovine endothelial cells [87]. Iwabayashi et al. also showed that 5-HT markedly stimulated the phosphorylation of ERK1/2. Moreover, after stimulation of human aortic endothelial cell with 5-HT on Matrigel-coated dishes, enhanced tubule formation was observed with increased tubule length in a dose-dependent manner. A role of the 5-HT4 receptor in angiogenic function is also reported [88]. Nishikawa et al. demonstrated an antiangiogenic effect of a selective 5-HT4 receptor agonist on HUVEC proliferation, cell cycle arrest, migration, and tube formation, suggesting that the 5-HT4 receptor has an opposite function to that of the other endothelial 5-HT receptors.

13.5 Induction of Angiogenesis in Diabetes

According to a previous study of proliferative effect of 5-HT in endothelial cells, it is suggested that stimulating endothelial 5-HT receptors might become a therapeutic strategy for treating ischemic disease by induction of therapeutic angiogenesis. A 5-HT1B receptor agonist, sumatriptan, has been reported; however, the 5-HT1B receptor is also expressed in VSMCs and mediates vasoconstriction, so sumatriptan cannot be used for selective stimulation of the endothelial 5-HT1B receptor [13, 89]. The 5-HT2B receptor is also known to mediate vasorelaxation, and a selective agonist, BW723C86, has been already reported. However, it is reported that the 5-HT2B receptor is critically involved in the pathogenesis of pulmonary hypertension [90], so this agonist might have adverse effects. This receptor-distribution problem was avoided by selective inhibition of the 5-HT2A receptor. Interestingly, Fujita et al. showed that selective blockade of 5-HT2A receptor by sarpogrelate increased coronary blood flow by induction of nitric oxide through the 5-HT1B receptor in the ischemic canine heart [91]. Since it was an acute hypoperfusion model, they did not assess the effect of blockade of 5-HT2A receptor on angiogenesis. However, it was clearly shown that addition of GR55562, a 5-HT1B receptor antagonist, to 5-HT2A receptor blocker administration significantly diminished the vasodilator effect of the 5-HT2A blocker. This implies that selective blockade of a vasoconstrictive 5-HT receptor in ischemic tissue might enhance not only vasodilation but also angiogenesis via stimulation of endothelial 5-HT receptors. Targeting this mechanism would bring about very rational therapy since serotonin concentration is known to increase under ischemic and hypoxic conditions [91–94], and as mentioned above, plasma serotonin level is increased in the pathological conditions of PAD and DM. It is noteworthy that insulin inhibits serotonin-induced Ca^{2+} influx in VSMCs [95], and insulin is reported to induce 5-HT2A receptor internalization [96]. Moreover, mRNA expression of 5-HT2A receptor is increased in ischemic limb and is further augmented in DM [97]. Thus, it is not hard to anticipate that adverse effects of 5-HT would be enhanced in DM or metabolic syndrome patients with insulin resistance. Therefore, inhibiting the 5-HT2A receptor in diabetes-related ischemic disease is a promising potent strategy. In fact, it is reported that administration of sarpogrelate restores ischemic limb blood flow [97–99] and capillary density in ischemic adductor muscle [97] in diabetic experimental animal models.

The effect of 5-HT on angiogenesis in a diabetic condition in vitro has been reported [99]. A high glucose condition is known to impair endothelial cell function and thereby depresses angiogenic function. A high glucose condition (33 mM) significantly inhibited the phosphorylation of Akt and the expression of eNOS in HAEC. However, 5-HT significantly increased Akt activity under a high glucose condition. Although a high glucose condition significantly decreased eNOS expression, 5-HT significantly restored eNOS expression. This protective effect of 5-HT on eNOS expression was significantly inhibited by LY294002, a specific PI3K inhibitor. High glucose also significantly inhibited tubule formation on Matrigel, while 5-HT significantly attenuated the decrease in tubule formation induced by high glucose. However, the decrease in Akt activity by LY294002 significantly diminished the protective effect of 5-HT on tubule formation. It is noteworthy that eNOS plays a critical role in angiogenesis [100]. Preservation of Akt/eNOS by 5-HT in a diabetic condition may be involved in restored tubule formation. Activation of Akt by 5-HT is also important for endothelial cell survival in a diabetic condition since the high glucose condition-induced increase in apoptotic cell number was significantly attenuated by 5-HT administration. These results suggest that 5-HT stimulates angiogenesis and protects endothelial cells through the Akt/eNOS pathway (Fig. 13.2).

13.6 Clinical Efficacy of 5-HT2A Receptor Antagonist

To date, using 5-HT modulation of the cardiovascular systems has produced only a few clinically useful drugs. A 5-HT modulator in widespread clinical use is the 5-HT2A receptor antagonist ketanserin, which also blocks α-adrenoceptors. Ketanserin was extensively studied in clinical and experimental studies in the 1980s and 1990s [14]. However, in clinical practice, ketanserin usage is restricted only to the treatment of preeclampsia, as serious side effects, such as prolonged QT intervals, may occur [101]. Instead of ketanserin, a novel 5-HT2A antagonist, sarpogrelate, has been developed. Sarpogrelate was first applied for treating patients with PAD [102, 103]. Blocking 5-HT2A by sarpogrelate has reportedly improved intermittent claudication, pain and cold sensation in the lower extremities, and ankle–brachial pressure index [103]. One of the mechanisms by which the beneficial effects of sarpogrelate operate is by enhancing vasodilation, as seen by sarpogrelate improving vascular function in PAD patients [104]. Although no direct evidence for involvement of enhanced collateral development by sarpogrelate in improvement of those symptoms is reported clinically, experimental studies suggest that 5-HT-induced angiogenesis plays a crucial role in ischemic diseases even in a severe diabetic state.

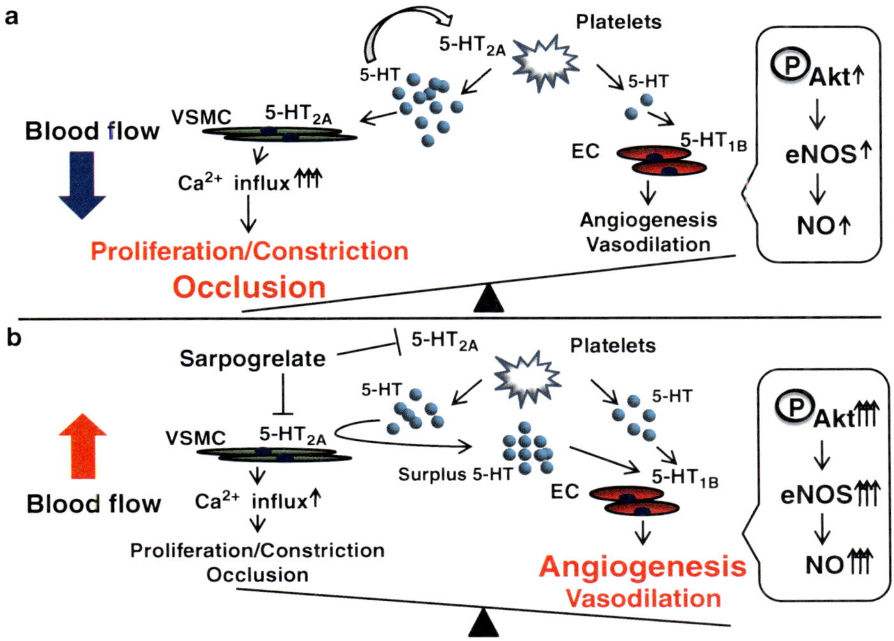

Fig. 13.2 Proposal mechanism of induction of angiogenesis by blockage of 5-HT2A under DM. In DM, platelets are aggregated and activated, leading to increased serum and tissue levels of 5-HT. Hypersensitivity of 5-HT2A to 5-HT of VSMC and platelets mediates vasoconstriction and atherosclerosis, while 5-HT1B-mediated vasodilation is impaired, leading to diminished peripheral blood flow. Selective blockade of 5-HT2A inhibits vasoconstriction of VSMC and activation and aggregation of platelets. Binding of 5-HT to 5-HT1B is relatively enhanced in the endothelium, leading to induction of angiogenesis and an increase in NO production via the Akt/eNOS pathway. Thus, blockade of 5-HT2A in DM restores peripheral blood flow via induction of angiogenesis and vasodilation

References

1. Barnes NM, Sharp T (1999) A review of central 5-HT receptors and their function. Neuropharmacology 38:1083–1152
2. Merritt JH, Chamness AF, Allen SJ (1978) Studies on blood-brain barrier permeability after microwave-radiation. Radiat Environ Biophys 15:367–377
3. Barter R, Pearse AG (1955) Mammalian enterochromaffin cells as the source of serotonin (5-hydroxytryptamine). J Pathol Bacteriol 69:25–31
4. Berger M, Gray JA, Roth BL (2009) The expanded biology of serotonin. Annu Rev Med 60:355–366
5. Leathwood PD (1987) Tryptophan availability and serotonin synthesis. Proc Nutr Soc 46:143–156
6. Markus CR, Olivier B, de Haan EH (2002) Whey protein rich in alpha-lactalbumin increases the ratio of plasma tryptophan to the sum of the other large neutral amino acids and improves cognitive performance in stress-vulnerable subjects. Am J Clin Nutr 75:1051–1056
7. Booij L, Merens W, Markus CR, Van der Does AJ (2006) Diet rich in alpha-lactalbumin improves memory in unmedicated recovered depressed patients and matched controls. J Psychopharmacol 20:526–535

8. Walther DJ, Peter JU, Bashammakh S et al (2003) Synthesis of serotonin by a second tryptophan hydroxylase isoform. Science 299:76
9. Walther DJ, Bader M (2003) A unique central tryptophan hydroxylase isoform. Biochem Pharmacol 66:1673–1680
10. Hoyer D, Hannon JP, Martin GR (2002) Molecular, pharmacological and functional diversity of 5-HT receptors. Pharmacol Biochem Behav 71:533–554
11. Hoyer D, Clarke DE, Fozard JR (1994) International Union of Pharmacology classification of receptors for 5-hydroxytryptamine (serotonin). Pharmacol Rev 46:157–203
12. Weiss O (1896) Ueber die Wirkungen von Blutserum-Injectionen ins Blut. Archiv Ges Physiol Menschen Thie LXV:215–230
13. Ramage AG, Villalon CM (2008) 5-Hydroxytryptamine and cardiovascular regulation. Trends Pharmacol Sci 29(9):472–481
14. Doggrell SA (2003) The role of 5-HT on the cardiovascular and renal systems and the clinical potential of 5-HT modulation. Expert Opin Investig Drugs 12:805–823
15. Mylecharane EJ (1990) Mechanisms involved in serotonin-induced vasodilatation. Blood Vessels 27:116–126
16. Haugen G, Mellembakken J, Stray-Pedersen S (1997) Characterization of the vasodilatatory response to serotonin in human umbilical arteries perfused in vitro. The influence of the endothelium. Early Hum Dev 47:185–193
17. Watts SW, Morrison SF, Davis RP, Barman SM (2012) Serotonin and blood pressure regulation. Pharmacol Rev 64:359–388
18. Nilsson T, Longmore J, Shaw D et al (1999) Characterisation of 5-HT receptors in human coronary arteries by molecular and pharmacological techniques. Eur J Pharmacol 372:49–56
19. Gilmore B, Michael M (2011) Treatment of acute migraine headache. Am Fam Physician 83:271–280
20. Vanhoutte PM (1998) Endothelial dysfunction and vascular disease. Verh K Acad Geneeskd Belg 60:251–266
21. Auch-Schwelk W, Paetsch I, Krackhardt F et al (2000) Modulation of contractions to ergonovine and methylergonovine by nitric oxide and thromboxane A2 in the human coronary artery. J Cardiovasc Pharmacol 36:631–639
22. Ellis ES, Byrne C, Murphy OE et al (1995) Mediation by 5-hydroxytryptamine2B receptors of endothelium-dependent relaxation in rat jugular vein. Br J Pharmacol 114:400–404
23. Centurion D, Sanchez-Lopez A, Ortiz MI et al (2000) Mediation of 5-HT-induced internal carotid vasodilatation in GR127935- and ritanserin-pretreated dogs by 5-HT7 receptors. Naunyn Schmiedebergs Arch Pharmacol 362:169–176
24. Golino P, Piscione F, Willerson JT et al (1991) Divergent effects of serotonin on coronary-artery dimensions and blood flow in patients with coronary atherosclerosis and control patients. N Engl J Med 324:641–648
25. Ishida T, Kawashima S, Hirata K et al (2001) Serotonin-induced hypercontraction through 5-hydroxytryptamine 1B receptors in atherosclerotic rabbit coronary arteries. Circulation 103:1289–1295
26. Criqui MH, Langer RD, Fronek A et al (1992) Mortality over a period of 10 years in patients with peripheral arterial disease. N Engl J Med 326:381–386
27. Criqui MH (2001) Peripheral arterial disease–epidemiological aspects. Vasc Med 6:3–7
28. Hirsch AT, Hartman L, Town RJ, Virnig BA (2008) National health care costs of peripheral arterial disease in the Medicare population. Vasc Med 13:209–215
29. Hirsch AT, Halverson SL, Treat-Jacobson D et al (2001) The Minnesota Regional Peripheral Arterial Disease Screening Program: toward a definition of community standards of care. Vasc Med 6:87–96
30. McDermott MM (2006) The magnitude of the problem of peripheral arterial disease: epidemiology and clinical significance. Cleve Clin J Med 73:S2–S7
31. Pell JP (1995) Impact of intermittent claudication on quality of life. The Scottish Vascular Audit Group. Eur J Vasc Endovasc Surg 9:469–472

32. Izquierdo-Porrera AM, Gardner AW, Bradham DD et al (2005) Relationship between objective measures of peripheral arterial disease severity to self-reported quality of life in older adults with intermittent claudication. J Vasc Surg 41:625–630
33. Tsai JC, Chan P, Wang CH et al (2002) The effects of exercise training on walking function and perception of health status in elderly patients with peripheral arterial occlusive disease. J Intern Med 252:448–455
34. Gornik HL, Creager MA (2006) Contemporary management of peripheral arterial disease: I. Cardiovascular risk-factor modification. Cleve Clin J Med 73:S30–S37
35. Zheng ZJ, Sharrett AR, Chambless LE et al (1997) Associations of ankle-brachial index with clinical coronary heart disease, stroke and preclinical carotid and popliteal atherosclerosis: the Atherosclerosis Risk in Communities (ARIC) Study. Atherosclerosis 131:115–125
36. Newman AB, Shemanski L, Manolio TA et al (1999) Ankle-arm index as a predictor of cardiovascular disease and mortality in the Cardiovascular Health Study. The Cardiovascular Health Study Group. Arterioscler Thromb Vasc Biol 19:538–545
37. Newman AB, Sutton-Tyrrell K, Vogt MT, Kuller LH (1993) Morbidity and mortality in hypertensive adults with a low ankle/arm blood pressure index. JAMA 270:487–489
38. Leng GC, Fowkes FG, Lee AJ et al (1996) Use of ankle brachial pressure index to predict cardiovascular events and death: a cohort study. Br Med J 313:1440–1444
39. Hirsch AT, Criqui MH, Treat-Jacobson D et al (2001) Peripheral arterial disease detection, awareness, and treatment in primary care. JAMA 286:1317–1324
40. Cimminiello C (2002) PAD. Epidemiology and pathophysiology. Thromb Res 106:V295–V301
41. Anon (1991) Second European consensus document on chronic critical leg ischemia. Circulation 84:IV1–IV26
42. Tapp RJ, Zimmet PZ, Harper CA et al (2004) Diabetes care in an Australian population: frequency of screening examinations for eye and foot complications of diabetes. Diabetes Care 27:688–693
43. Gregg EW, Sorlie P, Paulose-Ram R et al (2004) Prevalence of lower-extremity disease in the US adult population >=40 years of age with and without diabetes: 1999-2000 National Health and Nutrition Examination Survey. Diabetes Care 27:1591–1597
44. Dolan NC, Liu K, Criqui MH et al (2002) Peripheral artery disease, diabetes, and reduced lower extremity functioning. Diabetes Care 25:113–120
45. Prompers L, Schaper N, Apelqvist J et al (2008) Prediction of outcome in individuals with diabetic foot ulcers: focus on the differences between individuals with and without peripheral arterial disease. The EURODIALE Study. Diabetologia 51:747–755
46. Beckman JA, Creager MA, Libby P (2002) Diabetes and atherosclerosis: epidemiology, pathophysiology, and management. JAMA 287:2570–2581
47. Lanzer P (2001) Topographic distribution of peripheral arteriopathy in non-diabetics and type 2 diabetics. Z Kardiol 90:99–103
48. Faglia E, Clerici G, Clerissi J et al (2009) Long-term prognosis of diabetic patients with critical limb ischemia: a population-based cohort study. Diabetes Care 32:822–827
49. Kamalesh M, Shen J (2009) Diabetes and peripheral arterial disease in men: trends in prevalence, mortality, and effect of concomitant coronary disease. Clin Cardiol 32:442–446
50. Larsson J, Agardh CD, Apelqvist J, Stenstrom A (1998) Long-term prognosis after healed amputation in patients with diabetes. Clin Orthop Relat Res (350):149–158
51. Malmstedt J, Leander K, Wahlberg E et al (2008) Outcome after leg bypass surgery for critical limb ischemia is poor in patients with diabetes: a population-based cohort study. Diabetes Care 31:887–892
52. Currie CJ, Morgan CL, Peters JR (1998) The epidemiology and cost of inpatient care for peripheral vascular disease, infection, neuropathy, and ulceration in diabetes. Diabetes Care 21:42–48
53. De Vivo S, Palmer-Kazen U, Kalin B, Wahlberg E (2005) Risk factors for poor collateral development in claudication. Vasc Endovascular Surg 39:519–524
54. Waltenberger J (2001) Impaired collateral vessel development in diabetes: potential cellular mechanisms and therapeutic implications. Cardiovasc Res 49:554–560

55. Abaci A, Oguzhan A, Kahraman S et al (1999) Effect of diabetes mellitus on formation of coronary collateral vessels. Circulation 99:2239–2242
56. Yan J, Tie G, Park B, Yan Y et al (2009) Recovery from hind limb ischemia is less effective in type 2 than in type 1 diabetic mice: roles of endothelial nitric oxide synthase and endothelial progenitor cells. J Vasc Surg 50:1412–1422
57. Rivard A, Silver M, Chen D et al (1999) Rescue of diabetes-related impairment of angiogenesis by intramuscular gene therapy with adeno-VEGF. Am J Pathol 154:355–363
58. Taniyama Y, Morishita R, Hiraoka K et al (2001) Therapeutic angiogenesis induced by human hepatocyte growth factor gene in rat diabetic hind limb ischemia model: molecular mechanisms of delayed angiogenesis in diabetes. Circulation 104:2344–2350
59. Schiekofer S, Galasso G, Sato K et al (2005) Impaired revascularization in a mouse model of type 2 diabetes is associated with dysregulation of a complex angiogenic-regulatory network. Arterioscler Thromb Vasc Biol 25:1603–1609
60. van Golde JM, Ruiter MS, Schaper NC et al (2008) Impaired collateral recruitment and outward remodeling in experimental diabetes. Diabetes 57:2818–2823
61. Ruiter MS, van Golde JM, Schaper NC et al (2010) Diabetes impairs arteriogenesis in the peripheral circulation: review of molecular mechanisms. Clin Sci 119:225–238
62. Malyszko J, Urano T, Knofler R et al (1994) Daily variations of platelet aggregation in relation to blood and plasma serotonin in diabetes. Thromb Res 75:569–576
63. Barradas MA, Gill DS, Fonseca VA et al (1988) Intraplatelet serotonin in patients with diabetes mellitus and peripheral vascular disease. Eur J Clin Invest 18:399–404
64. Dallinger KJ, Jennings PE, Toop MJ et al (1987) Platelet aggregation and coagulation factors in insulin dependent diabetics with and without microangiopathy. Diabet Med 4:44–48
65. Pietraszek MH, Takada Y, Takada A et al (1992) Blood serotonergic mechanisms in type 2 (non-insulin-dependent) diabetes mellitus. Thromb Res 66:765–774
66. Robless PA, Okonko D, Lintott P et al (2003) Increased platelet aggregation and activation in peripheral arterial disease. Eur J Vasc Endovasc Surg 25:16–22
67. Hara K, Hirowatari Y, Yoshika M et al (2004) The ratio of plasma to whole-blood serotonin may be a novel marker of atherosclerotic cardiovascular disease. J Lab Clin Med 144:31–37
68. Maurer-Spurej E (2005) Serotonin reuptake inhibitors and cardiovascular diseases: a platelet connection. Cell Mol Life Sci 62:159–170
69. Tanaka N (2009) Pharmacological studies on responsiveness of 5-hydroxytryptamine in overcoming perioperative spasm of coronary artery bypass graft. Yakugaku Zasshi 129:1049–1053
70. Boston PC, Hodgson WC (1997) Changes in reactivity towards 5-hydroxytryptamine in the renal vasculature of the diabetic spontaneously hypertensive rat. J Hypertens 15:769–774
71. Brezun JM, Daszuta A (2000) Serotonin may stimulate granule cell proliferation in the adult hippocampus, as observed in rats grafted with foetal raphe neurons. Eur J Neurosci 12:391–396
72. Nebigil CG, Choi DS, Dierich A et al (2000) Serotonin 2B receptor is required for heart development. Proc Natl Acad Sci U S A 97:9508–9513
73. Seuwen K, Magnaldo I, Pouyssegur J (1988) Serotonin stimulates DNA synthesis in fibroblasts acting through 5-HT1B receptors coupled to a Gi-protein. Nature 335:254–256
74. Takuwa N, Ganz M, Takuwa Y et al (1989) Studies of the mitogenic effect of serotonin in rat renal mesangial cells. Am J Physiol 257:F431–F439
75. Tutton PJ, Barkla DH (1986) Serotonin receptors influencing cell proliferation in the jejunal crypt epithelium and in colonic adenocarcinomas. Anticancer Res 6:1123–1126
76. Hanley MR (1989) Mitogenic neurotransmitters. Nature 340:97
77. Ishizuka J, Beauchamp RD, Townsend CM et al (1992) Receptor-mediated autocrine growth-stimulatory effect of 5-hydroxytryptamine on cultured human pancreatic carcinoid cells. J Cell Physiol 150:1–7
78. Nemecek GM, Coughlin SR, Handley DA, Moskowitz MA (1986) Stimulation of aortic smooth muscle cell mitogenesis by serotonin. Proc Natl Acad Sci U S A 83:674–678
79. Kavanaugh WM, Williams LT, Ives HE, Coughlin SR (1988) Serotonin-induced deoxyribonucleic acid synthesis in vascular smooth muscle cells involves a novel, pertussis toxin-sensitive pathway. Mol Endocrinol 2:599–605

80. Araki S, Kawahara Y, Fukuzaki H, Takai Y (1990) Serotonin plays a major role in serum-induced phospholipase C-mediated hydrolysis of phosphoinositides and DNA synthesis in vascular smooth muscle cells. Atherosclerosis 83:29–34
81. Lee SL, Wang WW, Moore BJ, Fanburg BL (1991) Dual effect of serotonin on growth of bovine pulmonary artery smooth muscle cells in culture. Circ Res 68:1362–1368
82. Pakala R, Willerson JT, Benedict CR (1994) Mitogenic effect of serotonin on vascular endothelial cells. Circulation 90:1919–1926
83. Pakala R, Sheng WL, Benedict CR (1999) Serotonin fails to induce proliferation of endothelial cells preloaded with eicosapentaenoic acid and docosahexaenoic acid. Atherosclerosis 145:137–146
84. Pakala R, Benedict CR (1998) Effect of serotonin and thromboxane A2 on endothelial cell proliferation: effect of specific receptor antagonists. J Lab Clin Med 131:527–537
85. Asada M, Ebihara S, Yamanda S et al (2009) Depletion of serotonin and selective inhibition of 2B receptor suppressed tumor angiogenesis by inhibiting endothelial nitric oxide synthase and extracellular signal-regulated kinase 1/2 phosphorylation. Neoplasia 11:408–417
86. Tutton PJ, Steel GG (1979) Influence of biogenic amines on the growth of xenografted human colorectal carcinomas. Br J Cancer 40:743–749
87. McDuffie JE, Motley ED, Limbird LE, Maleque MA (2000) 5-Hydroxytryptamine stimulates phosphorylation of p44/p42 mitogen-activated protein kinase activation in bovine aortic endothelial cell cultures. J Cardiovasc Pharmacol 35:398–402
88. Nishikawa T, Tsuno NH, Shuno Y et al (2010) Antiangiogenic effect of a selective 5-HT4 receptor agonist. J Surg Res 159:696–704
89. Humphrey PP, Feniuk W, Perren MJ et al (1988) GR43175, a selective agonist for the 5-HT1-like receptor in dog isolated saphenous vein. Br J Pharmacol 94:1123–1132
90. Launay JM, Herve P, Peoc'h K et al (2002) Function of the serotonin 5-hydroxytryptamine 2B receptor in pulmonary hypertension. Nat Med 8:1129–1135
91. Fujita M, Minamino T, Sanada S et al (2004) Selective blockade of serotonin 5-HT2A receptor increases coronary blood flow via augmented cardiac nitric oxide release through 5-HT1B receptor in hypoperfused canine hearts. J Mol Cell Cardiol 37:1219–1223
92. Takano S, Hoshino Y, Li L et al (2004) Dual roles of 5-hydroxytryptamine in ischemia-reperfusion injury in isolated rat hearts. J Cardiovasc Pharmacol Ther 9:43–50
93. MacLean MR, Deuchar GA, Hicks MN et al (2004) Overexpression of the 5-hydroxytryptamine transporter gene: effect on pulmonary hemodynamics and hypoxia-induced pulmonary hypertension. Circulation 109:2150–2155
94. Esteve JM, Launay JM, Kellermann O, Maroteaux L (2007) Functions of serotonin in hypoxic pulmonary vascular remodeling. Cell Biochem Biophys 47:33–44
95. Kahn AM, Allen JC, Seidel CL, Song T (1994) Insulin inhibits serotonin-induced Ca2+ influx in vascular smooth muscle. Circulation 90:384–390
96. Ohkura M, Tanaka N, Kobayashi H et al (2005) Insulin induces internalization of the 5-HT2A receptor expressed in HEK293 cells. Eur J Pharmacol 518:18–21
97. Bir SC, Fujita M, Marui A et al (2008) New therapeutic approach for impaired arteriogenesis in diabetic mouse hindlimb ischemia. Circulation 72:633–640
98. Li TS, Furutani A, Takahashi M et al (2006) Impaired potency of bone marrow mononuclear cells for inducing therapeutic angiogenesis in obese diabetic rats. Am J Physiol Heart Circ Physiol 290:H1362–H1369
99. Iwabayashi M, Taniyama Y, Sanada F et al (2012) Role of serotonin in angiogenesis: induction of angiogenesis by sarpogrelate via endothelial 5-HT1B/Akt/eNOS pathway in diabetic mice. Atherosclerosis 220:337–342
100. Murohara T, Asahara T, Silver M et al (1998) Nitric oxide synthase modulates angiogenesis in response to tissue ischemia. J Clin Invest 101:2567–2578
101. Zehender M, Meinertz T, Hohnloser S et al (1989) Incidence and clinical relevance of QT prolongation caused by the new selective serotonin antagonist ketanserin. Am J Cardiol 63:826–832

102. Ito K, Notsu T (1991) Effect of sarpogrelate hydrochloride (MCI-9042) on peripheral circulation of chronic arterial occlusive disease. J Clin Ther Med 7:1243–1251
103. Norgren L, Jawien A, Matyas L et al (2006) Sarpogrelate, a 5-hT2A receptor antagonist in intermittent claudication. A Phase II European Study. Vasc Med 11:75–83
104. Miyazaki M, Higashi Y, Goto C et al (2007) Sarpogrelate hydrochloride, a selective 5-HT2A antagonist, improves vascular function in patients with peripheral arterial disease. J Cardiovasc Pharmacol 49:221–227

Chapter 14
Brain Angiogenesis After Stroke

Kazuhide Hayakawa, Ji Hae Seo, Nobukazu Miyamoto, Loc-Duyen D. Pham, Deepti Navaratna, Eng H. Lo, and Ken Arai

Abstract Stroke remains a major health problem worldwide, and is the leading cause of serious long-term disability. Although many advances have been made in terms of the basic molecular mechanisms underlying neuronal death, clinically effective neuroprotective drugs in stroke have not yet been discovered. Recent findings now suggest that strategies to enhance angiogenesis after focal cerebral ischemia may provide unique opportunities to improve clinical outcomes during stroke recovery. This chapter aims at summarizing current knowledge on mechanisms and potential targets for angiogenic therapies in brain after stroke. Crosstalk between cerebral endothelial cells and their neighboring cells may provide substrates for plasticity and remodeling in the recovering brain. A better understanding of the molecular interplay between all these complex pathways may lead to novel therapeutic approaches for this devastating disease.

Keywords Stroke • Angiogenesis • Cerebral endothelial cell • Neurovascular unit • Neurovascular niche • Oligovascular niche • Biphasic response • Brain remodeling

14.1 Introduction

Stroke is the second leading cause of death and a leading cause of adult disability worldwide. Under stroke conditions, brain function is perturbed due to cerebral ischemia caused by thrombosis or hemorrhage. In the central areas of ischemic regions, blood flow deficits are severe and brain cells die rapidly. In the peripheral penumbral areas, blood flow deficits are relatively mild, so that therapeutic salvage

K. Hayakawa • J.H. Seo • N. Miyamoto • L.-D.D. Pham • D. Navaratna • E.H. Lo • K. Arai (✉)
Neuroprotection Research Laboratory, Massachusetts General Hospital, Harvard Medical School, 149, 13th Street, MGH-East CNY149-2401, Charlestown, MA 02129, USA
e-mail: karai@partners.org

J.L. Mehta and N.S. Dhalla (eds.), *Biochemical Basis and Therapeutic Implications of Angiogenesis*, Advances in Biochemistry in Health and Disease 6,
DOI 10.1007/978-1-4614-5857-9_14,

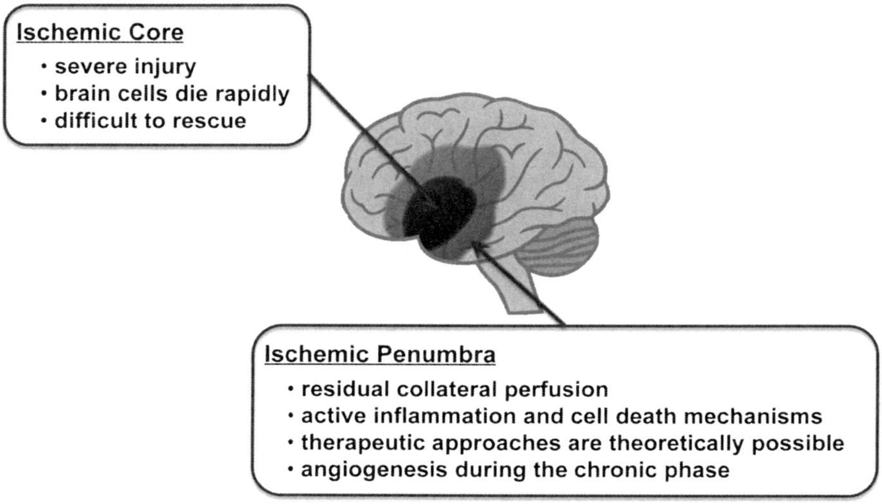

Fig. 14.1 Schematic of stroke brain. Under stroke conditions, brain function is perturbed due to cerebral ischemia (lack of blood supply to the brain). In central areas of ischemic regions (ischemic core), blood flow deficits are severe and brain cells die rapidly. In peripheral areas (ischemic penumbra), blood flow deficits are relatively mild and cell death progresses are slower. Hence, therapeutic salvage is theoretically possible, and angiogenesis may occur during the chronic phase

is theoretically possible (Fig. 14.1). However, therapeutic options for clinical management in stroke remain quite limited. In the acute phase, thrombolytic reperfusion with recombinant tissue plasminogen activator (t-PA) is still only used in less than 3% of all ischemic stroke patients worldwide [1]. During the chronic phase after stroke, standard treatments involving rehabilitation provide some support for recovering patients. However, many high-profile failures in a wide spectrum of pharmacologic neuroprotection trials have led to some pessimism in the field [2]. In recent years, accumulating data suggest that damaged brain can be surprisingly plastic, and intriguing mechanisms of neurogenesis and angiogenesis might provide novel substrates for brain repair [3, 4]. In this chapter, we will focus on key findings that emphasize interactions between growth factors, progenitor cells, and neurovascular/oligovascular signaling as potential mechanisms that may be augmented to stimulate cerebral angiogenesis and enhance stroke recovery.

14.2 Brain Angiogenesis After Stroke

Angiogenesis is a key restorative mechanism in response to ischemia in several non-CNS tissues. For example, myocardial infarction and limb ischemia can trigger endogenous angiogenesis in each organ, and therapeutics that enhanced angiogenesis can sometimes reduce injury in these disorders. Also in the brain, angiogenic responses may play important roles on brain remodeling after ischemic injury. Increasing evidence in both human stroke patients and animal stroke models suggests that the post-stroke penumbra is extremely resilient and is a site of intense remodeling and active angiogenesis. Autopsy studies show that brain ischemia stimulates angiogenesis in part via stereo-typed hypoxia-inducible factor (HIF-1) [5]. Proliferation of endothelial cells starts at several days after ischemic events [6]. Studies using mice with middle cerebral artery occlusion demonstrated that endothelial cell proliferation might begin as early as 12–24 h after ischemia and persist for up to several weeks thereafter [7, 8]. Studies using human brain samples also suggested that active angiogenesis takes place at 3–4 days after stroke, and the number of vessels appeared to be correlated longer survival times in ischemic stroke patients, suggesting that active angiogenesis may be beneficial [5, 9, 10]. In contrast, older patients who tend to do worse after stroke seem to have reduced new vessel formation after stroke [11, 12]. Furthermore, patients who develop dementia after stroke may suffer from reduced blood flow in adjacent cortical regions [13]. This raises the possibility that angiogenesis may improve cerebral perfusion and function as part of a network repair. However, the purpose of this angiogenic response remains speculative. Lyden and colleagues have proposed a "clean-up hypothesis," whereby newborn vessels serve to facilitate macrophage infiltration and clear up and remove cellular debris from pan-necrotic tissue [14, 15]. They demonstrated that microvessel density was always associated with increased numbers of macrophages. Ischemic brain areas without macrophages displayed no vascular changes compared with normal animals. This alternate hypothesis would suggest that post-stroke brain angiogenesis is only transient and not permanently involved in neuronal recovery. Nevertheless, the data in aggregate support a beneficial role for brain angiogenesis during recovery phase after ischemic stroke.

14.2.1 Growth Factors for Brain Angiogenesis

After focal cerebral ischemia, brain cells manufacture and secrete angiogenic peptides to stimulate angiogenesis. The precise regulatory mechanisms that underlie angiogenesis after ischemia still remain elucidated. But, several growth factors have been found to be upregulated after stroke to promote angiogenesis for brain remodeling. So far, at least 20 growth factors are known to induce angiogenesis, and here we will review the well-characterized growth factors.

14.2.1.1 Vascular Endothelial Growth Factor

Vascular endothelial growth factor (VEGF) is a major mediator for angiogenic responses in the brain. VEGF can accelerate angiogenesis and neurogenesis in the delayed stroke phase [16, 17]. VEGF can trigger remodeling responses in both endothelial cells and neurons. VEGF was increased within hours after ischemia and had a strong effect on the new vessel growing. Beginning minutes after stroke in rodents,

VEGF signals appear in neurons for days after ischemic onset, and can be found in astrocytes for up to a few weeks [18, 19]. In human, mRNA and proteins of VEGF165, VEGF189, and the receptor flk-1 were found to be upregulated in brain tissue or serum of patients following acute ischemic insult [20]. Elevated serum levels of VEGF were correlated to infarct volumes and clinical disabilities. An increase in angiogenesis by VEGF in rats was also associated with reduced neurological deficits after focal cerebral ischemia [19]. Boosting VEGF also seems to promote recovery. An intracerebroventricular injection of VEGF via osmotic pump, starting 24 h after onset of focal cerebral ischemia, stimulated angiogenesis and decreased infarct volume in rodent models of focal cerebral ischemia [21]. In addition to these biochemical and pharmacologic findings, genetic data have also been obtained. In transgenic mice overexpressing human VEGF165, brain microvessel density was significantly elevated compared with wild-type mice before ischemia, and the increase in microvessel density 3 days after stroke onset was improved [22]. These data show that VEGF promotes revascularization after stroke. Recently, encapsulated cell grafts overexpressing VEGF were implanted into rat striatum before induction of focal cerebral ischemia resulting in brain edema [23]. Angiogenesis was significantly increased around the area of the encapsulated graft after 24 h concomitant with a reduction in infarct size, but interestingly, there was no increase in cerebral blood flow at 1, 7, and 14 days compared with control untreated animals. These data suggest that the link between increased vascularization, increased blood flow, and recovery may not be interrelated or concomitant.

14.2.1.2 Fibroblast Growth Factor-2 (FGF-2/bFGF)

FGF-2 is a potent stimulator of endothelial cell migration, proliferation, sprouting, and tube formation. FGF-2 signaling also promotes mitogenesis and differentiation in neural progenitor cells in vivo [24]. In rat MCAO models, FGF-2 was elevated in neuron adjacent to infarct after 1 day following cerebral ischemia. Besides, endothelial cell, reactive astrocyte, and macrophage also expressed FGF-2 in the first 2 weeks after ischemia [6, 25]. The clinical findings were consistent with these experimental models. In human, mRNA and protein of FGF-2 were reported to be upregulated in the brains, serum, and cerebrospinal fluid (CSF) of patients who died of acute ischemic stroke [26]. The cellular localization of FGF-2 was found to be in endothelial cells in peri-infarct region of surviving patients after stroke, confirming its important role in angiogenesis in human as well as animal models [26]. Many studies examined the effect of FGF-2 administration on brain damage and recovery in animal stroke models. In rodent, FGF-2 administration 2 h after focal cerebral ischemia markedly reduced infarct volume [27]. Injection of FGF-2 enhanced recovery of sensorimotor function in rat stroke models [28, 29]. Later administration increased neuronal sprouting and enhanced neural recovery, though it could not reduce infarct volume [28, 30]. This effectiveness after later administration considered FGF-2 as a suitable therapy for human patients where treatment is often delayed; then human clinical trials have been carried out [31]. However, human clinical trials conducted in the USA were curtailed because of high dose toxicity.

14.2.1.3 Platelet-Derived Growth Factor-Beta

Platelet-derived growth factor-beta (PDGF-beta) and its receptor PDGFR-beta are essential for the recruitment of pericytes to cerebral endothelium. This step contributes to maintain the homeostasis of microvessel and the mechanical stability of capillary [32]. Thus, following cerebral ischemia, PDGF-beta is thought to support angiogenesis and vascular remodeling by mediating interactions of endothelium with pericytes. Analysis of the expression pattern following experimental cerebral ischemia showed that PDGFR-beta was specifically upregulated in vascular structures in the infarcted area mainly associates with pericytes 48 h after MCAO [33]. In human, PDGF-beta and its receptor can be detected on microvessel endothelial cells around cystic infarction for weeks after stroke following stroke [34].

14.2.1.4 Transforming Growth Factor-Beta

Transforming growth factor-beta (TGF-beta) controls proliferation, cellular differentiation, and other functions in most cells. TGF-beta contributes to angiogenesis by stabilizing newly formed capillary sprouts. Many studies of TGF-beta in mice and humans have demonstrated its pivotal role in modulation of angiogenesis. In rodent, TGF-beta was upregulated after hypoxia/ischemia, resulting in reduced infarct size and neuroprotection [35]. TGF-beta was detected to occur in astrocytes, activated microglial cells, and microvessels after cerebral ischemia. In the human brain, TGF-beta mRNA was upregulated in microvascular cells of the penumbra region of patients following ischemic stroke [36, 37]. Further human studies demonstrated that TGF-beta level was increased in CSF but not in serum [38]. Upregulation of TGF-beta seems to be associated with NOS1 in the serum and tissue of patients after stroke, suggesting that this interaction mediates pro-angiogenic function of TGF-beta [39, 40]. When TGF-beta was injected into animals prior to ischemia, it decreased infarct size, showing its neuroprotective role [41]. However, it had no beneficial effect if injected after the ischemic injury [42].

14.2.2 Biphasic Responses of Angiogenic Factors After Stroke

The biphasic nature of many mediators in neurobiology is now well known. For example, trophic factors such as NGF and BDNF promote cellular survival via their primary receptors TrkA and TrkB, respectively. But in contrast to these neuroprotective effects, both factors can also be neurotoxic via overactivation of the p75NTR receptor. As noted, the responses and regulatory mechanisms that underlie brain remodeling are highly complex. Growth factors seen above (and also other angiogenic factors) promote brain angiogenesis after stroke. Angiogenesis is an essential step for restoring brain function in injured brain, but those angiogenic factors may not be always supportive for the brain. As introduced, one major mediator in vascular responses after stroke is VEGF. VEGF is the prototypical biphasic mediator. VEGF can trigger remodeling responses in both endothelial cells and neurons [16, 17], and accelerate angiogenesis and neurogenesis responses in the delayed stroke phase. By contrast, in the acute phase in stroke, VEGF increases blood–brain barrier (BBB) permeability, which causes cerebral hemorrhage and brain edema [43, 44]. In fact, VEGF administration worsens BBB leakage by ischemic insults [19]. Within the context of the vascular remodeling, similar biphasic properties of many factors and mediators may also emerge. Therapies that can boost these endogenous signals and substrates of vascular remodeling might be a new direction for stroke treatments [45]. However, it remains to be fully elucidated how these approaches can be utilized in clinic. It is worth noting that most molecular targets for stroke therapy have biphasic roles in stroke pathophysiology [46, 47]. Here we will overview three major examples of the so-called biphasic responses in vascular remodeling after stroke.

14.2.2.1 Matrix Metalloproteinase

In recent years, dysregulation of neurovascular proteases has been implicated as central in neurovascular injury and remodeling after stroke. Hence, neurovascular proteases such as matrix metalloproteinases (MMPs) may have the biphasic properties after brain injury. The MMP family of extracellular proteases has been well studied in our field. MMPs comprise a family of zinc endopeptidases with major roles in the physiology and pathology of the mammalian CNS. To date, MMP-2 (gelatinase A), MMP-3 (stromelysin 1), MMP-7 (matrilysin), MMP-9 (gelatinase B), and MMP-13 (collagenase-3) are known to contribute to infarct extent and/or BBB disruption in the acute phase after stroke [48–52]. On the contrary, however, these same proteases may have a beneficial role during neurovascular repair. In a mouse stroke model, peri-infarct cortical areas demonstrate a secondary elevation in MMP-9 in endothelial and glial cells within networks of regrowing microvessels [53], and inhibition of MMPs during this delayed phase actually made outcomes worse with the development of hemorrhagic and malformed blood vessels and enlarged volumes of infarction and cavitation. Beyond the peri-infarct zone, other brain areas were also involved. Secondary MMP-9 signals co-localized with streams of migrating neuroblasts from the subventricular zone, and inhibition of these MMPs also blocked the movement of these neuroblasts, originally headed toward damaged brain [54].

14.2.2.2 High-Mobility Group Box 1

Besides VEGF and MMPs, attention is currently focused on the roles of high-mobility group box 1 (HMGB1) as well on brain remodeling after stroke. HMGB1, a highly conserved non-histone nuclear DNA-binding protein, is widely expressed in most eukaryotic cells including neural cells in several animal species including humans [55]. Traditionally, HMGB1 acts as a nuclear and cellular danger signal [56]. HMGB1 can exert different functions depending on its cellular localization.

It can be passively released from damaged cells or actively secreted from stimulated cells. Release of HMGB1 is observed after traumatic brain injury and ischemic stroke. In rodent middle cerebral artery occlusion models, levels of HMGB1 in the ischemic core are immediately decreased, and in turn, serum HMGB1 is rapidly increased [57–59]. In clinical stroke patients, HMGB1 is upregulated in serum of up to day 7 after stroke onset [60]. HMGB1 is also increased in CSF of subarachnoid hemorrhage patients on day 3, 7, and 14 after onset [60]. In addition, plasma HMGB1 in patients is acutely elevated 30 min after severe trauma in comparison to healthy subjects [61]. However, in contrast to the negative effects, HMGB1 may also possess beneficial actions. HMGB1 signaling can promote endothelial activation [62] and sprouting [63]. And it has also been reported that HMGB1 may increase neurite outgrowth and cell survival in neurons [63–66].

14.2.2.3 c-Jun N-Terminal Kinase (JNK)

The concept of biphasic angiogenic responses may apply more broadly to a large spectrum of other mediators such as intracellular signals. The stress-activated protein kinase JNK pathway is known to trigger many cell death pathways including caspases, and many studies have shown that JNK inhibitors are neuroprotective in rodent stroke models [67]. However, more recent data clearly support a beneficial role for JNK in CNS disease and repair [68]. JNK signaling is involved in neuronal precursor cell migration, microtubule assembly, and axonal guidance during brain development. After injury, this signal can contribute to dendritic sprouting and axonal regrowth. More recently, JNK has also been shown to mediate angiogenesis [69]. JNK mediates the regulation of both VEGF and MMPs, and blockade of JNK cascades with inhibitors can suppress angiogenesis [70, 71]. Whether similar pathways are activated in cerebral neurovascular repair and remodeling remains to be determined, but very recent paper reported that delayed JNK inhibition worsened vascular remodeling in rat stroke model PMID: 22699892.

14.2.3 Endothelial Progenitor Cell in Brain Angiogenesis

Interactions between angiogenesis and functional remodeling after stroke can be also manifested in terms of circulating endothelial progenitor cells (EPCs). EPCs are immature endothelial cells circulating in peripheral blood and are under maturation process to become endothelial cells [72]. Hence, EPCs possess functional and structural characteristics of both stem cells and mature endothelial cells. As discussed above, angiogenesis in the penumbra area is an important natural response to stroke. Although circulating EPCs represent only ~0.01% of cells in the blood under steady-state conditions, EPC numbers are highly affected by stroke onset. Emerging studies are beginning to elucidate the relationship between stroke outcome and the number of circulating

EPCs. In rodent models of focal cerebral ischemia, there was a strong correlation between the volume and severity of infarcts and the absolute number of circulating EPCs [73]. In clinical stroke patients, an increase in circulating EPCs after acute ischemic stroke was associated with good functional outcome and reduced infarct growth and maturation [74, 75]. Importantly, EPC levels were significantly lower in patients with severe neurological impairment compared with patients with less severe impairments at 48 h after ischemic stroke [75]. In mouse cerebral ischemia models, bone marrow-derived EPCs homed to the ischemic core and participated in cerebral neovascularization [39]. Recent experiments suggest that HMGB1 and interleukin-1beta can promote EPC homing and proliferation, respectively [76, 77]. Moreover, very recent study indicates that HMGB1 from reactive astrocytes recovers neurological function through EPC accumulation in the injured area after stroke [78]. These observations raise the possibility that EPCs can be used as a therapeutic approach for promoting repair [79]. However, the precise mechanisms of the EPC contribution to postnatal angiogenesis remain to be elucidated. It has been reported that bone marrow-derived EPCs did not incorporate into the adult growing vasculature [80, 81]. These reports suggest that EPCs support angiogenesis indirectly through growth factor release.

14.3 Neurovascular and Oligovascular Signaling for Brain Angiogenesis

Thus far, we have discussed the mechanisms of brain angiogenesis after stroke. But again, regulating mechanisms for brain angiogenesis after stroke are quite complex, and it may not be sufficient to focus on only endothelial cells (and their progenitor cells) to understand how new blood cells appear in the remodeling brains. In recent years, the concept of the "neurovascular unit" has emerged as a new paradigm for understanding the pathology in the CNS diseases including stroke [82–86]. This modular concept is defined at an intercellular level that comprises dynamic interactions between cerebral endothelial cells, glia, neurons, and other brain cell types (Fig. 14.2). Dysfunctional cross-talk between neurons, glia, and vascular compartments contributes to multiple aspects of acute pathophysiology in CNS disease. Impaired glutamate release–reuptake mechanisms in neurons and astrocytes can amplify excitotoxicity [87]. Perturbed signaling between cerebral endothelium, astrocytes, and pericytes can disrupt BBB integrity [85]. Dysfunctional coupling between neuronal activation and vascular responses can promote deleterious spreading depression [88]. Moreover, disordered signaling between all neurovascular and gliovascular elements can underlie the evolution of neuroinflammation and cell death [89]. In addition to the acute phase, the concept of the neurovascular unit has now been applied to discuss the mechanisms of the chronic phase

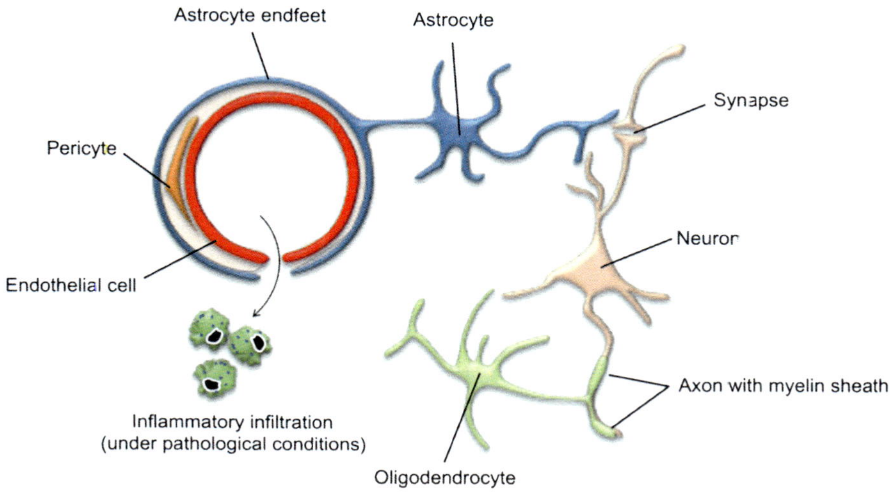

Fig. 14.2 Schematic of the neurovascular unit. This schematic depicts a cerebral blood vessel and surrounding brain cells such as astrocyte, neuron, pericyte, and oligodendrocyte

after stroke (Fig. 14.3). The evolution of brain injury and neurodegeneration comprises a dynamic balance and imbalance between initial triggers of injury and evolutionarily conserved responses of brain plasticity, remodeling, and compensation [90]. And just as cell–cell signaling in the entire neurovascular unit mediates acute injury, delayed recovery should also recruit analogous non-cell-autonomous mechanisms in the brain. Hence, dissecting these various signals and substrates within the neurovascular unit may reveal opportunities for developing novel therapeutic targets for stroke. In this section, we briefly outline the principles of the neurovascular unit and discuss recent data that may help us find common mechanisms of injury and repair after stroke, focusing on brain angiogenesis.

14.3.1 Neurovascular Damage in the Acute Phase

The fundamental mechanisms of brain cell death in the acute stroke phase are multifactorial. Accumulated data over the past two decades have implicated excitotoxicity, oxidative stress, and, in some circumstances, apoptotic-like pathways [82, 91]. When brain fails to generate sufficient ATP by reduction of blood flow supply, energy failure occurs and ionic gradients are lost. Glutamate reuptake processes are impaired, and accumulated glutamate promotes excessive calcium entry and release. Calcium-dependent synthases and proteases contribute to neuronal

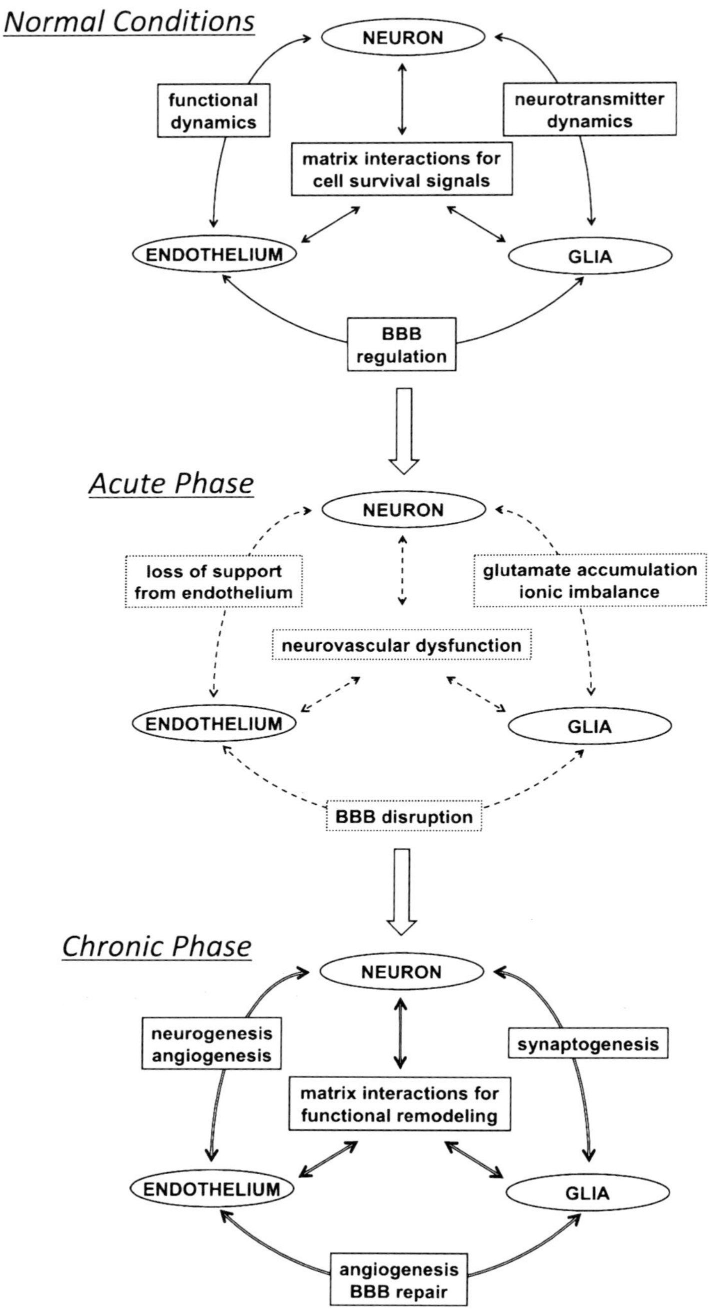

Fig. 14.3 Schematic of the neurovascular unit under normal conditions, acute and chronic phase after stroke. Homeostatic signaling in the neurovascular unit sustains normal brain function. Dysfunctional neurovascular signaling mediates injury after stroke. During the recovery phase after stroke, neurovascular signaling may also be critically important with repair mechanisms that may involve neurogenesis, synaptogenesis, and angiogenesis

death by degrading key cytoskeletal and enzymatic proteins. Abnormality of calcium homeostasis also generates nitric oxide and peroxynitrite, which directly strike neighboring cells. Moreover, mitochondrial functions such as oxidative phosphorylation fail and reactive oxygen radicals are released that further compromise cells by attacking proteins, lipids, and nucleic acids. In parallel with these ionic and free radical pathways, deleterious molecules such caspases may also promote cell death by suicidal endogenous mechanisms. However, most of the cell death pathways outlined here are well documented for neurons. Whether similar mechanisms should be targeted for glial and vascular compartments remains to be carefully assessed. Besides basic cell death mechanisms, one of the most important facets of early neurovascular damage is manifested as perturbations in BBB function. The BBB homeostasis is remarkably dependent on endothelial–astrocyte–matrix interactions [85, 92]. Perturbation of the neurovascular matrix (type IV collagen, heparan sulfate proteoglycan, laminin, fibronectin, etc.) disrupts the cell–matrix and cell–cell signaling that maintain neurovascular functions. Many proteinases might contribute to extracellular matrix proteolysis, and the extracellular protease systems become dysregulated under diseased conditions. As discussed, roles of the MMP family have been focused in the neurovascular damage after stroke [93]. MMP levels are increased in both experimental models of stroke [94–96] and stroke patients [49, 97]. Those excessive MMP activities might be deleterious. MMPs can degrade the extracellular matrix that comprises the basal lamina, thus damaging the BBB directly. In experimental stroke models, MMP inhibition reduces infraction and edema [98, 99]. In addition to BBB disruption, MMP-induced proteolysis of the neurovascular matrix might also promote programmed cell death by detachment of cells from the extracellular matrix (so-called anoikis) [100, 101]. These findings suggest that MMPs (and other extracellular proteases) mediate neurovascular damage during the acute stages of stroke. Ultimately, these neurovascular perturbations can also be interpreted as dysfunctional crosstalk between components of the neurovascular unit. However, none of above cell death pathways or neurovascular mechanisms have been successfully exploited for treating acute stroke patients. Although many translational barriers are involved, the heterogeneity of patients and tight timelines during acute pathology makes it difficult to block these early targets efficiently. Therefore, a recent emphasis in the field is beginning to assess opportunities for promoting neurovascular recovery (especially for angiogenesis) after stroke.

14.3.2 Neurovascular Repair in the Chronic Phase

Most stroke patients show some degree of recovery over time. For example, functional MRI studies demonstrate that peri-infarct areas are highly plastic [3, 102]. Representational areas shift as latent networks are unmasked, and parallel circuits are recruited adjacent to damaged regions [103]. One of the best early examples of cell–cell signaling in the neurovascular unit may be found in the original observations of the so-called neurovascular niche for neurogenesis and angiogenesis. From an evolutionary perspective, the underlying molecular mediators of neurogenesis and angiogenesis overlap and are highly conserved [104]. Molecular mechanisms of angiogenesis and neurogenesis have been evolutionarily conserved so that similar mediators and pathways are involved in both phenomena [105]. It is now accepted that cell–cell signaling between cerebral endothelium and neuronal precursor cells helps mediate and sustain pockets of ongoing angiogenesis and neurogenesis in adult brain [3, 84, 85, 106, 107]. Crosstalk between the vascular and neuronal compartments in the neurovascular niche is mediated by an exchange of soluble signals. This phenomenon is partly mediated by the ability of cerebral endothelium to secrete a rich repertoire of trophic factors [108–110]. In the normal brain, the neurovascular niche defines these complex mechanisms of cell–cell signaling between cerebral endothelium and neural precursors in the subventricular and subgranular zones of ongoing neurogenesis. In the context of post-stroke recovery, these close relationships between neurogenesis and angiogenesis are maintained. Neuroblasts migrate along perivascular routes [111]. Promotion of neurogenesis enhances vascular regrowth, and conversely, angiogenic stimulation enhances neurogenesis [112, 113]. Angiogenesis in peri-infarct regions has been detected in rodent models of cerebral ischemia [114] as well as in human stroke [36]. Hence, brain recovery after stroke comprises interdependent neurovascular plasticity and remodeling processes that recruit multiple common mediators and signals [105].

14.3.3 Cell–Cell Trophic Coupling in White Matter

For the most part, the concept of the neurovascular unit is used to guide investigation in gray matter. However, cell–cell trophic interactions are likely to be important in white matter as well. White matter is vulnerable to ischemic stress, and white matter damage is a clinically important part of stroke [82, 115]. Therefore, without considering white matter mechanisms, we may not be able to protect/recover the brain function against ischemic insults. Compared to the cellular mechanisms of neurovascular damage/repair in gray matter, white matter pathophysiology remains relatively understudied and poorly understood. However, the idea of the neurovascular unit is now applied to the white matter stroke research. The main components of white matter are the neuronal axon, oligodendrocyte (myelin), astrocyte, and endothelium. As in the neurovascular unit in gray matter, astrocytes and cerebral endothelial cells work together to maintain BBB in white matter [115]. In addition, astrocytes are in close apposition to OLGs within white matter [116], and couple with OLGs through gap junctions to maintain their functions [117]. Furthermore, astrocyte-derived soluble factors were also reported to nourish oligodendrocyte lineage cells [118, 119]. And, of course, myelin–axon interactions are essential for white matter homeostasis. OLGs not only myelinate axons but also maintain their functional integrity and survival through OLG-specific proteins and/or trophic factor release [120, 121]. Similar to gray matter, during the acute phase of stroke, several deleterious factors/cascades are activated. For example, MMPs are upregulated, and direct attack of MMPs on myelin components affects OLG survival and function [122]. Even if outright cell death does not occur, metabolic dysfunctions in OLGs might still affect the normal replenishment of myelin and synthesis of myelin-associated proteins, which eventually impair myelin–axon interactions. In the chronic phase, some endogenous responses might work for repairing white matter damage. However, it remains to be fully elucidated how angiogenesis and oligodendrogenesis occur during the chronic phase after stroke. As in the neurovascular niche, an oligovascular niche in the white matter may play an important role in supporting trophic interactions between brain endothelium and oligodendrocytes [123]. Brain endothelium-derived VEGF promotes OPC migration through focal adhesion kinase and reactive oxygen species-dependent mechanisms [124, 125]. On the contrary, after white matter injury, oligodendrocytes produced MMP-9 which may promote vascular remodeling [126]. Future studies should carefully examine the precise mechanisms of the cell–cell trophic coupling in white matter for better understanding the brain angiogenesis after stroke.

14.4 Therapeutic Implication

Therapeutic options for clinical management in stroke still remain quite limited. The treatment of only one FDA-approved drug t-PA is not easy due to its narrow therapeutic time window and related risks of brain hemorrhage [127]. Because recent preclinical studies have revealed that brain injury activates cellular signaling for angiogenesis and neurogenesis, strategies to promote angiogenesis are part of larger neurorestorative approaches in order to increase the diversity in therapeutic options for a variety of patients. Although there are no agents and manipulations in clinical use that can boost angiogenesis after stroke yet, we should discuss some key promising seeds here.

14.4.1 Combination Therapy with VEGF

As introduced, VEGF is the prototypical mediator for brain angiogenesis after stroke. Therefore, in theory, VEGF therapies could promote neurorestoration either directly as a neuroprotective agent or indirectly by inducing angiogenesis [21, 128, 129]. However, the function of VEGF as a vascular permeability factor also means that an untitrated response may lead to blood–brain barrier leakage, brain edema, vasodilation, and aberrant systemic hemodynamics [130–133]. In addition, VEGF-induced angiogenic vessels are hemorrhagic, aggravating inflammatory responses in the recovering penumbra [134, 135]. Nevertheless, recent findings now suggest that combinatorial therapy with other agents would be beneficial. The untoward side effects of VEGF were partially obviated by treatment with a combination of angiopoietins [136]. An alternate strategy might be to apply HIF prolyl hydroxylase inhibitors. These reagents may raise HIF-1 levels and increase expression of several hypoxia-response proteins that could avoid vascular leakage [137]. Moreover, several downstream effectors of VEGF have been tested as selective modulators for VEGF signaling [138]. The Roundabout (Robo) protein 4/Slit2 axis has been shown to selectively inhibit VEGF-165-induced migration, tube formation, and permeability in vitro and VEGF-165-stimulated vascular leak in vivo by blocking Src family kinase activation [139]. Thus, targeting the Robo4-Slit2 signaling or in recovering vessels may open newer therapeutic options along with VEGF to minimize tissue injury and maximize its beneficial effects [140].

14.4.2 Cell Junction Molecule

In addition to VEGF signaling, cell junction molecules would be interesting target for promoting brain angiogenesis. Cell junction molecules exist at the interface of multiple cellular decisions and play important roles in vascular permeability, quiescence, invasion, and differentiation [141–143]. Anti-integrin therapy for tumor angiogenesis has gained ground in the recent times [144]. GPIIb/IIIa and $\alpha v\beta 3$ integrins mediate endothelial–platelet interaction, and several antagonists targeted against these molecules have been found to have anti-angiogenic effects in vivo [145, 146]. Those findings suggest that modulation of cell adhesion molecules and their signaling might be a useful strategy in stroke therapy because of their ability to alter responsiveness to growth factors—either potentiate growth factor signaling or attenuate its effects where necessary [147–149]. Hence, the use of soluble adhesion molecules may deliver survival signals, alter growth factor responsiveness, and facilitate pertinent cell–cell communication within the remodeling penumbra. For example, soluble N-cadherin fragments were found to stimulate migration of endothelial cells through the FGF receptor [150, 151]. During vasculogenesis, N-cadherin mediates adhesion, recognition, and signaling between pericytes and endothelial cells and is required for normal vascular morphogenesis. The significant diversity in expression of cell junction molecules, the expression of tissue-specific isoforms, and their spatiotemporal functions in the CNS can be exploited for better vascular morphogenesis and neurogenerative outcomes.

14.4.3 Cell-Based Therapy

As noted in the section of EPCs, beyond the cell signaling targets, cell-based therapies would be promising approaches for the treatment of brain injury [152, 153]. Cell therapies using neural stem/progenitor cells (NSPCs) may replace lost brain cells, promote endogenous neurogenesis, and improve functional recovery [152]. NSPCs stabilize vasculature during ischemia, suggesting therapeutic application of NSPCs to promote revascularization and repair after brain injury [154]. Of course, there is little evidence to assess the applicability of NSPCs to stroke patients, and therefore, well-designed clinical trials are necessary to evaluate safety, toxicity, and efficacy as well as optimal cell type, route, and time of delivery for NSPCs [155, 156]. Another candidate for cell-based therapy would be mesenchymal stem cells (MSCs) isolated from bone marrow, adipose tissue, umbilical cord blood, placenta, and pancreas. MSCs exert powerful immunomodulatory effects, which include inhibition of proliferation and function of T cells, B cells, and natural killer cells. Those effects reduce immune reactions and increase tolerance of MSC recipients [157]. Moreover, MSCs secrete various growth factors including BDNF, VEGF, and FGF, which promote angiogenesis after stroke [158–162]. Hence, genetically engineered MSCs (and NSPCs) with overexpression of growth factors may be an improved source for cell therapy for stroke [163]. Finally, recent landmark experiments have shown that transient overexpression of a small number of transcription factors can reprogram differentiated cells into induced pluripotent stem (iPS) cells that resemble embryonic stem cells [164]. These iPS cells avoid the ethical issue inherent in embryonic tissues or oocytes and have the potential to generate patient-specific cell types for cell replacement therapy. iPS cells may offer promising opportunities for the treatment of brain injury. But again, pharmacological and cell-based therapies to induce rapid angiogenesis run the danger of leading to dysfunctional tissue architecture and exacerbating neuronal damage. How these promising experimental approaches can be tested long term in stroke patients remains to be carefully assessed.

14.5 Conclusion

The adult mammalian brain can be surprisingly plastic, especially after stroke and brain injury. Under normal conditions, newborn neurons in the subventricular and subgranular zones migrate to olfactory regions and the hippocampus. After brain injury, the birth rate of new cells seems to increase, and neuroblasts are rerouted toward damaged tissue. Along with neurogenesis, the recovering brain also exhibits complex patterns of vascular remodeling. This chapter provided an abbreviated summary and survey of major pathophysiological concepts in stroke, focusing on mechanisms of brain angiogenesis. Thus far, drugs that can be cyto-protective against stroke are not yet developed. Therefore, an emerging emphasis on promoting recovery after brain diseases is beginning to take shape in our field. Although there are many difficulties in translating findings in basic research into clinical applications, therapies that can boost endogenous angiogenic properties would be promising approaches for stroke patients in the future.

Acknowledgments Supported in part by the National Institutes of Health, the American Heart Association and the Deane Institute. Materials including figures in this chapter have been extensively drawn from previously published reviews including Lo et al., Nat Rev Neurosci 2003; Lo, Nat Med 2008; Arai et al., FEBS J 2009; Arai and Lo, Exp Transl Stroke Med 2009; Arai and Lo, Biol Pharm Bull 2009; Arai and Lo, FEBS J 2009; Navaratna et al., Cell Adh Migr 2009; Hayakawa et al., Ann N Y Acad Sci 2010; and Arai et al., J Child Neurol 2011.

References

1. Kleindorfer D, Lindsell CJ, Brass L, Koroshetz W, Broderick JP (2008) National us estimates of recombinant tissue plasminogen activator use: ICD-9 codes substantially underestimate. Stroke 39:924–928
2. Shuaib A, Lees KR, Lyden P et al (2007) NXY-059 for the treatment of acute ischemic stroke. N Engl J Med 357:562–571

3. Chopp M, Zhang ZG, Jiang Q (2007) Neurogenesis, angiogenesis, and MRI indices of functional recovery from stroke. Stroke 38:827–831
4. Chopp M, Li Y, Zhang J (2008) Plasticity and remodeling of brain. J Neurol Sci 265:97–101
5. Krupinski J, Kaluza J, Kumar P et al (1994) Role of angiogenesis in patients with cerebral ischemic stroke. Stroke 25:1794–1798
6. Chen HH, Chien CH, Liu HM (1994) Correlation between angiogenesis and basic fibroblast growth factor expression in experimental brain infarct. Stroke 25:1651–1657
7. Marti HJ, Bernaudin M, Bellail A et al (2000) Hypoxia-induced vascular endothelial growth factor expression precedes neovascularization after cerebral ischemia. Am J Pathol 156:965–976
8. Hayashi T, Noshita N, Sugawara T, Chan PH (2003) Temporal profile of angiogenesis and expression of related genes in the brain after ischemia. J Cereb Blood Flow Metab 23:166–180
9. Krupinski J, Kaluza J, Kumar P et al (1993) Some remarks on the growth-rate and angiogenesis of microvessels in ischemic stroke. Morphometric and immunocytochemical studies. Patol Pol 44:203–209
10. Szpak GM, Lechowicz W, Lewandowska E et al (1999) Border zone neovascularization in cerebral ischemic infarct. Folia Neuropathol 37:264–268
11. Allen CM (1984) Predicting outcome after acute stroke: role of computerised tomography. Lancet 2:464–465
12. Granger CV, Hamilton BB, Fiedler RC (1992) Discharge outcome after stroke rehabilitation. Stroke 23:978–982
13. Schmidt R, Schmidt H, Fazekas F (2000) Vascular risk factors in dementia. J Neurol 247:81–87
14. Manoonkitiwongsa PS, Jackson-Friedman C, McMillan PJ et al (2001) Angiogenesis after stroke is correlated with increased numbers of macrophages: the clean-up hypothesis. J Cereb Blood Flow Metab 21:1223–1231
15. Yu SW, Friedman B, Cheng Q, Lyden PD (2007) Stroke-evoked angiogenesis results in a transient population of microvessels. J Cereb Blood Flow Metab 27:755–763
16. Fagan SC, Hess DC, Hohnadel EJ et al (2004) Targets for vascular protection after acute ischemic stroke. Stroke 35:2220–2225
17. Hansen TM, Moss AJ, Brindle NP (2008) Vascular endothelial growth factor and angiopoietins in neurovascular regeneration and protection following stroke. Curr Neurovasc Res 5:236–245
18. Abe K, Setoguchi Y, Hayashi T, Itoyama Y (1997) Dissociative expression of adenoviral-mediated E. coli LacZ gene between ischemic and reperfused rat brains. Neurosci Lett 226:53–56
19. Zhang ZG, Zhang L, Jiang Q et al (2000) VEGF enhances angiogenesis and promotes blood–brain barrier leakage in the ischemic brain. J Clin Invest 106:829–838
20. Issa R, Krupinski J, Bujny T et al (1999) Vascular endothelial growth factor and its receptor, KDR, in human brain tissue after ischemic stroke. Lab Invest 79:417–425
21. Sun Y, Jin K, Xie L et al (2003) VEGF-induced neuroprotection, neurogenesis, and angiogenesis after focal cerebral ischemia. J Clin Invest 111:1843–1851
22. Wang Y, Kilic E, Kilic U et al (2005) VEGF overexpression induces post-ischaemic neuroprotection, but facilitates haemodynamic steal phenomena. Brain 128:52–63
23. Yano A, Shingo T, Takeuchi A et al (2005) Encapsulated vascular endothelial growth factor-secreting cell grafts have neuroprotective and angiogenic effects on focal cerebral ischemia. J Neurosurg 103:104–114
24. Wada K, Sugimori H, Bhide PG et al (2003) Effect of basic fibroblast growth factor treatment on brain progenitor cells after permanent focal ischemia in rats. Stroke 34:2722–2728
25. Speliotes EK, Caday CG, Do T et al (1996) Increased expression of basic fibroblast growth factor (bFGF) following focal cerebral infarction in the rat. Brain Res Mol Brain Res 39:31–42

26. Issa R, AlQteishat A, Mitsios N, Saka M, Krupinski J, Tarkowski E, Gaffney J, Slevin M, Kumar S, Kumar P (2005) Expression of basic fibroblast growth factor mRNA and protein in the human brain following ischaemic stroke. Angiogenesis 8:53–62
27. Jiang N, Finklestein SP, Do T et al (1996) Delayed intravenous administration of basic fibroblast growth factor (bFGF) reduces infarct volume in a model of focal cerebral ischemia/reperfusion in the rat. J Neurol Sci 139:173–179
28. Kawamata T, Dietrich WD, Schallert T et al (1997) Intracisternal basic fibroblast growth factor enhances functional recovery and up-regulates the expression of a molecular marker of neuronal sprouting following focal cerebral infarction. Proc Natl Acad Sci U S A 94:8179–8184
29. Li Q, Stephenson D (2002) Postischemic administration of basic fibroblast growth factor improves sensorimotor function and reduces infarct size following permanent focal cerebral ischemia in the rat. Exp Neurol 177:531–537
30. Kawamata T, Speliotes EK, Finklestein SP (1997) The role of polypeptide growth factors in recovery from stroke. Adv Neurol 73:377–382
31. Ren JM, Finklestein SP (2005) Growth factor treatment of stroke. Curr Drug Targets CNS Neurol Disord 4:121–125
32. Yamagishi S, Imaizumi T (2005) Pericyte biology and diseases. Int J Tissue React 27:125–135
33. Renner O, Tsimpas A, Kostin S et al (2003) Time- and cell type-specific induction of platelet-derived growth factor receptor-beta during cerebral ischemia. Brain Res Mol Brain Res 113:44–51
34. Krupinski J, Issa R, Bujny T et al (1997) A putative role for platelet-derived growth factor in angiogenesis and neuroprotection after ischemic stroke in humans. Stroke 28:564–573
35. Vivien D, Ali C (2006) Transforming growth factor-beta signalling in brain disorders. Cytokine Growth Factor Rev 17:121–128
36. Krupinski J, Kumar P, Kumar S, Kaluza J (1996) Increased expression of TGF-beta 1 in brain tissue after ischemic stroke in humans. Stroke 27:852–857
37. Ata AK, Funa K, Olsson Y (1997) Expression of various TGF-beta isoforms and type I receptor in necrotizing human brain lesions. Acta Neuropathol 93:326–333
38. Slevin M, Krupinski J, Slowik A, Kumar P, Szczudlik A, Gaffney J (2000) Serial measurement of vascular endothelial growth factor and transforming growth factor-beta1 in serum of patients with acute ischemic stroke. Stroke 31:1863–1870
39. Zhang ZG, Zhang L, Jiang Q, Chopp M (2002) Bone marrow-derived endothelial progenitor cells participate in cerebral neovascularization after focal cerebral ischemia in the adult mouse. Circ Res 90:284–288
40. Krupinski J, Vodovotz Y, Li C et al (1998) Inducible nitric oxide production and expression of transforming growth factor-beta1 in serum and CSF after cerebral ischaemic stroke in man. Nitric Oxide 2:442–453
41. Prehn JH, Backhauss C, Krieglstein J (1993) Transforming growth factor-beta 1 prevents glutamate neurotoxicity in rat neocortical cultures and protects mouse neocortex from ischemic injury in vivo. J Cereb Blood Flow Metab 13:521–525
42. Prehn JH, Peruche B, Unsicker K, Krieglstein J (1993) Isoform-specific effects of transforming growth factors-beta on degeneration of primary neuronal cultures induced by cytotoxic hypoxia or glutamate. J Neurochem 60:1665–1672
43. Le Cras TD, Spitzmiller RE, Albertine KH et al (2004) VEGF causes pulmonary hemorrhage, hemosiderosis, and air space enlargement in neonatal mice. Am J Physiol Lung Cell Mol Physiol 287:L134–L142
44. van Bruggen N, Thibodeaux H, Palmer JT et al (1999) VEGF antagonism reduces edema formation and tissue damage after ischemia/reperfusion injury in the mouse brain. J Clin Invest 104:1613–1620
45. Zhang ZG, Chopp M (2009) Neurorestorative therapies for stroke: underlying mechanisms and translation to the clinic. Lancet Neurol 8:491–500

46. Lo EH (2008) A new penumbra: transitioning from injury into repair after stroke. Nat Med 14:497–500
47. Navaratna D, Guo S, Arai K, Lo E (2009) Mechanisms and targets for angiogenic therapy after stroke. Cell Adh Migr 3:216–239
48. Anthony DC, Ferguson B, Matyzak MK et al (1997) Differential matrix metalloproteinase expression in cases of multiple sclerosis and stroke. Neuropathol Appl Neurobiol 23:406–415
49. Montaner J, Alvarez-Sabin J, Molina C et al (2001) Matrix metalloproteinase expression after human cardioembolic stroke: temporal profile and relation to neurological impairment. Stroke 32:1759–1766
50. Rosell A, Alvarez-Sabin J, Arenillas JF et al (2005) A matrix metalloproteinase protein array reveals a strong relation between MMP-9 and MMP-13 with diffusion-weighted image lesion increase in human stroke. Stroke 36:1415–1420
51. Alvarez-Sabin J, Delgado P, Abilleira S et al (2004) Temporal profile of matrix metalloproteinases and their inhibitors after spontaneous intracerebral hemorrhage: relationship to clinical and radiological outcome. Stroke 35:1316–1322
52. Horstmann S, Kalb P, Koziol J et al (2003) Profiles of matrix metalloproteinases, their inhibitors, and laminin in stroke patients: influence of different therapies. Stroke 34:2165–2170
53. Zhao BQ, Wang S, Kim HY et al (2006) Role of matrix metalloproteinases in delayed cortical responses after stroke. Nat Med 12:441–445
54. Lee SR, Kim HY, Rogowska J et al (2006) Involvement of matrix metalloproteinase in neuroblast cell migration from the subventricular zone after stroke. J Neurosci 26:3491–3495
55. Yang H, Wang H, Czura CJ, Tracey KJ (2005) The cytokine activity of HMGB1. J Leukoc Biol 78:1–8
56. Lotze MT, Tracey KJ (2005) High-mobility group box 1 protein (HMGB1): nuclear weapon in the immune arsenal. Nat Rev Immunol 5:331–342
57. Qiu J, Nishimura M, Wang Y et al (2008) Early release of HMGB-1 from neurons after the onset of brain ischemia. J Cereb Blood Flow Metab 28:927–938
58. Kim JB, Sig Choi J, Yu YM et al (2006) HMGB1, a novel cytokine-like mediator linking acute neuronal death and delayed neuroinflammation in the postischemic brain. J Neurosci 26:6413–6421
59. Hayakawa K, Mishima K, Nozako M et al (2008) Delayed treatment with minocycline ameliorates neurologic impairment through activated microglia expressing a high-mobility group box1-inhibiting mechanism. Stroke 39:951–958
60. Nakahara T, Tsuruta R, Kaneko T et al (2009) High-mobility group box 1 protein in CSF of patients with subarachnoid hemorrhage. Neurocrit Care 11:362–368
61. Cohen MJ, Brohi K, Calfee CS et al (2009) Early release of high mobility group box nuclear protein 1 after severe trauma in humans: role of injury severity and tissue hypoperfusion. Crit Care 13:R174
62. Treutiger CJ, Mullins GE, Johansson AS et al (2003) High mobility group 1 B-box mediates activation of human endothelium. J Intern Med 254:375–385
63. Schlueter C, Weber H, Meyer B et al (2005) Angiogenetic signaling through hypoxia: HMGB1: an angiogenetic switch molecule. Am J Pathol 166:1259–1263
64. Huttunen HJ, Kuja-Panula J, Sorci G et al (2000) Coregulation of neurite outgrowth and cell survival by amphoterin and S100 proteins through receptor for advanced glycation end products (RAGE) activation. J Biol Chem 275:40096–40105
65. Huttunen HJ, Kuja-Panula J, Rauvala H (2002) Receptor for advanced glycation end products (RAGE) signaling induces CREB-dependent chromogranin expression during neuronal differentiation. J Biol Chem 277:38635–38646
66. Passalacqua M, Patrone M, Picotti GB et al (1998) Stimulated astrocytes release high-mobility group 1 protein, an inducer of lan-5 neuroblastoma cell differentiation. Neuroscience 82:1021–1028
67. Kuan CY, Burke RE (2005) Targeting the JNK signaling pathway for stroke and Parkinson's diseases therapy. Curr Drug Targets CNS Neurol Disord 4:63–67
68. Waetzig V, Zhao Y, Herdegen T (2006) The bright side of JNKs-multitalented mediators in neuronal sprouting, brain development and nerve fiber regeneration. Prog Neurobiol 80:84–97

69. Uchida C, Gee E, Ispanovic E, Haas TL (2008) JNK as a positive regulator of angiogenic potential in endothelial cells. Cell Biol Int 32:769–776
70. Miura S, Matsuo Y, Saku K (2008) Jun N-terminal kinase inhibitor blocks angiogenesis by blocking VEGF secretion and an MMP pathway. J Atheroscler Thromb 15:69–74
71. Yoshino Y, Aoyagi M, Tamaki M et al (2006) Activation of p38 MAPK and/or JNK contributes to increased levels of VEGF secretion in human malignant glioma cells. Int J Oncol 29:981–987
72. Asahara T, Murohara T, Sullivan A et al (1997) Isolation of putative progenitor endothelial cells for angiogenesis. Science 275:964–967
73. Taguchi A, Matsuyama T, Moriwaki H et al (2004) Circulating CD34-positive cells provide an index of cerebrovascular function. Circulation 109:2972–2975
74. Sobrino T, Hurtado O, Moro MA et al (2007) The increase of circulating endothelial progenitor cells after acute ischemic stroke is associated with good outcome. Stroke 38:2759–2764
75. Yip HK, Chang LT, Chang WN et al (2008) Level and value of circulating endothelial progenitor cells in patients after acute ischemic stroke. Stroke 39:69–74
76. Chavakis E, Hain A, Vinci M et al (2007) High-mobility group box 1 activates integrin-dependent homing of endothelial progenitor cells. Circ Res 100:204–212
77. Rosell A, Arai K, Lok J et al (2009) Interleukin-1beta augments angiogenic responses of murine endothelial progenitor cells in vitro. J Cereb Blood Flow Metab 29:933–943
78. Hayakawa K, Pham L-D, Katusic Z et al (2012) Astrocytic high-mobility group box 1 promotes endothelial progenitor cell-mediated neurovascular remodeling during stroke recovery. Proc Natl Acad Sci U S A 109:7505–7515
79. Rouhl RP, van Oostenbrugge RJ, Damoiseaux J et al (2008) Endothelial progenitor cell research in stroke: a potential shift in pathophysiological and therapeutical concepts. Stroke 39:2158–2165
80. Ziegelhoeffer T, Fernandez B, Kostin S et al (2004) Bone marrow-derived cells do not incorporate into the adult growing vasculature. Circ Res 94:230–238
81. O'Neill TJT, Wamhoff BR, Owens GK, Skalak TC (2005) Mobilization of bone marrow-derived cells enhances the angiogenic response to hypoxia without transdifferentiation into endothelial cells. Circ Res 97:1027–1035
82. Lo EH, Dalkara T, Moskowitz MA (2003) Mechanisms, challenges and opportunities in stroke. Nat Rev Neurosci 4:399–415
83. del Zoppo GJ (2006) Stroke and neurovascular protection. N Engl J Med 354:553–555
84. Iadecola C (2004) Neurovascular regulation in the normal brain and in Alzheimer's disease. Nat Rev Neurosci 5:347–360
85. Zlokovic BV (2008) The blood–brain barrier in health and chronic neurodegenerative disorders. Neuron 57:178–201
86. Lok J, Gupta P, Guo S et al (2007) Cell-cell signaling in the neurovascular unit. Neurochem Res 32:2032–2045
87. Besancon E, Guo S, Lok J, Tymianski M, Lo EH (2008) Beyond NMDA and AMPA glutamate receptors: emerging mechanisms for ionic imbalance and cell death in stroke. Trends Pharmacol Sci 29:268–275
88. Dreier JP (2011) The role of spreading depression, spreading depolarization and spreading ischemia in neurological disease. Nat Med 17:439–447
89. Iadecola C, Anrather J (2011) The immunology of stroke: from mechanisms to translation. Nat Med 17:796–808
90. Lo EH (2010) Degeneration and repair in central nervous system disease. Nat Med 16:1205–1209
91. Lo EH, Moskowitz MA, Jacobs TP (2005) Exciting, radical, suicidal: how brain cells die after stroke. Stroke 36:189–192
92. Iadecola C, Nedergaard M (2007) Glial regulation of the cerebral microvasculature. Nat Neurosci 10:1369–1376
93. Zhao BQ, Tejima E, Lo EH (2007) Neurovascular proteases in brain injury, hemorrhage and remodeling after stroke. Stroke 38:748–752

94. Asahi M, Asahi K, Jung JC et al (2000) Role for matrix metalloproteinase 9 after focal cerebral ischemia: effects of gene knockout and enzyme inhibition with BB-94. J Cereb Blood Flow Metab 20:1681–1689
95. Gasche Y, Fujimura M, Morita-Fujimura Y et al (1999) Early appearance of activated matrix metalloproteinase-9 after focal cerebral ischemia in mice: a possible role in blood–brain barrier dysfunction. J Cereb Blood Flow Metab 19:1020–1028
96. Heo JH, Lucero J, Abumiya T et al (1999) Matrix metalloproteinases increase very early during experimental focal cerebral ischemia. J Cereb Blood Flow Metab 19:624–633
97. Kelly PJ, Morrow JD, Ning M et al (2008) Oxidative stress and matrix metalloproteinase-9 in acute ischemic stroke: The Biomarker Evaluation for Antioxidant Therapies in Stroke (BEAT-Stroke) study. Stroke 39:100–104
98. Asahi M, Wang X, Mori T et al (2001) Effects of matrix metalloproteinase-9 gene knockout on the proteolysis of blood–brain barrier and white matter components after cerebral ischemia. J Neurosci 21:7724–7732
99. Rosenberg GA, Estrada EY, Dencoff JE (1998) Matrix metalloproteinases and timps are associated with blood–brain barrier opening after reperfusion in rat brain. Stroke 29:2189–2195
100. Tagaya M, Haring HP, Stuiver I et al (2001) Rapid loss of microvascular integrin expression during focal brain ischemia reflects neuron injury. J Cereb Blood Flow Metab 21:835–846
101. Chen ZL, Strickland S (1997) Neuronal death in the hippocampus is promoted by plasmin-catalyzed degradation of laminin. Cell 91:917–925
102. Dijkhuizen RM, Singhal AB, Mandeville JB et al (2003) Correlation between brain reorganization, ischemic damage, and neurologic status after transient focal cerebral ischemia in rats: a functional magnetic resonance imaging study. J Neurosci 23:510–517
103. Murphy TH, Corbett D (2009) Plasticity during stroke recovery: from synapse to behaviour. Nat Rev Neurosci 10:861–872
104. Carmeliet P, Tessier-Lavigne M (2005) Common mechanisms of nerve and blood vessel wiring. Nature 436:193
105. Snapyan M, Lemasson M, Brill MS et al (2009) Vasculature guides migrating neuronal precursors in the adult mammalian forebrain via brain-derived neurotrophic factor signaling. J Neurosci 29:4172–4188
106. Zacchigna S, Lambrechts D, Carmeliet P (2008) Neurovascular signalling defects in neurodegeneration. Nat Rev Neurosci 9:169–181
107. Greenberg DA, Jin K (2005) From angiogenesis to neuropathology. Nature 438:954–959
108. Leventhal C, Rafii S, Rafii D et al (1999) Endothelial trophic support of neuronal production and recruitment from the adult mammalian subependyma. Mol Cell Neurosci 13:450–464
109. Shen Q, Goderie SK, Jin L et al (2004) Endothelial cells stimulate self-renewal and expand neurogenesis of neural stem cells. Science 304:1338–1340
110. Guo S, Kim WJ, Lok J et al (2008) Neuroprotection via matrix-trophic coupling between cerebral endothelial cells and neurons. Proc Natl Acad Sci U S A 105:7582–7587
111. Thored P, Wood J, Arvidsson A et al (2007) Long-term neuroblast migration along blood vessels in an area with transient angiogenesis and increased vascularization after stroke. Stroke 38:3032–3039
112. Ohab JJ, Fleming S, Blesch A, Carmichael ST (2006) A neurovascular niche for neurogenesis after stroke. J Neurosci 26:13007–13016
113. Taguchi A, Soma T, Tanaka H et al (2004) Administration of CD34+ cells after stroke enhances neurogenesis via angiogenesis in a mouse model. J Clin Invest 114:330–338
114. Ding G, Jiang Q, Li L et al (2008) Angiogenesis detected after embolic stroke in rat brain using magnetic resonance T2*WI. Stroke 39:1563–1568
115. Arai K, Lo EH (2009) Oligovascular signaling in white matter stroke. Biol Pharm Bull 32:1639–1644
116. Butt AM, Ibrahim M, Ruge FM, Berry M (1995) Biochemical subtypes of oligodendrocyte in the anterior medullary velum of the rat as revealed by the monoclonal antibody Rip. Glia 14:185–197

117. Orthmann-Murphy JL, Abrams CK, Scherer SS (2008) Gap junctions couple astrocytes and oligodendrocytes. J Mol Neurosci 35:101–116
118. Corley SM, Ladiwala U, Besson A, Yong VW (2001) Astrocytes attenuate oligodendrocyte death in vitro through an alpha(6) integrin-laminin-dependent mechanism. Glia 36:281–294
119. Arai K, Lo EH (2010) Astrocytes protect oligodendrocyte precursor cells via MEK/ERK and PI3K/Akt signaling. J Neurosci Res 88:758–763
120. Nave KA, Trapp BD (2008) Axon-glial signaling and the glial support of axon function. Annu Rev Neurosci 31:535–561
121. Rosenberg SS, Ng BK, Chan JR (2006) The quest for remyelination: a new role for neurotrophins and their receptors. Brain Pathol 16:288–294
122. Chandler S, Coates R, Gearing A et al (1995) Matrix metalloproteinases degrade myelin basic protein. Neurosci Lett 201:223–226
123. Arai K, Lo EH (2009) An oligovascular niche: cerebral endothelial cells promote the survival and proliferation of oligodendrocyte precursor cells. J Neurosci 29:4351–4355
124. Hayakawa K, Seo JH, Pham LD, Miyamoto N, Som AT, Guo S, Kim KW, Lo EH, Arai K (2012) Cerebral endothelial derived vascular endothelial growth factor promotes the migration but not the proliferation of oligodendrocyte precursor cells in vitro. Neurosci Lett 513:42–56
125. Hayakawa K, Pham LD, Som AT et al (2011) Vascular endothelial growth factor regulates the migration of oligodendrocyte precursor cells. J Neurosci 31:10666–10670
126. Pham LD, Hayakawa K, Seo JH et al (2012) Crosstalk between oligodendrocytes and cerebral endothelium contributes to vascular remodeling after white matter injury. Glia 60:875–881
127. Weintraub MI (2006) Thrombolysis (tissue plasminogen activator) in stroke: a medicolegal quagmire. Stroke 37:1917–1922
128. Carmeliet P, Storkebaum E (2002) Vascular and neuronal effects of VEGF in the nervous system: implications for neurological disorders. Semin Cell Dev Biol 13:39–53
129. Ferrara N, Gerber HP (2001) The role of vascular endothelial growth factor in angiogenesis. Acta Haematol 106:148–156
130. Weis SM, Cheresh DA (2005) Pathophysiological consequences of VEGF-induced vascular permeability. Nature 437:497–504
131. Croll SD, Goodman JH, Scharfman HE (2004) Vascular endothelial growth factor (VEGF) in seizures: a double-edged sword. Adv Exp Med Biol 548:57–68
132. Bates DO, Harper SJ (2002) Regulation of vascular permeability by vascular endothelial growth factors. Vascul Pharmacol 39:225–237
133. Forstreuter F, Lucius R, Mentlein R (2002) Vascular endothelial growth factor induces chemotaxis and proliferation of microglial cells. J Neuroimmunol 132:93–98
134. Schoch HJ, Fischer S, Marti HH (2002) Hypoxia-induced vascular endothelial growth factor expression causes vascular leakage in the brain. Brain 125:2549–2557
135. Dobrogowska DH, Lossinsky AS, Tarnawski M, Vorbrodt AW (1998) Increased blood–brain barrier permeability and endothelial abnormalities induced by vascular endothelial growth factor. J Neurocytol 27:163–173
136. Thurston G, Rudge JS, Ioffe E, Zhou H, Ross L, Croll SD, Glazer N, Holash J, McDonald DM, Yancopoulos GD (2000) Angiopoietin-1 protects the adult vasculature against plasma leakage. Nat Med 6:460–463
137. Mu D, Jiang X, Sheldon RA et al (2003) Regulation of hypoxia-inducible factor 1alpha and induction of vascular endothelial growth factor in a rat neonatal stroke model. Neurobiol Dis 14:524–534
138. Eliceiri BP, Paul R, Schwartzberg PL et al (1999) Selective requirement for Src kinases during VEGF-induced angiogenesis and vascular permeability. Mol Cell 4:915–924
139. Jones CA, London NR, Chen H et al (2008) Robo4 stabilizes the vascular network by inhibiting pathologic angiogenesis and endothelial hyperpermeability. Nat Med 14:448–453
140. Acevedo LM, Weis SM, Cheresh DA (2008) Robo4 counteracts VEGF signaling. Nat Med 14:372–373

141. Wallez Y, Huber P (2008) Endothelial adherens and tight junctions in vascular homeostasis, inflammation and angiogenesis. Biochim Biophys Acta 1778:794–809
142. Vestweber D (2008) VE-cadherin: the major endothelial adhesion molecule controlling cellular junctions and blood vessel formation. Arterioscler Thromb Vasc Biol 28:223–232
143. Liebner S, Cavallaro U, Dejana E (2006) The multiple languages of endothelial cell-to-cell communication. Arterioscler Thromb Vasc Biol 26:1431–1438
144. Tucker GC (2006) Integrins: molecular targets in cancer therapy. Curr Oncol Rep 8:96–103
145. Trikha M, Zhou Z, Timar J et al (2002) Multiple roles for platelet GPIIb/IIIa and alphavbeta3 integrins in tumor growth, angiogenesis, and metastasis. Cancer Res 62:2824–2833
146. Varner JA, Nakada MT, Jordan RE, Coller BS (1999) Inhibition of angiogenesis and tumor growth by murine 7E3, the parent antibody of c7E3 Fab (abciximab; ReoPro). Angiogenesis 3:53–60
147. Carmeliet P, Collen D (2000) Molecular basis of angiogenesis. Role of VEGF and VE-cadherin. Ann N Y Acad Sci 902:249–262; discussion 262–264
148. Grazia Lampugnani M, Zanetti A, Corada M et al (2003) Contact inhibition of VEGF-induced proliferation requires vascular endothelial cadherin, beta-catenin, and the phosphatase dep-1/cd148. J Cell Biol 161:793–804
149. Francavilla C, Loeffler S, Piccini D et al (2007) Neural cell adhesion molecule regulates the cellular response to fibroblast growth factor. J Cell Sci 120:4388–4394
150. Skaper SD, Facci L, Williams G et al (2004) A dimeric version of the short N-cadherin binding motif HAVDI promotes neuronal cell survival by activating an N-cadherin/fibroblast growth factor receptor signalling cascade. Mol Cell Neurosci 26:17–23
151. Williams EJ, Furness J, Walsh FS, Doherty P (1994) Activation of the FGF receptor underlies neurite outgrowth stimulated by L1, N-CAM, and N-cadherin. Neuron 13:583–594
152. Miljan EA, Sinden JD (2009) Stem cell treatment of ischemic brain injury. Curr Opin Mol Ther 11:394–403
153. Richardson RM, Singh A, Sun D et al (2010) Stem cell biology in traumatic brain injury: effects of injury and strategies for repair. J Neurosurg 112:1125–1138
154. Roitbak T, Li L, Cunningham LA (2008) Neural stem/progenitor cells promote endothelial cell morphogenesis and protect endothelial cells against ischemia via hif-1alpha-regulated VEGF signaling. J Cereb Blood Flow Metab 28:1530–1542
155. Locatelli F, Bersano A, Ballabio E et al (2009) Stem cell therapy in stroke. Cell Mol Life Sci 66:757–772
156. Bersano A, Ballabio E, Lanfranconi S et al (2010) Clinical studies in stem cells transplantation for stroke: a review. Curr Vasc Pharmacol 8:29–34
157. Siegel G, Schafer R, Dazzi F (2009) The immunosuppressive properties of mesenchymal stem cells. Transplantation 87:S45–S49
158. Chopp M, Li Y (2002) Treatment of neural injury with marrow stromal cells. Lancet Neurol 1:92–100
159. Chen J, Zhang ZG, Li Y et al (2003) Intravenous administration of human bone marrow stromal cells induces angiogenesis in the ischemic boundary zone after stroke in rats. Circ Res 92:692–699
160. Kinnaird T, Stabile E, Burnett MS et al (2004) Marrow-derived stromal cells express genes encoding a broad spectrum of arteriogenic cytokines and promote in vitro and in vivo arteriogenesis through paracrine mechanisms. Circ Res 94:678–685
161. Zacharek A, Chen J, Cui X et al (2007) Angiopoietin1/Tie2 and VEGF/Flk1 induced by MSC treatment amplifies angiogenesis and vascular stabilization after stroke. J Cereb Blood Flow Metab 27:1684–1691
162. Pavlichenko N, Sokolova I, Vijde S et al (2008) Mesenchymal stem cells transplantation could be beneficial for treatment of experimental ischemic stroke in rats. Brain Res 1233:203–213
163. van Velthoven CT, Kavelaars A, van Bel F, Heijnen CJ (2009) Regeneration of the ischemic brain by engineered stem cells: fuelling endogenous repair processes. Brain Res Rev 61:1–13
164. Yamanaka S (2007) Strategies and new developments in the generation of patient-specific pluripotent stem cells. Cell Stem Cell 1:39–49

Chapter 15
Angiogenesis in Myocardial Ischemia

Yasuyuki Fujita, Takayuki Asahara, and Atsuhiko Kawamoto

Abstract Progress in cardiovascular biology including the identification of various angiogenic growth factors and the discovery of somatic stem/progenitor cells including bone marrow-derived endothelial progenitor cells, mesenchymal stem cells, skeletal myoblasts, adipose tissue-derived regenerative cells, and resident cardiac stem/progenitor cells has drastically developed the field of therapeutic angiogenesis for refractory ischemic heart disease. Accumulating evidences in both animal studies and human clinical trials support the notion that protein or gene therapy of various angiogenic growth factors and transplantation of the somatic stem/progenitor cells may have significant potency of enhancing new blood vessel formation in ischemic myocardium. This chapter provides an overview of the preclinical and clinical reports to demonstrate the usefulness and the current limitations of the therapeutic strategy.

Keywords Angiogenesis • Arteriogenesis • Cell-based therapy • Gene therapy • Growth factors • Myocardial ischemia • Protein therapy • Stem/progenitor cells • Vasculogenesis

15.1 Introduction

Ischemic heart diseases (IHD) including myocardial infarction (MI) and ischemia is the leading cause of morbidity and mortality in the developed countries, and WHO has estimated that cardiovascular deaths will increase from 17.3 million in

Y. Fujita • A. Kawamoto, M.D. (✉)
Division of Vascular Regeneration, Unit of Regenerative Medicine, Institute of Biomedical Research and Innovation, 2-2 Minatojima Minamimachi, Chuo-Ku, Kobe 650-0047, Japan
e-mail: kawamoto@fbri.org

T. Asahara
Department of Regenerative Medicine Science, Tokai University School of Medicine, Isehara, Japan

J.L. Mehta and N.S. Dhalla (eds.), *Biochemical Basis and Therapeutic Implications of Angiogenesis*, Advances in Biochemistry in Health and Disease 6,
DOI 10.1007/978-1-4614-5857-9_15,

2008 to 23.6 million in 2030 due to aging population [1]. MI induces the necrosis or apoptosis of millions of cardiomyocytes which are replaced with noncontractile scar tissue in several weeks following the infarction. Myocardial scar formation results in an increased mechanical load on surviving myocardial cells, inducing a remodeling process which eventually progresses to clinical heart failure.

The standard of therapy for MI is early reperfusion of the occluded coronary artery with percutaneous coronary intervention (PCI), thrombolytic therapy, or coronary artery bypass graft (CABG) to reverse ischemia and rescue the number of surviving myocardial cells [2]. In addition to the early reperfusion therapy, pharmaceutical drugs including angiotensin converting enzyme (ACE) inhibitors and beta-blockers are used to prevent remodeling after MI and progression of heart failure. Both ACE inhibitors and beta-blockers improve long-term survival in MI patients. On the other hand, a certain number of patients with coronary artery disease (CAD) are resistant to the conventional therapies. Typically, revascularization therapy is no longer available in such patients because of a lack of suitable conduit vessels or diffuse coronary lesions. Therefore, new strategies for refractory CAD are urgently required. The challenge to improve blood flow to the ischemic heart in severe IHD patients has provoked extensive research programs and numerous innovative approaches in the fields of molecular biology and pharmacology. One of these approaches is to promote the growth of new blood vessels in the myocardium using several potential compounds, delivery vectors, and delivery systems to the ischemic myocardium. Accumulating evidences in both animal studies and clinical trials support the notion that various angiogenic growth factors and stem/progenitor cells can enhance formation of new blood vessels. Progress in understanding the molecular and cellular mechanisms of new blood vessel formation including the identification of a number of angiogenic growth factors, successful preclinical studies and promising results of early clinical trials have promoted the recent research of therapeutic angiogenesis.

In this chapter, we provide an overview of the biology of angiogenesis, the basic characteristics of main angiogenic factors, and the mechanism of cell therapy. Later, we introduce some of the clinical experiences with discussion regarding the problem and the future perspective of the therapeutic angiogenesis.

15.2 Molecular and Cellular Mechanism of Coronary Vasculature Formation

Deprivation of oxygen and nutrition due to ischemia brings about a serious threat for tissue viability. In response to the development of tissue ischemia, angiogenic growth factors are upregulated and circulating cellular elements are mobilized to induce neovascular formation (angiogenesis, vasculogenesis, and arteriogenesis) in a coordinated manner [3].

15.2.1 *Definition of Angiogenesis, Vasculogenesis, and Arteriogenesis*

Angiogenesis refers to the extension of preexisting blood vessels through migration and proliferation of terminally differentiated endothelial cells [4, 5]. On the other hand, vasculogenesis, which is commonly known as a crucial paradigm for embryonic neovascularization, refers to the de novo formation of blood vessels derived from stem/progenitor cells. The embryonic process is initiated by forming a blood island consisting of hemangioblasts, surrounded by endothelial cells. The growth and fusion of blood islands in the yolk sac forms the primary capillary plexus and the coalescence of angioblasts subsequently causes the vessel to elongate [6]. Historically, vasculogenesis was thought to be restricted to embryonic development, whereas postnatal capillary growth was thought to occur solely through angiogenesis. However, recent findings revealed that neocapillaries can also develop in previously avascular tissue. This process is initiated by the recruitment of endothelial progenitor cells (EPCs) to foci of neovascularization [7, 8]. Finally, arteriogenesis refers to the process of the transformation of small arterioles into larger conductance arteries with a sufficient diameter to be visualized by angiography. Collectively, angiogenesis, vasculogenesis, and arteriogenesis are referred to as neovascularization [9].

15.2.2 *Angiogenic Growth Factors*

15.2.2.1 Vascular Endothelial Growth Factor Family

The formation of new blood vessels causes in response to the stimulation of angiogenic growth factors, which regulate endothelial migration, proliferation, survival, and proteolytic activity. Among the first growth factors identified to improve angiogenesis, a member of vascular endothelial growth factor (VEGF) has been identified as a critical regulator of pro-angiogenic process [10–12]. VEGF-1 (VEGF-A) is a member of a family consisting of six related growth factors, i.e., VEGF-2 (VEGF-B), VEGF-3 (VEGF-C), VEGF-D, VEGF-E, and placenta growth factor (PIGF) that bind to three specific tyrosine kinase receptors, VEGFR-1, VEGFR-2, and VEGFR-3 [8, 13, 14], and two co-receptors, neurophilin-1 (Nrp-1) and Nrp-2. VEGF-A consists of five isoforms resulting from alternate splicing of mRNA from a single gene, such as $VEGF_{121}$, $VEGF_{145}$, $VEGF_{165}$, $VEGF_{189}$, and $VEGF_{206}$. VEGF-A and its splicing variants and VEGF-B preferentially bind to VEGFR-1 and VEGFR-2 associated with angiogenesis, while VEGF-C and VEGF-D bind to VEGFR-2 and VEGFR-3, contributing to lymphangiogenesis. Among the primary receptors, VEGFR-2 is expressed on almost all endothelial cells, while VEGFR-1 and VEGFR-3 are selectively expressed in distinct vascular beds. Two co-receptors, Nrp-1 and Nrp-2, are thought to raise the binding affinity of VEGF ligands to VEGF receptors; however, the roles of Nrp-1 and Nrp-2 in angiogenesis remain unclear [15–18]. Although there are several isoforms and variants of VEGF and their corresponding receptors, the angiogenic effects are mainly mediated through the interaction of VEGF-A with VEGFR-2, while the other VEGFs secondarily contribute to angiogenesis. The binding of VEGF to its receptor on the surface of endothelial cells activates intracellular receptor tyrosine kinases and thereby triggers multiple downstream signals that promote vascular endothelial cell growth, survival, and proliferation [19]. VEGF production can be triggered by several conditions including hypoxia, tumor, and inflammation [20, 21]. Accumulated data indicate that VEGF is an important factor for the mobilization of EPCs from bone marrow (BM) [22, 23].

15.2.2.2 Fibroblast Growth Factor Family

Fibroblast growth factor (FGF) family is composed of 23 members, FGF-1 through FGF-23, which are multifunctional proteins binding to the alternative splice isoforms of the seven FGF receptors (FGFRs) (FGFR1b, FGFR1c, FGFR2b, FGFR2c, FGFR3b, FGFR3c, and FGFR4). Activation of the FGFRs, which are expressed on endothelial cells and smooth muscle cells, promotes angiogenesis and arteriogenesis by modulating the phenotypes of endothelial cells and vascular smooth muscle cells [24]. The FGF homologous factors (FGF-11–FGF-14) have high sequence identity with the FGF family, but do not activate FGFRs. Therefore, these factors are not generally considered to be the members of FGF family. The FGF ligands perform their diverse functions by binding and activating FGFR family of tyrosine kinase receptors in heparan sulfate glycosaminoglycan (HSGAG)-dependent manner. In particular, FGF-1 (acidic or aFGF), FGF-2 (basic or bFGF), and FGF-4 are highly potent endothelial cell mitogens [25].

15.2.2.3 Granulocyte-Macrophage Colony-Stimulating Factor

Granulocyte-macrophage colony-stimulating factor (GM-CSF) stimulates the growth and differentiation of granulocyte and macrophage precursor cells. The GM-CSF induces peripheral monocytosis and reduces monocyte apoptosis [26, 27]. GM-CSF can also effectively mobilize EPCs from BM to peripheral blood, which may then participate in regenerative activity [28].

15.2.2.4 Granulocyte Colony-Stimulating Factor

Granulocyte colony-stimulating factor (G-CSF) is a potent hematopoietic protein that promotes the development and function of granulocytes and mobilizes progenitor cells from the BM [29]. Mobilization of progenitor cells is initiated when G-CSF binds to the receptors on the cell surface, leading to the release of enzymes that digest adhesion molecules [30]. G-CSF also directly influences the activity of some nonhematopoietic cells, such as cardiomyocytes and endothelial cells [31]. When administered immediately after MI, G-CSF activates the Janus kinase/signal transducer and activator of transcription (JAK-STAT) pathway, which promotes the production of several anti-apoptosis-related proteins, decreases cardiomyocyte death, and limits infarct size [32]. In a murine model of MI, G-CSF treatment was associated with improvement of left ventricular (LV) function and enhancement of arteriogenesis [33]. On the other hand, G-CSF can also stimulate the differentiation of lineage-committed progenitor cells into neutrophils and macrophages [34], which could worsen inflammation and cardiac remodeling [35].

15.2.2.5 Other Angiogenic Growth Factors

Several other angiogenic growth factors have been identified such as erythropoietin (EPO), growth hormone (GH), angiopoietin (Ang), hepatocyte growth factor (HGF), placental growth factor (PIGF), stem cell factor (SCF), and more [36–52]. However, few of these factors have been investigated in clinical trials. Early clinical trials have focused mainly on VEGF and FGF.

15.2.3 Stem/Progenitor Cells

With the exception of administration of angiogenic growth factors and gene therapy, the principal alternative approach is cell-based therapy. Among a variety of stem/ progenitor cells, EPCs and BM-derived mononuclear cells (BM-MNCs) are most widely researched for their ability to induce cardiac neovascularization.

15.2.3.1 Endothelial Progenitor Cells

EPCs were first isolated from the MNC fraction of human peripheral blood [7]. They are phenotypically characterized by expression of antigens associated with hematopoietic stem cells (HSCs) including CD133, CD34, c-kit, VEGFR-2, CD144 (vascular endothelial (VE)-cadherin), and Sca-1. The discovery of circulating EPCs changed the traditional paradigm that "vasculogenesis" occurs exclusively in the developing embryo. EPC levels in the peripheral blood are low under normal conditions; however, EPCs residing in the BM are mobilized into peripheral blood in response to physiological and pathological stimuli, such as myocardial and peripheral ischemia [53, 54]. Mobilized EPCs recruit to the foci of neovascularization where they form structural components of the growing vasculature [55]. Accumulated recent insights into the mechanism of EPC-mediated neovascularization reveal that EPCs secrete paracrine factors including VEGF-A, VEGF-B, stromal cell-derived factor-1 (SDF-1), and insulin-like growth factor-1 (IGF-1) in addition to the direct contribution to vasculogenesis [56]. The paracrine effect of EPCs inhibits cell death, enhances cell proliferation, activates resident stem/progenitor cells in the ischemic tissue, and recruits additional stem/progenitor cells to the ischemic site [57–61]. In preclinical studies, an EPC-enriched fraction, $CD34^+$ human cells were administrated, by either intravenous infusion [62] or intramyocardial injection [63] into rats following MI. In both studies, improvement of LV function and myocardial neovascularization was confirmed. Furthermore, in the study of intramyocardial injection, a subpopulation of cardiomyocytes and smooth muscle cells in the rat heart were shown to be of human origin [63]. This finding indicates that both neovascularization and cardiomyogenesis following EPC transplantation may contribute to improvement of LV function. However, the proportional contribution of both mechanisms remains unknown.

15.2.3.2 Mesenchymal Stem Cells

Mesenchymal stem cells (MSCs) are a subset of stem cells that exist in the BM stroma and can differentiate into the mesenchymal lineages, including osteoblasts, chondrocytes, and adipocytes [64, 65]. MSCs can be distinguished from BM hematopoietic cells by their ability to adhere to the culture dish [66]. MSCs can also be induced to differentiate in vitro into cardiomyocytes, which has encouraged a large number of animal and clinical studies of cardiac repair and regeneration [67–69]. Several preclinical studies of transplantation of MSCs in mice model of MI demonstrated improvement of LV function and reduction of infarct size [69–71]. However, in vivo differentiation capacity of MSCs into cardiomyocytes in the ventricular tissue is still controversial. A preclinical study revealed that implanted MSCs could differentiate into osteoblasts and form calcification in the ventricular tissue [72, 73].

15.2.3.3 Skeletal Myoblasts

Skeletal myoblasts (MBs), which are present in the basal lamina of adult skeletal muscle, are functionally indistinguishable from embryonic MBs and are committed to multiply after injury [74]. Skeletal MBs show strong resistance against ischemia, and have the capability to differentiate into myotubes in the body [75]. Therefore, MBs have been recognized as a candidate of cell-based therapeutic agent for cardiomyocyte regeneration. Several experimental studies revealed the improvement of LV function post skeletal MB administration [76, 77].

15.2.3.4 Adipose Tissue-Derived Regenerative Cells

A multipotent stem cell population with highly proliferative potential was isolated from human adipose tissue [78]. Adipose tissue-derived regenerative cells (ADRCs) have been shown to differentiate into endothelial cells, incorporate into vessels, and promote post-ischemic neovascularization [79, 80]. However, the main mechanism underlying the angiogenic effect of ADRCs on ischemic tissue is believed to be a paracrine effect. Several preclinical studies of acute MI (AMI) demonstrated that administration of ADRCs immediately after AMI contributed to improvement of LV function [81–83].

15.2.3.5 Resident Cardiac Stem Cells

Resident cardiac stem cells (CSCs) refer to the cells that exist in cardiac tissue and can differentiate into endothelial cells, smooth muscle cells, and cardiomyocytes.

CSCs were first identified as an elevated number of proliferative and immature cardiomyocytes, Lin^- $c\text{-}kit^+$ cells in the infarct border zone of AMI patients [84]. They are self-renewing, clonogenic, and multipotent cells, which are capable of differentiating into cardiomyocytes, smooth muscle cells, and endothelial cells. When injected into a rat ischemic heart, these cells or their clonal progeny reconstitutes well-differentiated myocardium, formed by new blood vessels and cardiomyocytes. Subsequently, undifferentiated cells that grow as self-adherent clusters were isolated from subcultures of postnatal atrial or ventricular human biopsy specimens and from murine hearts. The cells were termed "cardiospheres" because they form sphenoid aggregates in culture. The cardiosphere-derived cardiac cells (CDCs) can also differentiate into cardiomyocytes, smooth muscle cells, and endothelial cells [85]. Several studies suggest that transplantation of CDCs contributes to improvement of LV function [86, 87].

15.3 Clinical Trials

15.3.1 Protein and Gene Therapy for Ischemic Heart Disease (Tables 15.1 and 15.2)

15.3.1.1 Vascular Endothelial Growth Factor Family

Successful therapeutic angiogenesis has been reported with VEGF protein or genes that encode VEGF in preclinical models [88–91]. Safety and feasibility of VEGF therapy for myocardial ischemia has been assessed in several phase I studies [92–97].

The VIVA (Vascular Endothelial Growth Factor in Ischemia for Vascular Angiogenesis) trial was the first randomized, double-blind, placebo-controlled phase II trial. One hundred and seventy-eight patients with stable exertional angina, who were unsuitable for standard revascularization, were randomized to receive placebo, low-dose recombinant human vascular endothelial growth factor (rhVEGF), or high-dose rhVEGF by intracoronary infusion, followed by intravenous infusion on

Table 15.1 Randomized controlled trials of protein therapy for myocardial angiogenesis

Trial	Design	Therapy	Route of administration	Case no.	Disease	Results
VIVA [15]	RCT (double-blind)	RhVEGF-A_{165}	IC+IV	178	Chronic CAD	Symptom↑
Laham et al. [16]	RCT (double-blind)	bFGF	Epicardial+CABG	24	Chronic CAD	Myocardial perfusion↑
FIRST [17]	RCT (open label)	rFGF-2	IC	337	Chronic CAD	No effect
Seiler et al. [26]	RCT (double-blind)	GM-CSF	IC+SC	21	Chronic CAD	Collateral flow↑
Zbinden et al. [103]	RCT (double-blind)	GM-CSF	SC	14	Chronic CAD	Collateral flow↑
Valgimigli et al. [107]	RCT	G-CSF	SC	20	STEMI	No effect
FIRSTLINE-AMI [105]	RCT (open label)	G-CSF	SC	50	STEMI	EF↑
STEMMI [109]	RCT (double-blind)	G-CSF	SC	78	STEMI	No effect
REVIVAL-2 [110]	RCT (double-blind)	G-CSF	SC	114	STEMI	No effect
G-CSF-STEMI [108]	RCT (double-blind)	G-CSF	SC	44	STEMI	No effect

CABG coronary artery bypass grafting; *CAD* coronary artery disease; *EF* ejection fraction; *ETT* exercise tolerance test; *FGF* fibroblast growth factor; *G-CSF* granulocyte colony-stimulating factor; *GM-CSF* granulocyte-macrophage colony-stimulating factor; *IC* intracoronary; *IV* intravenous; *SC* subcutaneous; *STEMI* ST-segment elevation myocardial infarction; *VEGF* vascular endothelial growth factor

days 3, 6, and 9 [15]. This study demonstrated that rhVEGF therapy was safe and well tolerated in short term; however, changes in exercise treadmill time (the primary end point) and quality of life did not differ significantly between groups by day 120. By day 120 after treatment, only the angina class was significantly improved in patients receiving high-dose rhVEGF by at least one Canada Cardiovascular Society (CCS) class ($p=0.05$ vs. placebo). In the EUROINJECT One phase II trial [12], 80 no-option patients with CCS class III or IV angina were assigned randomly to receive injections of either 0.5-mg VEGF-A_{165} plasmid or a placebo plasmid into myocardial lesions identified as stress-induced perfusion defect areas. At the 3-month follow-up visit, the perfusion defects did not differ between the treatment groups; however, VEGF treatment was associated with improvement of regional wall motion. Catheter-based intracoronary VEGF gene transfer was examined in the KAT (Kuopio Angiogenesis Trial) (phase II trial) [98]. Following PCI, patients with CCS class II or III angina (90 % received stent implantation) were randomized to receive intracoronary injections of adenovirus-encoded $VEGF_{165}$, $VEGF_{165}$ plasmid liposome, or Ringer's lactate. Six months after treatment, there were no significant differences among the three treatment groups in functional status; however, myocardial perfusion improved in patients treated with adenoviral VEGF.

Table 15.2 Randomized controlled trials of gene therapy for myocardial angiogenesis

Trial	Design	Therapy	Route of administration	Case no.	Disease	Results
KAT [98]	RCT (double-blind)	Adenoviral or plasmid $hVEGF_{165}$	IC	103	Chronic CAD	Myocardial perfusion↑
REVASC [128]	RCT	$AdVEGF_{121}$	IM (mini-thoracotomy)	67	Chronic CAD	ETT↑
EUROINJECT ONE [12, 129]	RCT (double-blind)	$phVEGF\text{-}A_{165}$	IM (TED)	80	Chronic CAD	Regional wall motion↑
NORTHERN [130]	RCT (double-blind)	Plasmid $VEGF_{165}$	IM (TED)	93	Chronic CAD	No effect
NOVA [131]	RCT (double-blind)	$AdGV\text{-}VEGF_{121}$	IM (TED)	17	Chronic CAD	No effect
VIF-CAD [132]	RCT (double-blind)	Plasmid VEGF-A165/bFGF	IM	52	Chronic CAD	No effect
AGENT [99]	RCT (double-blind)	Ad5FGF-4	IC	79	Chronic CAD	ETT↑
AGENT-2 [101]	RCT (double-blind)	Ad5FGF-4	IC	52	Chronic CAD	Myocardial perfusion↑
AGENT-3, AGENT-4 [102]	RCT (double-blind)	Ad5FGF-4	IC	532	Chronic CAD	ETT↑ (gender-specific)

CAD coronary artery disease; *EF* ejection fraction; *ETT* exercise tolerance test; *FGF* fibroblast growth factor; *IC* intracoronary; *IM* intramyocardial; *RCT* randomized controlled trial; *TED* transendocardial delivery; *VEGF* vascular endothelial growth factor

15.3.2 Fibroblast Growth Factor Family

FIRST (Initiating RevaScularization Trial) is the first study of rFGF-2 treatment for chronic myocardial ischemia [17]. Three hundred and thirty-seven patients were assigned randomly to receive a single intracoronary infusion of rFGF-2 at 0, 0.3, 3, or 30 μg/kg rFGF-2. Significant improvement of exercise tolerance and myocardial nuclear perfusion was reported at 90 days, but not 180 days, because of continued improvement in the placebo group. Hypotension was noted with increased frequency in the 30 μg/kg FGF-2 group. The trial confirmed that administration of recombinant FGF is safe and feasible, however, revealed that the efficacy of recombinant FGF was questionable.

The AGENT (Angiogenic Gene Therapy) trial examined the effects of a replication defective adenovirus (Ad) containing a human FGF gene in patients with CAD and medically refractory angina. In the first AGENT trial, which is a randomized, double-blind, placebo-controlled trial, intracoronary delivery of an adenovirus coding for FGF-4 transcription (Ad5FGF-4) was safe and shown to improve exercise treadmill times [99]. The phase II AGENT 2 trial investigated whether a further dose of Ad5FGF-4 treatment improved regional myocardial perfusion [100]. Intracoronary injection of Ad5FGF-4 in 35 patients resulted in an encouraging trend for improved myocardial perfusion; however, the outcomes were not statistically significant. The phase III AGENT 3 and 4 trials [101, 102] enrolled patients with CCS class II to IV angina, who were unsuitable for revascularization. Both trials were stopped prematurely when a planned interim analysis of the AGENT 3 cohort indicated that change in total treadmill exercise duration between baseline and 12 weeks after treatment (the primary end point) would not reach statistical significance. Differences in secondary outcomes (such as change in CCS class or other clinical variables) were not also significant.

15.3.3 Granulocyte-Macrophage Colony-Stimulating Factor

In the first clinical study of GM-CSF for patients with CAD, 21 patients were randomized to receive a single intracoronary injection of 40 μg of GM-CSF, followed by subcutaneous injections (10 μg/kg) every other day for 2 weeks. GM-CSF was associated with significant improvement in electrocardiographic signs of myocardial ischemia during coronary balloon occlusion and collateral flow index (CFI). CFI did not significantly change in patients who received placebo injections [26]. Improvements in CFI were also reported in a subsequent study of GM-CSF (10 μg/kg/day subcutaneous injection for 2 weeks) therapy, but 2 of 7 GM-CSF-treated patients experienced acute coronary syndrome during the 2-week course of therapy [103].

15.3.4 *Granulocyte Colony-Stimulating Factor*

In phase I trials, G-CSF administration after PCI in patients with AMI revealed improvement of cardiac function [31, 104–107]. In the FIRSTLINE-AMI (Front-Integrated Revascularization and Stem Cell Liberation in Evolving Acute Myocardial Infarction) randomized, open-label trial, 25 patients with ST-segment elevation MI (STEMI) were randomly assigned to receive subcutaneous injection of G-CSF at 10 μg/kg/day for 6 days starting within 90 min after primary PCI for STEMI; the control group ($n=25$) did not receive placebo injections but had standard post-interventional care. Four months [105] and 1 year [104] after the PCI procedure, improvement of LV ejection fraction (LVEF) was significantly greater in G-CSF-treated patients than in the control group. However, these promising results were not reproduced in subsequent double-blind, placebo-controlled trials, such as the STEMMI (Stem Cells in Myocardial Infarction), REVIVAL-2 (Regenerate Vital Myocardium by Vigorous Activation of Bone Marrow Stem Cells), and G-CSF-STEMI (Granulocyte Colony-Stimulating Factor ST-Segment Elevation Myocardial Infarction) trials [108–110].

15.3.5 *Cell-Based Therapy for Ischemic Heart Disease*

The discovery of EPCs [7] guided to the development of stem/progenitor cell-based strategies for ischemic cardiovascular disease. Since then various stem/progenitor cell population from a variety of sources have been proposed for cell-based therapy for IHD, and the promising results from experimental studies in rodents promoted the initiation of clinical pilot trials. This section provides a summary of the clinical evidence of the role of stem/progenitor cells in blood vessel formation and tissue repair in IHD (Table 15.3).

15.3.5.1 Bone Marrow-Derived Mononuclear Cells

The first clinical trial of BM-MNC therapy for IHD was reported in 2002 [111]. Following emergent balloon angioplasty and stent implantation, ten patients received an intracoronary infusion of autologous BM-MNCs 5–9 days after AMI. Ten non-randomized control patients with AMI were treated by standard therapy alone. At the 3-month follow-up, the infarct region size significantly decreased in the cell treatment group. Furthermore, stroke volume, end-systolic volume, and contractility were improved after the cell treatment. In the TOPCARE-AMI (Transplantation of Progenitor Cells and Regeneration Enhancement in Acute Myocardial Infarction) trial, 20 patients with reperfused AMI were randomly allocated to receive intracoronary infusion of either BM-MNCs or circulating blood-derived progenitor cells (CPCs) into infarct artery 4.3 ± 1.5 days after AMI.

Table 15.3 Randomized controlled trials of cell therapy for myocardial angiogenesis

Trial	Design	Cell type	Route of administration	Case no.	Disease	Results
TOPCARE-AMI [113]	RCT	CPCs or BM-MNCs	IC day 3–7 after PCI	59	STEMI	EF↑
BOOST [114, 133]	RCT	BM-MNCs	IC	60	STEMI	EF↑
ASAMI [134]	RCT (double-blind)	BM-MNCs	IC	100	STEMI	No effect
REPAIR-AMI [115]	RCT (double-blind)	BM-MNCs	IC	204	STEMI	EF↑
PROTECT-CAD [135]	RCT	BM-MNCs	IM (TED)	28	Chronic CAD	Regional function↑, symptom↑
MYSTAR [136]	RCT	BM-MNCs	IM and IC	60	AMI	EF↑
TOPCARE-CHD [117]	RCT	CPC or BM-MNCs	IC	75	OMI	Regional function↑
LateTIME [119]	RCT (double-blind)	BM-MNCs	IC	87	Recent MI	No effect
Focus-CCTRN [118]	RCT (double-blind)	BM-MNCs	IM (TED)	92	Chronic CAD	No effect
MAGIC-Cell [121]	RCT	G-CSF mobilized PB-MNC or G-CSF alone	IC	27	Recent MI	EF↑, ETT↑
Losordo et al. [122]	RCT (double-blind)	G-CSF mobilized CD34 cells	IM (TED)	24	Chronic CAD	ETT↑, symptom↑
REGENT [137]	RCT	BM-MNC or $CD34^+CXCR4^+$ cells	IC	200	AMI	No effect
ACT34-CMI [123]	RCT (double-blind)	G-CSF mobilized PB-$CD34^+$ cells	IM (TED)	167	Chronic CAD	Symptom↑, ETT↑
Chen et al. [124]	RCT (open label)	MSCs	IC	69	AMI	EF↑
MAGIC [125]	RCT (double-blind)	Skeletal MBs	IM+CABG	97	OMI	No effect
APOLLO [126] (in progress)	RCT (double-blind)	ADRCs	IC	14	STEMI	Infarct size↑, perfusion↑
SCIPIO [127] (in progress)	RCT	CSCs (lin-c^-kit^+ cells)	IC	23	Chronic CAD	EF↑, infarct size↑
CADUCEUS [138]	RCT	CSCs (CDCs)	IC	25	Recent MI	EF↑, scar mass↑

ADRCs adipose tissue-derived regenerative cells; *AMI* acute myocardial infarction; *BM-, PB-MNC* bone marrow-derived, peripheral blood-derived mononuclear cells; *CDCs* cardiosphere-derived cardiac cells; *CSCs* cardiac stem cells; *CPCs* circulating progenitor cells; *EF* ejection fraction; *MBs* myoblasts; *MSCs* mesenchymal stem cells; *OMI* old myocardial infarction

BM-MNCs and CPCs significantly improved the LVEF, perfusion, and infarct size compared with a matched reference group [112]. There were no adverse events related to cell transplantation during 1-year follow-up, indicating that intracoronary infusion of either BM-MNCs or CPCs is safe and feasible in patients after AMI [113]. The first randomized trial was the BOOST (Bone marrow transfer to enhance ST-elevation infarct regeneration) trial [114]. In this study, 60 patients with STEMI were randomly assigned to receive intracoronary infusion of BM-MNCs 4.8 ± 1.3 days after PCI or were not treated (a control group received optimum post-infarction medical treatment). BM-MNCs significantly increased LVEF and reduced end-systolic volumes as assessed by cardiac MRI at 6 months but not after 18 months after treatment. The REPAIR-AMI (Reinfusion of Enriched Progenitor Cells and Infarct Remodeling in Acute Myocardial Infarction) group [115] enrolled 204 patients with STEMI who were randomized to receive an intra-coronary infusion of BM-MNCs or placebo after successful reperfusion therapy. At 4 months, LVEF was significantly improved in the treatment group than the placebo group. Intracoronary infusion of BM-MNCs was associated with a reduction in the combined clinical end point of death, recurrence of myocardial infarction, and any revascularization procedure at 1 year. These trials suggested that cell-based therapy might be attributed to the improvement of patient prognosis. However, the ASTAMI (Autologous Stem cell Transplantation in Acute Myocardial Infarction) study could not show significant benefit of an intracoronary injection of BM-MNC on STEMI during 3 years follow-up [116].

On the other hand, BM-MNC transplantation was also applied to chronic CAD including old MI and chronic severe myocardial ischemia. TOPCARE-CHD (Transplantation of Progenitor Cells and Recovery of LV [Left Ventricular] Function in Patients with Chronic Ischemic Heart Disease) group accomplished a crossover study [117]. Seventy-five patients with stable IHD, who had had a MI at least 3 months previously, were randomly assigned to receive either no cell infusion or infusion of CPCs or BM-MNCs into the patent coronary artery. Assessment of cardiac functions at 3 months after each cell infusion revealed that transplantation of BM-MNCs was associated with moderate but significant improvement in LVEF. However, the recent clinical trials revealed that intracoronary infusion of BM-MNCs demonstrated no functional benefit in chronic CAD or recent MI [118, 119]. One of the reasons of the low efficacy may be that BM-MNCs contain higher number of inflammatory and mesenchymal cells, which may accelerate tissue damage in ischemic myocardium [73, 120].

15.3.5.2 G-CSF Mobilized Peripheral Blood Mononuclear Cells

In the MAGIC-Cell (Myocardial Regeneration and Angiogenesis in Myocardial Infarction with G-CSF and Intra-Coronary Stem Cell Infusion) trial, 27 patients with revascularized AMI were randomly assigned to G-CSF mobilized cell infusion group, G-CSF alone group, or control group. Among three groups, cell infusion group showed significant improvement of cardiac function at 6 months.

However, the trial was halted prematurely because of high rate of in-stent restenosis at the culprit lesion in patients who received G-CSF [121].

15.3.5.3 Endothelial Progenitor Cells

The first pilot (phase I/IIa) study of intramyocardial injection of autologous $CD34^+$ stem cells in patients with intractable angina provided early evidence of feasibility, safety, and bioactivity of the stem cells [122]. The promising results encouraged the ACT34-CMI study, a prospective, double-blind, randomized, controlled phase II trial [123]. In this study, 167 patients with refractory angina were randomized to receive intramyocardial injection of 1×10^5 or 5×10^5 cells/kg of mobilized autologous $CD34^+$ cells or an equal volume of diluent (placebo). Improvement of weekly angina frequency and exercise tolerance was significantly greater in low-dose patients, but not high-dose group, than in placebo-treated patients at 6 and 12 months. Three patients in the control group died in this study. No major adverse cardiovascular events were causatively related to the cell therapy. The pivotal, phase III study of the $CD34^+$ cell therapy is ongoing.

15.3.5.4 Mesenchymal Stem Cells

Sixty-nine patients with revascularized AMI were enrolled in a randomized and controlled clinical trial. BM-MNCs were harvested from the patients and cultured for 10 days to obtain MSCs. The patients received the infusion of MSCs or saline into the infarct-related artery. MSCs infusion resulted in improvement of LV function [124].

15.3.5.5 Skeletal Myoblasts

Skeletal MBs can also be a useful tool to treat IHD. The first randomized, placebo-controlled trial is the MAGIC (Myoblast Autologous Grafting in Ischemic Cardiomyopathy) phase II study [125]. The study enrolled 97 patients with old MI randomly assigned to the treatment group (high-dose; 400 million or low-dose; 800 million MBs) or the placebo group. All patients received a cardioverter-defibrillator due to concerns about a possible arrhythmogenic effect of cell transplantation. All treated patients underwent a skeletal muscle biopsy. The muscle-derived cells were cultured and expanded for 3 weeks to obtain the skeletal MBs. Skeletal MBs were injected into myocardium after CABG. The study was stopped early because the cell therapy did not show superior effect on the primary efficacy end point, improvement of LV regional contractility and grovel function compared with placebo treatment. However, the high-dose cell group demonstrated a significant decrease in LV volumes compared with the placebo group. Although a higher number of arrhythmic events occurred in the MB-treated patients than the control subjects, the rates of major cardiac adverse events and of ventricular arrhythmias did not differ significantly between the pooled treatment and placebo groups at 6 months.

15.3.5.6 Adipose Tissue-Derived Regenerative Cells

The first double-blind, randomized controlled trial of ADRCs was the APOLLO (AdiPOse-derived Stem ceLLs in the Treatment of Patients With ST-elevation myOcardial Infarction) trial. A total of 14 patients with reperfused STEMI were enrolled and randomized to receive an intracoronary infusion of either ADRCs or placebo solution. After a liposuction procedure of periumbilical region, ADRCs were isolated. Within 24 h after primary PCI, ADRCs were infused into the culprit coronary artery. LVEF significantly increased when assessed by sestamibi single photon emission computed tomography (MIBI-SPECT) and MRI at 6 months after ADRCs infusion in a treatment group. In addition, the perfusion defect and infarct size of LV were significantly reduced when assessed by MIBI-SPECT and MRI, respectively, at 6 months after the cell therapy. No unanticipated adverse effects related to the ADRC therapy were reported [126]. The phase IIb/III clinical trial is ongoing.

15.3.5.7 Resident Cardiac Stem Cells

To date, two research groups reported the results of clinical trials for IHD using resident CSCs including $Lin^-c\text{-}kit^+$ cells and CDCs.

The first randomized controlled clinical trial is the SCIPIO (Stem Cell Infusion in Patients with Ischemic cardiomyopathy) trial, which is in progress. Sixteen patients were assigned to the treatment group and seven to the control group. In the treatment group, one million autologous CSCs ($Lin^-c\text{-}kit^+$ cells) were administered by intracoronary infusion at a mean of 113 ± 4 days after CABG surgery; controls received standard of care post CABG. In 14 CSC-treated patients, LVEF significantly increased at 4 months after infusion. In contrast, in seven control patients, LVEF did not change by 8 months after CABG. Furthermore, the favorable effects of CSCs were even more pronounced at 1 year in eight patients. In the seven treated patients in whom cardiac MRI could be done, infarct size significantly decreased at 1 year [127].

As for CDCs, in the prospective, randomized CADUCEUS (CArdiosphere-Derived aUtologous stem CElls to reverse ventricUlar dysfunction) trial, 31 patients with recent MI were randomly allocated to the CDCs group or the control group. Twenty-five out of 31 eligible patients were included in this study (17 to CDC group and 8 to control group). For patients assigned to receive CDCs, autologous cells grown from endocardial biopsy specimens were infused into the infarct-related artery 1.5–3 months after MI. MRI analysis at 6 months revealed significant reduction in scar mass, significant increase in viable heart mass, and significant improvement of regional contractility and regional systolic wall thickening in patients treated with CDCs compared with controls. No complications were reported within 24 h post CDC infusion. By 6 months, no patients in both groups had died, developed cardiac tumors, or experienced MACE.

The initial favorable results in these two clinical trials support that implantation of CSCs may be a promising therapeutic modality for IHD.

15.4 Conclusions

Summarizing the results of the clinical trials regarding therapeutic myocardial angiogenesis, both protein and gene therapies are feasible and relatively safe. However, efficacy of these therapies may be modest or still controversial. Especially, G-CSF therapy did not significantly improve LV function post MI.

Theoretically, stem/progenitor cell therapies may be superior over protein or gene therapy due to not only direct vasculogenic properties but also paracrine action by secreting multiple growth factors besides a single angiogenic factor. Early phase clinical trials revealed safety and feasibility of the cell therapies for IHD. However, effect of most cell-based strategies on LV perfusion or function has not yet been confirmed in large-scale, randomized, clinical trials. Among various cell types, intramyocardial injection of $CD34^{+}$ cells resulted in significant improvement of angina symptom and exercise tolerance in a phase II trial, in which relatively large number of patients were enrolled. Improvement of myocardial ischemia but not LV dysfunction may be more realistic target for therapeutic angiogenesis. Novel approach using resident CSCs may be a hopeful strategy to treat severe heart failure by inducing both angiogenesis and cardiomyogenesis. In either approach, further investigation of more suitable cell types and more efficient cell delivery methods would be warranted for the future clinical application.

References

1. WHO. World Health Organization, Cardiovascular Diseases. http://www.who.int/cardiovascular_diseases/en/
2. White HD, Chew DP (2008) Acute myocardial infarction. Lancet 372:570–584
3. Folkman J (1995) Angiogenesis in cancer, vascular, rheumatoid and other disease. Nat Med 1:27–31
4. Goussetis E, Manginas A, Koutelou M et al (2006) Intracoronary infusion of CD133+ and CD133-CD34+ selected autologous bone marrow progenitor cells in patients with chronic ischemic cardiomyopathy: cell isolation, adherence to the infarcted area, and body distribution. Stem Cells 24:2279–2283
5. Askari AT, Unzek S, Popovic ZB et al (2003) Effect of stromal-cell-derived factor 1 on stem-cell homing and tissue regeneration in ischaemic cardiomyopathy. Lancet 362:697–703
6. Fadini GP, Losordo D, Dimmeler S (2012) Critical reevaluation of endothelial progenitor cell phenotypes for therapeutic and diagnostic use. Circ Res 110:624–637
7. Asahara T, Murohara T, Sullivan A et al (1997) Isolation of putative progenitor endothelial cells for angiogenesis. Science 275:964–967

8. Asahara T, Masuda H, Takahashi T et al (1999) Bone marrow origin of endothelial progenitor cells responsible for postnatal vasculogenesis in physiological and pathological neovascularization. Circ Res 85:221–228
9. Simons M (2005) Angiogenesis: where do we stand now? Circulation 111:1556–1566
10. Lewis BS, Flugelman MY, Weisz A et al (1997) Angiogenesis by gene therapy: a new horizon for myocardial revascularization? Cardiovasc Res 35:490–497
11. Ripa RS, Wang Y, Jorgensen E et al (2006) Intramyocardial injection of vascular endothelial growth factor-A165 plasmid followed by granulocyte-colony stimulating factor to induce angiogenesis in patients with severe chronic ischaemic heart disease. Eur Heart J 27:1785–1792
12. Kastrup J, Jorgensen E, Ruck A et al (2005) Direct intramyocardial plasmid vascular endothelial growth factor-A165 gene therapy in patients with stable severe angina pectoris A randomized double-blind placebo-controlled study: the Euroinject One trial. J Am Coll Cardiol 45:982–988
13. The Editors (2012) Circulation research thematic synopsis: stem cells and cardiac progenitor cells. Circ Res 111:e1–e14
14. Losordo DW, Dimmeler S (2004) Therapeutic angiogenesis and vasculogenesis for ischemic disease. Part I: angiogenic cytokines. Circulation 109:2487–2491
15. Henry TD, Annex BH, McKendall GR et al (2003) The VIVA trial: vascular endothelial growth factor in ischemia for vascular angiogenesis. Circulation 107:1359–1365
16. Laham RJ, Sellke FW, Edelman ER et al (1999) Local perivascular delivery of basic fibroblast growth factor in patients undergoing coronary bypass surgery: results of a phase I randomized, double-blind, placebo-controlled trial. Circulation 100:1865–1871
17. Simons M, Annex BH, Laham RJ et al (2002) Pharmacological treatment of coronary artery disease with recombinant fibroblast growth factor-2: double-blind, randomized, controlled clinical trial. Circulation 105:788–793
18. Kawasaki T, Kitsukawa T, Bekku Y et al (1999) A requirement for neuropilin-1 in embryonic vessel formation. Development 126:4895–4902
19. Eichmann A, Simons M (2012) VEGF signaling inside vascular endothelial cells and beyond. Curr Opin Cell Biol 24:188–193
20. Costa C, Soares R, Reis-Filho JS et al (2002) Cyclo-oxygenase 2 expression is associated with angiogenesis and lymph node metastasis in human breast cancer. J Clin Pathol 55:429–434
21. Waltenberger J, Mayr U, Pentz S, Hombach V (1996) Functional upregulation of the vascular endothelial growth factor receptor KDR by hypoxia. Circulation 94:1647–1654
22. Kalka C, Masuda H, Takahashi T et al (2000) Vascular endothelial growth factor(165) gene transfer augments circulating endothelial progenitor cells in human subjects. Circ Res 86:1198–1202
23. Asahara T, Takahashi T, Masuda H et al (1999) VEGF contributes to postnatal neovascularization by mobilizing bone marrow-derived endothelial progenitor cells. EMBO J 18:3964–3972
24. Itoh N, Ornitz DM (2004) Evolution of the Fgf and Fgfr gene families. Trends Genet 20:563–569
25. Beenken A, Mohammadi M (2009) The FGF family: biology, pathophysiology and therapy. Nat Rev Drug Discov 8:235–253
26. Seiler C, Pohl T, Wustmann K et al (2001) Promotion of collateral growth by granulocyte-macrophage colony-stimulating factor in patients with coronary artery disease: a randomized, double-blind, placebo-controlled study. Circulation 104:2012–2017
27. Maekawa Y, Anzai T, Yoshikawa T et al (2004) Effect of granulocyte-macrophage colony-stimulating factor inducer on left ventricular remodeling after acute myocardial infarction. J Am Coll Cardiol 44:1510–1520
28. Deng Z, Yang C, Deng H et al (2006) Effects of GM-CSF on the stem cells mobilization and plasma C-reactive protein levels in patients with acute myocardial infarction. Int J Cardiol 113:92–96

29. Anderlini P, Donato M, Chan KW et al (1999) Allogeneic blood progenitor cell collection in normal donors after mobilization with filgrastim: the M.D. Anderson Cancer Center experience. Transfusion 39:555–560
30. Levesque JP, Hendy J, Takamatsu Y et al (2003) Disruption of the CXCR4/CXCL12 chemotactic interaction during hematopoietic stem cell mobilization induced by GCSF or cyclophosphamide. J Clin Invest 111:187–196
31. Kuethe F, Figulla HR, Herzau M et al (2005) Treatment with granulocyte colony-stimulating factor for mobilization of bone marrow cells in patients with acute myocardial infarction. Am Heart J 150:115
32. Harada M, Qin Y, Takano H et al (2005) G-CSF prevents cardiac remodeling after myocardial infarction by activating the Jak-Stat pathway in cardiomyocytes. Nat Med 11:305–311
33. Deindl E, Zaruba MM, Brunner S et al (2006) G-CSF administration after myocardial infarction in mice attenuates late ischemic cardiomyopathy by enhanced arteriogenesis. FASEB J 20:956–958
34. Suda T, Suda J, Kajigaya S et al (1987) Effects of recombinant murine granulocyte colony-stimulating factor on granulocyte-macrophage and blast colony formation. Exp Hematol 15:958–965
35. Vandervelde S, van Luyn MJ, Tio RA, Harmsen MC (2005) Signaling factors in stem cell-mediated repair of infarcted myocardium. J Mol Cell Cardiol 39:363–376
36. Heeschen C, Aicher A, Lehmann R et al (2003) Erythropoietin is a potent physiologic stimulus for endothelial progenitor cell mobilization. Blood 102:1340–1346
37. Lipsic E, van der Meer P, Voors AA et al (2006) A single bolus of a long-acting erythropoietin analogue darbepoetin alfa in patients with acute myocardial infarction: a randomized feasibility and safety study. Cardiovasc Drugs Ther 20:135–141
38. Ruifrok WP, de Boer RA, Westenbrink BD et al (2008) Erythropoietin in cardiac disease: new features of an old drug. Eur J Pharmacol 585:270–277
39. Fazio S, Sabatini D, Capaldo B et al (1996) A preliminary study of growth hormone in the treatment of dilated cardiomyopathy. N Engl J Med 334:809–814
40. Genth-Zotz S, Zotz R, Geil S et al (1999) Recombinant growth hormone therapy in patients with ischemic cardiomyopathy: effects on hemodynamics, left ventricular function, and cardiopulmonary exercise capacity. Circulation 99:18–21
41. Isgaard J, Bergh CH, Caidahl K et al (1998) A placebo-controlled study of growth hormone in patients with congestive heart failure. Eur Heart J 19:1704–1711
42. O'Driscoll JG, Green DJ, Ireland M et al (1997) Treatment of end-stage cardiac failure with growth hormone. Lancet 349:1068
43. Osterziel KJ, Strohm O, Schuler J et al (1998) Randomised, double-blind, placebo-controlled trial of human recombinant growth hormone in patients with chronic heart failure due to dilated cardiomyopathy. Lancet 351:1233–1237
44. Volterrani M, Desenzani P, Lorusso R et al (1997) Haemodynamic effects of intravenous growth hormone in congestive heart failure. Lancet 349:1067–1068
45. Kim KL, Shin IS, Kim JM et al (2006) Interaction between Tie receptors modulates angiogenic activity of angiopoietin2 in endothelial progenitor cells. Cardiovasc Res 72:394–402
46. Takahashi K, Ito Y, Morikawa M et al (2003) Adenoviral-delivered angiopoietin-1 reduces the infarction and attenuates the progression of cardiac dysfunction in the rat model of acute myocardial infarction. Mol Ther 8:584–592
47. Bussolino F, Di Renzo MF, Ziche M et al (1992) Hepatocyte growth factor is a potent angiogenic factor which stimulates endothelial cell motility and growth. J Cell Biol 119:629–641
48. Yasuda S, Goto Y, Baba T et al (2000) Enhanced secretion of cardiac hepatocyte growth factor from an infarct region is associated with less severe ventricular enlargement and improved cardiac function. J Am Coll Cardiol 36:115–121
49. Khurana R, Moons L, Shafi S et al (2005) Placental growth factor promotes atherosclerotic intimal thickening and macrophage accumulation. Circulation 111:2828–2836

50. Kolakowski S Jr, Berry MF, Atluri P et al (2006) Placental growth factor provides a novel local angiogenic therapy for ischemic cardiomyopathy. J Card Surg 21:559–564
51. Luttun A, Tjwa M, Moons L et al (2002) Revascularization of ischemic tissues by PlGF treatment, and inhibition of tumor angiogenesis, arthritis and atherosclerosis by anti-Flt1. Nat Med 8:831–840
52. Kuang D, Zhao X, Xiao G et al (2008) Stem cell factor/c-kit signaling mediated cardiac stem cell migration via activation of p38 MAPK. Basic Res Cardiol 103:265–273
53. Masuda H, Kalka C, Takahashi T et al (2007) Estrogen-mediated endothelial progenitor cell biology and kinetics for physiological postnatal vasculogenesis. Circ Res 101:598–606
54. Takahashi T, Kalka C, Masuda H et al (1999) Ischemia- and cytokine-induced mobilization of bone marrow-derived endothelial progenitor cells for neovascularization. Nat Med 5:434–438
55. Orlic D, Kajstura J, Chimenti S et al (2001) Bone marrow cells regenerate infarcted myocardium. Nature 410:701–705
56. Narmoneva DA, Vukmirovic R, Davis ME et al (2004) Endothelial cells promote cardiac myocyte survival and spatial reorganization: implications for cardiac regeneration. Circulation 110:962–968
57. Kamihata H, Matsubara H, Nishiue T et al (2002) Improvement of collateral perfusion and regional function by implantation of peripheral blood mononuclear cells into ischemic hibernating myocardium. Arterioscler Thromb Vasc Biol 22:1804–1810
58. Kinnaird T, Stabile E, Burnett MS et al (2004) Marrow-derived stromal cells express genes encoding a broad spectrum of arteriogenic cytokines and promote in vitro and in vivo arteriogenesis through paracrine mechanisms. Circ Res 94:678–685
59. Kinnaird T, Stabile E, Burnett MS et al (2004) Local delivery of marrow-derived stromal cells augments collateral perfusion through paracrine mechanisms. Circulation 109:1543–1549
60. Rehman J, Li J, Orschell CM, March KL (2003) Peripheral blood "endothelial progenitor cells" are derived from monocyte/macrophages and secrete angiogenic growth factors. Circulation 107:1164–1169
61. Wollert KC, Drexler H (2005) Clinical applications of stem cells for the heart. Circ Res 96:151–163
62. Kocher AA, Schuster MD, Szabolcs MJ et al (2001) Neovascularization of ischemic myocardium by human bone-marrow-derived angioblasts prevents cardiomyocyte apoptosis, reduces remodeling and improves cardiac function. Nat Med 7:430–436
63. Iwasaki H, Kawamoto A, Ishikawa M et al (2006) Dose-dependent contribution of CD34-positive cell transplantation to concurrent vasculogenesis and cardiomyogenesis for functional regenerative recovery after myocardial infarction. Circulation 113:1311–1325
64. Jiang Y, Jahagirdar BN, Reinhardt RL et al (2002) Pluripotency of mesenchymal stem cells derived from adult marrow. Nature 418:41–49
65. Pittenger MF, Mackay AM, Beck SC et al (1999) Multilineage potential of adult human mesenchymal stem cells. Science 284:143–147
66. Alhadlaq A, Mao JJ (2004) Mesenchymal stem cells: isolation and therapeutics. Stem Cells Dev 13:436–448
67. Makino S, Fukuda K, Miyoshi S et al (1999) Cardiomyocytes can be generated from marrow stromal cells in vitro. J Clin Invest 103:697–705
68. Shiota M, Heike T, Haruyama M et al (2007) Isolation and characterization of bone marrow-derived mesenchymal progenitor cells with myogenic and neuronal properties. Exp Cell Res 313:1008–1023
69 Tomita S, Li RK, Weisel RD et al (1999) Autologous transplantation of bone marrow cells improves damaged heart function. Circulation 100:II247–II256
70. Amado LC, Saliaris AP, Schuleri KH et al (2005) Cardiac repair with intramyocardial injection of allogeneic mesenchymal stem cells after myocardial infarction. Proc Natl Acad Sci USA 102:11474–11479
71. Dai W, Hale SL, Martin BJ et al (2005) Allogeneic mesenchymal stem cell transplantation in postinfarcted rat myocardium: short- and long-term effects. Circulation 112:214–223

72. Breitbach M, Bostani T, Roell W et al (2007) Potential risks of bone marrow cell transplantation into infarcted hearts. Blood 110:1362–1369
73. Yoon YS, Park JS, Tkebuchava T et al (2004) Unexpected severe calcification after transplantation of bone marrow cells in acute myocardial infarction. Circulation 109:3154–3157
74. Murry CE, Wiseman RW, Schwartz SM, Hauschka SD (1996) Skeletal myoblast transplantation for repair of myocardial necrosis. J Clin Invest 98:2512–2523
75. Allen RE, Rankin LL (1990) Regulation of satellite cells during skeletal muscle growth and development. Proc Soc Exp Biol Med 194:81–86
76. Ghostine S, Carrion C, Souza LC et al (2002) Long-term efficacy of myoblast transplantation on regional structure and function after myocardial infarction. Circulation 106:I131–I136
77. Pagani FD, DerSimonian H, Zawadzka A et al (2003) Autologous skeletal myoblasts transplanted to ischemia-damaged myocardium in humans. Histological analysis of cell survival and differentiation. J Am Coll Cardiol 41:879–888
78. Zuk PA, Zhu M, Mizuno H et al (2001) Multilineage cells from human adipose tissue: implications for cell-based therapies. Tissue Eng 7:211–228
79. Miranville A, Heeschen C, Sengenes C et al (2004) Improvement of postnatal neovascularization by human adipose tissue-derived stem cells. Circulation 110:349–355
80. Planat-Benard V, Silvestre JS, Cousin B et al (2004) Plasticity of human adipose lineage cells toward endothelial cells: physiological and therapeutic perspectives. Circulation 109:656–663
81. Madonna R, Geng YJ, De Caterina R (2009) Adipose tissue-derived stem cells: characterization and potential for cardiovascular repair. Arterioscler Thromb Vasc Biol 29:1723–1729
82. Meliga E, Strem BM, Duckers HJ, Serruys PW (2007) Adipose-derived cells. Cell Transplant 16:963–970
83. Valina C, Pinkernell K, Song YH et al (2007) Intracoronary administration of autologous adipose tissue-derived stem cells improves left ventricular function, perfusion, and remodelling after acute myocardial infarction. Eur Heart J 28:2667–2677
84. Beltrami AP, Barlucchi L, Torella D et al (2003) Adult cardiac stem cells are multipotent and support myocardial regeneration. Cell 114:763–776
85. Messina E, De Angelis L, Frati G et al (2004) Isolation and expansion of adult cardiac stem cells from human and murine heart. Circ Res 95:911–921
86. Dawn B, Stein AB, Urbanek K et al (2005) Cardiac stem cells delivered intravascularly traverse the vessel barrier, regenerate infarcted myocardium, and improve cardiac function. Proc Natl Acad Sci U S A 102:3766–3771
87. Smith RR, Barile L, Cho HC et al (2007) Regenerative potential of cardiosphere-derived cells expanded from percutaneous endomyocardial biopsy specimens. Circulation 115:896–908
88. Banai S, Jaklitsch MT, Shou M et al (1994) Angiogenic-induced enhancement of collateral blood flow to ischemic myocardium by vascular endothelial growth factor in dogs. Circulation 89:2183–2189
89. Harada K, Friedman M, Lopez JJ et al (1996) Vascular endothelial growth factor administration in chronic myocardial ischemia. Am J Physiol 270:H1791–H1802
90. Pearlman JD, Hibberd MG, Chuang ML et al (1995) Magnetic resonance mapping demonstrates benefits of VEGF-induced myocardial angiogenesis. Nat Med 1:1085–1089
91. Lopez JJ, Laham RJ, Stamler A et al (1998) VEGF administration in chronic myocardial ischemia in pigs. Cardiovasc Res 40:272–281
92. Henry TD, Rocha-Singh K, Isner JM et al (2001) Intracoronary administration of recombinant human vascular endothelial growth factor to patients with coronary artery disease. Am Heart J 142:872–880
93. Losordo DW, Vale PR, Symes JF et al (1998) Gene therapy for myocardial angiogenesis: initial clinical results with direct myocardial injection of phVEGF165 as sole therapy for myocardial ischemia. Circulation 98:2800–2804
94. Rosengart TK, Lee LY, Patel SR et al (1999) Angiogenesis gene therapy: phase I assessment of direct intramyocardial administration of an adenovirus vector expressing VEGF121

cDNA to individuals with clinically significant severe coronary artery disease. Circulation 100:468–474
95. Symes JF, Losordo DW, Vale PR et al (1999) Gene therapy with vascular endothelial growth factor for inoperable coronary artery disease. Ann Thorac Surg 68:830–836; discussion 836–837
96. Vale PR, Losordo DW, Milliken CE et al (2000) Left ventricular electromechanical mapping to assess efficacy of phVEGF(165) gene transfer for therapeutic angiogenesis in chronic myocardial ischemia. Circulation 102:965–974
97. Vale PR, Losordo DW, Milliken CE et al (2001) Randomized, single-blind, placebo-controlled pilot study of catheter-based myocardial gene transfer for therapeutic angiogenesis using left ventricular electromechanical mapping in patients with chronic myocardial ischemia. Circulation 103:2138–2143
98. Hedman M, Hartikainen J, Syvanne M et al (2003) Safety and feasibility of catheter-based local intracoronary vascular endothelial growth factor gene transfer in the prevention of postangioplasty and in-stent restenosis and in the treatment of chronic myocardial ischemia: phase II results of the Kuopio Angiogenesis Trial (KAT). Circulation 107:2677–2683
99. Grines CL, Watkins MW, Helmer G et al (2002) Angiogenic Gene Therapy (AGENT) trial in patients with stable angina pectoris. Circulation 105:1291–1297
100. Grines C, Rubanyi GM, Kleiman NS et al (2003) Angiogenic gene therapy with adenovirus 5 fibroblast growth factor-4 (Ad5FGF-4): a new option for the treatment of coronary artery disease. Am J Cardiol 92:24N–31N
101. Grines CL, Watkins MW, Mahmarian JJ et al (2003) A randomized, double-blind, placebo-controlled trial of Ad5FGF-4 gene therapy and its effect on myocardial perfusion in patients with stable angina. J Am Coll Cardiol 42:1339–1347
102. Henry TD, Grines CL, Watkins MW et al (2007) Effects of Ad5FGF-4 in patients with angina: an analysis of pooled data from the AGENT-3 and AGENT-4 trials. J Am Coll Cardiol 50:1038–1046
103. Zbinden S, Zbinden R, Meier P et al (2005) Safety and efficacy of subcutaneous-only granulocyte-macrophage colony-stimulating factor for collateral growth promotion in patients with coronary artery disease. J Am Coll Cardiol 46:1636–1642
104. Ince H, Petzsch M, Kleine HD et al (2005) Prevention of left ventricular remodeling with granulocyte colony-stimulating factor after acute myocardial infarction: final 1-year results of the Front-Integrated Revascularization and Stem Cell Liberation in Evolving Acute Myocardial Infarction by Granulocyte Colony-Stimulating Factor (FIRSTLINE-AMI) trial. Circulation 112:I73–I80
105. Ince H, Petzsch M, Kleine HD et al (2005) Preservation from left ventricular remodeling by front-integrated revascularization and stem cell liberation in evolving acute myocardial infarction by use of granulocyte-colony-stimulating factor (FIRSTLINE-AMI). Circulation 112:3097–3106
106. Suarez de Lezo J, Torres A, Herrera I et al (2005) Effects of stem-cell mobilization with recombinant human granulocyte colony stimulating factor in patients with percutaneously revascularized acute anterior myocardial infarction. Rev Esp Cardiol 58:253–261
107. Valgimigli M, Rigolin GM, Cittanti C et al (2005) Use of granulocyte-colony stimulating factor during acute myocardial infarction to enhance bone marrow stem cell mobilization in humans: clinical and angiographic safety profile. Eur Heart J 26:1838–1845
108. Engelmann MG, Theiss HD, Hennig-Theiss C et al (2006) Autologous bone marrow stem cell mobilization induced by granulocyte colony-stimulating factor after subacute ST-segment elevation myocardial infarction undergoing late revascularization: final results from the G-CSF-STEMI (Granulocyte Colony-Stimulating Factor ST-Segment Elevation Myocardial Infarction) trial. J Am Coll Cardiol 48:1712–1721
109. Ripa RS, Jorgensen E, Wang Y et al (2006) Stem cell mobilization induced by subcutaneous granulocyte-colony stimulating factor to improve cardiac regeneration after acute ST-elevation myocardial infarction: result of the double-blind, randomized, placebo-controlled stem cells in myocardial infarction (STEMMI) trial. Circulation 113:1983–1992
110. Zohlnhofer D, Ott I, Mehilli J et al (2006) Stem cell mobilization by granulocyte colony-stimulating factor in patients with acute myocardial infarction: a randomized controlled trial. JAMA 295:1003–1010

111. Strauer BE, Brehm M, Zeus T et al (2002) Repair of infarcted myocardium by autologous intracoronary mononuclear bone marrow cell transplantation in humans. Circulation 106:1913–1918
112. Assmus B, Schachinger V, Teupe C et al (2002) Transplantation of progenitor cells and regeneration enhancement in acute myocardial infarction (TOPCARE-AMI). Circulation 106:3009–3017
113. Schachinger V, Assmus B, Britten MB et al (2004) Transplantation of progenitor cells and regeneration enhancement in acute myocardial infarction: final one-year results of the TOPCARE-AMI trial. J Am Coll Cardiol 44:1690–1699
114. Wollert KC, Meyer GP, Lotz J et al (2004) Intracoronary autologous bone-marrow cell transfer after myocardial infarction: the BOOST randomised controlled clinical trial. Lancet 364:141–148
115. Schachinger V, Erbs S, Elsasser A et al (2006) Intracoronary bone marrow-derived progenitor cells in acute myocardial infarction. N Engl J Med 355:1210–1221
116. Beitnes JO, Gjesdal O, Lunde K et al (2011) Left ventricular systolic and diastolic function improve after acute myocardial infarction treated with acute percutaneous coronary intervention, but are not influenced by intracoronary injection of autologous mononuclear bone marrow cells: a 3 year serial echocardiographic sub-study of the randomized-controlled ASTAMI study. Eur J Echocardiogr 12:98–106
117. Assmus B, Honold J, Schachinger V et al (2006) Transcoronary transplantation of progenitor cells after myocardial infarction. N Engl J Med 355:1222–1232
118. Perin EC, Willerson JT, Pepine CJ et al (2012) Effect of transendocardial delivery of autologous bone marrow mononuclear cells on functional capacity, left ventricular function, and perfusion in chronic heart failure: the FOCUS-CCTRN trial. JAMA 307:1717–1726
119. Traverse JH, Henry TD, Ellis SG et al (2011) Effect of intracoronary delivery of autologous bone marrow mononuclear cells 2 to 3 weeks following acute myocardial infarction on left ventricular function: the LateTIME randomized trial. JAMA 306:2110–2119
120. Kawamoto A, Iwasaki H, Kusano K et al (2006) CD34-positive cells exhibit increased potency and safety for therapeutic neovascularization after myocardial infarction compared with total mononuclear cells. Circulation 114:2163–2169
121. Kang HJ, Kim HS, Zhang SY et al (2004) Effects of intracoronary infusion of peripheral blood stem-cells mobilised with granulocyte-colony stimulating factor on left ventricular systolic function and restenosis after coronary stenting in myocardial infarction: the MAGIC cell randomised clinical trial. Lancet 363:751–756
122. Losordo DW, Schatz RA, White CJ et al (2007) Intramyocardial transplantation of autologous CD34+ stem cells for intractable angina: a phase I/IIa double-blind, randomized controlled trial. Circulation 115:3165–3172
123. Losordo DW, Henry TD, Davidson C et al (2011) Intramyocardial, autologous CD34+ cell therapy for refractory angina. Circ Res 109:428–436
124. Chen SL, Fang WW, Ye F et al (2004) Effect on left ventricular function of intracoronary transplantation of autologous bone marrow mesenchymal stem cell in patients with acute myocardial infarction. Am J Cardiol 94:92–95
125. Menasche P, Alfieri O, Janssens S et al (2008) The Myoblast Autologous Grafting in Ischemic Cardiomyopathy (MAGIC) trial: first randomized placebo-controlled study of myoblast transplantation. Circulation 117:1189–1200
126. Houtgraaf JH, den Dekker WK, van Dalen BM et al (2012) First experience in humans using adipose tissue-derived regenerative cells in the treatment of patients with ST-segment elevation myocardial infarction. J Am Coll Cardiol 59:539–540
127. Bolli R, Chugh AR, D'Amario D et al (2011) Cardiac stem cells in patients with ischaemic cardiomyopathy (SCIPIO): initial results of a randomised phase 1 trial. Lancet 378:1847–1857
128. Stewart DJ, Hilton JD, Arnold JM et al (2006) Angiogenic gene therapy in patients with nonrevascularizable ischemic heart disease: a phase 2 randomized, controlled trial of AdVEGF(121) (AdVEGF121) versus maximum medical treatment. Gene Ther 13:1503–1511

129. Gyongyosi M, Khorsand A, Zamini S et al (2005) NOGA-guided analysis of regional myocardial perfusion abnormalities treated with intramyocardial injections of plasmid encoding vascular endothelial growth factor A-165 in patients with chronic myocardial ischemia: subanalysis of the EUROINJECT-ONE multicenter double-blind randomized study. Circulation 112:I157–I165
130. Stewart DJ, Kutryk MJ, Fitchett D et al (2009) VEGF gene therapy fails to improve perfusion of ischemic myocardium in patients with advanced coronary disease: results of the NORTHERN trial. Mol Ther 17:1109–1115
131. Kastrup J, Jorgensen E, Fuchs S et al (2011) A randomised, double-blind, placebo-controlled, multicentre study of the safety and efficacy of BIOBYPASS (AdGVVEGF121.10NH) gene therapy in patients with refractory advanced coronary artery disease: the NOVA trial. EuroIntervention 6:813–818
132. Kukula K, Chojnowska L, Dabrowski M et al (2011) Intramyocardial plasmid-encoding human vascular endothelial growth factor A165/basic fibroblast growth factor therapy using percutaneous transcatheter approach in patients with refractory coronary artery disease (VIF-CAD). Am Heart J 161:581–589
133. Meyer GP, Wollert KC, Lotz J et al (2006) Intracoronary bone marrow cell transfer after myocardial infarction: eighteen months' follow-up data from the randomized, controlled BOOST (BOne marrOw transfer to enhance ST-elevation infarct regeneration) trial. Circulation 113:1287–1294
134. Lunde K, Solheim S, Aakhus S et al (2006) Intracoronary injection of mononuclear bone marrow cells in acute myocardial infarction. N Engl J Med 355:1199–1209
135. Tse HF, Thambar S, Kwong YL et al (2007) Prospective randomized trial of direct endomyocardial implantation of bone marrow cells for treatment of severe coronary artery diseases (PROTECT-CAD trial). Eur Heart J 28:2998–3005
136. Gyongyosi M, Lang I, Dettke M et al (2009) Combined delivery approach of bone marrow mononuclear stem cells early and late after myocardial infarction: the MYSTAR prospective, randomized study. Nat Clin Pract Cardiovasc Med 6:70–81
137. Tendera M, Wojakowski W, Ruzyllo W et al (2009) Intracoronary infusion of bone marrow-derived selected CD34+CXCR4+ cells and non-selected mononuclear cells in patients with acute STEMI and reduced left ventricular ejection fraction: results of randomized, multicentre Myocardial Regeneration by Intracoronary Infusion of Selected Population of Stem Cells in Acute Myocardial Infarction (REGENT) trial. Eur Heart J 30:1313–1321
138. Makkar RR, Smith RR, Cheng K et al (2012) Intracoronary cardiosphere-derived cells for heart regeneration after myocardial infarction (CADUCEUS): a prospective, randomised phase 1 trial. Lancet 379:895–904

Chapter 16
Adipose Tissue-Derived Mesenchymal Stem Cell and Angiogenesis in Ischemic Heart Disease

Lina Badimon, Blanca Oñate, and Gemma Vilahur

Abstract Acute myocardial infarction is one of the most important causes of death and disability worldwide. The limited capacity of the adult heart to self-regenerate and revascularize the ischemic damaged tissue leads to tissue loss, ventricular remodeling, and persistent deterioration in cardiac performance increasing the frequency of heart failure. Over the last several years, adult stem cells have appeared as one of the novel promising therapeutic approaches for the treatment of ischemic heart disease. However, the quest for the best cell type is still ongoing. This ideal cell type should be capable of differentiating into functional cardiomyocytes and of forming new vessels to nourish the damaged area. Recent studies have shown that adipose tissue contains multipotent stem cells (the so-called adipose tissue-derived stem cells or ASC) that are capable of regenerating injured myocardium by differentiating into cardiac resident cells or by secreting multiple angiogenic growth factors (paracrine effects). Moreover, due to ease of harvesting these cells in large numbers and low immunogenicity, white adipose tissue has become an attractive stem cell source. In this chapter, we review the principal characteristics of ASC as well as their capacity to repair cardiac damage in the setting of ischemic heart disease as compared with other adult stem cells, with special attention to their pro-angiogenic mechanisms of action.

L. Badimon (✉)
Cardiovascular Research Center, CSIC-ICCC, Hospital de la Santa Creu i Sant Pau, IIB-Sant Pau, c/Sant Antoni MªClaret 167, 08025 Barcelona, Spain

CIBEROBN-Pathophysiology of Obesity and Nutrition, Barcelona, Spain

Cardiovascular Research Chair, UAB, Barcelona, Spain
e-mail: lbadimon@csic-iccc.org

B. Oñate • G. Vilahur
Cardiovascular Research Center, CSIC-ICCC, Hospital de la Santa Creu i Sant Pau, IIB-Sant Pau, c/Sant Antoni MªClaret 167, 08025 Barcelona, Spain

CIBEROBN-Pathophysiology of Obesity and Nutrition, Barcelona, Spain

J.L. Mehta and N.S. Dhalla (eds.), *Biochemical Basis and Therapeutic Implications of Angiogenesis*, Advances in Biochemistry in Health and Disease 6,
DOI 10.1007/978-1-4614-5857-9_16, © Springer Science+Business Media New York 2013

Keywords Cell therapy • Adipose tissue-derived stem cells • Myocardial infarction • Angiogenesis • Cytokines

16.1 Introduction

Stem cells are unspecialized cells in the human body that, upon receiving a signal, are capable of becoming specialized cells with new specialized cell functions [1]. Very broadly they comprise two major classes: embryonic stem cells and adult stem cells. Embryonic stem cells exhibit strong proliferation and differentiation potential both in vitro and in vivo. However, several major issues, including ethical concerns, immune rejection, and teratoma formation, limit the practical use of these cells. In contrast, autologous adult stem cells do not come with any ethical, immunorejection, or oncological concern and can be isolated from numerous organs and tissues including the bone marrow, skeletal muscle, blood, epidermis, brain, liver, myocardium, perinatal organs, and, more recently, the adipose tissue.

Adipose tissue is a mesodermal tissue that contains several cell types, with the highest percentage of cells (80%) being adipocytes, cells specialized in storing energy as fat. Other cell types include endothelial and smooth muscle cells, leukocytes, mast cell precursors, cells with hematopoietic progenitor activity, and stem cells, all of which compose the stromal vascular fraction (SVF). After a simple harvest and collagenase digestion, adipose tissue segregates the mature adipocytes from the SVF by the floating ability of the formers. Upon adherent culture, the multipotent fraction of SVF is enriched and homogenized as cells with fibroblast-like morphology. Although a large amount of acronyms have been employed to name these cells, the *International Fat Applied Technology Society* standardized the nomenclature in 2004 by adopting the term adipose-derived stem cells (ASC) to identify them [2].

In this chapter, we review the principal characteristics of ASC as well as their capacity to repair cardiac damage in the setting of ischemic heart disease as compared with other adult stem cells, with special attention to their pro-angiogenic mechanisms of action.

16.2 Adipose Tissue-Derived Mesenchymal Stem Cells Characteristics

Few cell types have captivated so many biomedical researchers over the last 10 years as have adult mesenchymal stem cells (MSC) [3–5]. MSC are characterized by the cell surface expression of CD105 (endoglin, SH2), CD73 (ecto-5′-nucleotidase), and CD90 (Thy1) and the absence of the hematopoietic markers, CD45, CD19, CD19 or CD79, CD14 or CD11b, and HLA-DR. MSC readily adhere to cell culture surfaces, are easily expandable, and are capable (upon induction) to suffer in vitro differentiation to adipogenic, chondrogenic, and osteogenic cells [6].

MSC are also highly active metabolically, secreting not only components of the extracellular matrix but also a vast array of cytokines [7]. Recent work by Ranganath and coworkers has nicely and extensively documented the secretome and proteome of MSC [5]. Besides, MSC that present low immunogenicity appearing well suited for both autologous and allogeneic transplantation settings can be harvested from bone marrow (BM), perinatal organs (e.g., cord blood, placenta, amniotic fluid), skin, and white adipose tissue (WAT) [8–11]. As to the later, this mesodermal tissue stores excess energy in the form of fat. However, WAT not only serves as a reservoir tissue but also provides mechanical support and thermal insulation and works as an efficient endocrine organ, secreting molecules capable of regulating immune response, blood pressure, angiogenesis, or bone mass among others [12]. At the histological level, WAT is composed mostly of adipocytes and the SVF. As mentioned above, the SVF includes endothelial and smooth muscle cells, leukocytes, mast cell precursors, cells with hematopoietic progenitor activity, and ASC [2, 13].

For many years it was thought that the hyperplastic response of fat tissue was solely derived from a population of unipotent tissue progenitor cells, the so-called pre-adipocytes [14]. However, in 2001 Zuk et al. [15] reported that cells with potential beyond that of the adipocytic lineage were present within the adipose tissue. Using in vitro clonogenic assays, the investigators demonstrated the presence of cells capable of undergoing extensive self-renewal and multilineage differentiation, ASC.

ASC phenotypic characterization remains to be fully described. To date, the *Mesenchymal and Tissue Stem Cell Committee of the International Society for Cellular Therapy* has proposed a minimum set of four criteria to define MSC [6]:

1. These cells have to be plastic-adherent when maintained under human standard culture conditions.
2. These cells must have the ability for osteogenic, adipogenic, and chondrogenic in vitro differentiation.
3. These cells (>95%) must express CD73, CD90, and CD105.
4. These cells must lack (<2%) expression of the hematopoietic lineage markers c-kit, CD14, CD11b, CD34, CD45, CD19, CD79α, and human leukocyte antigen HLA-DR.

Given the similarities between ASC and bone-marrow mesenchymal stem cells (BM-MSC) and the good characterization of the later, comparisons between both stem cell populations are commonly used to describe ASC characteristics. Table 16.1 details the immunophenotypic characterization of ASC and compares it to that of BM-MSC. As observed, both cell types express CD105, CD13 (aminopeptidase), CD29 (β1 integrin), CD90, SH-3 (ecto-5′-nucleotidase: an epitope present on CD73), HLA-1 (human leukocyte antigen class I), and Sca-1 (stem cell antigen 1). Conversely, neither of the cells expresses the hematopoietic cell marker CD45 (leukocyte common antigen), the endothelial cell markers CD31 and CD144 (VE-cadherin), c-Kit (CD117), CD62P (P-selectin) nor HLA-DR [9, 13, 20, 21, 25, 27]. ADSC and BM-MSC differ in the expression of adhesion molecules involved in homing and mobilization of hematopoietic stem cells (HSC) including CD49d (α4 integrin), CD106 (component of VCAM-1), and CD54 (ICAM-1) [9, 19]. In other words, CD49d, which forms a heterodimer with CD29 to create very late activation

Table 16.1 Molecular phenotype of ADSC and BM-MSC

Markers	Other names	ADSC	BM-MSC	References
αSMA	Smooth muscle actin	+		[16, 17]
CD10	Neutral endopeptidase (NEP)	+	+	[16, 18]
CD104	β4 integrin	–	–	[9, 19]
CD105	Endoglin (SH2)	+	+	[9, 13, 16, 18–23]
CD106	Component of vascular cell adhesion molecule-1 (VCAM-1)	–	+	[9, 13, 19, 22, 23]
CD117	c-Kit	–	–	[16, 21, 24]
CD11b	αM integrin	–	–	[16, 24]
CD11c	αX integrin	–	–	[19]
CD13	Alanine aminopeptidase	+	+	[9, 16, 18–21, 25]
CD133	Prominin 1	–	–	[16, 21]
CD14	INFγ R	–	–	[9, 16, 19, 20, 25]
CD144	VE-cadherin	–	–	[16, 19]
CD146	Melanoma cell adhesion molecule (MCAM)	+	+	[16, 18]
CD15	Stage-specific embryonic antigen 1 (SSEA-1)	–	–	[19, 21]
CD16	Fc receptor for IgG	–	–	[9, 19]
CD166	Activated leukocyte cell adhesion molecule (ALCAM)	+	+	[16, 18, 19, 22]
CD19	B-lymphocyte surface antigen B4	–	–	[16, 19]
CD24	Heat stable antigen (HSA)	+	+	[17, 19]
CD29	β1 integrin	+	+	[9, 16–19]
CD3	T cell receptor (TCR)	–	–	[19]
CD31	PECAM-1	–	–	[9, 13, 16, 19, 20, 22–24]
CD33		–	–	[17, 19]
CD34		–	–	[9, 13, 16, 17, 21, 23, 24]
CD38		–	–	[19]
CD4	MHC-II coreceptor	–	–	[19]
CD44	PGP-1, hyaluronic acid/fibronectin receptor	+	+	[9, 13, 16, 18–21, 23, 24]
CD45	Leukocyte common antigen (LCA)	–	–	[9, 13, 16, 17, 19, 20, 23–25]
CD49d	α4 integrin	+	–	[9, 13, 16, 18, 19, 21, 23]
CD49e	α5 integrin	+	+	[16, 18]
CD54	Intracellular adhesion molecule-1 (ICAM-1)	+	Minimum levels	[16, 18, 19, 22]
CD55	Complement decay-accelerating factor (DAF)	+	+	[16, 18]
CD56	Neural cell adhesion molecule (NCAM)	–	–	[9, 19]
CD58		+	+	[19]

(continued)

Table 16.1 (continued)

Markers	Other names	ADSC	BM-MSC	References
CD59	Membrane attack complex inhibition factor (MACIF)	+	+	[16, 18]
CD61	β3 integrin	–	–	[9]
CD62E	E-selectin	–	–	[9]
CD62P	P-selectin	–	–	[19]
CD71	Transferrin receptor	+	+	[9, 17]
CD73	Ecto-5′-nucleotidase (SH3)	+	+	[9, 13, 16, 19, 20, 23]
CD79α	Immunoglobulin-associated α	–		[16]
CD80	B7.1	–		[16]
CD9		+	+	[16, 18]
CD90	Thymus cell antigen-1 (Thy-1)	+	+	[9, 13, 16, 17, 19–23]
gly-A	Glycophorin A	–		[21]
HLA-I	Human leukocyte antigen class I (A, B, C)	+	+	[16, 20, 22]
HLA-DR	Human leukocyte antigen class II (DR, DP, DQ)	–	–	[15, 20–22, 26]
Lin		–	–	[16, 24]
MyD88	Myeloid differentiation primary response gene (88)	–		[16]
Sca-1	Stem cell antigen 1, ly-6A/E	+	+	[24]
Stro-1		Low levels	+	[9, 13, 16, 19, 21, 23]
VEGFR2	Flk-1, KDR	–		[21]

+ present; – absent

antigen 4 (VLA-4), is expressed in ADSC but not in BM-MSC. On the contrary, CD106, which is a component of VCAM-1, a VLA-4 ligand, is expressed in BM-MSC and not in ASC. Finally, CD54 (ICAM-1), an adhesion molecule involved in hematopoietic cell traffic across the endothelium, is expressed in ASC but not in BM-MSC [19].

ASC CD34 expression, a marker of hematopoietic and progenitor stem cells, remains controversial [28, 29]. While some authors detect high levels of CD34 expression in ASC [22, 30, 31], others do not [9, 21]. Alternatively, some authors describe that CD34+ cells are only present in freshly isolated ASC and CD34 expression notably decreases after 1-week culture. The same pattern occurs in other hematopoietic lineage and endothelial cell markers such as CD45, CD144, and CD14 [20, 24, 32]. Overall, these results suggest that ASC might not have the capability to differentiate into endothelial progenitor cells (EPC) or mature endothelial cells.

So far, no differences in morphology, immune phenotype, and plastic adherence have been observed between ASC and BM-MSC [13, 20, 33]. ASC also present a normal diploid karyotype, display extensive self-renewal capacity in vitro, and can be cryopreserved and expanded easily [21]. Moreover, ASC have shown to possess

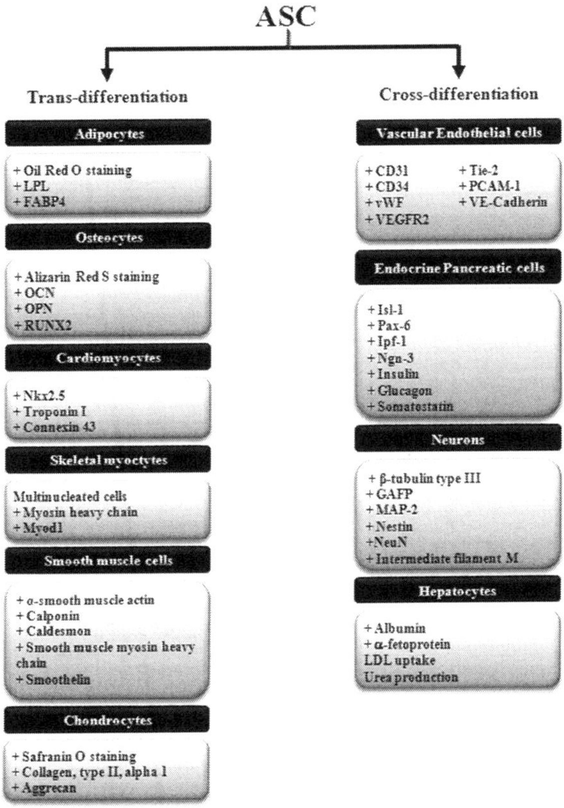

Fig. 16.1 Adipose-derived stem cells potential differentiation. ASC can differentiate into cells from the same germ layer, mesoderm, termed "transdifferentiation." Additionally, ASC can also differentiate into cells from non-mesodermal origin, named "cross-differentiation." Each differentiated cell acquires its characteristic phenotype which can be detected using different markers. *FABP4* fatty acid binding protein 4; *GAFP* glial acidic fibrillar protein; *Ipf-1* insulin promoter factor 1; *Isl-1* insulin gene enhancer protein; *LDL* low-density lipoprotein; *LPL* lipoprotein lipase; *MAP-2* microtubule-associated protein 2; *Myod1* myogenic differentiation 1; *NeuN* neuronal nuclei; *Ngn-3* neurogenin 3; *Nkx2.5* NK2 homeobox 5; *OCN* osteocalcin; *OPN* osteopontin; *Pax-6* paired box protein; *PECAM-1* platelet endothelial cell adhesion molecule 1; *RUNX2* Runt-related transcription factor 2; *VEGFR2* vascular endothelial growth factor receptor 2; *vWF* von Willebrand factor

an osteogenic, chondrogenic, and adipogenic differentiation capacity. This plasticity process is known as "transdifferentiation" (i.e., lineage-specific differentiation) [34]. On the other hand, ASC plasticity is not just confined to differentiate into cells from the same germ layer but can also differentiate into cells from non-mesodermal origin such as neurons, endocrine pancreatic cells, hepatocytes, and endothelial cells in a process known as "cross-differentiation" (Figs. 16.1 and 16.2) [16].

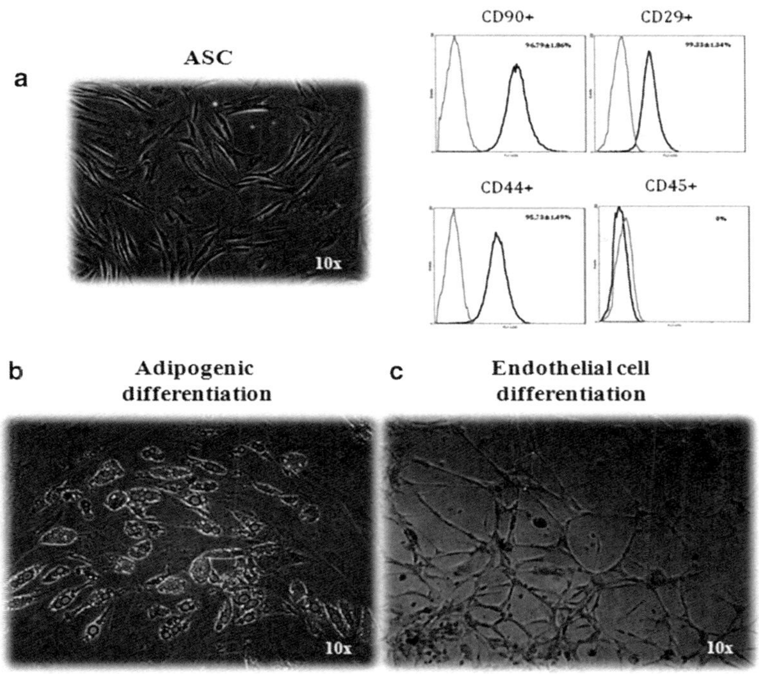

Fig. 16.2 Non-differentiated and differentiated ASC. (**a**) Typical image of cultured ASC in non-differentiating standard medium (*left*) and flow cytometric characterization of ASC (*right*). (**b**) Adipogenic differentiation of ASC after 15 days culture in specific supplemented medium. (**c**) Tubular structures after culturing ASC on Matrigel for a total of 7 days (data from original experiments by the authors)

Finally, Class I and II major histocompatibility complexes MHC are expressed in only 1% of ASC, and therefore it has been hypothesized that these could behave as universal donor cells and could be used for autologous and allogeneic transplantation because of their low immunogenic potential [35].

16.3 ASC Isolation

Compared with BM, adipose tissue is abundant in most individuals and can be harvested using a less invasive method such as liposuction [13, 36–38]. Indeed, traditional BM isolation procedures are painful, frequently requiring general or spinal anesthesia, thereby limiting the amount of marrow that can be obtained. In contrast, adipose tissue can be easily obtained with minimally invasive procedures. Adipose tissue also offers a significantly higher stem cell density than does bone marrow (5% vs. 0.01%) [13]. Moreover, low BM-MSC numbers are usually harvested (approximately 1 BM-MSC per 10^5 adherent stromal cells) requiring ex vivo cell expansion during long periods to obtain therapeutic cell doses. In contrast, adipose tissue can be largely obtained avoiding ex vivo expansion in order to obtain a clinically relevant therapeutic amount and, if required, ASC expansion occurs more rapidly (ten times within 1 week) without modifying ASC pluripotency potential [36, 37]. All these favorable features make adipose tissue an attractive stem cell source for tissue repair [39].

SVF isolation from the adipose tissue is based on a multistep laboratory process for which several protocols have been proposed [40]. Yet, the majority of laboratories use collagenases that generate SVF (cell pellet) following centrifugation [9, 41]. One of the most significant issues limiting the interpretation of results and clinical progression relies that some authors use the entire unpurified SVF in their experiments [42] whereas others use isolated/purified ASC. ASC can be isolated from the SVF through culturing on plastic dish or by means of fluorescence-activated cell sorting for their specific surface markers (e.g., CD73+, CD90+, CD105+, CD45−, CD34−, CD79−). Therefore, ASC standardization is required to allow direct comparison of results across all research groups and better define the clinical potential for ASC vs. any of the other cell populations present in the SVF [39].

16.4 Adult Stem Cells in Tissue Repair Following Myocardial Infarction

Despite recent advances in medical therapy and interventional techniques, ischemic heart disease remains one of the major causes of morbidity and mortality in developed countries [43, 44]. Because of the limited potential of native cardiomyocytes to regenerate injured mature heart tissue and/or to restore cardiac vascular network, heart transplantation is the only effective treatment of end-stage ischemic heart disease. Unfortunately, the insufficient number of available donor hearts and immunorejection limit the use of transplantation. Thus, new therapeutic approaches aimed at stimulating cardiac repair are a primary focus of attention. Cardiomyoplasty, a process to replace or regenerate cardiomyocytes through cell transplantation, has been an area of intense research during the last years. Several types of adult stem cells have been used for cardiovascular cell-based therapies such as bone marrow-derived cells, skeletal myoblasts, resident cardiac stem cells, stem cells from perinatal organs/tissues, and, last but not least, ASC (Fig. 16.3) [45–48]. Following, we briefly summarize the most important advantages and disadvantages associated to the use of the different adult stem cells sources.

16.4.1 Bone Marrow-Derived Cells

BM-derived cells have been extensively investigated because of the diversity of stem cell populations that can be harvested including HSC, EPC, and BM-MSC. However, the potential disadvantage of delivering a mixture of cells within the injured heart is

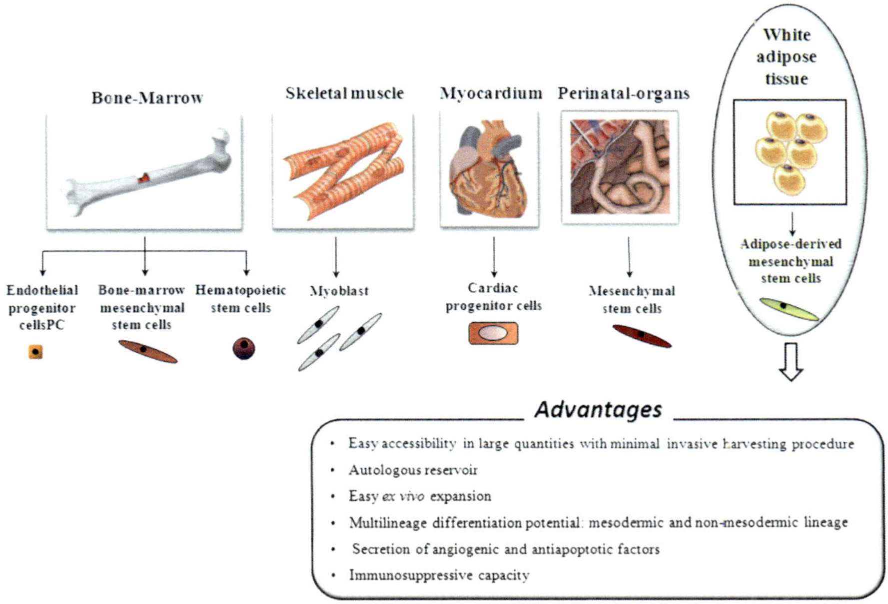

Fig. 16.3 Adult stem cell sources most commonly used for in vivo experimental assays and ASC advantages

that the eventual percentage of cells that are therapeutically useful may be small. To overcome this limitation, stem cell markers have proved useful in order to isolate a pure population from the unfractionated bone marrow population [49, 50].

1. HSC are the best characterized adult stem cell population expressing CD34 surface antigen. HSC are able to self-renew and differentiate into different precursors of specialized hematopoietic cells. It is still a matter of debate whether HSC can differentiate into cardiomyocytes. Indeed, a pioneering study in mice suggested HSC potential to differentiate into cardiomyocytes [50], whereas subsequent studies with genetically labeled cells showed no such differentiation [51, 52]. Despite all this, HSC are the only stem cell population being routinely used in the clinical setting.
2. EPC are a specialized subset of hematopoietic cells found in the bone marrow and peripheral circulation that express the cell surface marker CD133 [29]. EPC are mobilized from bone marrow and recruited to foci of neovascularization where they form neovessels in situ [53]. Although they have not been shown to differentiate into cardiomyocytes, EPC appear to promote angiogenesis [54] and likely provide paracrine survival signals to cardiomyocytes [55]. Despite such great potential to promote angiogenesis, it is technically challenging to isolate EPC in significant quantities.

3. BM-MSC represents a rare population of bone marrow-derived cells (0.01–0.001%) that, as already mentioned, neither express CD34 nor CD133 in their surface. In fact, these cells lack specific surface markers and as detailed in Table 16.1 have an immunophenotype positive for many adhesion proteins. BM-MSC, besides differentiating into bone, cartilage, and adipocytes, under certain in vitro conditions may differentiate into cardiac myocytes [56]. Clones of these cells can easily be expanded in vitro, exhibit relatively low immunogenicity, and might be particularly useful when autologous stem cells are not readily available [57].

16.4.2 Skeletal Myoblasts

Skeletal myoblasts are resident satellite stem cells in the muscle. These were the first cell type tested in the context of cardiac regeneration because of their resistance to ischemia-induced apoptosis [58] and their ability to regenerate after cardiac injury [59, 60]. Skeletal myoblasts were firstly thought to be an ideal stem cell source for cell-based therapies because of their in vitro expansion potential, and in vivo capacity to differentiate into multinucleated myotubes avoiding the risk of tumor formation. Indeed, myotubes exhibit skeletal muscle phenotype after transplantation leading to improvements in left ventricular systolic and diastolic function [61]. However, their inability to differentiate into cardiomyocytes or to integrate electrically with surviving cardiomyocytes accounts for the greater risk of arrhythmias after skeletal myoblasts transplantation [62]. Additionally, these cells are not able to extravasate and migrate to ischemic areas; this eventually leads to occlusion of the distal microcirculation after intracoronary administration resulting in additional myocardial injury [63, 64]. Furthermore, to obtain a sufficient number of cells, myoblasts need to be cultured and expanded for approximately 3 weeks before transplantation which is an important limitation in acute interventions [65]. This approach is not used presently after failure of this therapy in a major clinical trial [66].

16.4.3 Resident Cardiac Stem Cells

Resident cardiac stem cells obtained from cardiac biopsies have the capability to differentiate into cardiomyocytes or vascular lineages [67]. These features make these cells a suitable source for cardiac tissue repair [68]. However, their limited number and further requirement of ex vivo separation and expansion over several weeks before implantation represent important drawbacks for their routine use [69].

In summary, among adult stem cell types, BM-MSC seems to be highly advantageous for cellular therapy accounting to its inherent characteristics (multipotent capabilities, rapid proliferation, angiogenesis induction, and myogenic differentiation) [70, 71]. However, the fact that ASC have shown to possess many of the same characteristics/qualities of bone BM-MSC while offering more easy/less invasive obtention, higher harvest and yield parameters make ASC an ideal stem cell approach for cardiac repair [13, 36, 37].

16.5 ASC Reparative Potential in the Setting of Ischemic Heart Disease

Several experimental studies have supported the use of ASC for many purposes in regenerative medicine (neuronal, adipogenic, skeletal myogenic. hematopoietic, chondrogenic and osteogenic, hepatic, cardiac, and vascular regeneration), immunomodulation (e.g., graft vs. host disease), to obtain hormone-secreting cells, or even in plastic and reconstructive surgery. Over the last several years, a number of animal studies have explored the use, behavior, and effect of ASC in the setting of ischemic heart disease [56, 57, 72, 73]. Table 16.2 shows the studies performed so far in rabbits and rodents, and Table 16.3 shows those performed in swine, a more human-like animal model. Overall, the studies have shown the in vivo capability of ASC either to differentiate into different cardiac resident cells or to exert paracrine effects by secreting different growth factors with important pro-angiogenic and anti-apoptotic properties, eventually leading to left ventricular functional improvement and infarct size reduction with the absence of potentially fatal arrhythmias.

16.5.1 Target-Cell Differentiation of ASC

As depicted in Fig. 16.1, ASC may give rise to cell types of different lineages including vascular endothelial cells, smooth muscle cells, and cardiomyocytes (i.e., cardiac resident cells).

The first study in the setting of acute myocardial infarction investigated the effect of freshly isolated ASC delivered into the left ventricular chamber (an approximation of intracoronary delivery) in a cryoinjury-related myocardial damage rat model [74]. This study demonstrated, for the first time, myocardial engraftment of ASC with expression of cardiac-specific markers. In addition, functional and pathological assessments revealed significant improvements in global ejection fraction and a marked increase in capillary density in the myocardial border area of treated animals compared with controls [75]. Since then, the potential to generate cardiomyocytes and vascular cells (endothelial cells and smooth muscle cells) from ASC became a focus of attention. However, subsequent studies reported conflicting data; some showed engrafted ASC expression of specific cardiac markers (troponin I and myosin light chain 2) [39, 74, 113–115] and/or von Willebrand factor and smooth muscle actin expression whereas others showed the absence of such ASC differentiation capacity (Tables 16.2 and 16.3) [77, 111]. Such discrepancies in ASC in vivo differentiation potential may have resulted from different tissue

Table 16.2 Studies performed in rodents and rabbits to assess the efficacy and safety of ASC in the setting of acute myocardial infarction

References	Cell source	Animal model	Injury	Outcome
Strem et al. [74, 75]	Subcutaneous ASC from mice	Mice	Myocardial cryoinjury	Engrafted ASC expresses specific cardiomyocytes markers
Miyahara et al. [17]	Subcutaneous ASC sheets from rat	Rats	MI by ligation of the left coronary	Improved scar and cardiac function
Mazo et al. [76]	Subcutaneous ASC, AD-CMG, or BM-MNC from GFP mice	SD rats	Left coronary artery ligation	ASC improved heart function, tissue viability, and increased angiogenesis and decreased fibrosis
Cai et al. [77]	Human subcutaneous ASC	Athymic rats	MI by LAD ligation	ASC improved heart function, tissue viability, and increased angiogenesis and decreased fibrosis
Léobon et al. [78]	Subcutaneous AD-CMG	Mice	AMI following AD-CMG injection	At 4 weeks, AD-CMG-treated group showed a reduction of remodeling and stability of LVEF and a stimulation of angiogenesis in peri-infarct areas
Schenke-Layland et al. [79]	Subcutaneous ASC from rats	Rats	AMI by LAD occlusion following reperfusion	Improvement in cardiac performance despite low rate of absolute engraftment
van der Bogt et al. [80]	Subcutaneous ASC and BM-MSC from mice	FVB mice	MI	No amelioration detected. Higher apoptosis
Wang et al. [81]	Subcutaneous ASC from rats	Rats	MI by occlusion of the LAD	At 1 month, greater LVEF, wall thickening, and a higher capillary density. Only 0.5% of the engrafted ASC were positive for cardiac-specific markers
Zhu et al. [82]	Human subcutaneous ASC expressing hepatocyte growth factor	SD rats	AMI	ASC improved heart function and decreased fibrosis
Bai et al. [83]	Human subcutaneous ASC	SCID mice	MI	Improvement in myocardial performance, cardiomyogenic differentiation, and higher angiogenesis
Bayes-Genis et al. [84]	Human ASC from cardiac adipose tissue	Mouse and rats	MI	Improvement in myocardial performance, cardiomyogenic differentiation, and higher vasculogenesis
Danoviz et al. [85]	Rat subcutaneous ASC using fibrin-a, collagen, or culture medium as vehicle	Rats	MI	Mitigation of the negative cardiac remodeling process

Hwangbo et al. [149]	Human ASC	SD rats	MI by permanent ligation of the LAD	At 4 weeks, improved cardiac performance, cardiomyogenic differentiation, and higher capillary density
Lin et al. [86]	Combined ASC from rat adipose tissue surrounding the epididymis and sildenafil therapy	Lewis rats	Dilated cardiomyopathy	Lower apoptosis and fibrosis, improved cardiac performance, and higher angiogenesis
Okura et al. [87]	Human omental ASC differentiated into cardiomyoblast-like cells	Nude rats	MI	Improvement in myocardial performance and cardiomyogenic differentiation
Zhang et al. [88]	Rat subcutaneous ASC co-injected with fibrin glue	Rats	MI	Fibrin + ASC produced higher cell retention and 4-week graft size than ASC group alone. Heart function and arteriole densities within the infarcted area improved in fibrin + ASC compared to ASC group
Bai et al. [89]	Human subcutaneous ASC	Immuno-deficient mice	MI by permanent ligation of the LAD	Engrafted ASC, 3.5% differentiated into cardiomyocytes or endothelial cells
Berardi et al. [90]	Human ASC treated with SNAP	Rats	MI	SNAP-treated ASC improved the cardiac performance compared to ASC untreated cells and was associated with increased expression of troponin T and von Willebrand factor
Cai et al. [91]	Rat subcutaneous ASC cocultured with cardiomyocytes	Rats	AMI	Pre-specification of ASC was beneficial for enhancing ASC's therapeutic efficacy on cardiac function after AMI
Gaebel et al. [92]	Human BM-MSC, ASC, and MSC from umbilical cord blood	SCID mice	LAD ligation	Human MSC originating from different sources showed a different healing performance in cardiac regeneration, and CD105(+) MSC exhibited a favorable survival pattern in infarcted hearts
Hamdi et al. [93]	Trilayered ASC sheet from rat subcutaneous adipose tissue	Rats	Coronary ligation	Cell sheet-treated rats survive better than cell-injected, were absent of LV remodeling, improved end-diastolic volume, and showed greater cell engraftment

(continued)

Table 16.2 (continued)

References	Cell source	Animal model	Injury	Outcome
Ii et al. [94]	Human ASC	Nude rats	MI	Improved cardiac performance and increased capillary density. Cardiac or vascular lineage transdifferentiation was absent
Paul et al. [95]	Human ASC and BM-MSC	Rats	AMI	Improved cardiac function and reduced infarction vs. BM-MSC and untreated control groups
van Dijk et al. [96]	Freshly isolated SVF cells of adipose tissue and cultured ASC from rat subcutaneous adipose tissue	Rats	AMI	Intravenous delivery of both SVF cells and ASC significantly reduced infarct size when infused at day 7 post-MI (not at day 1 post-MI)
Bagno et al. [97]	Rat subcutaneous ASC + Matrigel	Wistar rats		At 6 weeks, cardiac function improvement and reduced scar size
Otto Beitnes et al. [98]	Human subcutaneous ASC and MSC from skeletal muscle	Nude rats	MI	Cardiac function improvement and reduced scar size
Fang et al. [99]	Human amniotic epithelial cells, cord blood-derived MSC, and subcutaneous ASC	Athymic nude rats	MI	Cardiac function improvement and reduced scar size comparable to that of cord blood-derived MSC and ASC
Hoke et al. [100]	Human epicardial ASC preconditioning by inhibition of phosphodiesterase-5	CD-1 mice	MI	Preconditioned ASC-treated hearts showed consistently superior cardiac function when compared with nonpreconditioned ASC: reduced fibrosis, increased vascular density, decreased resident myocyte apoptosis and VEGF, bFGF, and Ang-1 were elevated
Li et al. [101]	Human CDC, BM-MSC, ASC, and BM-MNC	Mice	MI	At 3 weeks, CDC resulted in superior improvement of cardiac function, highest cell engraftment, and myogenic differentiation rates
Liu et al. [102]	Rat subcutaneous ASC in chitosan hydrogel	SD rats	MI by LAD ligation	ASC in chitosan hydrogel enhance stem cell engraftment, survival, and homing

Paul et al. [103]	Human ASC delivered in fluorogenic genipin-crosslinked alginate chitosan microcapsules	Lewis rats	MI by occlusion of the LAD	Microencapsulated ASC showed a greater implant retention, attenuation of infarct size, enhanced vasculogenesis, and improvement of cardiac function compared to free ASC
Paul et al. [103]	Human ASC expressing Ang-1	Rats	AMI	The transplanted hASC-Ang1 group showed higher cell retention, capillary density, reduction in infarct size, and higher cardiac performance compared to animals treated with ASC or the untreated group
Shi et al. [104]	Rat subcutaneous ASC overexpressing eNOS	Rats	AMI	Transplantation of ASC with eNOS embedded reduced infarcted size compared with normal ASC
Zhang et al. [105]	Rabbit ASC	New Zealand white rabbits	MI by ligation of left anterior descending coronary artery	Improvement in myocardial performance in ASC-implanted hearts relative to the control hearts
Yang et al. [106]	Rabbit subcutaneous HO-1-ASC	Rabbits	MI	HO-1-ASC-treated animals showed improved cardiac functions, LV dimension, and cardiomyogenic and angiogenic differentiation

AD-CMG ASC differentiated to cardiomyogenic cells; *AMI* acute myocardial infarction; *Ang-1* angiopoietin-1; *ASC* adipose-derived stem cells; *bFGF* basic fibroblast growth factor; *BM-MNC* bone marrow-derived mononuclear cells; *BM-MSC* bone marrow-derived mesenchymal stem cells; *CDC* cardiosphere-derived cells; *eNOS* endothelial nitric oxide syntheses; *GFP* green fluorescence protein; *HO-1-ASC* ASC transduced with heme oxygenase-1; *LAD* left anterior descending coronary artery; *LV* left ventricle; *LVEF* left ventricular ejection fraction; *MI* myocardial infarction; *MSC* mesenchymal stem cells; *SCID* severe combined immunodeficiency; *SD* Sprague Dawley; *SNAP* *S*-nitroso-*N*-acetyl-d,l-penicillamine; *SVF* stroma vascular fraction; *VEGF* vascular endothelial growth factor sources, procedures for isolating ASC, culture media, animal models, delivery routes, or limits of histological analysis. Whatsoever, various independent groups have commonly observed a low ASC in vivo differentiation efficiency [77, 83, 111]. In fact, these observations have raised the question if the benefits observed after ASC transplantation are directly related to stem cell transdifferentiation or to ASC paracrine-related effects [34, 116–118].

16.5.2 Therapeutic Angiogenic Potential of ASC

Judah Folkman, one of the pioneers of angiogenesis research, defined the concept of therapeutic angiogenesis a strategy to treat tissue ischemia by promoting vessels proliferation in the early 1970s [119]. After the identification of angiogenic growth factors including vascular endothelial growth factor (VEGF) and basic fibroblast growth factor (bFGF), investigators in the cardiovascular field started testing the hypothesis that facilitating angiogenesis with growth factor genes and/or cytokines would improve tissue perfusion and function in severe ischemic heart disease [120, 121]. A number of preclinical studies using recombinant proteins or by gene transfer supported the safety and feasibility of therapeutic angiogenesis. These studies supported the development of several clinical trials [122, 123]. Initial uncontrolled small clinical trials showed salutary effects of this approach; however, subsequent larger randomized placebo-controlled trials failed to show consistent and significant improvement of angiogenesis, cardiac function, or patient symptomatology [122, 124–126]. These observations suggested that the clinical use of therapeutic angiogenesis using single factors may have limited efficacy. In fact, it has been postulated that the life span of cytokines, their small size—which restricts their retention within a tissue for prolonged periods—and difficulties in their administration may have contributed to such unexpected results. With this in mind, delivery/administration of stem cells capable of directly inducing vessel formation or secrete pro-angiogenic factors appeared as a promising therapeutic approach for ischemic tissue repair. In this regard, EPCs have shown to strongly contribute to neovascularization by directly participating in blood vessel formation [29]. Indeed, mobilized bone marrow EPCs circulate in the peripheral blood and neovascularize and re-endothelialize injured tissues [127]. However, although transplantation of culture expanded EPCs successfully augmented angiogenesis and tissue blood supply in experimental animal models with ischemia, this approach has not developed into clinical trials because of difficulties in culture expansion techniques, and limited number of cells available for implant [128].

On the other hand, MSC, either isolated from bone marrow or adipose tissue, were also found to induce angiogenesis although through paracrine secretion of angiogenic growth factors [77, 118]. Although previous studies failed to show the differentiation of ASC into endothelial cell lineage, ASC have shown to secrete a large number of angiogenesis-related cytokines and cytoprotective growth factors including VEGF, hepatocyte growth factor (HGF), insulin-like growth factor 1

(IGF-1), placental growth factor (PGF), transforming growth factor-β (TGFβ), granulocyte/macrophage growth factor (GM-CSF), monocyte chemotactic protein-1 (MCP-1), and stromal-derived factor-1α (SDF-1α) and moderately express fibroblast growth factor (FGF) and angiopoietin-1 (Ang-1) [27, 32, 129]. In contrast ASC express low levels of angiopoietin-2 (Ang-2) [27, 32, 129–133]. Thus, it seems that ASC may be useful for therapeutic angiogenesis and apoptosis inhibition through a growth factor-dependent pathway [24, 130–134].

In vitro human studies with ASC have demonstrated the capability of ASC to secrete both VEGF and IGF-I at bioactive levels able to prevent cardiomyocytes apoptosis and promote endothelial cells tube formation [134]. Further, ASC-secreted IGF-I markedly diminished apoptosis execution whereas VEGF was capable of preventing apoptosis (although to a lesser extent) and promoting endothelial cell tube formation. Moreover, an in vivo study demonstrated that intramyocardial-injected human ASC significantly promoted angiogenesis and inhibited apoptosis of cells in the injured hearts 4 weeks after cell injection [94]. In this study, vWF-positive blood vessel counts revealed that the vascular densities were higher in the peri-infarct regions of cell-injected myocardium than in PBS-injected myocardium. Additionally, significantly smaller percentages of apoptotic cells were observed in the peri-infarct regions of cell-injected hearts than in PBS-injected hearts. Interestingly, ASC showed an upregulation and co-expression of VEGF, bFGF, and SDF-1α [94]. SDF-1α/CXCR4/SDF-1α receptor interaction has shown to induce a rapid and effective mobilization of stem/progenitor cells from bone marrow [135] and to be essential for organ system vascularization [136]. Altogether, these data suggest a cooperative effect of transplanted ASC with endogenous BM-derived stem/progenitors via an SDF-1α-dependent cell homing mechanism likely promoting angio/vasculogenesis in the ischemic myocardium [94].

The amount of oxygen levels has also shown to affect ASC functional response [32, 137]. Interestingly, whereas severe ischemia/hypoxia has shown to induce degenerative changes in adipose tissue with subsequent adipocytes death, it has conversely shown to trigger ASC activation [138]. Rehman et al. [32] found that VEGF secretion in ASC increased fivefolds when cultured under hypoxic conditions. Conditioned media obtained from hypoxic ASC significantly increased endothelial cell growth and reduced endothelial cell apoptosis, suggesting that ASC could respond to ischemia and promote angiogenesis by modulating VEGF production. In addition, based on these positive effects of hypoxia on ASC potential, a recent study has demonstrated that preconditioning of ASC with hypoxia increased cell survival and ASC paracrine effects in vitro by promoting endothelial cell survival and pro-angiogenic responses via the hypoxia-inducible factor 1, α subunit (HIF-1α) downstream target gene VEGF [139].

As already stated, ASC safety and efficacy has been assessed in multiple preclinical MI-experimental animal models (Table 16.3). As shown in Table 16.3, so far, two studies have supported higher capillarity density in the border infarcted cardiac zone and heart function improvement in ASC-treated animals 1 month after myocardial infarction, similar to that detected after BM-MSC administration [109, 110]. A recent longer-term follow-up study showed that, despite the lack of cell

Table 16.3 Studies performed in swine to assess the efficacy and safety of ASC in the setting of acute myocardial infarction

References	Cell source	Animal model	Injury	Outcome
Watanabe et al. [107]	ASC	Swine	AMI through LAD balloon occlusion	At 6 months, 3% increase in LVEF in ASC-treated pigs vs. 9% decline in controls
Fotuhi et al. [108]	Subcutaneous ASC from swine	Swine	MI by ligation of the left coronary	Lower arrhythmogenesis in ASC-treated animals
Valina et al. [109]	Subcutaneous ASC or BM-MSC from swine	Swine	AMI induced by an angioplasty balloon in the mid-left anterior descending artery	At 4 weeks ASC improve LVEF after MI associated with greater capillary density and wall thickness
Alt et al. [110]	Swine subcutaneous uncultured ASC	Swine	AMI by LAD occlusion following reperfusion	Higher perfusion, myocardial salvage, LVEF, and capillary density compared to control animals
Rigol et al. [111]	Swine subcutaneous ASC via intracoronary or transendo-cardial administration	Swine	MI	Intracoronary administration was more effective in increasing the number of small vessels than transendocardial administration. Ejection fraction was not modified
Mazo et al. [112]	ASC from pigs	Gottingen minipigs	Ischemia/reperfusion	Despite the lack of cell engraftment at 3 months there was an improvement in cardiac function and geometry, increase angiogenesis and vasculogenesis, and decrease fibrosis and cardiac hypertrophy
Yang et al. [106]	Human ASC immobilized in agarose gel patches	Swine	Cryoinjury infarction	At 4 weeks, better perfusion, lower infarction, and enhanced kinetics

MI acute myocardial infarction; *ASC* adipose-derived stem cells; *BM-MSC* bone marrow-derived mesenchymal stem cells; *LAD* left anterior descending coronary artery; *LVEF* left ventricular ejection fraction; *MI* myocardial infarction engraftment after 3 months, ASC transplantation is associated with a marked improvement in cardiac function, positive remodeling, and an increase in angiogenesis and vasculogenesis, further supporting long-term ASC-related paracrine effects [112]. However, these observations also suggest that, despite ASC several potential advantages, low engraftment in the ischemic tissue may present an obstacle in the clinical setting [140]. Therefore, several strategies for improving ASC survival and engraftment are being developed including ASC administration in combination with growth factor delivery [141], genetic modification of ASC [142], and/or the use of tissue engineering scaffolds [143, 144].

Finally, it needs to be acknowledged that several reports have suggested that ASC are able, both in vitro and in vivo, to increase vascularity by differentiation into pericytes, cells capable of stabilizing the microvasculature, preventing vascular regression, and promoting long-term microvessel maintenance [145, 146].

16.6 Clinical Use of ASC

The evidence that stem cells may reconstitute ischemic myocardium and improve cardiac function in animal models (Tables 16.2 and 16.3) has led to the initiation of clinical trials. However, moving towards the clinical arena is currently a matter of hot debate. Whereas some basic scientists and clinicians believe that within the next years human stem cells will be used to replace dead or dying cells within organs, others argue that more information is needed before assessing their therapeutic potential in humans.

Up to date, ASC use in humans has been successfully employed in the treatment of a variety of diseases including the treatment of Crohn's disease fistulas, osteogenesis imperfecta, and breast augmentation and reconstruction after partial mastectomy. However, ASC use for the treatment of ischemic heart disease is still in a phase I–II clinical trials. Other ongoing trials are exploring the safety, feasibility, and efficacy of freshly isolated ASC in patients who have suffered acute myocardial infarction (APOLLO and ADI-ME-CHF-002), chronic myocardial ischemia (MyStromalCell trial), non-revascularizable ischemia (PRECISE trial), and non-ischemic congestive heart failure (ADI-ME-CHF-001 trial).

16.7 Conclusions

The last few years have witnessed growing interest in regenerative therapy for the failing heart by cell transplantation. Among all stem cells, adult stem cells are an attractive source of cells for therapy, especially in view of the recent claims that they are remarkably plastic in their developmental potential when exposed to new environments. Although initial studies with skeletal myoblasts are more than 10 years old, the potential of other stem cells has led to a flurry of experimental studies generating generally positive but occasionally contradictory results. Recently it had been shown that adipose tissue, in addition to committed adipogenic, endothelial progenitor cells and pluripotent vascular progenitor cells, also contains the multipotent cell types, ASC, that have extensive proliferative capacity, and are able to differentiate into several lineages (mesenchymal or not) and secrete several factors with pro-angiogenic and anti-apoptotic effects. These observations have generated much interest because adipose tissue can be easily and safely harvested in large quantities and with minimal morbidity making it an appealing source for cell therapy. The true in vivo differentiation capacity of ASC and their possible contribution to cardiac tissue regeneration remain controversial. Although the efficacy of autologous ASC implantation for therapeutic angiogenesis has been established in animal studies, it is still unknown how many cells should be used for optimal angiogenesis. In addition, optimal cell isolation method remains to be determined. Finally, it needs to be determined if the number or function of ASC may be affected by patients' concomitant conditions or atherogenic risk factors. Previous studies demonstrated that ASC isolated from older animals have impaired proliferation and angiogenic potential [147]. We have recently provided evidence that the presence of obesity, a cardiovascular disease risk factor, affects properties of ASC; these cells exhibit a reduced differentiation and angiogenic potential as compared to ASC from nonobese metabolically normal individuals [148]. These observations suggest that subcutaneous fat depots from obese patients, who are likely to suffer a cardiovascular event, might present autologous ASC with impaired therapeutic capacity. Further studies in animal models with clinical translation are warranted to determine the impact of cardiovascular risk factors on ASC functional capability. Indeed, results of ongoing human trials will provide an understanding of the real safety and efficacy or these cells in patients with ischemic heart disease, giving new insights into the possible role of ASC in the treatment of cardiovascular disease. However, before moving to the clinical setting, further basic and translational research seems warranted in order to better characterize the surviving cells and to answer questions like how long these injected cells can function in the heart and whether stem cell transplantation is "regenerative cell therapy" or "cell-based paracrine therapy."

Acknowledgements This work was supported by grants TERCEL, SAF 2010-16549, CIBEROBN CB06/03 and by the ENCITE (European Network for Cell Imaging and Tracking Expertise) project Cooperation Health-2007-1.2-4 In Vivo Image-guidance for Cell Therapy, Large-scale Integrating Project (to LB). We thank Fundacion Jesus Serra-FIC, Barcelona, for their continuous support. BO is a recipient of a predoctoral fellowship from Instituto Salud Carlos III, Madrid, and GV is a recipient of a contract from the Innovation and Science Spanish Ministry (RyC-2009-5495).

Disclosure No competing financial interests exist.

References

1. Armstrong L, Lako M, Buckley N et al (2012) Editorial: our top 10 developments in stem cell biology over the last 30 years. Stem Cells 30:2–9
2. Daher SR, Johnstone BH, Phinney DG, March KL (2008) Adipose stromal/stem cells: basic and translational advances: the IFATS collection. Stem Cells 26:2664–2665

3. Uccelli A, Moretta L, Pistoia V (2008) Mesenchymal stem cells in health and disease. Nat Rev Immunol 8:726–736
4. Bianco P, Robey PG, Simmons PJ (2008) Mesenchymal stem cells: revisiting history, concepts, and assays. Cell Stem Cell 2:313–319
5. Ranganath SH, Levy O, Inamdar MS, Karp JM (2012) Harnessing the mesenchymal stem cell secretome for the treatment of cardiovascular disease. Cell Stem Cell 10:244–258
6. Dominici M, Le Blanc K, Mueller I et al (2006) Minimal criteria for defining multipotent mesenchymal stromal cells. The International Society for Cellular Therapy position statement. Cytotherapy 8:315–317
7. Horwitz EM, Dominici M (2008) How do mesenchymal stromal cells exert their therapeutic benefit? Cytotherapy 10:771–774
8. Nadri S, Soleimani M (2007) Comparative analysis of mesenchymal stromal cells from murine bone marrow and amniotic fluid. Cytotherapy 9:729–737
9. Zuk PA, Zhu M, Ashjian P et al (2002) Human adipose tissue is a source of multipotent stem cells. Mol Biol Cell 13:4279–4295
10. Shih DT, Lee DC, Chen SC et al (2005) Isolation and characterization of neurogenic mesenchymal stem cells in human scalp tissue. Stem Cells 23:1012–1020
11. Sarugaser R, Lickorish D, Baksh D et al (2005) Human umbilical cord perivascular (HUCPV) cells: a source of mesenchymal progenitors. Stem Cells 23:220–229
12. Mazo M, Gavira JJ, Pelacho B, Prosper F (2011) Adipose-derived stem cells for myocardial infarction. J Cardiovasc Transl Res 4:145–153
13. Fraser JK, Wulur I, Alfonso Z, Hedrick MH (2006) Fat tissue: an underappreciated source of stem cells for biotechnology. Trends Biotechnol 24:150–154
14. Strawford A, Antelo F, Christiansen M, Hellerstein MK (2004) Adipose tissue triglyceride turnover, de novo lipogenesis, and cell proliferation in humans measured with 2H2O. Am J Physiol 286:E577–E588
15. Zuk PA, Zhu M, Mizuno H et al (2001) Multilineage cells from human adipose tissue: implications for cell-based therapies. Tissue Eng 7:211–228
16. Schaffler A, Buchler C (2007) Concise review: adipose tissue-derived stromal cells—basic and clinical implications for novel cell-based therapies. Stem Cells 25:818–827
17. Miyahara Y, Nagaya N, Kataoka M et al (2006) Monolayered mesenchymal stem cells repair scarred myocardium after myocardial infarction. Nat Med 12:459–465
18. Williams KJ, Picou AA, Kish SL et al (2008) Isolation and characterization of porcine adipose tissue-derived adult stem cells. Cells Tissues Organs 188:251–258
19. De Ugarte DA, Alfonso Z, Zuk PA et al (2003) Differential expression of stem cell mobilization-associated molecules on multi-lineage cells from adipose tissue and bone marrow. Immunol Lett 89:267–270
20. Puissant B, Barreau C, Bourin P et al (2005) Immunomodulatory effect of human adipose tissue-derived adult stem cells: comparison with bone marrow mesenchymal stem cells. Br J Haematol 129:118–129
21. Rodriguez AM, Pisani D, Dechesne CA et al (2005) Transplantation of a multipotent cell population from human adipose tissue induces dystrophin expression in the immunocompetent mdx mouse. J Exp Med 201:1397–1405
22. Jurgens WJ, Oedayrajsingh-Varma MJ, Helder MN et al (2008) Effect of tissue-harvesting site on yield of stem cells derived from adipose tissue: implications for cell-based therapies. Cell Tissue Res 332:415–426
23. Fraser JK, Schreiber R, Strem B et al (2006) Plasticity of human adipose stem cells toward endothelial cells and cardiomyocytes. Nat Clin Pract Cardiovasc Med 3(suppl 1):S33–S37
24. Nakagami H, Morishita R, Maeda K et al (2006) Adipose tissue-derived stromal cells as a novel option for regenerative cell therapy. J Atheroscler Thromb 13:77–81
25. Planat-Benard V, Silvestre JS, Cousin B et al (2004) Plasticity of human adipose lineage cells toward endothelial cells: physiological and therapeutic perspectives. Circulation 109:656–663

26. Djouad F, Plence P, Bony C et al (2003) Immunosuppressive effect of mesenchymal stem cells favors tumor growth in allogeneic animals. Blood 102:3837–3844
27. Nakagami H, Maeda K, Morishita R et al (2005) Novel autologous cell therapy in ischemic limb disease through growth factor secretion by cultured adipose tissue-derived stromal cells. Arterioscler Thromb Vasc Biol 25:2542–2547
28. Bensinger W, Singer J, Appelbaum F et al (1993) Autologous transplantation with peripheral blood mononuclear cells collected after administration of recombinant granulocyte stimulating factor. Blood 81:3158–3163
29. Asahara T, Murohara T, Sullivan A et al (1997) Isolation of putative progenitor endothelial cells for angiogenesis. Science 275:964–967
30. Gronthos S, Franklin DM, Leddy HA et al (2001) Surface protein characterization of human adipose tissue-derived stromal cells. J Cell Physiol 189:54–63
31. Festy F, Hoareau L, Bes-Houtmann S et al (2005) Surface protein expression between human adipose tissue-derived stromal cells and mature adipocytes. Histochem Cell Biol 124: 113–121
32. Rehman J, Traktuev D, Li J et al (2004) Secretion of angiogenic and antiapoptotic factors by human adipose stromal cells. Circulation 109:1292–1298
33. Kern S, Eichler H, Stoeve J et al (2006) Comparative analysis of mesenchymal stem cells from bone marrow, umbilical cord blood, or adipose tissue. Stem Cells 24:1294–1301
34. Hombach-Klonisch S, Panigrahi S, Rashedi I et al (2008) Adult stem cells and their transdifferentiation potential-perspectives and therapeutic applications. J Mol Med 86(12): 1301–1314
35. Sanz-Ruiz R, Fernandez-Santos E, Dominguez-Munoa M et al (2009) Early translation of adipose-derived cell therapy for cardiovascular disease. Cell Transplant 18:245–254
36. Casteilla L, Planat-Benard V, Cousin B et al (2005) Plasticity of adipose tissue: a promising therapeutic avenue in the treatment of cardiovascular and blood diseases? Arch Mal Coeur Vaiss 98:922–926
37. Oedayrajsingh-Varma MJ, van Ham SM, Knippenberg M et al (2006) Adipose tissue-derived mesenchymal stem cell yield and growth characteristics are affected by the tissue-harvesting procedure. Cytotherapy 8:166–177
38. Martínez-González J, Viñals M et al (1997) Mevalonate deprivation impairs IGF-I/insulin signaling in human vascular smooth cells. Atherosclerosis 135:213–223
39. Bai X, Alt E (2010) Myocardial regeneration potential of adipose tissue-derived stem cells. Biochem Biophys Res Commun 401:321–326
40. Sanz-Ruiz R, Santos ME, Munoa MD et al (2008) Adipose tissue-derived stem cells: the friendly side of a classic cardiovascular foe. J Cardiovasc Transl Res 1:55–63
41. Rodriguez AM, Elabd C, Amri E-Z et al (2005) The human adipose tissue is a source of multipotent stem cells. Biochimie 87:125–128
42. Locke M, Feisst V, Dunbar PR (2011) Concise review: human adipose-derived stem cells: separating promise from clinical need. Stem Cells 29:404–411
43. Fuster V, Badimon L, Badimon JJ, Chesebro JH (1992) The pathogenesis of coronary artery disease and the acute coronary syndromes (1). N Engl J Med 326:242–250
44. Fuster V, Badimon L, Badimon JJ, Chesebro JH (1992) The pathogenesis of coronary artery disease and the acute coronary syndromes (2). N Engl J Med 326:310–318
45. Beltrami AP, Urbanek K, Kajstura J et al (2001) Evidence that human cardiac myocytes divide after myocardial infarction. N Engl J Med 344:1750–1757
46. Condorelli G, Borello U, De Angelis L et al (2001) Cardiomyocytes induce endothelial cells to trans-differentiate into cardiac muscle: implications for myocardium regeneration. Proc Natl Acad Sci U S A 98:10733–10738
47. Kudo M, Wang Y, Wani MA et al (2003) Implantation of bone marrow stem cells reduces the infarction and fibrosis in ischemic mouse heart. J Mol Cell Cardiol 35:1113–1119
48. Smart N, Riley PR (2008) The stem cell movement. Circ Res 102:1155–1168
49. Sanchez PL, Villa A, Sanz R et al (2007) Present and future of stem cells for cardiovascular therapy. Ann Med 39:412–427

50. Orlic D, Kajstura J, Chimenti S et al (2001) Bone marrow cells regenerate infarcted myocardium. Nature 410:701–705
51. Murry CE, Soonpaa MH, Reinecke H et al (2004) Haematopoietic stem cells do not transdifferentiate into cardiac myocytes in myocardial infarcts. Nature 428:664–668
52. Balsam LB, Wagers AJ, Christensen JL et al (2004) Haematopoietic stem cells adopt mature haematopoietic fates in ischaemic myocardium. Nature 428:668–673
53. Miller-Kasprzak E, Jagodzinski PP (2007) Endothelial progenitor cells as a new agent contributing to vascular repair. Arch Immunol Ther Exp 55:247–259
54. Rubart M, Field LJ (2006) Cardiac regeneration: repopulating the heart. Annu Rev Physiol 68:29–49
55. Narmoneva DA, Vukmirovic R, Davis ME et al (2004) Endothelial cells promote cardiac myocyte survival and spatial reorganization: implications for cardiac regeneration. Circulation 110:962–968
56. Makino S, Fukuda K, Miyoshi S et al (1999) Cardiomyocytes can be generated from marrow stromal cells in vitro. J Clin Invest 103:697–705
57. Pittenger MF, Mackay AM, Beck SC et al (1999) Multilineage potential of adult human mesenchymal stem cells. Science 284:143–147
58. Murry CE, Field LJ, Menasche P (2005) Cell-based cardiac repair: reflections at the 10-year point. Circulation 112:3174–3183
59. Menasche P (2007) Skeletal myoblasts as a therapeutic agent. Prog Cardiovasc Dis 50:7–17
60. Shi X, Garry DJ (2006) Muscle stem cells in development, regeneration, and disease. Genes Dev 20:1692–1708
61. Dowell JD, Rubart M, Pasumarthi KBS et al (2003) Myocyte and myogenic stem cell transplantation in the heart. Cardiovasc Res 58:336–350
62. Laflamme MA, Murry CE (2005) Regenerating the heart. Nat Biotechnol 23:845–856
63. Aicher A, Brenner W, Zuhayra M et al (2003) Assessment of the tissue distribution of transplanted human endothelial progenitor cells by radioactive labeling. Circulation 107:2134–2139
64. Dimmeler S, Zeiher AM, Schneider MD (2005) Unchain my heart: the scientific foundations of cardiac repair. J Clin Invest 115:572–583
65. Menasche P, Hagege AA, Vilquin JT et al (2003) Autologous skeletal myoblast transplantation for severe postinfarction left ventricular dysfunction. J Am Coll Cardiol 41:1078–1083
66. Menasche P, Alfieri O, Janssens S et al (2008) The myoblast autologous grafting in ischemic cardiomyopathy (MAGIC) trial: first randomized placebo-controlled study of myoblast transplantation. Circulation 117:1189–1200
67. Barile L, Messina E, Giacomello A, Marban E (2007) Endogenous cardiac stem cells. Prog Cardiovasc Dis 50:31–48
68. Oh H, Bradfute SB, Gallardo TD et al (2003) Cardiac progenitor cells from adult myocardium: homing, differentiation, and fusion after infarction. Proc Natl Acad Sci U S A 100:12313–12318
69. Oettgen P (2006) Cardiac stem cell therapy: need for optimization of efficacy and safety monitoring. Circulation 114:353–358
70. Wang JS, Shum-Tim D, Galipeau J et al (2000) Marrow stromal cells for cellular cardiomyoplasty: feasibility and potential clinical advantages. J Thorac Cardiovasc Surg 120:999–1005
71. Wakitani S, Saito T, Caplan AI (1995) Myogenic cells derived from rat bone marrow mesenchymal stem cells exposed to 5-azacytidine. Muscle Nerve 18:1417–1426
72. Prockop DJ (1997) Marrow stromal cells as stem cells for nonhematopoietic tissues. Science 276:71–74
73. Nagaya N, Fujii T, Iwase T et al (2004) Intravenous administration of mesenchymal stem cells improves cardiac function in rats with acute myocardial infarction through angiogenesis and myogenesis. Am J Physiol Heart Circ Physiol 287:H2670–H2676
74. Strem BM, Zhu M, Alfonso Z et al (2005) Expression of cardiomyocytic markers on adipose tissue-derived cells in a murine model of acute myocardial injury. Cytotherapy 7:282–291

75. Strem BM, Jordan M, Kim J, Yang J, Anderson CD, Daniels E (2005) Adipose tissue-derived stem cells enhance cardiac function following surgically-induced myocardial infarction. Circulation 112(suppl II):274
76. Mazo M, Planat-Bénard V, Abizanda G et al (2008) Transplantation of adipose derived stromal cells is associated with functional improvement in a rat model of chronic myocardial infarction. Eur J Heart Fail 10:454–462
77. Cai L, Johnstone BH, Cook TG et al (2009) IFATS collection: human adipose tissue-derived stem cells induce angiogenesis and nerve sprouting following myocardial infarction, in conjunction with potent preservation of cardiac function. Stem Cells 27:230–237
78. Léobon B, Roncalli J, Joffre C et al (2009) Adipose-derived cardiomyogenic cells: in vitro expansion and functional improvement in a mouse model of myocardial infarction. Cardiovasc Res 83:757–767
79. Schenke-Layland K, Strem BM, Jordan MC et al (2009) Adipose tissue-derived cells improve cardiac function following myocardial infarction. J Surg Res 153:217–223
80. van der Bogt KE, Schrepfer S, Yu J et al (2009) Comparison of transplantation of adipose tissue- and bone marrow-derived mesenchymal stem cells in the infarcted heart. Transplantation 87:642–652
81. Wang L, Deng J, Tian W et al (2009) Adipose-derived stem cells are an effective cell candidate for treatment of heart failure: an MR imaging study of rat hearts. Am J Physiol Heart Circ Physiol 297:H1020–H1031
82. Zhu XY, Zhang XZ, Xu L et al (2009) Transplantation of adipose-derived stem cells overexpressing hHGF into cardiac tissue. Biochem Biophys Res Commun 379:1084–1090
83. Bai X, Yan Y, Song YH et al (2010) Both cultured and freshly isolated adipose tissue-derived stem cells enhance cardiac function after acute myocardial infarction. Eur Heart J 31:489–501
84. Bayes-Genis A, Soler-Botija C, Farré J et al (2010) Human progenitor cells derived from cardiac adipose tissue ameliorate myocardial infarction in rodents. J Mol Cell Cardiol 49:771–780
85. Danoviz ME, Nakamuta JS, Marques FL et al (2010) Rat adipose tissue-derived stem cells transplantation attenuates cardiac dysfunction post infarction and biopolymers enhance cell retention. PLoS One 5:e12077
86. Lin YC, Leu S, Sun CK et al (2010) Early combined treatment with sildenafil and adiposederived mesenchymal stem cells preserves heart function in rat dilated cardiomyopathy. J Transl Med 8:88
87. Okura H, Matsuyama A, Lee CM et al (2010) Cardiomyoblast-like cells differentiated from human adipose tissue-derived mesenchymal stem cells improve left ventricular dysfunction and survival in a rat myocardial infarction model. Tissue Eng Part C Methods 16:417–425
88. Zhang X, Wang H, Ma X et al (2010) Preservation of the cardiac function in infarcted rat hearts by the transplantation of adipose-derived stem cells with injectable fibrin scaffolds. Exp Biol Med (Maywood) 235:1505–1515
89. Bai X, Yan Y, Caleman M et al (2011) Tracking long-term survival of intramyocardially delivered human adipose tissue-derived stem cells using bioluminescence imaging. Mol Imaging Biol 13:633–645
90. Berardi GR, Rebelatto CK, Tavares HF et al (2011) Transplantation of SNAP-treated adipose tissue-derived stem cells improves cardiac function and induces neovascularization after myocardium infarct in rats. Exp Mol Pathol 90:149–156
91. Cai A, Zheng D, Dong Y et al (2011) Efficacy of Atorvastatin combined with adipose-derived mesenchymal stem cell transplantation on cardiac function in rats with acute myocardial infarction. Acta Biochim Biophys Sin (Shanghai) 43:857–866
92. Gaebel R, Furlani D, Sorg H et al (2011) Cell origin of human mesenchymal stem cells determines a different healing performance in cardiac regeneration. PLoS One 6:e15652
93. Hamdi H, Planat-Benard V, Bel A et al (2011) Epicardial adipose stem cell sheets results in greater post-infarction survival than intramyocardial injections. Cardiovasc Res 91:483–491
94. Ii M, Horii M, Yokoyama A et al (2011) Synergistic effect of adiposederived stem cell therapy and bone marrow progenitor recruitment in ischemic heart. Lab Invest 91:539–552

95. Paul A, Srivastava S, Cheng G et al (2011) Functional assessment of adipose stem cells for xenotransplantation using myocardial. Infarction immunocompetent models: comparison with bone marrow stem cells. Cell Biochem Biophys [Epub ahead of print]
96. van Dijk A, Naaijkens BA, Jurgens WJ et al (2011) Reduction of infarct size by intravenous injection of uncultured adipose derived stromal cells in a rat model is dependent on the time point of application. Stem Cell Res 7:219–229
97. Bagno LL, Werneck-de-Castro JP, Oliveira PF et al (2012) Adipose-derived stromal cell therapy improves cardiac function after coronary occlusion in rats. Cell Transplant 21 (9):1985–1996
98. Otto Beitnes J, Oie E, Shahdadfar A et al (2012) Intramyocardial injections of human mesenchymal stem cells following acute myocardial infarction modulate scar formation and improve left ventricular function. Cell Transplant 21:1697–1709
99. Fang CH, Jin J, Joe JH et al (2012) In vivo differentiation of human amniotic epithelial cells into cardiomyocyte-like cells and cell transplantation effect on myocardial infarction in rats: comparison with cord blood and adipose tissue-derived mesenchymal stem cells. Cell Transplant 21:1687–1696
100. Hoke NN, Salloum FN, Dakks DA et al (2012) Preconditioning by phosphodiesterase-5 inhibition improves therapeutic efficacy of adipose-derived stem cells following myocardial infarction in mice. Stem Cells 30:326–335
101. Li TS, Cheng K, Malliaras K et al (2012) Direct comparison of different stem cell types and subpopulations reveals superior paracrine potency and myocardial repair efficacy with cardiosphere-derived cells. J Am Coll Cardiol 59:942–953
102. Liu Z, Wang H, Wang Y et al (2012) The influence of chitosan hydrogel on stem cell engraftment, survival and homing in the ischemic myocardial microenvironment. Biomaterials 33:3093–3106
103. Paul A, Cheng G, Khan A et al (2012) Genipin-cross-linked microencapsulated human adipose stem cells augment transplant retention resulting in attenuation of chronically infarcted rat heart fibrosis and cardiac dysfunction. Cell Transplant 21:2735–2751
104. Shi CZ, Zhang XP, Lv ZW et al (2012) Adipose tissue-derived stem cells embedded with eNOS restore cardiac function in acute myocardial infarction model. Int J Cardiol 154:2–8
105. Zhang DZ, Gai LY, Liu HW et al (2007) Transplantation of autologous adipose-derived stem cells ameliorates cardiac function in rabbits with myocardial infarction. Chin Med J (Engl) 120:300–307
106. Yang JJ, Yang X, Liu ZQ et al (2012) Transplantation of adipose tissue-derived stem cells overexpressing heme oxygenase-1 improves functions and remodeling of infarcted myocardium in rabbits. Tohoku J Exp Med 226:231–241
107. Watanabe C (2004) Intracoronary adipose tissue derived stem cells therapy preserves left ventricular function in a porcine infarct model. Paper presented at Transvascular Cardiovascular Therapeutics Annual Meeting, September 2004, Washington DC, USA
108. Fotuhi P, Song YH, Alt E (2007) Electrophysiological consequence of adipose-derived stem cell transplantation in infarcted porcine myocardium. Europace 9:1218–1221
109. Valina C, Pinkernell K, Song YH et al (2007) Intracoronary administration of autologous adipose tissue-derived stem cells improves left ventricular function, perfusion, and remodelling after acute myocardial infarction. Eur Heart J 28:2667–2677
110. Alt E, Pinkernell K, Scharlau M et al (2010) Effect of freshly isolated autologous tissue resident stromal cells on cardiac function and perfusion following acute myocardial infarction. Int J Cardiol 144:26–35
111. Rigol M, Solanes N, Farre J et al (2010) Effects of adipose tissue-derived stem cell therapy after myocardial infarction: impact of the route of administration. J Card Fail 16:357–366
112. Mazo M, Hernández S, Gavira JJ et al (2012) Treatment of reperfused ischemia with adiposederived stem cells in a preclinical Swine model of myocardial infarction. Cell Transplant 21:2723–2733
113. Kim YM, Jeon ES, Kim MR et al (2008) Angiotensin II-induced differentiation of adipose tissue-derived mesenchymal stem cells to smooth muscle-like cells. Int J Biochem Cell Biol 40:2482–2491

114. Rodriguez LV, Alfonso Z, Zhang R et al (2006) Clonogenic multipotent stem cells in human adipose tissue differentiate into functional smooth muscle cells. Proc Natl Acad Sci U S A 103:12167–12172
115. Ning H, Liu G, Lin G et al (2009) Fibroblast growth factor 2 promotes endothelial differentiation of adipose tissue-derived stem cells. J Sex Med 6:967–979
116. Jackson KA, Majka SM, Wang H et al (2001) Regeneration of ischemic cardiac muscle and vascular endothelium by adult stem cells. J Clin Invest 107:1395–1402
117. Laflamme MA, Myerson D, Saffitz JE, Murry CE (2002) Evidence for cardiomyocyte repopulation by extracardiac progenitors in transplanted human hearts. Circ Res 90:634–640
118. Gnecchi M, He H, Liang OD et al (2005) Paracrine action accounts for marked protection of ischemic heart by Akt-modified mesenchymal stem cells. Nat Med 11:367–368
119. Folkman J (1971) Tumor angiogenesis: therapeutic implications. N Engl J Med 285: 1182–1186
120. Yanagisawa-Miwa A, Uchida Y, Nakamura F et al (1992) Salvage of infarcted myocardium by angiogenic action of basic fibroblast growth factor. Science 257:1401–1403
121. Takeshita S, Zheng LP, Brogi E et al (1994) Therapeutic angiogenesis. A single intraarterial bolus of vascular endothelial growth factor augments revascularization in a rabbit ischemic hind limb model. J Clin Invest 93:662–670
122. Losordo DW, Dimmeler S (2004) Therapeutic angiogenesis and vasculogenesis for ischemic disease. Part I: angiogenic cytokines. Circulation 109:2487–2491
123. Freedman SB, Isner JM (2002) Therapeutic angiogenesis for coronary artery disease. Ann Intern Med 136:54–71
124. Losordo DW, Vale PR, Hendel RC et al (2002) Phase 1/2 placebo-controlled, double-blind, dose-escalating trial of myocardial vascular endothelial growth factor 2 gene transfer by catheter delivery in patients with chronic myocardial ischemia. Circulation 105:2012–2018
125. Henry TD, Annex BH, McKendall GR et al (2003) The VIVA trial: vascular endothelial growth factor in ischemia for vascular angiogenesis. Circulation 107:1359–1365
126. Lazarous DF, Shou M, Scheinowitz M et al (1996) Comparative effects of basic fibroblast growth factor and vascular endothelial growth factor on coronary collateral development and the arterial response to injury. Circulation 94:1074–1082
127. Takahashi T, Kalka C, Masuda H et al (1999) Ischemia- and cytokine-induced mobilization of bone marrow-derived endothelial progenitor cells for neovascularization. Nat Med 5:434–438
128. Yamaguchi J, Kusano KF, Masuo O et al (2003) Stromal cell-derived factor-1 effects on ex vivo expanded endothelial progenitor cell recruitment for ischemic neovascularization. Circulation 107:1322–1328
129. Wang M, Crisostomo PR, Herring C et al (2006) Human progenitor cells from bone marrow or adipose tissue produce VEGF, HGF, and IGF-I in response to TNF by a p38 MAPKdependent mechanism. Am J Physiol Regul Integr Comp Physiol 291:R880–R884
130. Dernbach E, Urbich C, Brandes RP et al (2004) Antioxidative stress-associated genes in circulating progenitor cells: evidence for enhanced resistance against oxidative stress. Blood 104:3591–3597
131. Kocher AA, Schuster MD, Szabolcs MJ et al (2001) Neovascularization of ischemic myocardium by human bone-marrow-derived angioblasts prevents cardiomyocyte apoptosis, reduces remodeling and improves cardiac function. Nat Med 7:430–436
132. Schuster MD, Kocher AA, Seki T et al (2004) Myocardial neovascularization by bone marrow angioblasts results in cardiomyocyte regeneration. Am J Physiol Heart Circ Physiol 287:H525–H532
133. Murohara T (2009) Autologous adipose tissue as a new source of progenitor cells for therapeutic angiogenesis. J Cardiol 53:155–163
134. Sadat S, Gehmert S, Song YH et al (2007) The cardioprotective effect of mesenchymal stem cells is mediated by IGF-I and VEGF. Biochem Biophys Res Commun 363:674–679
135. Liles WC, Broxmeyer HE, Rodger E et al (2003) Mobilization of hematopoietic progenitor cells in healthy volunteers by AMD3100, a CXCR4 antagonist. Blood 102:2728–2730

136. Tachibana K, Hirota S, Iizasa H et al (1998) The chemokine receptor CXCR4 is essential for vascularization of the gastrointestinal tract. Nature 393:591–594
137. Rasmussen JG, Frobert O, Pilgaard L et al (2011) Prolonged hypoxic culture and trypsinization increase the pro-angiogenic potential of human adipose tissue-derived stem cells. Cytotherapy 13:318–328
138. Suga H, Eto H, Aoi N et al (2010) Adipose tissue remodeling under ischemia: death of adipocytes and activation of stem/progenitor cells. Plast Reconstr Surg 126:1911–1923
139. Stubbs SL, Hsiao ST, Peshavariya HM et al (2012) Hypoxic preconditioning enhances survival of human adipose-derived stem cells and conditions endothelial cells in vitro. Stem Cells Dev 21:1887–1896
140. Tateishi-Yuyama E, Matsubara H, Murohara T et al (2002) Therapeutic angiogenesis for patients with limb ischaemia by autologous transplantation of bone-marrow cells: a pilot study and a randomised controlled trial. Lancet 360:427–435
141. Jay SM, Shepherd BR, Bertram JP et al (2008) Engineering of multifunctional gels integrating highly efficient growth factor delivery with endothelial cell transplantation. FASEB J 22:2949–2956
142. Deuse T, Peter C, Fedak PW et al (2009) Hepatocyte growth factor or vascular endothelial growth factor gene transfer maximizes mesenchymal stem cell-based myocardial salvage after acute myocardial infarction. Circulation 120:S247–S254
143. Fitzpatrick JR III, Frederick JR, McCormick RC et al (2010) Tissue-engineered proangiogenic fibroblast scaffold improves myocardial perfusion and function and limits ventricular remodeling after infarction. J Thorac Cardiovasc Surg 140:667–676
144. Bhang SH, Cho SW, La WG et al (2011) Angiogenesis in ischemic tissue produced by spheroid grafting of human adipose-derived stromal cells. Biomaterials 32:2734–2747
145. Amos PJ, Shang H, Bailey AM et al (2008) IFATS collection: the role of human adiposederived stromal cells in inflammatory microvascular remodeling and evidence of a perivascular phenotype. Stem Cells 26:2682–2690
146. Zannettino AC, Paton S, Arthur A et al (2008) Multipotential human adipose-derived stromal stem cells exhibit a perivascular phenotype in vitro and in vivo. J Cell Physiol 214:413–421
147. Efimenko A, Starostina E, Kalinina N, Stolzing A (2011) Angiogenic properties of aged adipose derived mesenchymal stem cells after hypoxic conditioning. J Transl Med 9:10
148. Onate B, Vilahur G, Ferrer-Lorente R et al (2012) The subcutaneous adipose tissue reservoir of functionally active stem cells is reduced in obese patients. FASEB J 26(10):4327–4336
149. Hwangbo S, Kim J, Her S et al (2010) Therapeutic potential of human adipose stem cells in a rat myocardial infarction model. Yonsei Med J. 51:69–76

Chapter 17
Trials of Angiogenesis Therapy in Patients with Ischemic Heart Disease

Sandeep Singla and Jawahar L. Mehta

Abstract After initial success in in vitro and in vivo studies showing tremendous potential of angiogenic therapy, therapeutic angiogenesis has been studied in humans with ischemic heart disease (IHD) not responding to conventional treatments. Vascular endothelial growth factor, fibroblast growth factor, and granulocyte colony stimulating factor are among the most commonly tested cytokines in human trials. Delivery as a protein or vector with gene encoding for specific protein have been tested in these trials. In addition, multitude of delivery routes ranging from direct intramyocardial transfer to intracoronary infusion to systemic administration via subcutaneous route have been tried. Clinical, radiographic, and angiographic parameters have been used to gauge the effect of these therapies in human trials. While promising results were attained in small phase I studies, these salutary results have not been replicated in large phase II and III studies. Multitude of variables including duration of exposure, type of vector, and need for co-factor influence the process of angiogenesis. Further studies accounting for these variables are needed to fully determine the potential of therapeutic angiogenesis in IHD.

Keywords Angiogenesis • Vascular endothelial growth factor • Fibroblast growth factor • Granulocyte colony stimulating factor

---

S. Singla, M.D. (✉) • J.L. Mehta
Division of Cardiovascular Medicine, University of Arkansas for Medical Sciences and the Central Arkansas Veterans Healthcare System, 4301 West Markham, Little Rock, AR 72205, USA
e-mail: tanusandeepsingla@gmail.com

J.L. Mehta and N.S. Dhalla (eds.), *Biochemical Basis and Therapeutic Implications of Angiogenesis*, Advances in Biochemistry in Health and Disease 6,
DOI 10.1007/978-1-4614-5857-9_17, © Springer Science+Business Media New York 2013

17.1 Introduction

Ischemic heart disease (IHD) is the most common clinical condition associated with significant morbidity and mortality throughout the world. The most common clinical manifestation of IHD is angina or its equivalents—shortness of breath or fatigue. While relief of angina by medical therapies (nitrates, beta blockers, calcium channel blockers, ranolazine, etc.) is accepted as first-line therapy, as stand-alone modality it usually provides partial relief. Conventional approach for relief of angina in such cases involves revascularization, either by percutaneous or surgical approaches. Both these revascularization approaches are associated with significant peri-procedural and post-procedural morbidity and mortality. Furthermore, both these techniques exclude a significant proportion of cases that are ineligible for these procedures either because of risk involved or poor target arteries (e.g., small vessel disease) or associated medical comorbidities.

Therapeutic angiogenesis in the ischemic myocardium represents another strategy that holds promise in providing relief in this disabling disease. Since its recognition almost three decades ago, data have been generated from in vitro and in vivo animal models supporting the feasibility of therapeutic angiogenesis in IHD. Experimental neovascularization of the myocardium has been achieved with the use of agents that promote the formation of blood vessels. These agents included basic fibroblast growth factor (bFGF), vascular endothelial growth factor (VEGF) and granulocyte colony stimulating factor (G-CSF), and gene therapy by encoding a specific protein in vectors which are then directly delivered to host cells. Proteins being active agents ensure immediate exposure to active agents and this is most conventional and practical means for delivery of angiogenic cytokines. Genes can be transferred using transgenes incorporated in viral vectors and via nonviral methods using naked plasmid DNA. Gene therapy has an advantage of sustained exposure to angiogenic factors as growth factors are secreted in vivo for a given period of time. Moreover, genes can be transferred directly to specific biological site, limiting the side effects. However introduction of foreign genetic material and exposure to viral vectors may mediate an immune and inflammatory response and there is potential for long-term, low-level systemic exposure to secreted angiogenic factors. Moreover, duration and level of gene expression are unpredictable.

17.2 Human Trials of Cardiovascular Protein Therapy

Human trials of angiogenic therapy have been conducted since late 1990s. In this section, we review major trials of therapeutic angiogenesis reported in English literature. Biology of therapeutic angiogenesis including role of growth factors has been discussed in previous sections.

17.2.1 Fibroblast Growth Factor Trials (Table 17.1)

In a phase I study of 20 patients with 3-vessel coronary artery disease undergoing coronary artery bypass grafting (CABG), FGF-1 was injected into the myocardium close to the internal mammary artery (IMA)-left anterior descending (LAD) anastomosis. As control, 20 other patients undergoing CABG received heat denatured FGF-1 [1]. Twelve weeks later, the IMA bypass grafts were selectively imaged by intra-arterial digital subtraction angiography and quantitatively evaluated. The formation of capillaries could be demonstrated in all cases around the site of injection; this was not seen in the 20 control subjects. This was the first human study providing evidence for proof of concept of therapeutic angiogenesis in treatment of IHD. This study was limited insofar that it did not provide clinical outcomes and provided data only over 12 week time period.

Sellke et al. [2] reported a similar study in 1998 where bFGF was administered using heparin-alginate slow-release device implanted in the epicardial fat in patients undergoing CABG. Eight patients were enrolled in this study. There was no mortality. At 3-month follow-up all patients remained free of angina. Seven patients were examined with stress perfusion scans. Three patients had clear enhancement of perfusion to the unrevascularized myocardium, 1 patient had a new fixed defect, and 3 had minimal overall change but had evidence of new small, fixed perfusion defects. Seven patients had improved or similar myocardial contractile function (ejection fraction at 3-month follow-up—0.53 ± 0.22 vs. 0.47 ± 0.14 preoperatively). Similar to the study by Schumacher et al. [1], this study also demonstrated the safety and technical feasibility of therapeutic angiogenesis with bFGF delivered by heparin-alginate slow-release device, in addition to providing data on clinical outcomes. These investigators later in 1999 reported a placebo-controlled randomized, double-blind study with the use of bFGF. In this trial, 24 patients with viable and ischemic myocardium supplied by a major coronary artery that could not receive graft were randomized to receive heparin-alginate pellets containing 10 μg of bFGF (n-8) or 100 μg of bFGF (n-8) or placebo (n-8). One patient assigned to the 10-μg bFGF group had angina, whereas all patients in the 100-μg bFGF group remained angina-free. Stress nuclear perfusion imaging at baseline and 3 months after CABG showed a trend toward worsening of the defect size in the placebo group (20.8–23.9%, $P=0.06$), no significant change in the 10-μg group, and significant improvement in the 100-μg group (19.2 to 9.1, $P=0.01$). Magnetic resonance assessment of the target ischemic zone in a subset of patients showed a trend toward a reduction in the target ischemic area in the 100-μg group (10.7% to 3.7% $P=0.06$) [3]. This study provided further evidence for technical feasibility and safety of this procedure as there was no excess morbidity or mortality associated with the administration of bFGF. More importantly, this study provided data on critical dose to have therapeutic effect since there was significant benefit of 100 μg dose and minimal benefit at 10 μg dose.

Administration of bFGF via intramyocardial route carries the inherent risk associated with thoracotomy. This makes this procedure suitable only for a small

Table 17.1 Trials of angiogenesis with FGF protein therapy

References	Cytokine	Delivery mode	*N*	Control Y/N	Follow up	Outcome	Adverse events	Comments
[1]	FGF-1	Intramyocardial	20	Y	12 weeks	Capillary density	None	First study No clinical outcomes reported Long-term data was not available
[2]	B-FGF	Epicardial heparin-alginate beads	8	N	6 months	Angina, SPECT-MPI	Perioperative MI	First study to report clinical endpoints
[3]	B-FGF	Epicardial heparin-alginate beads	16	Y	16.0±6.8 months	Angina, SPECT-MPI, MRI	Clinical (none)	Evidence for dose effect
[4]	B-FGF	Intracoronary injection	17	Y	29 days	Exercise time and exercise-induced ECG changes	Hypotension Bradycardia Atrial fibrillation	No systemic angiogenesis
[5]	FGF-2	Intracoronary infusion	52	N	6 months	Clinical TMST angina MRI	Hypotension	Dose-dependent hypotension
[6]	r FGF2	Intracoronary infusion	337	Y	180 days	Exercise tolerance Seattle angina Question MRI	Hypotension	Largest trial to date for FGF-induced myocardial angiogenesis No benefit was noted with FGF

fraction of IHD patients. In addition, repeated bFGF injections would not be possible if thoracotomy is required each time. To overcome this issue, multiple parenteral routes have been explored. Unger et al. [4] reported the first human trial of intracoronary delivery of bFGF. This was a phase I, randomized trial involving 25 patients with CAD and stable angina. This trial was conducted to evaluate the safety, tolerability, pharmacokinetics, and pharmacodynamics of intracoronary bFGF administration. Subjects were randomized in a 2:1 ratio to a single dose of bFGF or placebo. bFGF was injected into the left main coronary artery in doses ranging from 3 to 100 μg/kg. Sustained hypotension and bradycardia were more common with doses of 30–100 μg/kg. The 100 μg/kg dose caused acute hypotension (~10%) that did not appear to be dose-related. Importantly, there was nothing to suggest systemic angiogenesis. Though not designed primary to assess therapeutic efficacy, this study did report exercise time and electrocardiographic indexes of ischemia which were not different between two groups. In comparison to previous studies of intramyocardial injection, lack of therapeutic efficacy of intracoronary bFGF administration may reflect need for prolonged exposure to bFGF to attain therapeutic angiogenesis in target tissue.

Laham et al. [5] further tested the concept of intracoronary administration of FGF-2 with longer duration of infusion to overcome the issue of transient exposure to growth factor via intracoronary route. Fifty-two patients received intracoronary infusions of FGF-2 ranging in dose from 0.33 to 48 μg/kg. FGF-2 infusions were well tolerated except for systemic hypotension which was dose-related. Over a 6 month period there were 4 deaths and 4 non-Q-wave myocardial infarctions. There was significant improvement in quality of life as assessed by Seattle Angina Questionnaire. There was also significant improvement in exercise tolerance as assessed by treadmill exercise testing. Significant improvement was seen in target wall thickening with MRI. This study in addition to adding on previous data on feasibility and safety also provided evidence for dose effect of FGF therapy in myocardial angiogenesis.

Based on the results of above mentioned phase I studies, Simons et al. [6] conducted a phase II FGF Initiating RevaScularization Trial (FIRST) to evaluate further the efficacy and safety of recombinant FGF-2. This was a multicenter, randomized, double-blind, placebo-controlled trial of a single intracoronary infusion of rFGF-2 at 0, 0.3, 3, or 30 μg/kg dose ($n=337$ patients). Efficacy was evaluated at 90 and 180 days by exercise tolerance test, myocardial nuclear perfusion imaging, Seattle Angina Questionnaire, and Short-Form 36 questionnaire. Exercise tolerance was increased at 90 days in all groups and was not significantly different between placebo and FGF-treated groups. FGF-2 reduced angina symptoms as measured by the angina frequency score of the Seattle Angina Questionnaire (overall $P=0.035$) and the physical component summary scale of the Short-Form 36 (pairwise $P=0.033$, all FGF groups vs. placebo). These differences were more pronounced in highly symptomatic patients (baseline angina frequency score ≤40 or Canadian Cardiovascular Society score of III or IV). None of the differences were significant at 180 days because of continued improvement in the placebo group as well. Adverse events were similar across all groups, except for hypotension, which occurred with higher frequency in the 30 μg/kg rFGF2 group. The authors concluded that single intracoronary infusion of rFGF2 does not improve exercise tolerance or myocardial perfusion, but does show trends toward symptomatic improvement at 90 (but not 180) days. Lack of improvement with FGF despite it being a potent angiogenic factor may represent variability in effect secondary to delivery route. Intracoronary delivery may result in transient exposure of myocardium to FGF and thus a non-sustained effect. In comparison, previous studies with intramyocardial administration of FGF had resulted in significant clinical improvement.

Recent work by Jang et al. [7] on improving the "bioefficacy" of FGF-2 holds promise for the field of therapeutic angiogenesis. In their animal model of limb ischemia FGF-2 was co-delivered with a proteoglycan syndecan-4 in proteoliposomal preparation. This co-delivery markedly enhanced neovascularization in animal model. Another strategy tested in animal model of myocardial infarction involves co-delivery of growth factors including FGF-2 with self-assembling peptide fibers that provide microenvironment to help recruit endothelial cells [8]. These newer strategies need to be tested in the IHD patients.

17.2.2 Vascular Endothelial Growth Factor Trials (Table 17.2)

Hendel et al. [9] first reported the impact of recombinant human studies VEGF (rhVEGF) on myocardial perfusion in humans. In a small ($n=14$) phase I study, they demonstrated safety and efficacy of intracoronary rhVEGF infusion. Though this trial was not powered to test therapeutic efficacy, rhVEGF improved myocardial perfusion in serial SPECT studies. Additionally, similar to studies with FGF [4, 5], only higher doses were associated with improved perfusion.

In another phase I clinical study by Henry et al. [10], 15 patients were given intracoronary rhVEGF by two sequential infusions. This trial was also designed to primarily assess safety and tolerability of intracoronary rhVEGF. rhVEGF was well tolerated at infusion rates up to 0.005 μg/kg/min with dose-related hypotension being the most common side effect as seen with FGF therapy [6]. In addition, this study demonstrated improved perfusion on SPECT-MPI and increased collateral density score on follow-up angiograms. These initial clinical trials played pivotal role in helping design the subsequent large clinical trials.

VEGF in Ischemia for Vascular Angiogenesis (VIVA) was a double-blind placebo control trial [11]. In this trial 178 patients with stable IHD unsuitable for standard revascularization were randomized to receive placebo, or low-dose or high-dose intracoronary rhVEGF protein followed by interrupted intravenous infusions. Endpoints included exercise treadmill testing, quality of life assessments, and nuclear perfusion imaging at 60 days. There was no improvement beyond placebo in all measurements by day 60. By day 120, high-dose rhVEGF resulted in significant improvement in angina and favorable trends in treadmill exercise time and angina frequency. Similar to previous phase I trials, this study

Table 17.2 Trials of angiogenesis with VEGF protein therapy

Trial	Agent	Delivery mode	N	Control Y/N	Duration of follow-up	Endpoints	Adverse events	Comments
[9]	rhVEGF	Intracoronary	14	N	60 days	Safety endpoints, SPECT-MPI	Not reported	Dose-dependent effect, higher dose = more efficacy
[10]	rhVEGF	Intracoronary	20	N	60 days	Safety endpoints, SPECT-MPI, coronary angiogram	Hypotension	
[11]	rhVEGF	Intracoronary + intravenous	178	Y	120 days	Exercise treadmill test, angina class, quality of life, SPECT-MPI	Transient hypotension	Higher dose demonstrated benefit at 120 days but not at 60 days

also demonstrated excellent short-term safety of using intracoronary rhVEGF. Benefit seen at longer follow-up and with higher dose brings into question of carrying out studies with higher doses and for longer duration to assess the true therapeutic efficacy of this approach.

17.2.3 Granulocyte Colony Stimulating Factor Trials (Table 17.3)

In addition to VEGF and FGF, G-CSF has also been used in human studies to promote neovascularization in IHD. In contrast to VEGF and FGF which promote local angiogenesis, G-CSF injection promotes mobilization of progenitor cells from bone marrow into ischemic myocardium. These cells in turn once localized to tissue of interest lead to secretion of cytokines which promote local angiogenesis. An important difference between G-CSF therapy and treatment with FGF and VEGF is that in addition to neovascularization, G-CSF also leads to replacement of necrotic myocytes by stem cells. Thus G-CSF therapy may really represent a dual-pronged strategy in management of IHD patients.

This concept was first tested in humans by Kuethe et al. [12] in 2005 in 14 patients against 9 matching controls with acute myocardial infarction. Patients received G-CSF subcutaneously for a mean of 7.0 ± 1.0 days, 48 h after successful recanalization and stent implantation. Treatment effect was evaluated by SPECT-MPI-based regional wall motion and perfusion, and ejection fraction estimation by radioventriculography before discharge and at 3 months. There was a significant improvement in regional wall motion and perfusion with G-CSF therapy. This initial trial provided proof of concept for utility of G-CSF for promoting angiogenesis following recanalization of blocked coronary arteries.

In 2001, Seiler et al. [13] reported in 21 patients with extensive coronary artery disease not eligible for CABG, effect of granulocyte-macrophage colony stimulating factor (GM-CSF, Molgramostim) on quantitatively assessed collateral flow index (CFI) in a randomized, double-blind, placebo-controlled fashion. The study protocol involved both intracoronary and intravenous infusion of GM-CSF. The treatment-induced difference in CFI was $+0.11 \pm 0.12$ in the GM-CSF group and -0.07 ± 0.12 in the placebo group ($P = 0.01$) which was suggestive of improvement. This was the first clinical study investigating the potential of GM-CSF to promote angiogenesis in IHD patients.

Same group reported in 2005 on the safety and efficacy of short-term subcutaneous GM-CSF therapy for promoting coronary collateral growth [14]. This was a randomized, double-blind, placebo-controlled trial of a 2-week period with subcutaneous GM-CSF (10 μg/kg; $n = 7$) or placebo ($n = 7$). In 14 men with chronic stable IHD (Table 17.3), efficacy measurements were made in a stenotic as well as a normal coronary artery before and after GM-CSF administration. There was significant improvement in CFI in the treatment group. Among 11 determined cytokines, chemokines, and their monocytic receptor concentrations, the treatment-induced change in CFI was predicted by the respective change in tumor necrosis factor-alpha concentration.

Table 17.3 Trials of angiogenesis with G-CSF or GM-CSF protein therapy

Trial	Factor	Delivery mode	N	Control Y/N	Follow-up period	Endpoints	Adverse events	Comments
[13]	GM-CSF	Intracoronary + subcutaneous	21	Y	2 weeks	Collateral flow index	Low grade fever, Skin rash	
[14]	GM-CSF	Subcutaneous	14	Y	2 weeks	Collateral flow index	Acute coronary syndrome	
[12]	G-CSF	Subcutaneous	14	Y	3 months	Wall motion and perfusion on SPECT-MPI	NA	Marked improvement in treatment arm
[15]	G-CSF	Subcutaneous	20	N	6 months	SPECT-MPI perfusion and LV function	Transient transaminitis	
[16]	G-CSF	Subcutaneous	114	Y	6 months	SPECT-MPI, MRI	Bone pain and muscle discomfort	
[17, 18]	G-CSF	Subcutaneous G-CSF followed by intracoronary stem cells	27	Y	6 months, 2 years		In-stent restenosis at culprit lesions	2-Year follow-up results showed consistent improvement in cell infusion group as compared to G-CSF alone group, though not statistically significant when compared to control

In this trial, 2 of the 7 patients in the GM-CSF group and none in the placebo group suffered an acute coronary syndrome during the treatment period raising questions about its safety. It was speculated that the precipitation of coronary syndromes with GM-CSF might have been related to its pro-atherogenic effects.

Valgimigli et al. [15] randomized 14 patients with ST elevation myocardial infarction who went for primary percutaneous coronary intervention (PCI) to receive G-CSF or placebo. G-CSF therapy induced a significant increase of white blood count, CD34(+) cells, and CD34(+) cells co-expressing AC133 and VEGFR-2. At follow-up, treated and placebo groups did not differ for the angiographic coronary late loss and showed a similar pattern of perfusion defect recovery, whereas in the treatment group at 6 months left ventricular ejection fraction tended to be higher ($P=0.068$) and left ventricular end-diastolic volume relatively lower ($P=0.054$). Similar to the study by Kuethe et al. [12], this study demonstrated efficacy of subcutaneous approach only for G-CSF therapy.

Based on the initial results of small phase I human trials of G-CSF, a larger phase II trial was conducted. REgenerate VItal myocardium by Vigorous Activation of bone marrow stem celLs (REVIVAL-2) trial was one such, double-blind, randomized placebo-controlled trial to assess the efficacy of stem cell mobilization with G-CSF in patients with myocardial infarction [16]. One hundred and fourteen patients diagnosed with ST-segment elevation myocardial infarction who had successful reperfusion by PCI within 12 h after onset of symptoms were randomized after 5 days to receive subcutaneously either a daily dose of 10 μg/kg of G-CSF ($n=56$) or placebo ($n=58$) for 5 days. Treatment with G-CSF produced a significant mobilization of stem cells. Between baseline and follow-up, left ventricular infarct size according to scintigraphy was reduced by a mean (SD) of 6.2% (9.1%) in the G-CSF group and 4.9% (8.9%) in the placebo group ($P=0.56$) and left ventricular ejection fraction was improved by 0.5% (3.8%) in the G-CSF group and 2.0% (4.9%) in the placebo group ($P=0.14$). Angiographic restenosis occurred in 19 (35.2%) of 54 patients in the G-CSF group and in 17 (30.9%) of 55 patients in the placebo group ($P=0.79$). The most common adverse event among patients assigned to G-CSF was mild to moderate bone pain and muscle discomfort. Thus stem cell mobilization by G-CSF therapy in patients with acute myocardial infarction and successful mechanical reperfusion had no influence on infarct size, left ventricular function, or coronary restenosis.

In a slightly different version of this strategy, Kang et al. [17] examined the efficacy of intracoronary infusion of peripheral blood stem cells collected after G-CSF therapy in the MAGIC trial. This trial involved 27 patients with myocardial infarction who underwent coronary stenting for the culprit lesion and were prospectively randomized into three groups; cell infusion ($n=10$), G-CSF alone ($n=10$), and control group ($n=7$). At 6 month follow-up exercise capacity as assessed by treadmill exercise time, myocardial perfusion and systolic function improved significantly in patients who received cell infusion. However, an unexpectedly high rate of in-stent restenosis at culprit lesion in patients who received G-CSF was noted and thus the trial was terminated prematurely. Two-year follow-up data from this study supports a persistent improvement in group which received intracoronary infusion of stem cells, though it was not statistically significant when compared with control group [18].

Therapeutic success with MAGIC trial strategy in comparison to strategy used with REVIVAL trial may be related to more direct availability of stem cells to injured myocardium. Moreover, waning of significant benefit at 2 years in MAGIC trial raises the question of performing repeated injection of stem cells/G-CSF therapy to promote myocardial angiogenesis and regeneration.

17.3 Human Trial of Dipyridamole Therapy

Belardinelli et al. [19] evaluated the effect of oral dipyridamole in inducing coronary collateral growth in IHD. They randomized 30 male patients with coronary artery disease and left ventricular systolic dysfunction (ejection fraction >40%) into three matched groups to receive dipyridamole alone ($n=10$) orally for 8 weeks or exercise training at 60% of peak $VO_{(2)}$ three times a week for 8 weeks and receive dipyridamole ($n=10$), or neither exercise testing nor dipyridamole ($n=10$). Thallium uptake of the collateral-dependent myocardium, coronary collateral score, and wall thickening score increased significantly only in groups receiving dipyridamole. There was significant correlations between changes in adenosine levels and increases of both thallium uptake ($r=-0.70$; $P=0.001$) and collateralization ($r=0.72$; $P=0.001$).

Further studies are needed to better define the role of oral dipyridamole as sole therapy or in combination with growth factors in treatment of IHD.

17.4 Human Trials of Gene Therapy

17.4.1 VEGF Gene Trials (Table 17.4)

The first investigation to use VEGF165 plasmid for myocardial gene therapy was an open-label study conducted by Losordo et al. [20]. Naked plasmid DNA encoding VEGF (phVEGF165) was injected directly into ischemic myocardium of five patients with angina via mini left anterior thoracotomy who failed conventional therapy. All patients had significant improvement in angina frequency and severity which correlated with increased myocardial perfusion on SPECT-MPI imaging. Coronary angiography also showed improved Rentrop score. This study provided the first evidence for favorable clinical effects of direct myocardial injection of naked DNA encoding VEGF [20].

Following this landmark study, there were four more phase I clinical studies where phVEGF165 was delivered directly into ischemic myocardium via mini left anterior thoracotomy in patients with medically intractable angina leading to reduced clinical

Table 17.4 Trials of angiogenesis with VEGF gene therapy

Trial	Factor	Delivery route	*N*	Control Y/N	Follow-up	Endpoints	Adverse effects	Comments
[20]	phVEGF	Intramyocardial	5	N	60 days	SPECT myocardial perfusion, coronary angiography	None	First human study with viral vector
[22]	ph-VEGF 165	Intramyocardial	20	N	60 days	SPECT myocardial perfusion, coronary angiography	None	
[23]	ph-VEGF 165	Intramyocardial	13	N	60 days	NOGA electromechanical mapping	None	
[24]	ph-VEGF A165	Intramyocardial	7	N	12 months	SPECT-MPI, tissue velocity imaging	None	Longest follow-up
[21, 25]	AdVEGF121	Intramyocardial	21	N	30 days	Coronary angiography, SPECT-MPI, and treadmill exercise	None	Continued improvement in angina and exercise capacity at 6 months [25]
[26]	AdVEGF121	Intramyocardial	67	Y	26 weeks	Treadmill exercise stress test, Seattle angina Questionnaire	Similar between control and treatment group	
[28]	VEGF	Intracoronary	15	Y	6 months	Coronary angiography	No adverse events	Established safety and efficacy of intracoronary route for viral vector delivery
[29]	PhVEGF-2	Intramyocardial	6	Y	365 days	NOGA electromechanical mapping, SPECT-MPI	None	
[30]	PhVEGF-2	Intramyocardial			12 weeks	CCS angina class, Seattle angina questionnaire	None	
[31]	VEGF	Intracoronary	103	Y	6 months	Myocardial perfusion	No serious adverse events	

[32]	phVEGF-A165	Intramyocardial	80	Y	3 months	Myocardial perfusion, wall motion abnormality, CCS angina class	Five adverse events in treatment group	In a sub-study improvement in perfusion with VEGF therapy in NOGA-defined ischemic areas [33]
[34]	VEGF165 + G-CSF	Intramyocardial + subcutaneous G-CSF	32	Y	3 months	SPECT-MPI	No major side effects	
[35]	VEGF165	Intramyocardial	93	Y	6 months	SPECT-MPI, exercise treadmill time, angina symptoms	No difference in major events	First multicenter trial of VEGF gene therapy via Intramyocardial route
[36]	VEGF121	Intramyocardial	17	Y	52 weeks	Exercise capacity, myocardial perfusion	none	
[41]	VEGF + l-arginine	Intramyocardial VEGF + oral dipyridamole	19	I	Y	3 months	PET—perfusion and wall motion	

symptoms, reduced evidence of ischemia on SPECT imaging, and improved collateral filling of at least one occluded vessel on angiography [21–24].

In a study done by Rosengart et al. [21], the viral vector expressing the 121-amino-acid form of human VEGF was administered to individuals with clinically significant coronary artery disease. It was administered directly to an ischemic area of the myocardium as an adjunct to conventional CABG surgery in a region that could not be bypassed in 15 patients (group A) or through a mini-thoracotomy as sole therapy in 6 patients (group B). There was no evidence of systemic or cardiac adverse events related to vector administration. In both groups, coronary angiography and stress sestamibi scan assessment of wall motion 30 days after therapy suggested improvement in the area of vector administration but no improvement in relative blood flow. All patients reported improvement in angina class after therapy, but in group A patients this could not be solely attributed to vector administration (had concomitant CABG). In group B, in which gene transfer was the only therapy, treadmill exercise assessment suggested improvement in most individuals assessed 30 days after therapy and at 6 month follow-up. Trends toward improvement in angina class and exercise treadmill testing at 6-month follow-up in the sole therapy group suggest that the effects of this therapy are persistent for $\geq$6 months [25].

The other large trial was Randomized evaluation of VEGF for Angiogenesis (REVASC) that enrolled 67 patients with severe angina pectoris with no option for revascularization to compare the efficacy and safety of AdVEGF121 with epicardial injection at mini-thoracotomy with maximum medical therapy [26]. Of the 67 patients enrolled, 35 continued maximum medical treatment and 32 received AdVEGF121. Exercise time to 1 mm ST-segment depression, the predefined endpoint, was increased in the AdVEGF121 group compared to control at 26 weeks ($P=0.026$), but not at 12 weeks. Total exercise duration and time to level 2 angina were also significantly improved in the AdVEGF121 group compared to control at weeks 12 (P=0.008 and 0.006) and 26 (P=0.015 and 0.003). Significant improvement in the Canadian Cardiovascular Society class score was evident in the AdVEGF121 group as compared to control as early as 6 weeks and continued to improve at 12 and 26 weeks ($P=0.001$ at all time points). Despite the inability to blind for treatment assignment, this study provided the first large scale objective data for the efficacy of therapeutic angiogenesis in patients with severe symptoms who were not candidates for traditional revascularization procedures.

In studies mentioned above, efficacy and safety of gene transfer in ischemic myocardium was established but the use of operative thoracotomy to deliver DNA precluded use of randomization against placebo effect. Catheter-based intracoronary infusion of viral vectors encoding angiogenic growth factors initiated nonoperative gene transfer in ischemic myocardium [27].

Laitinen et al. studied the safety and feasibility of catheter-mediated VEGF plasmid/liposome (P/L) gene transfer in human coronary arteries after PCI in a randomized, double-blinded, placebo-controlled study. Ten patients received VEGF P/L, three patients received beta-galactosidase P/L, and two patients received Ringer lactate. Catheter-mediated intracoronary gene transfer performed after angioplasty was found to be safe and well tolerated. However, there were no differences in the degree of coronary stenosis between treatment and control groups on angiography after 6 months [28].

In a single blind placebo-controlled pilot study of catheter-based myocardial gene transfer, patients were randomized to receive 200 μg of naked plasmid DNA encoding phVEGF-2 or placebo. Injections were guided by NOGA left ventricular electromechanical mapping. phVEGF-2 transfected patients had reduced angina and improved myocardial perfusion [29].

Another double-blind placebo control trial involving injection of naked plasmid DNA encoding ph VEGF-2 into left ventricular myocardium via catheter in escalating doses of 200 μg ($n=9$), 800 μg ($n=9$), or 2,000 μg ($n=1$) was reported by Losordo et al. [30]. At 12-week follow-up, angina class was decreased and mean duration of exercise was increased significantly in transfected patients. In addition, there was functional improvement by ≥2 Canadian Cardiovascular Society classes and Seattle Angina questionnaire data showed strong trend toward efficacy of phVEGF.

Kupio Angiogenesis Trial (KAT) was the first randomized double-blind, placebo-controlled trial studying the use of intracoronary infusion of VEGF gene for promoting myocardial angiogenesis [31]. Following PCI, 37 patients received VEGF-adenovirus, 28 patients received VEGF plasmid liposome, and 38 patients received ringer's lactate. There were no differences in clinical restenosis rate or minimal lumen diameter as measured by quantitative coronary angiography but significant improvement was detected in myocardial perfusion in the VEGF-Adv-treated patients at 6-month follow-up. Myocardial perfusion improvement was also associated with improvement in clinical parameters of functional capacity and exercise time in patients who received VEGF-adenovirus.

The EUROINJECT-ONE phase II multicenter trial randomized 80 patients with severe stable IHD with no other treatment option to direct intramyocardial phVEGF-A165 or placebo plasmid injection [32]. This study was the first to use VEGF gene transfer, but failed to show significant improvement in stress-induced myocardial perfusion abnormalities compared with placebo (38 ± 3% and 44 ± 2%, respectively); however, improved regional wall motion, as assessed both by NOGA ($P=0.04$) and by ventriculography ($P=0.03$), was evident. There were no adverse events attributable to phVEGF-A165 therapy. Further sub-study of this randomized trial was done to analyze changes in myocardial perfusion in NOGA-defined regions with intramyocardial injections of plasmid encoding human phVEGF-A165. The injected area (region of interest, ROI) was delineated as a best polygon by connecting the injection points marked on NOGA polar maps. The extents of the ROI were similar in the VEGF and placebo groups (19.4 ± 4.2% vs. 21.5 ± 5.4% of entire myocardium). No differences were found between VEGF and placebo groups at baseline with regard to the perfusion defect severity. At follow-up, a trend toward improvement in perfusion defect severity at stress was observed in VEGF group as compared with placebo (68.5 ± 11.9% vs. 62.5 ± 13.5%, $P=0.072$) without reaching normal values [33].

Ripa et al. [34] performed a pilot study of combined VEGF165 gene therapy and stem cell mobilization in patients with IHD who were symptomatic but were not candidates for revascularization. Sixteen patients received intramyocardial injections of

VEGF165 plasmid followed 1 week later by administration of G-CSF to mobilize progenitor cells from the bone marrow; the historical control groups consisted of 16 VEGF plasmid-treated patients and 16 control plasmid-treated patients from the Euroinject-One trial. The number of circulating progenitor cells (identified via CD34+ cells) increased significantly after G-CSF treatment, but there was no improvement in the primary endpoint of change in myocardial stress perfusion. The authors speculated that the homing of mobilized stem cells to the ischemic area may have been inadequate and suggested that co-transfer of a plasmid encoding stromal cell-derived factor 1, a progenitor cell-homing factor, or higher doses of VEGF plasmid may be required to get full therapeutic benefit. Other explanations for a lack of benefit must be considered, however, including the fact that SPECT scanning has not been validated for documenting changes in perfusion that may occur following local therapy, which could potentially result in sub-segmental, non-transmural alterations in flow. In addition, the timing of G-CSF administration may not have coincided with the peak of VEGF gene expression, thereby diminishing the possibility of synergy.

NOGA angiogenesis Revascularization Therapy assessment by radioNucleotide imaging (NORTHERN) trial was a double-blind, placebo-controlled trial in which VEGF165 DNA was delivered to left ventricular myocardium [35]. Seventy-two no-option and 21 patients with single vessel coronary occlusion or diffuse in-stent restenosis patients were randomized to receive gene therapy ($n=48$) or saline ($n=45$). Although there was no difference between the VEGF-treated and the placebo groups in the primary endpoint of change in myocardial perfusion from baseline at 3 or 6 months, assessed by SPECT imaging, a significant reduction in the ischemic area and improvement in perfusion scores were seen in both groups over time. Also, there was no difference between placebo and treatment arms with similar improvements in exercise treadmill time and angina symptoms in both groups at 3 and 6 months.

NOVA, another randomized double-blind placebo-controlled multicenter gene therapy trial, was conducted to study the efficacy of adenovirus carrying VEGF121 (BIOBYPASS) in patients with refractory advanced coronary artery disease [36]. Seventeen patients with severe CAD were randomized to receive BIOBYPASS ($n=12$) or placebo ($n=5$) as 12 intramyocardial injections into the ischemic area using the NOGA mapping. Direct intramyocardial injection of BIOBYPASS was safe but did not improve exercise capacity, time to ischemia threshold, or myocardial perfusion compared to sham injection in patients with refractory myocardial ischemia.

17.4.2 FGF Gene Trials (Table 17.5)

Angiogenic GENe Therapy (AGENT) trial was a double-blind controlled trial which evaluated the safety and anti-ischemic effects of five ascending doses of adenovirus (Ad) containing a human FGF gene in patients with angina [37]. Single intracoronary infusion of Ad5-FGF4 was found to be safe and well tolerated.

Table 17.5 Trials of angiogenesis with FGF gene therapy

Trial	Agent	Route of delivery	*N*	Control Y/N	Follow-up	Endpoints	Adverse events	Comments
[37]	FGF	Intracoronary infusion	79	Y	311 days (mean)	Exercise treadmill time	No difference in adverse event rate	
[38]	FGF	Intracoronary infusion	52	Y	8 weeks	Myocardial perfusion		
[39]	FGF	Intracoronary infusion	532	Y	12 weeks	Exercise treadmill time	No significant difference	No significant benefit with treatment. Upon subgroup analysis, women had benefit with treatment
	FGF/VEGF	Intramyocardial	52	Y	12 months	SPECT-MPI, exercise capacity, angina control		

Patients who received Ad5-FGF4 tended to have greater improvements in exercise time at 4 weeks (1.3 min vs. 0.7 min, NS). Serious adverse events during follow-up (mean, 311 days) were not different between placebo and treatment group. A protocol-specified, subgroup analysis showed the greatest improvement in patients with baseline ETT <10 min (1.6 min vs. 0.6 min, $P=0.01$).

AGENT 2 was another randomized, double blind trial in which 35 patients with stable angina and reversible ischemia were given intracoronary injection of Ad5-FGF4. At 8 weeks, Ad5-FGF4 resulted in significant reduction in ischemic defect on SPECT. Additionally, there was significant change in reversible perfusion defect size compared to placebo after removing an outlier [38]. AGENT 3 and 4 trials were parallel randomized placebo-controlled double blind trials designed to evaluate the efficacy of low and high dose of Ad5-FGF4 for therapeutic angiogenesis in myocardial ischemia [39]. Patients who remained symptomatic despite anti-anginal medications and who did not require immediate PCI or CABG were enrolled in AGENT-3, whereas those who were unsuitable for PCI or CABG were enrolled in AGENT 4 study. The primary endpoint was change from baseline in total exercise time at 12 weeks, and at secondary time points of 4 weeks and 6 months. Pre-specified secondary endpoints in both studies included time to 1 mm ST-segment depression, time to onset of angina, change in Canadian Cardiovascular Society class, percentage of patients with ≥30% increase in exercise time, coronary events or death at 1 year, change in angina incidence and nitroglycerin use by patient diary, and change in quality of life. There was no significant difference between the active groups and the placebo for the primary endpoint in both trials. Subgroup analysis by gender at 12 weeks of follow-up revealed a significant effect in women in the primary endpoint, total exercise time, for both dose groups but no effect in men. This was due to a negligible placebo effect in females compared with large placebo effect in males. There were no differences between either dose group and placebo for any of the secondary endpoints except for Canadian Cardiovascular Society class in high-dose group, in which there was significant improvement in all patients over placebo at week 12, month 6, and month 12 ($P<0.05$). The latter effect was again shown to be driven by the female population as in the male only subgroup; there was no significant change in Canadian Cardiovascular Society class. Other secondary endpoints of time to 1 mm ST-segment depression and time to angina also showed gender-specific significant or nearly so effects in females at 12 weeks and 6 months in both dose groups. Though these trials were stopped prematurely, subgroup analysis on pooled data revealed gender differences in angiogenic response which may suggest effect of differential hormonal milieu on the biology of angiogenesis.

17.4.3 Other Gene Therapy Trials

VIF-CAD, a randomized, placebo-controlled, double-blind trial, was designed to study therapeutic angiogenesis by percutaneous intramyocardial transfer of bicistronic (vascular endothelial growth factor/fibroblast growth factor [VEGF/FGF]) plasmid (Pvif) in patients with refractory heart ischemia [40]. Fifty-two patients with refractory angina were randomized to receive VEGF/FGF plasmid ($n=33$) or placebo plasmid ($n=19$) into myocardial region showing stress-induced perfusion defect. Stress and rest perfusion defect at 5 months did not differ in groups. Canadian Cardiovascular Society functional class improved after 5 ($P=0.02$) and 12 months ($P=0.0607$) in the treatment group.

Similarly, Endothelial Modulation in Angiogenic Therapy (EMAT) phase I double blind, placebo-controlled trial tested safety and efficacy intramyocardial VEGF165 combined with l-arginine, a nitric oxide donor, in patients undergoing CABG. Nineteen patients with surgical 3-vessel coronary disease and a severely diffusely diseased LAD artery were randomized to receive placebo injection + placebo oral supplementation, placebo injection + oral l-arginine supplementation, intramuscular VEGf165 + placebo oral supplementation, or intramuscular VEGf165 + oral l-arginine supplement. Patients who received the combination of VEGF and l-arginine had improved anterior wall perfusion on positron emission tomography ($P=0.02$), a trend toward smaller rest and stress perfusion defects ($P=0.10$), and better anterior wall contractility ($P=0.02$, Kruskal-Wallis) at 3 months vs. baseline. Thus concomitant endothelial modulation with l-arginine not only has the potential to make angiogenesis effective but also may have implications for cell therapy trials [41]. These kind of novel combined approaches may help overcome the hurdles realized in last decade and a half of human trials of neoangiogenesis in patients with IHD.

17.5 Conclusion

The success of therapeutic angiogenesis in in vitro and animal models has not been realized in human studies. This reflects the complex biology of IHD in humans and helps us realize that simplistic preclinical models of therapeutic approach may not always be successful in real life situations. Future research in co-delivery of growth factors, prolonging time of exposure of ischemic myocardium to a given growth factor while at the same time limiting systemic exposure and discovering more selective and efficacious angiogenic factors are some of the few approaches which will help define this novel field in coming years.

References

1. Schumacher B, Pecher P, von Specht BU, Stegmann T (1998) Induction of neoangiogenesis in ischemic myocardium by human growth factors: first clinical results of a new treatment of coronary heart disease. Circulation 97:645–650
2. Sellke FW, Laham RJ, Edelman ER et al (1998) Therapeutic angiogenesis with basic fibroblast growth factor: technique and early results. Ann Thorac Surg 65:1540–1544
3. Laham RJ, Sellke FW, Edelman ER et al (1999) Local perivascular delivery of basic fibroblast growth factor in patients undergoing coronary bypass surgery: results of a phase I randomized, double-blind, placebo-controlled trial. Circulation 100:1865–1871

4. Unger EF, Goncalves L, Epstein SE et al (2000) Effects of a single intracoronary injection of basic fibroblast growth factor in stable angina pectoris. Am J Cardiol 85(12):1414–1419
5. Laham RJ, Chronos NA, Pike M et al (2000) Intracoronary basic fibroblast growth factor (FGF-2) in patients with severe ischemic heart disease: results of a phase I open-label dose escalation study. J Am Coll Cardiol 36:2132–2139
6. Simons M, Annex BH, Laham RJ, Kleiman N, Henry T, Dauerman H et al (2002) Pharmacological treatment of coronary artery disease with recombinant fibroblast growth factor-2: double-blind, randomized, controlled clinical trial. Circulation 105:788–793
7. Jang E, Albadawi H, Watkins MT et al (2012) Syndecan-4 proteoliposomes enhance fibroblast growth factor-2 (FGF-2)-induced proliferation, migration, and neovascularization of ischemic muscle. Proc Natl Acad Sci U S A 109:1679–1684
8. Kim JH, Jung Y, Kim SH et al (2011) The enhancement of mature vessel formation and cardiac function in infarcted hearts using dual growth factor delivery with self-assembling peptides. Biomaterials 32:6080–6088
9. Hendel RC, Henry TD, Rocha-Singh K et al (2000) Effect of intracoronary recombinant human vascular endothelial growth factor on myocardial perfusion: evidence for a dose-dependent effect. Circulation 101:118–121
10. Henry TD, Rocha-Singh K, Isner JM et al (2001) Intracoronary administration of recombinant human vascular endothelial growth factor to patients with coronary artery disease. Am Heart J 142:872–880
11. Henry TD, Annex BH, McKendall GR et al (2003) The VIVA trial: vascular endothelial growth factor in ischemia for vascular angiogenesis. Circulation 107:1359–1365
12. Kuethe F, Figulla HR, Herzau M et al (2005) Treatment with granulocyte colony-stimulating factor for mobilization of bone marrow cells in patients with acute myocardial infarction. Am Heart J 150:115
13. Seiler C, Pohl T, Wustmann K et al (2001) Promotion of collateral growth by granulocyte-macrophage colony-stimulating factor in patients with coronary artery disease: a randomized, double-blind, placebo-controlled study. Circulation 104:2012–2017
14. Zbinden S, Zbinden R, Meier P et al (2005) Safety and efficacy of subcutaneous-only granulocyte-macrophage colony-stimulating factor for collateral growth promotion in patients with coronary artery disease. J Am Coll Cardiol 46:1636–1642
15. Valgimigli M, Rigolin GM, Cittanti C et al (2005) Use of granulocyte-colony stimulating factor during acute myocardial infarction to enhance bone marrow stem cell mobilization in humans: clinical and angiographic safety profile. Eur Heart J 26:1838–1845
16. Zohlnhofer D, Ott I, Mehilli J et al (2006) Stem cell mobilization by granulocyte colony-stimulating factor in patients with acute myocardial infarction: a randomized controlled trial. JAMA 295:1003–1010
17. Kang HJ, Kim HS, Zhang SY et al (2004) Effects of intracoronary infusion of peripheral blood stem-cells mobilised with granulocyte-colony stimulating factor on left ventricular systolic function and restenosis after coronary stenting in myocardial infarction: the MAGIC cell randomised clinical trial. Lancet 363:751–756
18. Kang HJ, Kim HS, Koo BK et al (2007) Intracoronary infusion of the mobilized peripheral blood stem cell by G-CSF is better than mobilization alone by G-CSF for improvement of cardiac function and remodeling: 2-year follow-up results of the myocardial regeneration and angiogenesis in myocardial infarction with G-CSF and intra-coronary stem cell infusion (MAGIC cell) 1 trial. Am Heart J 153:237.e1–e8
19. Belardinelli R, Belardinelli L, Shryock JC (2001) Effects of dipyridamole on coronary collateralization and myocardial perfusion in patients with ischaemic cardiomyopathy. Eur Heart J 22:1205–1213
20. Losordo DW, Vale PR, Symes JF et al (1998) Gene therapy for myocardial angiogenesis: Initial clinical results with direct myocardial injection of phVEGF165 as sole therapy for myocardial ischemia. Circulation 98:2800–2804
21. Rosengart TK, Lee LY, Patel SR et al (1999) Angiogenesis gene therapy: phase I assessment of direct intramyocardial administration of an adenovirus vector expressing VEGF121 cDNA to individuals with clinically significant severe coronary artery disease. Circulation 100:468–474

22. Symes JF, Losordo DW, Vale PR et al (1999) Gene therapy with vascular endothelial growth factor for inoperable coronary artery disease. Ann Thorac Surg 68:830–836; discussion 836–837
23. Vale PR, Losordo DW, Milliken CE et al (2000) Left ventricular electromechanical mapping to assess efficacy of phVEGF(165) gene transfer for therapeutic angiogenesis in chronic myocardial ischemia. Circulation 102:965–974
24. Sarkar N, Ruck A, Kallner G et al (2001) Effects of intramyocardial injection of phVEGF-A165 as sole therapy in patients with refractory coronary artery disease—12-month follow-up: angiogenic gene therapy. J Intern Med 250:373–381
25. Rosengart TK, Lee LY, Patel SR et al (1999) Six-month assessment of a phase I trial of angiogenic gene therapy for the treatment of coronary artery disease using direct intramyocardial administration of an adenovirus vector expressing the VEGF121 cDNA. Ann Surg 230:466–470; discussion 470–472
26. Stewart DJ, Hilton JD, Arnold JM, Gregoire J et al (2006) Angiogenic gene therapy in patients with nonrevascularizable ischemic heart disease: a phase 2 randomized, controlled trial of AdVEGF(121) (AdVEGF121) versus maximum medical treatment. Gene Ther 13:1503–1511
27. Giordano FJ, Ping P, McKirnan MD et al (1996) Intracoronary gene transfer of fibroblast growth factor-5 increases blood flow and contractile function in an ischemic region of the heart. Nat Med 2:534–539
28. Laitinen M, Hartikainen J, Hiltunen MO et al (2000) Catheter-mediated vascular endothelial growth factor gene transfer to human coronary arteries after angioplasty. Hum Gene Ther 11:263–270
29. Vale PR, Losordo DW, Milliken CE et al (2001) Randomized, single-blind, placebo-controlled pilot study of catheter-based myocardial gene transfer for therapeutic angiogenesis using left ventricular electromechanical mapping in patients with chronic myocardial ischemia. Circulation 103:2138–2143
30. Losordo DW, Vale PR, Hendel RC et al (2002) Phase 1/2 placebo-controlled, double-blind, dose-escalating trial of myocardial vascular endothelial growth factor 2 gene transfer by catheter delivery in patients with chronic myocardial ischemia. Circulation 105:2012–2018
31. Hedman M, Hartikainen J, Syvanne M et al (2003) Safety and feasibility of catheter-based local intracoronary vascular endothelial growth factor gene transfer in the prevention of postangioplasty and in-stent restenosis and in the treatment of chronic myocardial ischemia: phase II results of the kuopio angiogenesis trial (KAT). Circulation 107:2677–2683
32. Kastrup J, Jorgensen E, Ruck A et al (2005) Direct intramyocardial plasmid vascular endothelial growth factor-A165 gene therapy in patients with stable severe angina pectoris A randomized double-blind placebo-controlled study: the euroinject one trial. J Am Coll Cardiol 45:982–988
33. Gyongyosi M, Khorsand A, Zamini S et al (2005) NOGA-guided analysis of regional myocardial perfusion abnormalities treated with intramyocardial injections of plasmid encoding vascular endothelial growth factor A-165 in patients with chronic myocardial ischemia: subanalysis of the EUROINJECT-ONE multicenter double-blind randomized study. Circulation 112(9 suppl):I157–I165
34. Ripa RS, Jorgensen E, Wang Y et al (2006) Stem cell mobilization induced by subcutaneous granulocyte-colony stimulating factor to improve cardiac regeneration after acute ST-elevation myocardial infarction: result of the double-blind, randomized, placebo-controlled stem cells in myocardial infarction (STEMMI) trial. Circulation 113:1983–1992
35. Stewart DJ, Kutryk MJ, Fitchett D et al (2009) VEGF gene therapy fails to improve perfusion of ischemic myocardium in patients with advanced coronary disease: results of the NORTHERN trial. Mol Ther 17:1109–1115
36. Kastrup J, Jorgensen E, Fuchs S et al (2011) A randomised, double-blind, placebo-controlled, multicentre study of the safety and efficacy of BIOBYPASS (AdGVVEGF121.10NH) gene therapy in patients with refractory advanced coronary artery disease: the NOVA trial. EuroIntervention 6:813–818
37. Grines CL, Watkins MW, Helmer G et al (2002) Angiogenic gene therapy (AGENT) trial in patients with stable angina pectoris. Circulation 105:1291–1297
38. Grines CL, Watkins MW, Mahmarian JJ et al (2003) A randomized, double-blind, placebo-controlled trial of Ad5FGF-4 gene therapy and its effect on myocardial perfusion in patients with stable angina. J Am Coll Cardiol 42:1339–1347

39. Henry TD, Grines CL, Watkins MW et al (2007) Effects of Ad5FGF-4 in patients with angina: an analysis of pooled data from the AGENT-3 and AGENT-4 trials. J Am Coll Cardiol 50:1038–1046
40. Kukula K, Chojnowska L, Dabrowski M et al (2011) Intramyocardial plasmid-encoding human vascular endothelial growth factor A165/basic fibroblast growth factor therapy using percutaneous transcatheter approach in patients with refractory coronary artery disease (VIF-CAD). Am Heart J 161:581–589
41. Ruel M, Beanlands RS, Lortie M et al (2008) Concomitant treatment with oral L-arginine improves the efficacy of surgical angiogenesis in patients with severe diffuse coronary artery disease: the endothelial modulation in angiogenic therapy randomized controlled trial. J Thorac Cardiovasc Surg 135:762–770, 770.e1

Chapter 18
Angiogenesis in Cancer

Gerald W. Prager and Christoph C. Zielinski

Abstract Intensive research on molecular mechanisms responsible for tumor angiogenesis has opened an abundance of potential therapeutic options, whereby first anti-angiogenic strategies have already led to improved prognosis in certain cancers. Thereby, the main angiogenic growth factor, the vascular endothelial growth factor (VEGF), has first become the focus of therapeutic interventions in advanced colorectal cancer, non-small cell lung cancer, breast cancer, ovarian cancer, glioblastoma, or renal cell cancer. However, not all patients benefit from anti-VEGF treatment, and those who do respond to such treatment may finally become resistant to anti-angiogenic therapies and relapse. This might lie in the fact that tumor angiogenesis is not only induced by VEGF but by many other VEGF-independent mechanisms. Furthermore, anti-VEGF therapies provoke hypoxia-induced upregulation of pro-angiogenic molecules. This book chapter covers main molecular mechanisms involved in tumor angiogenesis and gives insights in future therapeutic strategies.

Keywords Tumor angiogenesis • Endothelial cells • Integrins • Urokinase • VEGF

18.1 Introduction

The involvement of the vascular system in tumor growth was first described more than 100 years ago by Goldman [1]. That the formation of new blood vessels form preexisting vasculature, the so-called angiogenesis, is essential for tumor growth and metastasis formation, and interfering with this process disrupts progression of

---

G.W. Prager, M.D. (✉) • C.C. Zielinski
Clinical Division of Oncology, Department of Medicine I, Medical University of Vienna, Comprehensive Cancer Center Vienna, Waehringer Guertel 18-20, 1090 Vienna, Austria
e-mail: gerald.prager@meduniwien.ac.at

J.L. Mehta and N.S. Dhalla (eds.), *Biochemical Basis and Therapeutic Implications of Angiogenesis*, Advances in Biochemistry in Health and Disease 6,
DOI 10.1007/978-1-4614-5857-9_18,

disease, was first proposed by Folkman in 1971 [2]. After extensive research on basic mechanisms involved in tumor angiogenesis, the major pro-angiogenic growth factor, the vascular endothelial growth factor (VEGF), has soon become the focus of therapeutic interventions. While under physiological conditions VEGF is regulated as a hypoxia-regulated gene via binding of hypoxia-inducible factor (HIF) to its promoter, VEGF is highly overexpressed by a variety of different cancers [3]. VEGF-A induces every single angiogenic step: (1) plasma protein transudation into the extracellular space via increase in endothelial permeability, which offers a provisional matrix as a scaffold for invaded endothelial cells; (2) cell survival, which compensates for anoikis, the so-called apoptosis upon detachment. Cell detachment is observed when endothelial cells emigrate from the basal membrane into surrounding tissues; (3) endothelial cell migration and invasion; (4) endothelial cell proliferation; and finally (5) capillary-tube formation. In endothelial cells VEGF-A mainly binds to its receptor-1 (flt-1) as well as to VEGFR-2 (flk-1). Recent reports have also demonstrated neuropilin-1 as well as neuropilin-2 to be involved in VEGF-induced endothelial cell migration [4]. In this context, recent clinical reports have suggested that neuropilin levels of tumor tissues predict for anti-VEGF therapies in gastric cancer [5]. Bevacizumab (Genentech/Roche) is an FDA- and an EMEA-approved monoclonal antibody targeting VEGF-A for the treatment of certain cancers. The fusion protein aflibercept (Regeneron Pharmaceuticals and Sanofi-Aventis Pharmaceuticals) is a soluble recombinant decoy VEGF receptor composed of domain 2 of VEGFR-1 and domain 3 of VEGFR-2 fused to the Fc of IgG1. It blocks all VEGF-A isoforms, VEGF-B, and the placental-like growth factor (PlGF). Both anti-VEGF molecules have been shown to be effective after progression upon a bevacizumab containing first-line regiment in colorectal cancer, but aflibercept revealed an increased toxicity profile when crossover compared to VEGF [6]. In addition, small-molecule inhibitors of the VEGFR-2 tyrosine kinase such as sorafenib, sunitinib, afatinib, and pazopanib are approved for certain tumor types. However, not all patients benefit from an anti-VEGF approach, and patients who do benefit might become resistant. In fact, besides angiogenesis tumors are able to induce a variety of mechanisms to increase the vascular network: (1) Tumors can abduct the existing vasculature, a process which is called vessel co-option; (2) preexisting blood vessels can split into two vessels, which is called intussusception; (3) tumor cells can align along preexisting vessels, a phenomenon called vascular mimicry; and (4) the involvement of endothelial progenitor cells (EPCs) derived from the bone marrow might also participate in the formation of tumor vessels. Finally, tumor stem cells can also generate tumor endothelium [7]. In addition, tumor angiogenesis does not only depend on VEGF but is induced by a variety of different growth factors, proteolytic enzymes, and adhesion molecules. Furthermore, hypoxia induced by anti-VEGF treatments leads to upregulation of hypoxia-induced proteins, such as proteolytic enzymes, which can lead to tumor propagation and metastasis formation [8]. The following subchapters will focus on angiogenic mechanisms beyond VEGF, which provide potential novel therapeutic strategies in cancer, and we apologize for any omissions that this focus engenders.

18.2 Growth Factors Beyond VEGF

18.2.1 Fibroblast Growth Factor

The family of the fibroblast growth factor consists of 18 members binding to four homologous high-affinity FGF receptors [9]. Among them, FGF-2 via binding to its receptor FGFR-1 has been shown to promote angiogenesis in a VEGF-independent manner [10]. While FGF-2 is highly overexpressed in different types of tumors [11], gene silencing by antisense mRNA led to a dramatic reduction in tumor size [12]. That FGF-1 and FGF-2 expression becomes highly upregulated whenever VEGF/VEGFR pathway is targeted has been demonstrated in a RIP-tag model, and co-targeting FGF-2 via an FGF trap together with an anti-VEGF treatment was able to decrease tumor growth by a reduction of tumor vascularization [10, 13]. A therapeutic benefit by reduction in FGF-2/FGFR-1 signaling was so far demonstrated by the use of thalidomide or suramin when used in advanced prostate or renal cell cancer [12]. Another approach is to interfere with FGF/FGFR signaling represented by FGF-traps. The FGF trap FP-1039 (Five Prime Therapeutics) is thought to interfere with FGFR-induced proliferation and angiogenesis. It is currently tested in clinical Phase II trials for activity and safety in advanced and recurrent endometrial cancer (http://ClinicalTrials.gov).

18.2.2 Platelet-Derived Growth Factor

The growth factor family of platelet-derived growth factor (PDGF) contributes to mature newly formed blood vessels: Activated endothelial cells release PDGF-B to chemo-attract PDGF receptor-β pericytes [7]. Interfering with PDGF-B binding to its receptor diminished tumor growth by pericyte detachment, which resulted to immature vessel formation, and vessel regression [14]. Paradoxically, overexpression of PDGF-B inhibited tumor growth by promoting pericyte recruitment, which induced downregulation of endothelial cell proliferation [15]. Another dual role of PDGF-B/PDGFR family has been demonstrated in metastasis formation, while in primary tumors pericyte recruitment blocked tumor cell intravasation by reduction of tumor-vessel leakage and increasing the barrier function; pericytes at pre- and micrometastatic niches are thought to promote tumor colonization by expression of angiogenic growth factors [7]. In fact, FGF-2 was shown to overexpress PDGFR-α and PDGFR-β on transcriptional levels in endothelial cells, which increased the responsiveness to PDGF-BB [16]. The role of PDGF-C in tumorigenesis is discussed below. Taken together, future clinical studies are required to examine benefit for specific PDGF/PDGFR targeting, while less specific tyrosine kinase inhibitors such as pazopanib, which targets not only PDGFR-α/β but also VEGFR family as well as c-kit, have been shown to be efficient in certain cancers.

18.2.3 Delta/Jagged Family

Delta 1-like ligand 4 (Dll4), a member of the Delta/Jagged family of transmembrane ligand of the Notch receptor, mediates cell–cell communication and controls cell determination. Thus, Delta–Notch pathway is regulating proper vascular development which is crucial for artery–vein differentiation but also enables blood vessel formation [17]. In detail, Notch signaling blocks endothelial cell proliferation by downregulation of VEGFR-2, inducing cell cycle arrest in the G0/G1 phase as well as regulation of Erk1/2 and Akt phosphorylation. It also regulates pericyte recruitment and regulates vessel branches (summarized by Sainson and Harris [18]). Genetic knockout led to abnormal vessel sprouting, while blockage of the Dll4-Notch pathway in tumors resulted in excessive but hypoperfused nonproductive angiogenesis. This resulted in inhibitory effects on tumor growth even after progress upon anti-VEGF treatment [19]. Inhibition of Dll4 or Jagged1 may prove to be effective in the treatment of cancer by interfering with functional vascular network formation.

18.2.4 Wnt/Frizzled

Wnts are a family of 19 secreted glycoproteins that accumulate in the extracellular matrix to activate pathways in adjacent cells [20]. They bind to their frizzled (FZD) receptors, which are G-proteins, and stimulate endothelial cell proliferation and cell survival. Furthermore, the wnt/β-catenin pathway was demonstrated to regulate VEGF-A gene transcription [21]. Antagonizing wnt by soluble frizzled-related protein-4 (sFRP-4) has been demonstrated to inhibit endothelial cell migration and sprout formation leading to restricted tumor growth in an in vivo tumor model [22]. Because some members of the Wnt family have distinct functions, such as preservation of the brain-barrier function, specific inhibitors for therapeutic interference with wnt/frizzled have to be synthesized. That the wnt-signaling pathway is a promising target in cancer is reflected by the fact that aberrant regulation of the wnt pathway leads to neoplastic proliferation of various cancer types [23]. Notch stimulates wnt/frizzled-induced endothelial cell proliferation in stalk cells, while wnt can also activate the Notch pathway in a positive feedback loop. This leads to a Notch-like phenotype induced by the wnt/frizzled pathway, which is characterized by branching defects, aberrant vascular remodeling, and loss of venous identity.

18.2.5 Ang and Tie Signaling System

The human Ang–tie system consists of three secreted ligands (Ang-1, Ang-2, and Ang-4), binding to two tyrosine kinase receptors (tie-1 and tie-2). Ang-1 functions as a tie-2 receptor agonist [24], and Ang-2 normally functions as an Ang-1 antagonist [25].

Thus, the angiopoietin system has a dual function in angiogenesis, while Ang-1 induces tie-2 clustering in trans at cell–cell junctions, thereby promoting vessel integrity and quiescence, leading to inhibition of tumor cell extravasation; it can also promote tumor growth by inducing endothelial cell survival [26]. These conflicting biologic activities might reflect difficulties in targeting Ang-1 for anticancer treatment. In contrast to Ang-1, Ang-2 might be the more promising target to interfere with tumor angiogenesis as it promotes tumor angiogenesis and recruits tie-2 expressing monocytes (TEMs), which are thought to produce growth factors and matrix-remodeling enzymes that may activate ECs and facilitate the local growth of new blood vessels [27]. Targeting Ang-2 has been demonstrated to lead to vessel regression and double targeting of VEGF-A and Ang-2 led to reduction in tumor angiogenesis, metastasis, and vascular leakage. Thereby, Ang-2 affected early stages of tumor development and vessel maturation but seemed to be dispensable for later stages of tumor growth. Tie-2 presents an attractive opportunity for therapeutic intervention as it is not only crucial for angiogenesis and vascular homeostasis but it also provides an important link between angiogenic and inflammatory pathways. Currently, several early clinical trials are testing efficacy of anti-angiopoietin system therapies [28].

18.3 Tumor Microenvironment

Initial research on tumor angiogenesis has focused on growth factors release by tumor cells, which stimulate pro-angiogenic endothelial cell behavior [3, 29]. It is now accepted that cells of the tumor microenvironment such as fibroblasts, pericytes, mesenchymal stem cells, as well as cells of the immune system also participate in angiogenesis via secretion of angiogenic cytokines or modulation of the immune system [30]. Hypoxia recruits bone marrow-derived cells such as TEMs, tumor-associated macrophages (TAM), neutrophils, mast cells, and CD11b+Gr-1+ myeloid-derived suppressor cells, which release pro-angiogenic molecules such as VEGF, MMPs, and BV8 (also known as prokineticin-2). Especially in anti-VEGF refractory cancer, BV8 secretion by CD11b+Gr-1+ cells is frequently observed. In detail, CD11b+Gr-1+ cells, which are highly recruited by tumors, secrete BV8, which results in mobilization of hematopoietic stem cells into the peripheral blood and local stimulation of angiogenesis [31]. Interestingly, the well-known inducer of granulocyte-progenitor cell differentiation and mobilization, G-CSF, highly upregulates expression of BV8. Thus, functional blocking anti-G-CSF treatment has been shown to abolish the increase in BV8 secretion in CD11b+Gr-1+ cells after tumor transplantation in a mouse model [32], leading to a reduced number of circulating as well as tumor-associated CD11b+Gr-1+ cells. Consistently, targeting of either G-CSF or BV8 by functional blocking antibodies reduced tumor growth by blocking angiogenesis. Notably, the effect was independent and additive to anti-VEGF mAB as tumor refractory to anti-VEGF treatment responded to anti-BV8 approaches. Moreover, G-CSF treatment led to a reduced responsiveness to anti-VEGF therapy in cancer [33]. Although the clinical relevance is discussed controversial, administration of G-CSF has been correlated with enhanced angiogenesis and tumor growth [34, 35]. Further studies are required to evaluate a therapeutic role of targeting BV8.

18.3.1 Fibroblasts

Tumor-associated fibroblasts (TAFs) promote tumor angiogenesis via secretion of the stroma-derived growth factor-1 (SDF-1), which recruits EPCs to the tumor, thereby promoting tumor angiogenesis. Although there is a considerable controversy about the role of bone marrow-derived EPCs in tumor angiogenesis, there is some evidence that the stem-cell-like EPCs are able to differentiate into tumor endothelium [36], which might be essential especially in early stages of tumor development [37]. Another pro-angiogenic mechanism to induce tumor vascularization by TAFs was recently reported by Crawford et al. [38], who demonstrated that TAFs in anti-VEGF resistant tumors highly express platelet-derived growth factor (PDGF-C). PDGF-C induced angiogenic cell behavior in endothelial cells, pericytes, as well as in smooth muscle cells, which led to enhanced cell proliferation, migration, and cell survival. In VEGF-deficient fibrosarcomas, tumor angiogenesis developed in the absence of VEGF, and the PDGFR-α-mediated recruitment of stromal fibroblasts, was believed to be responsible for the persistent tumor angiogenesis [39], further affirming a VEGF-independent tumor angiogenesis induced by PDGF-C. Thus, a combination therapy against VEGF and PDGF-C using functional blocking antibodies was more effective than anti-VEGF treatment alone. Consistently, tumors, which highly express PDGF-C, such as pancreatic cancer, do not respond towards anti-VEGF targeting treatment regiments [40, 41]. However, as neither sorafenib nor imatinib have been shown to be effective in pancreatic cancer, secondary mechanisms of PDGF-C—such as pericyte recruitment to avoid metastasis formation—might be involved [42–44].

18.3.2 The Proteolytic Systems

For efficient endothelial cell invasion, proteolytic enzymes for matrix degradation are required. Among them, matrix metalloproteinases (MMPs) and the urokinase/plasminogen system have been shown to be the most effective families of proteases during angiogenesis.

18.3.2.1 Plasminogen Activation System

First described as the main system for fibrinolysis for clot degradation in hemostasis, the urokinase/plasminogen system is now accepted as a major contributor for angiogenic endothelial cell behavior. The inactive proenzyme plasminogen is converted into active plasmin by one of the two serine proteases, either the urokinase plasminogen activator (uPA) or the tissue-type plasminogen activator (tPA). The plasminogen activation inhibitors-1 and -2 (PAI-1 and PAI-2) inhibit tPA as well as uPA, thereby tightly regulating pro- and anticoagulative activities. While by genetically knockout studies have revealed that tPA is primarily involved in intravascular fibrinolysis, the uPA system is mainly involved in cell migration and tissue remodeling [45]. Furthermore, uPA also contributes to many important physiological processes such as ovulation, inflammation, and wound healing [46, 47].

In angiogenesis, the urokinase/plasminogen system has been shown to play a central role: Mice which were deficient in any of the key players of the fibrinolytic system, uPA, uPAR, or PAI 1, revealed impaired VEGF-induced angiogenesis, which led to defects in wound healing and tumor progression [48–51].

18.3.2.2 Urokinase Plasminogen Activator/Urokinase Plasminogen Activator Receptor/Plasminogen Activation Inhibitor-1

uPA is secreted as an inactive single chain zymogen, pro-urokinase (pro-uPA), or single chain-urokinase (sc-uPA), which is cleaved into the active two-chain form, uPA. It consists of three domains: (1) an amino-terminal growth factor domain, (2) a kringle domain, and (3) a carboxy terminal domain, whereby the latter one is catalytically active. The amino-terminal growth factor domain is necessary for binding to its cell surface receptor, the urokinase plasminogen activator receptor (uPAR, CD87) [52]. uPAR is a glycophosphatidylinositol (GPI)-anchored protein, which is enriched in glycosphingolipid rafts. It binds the inactive pro-uPA and the active uPA with high affinity (1 nM). To interact with uPA, uPAR forms a pocket-like binding site, which involves all three domains of uPAR. Thus, uPAR can concomitantly interact with other molecules via lateral binding motifs [53]. Thereby, besides focusing the proteolytic activity of uPA to the cell surface of activated cells, uPAR also serves as a signaling receptor because uPAR, which lacks a transmembrane domain cis-interacts with transmembrane proteins, such as integrin adhesion receptors, receptor tyrosine kinases (RTKs) (EGF receptor), G-protein coupled receptors, or M6P receptor. uPAR interaction with other molecules is thereby regulated by several mechanisms: (1) its expression, which is induced by major pro-angiogenic growth factors such as VEGF-A or FGF-2 [54] or by hypoxia; (2) by uPA binding, which is thought to induce a conformational change, thereby decrypting binding sites on the surface of uPAR; and (3) by internalization and recycling of uPAR, which is a tightly regulated process. Especially the subcellular distribution is thought to be essential for effective endothelial cell migration and invasion. In detail, whenever endothelial cells become activated by angiogenic growth factors, uPAR as well as uPA and PAI-1 become transcriptionally upregulated. uPA is expressed and secreted as its inactive zymogen pro-uPA, which binds to the cell surface via uPAR where it can become activated via enzymes, among them most prominently plasmin. Active uPA activates plasmin in a positive feedback loop and cleaves a range of extracellular matrix (ECM) proteins. It leads to degradation and clearance of fibrin blood clots (fibrinolysis) and activates matrix metalloproteases (MMPs), such as MMP3, MMP9, MMP12, and MMP13. Enzymatic degradation of the extracellular matrix by these enzymes is thought to facilitate endothelial cell migration and invasion during angiogenesis. That the enzymatic activity is thereby focused to the leading edge of migrating cells was described more than 20 years ago [55]. The subcellular distribution mechanisms which guarantees localized proteolytic activity, however, was described recently: VEGF activation of endothelial cells leads to a PI3kinase-dependent change in integrin beta-1 activity in endothelial cells, which leads to MT1-MMP as well as MMP-2 activation. As a consequence, pro-uPA becomes activated on the surface of endothelial cells [56]. Eventually, uPAR-bound active uPA becomes inhibited by the serpin plasminogen activator inhibitor-1, and a ternary complex is formed. The uPAR/uPA/PAI-1 complex becomes internalized via clathrin-coated pits, which is mediated by members of the LDLR-system. In endothelial cells VLDL and ApoER-2 are the most common representatives of the LDLR family. While uPA/PAI-1 become degraded by the lysosome, naked uPAR can recycle back to the cell surface [57], whereby recycled uPAR is focused to newly formed focal contacts at the leading edge of migrating endothelial cells [48]. Interfering with uPAR-complex binding to members of the LDLR family by the receptor-associated protein (RAP), cleavage of the GPI-anchor of uPAR by phosphatidylinositol-specific phospholipase C (PI-PLC), or genetically knockout of uPAR led to abrogation of VEGF-induced endothelial cell migration in vitro and in vivo [48]. Finally, in confluent grown endothelial cells, the density-enhanced phosphatase-1 (DEP-1), which negatively regulates the VEGF-dependent activation of ERK1/2, led to downregulation of uPAR expression. Overexpression of active ERK1 rescued the DEP-1 effect on uPAR. Notably, high DEP-1 expression was capable to interfere with angiogenic endothelial cell behavior, which was reflected by diminished endothelial cell migration, proliferation, and capillary-like tube formation in vitro. The biological role in regulating angiogenesis was further supported by the fact that a specific blocking antibody against DEP-1 was capable to interfere with angiogenesis [58]. However, intermediate levels of DEP-1 expression were shown to induce src-family kinase activation [59], a process which is thought to support angiogenic cell functions. Thus, different levels of DEP-1 expression may lead to a biphasic function in angiogenesis. Besides its proteolytic activities, uPAR is also known as an inducer of intracellular signal transduction, thereby mediating cell adhesion, migration, proliferation, and differentiation [60]. As uPAR is a GPI-anchored protein, cis-interactions with transmembrane proteins is a prerequisite for induction of intracellular signaling events.

18.3.2.3 Urokinase Plasminogen Activator Receptor and Low-Density Lipoprotein Receptor-Related Protein 1

Low-density lipoprotein receptor-related protein 1 (LRP1) belongs to the LDL receptor-related protein gene family. It is a large transmembrane protein comprising a heavy chain of 515 kDa which is a ligand-binding subunit and a light chain of 85 kDa which is the transmembrane domain. This scavenger receptor is capable of bind and endocytose multiple ligands. It recognizes more than 30 extracellular ligands including uPA/tPA, PAI 1, apolipoprotein E, α2-macroglobulin, LDL, lactoferrin, and VLDL. Besides its role in internalization of uPA/PAI-1/uPAR-complexes as described above, the internalization via LRP to the endosomes also activates the MAP kinase pathway, thereby initiating cell migration on vitronectin-coated plates and invasion into matrigel. It has further been reported that LRP 1 regulates the activation of Rac 1 in an uPAR-dependent manner [61]. Rac1 activation was reduced in uPAR$^{-/-}$ MEFs leading to decreased cell migration on vitronectin [62]. Interference with uPAR-complex formation affects not only plasminogen activation and uPA-mediated ECM degradation but also interferes with cell migration and cell signaling [63, 64].

Recently, the simultaneous activation of LRP and tyrosine kinase receptor (e.g., EGFR) were shown to initiate sustained ERK activation, which led to enhanced adhesion of HT080 fibrosarcoma cells to vitronectin and an increased cell migration. This effect was dependent on the presence of uPA, uPAR, and LRP 1 as sustained ERK activation was absent in uPA ($^{-/-}$), uPAR ($^{-/-}$), and LRP 1($^{-/-}$) fibroblasts [65]. Consistently, Carroll et al. described that high PAI 1 expressing HT1080 cells induced more lung metastasis in mice due to PAI-1-mediated cell signaling. This was explained by the observation that PAI-1 downregulated dual-specific phosphatases that in turn led to sustained activation of ERK1/2 and enhanced surface expression of β3 and β5 integrins.

18.3.2.4 Interaction of uPAR with Integrins

Besides its scaffold function as a linker between the extracellular matrix and the cytoskeleton, integrin adhesion receptors also induce intracellular signal transduction upon cell adhesion (see below). While adhesion-induced signal transduction is also induced in the absence of the urokinase receptor system, recent studies have demonstrated that for efficient uPAR signaling, the integrin family is required. Consistently, uPAR co-localizes with integrins in focal adhesions [66–68]. A direct interaction of purified uPAR and integrin α5β1 was suggested [69]; thereby binding of uPAR to the beta-propeller of α3β1 empowers vitronectin adhesion by integrins. How uPAR modifies other β1 integrins was recently suggested by Dr. Chapmans group, who found that uPAR resulted in RGD-independent binding of α5β1-uPAR to Fn type III repeats 12–15 in addition to type III repeats 9–11 bound by α5β1. The mechanism suggested an uPAR-induced modified "bent" integrin conformation [69].

In uPAR overexpressing tumor cells, uPAR was demonstrated to interact with α5β1 integrins, which led to activation of ERK, a central kinase in tumorigenicity. When uPAR/α5β1 integrin interaction was blocked, e.g., by inhibitory peptides, ERK activation was reduced and forced tumor cells into dormancy [70]. When uPAR was overexpressed in human embryonic kidney (HEK) cells, uPAR/integrin β1 complex induced an increased cell adhesion of these cells on to vitronectin through the activation of Src kinase and focal adhesion kinase (FAK). In many studies, uPAR was observed to mediate signaling via integrins; this leads to the activation of FAK, Src, and MAP kinase, which orchestrate cell adhesion, migration, invasion, and tumor metastasis [67, 71–73].

18.3.2.5 uPAR Interaction with M6P Receptor and GPCRs

uPAR interacts directly with the N-terminus of the M6P receptor enhancing the proteolytic cleavage of uPAR and interferes the plasminogen activation, cell adhesion on various matrices, and cell invasion. Lentiviral-mediated silencing of M6P receptor in human endothelial cells upregulated uPA, uPAR, and αvβ3 integrins, which led to enhanced cell adhesion and cell migration on various matrix proteins [74]. When uPA/uPAR complex interacts with the M6P receptor, uPA mediates the proteolytic cleavage of uPAR and the soluble domain 1 of uPAR is released. Cleaved uPAR cannot participate in plasminogen activation but interacts with G-protein coupled receptor FPRL1, a homologue of formyl peptide receptor (FPRL1), thereby enhancing cell migration [75].

18.3.2.6 Interaction of uPAR with Receptor Tyrosine Kinase

The "uPAR interactome," the so-called multimolecular interaction partners of uPAR, also includes interaction of uPAR with RTKs such as PDGF receptor, VEGF receptor, and EGF receptor. Thus, elevated levels of uPAR led to a ligand-independent EGF-receptor signaling. uPAR further potentiated the association of EGFR and αvβ1 integrin, activating FAK and ERK signaling pathways, which resulted in enhanced cell proliferation in an MAPK-dependent manner [76]. Antagonizing EGFR blocked uPAR-initiated cell proliferation, thus, suggesting that EGFR is required for the ERK activation and cell proliferation initiated by uPAR/α5β1 integrin association.

In another study by Cavallaro et al., it was demonstrated that FGF-2-stimulated cell migration was blocked by antibodies against uPA or uPAR [77], which is consistent with our observation that a variety of angiogenic growth factors induce uPAR-associated fibrinolytic activity and the distribution of the uPAR system to the leading edge of migrating endothelial cells to promote cell migration in vitro and in vivo.

LRP also functions as a co-receptor for PDGFR, thereby mediating signal transduction, such as ERK phosphorylation [78]. Notably, this was dependent on uPA, and uPAR, as the $uPA^{-/-}$ and $uPAR^{-/-}$ cells could not induce sustained ERK activation [65].

18.3.2.7 Plasminogen Activation System and Cancer

High levels of uPA and uPAR in tumors have been correlated with aggressiveness of cancer. Paradoxically, the uPA inhibitor PAI 1 was described as a negative predictor of disease outcome in a variety of tumors. uPA and PAI-1 have become the only level of evidence-1 tumor marker when predicting outcome in node-negative breast cancer [79]. Also in other cancers such as esophageal adenocarcinoma [80], gastric cancer [81], lung cancer [82] (Oka, 1991 3/id), ovarian cancer [83], or colorectal cancer [84, 85], the uPA system was suggested as a prognostic or predictive marker. Consistently, we observed that high PAI-1 expressing fibrosarcoma cells were prone for lung metastasis formation in an in vivo model (Carroll, 2000 34/id).

18.4 Integrin Adhesion Receptors

Recent advances in cancer research have identified the important contribution of the "tumor microenvironment" to tumorigenesis. Thus, tumor biology cannot be restricted to enumerating the traits of tumor cells but encompasses somatic cells recruited by the tumors, which form the tumor-associated stroma [86]. To interact with the tumor stroma or other cell types, integrin adhesion receptors are the major adaptors. Cell adhesion via integrins, furthermore, mediates intracellular signaling, thereby affecting cell behavior such as cell migration, invasion, cell survival, and cell proliferation. Especially in endothelial cells, integrin-mediated cell activation is thought to be essential for every step in angiogenesis [87].

Integrin adhesion receptors are heterodimeric molecules, which consist of 1 of 18 alpha chains and 1 of 8 beta chains, forming 24 different heterodimers, which are characterized by respective ligand specificities [88]. Integrin subunits are formed by large extracellular domain, a single transmembrane domain and a short cytoplasmic domain [89]. The α and β subunits of integrins are linked via non-covalent bonds.

β1 subunits are the major integrins expressed in resting endothelial cells, where they recognize matrix proteins of the basal membrane or interact in differentiated blood vessels with pericytes via their cell surface receptor vascular cell adhesion molecule-1 (VCAM-1) [90].

During angiogenesis, activated endothelial cells express the broad-ligand-specific αvβ3 and αvβ5 integrins, which bind the Arg–Gly–Asp (RGD) motif of ECM proteins, among them vitronectin, fibronectin, fibrinogen, osteopontin, and Cyr61. The switch in class of integrin expression during angiogenesis to the broad-ligand-specific αvβ3 and αvβ5 integrins supports invaded endothelial cells for angiogenic endothelial cells behavior by guaranteeing cell attachment, adhesion-induced signal transduction, and cell survival (Fig. 18.1).

When endothelial cells lose cell–matrix contacts during invasion into surrounding tissue, they become prone to apoptosis, the so-called anoikis upon detachment. Thus, anti-apoptotic signals, which are provoked by angiogenic growth factors as well as adhesion-induced signal transduction via integrins, are essential to guarantee any further steps in angiogenesis [91].

Since they have no intrinsic kinase activity, integrin cytoplasmic tails interact with several kinases and related adaptor proteins. Thereby, the β subunit is adjuvant as the major site for recruiting signaling molecules. In detail, upon cell adhesion to

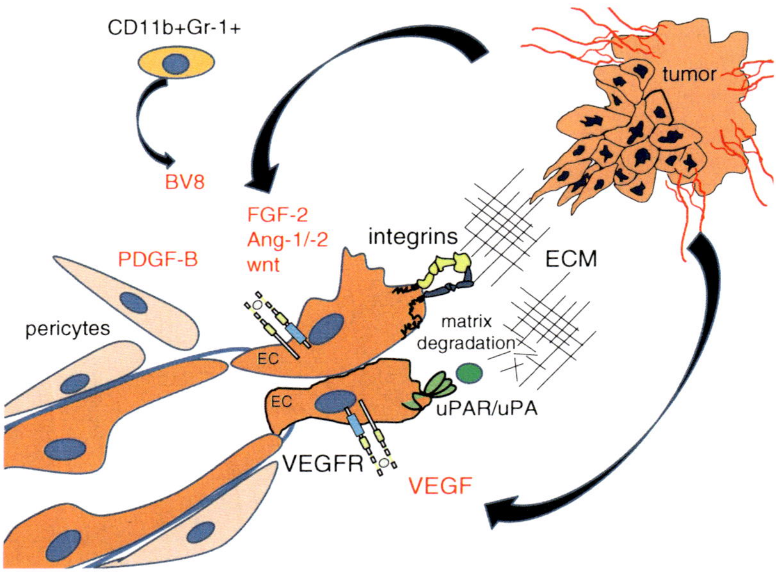

Fig. 18.1 Cartoon illustrating tumor-induced angiogenesis: growth factors secreted by the tumor activate endothelial cells. Activated endothelial cells become polarized and start to transmigrate through the basal membrane into the surrounding matrix. On the leading edge, endothelial cells express integrins for cell adhesion as well as proteolytic enzymes for matrix degradation

a integrin ligand matrix protein, pp125 FAK becomes phosphorylated at tyrosine 397, leading to activation of src-family kinases (SFK), which controls not only the activity of Rho GTPases but also "superactivates" FAK, which leads to p130cas phosphorylation. The so formed signaling complex induces downstream signaling cascades such as PI3kinase/AKT, which affects endothelial cell proliferation, migration, and survival, or Mek/Erk, which induces cell proliferation [92, 93, 94]. Besides transducing signals from the outside to the inside (outside-in), integrins also operate as bidirectional transducer molecules by matching signals from the inside to the outside (inside-out), which induces activation of integrins. Thereby, the cytoplasmic tail of β-integrins interacts with the FERM domain of talin, which leads to high-affinity conformation of integrins and enables ligand binding, thereby linking the ECM to the cytoskeleton [95]. Alternatively, kindling binding to the beta tails activates integrins [96]. Both signaling directions are tightly regulated in focal adhesion to support cell adhesion, spreading, and motility [97].

18.4.1 Integrins and Growth Factors

Integrins are able to affect cell behavior by interfering with signal transduction pathways induced by RTKs receptors for FGF-2 or VEGF-A [98]. While FGF-2 induced angiogenic endothelial cell behavior requires integrin αvβ3, VEGF, and αvβ5, integrins are thought to contribute for endothelial cell activation [99]. The distinction of these different mechanisms is given by the fact that integrins β3 and β5 affect activation of the Ras/Raf/Mek/Erk pathways in the vasculature in different ways. While αvβ5 integrins induce activation of focal adhesion kinases (FAK) and Src kinases, αvβ3 integrins activate p21 kinases [100]. In addition αvβ3 induces activation of Raf at tyrosine 340/341 and Mek-dependent protection from extrinsic-mediated apoptosis, which is initiated by pro-apoptotic ligand binding to receptors, including TNF-α and Fas. In contrast, αvβ5-affected pathways activate Raf on serine 338/339, which leads to Raf-1 mitochondrial translocation and shelter from the intrinsic pathway of apoptosis.

18.4.2 Interfering with Angiogenesis via Integrins

Various integrins have been functionally related to tumor growth and tumor-vessel formation, such as α1β1, α2β1, α4β1, α5β1, α6β1, α6β4, α9β1, αvβ3, and αvβ5 [101]. A major regulator for developmental angiogenesis and blood vessel formation is resembled by fibronectin [102]. Its receptor, the β1 integrin, is an essential mediator in angiogenesis, as the genetically knockout of β1 integrins led to embryonic lethality or blocked blood vessel development in teratomas [103, 104]. Integrins α1β1 and α2β1, which bind to laminin and collagen, seemed to be indispensable for angiogenesis [105], because functional blocking α1β1 and α2β1 integrins inhibited angiogenesis in vitro, as well as it reduced tumor growth and angiogenesis in vivo [106–108]. Another fibronectin receptor, α5 integrin, is also essential, because its loss led to vascular deformation and developmental deviations [109]. α5β1 integrins have been described to be overexpressed in endothelial cells of malignant tumors, suggesting their functional role in tumor angiogenesis [110, 111], which was supported by the observation that targeting α5β1 was able to block tumor angiogenesis [110]. That α5 is also required for embryonic vasculature development was demonstrated by its genetically knockout, which was embryonic lethal on day E10–E11 [112]. α5β1 integrin antagonists are currently under development or tested in clinical trials for the treatment of cancer. Volociximab, originally known as Eos200-4, is a humanized monoclonal antibody against α5β1 integrins [113] and is able to inhibit signaling cascades, thereby inducing cell death and preventing capillary-tube formation in vitro. Because no effects on cell proliferation were observed when volociximab was combined with an anti-VEGF antibody, it was suggested that α5β1 integrins are downstream of the VEGF signaling pathway. In vivo, volociximab revealed antitumor as well as anti-angiogenic capacities [114] without any dose-limiting toxicities in a Phase I dose escalating trial. In a Phase II study, volociximab showed promising potential efficacy in prolonging disease progression and overall survival. In clear cell renal cell carcinoma, volociximab delayed disease progression up to 22 months. Additional Phase II and III trials, in which volociximab is used as single agent as well as in combination regiments, are currently ongoing to test its effect in advanced melanoma, non-small cell lung cancer, and peritoneal cancer [113].

Although β3 integrins are highly upregulated in activated endothelial cells, patients suffering from the Glanzmann thrombasthenia, which is characterized by null mutation in integrin β3 subunit, do not have any aberrations in angiogenesis and vascular development. However, knockout studies of αv, including αvβ3, αvβ5, and all other αv combinations, revealed a deficiency in vascular integrity [115]. It was hypothesized that αvβ8 is the key αv integrin as vasculature abnormalities aroused only in β8 integrin knockouts, but not in β5 or β3 knockout mice. Paradoxically, integrin β3 knockout as well as integrin β3/β5 double knockout mice revealed enhanced tumor angiogenesis [116], which was explained by an observed upregulation of receptor tyrosine kinases, such as VEGFR-2.

Low toxicity has been recognized for various antagonists of αvβ3 integrins as they are mainly expressed in activated or remodeling tissues such as in tumor cells or its microenvironment. Therefore, they are thought to be a central target in tumor angiogenesis [117]. Abergrin (vitaxin, Medi-522) is a humanized antibody against αvβ3 integrins and was the first anti-integrin approach introduced into clinical trials for malignant tumors [118]. This antibody blocks binding to vitronectin and fibrinogen and prevents cell adhesion, migration, proliferation, and integrin-mediated cell signaling in a dose-dependent manner [119]. Tumors of patients with a high expression of αvβ3 had a poorer prognosis than patients with lower expression levels. Consistently, high expression of αvβ3 in tumors was accompanied with an aggressive phenotype and augmented metastatic potential and invasive traits. Furthermore, association of αvβ3 with Src promoted attachment-independent autonomous cell growth, while αvβ3 integrin-mediated FAK signaling fostered the survival of cancer stem cells [91].

αvβ3 integrin expression and activation is induced by angiogenic growth factors, and blocking of αvβ3 integrins restrains tumor angiogenesis and tumor growth in in vivo models [99, 120, 121]. Intetumumab (formerly named CNTO 95) is a fully human monoclonal antibody that recognizes all members of the α_v integrin family, which revealed anti-angiogenic and antitumor properties. Its functional blocking activities against α_v integrins resulted in inhibition of cell adhesion, migration, proliferation, and invasion of both tumor and endothelial cells in vitro and in vivo [122]. Intetumumab was recently tested in a clinical Phase I/II trial for the treatment of melanoma [123].

The small-molecule cilengitide (EMD 121974), a cyclic RGD-peptide inhibitor, which mimics the ligand-binding site for αvβ3 and αvβ5 integrins, has been tested in clinical Phase I and II trials and revealed enhanced progression-free survival in glioblastoma patients [124]. Cilengitide is currently examined in a clinical Phase III trial. Also for several other tumor identities, cilengitide is now in clinical Phase II studies including breast cancer, squamous cell cancer, non-small cell lung cancer, and melanoma [125–127]. Cilengitide is able to promote detachment in cells already attached to the matrix; it diminishes cell growth, cell proliferation, and cell survival in vitro. Moreover, it inhibits FAK-Src-Akt and Erk signaling pathways [128, 129]. However, role of interfering with αv integrins in tumor cells or endothelial cells is so far not fully understood, which is reflected by the preclinical observations of Reynolds et al., who described a biphasic effect of cilengitide-promoting angiogenesis and tumor growth at nanomolar levels [130].

Current concepts of anti-integrin therapeutics are focused on targeting the receptor-binding sites directly or indirectly; new functional insights, however, propose alternative approaches such as targeting most upstream integrin signaling molecules, such as FAK or SFK [131].

18.5 Clinical Perspectives and Challenges

During adulthood, the majority of blood vessels remain quiescent, and angiogenesis occurs only during embryonic development. Whereas in many diseases, angiogenesis processes are activated, resembling tumor growth and metastasis as well as ocular and inflammatory diseases. Huge advances have been made over the last two decades in finding and engineering anti-angiogenic therapeutics, but so far most emphasis was given on anti-VEGF approaches. These concepts, however, were only effective in certain types of cancer, and not every patient respondis on such anti-VEGF approach or become finally resistant towards these treatments. The characterization of biomarkers predicting therapeutic response towards anti-VEGF treatment is urged. As anti-VEGF-induced hypoxia leads to upregulation of HIF-dependent proteolytic systems, enhanced tumor cell invasion and metastasis formation were observed. In detail, Ebos et al. [132] recently described a phenomenon of accelerated metastasis formation during treatment of the VEGFR/PDGFR kinase inhibitor sunitinib in an in vivo model. This was explained by the fact that pro-angiogenic cytokines and growth factors such as SDF-1 or G-CSF were upregulated under treatment; furthermore, bone marrow-derived circulating cells forming a pre-metastatic niche and the unspecific effect of tyrosine kinase inhibitors were suspected to be responsible for the "metastatic conditioning." These observations were consistent with the study of Paez-Ribes et al. [133], who found in several tumor models a more invasive phenotype of tumor cells during VEGF-targeting treatments. The switch to an increased invasive phenotype of tumor cells is reflected in clinical observations in a subset of glioblastoma multiforme (GBM) patients that developed multifocal recurrence of their disease during the course of their disease with bevacizumab [134]. This was accompanied by vessel co-option, a phenomenon in which tumor cells grow along preexisting vessels and invade healthy tissue. Especially GBM is characterized by a high degree of microvascular proliferation and overexpression of pro-angiogenic growth factors such as VEGF and disease progression depends on angiogenesis. Although this has led to the FDA approval of single-agent bevacizumab for patients with pretreated GBM, the above-mentioned escape mechanisms leading to progression of the disease are currently in the focus of new anti-angiogenic concepts. Thus, a variety of different anti-angiogenic agents are tested in clinical trials for treatment efficacy in GBM (an overview is given in Table 18.1). In summary, as tumor growth might eventually escape

Table 18.1 An overview about angiogenic molecules, which therapeutic role in glioblastoma multiforme patients are currently evaluated in clinical trials

Target	Agents
VEGF-A	Aflibercept, bevacizumab
PIGF	Aflibercept
VEGFR-1	Cediranib, pazopanib, sorafenib
VEGFR-2	Angiocept, cediranib, pazopanib, sorafenib, sunitinib, vandetanib, vatalanib
VEGFR-3	Cediranib, pazopanib, sorafenib
PDGFR	MLN-518, pazopanib, sorafenib, sunitinib, vatalanib
Thrombospondin receptor (CD36)	ABT-510
Kit	MLN-518, pazopanib, vatalanib
Integrins	Cilengitide
Y-secretase/Notch	Ro4929097
Tie-2	Sorafenib

Agents are in alphabetic order. We apologize for any omissions that this table engenders

anti-VEGF treatments, novel anti-angiogenic strategies should consider other angiogenic growth factors, the role of proteolytic enzymes, and/or integrin adhesion receptors as promising concepts in the treatment of cancer.

References

1. Goldman E (1907) The growth of malignant disease in man and the lower animals with special reference to the vascular system. Lancet ii:1236–1240
2. Folkman J (1971) Tumor angiogenesis: therapeutic implications. N Engl J Med 285:1182–1186
3. Ellis LM, Hicklin DJ (2008) VEGF-targeted therapy: mechanisms of anti-tumour activity. Nat Rev Cancer 8:579–591
4. Schwarz Q, Ruhrberg C (2010) Neuropilin, you gotta let me know: should I stay or should I go? Cell Adh Migr 4:61–66
5. Shah MA, van Cutsem E, Kang YK et al (2012) Survival analysis according to disease subtype in AVAGAST: first-line capecitabine and cisplatin plus bevacizumab or placebo in patients with advanced gastric cancer. J Clin Oncol 30, 2012 (Suppl 4; Abstr 5)
6. Van Cutsem E, Tabernero J, Lakomy R, Prenen H, Prausová J, Macarulla T, Ruff P, van Hazel GA, Moiseyenko V, Ferry D, McKendrick J, Polikoff J, Tellier A, Castan R, Allegra C. (2012) Addition of aflibercept to fluorouracil, leucovorin, and irinotecan improves survival in a phase III randomized trial in patients with metastatic colorectal cancer previously treated with an oxaliplatin-based regimen. J Clin Oncol. 2012 Oct 1;30(28):3499–506. (Epub 2012 Sep 4)
7. Carmeliet P, Jain RK (2011) Molecular mechanisms and clinical applications of angiogenesis. Nature 473:298–307
8. Loges S, Mazzone M, Hohensinner P, Carmeliet P (2009) Silencing or fueling metastasis with VEGF inhibitors: antiangiogenesis revisited. Cancer Cell 15:167–170
9. Sakurai T, Kudo M (2011) Signaling pathways governing tumor angiogenesis. Oncology 81(Suppl 1):24–29
10. Cao Y, Cao R, Hedlund EM (2008) R Regulation of tumor angiogenesis and metastasis by FGF and PDGF signaling pathways. J Mol Med (Berl) 86(7):785–789

11. Korc M, Friesel RE (2009) The role of fibroblast growth factors in tumor growth. Curr Cancer Drug Targets 9:639–651
12. Beenken A, Mohammadi M (2008) The FGF family: biology, pathophysiology and therapy. Nat Rev Drug Discov 8:235–253
13. Casanovas O, Hicklin DJ, Bergers G, Hanahan D (2005) Drug resistance by evasion of anti-angiogenic targeting of VEGF signaling in late-stage pancreatic islet tumors. Cancer Cell 8:299–309
14. Bergers G, Benjamin LE (2003) Tumorigenesis and the angiogenic switch. Nat Rev Cancer 3:401–410
15. McCarty MF, Somcio RJ, Stoeltzing O et al (2007) Overexpression of PDGF-BB decreases colorectal and pancreatic cancer growth by increasing tumor pericyte content. J Clin Invest 117:2114–2122
16. Nissen LJ, Cao R, Hedlund EM et al (2007) Angiogenic factors FGF2 and PDGF-BB synergistically promote murine tumor neovascularization and metastasis. J Clin Invest 117:2766–2777
17. Dufraine J, Funahashi Y, Kitajewski J (2008) Notch signaling regulates tumor angiogenesis by diverse mechanisms. Oncogene 27:5132–5137
18. Sainson RC, Harris AL (2008) Regulation of angiogenesis by homotypic and heterotypic notch signalling in endothelial cells and pericytes: from basic research to potential therapies. Angiogenesis 11:41–51
19. Kuhnert F, Kirshner JR, Thurston G (2011) Dll4-Notch signaling as a therapeutic target in tumor angiogenesis. Vasc Cell 3:20
20. Zerlin M, Julius MA, Kitajewski J (2008) Wnt/Frizzled signaling in angiogenesis. Angiogenesis 11:63–69
21. Easwaran V, Lee SH, Inge L et al (2003) beta-Catenin regulates vascular endothelial growth factor expression in colon cancer. Cancer Res 63:3145–3153
22. Muley A, Majumder S, Kolluru GK et al (2010) Secreted frizzled-related protein 4: an angiogenesis inhibitor. Am J Pathol 176:1505–1516
23. Takahashi-Yanaga F, Kahn M (2010) Targeting Wnt signaling: can we safely eradicate cancer stem cells? Clin Cancer Res 16:3153–3162
24. Davis S, Aldrich TH, Jones PF et al (1996) Isolation of angiopoietin-1, a ligand for the TIE2 receptor, by secretion-trap expression cloning. Cell 87:1161–1169
25. Maisonpierre PC, Suri C, Jones PF et al (1997) Angiopoietin-2, a natural antagonist for Tie2 that disrupts in vivo angiogenesis. Science 277:55–60
26. Augustin HG, Koh GY, Thurston G, Alitalo K (2009) Control of vascular morphogenesis and homeostasis through the angiopoietin-Tie system. Nat Rev Mol Cell Biol 10:165–177
27. De PM, Naldini L (2009) Tie2-expressing monocytes (TEMs): novel targets and vehicles of anticancer therapy? Biochim Biophys Acta 1796:5–10
28. Huang H, Bhat A, Woodnutt G, Lappe R (2010) Targeting the ANGPT-TIE2 pathway in malignancy. Nat Rev Cancer 10:575–585
29. Ferrara N (2005) VEGF as a therapeutic target in cancer. Oncology 69(Suppl 3):11–16
30. Zhong C, Qu X, Tan M et al (2009) Characterization and regulation of bv8 in human blood cells. Clin Cancer Res 15:2675–2684
31. LeCouter J, Zlot C, Tejada M et al (2004) Bv8 and endocrine gland-derived vascular endothelial growth factor stimulate hematopoiesis and hematopoietic cell mobilization. Proc Natl Acad Sci USA 101:16813–16818
32. Shojaei F, Wu X, Zhong C et al (2007) Bv8 regulates myeloid-cell-dependent tumour angiogenesis. Nature 450:825–831
33. Shojaei F, Wu X, Qu X et al (2009) G-CSF-initiated myeloid cell mobilization and angiogenesis mediate tumor refractoriness to anti-VEGF therapy in mouse models. Proc Natl Acad Sci USA 106:6742–6747
34. Okazaki T, Ebihara S, Asada M et al (2006) Granulocyte colony-stimulating factor promotes tumor angiogenesis via increasing circulating endothelial progenitor cells and Gr1+CD11b+ cells in cancer animal models. Int Immunol 18:1–9

35. Natori T, Sata M, Washida M et al (2002) G-CSF stimulates angiogenesis and promotes tumor growth: potential contribution of bone marrow-derived endothelial progenitor cells. Biochem Biophys Res Commun 297:1058–1061
36. Shojaei F, Zhong C, Wu X et al (2008) Role of myeloid cells in tumor angiogenesis and growth. Trends Cell Biol 18:372–378
37. Nolan DJ, Ciarrocchi A, Mellick AS et al (2007) Bone marrow-derived endothelial progenitor cells are a major determinant of nascent tumor neovascularization. Genes Dev 21:1546–1558
38. Crawford Y, Kasman I, Yu L et al (2009) PDGF-C mediates the angiogenic and tumorigenic properties of fibroblasts associated with tumors refractory to anti-VEGF treatment. Cancer Cell 15:21–34
39. Dong J, Grunstein J, Tejada M et al (2004) VEGF-null cells require PDGFR alpha signaling-mediated stromal fibroblast recruitment for tumorigenesis. EMBO J 23:2800–2810
40. Kindler HL, Niedzwiecki D, Hollis D et al (2010) Gemcitabine plus bevacizumab compared with gemcitabine plus placebo in patients with advanced pancreatic cancer: phase III trial of the Cancer and Leukemia Group B (CALGB 80303). J Clin Oncol 28:3617–3622
41. Kerbel RS (2008) Tumor angiogenesis. N Engl J Med 358:2039–2049
42. Kindler HL, Wroblewski K, Wallace JA et al (2012) Gemcitabine plus sorafenib in patients with advanced pancreatic cancer: a phase II trial of the University of Chicago Phase II Consortium. Invest New Drugs 30:382–386
43. Gharibo M, Patrick-Miller L, Zheng L et al (2008) A phase II trial of imatinib mesylate in patients with metastatic pancreatic cancer. Pancreas 36:341–345
44. Xian X, Hakansson J, Stahlberg A et al (2006) Pericytes limit tumor cell metastasis. J Clin Invest 116:642–651
45. Crippa MP (2007) Urokinase-type plasminogen activator. Int J Biochem Cell Biol 39:690–694
46. Carmeliet P, Kieckens L, Schoonjans L et al (1993) Plasminogen activator inhibitor-1 gene-deficient mice. I. Generation by homologous recombination and characterization. J Clin Invest 92:2746–2755
47. Binder BR, Mihaly J, Prager GW (2007) uPAR-uPA-PAI-1 interactions and signaling: a vascular biologist's view. Thromb Haemost 97:336–342
48. Prager GW, Breuss JM, Steurer S et al (2004) Vascular Endothelial Growth Factor Receptor-2-Induced Initial Endothelial Cell Migration Depends on the Presence of the Urokinase Receptor. Circ Res 94:1562–1570
49. D'Alessio S, Gerasi L, Blasi F (2008) uPAR-deficient mouse keratinocytes fail to produce EGFR-dependent laminin-5, affecting migration in vivo and in vitro. J Cell Sci 121:3922–3932
50. Lund LR, Green KA, Stoop AA et al (2006) Plasminogen activation independent of uPA and tPA maintains wound healing in gene-deficient mice. EMBO J 25:2686–2697
51. Bajou K, Masson V, Gerard RD et al (2001) The plasminogen activator inhibitor PAI-1 controls in vivo tumor vascularization by interaction with proteases, not vitronectin. Implications for antiangiogenic strategies. J Cell Biol 152:777–784
52. Appella E, Blasi F (1987) The growth factor module of urokinase is the binding sequence for its receptor. Ann N Y Acad Sci 511:192–195
53. Huai Q, Mazar AP, Kuo A et al (2006) Structure of human urokinase plasminogen activator in complex with its receptor. Science 311:656–659
54. Pepper MS, Ferrara N, Orci L, Montesano R (1991) Vascular endothelial growth factor (VEGF) induces plasminogen activators and plasminogen activator inhibitor-1 in microvascular endothelial cells. Biochem Biophys Res Commun 181:902–906
55. Estreicher A, Muhlhauser J, Carpentier JL et al (1990) The receptor for urokinase type plasminogen activator polarizes expression of the protease to the leading edge of migrating monocytes and promotes degradation of enzyme inhibitor complexes. J Cell Biol 111:783–792

56. Prager GW, Breuss JM, Steurer S et al (2004) Vascular endothelial growth factor (VEGF) induces rapid prourokinase (pro-uPA) activation on the surface of endothelial cells. Blood 103:955–962
57. Nykjaer A, Conese M, Christensen EI et al (1997) Recycling of the urokinase receptor upon internalization of the uPA:serpin complexes. EMBO J 16:2610–2620
58. Takahashi T, Takahashi K, Mernaugh RL et al (2006) A monoclonal antibody against CD148, a receptor-like tyrosine phosphatase, inhibits endothelial-cell growth and angiogenesis. Blood 108:1234–1242
59. Pera IL, Iuliano R, Florio T, Susini C, Trapasso F, Santoro M et al (2005) The rat tyrosine phosphatase eta increases cell adhesion by activating c-Src through dephosphorylation of its inhibitory phosphotyrosine residue. Oncogene 24:3187–3195
60. Blasi F, Carmeliet P (2002) uPAR: a versatile signalling orchestrator. Nat Rev Mol Cell Biol 3:932–943
61. Webb DJ, Nguyen DH, Gonias SL (2000) Extracellular signal-regulated kinase functions in the urokinase receptor-dependent pathway by which neutralization of low density lipoprotein receptor-related protein promotes fibrosarcoma cell migration and matrigel invasion. J Cell Sci 113:123–134
62. Ma Z, Thomas KS, Webb DJ et al (2002) Regulation of Rac1 activation by the low density lipoprotein receptor-related protein. J Cell Biol 159:1061–1070
63. Nykjaer A, Petersen CM, Moller B et al (1992) Purified alpha 2-macroglobulin receptor/LDL receptor-related protein binds urokinase.plasminogen activator inhibitor type-1 complex. Evidence that the alpha 2-macroglobulin receptor mediates cellular degradation of urokinase receptor-bound complexes. J Biol Chem 267:14543–14546
64. Nguyen DH, Webb DJ, Catling AD et al (2000) Urokinase-type plasminogen activator stimulates the Ras/Extracellular signal-regulated kinase (ERK) signaling pathway and MCF-7 cell migration by a mechanism that requires focal adhesion kinase, Src, and Shc. Rapid dissociation of GRB2/Sps-Shc complex is associated with the transient phosphorylation of ERK in urokinase-treated cells. J Biol Chem 275:19382–19388
65. Geetha N, Mihaly J, Stockenhuber A et al (2011) Signal integration and coincidence detection in the mitogen-activated protein kinase/extracellular signal-regulated kinase (ERK) cascade: concomitant activation of receptor tyrosine kinases and of LRP-1 leads to sustained ERK phosphorylation via down-regulation of dual specificity phosphatases (DUSP1 and -6). J Biol Chem 286:25663–25674
66. Wei Y, Eble JA, Wang Z et al (2001) Urokinase receptors promote beta1 integrin function through interactions with integrin alpha3beta1. Mol Biol Cell 12:2975–2986
67. Aguirre Ghiso JA, Kovalski K, Ossowski L (1999) Tumor dormancy induced by downregulation of urokinase receptor in human carcinoma involves integrin and MAPK signaling. J Cell Biol 147:89–104
68. Carriero MV, Del VS, Capozzoli M et al (1999) Urokinase receptor interacts with alpha(v) beta5 vitronectin receptor, promoting urokinase-dependent cell migration in breast cancer. Cancer Res 59:5307–5314
69. Wei Y, Czekay RP, Robillard L et al (2005) Regulation of alpha5beta1 integrin conformation and function by urokinase receptor binding. J Cell Biol 168:501–511
70. Chaurasia P, Guirre-Ghiso JA, Liang OD et al (2006) Region in urokinase plasminogen receptor domain III controlling a functional association with alpha5beta1 integrin and tumor growth. J Biol Chem 281:14852–14863
71. Aguirre-Ghiso JA, Estrada Y, Liu D, Ossowski L (2003) ERK(MAPK) activity as a determinant of tumor growth and dormancy; regulation by p38(SAPK). Cancer Res 63:1684–1695
72. Smith HW, Marra P, Marshall CJ (2008) uPAR promotes formation of the p130Cas-Crk complex to activate Rac through DOCK180. J Cell Biol 182:777–790
73. Smith HW, Marshall CJ (2010) Regulation of cell signalling by uPAR. Nat Rev Mol Cell Biol 11:23–36

74. Leksa V, Godar S, Cebecauer M et al (2002) The N-terminus of mannose 6-phosphate/ Insulin-like growth factor 2 receptor in regulation of fibrinolysis and cell migration. J Biol Chem 280:14811–14818
75. Resnati M, Pallavicini I, Wang JM et al (2002) The fibrinolytic receptor for urokinase activates the G protein-coupled chemotactic receptor FPRL1/LXA4R. Proc Natl Acad Sci U S A 99:1359–1364
76. Liu D, Aguirre GJ, Estrada Y, Ossowski L (2002) EGFR is a transducer of the urokinase receptor initiated signal that is required for in vivo growth of a human carcinoma. Cancer Cell 1:445–457
77. Cavallaro U, Wu Z, Di Palo A et al (1998) FGF-2 stimulates migration of Kaposi's sarcoma-like vascular cells by HGF-dependent relocalization of the urokinase receptor. FASEB J 12:1027–1034
78. Loukinova E, Ranganathan S, Kuznetsov S et al (2002) Platelet-derived growth factor (PDGF)-induced tyrosine phosphorylation of the low density lipoprotein receptor-related protein (LRP). Evidence for integrated co-receptor function between LRP and the PDGF. J Biol Chem 277:15499–15506
79. Schmitt M, Mengele K, Napieralski R et al (2010) Clinical utility of level-of-evidence-1 disease forecast cancer biomarkers uPA and its inhibitor PAI-1. Expert Rev Mol Diagn 10:1051–1067
80. Nekarda H, Schlegel P, Schmitt M et al (1998) Strong prognostic impact of tumor-associated urokinase-type plasminogen activator in completely resected adenocarcinoma of the esophagus. Clin Cancer Res 4:1755–1763
81. Kaneko T, Konno H, Baba M et al (2003) Urokinase-type plasminogen activator expression correlates with tumor angiogenesis and poor outcome in gastric cancer. Cancer Sci 94:43–49
82. Ishida T, Sugio K, Yokoyama K et al (1991) Immunohistochemical evidences of prognostic parameters associated with tumor development of pulmonary adenocarcinoma. Nihon Geka Gakkai Zasshi 92:1107–1110
83. Kuhn W, Pache L, Schmalfeldt B et al (1994) Urokinase (uPA) and PAI-1 predict survival in advanced ovarian cancer patients (FIGO III) after radical surgery and platinum-based chemotherapy. Gynecol Oncol 55:401–409
84. Mulcahy HE, Duffy MJ, Gibbons D et al (1994) Urokinase-type plasminogen activator and outcome in Dukes' B colorectal cancer. Lancet 344:583–584
85. Berger DH (2002) Plasmin/plasminogen system in colorectal cancer. World J Surg 26:767–771
86. Hanahan D, Weinberg RA (2011) Hallmarks of cancer: the next generation. Cell 144:646–674
87. Weis SM, Cheresh DA (2011) Tumor angiogenesis: molecular pathways and therapeutic targets. Nat Med 17:1359–1370
88. van der Flier A, Sonnenberg A (2001) Function and interactions of integrins. Cell Tissue Res 305:285–298
89. Colombatti A, Bonaldo P, Doliana R (1993) Type A modules: interacting domains found in several non-fibrillar collagens and in other extracellular matrix proteins. Matrix 13:297–306
90. Gamble LJ, Borovjagin AV, Matthews QL (2010) Role of RGD-containing ligands in targeting cellular integrins: Applications for ovarian cancer virotherapy (Review). Exp Ther Med 1:233–240
91. Desgrosellier JS, Cheresh DA (2010) Integrins in cancer: biological implications and therapeutic opportunities. Nat Rev Cancer 10:9–22
92. Bolos V, Gasent JM, Lopez-Tarruella S, Grande E (2010) The dual kinase complex FAK-Src as a promising therapeutic target in cancer. Onco Targets Ther 3:83–97
93. Hehlgans S, Haase M, Cordes N (2007) Signalling via integrins: implications for cell survival and anticancer strategies. Biochim Biophys Acta 1775:163–180

94. Stupack DG, Cheresh DA (2002) ECM remodeling regulates angiogenesis: endothelial integrins look for new ligands. Sci STKE 2002:E7
95. Critchley DR (2004) Cytoskeletal proteins talin and vinculin in integrin-mediated adhesion. Biochem Soc Trans 32(Pt 5):831–836
96. Shattil SJ, Kim C, Ginsberg MH (2010) The final steps of integrin activation: the end game. Nat Rev Mol Cell Biol 11:288–300
97. Hynes RO (2002) Integrins: bidirectional, allosteric signaling machines. Cell 110:673–687
98. Walker JL, Fournier AK, Assoian RK (2005) Regulation of growth factor signaling and cell cycle progression by cell adhesion and adhesion-dependent changes in cellular tension. Cytokine Growth Factor Rev 16:395–405
99. Friedlander M, Brooks PC, Shaffer RW et al (1995) Definition of two angiogenic pathways by distinct alpha v integrins. Science 270:1500–1502
100. Hood JD, Frausto R, Kiosses WB et al (2003) Differential alphav integrin-mediated Ras-ERK signaling during two pathways of angiogenesis. J Cell Biol 162:933–943
101. Avraamides CJ, Garmy-Susini B, Varner JA (2008) Integrins in angiogenesis and lymphangiogenesis. Nat Rev Cancer 8:604–617
102. George EL, Georges-Labouesse EN, Patel-King RS et al (1993) Defects in mesoderm, neural tube and vascular development in mouse embryos lacking fibronectin. Development 119:1079–1091
103. Fassler R, Meyer M (1995) Consequences of lack of beta 1 integrin gene expression in mice. Genes Dev 9:1896–1908
104. Fassler R, Pfaff M, Murphy J et al (1995) Lack of beta 1 integrin gene in embryonic stem cells affects morphology, adhesion, and migration but not integration into the inner cell mass of blastocysts. J Cell Biol 128:979–988
105. Senger DR, Claffey KP, Benes JE et al (1997) Angiogenesis promoted by vascular endothelial growth factor: regulation through alpha1beta1 and alpha2beta1 integrins. Proc Natl Acad Sci USA 94:13612–13617
106. Whelan MC, Senger DR (2003) Collagen I initiates endothelial cell morphogenesis by inducing actin polymerization through suppression of cyclic AMP and protein kinase A. J Biol Chem 278:327–334
107. Sweeney SM, DiLullo G, Slater SJ et al (2003) Angiogenesis in collagen I requires alpha2beta1 ligation of a GFP*GER sequence and possibly p38 MAPK activation and focal adhesion disassembly. J Biol Chem 278:30516–30524
108. Senger DR, Perruzzi CA, Streit M et al (2002) The alpha(1)beta(1) and alpha(2)beta(1) integrins provide critical support for vascular endothelial growth factor signaling, endothelial cell migration, and tumor angiogenesis. Am J Pathol 160:195–204
109. Goh KL, Yang JT, Hynes RO (1997) Mesodermal defects and cranial neural crest apoptosis in alpha5 integrin-null embryos. Development 124:4309–4319
110. Kim S, Bell K, Mousa SA, Varner JA (2000) Regulation of angiogenesis in vivo by ligation of integrin alpha5beta1 with the central cell-binding domain of fibronectin. Am J Pathol 156:1345–1362
111. Boudreau NJ, Varner JA (2004) The homeobox transcription factor Hox D3 promotes integrin alpha5beta1 expression and function during angiogenesis. J Biol Chem 279:4862–4868
112. Yang JT, Rayburn H, Hynes RO (2003) Embryonic mesodermal defects in alpha 5 integrin-deficient mice. Development 119:1093–1105
113. Kuwada SK (2007) Drug evaluation: Volociximab, an angiogenesis-inhibiting chimeric monoclonal antibody. Curr Opin Mol Ther 9:92–98
114. Ricart AD, Tolcher AW, Liu G, Holen K et al (2008) Volociximab, a chimeric monoclonal antibody that specifically binds alpha5beta1 integrin: a phase I, pharmacokinetic, and biological correlative study. Clin Cancer Res 14:7924–7929
115. Bader BL, Rayburn H, Crowley D, Hynes RO (1998) Extensive vasculogenesis, angiogenesis, and organogenesis precede lethality in mice lacking all alpha v integrins. Cell 95:507–519

116. Reynolds LE, Wyder L, Lively JC et al (2002) Enhanced pathological angiogenesis in mice lacking beta3 integrin or beta3 and beta5 integrins. Nat Med 8:27–34
117. Brooks PC, Clark RA, Cheresh DA (1994) Requirement of vascular integrin alpha v beta 3 for angiogenesis. Science 264:569–5671
118. Zhang D, Pier T, McNeel DG et al (2007) Effects of a monoclonal anti-alphavbeta3 integrin antibody on blood vessels—a pharmacodynamic study. Invest New Drugs 25:49–55
119. Cai W, Wu Y, Chen K et al (2006) In vitro and in vivo characterization of 64Cu-labeled Abegrin, a humanized monoclonal antibody against integrin alpha v beta 3. Cancer Res 66:9673–9681
120. Tonnesen MG, Feng X, Clark RA (2000) Angiogenesis in wound healing. J Investig Dermatol Symp Proc 5:40–46
121. Drake CJ, Cheresh DA, Little CD (1995) An antagonist of integrin alpha v beta 3 prevents maturation of blood vessels during embryonic neovascularization. J Cell Sci 108:2655–2661
122. Trikha M, Zhou Z, Nemeth JA et al (2004) CNTO 95, a fully human monoclonal antibody that inhibits alphav integrins, has antitumor and antiangiogenic activity in vivo. Int J Cancer 110:326–335
123. O'Day SJ, Pavlick AC, Albertini MR et al (2011) Clinical and pharmacologic evaluation of two dose levels of intetumumab (CNTO 95) in patients with melanoma or angiosarcoma. Invest New Drugs 30:1074–1081
124. Reardon DA, Neyns B, Weller M et al (2011) Cilengitide: an RGD pentapeptide alphanubeta3 and alphanubeta5 integrin inhibitor in development for glioblastoma and other malignancies. Future Oncol 7:339–354
125. Weller M (2011) Novel diagnostic and therapeutic approaches to malignant glioma. Swiss Med Wkly 141:w13210
126. Vermorken JB, Guigay J, Mesia R et al (2011) Phase I/II trial of cilengitide with cetuximab, cisplatin and 5-fluorouracil in recurrent and/or metastatic squamous cell cancer of the head and neck: findings of the phase I part. Br J Cancer 104:1691–1696
127. Cheng C, Komljenovic D, Pan L et al (2011) Evaluation of treatment response of cilengitide in an experimental model of breast cancer bone metastasis using dynamic PET with (18) F-FDG. Hell J Nucl Med 14:15–20
128. Oliveira-Ferrer L, Hauschild J, Fiedler W et al (2008) Cilengitide induces cellular detachment and apoptosis in endothelial and glioma cells mediated by inhibition of FAK/src/AKT pathway. J Exp Clin Cancer Res 27:86
129. Nisato RE, Tille JC, Jonczyk A et al (2003) alphav beta 3 and alphav beta 5 integrin antagonists inhibit angiogenesis in vitro. Angiogenesis 6:105–119
130. Reynolds AR, Hart IR, Watson AR et al (2009) Stimulation of tumor growth and angiogenesis by low concentrations of RGD-mimetic integrin inhibitors. Nat Med 15:392–400
131. Millard M, Odde S, Neamati N (2011) Integrin targeted therapeutics. Theranostics 1:154–188
132. Ebos JM, Lee CR, Cruz-Munoz W et al (2009) Accelerated metastasis after short-term treatment with a potent inhibitor of tumor angiogenesis. Cancer Cell 15:232–239
133. Paez-Ribes M, Allen E, Hudock J et al (2009) Antiangiogenic therapy elicits malignant progression of tumors to increased local invasion and distant metastasis. Cancer Cell 15:220–231
134. Norden AD, Young GS, Setayesh K et al (2008) Bevacizumab for recurrent malignant gliomas: efficacy, toxicity, and patterns of recurrence. Neurology 70:779–787

Chapter 19
Potential Mechanisms Linking Oxidized LDL to Susceptibility to Cancer

Magomed Khaidakov and Jawahar L. Mehta

Abstract Last decades have witnessed an explosive growth of obesity in developed countries. This condition is associated with not only increased risk for atherosclerosis and resultant cardiovascular pathologies but also confers significant increase in probability for the development of various forms of cancer. Experimental and epidemiological evidence accumulated in respective fields highlights multiple overlaps in pathobiology of atherosclerosis and cancer including signaling pathways, inflammation, cytokine involvement, proliferation, and angiogenesis. Recent discovery of additional layer of regulation of gene expression based on miRNAs also revealed multiple similarities in miRNA profiles. This suggests that atherogenesis and obesity-mediated increase in susceptibility to cancers may have common etiological roots. Recent advances in cardiovascular research identified oxidatively modified LDL (and not LDL itself) as a primary factor responsible for initiation and progression of atherogenesis, and the analysis of potential involvement of ox-LDL in carcinogenesis is the focus of the present review.

Keywords Ox-LDL • Cancer • Inflammation • Angiogenesis • miRNAs

19.1 Introduction

Epidemiological and experimental data link obesity to elevated risk of variety of cancers. Recent meta-analysis of 218 datasets including close to 300,000 cases [1] revealed strong positive association between increasing BMI and esophageal

M. Khaidakov (✉) • J.L. Mehta
Division of Cardiovascular Medicine, University of Arkansas for Medical Sciences and the Central Arkansas Veterans Healthcare System, 4301 West Markham, Little Rock, AR 72205, USA
e-mail: mkhaidakov@uams.edu

J.L. Mehta and N.S. Dhalla (eds.), *Biochemical Basis and Therapeutic Implications of Angiogenesis*, Advances in Biochemistry in Health and Disease 6,

adenocarcinomas as well as thyroid, colon, and renal cancers in men. In women, significant associations were found with endometrial, gallbladder and renal cancers and esophageal adenocarcinoma. Overall, incidence of 8.6% (5.6–11.5%) of cancers is attributed to obesity [2]. Obesity also accounts for 52 and 88% higher mortality rate across all cancers in men and women, respectively [3].

Several lines of experimental evidence suggest that obesity enhances susceptibility to cancer via multiple mechanisms. Among them are oxidative damage [4], inflammation [5], paracrine interactions between epithelial cells and adipocytes [6], metabolic syndrome-mediated alterations in insulin signaling [7], and conversion of androgens into estrogens in obese postmenopausal women [8].

One of the crucial components of obesity—extensively studied in the cardiovascular field but almost overlooked in cancer research—is oxidized LDL (ox-LDL). In relation to atherosclerosis, ox-LDL has been shown to influence practically all key steps of atherogenesis including oxidative stress, inflammatory reactions, proliferation, and angiogenesis—all of which are relevant to pathobiology of tumor initiation and progression. The current review is an attempt to consolidate available data on ox-LDL as a possible link between dyslipidemia and carcinogenesis.

19.2 Lipoprotein Modifications, Change of Destination, and Mechanisms of Entry

Intracellular lipid concentration is strictly controlled, and excess of available LDL translates into downregulation of LDL receptor [9]. Dyslipidemia, however, is associated with global increase in NFκB activity [10], downregulation of antioxidant defense [11], and, as a result, enhanced oxidation of LDL [12–15]. Elevated circulating concentrations of ox-LDL are a typical feature of metabolic syndrome [13–15] with high CVD predictive value as a risk factor [16]. Another fundamentally important feature of ox-LDL is a shift in its identity as the process of LDL oxidation results in increasing proportion of free modified phospholipids or those covalently linked to lysins [17]. It is poorly recognized by LDLr [18] and, instead, becomes a ligand for scavenger receptors [19] which, unlike LDLr, are upregulated by ox-LDL [20, 21]. As a bypass for restricted entry through LDLr, ox-LDL/SR interaction presents one of the mechanisms for smuggling excessive amounts of lipoproteins into the cells.

19.3 DNA Damage and Repair

Several constituents of ox-LDL are mutagenic and carcinogenic including typical products of lipid peroxidation malonyldialdehyde (MDA) and 4-hydroxynonenal (4HNE). MDA preferentially forms adducts to deoxyguanosines and deoxyadenosines [22]. 4-HNE is a major product of lipid peroxidation involved in formation of 4-HNE adducts, double strand breaks, DNA cross-linking, and mutagenesis [23]. In addition, treatment with 4HNE compromises nucleotide excision repair and sensitizes cells to other genotoxic exposures [24].

Apart from delivering products of lipid oxidation into the cells, ox-LDL triggers a secondary wave of oxidant stress by stimulating NADPH oxidase (see below). The content of 8-hydroxyguanine, a marker of mutagenic oxidative DNA damage, significantly increases in vascular cells response to ox-LDL [25, 26]. Similarly, circulating concentrations of 8-hydroxyguanine correlate with serum titers of antibodies to ox-LDL (but not native LDL) in dyslipidemic patients [27]. Moreover, even at nontoxic concentrations, ox-LDL (but not native LDL) was shown to significantly suppress base excision repair activity via downregulation of its key enzyme DNA polymerase β [25].

19.4 ROS Production

Generation of ROS is an intrinsic part of ox-LDL-mediated signaling and involves various oxidant-producing entities including NADPH oxidase [28–30] and mitochondria [31–34]. NADPH oxidase is systemically upregulated in obesity [35], and overexpression of its various subunits has been reported for many cancers (reviewed in [36]). In addition to causing mutations, the oncogenic effects of ROS originating from NADPH oxidase appear to be based on transformation-specific mechanisms. Overexpression of NADPH oxidase is mitogenic by itself [37] but is also mechanistically linked to transformation initiated by mutated Ras [38]. Nox1 transcription is Ras dependent, and Ras-mediated transformation is accompanied with overproduction of superoxide while its prevention, in turn, inhibits proliferation of Ras-transformed cells and cancer cell lines [38, 39]. Increased NOX1 activity alone also appears to be oncogenic as its overexpression in NIH3T3 cells resulted in anchorage-independent growth and increased tumorigenic potential in immunodeficient mice [40]. In absence of Ras mutation, the oncogenic potential of Nox1 may be linked to inactivation of p53 via inhibition of SIRT1-mediated p53 Lys382 acetylation [41]. Similar effects are reported for Nox4 which, upon overexpression, induced tumorigenic transformation of normal breast epithelial cells [42].

Significant portion of ROS generated in the cells exposed to ox-LDL was shown to co-localize to mitochondria, and their production was significantly impaired in cells with depleted mitochondrial DNA [32]. This process also required activity of NO synthase as NOS inhibition abrogated generation of ROS and reintroduction of NO donors restored it. In endothelial cells, ROS stimulating effects of ox-LDL exposure are attributed to dose and time-dependent increase in activity of c-Jun NH_2-terminal kinase followed by increase in mitochondrial membrane potential and enhanced ubiquination of MnSOD [33]. The accounts on effects of ox-LDL treatment of electron transport chain complexes vary. In HUVECs, upregulation of Complex I activity cells was evident within minutes of exposure to moderate (50 μg/mL) ox-LDL concentration [34]. In contrast, in another study on porcine endothelial cells, mildly oxidized ox-LDL utilized at higher concentration inhibited NADH-ubiquinone dehydrogenase (complex I), cytochrome *c* reductase (complex III), and cytochrome *c* oxidase (complex IV) activities to approximately the same extent as native LDL [43].

19.5 Inflammation

Inflammation-mediated oxidative stress stimulates site-specific production of several cytokines, chemokines, transcription, and growth factors, many of which display pro-oncogenic properties (reviewed in [5, 44]). A connection between chronic inflammation and increased incidence of cancer has been first noticed by Galen [45] and rediscovered by Virchow in nineteenth century [46]. Persistent low-grade inflammation is also characteristic for obesity [47] and several aspects of this phenomenon are often viewed as a mechanistic bridge between lipid metabolism perturbations and cancer. What components of dyslipidemia are responsible for inflammation is not completely elucidated, although epidemiological, clinical, and experimental data increasingly point toward ox-LDL as one of the possible culprits.

Elevated levels of ox-LDL are typical for obese individuals [12–15]. Similarly, considerably higher concentrations of circulating ox-LDL accompanied with concomitant elevation of inflammatory molecules including circulating CRP and IL-6 were found in extremely obese children [47]. In another study on a group of clinically healthy 58-year old men ($n=391$), circulating ox-LDL have been shown to correlate with TNF-α and CRP whereas IL-6 did not show significant association [48]. Progression of rheumatoid arthritis is fueled by inflammatory synovitis, and a recent study has shown that ox-LDL was overabundant in synovial liquid of RA patients compared to healthy individuals [49]. Unlike native LDL, ox-LDL, when injected into the joints of mice, caused swelling, upregulation of LOX-1, synovial hyperplasia, and increased inflammatory cell infiltration [49].

19.5.1 NF-κB

One of the typical downstream consequences of oxidant stress is activation of redox-sensitive NFκB [10, 50, 51]. NF-κB family of transcription factors controls numerous functions including immunity, inflammation, proliferation, and apoptosis. It reacts to multitude of stimuli, and its persistent activation has been implicated in pathobiology of a range of condition from arthritis to Alzheimer (see Gilmor Lab database, ref. [52]). NFκB has been shown to be implicated in the development of many forms of cancer and abrogation; its downstream signaling is currently a focus of efforts for many pharmaceutical companies [53, 54]. Recent microarray studies reported significant transcriptional overlaps between oncogenic transformation and obesity, atherosclerosis, and metabolic syndrome [55]. Transformation transcriptomes of mammary epithelial MCF10a cells and primary fibroblasts revealed involvement of ox-LDL specific scavenger receptor LOX-1 in NF-κB-dependent pro-inflammatory and hypoxic signaling. Moreover, treatment of MCF-10A cells with low concentration of ox-LDL induced cellular transformation and colony formation in NF-κB-dependent manner. These data are compatible with the analysis of LOX-1 knockout transcriptome from our group that showed downregulation of a quarter of known NF-κB target genes compared to WT animals [56, 57].

Ox-LDL has been shown to activate NFκB in various cell types [58]. In vivo, injection of native human LDL resulted in gradual accumulation of ox-LDL epitopes within endothelial lining and media concomitant with activation of NFκB along with upregulation of its target gene ICAM-1 [59]. Interestingly, in contrast to native LDL, injection of ox-LDL had no effect on NFκB activity in arteries after 24 h postinjection. This is probably indicative of extremely fast clearance of ox-LDL from circulation. This observation also suggests that LDL oxidation occurs either after internalization of LDL (LDL receptor route) or immediately before capture by scavenger receptors on the surface of the cells.

Ox-LDL (as well as other oxidatively modified lipoproteins)-mediated stimulation of NFκB is to large extent driven by stimulation of ROS production. In in vitro studies on bovine chondrocytes and human endothelial cells, generation of ROS was a prerequisite to activation of NFκB [51, 60]. In chondrocytes, pretreatment of cells with antioxidant reduced NFκB response [60]. Similarly, HUVECs exposed to ox-HDL in presence of probucol or inhibitor of NADPH oxidase diphenylene iodonium displayed reduced degradation of IkBa [51]. Both studies highlighted importance of scavenger receptors as application of neutralizing antibodies to LOX-1 attenuated ox-LDL effects.

Another possible mechanism of NFκB activation may involve regulation of PI3K/Akt cascade. In our studies (unpublished), we observed that exposure of human mammary epithelial cells (MCF10A) to nontoxic concentrations of highly oxidized LDL (5 μg/mL) results in stimulation of oncomiR-21 accompanied with inhibition of its target gene PTEN and corresponding upregulation of PI3K and Akt.

19.5.2 Cytokines

NF-κB targets a number of cytokines with pro-inflammatory and anti-inflammatory properties including TNFα, IL1β, IL-6, IL-8, IL-10, and IL-12. TNFα is a pleiotropic cytokine that has been shown to promote proliferation/angiogenesis and anchorage-independent growth [61]. Being a target of NFκb, TNFα generates a positive loop for continuous NFκB activation via phosphorylation and degradation of IKB proteins [62], and abrogation of TNFα reduces susceptibility to cancer development [63]. IL1β was found to be highly expressed in microenvironments of murine lung, pancreatic, and breast tumors, and its inhibition was detrimental to tumor inflammation and growth [64]. In multiple myeloma patients, circulating concentrations of IL1β, IL-6, and IL-8 were higher at advanced stages of disease and positively correlated with proliferation index [65]. Connection of IL-6 to tumorigenesis is well documented for variety of cancers [66]. Its primary signaling involves STAT3 via activation of Janus kinases [67], and it is mitogenic and protective to tumor cells exposed to DNA-damaging agents [68, 69]. The action of IL-8 is based on interaction with CXCR1 and CXCR2 receptors triggering multiple pathways including PI3K/Akt, MAPK, and PKC resulting in enhanced tumorigenicity, metastatic potential, and angiogenesis in variety of tumors (reviewed in [70]).

Ox-LDL has been shown to stimulate a range of pro-oncogenic cytokines in different cell types (Table 19.1) including adipocytes [71], monocytes and macrophages [72–79], endothelial [80–83], mast [84], and epithelial cells [85, 86]. The key role of ox-LDL in dyslipidemia-induced expression of cytokines has been highlighted in a recent in vivo study, where injections of ox-LDL to mice maintained on high-fat diet led to a very significant exacerbation of liver inflammation accompanied with increased expression of IL-6 and TNFα [87].

Mechanistically, a number of studies highlighted a central role of NFκB and MAPK activation in ox-LDL signaling across different cell types [71, 74, 83, 85, 86]. It is interesting to note, however, that in the experiment that utilized the lowest concentration of ox-LDL [79], activation of NFκB was not evident and, instead, significant stimulation of AP-1 was reported—probably suggesting "peace-time" and more forceful "war-time" modes of ox-LDL signaling.

Recent studies on endothelial cells also suggest an independent contribution of LOX-1 per se (a predominant scavenger receptor in this cell type which was considered to function as a mere shuttle for its ligands) to ox-LDL-mediated effects [83]. Affinity proteomic analysis has identified ligand-dependent association of LOX-1 with ROCK2 which was essential for NFκB activation and cytokine production.

19.5.3 Adipocytes

Adipocytes have been shown to be involved in carcinogenesis as stromal components critically contributing to tumor microenvironment. Obesity is associated with dysfunction of adipocytes and substantial changes in their paracrine/endocrine output that is comprised of a panel of adipocyte-specific (leptin and adiponectin) and more common (TNFα, IL-6) cytokines (reviewed in [88]). Leptin is involved in regulation of energy balance and appetite control and positively correlates with obesity [89]. In the cells, leptin has been shown to activate JAK, STAT, and MAPK signaling pathways [90] and promote a switch toward preferential use of fatty acids, inflammatory reactions [91], proliferation [92], and reduction of apoptosis [93], all of which are consistent with increased susceptibility to cancer. Adiponectin, on the other hand, exerts anti-inflammatory effects [94]. It generally inversely correlates with obesity [95], possibly due to inhibitory effects of IL-6 and TNFα on its expression [96]. On the cellular level, adiponectin has been shown to inhibit proliferation in many cell types [97, 98], partially, by the virtue of its affinity to various growth factors [98]. Adiponectin also exerts similar effects on cancer cells and prevents

Table 19.1 Effects of ox-LDL on cytokine production

Cell type	Exposure [mg/ml] × [hours]	Parameter	Ox-LDL	LDL	Additional comments	Ref
HUVEC	[100] × [24]	IL-6	↑	N/A		[80]
		TNF-α	↑			
HUVEC	[100] × [24]	IL-6	↑	N/A		[82]
HAECT	[35] × [0.3]	IL-8	↑	N/A	• Ox-LDL/LOX-1 was dynamically associated with ROCK2 and ROCK2 inhibition attenuated IL-8 and NFkB responses to ox-LDL	[83]
		NFkB activity	↑			
Mast cells	Ox-LDL/IgG ICs [50] × [3–24]	IL-8	↑	→		[84]
		TNF-α	↑	→		
		MCP-1	↑	→		
Human monocytes	[10–150] × [24]	IL-1β	↑ →	→		[72]
		IL-8	↑ ↑	→		
THP-1 cells		TNF-α	→	→		
		MIP-1α	→			
Human monocytes	[10–100] × [24]	IL-10	↑	N/A	• mmLDL failed to induce IL-10 and IL-12 • Changes in IL-10 and IL-12 were reciprocal and addition of IL10 prevented ox-LDL mediated stimulation of IL12	[78]
		IL-12 p40	↑			
		IL-12 p70	↑			
Human M1 macrophages	[50] × [24]	IL-6	↑ ↑	N/A		[73]
		IL-8	↑ ↑			
Human M2 macrophages		IL-10	→ ↓			
		MCP-1	→ ↓			
		PPAR-γ	↓ →			
		LXR-α	→ ↑			
Murine macrophages Raw264.7	[40] × [24]	MCP-1	↑	N/A		[74]
		TNF-α	↑			
		COX2	↑			
		IκBα	↓			
		p-ERK1/2	↑			

(continued)

Table 19.1 (continued)

Cell type	Exposure [mg/ml] × [hours]	Parameter	Ox-LDL	LDL	Additional comments	Ref
Monocytes Macrophages	mmLDL [50–70] × [24]	IL-1β	↑↑	N/A	• Secretion of ILs in response to mmLDL was attenuated by inhibition of CD14, TLR2 and TLR4	[76]
		IL-6	↑↑			
		IL-10	↑↑			
Murine macrophages	[5] × [18]	TNF-α			• Stimulation was completely prevented by NFkB inhibitor and partially by MEK1/2 inhibitor.	[75]
		RANTES				
Human monocytes	[0–15] × [2–48]	TNF-α	↑	→	• Conditioned medium from ox-LDL exposed monocytes stimulated SMC proliferation • Addition of TNF-α antibody partially canceled this effect. • High basal levels of IL-1β and activity of NFkB.	[79]
		IL-1β	→	→		
		NFkB activity	→	→		
		AP-1 activity	↑	→		
Gingival epithelial Ca9-22 Oral epithelial HO-1-u-1 Primary oral keratinocytes	[10–50] × [24]	IL-8	↑↑↑	→→→	• Of several constituents of ox-LDL only 7-ketocholesterol stimulated IL-8 • Stimulation of IL-8 by ox-LDL was suppressed by MAPK inhibitors	[86]
		PGE2	→→→	→→→		
Gingival epithelial Ca9-22 Primary oral keratinocytes	[mild oxidation] [50] × [6]	IL-1β	↑↑	→	Responses suppressed by NFkB inhibitor	[85]
		IL-8	↑↑	→		
		PGES1	→↑	→		
		PGES2	→→	→		
		COX2	↑↑	→		
		NFkB activity	↑↑	→		
Adipocytes	[0–50] × [6–48]	PAI-1	↑	→	• Antioxidant treatment prevented both ROS and PAI-1 upregulation • PAI expression was also attenuated by Inhibition of 44/42 MAPK but not p38 MAPK and JNK	[71]

tumor growth via inhibiting Akt and Wnt/β-catenin signaling [99, 100]. Anti-proliferative effects of adiponectin appear to be context dependent as it has been reported to be mitogenic in other studies [101, 102]. In endothelial cell, however, adiponectin was reported to counteract ox-LDL-mediated release of superoxide, stimulation of proliferation, and activation of p44/42 MAPK pathway [97].

Ox-LDL was found to positively correlate with leptin and negatively correlate with adiponectin in a number of studies [103–106], which is likely a reflection of dyslipidemia as their common denominator. Adipocytes express scavenger receptors and internalize ox-LDL [71]. In vivo, high-fat diet significantly upregulates LOX-1 expression in adipose tissues [107]. In pre-adipocytes, ox-LDL has been shown to delay their differentiation by increasing their proliferative potential via preventing prerequisite silencing of pre-adipocyte factor 1 (Pref-1) and thus consequent activation of PPARγ [108]. Moreover, effects of ox-LDL are likely to be even more pronounced in mature adipocytes since LOX-1 is one of the target genes for PPARγ and is dramatically upregulated with its stimulation [109].

As in other cell types, ox-LDL stimulates ROS production and NFκB activation in adipocytes as well [71, 110] which has been shown to be ultimately responsible for expression pro-oncogenic cytokine plasminogen activator inhibitor 1 (PAU-1). Although experimental proof is lacking, it can be assumed by extension that other cytokines targeted by NFκB are upregulated as well. This assumption is partially corroborated by in vivo studies in mice where deletion of LOX-1 in mice fed with high-fat diet largely prevented upregulation of several cytokines including MCP-1, MIP1-alpha, and IL-6 [107].

19.6 Proliferation/Angiogenesis

Continuing stimulation of proliferation of cancer cells supported by formation of adequate vascular network are main determinants of tumor aggressiveness. Escape from internal mechanisms of proliferation control is compulsory for carcinogenesis, and dyslipidemic environment creates additional incentives for transformation and accelerated growth. We already mentioned of ability of ox-LDL to cause transformation of mammary epithelial cells [55]. In cancer cells, even very small concentrations of ox-LDL (as low as 0.1 μg/mL) were also reported to stimulate proliferation [111]. Similarly, small doses of ox-LDL (but not native LDL) were shown to exert pro-proliferative and pro-angiogenic effects in vascular cells through activation of NADPH oxidase, MAPK signaling, NF-κB, and EGF expression [112, 113].

The mechanisms by which ox-LDL stimulates proliferation and angiogenesis are likely very complex due to its composition and a mixture of autocrine responses with signaling forced by paracrine interactions (Fig. 19.1). For example, LDL oxidation generates multiple species of modified molecules with competing signaling properties. While lysophosphatidyl choline and 4HNE stimulate DNA synthesis in smooth muscle cells [114, 115], oxysterols are inhibitory [114]. Oxidized phospholipids stimulate formation of sprouts from endothelial spheroids and promote

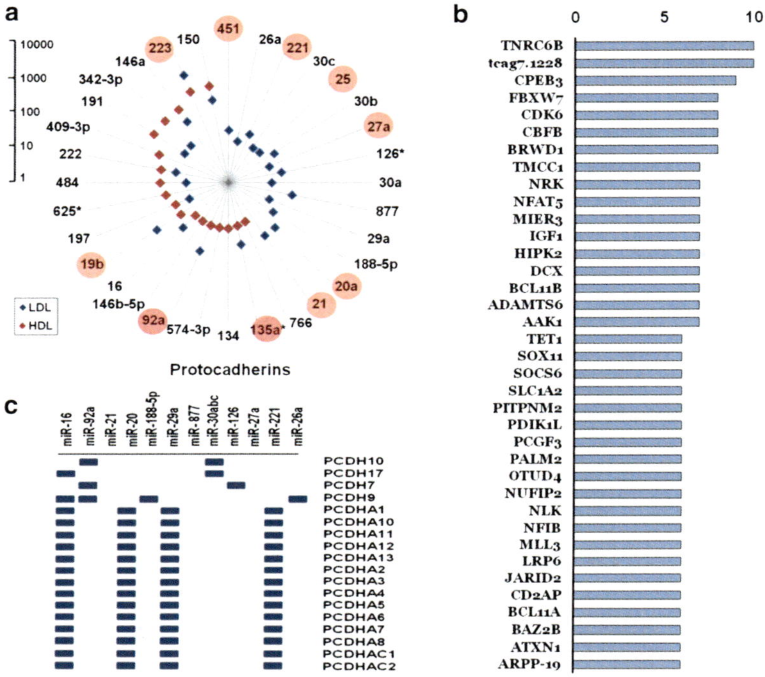

Fig. 19.1 A diagram of a fraction of pro-oncogenic effects of ox-LDL. Ox-LDL *black arrows* indicate stimulation. *Red connectors* indicate inhibition

neovascularization via upregulation of cyclooxygenase-2 (COX-2)—a primary effector of VEGF angiogenic action [116]. On the other hand, some oxidation products of phospholipids act as PPARγ agonists causing activation of PPRE-responsive genes including CD36 (receptor for both ox-LDL and trombospondin) that may have anti-angiogenic consequences [117].

Most cytokines stimulated by ox-LDL have been shown to be pro-proliferative and angiogenic. IL-1β was shown to be a powerful stimulator of angiogenesis acting through activation of NFκB via its primary signal transducer IL-1-associated receptor kinase [118, 119] and stimulation of VEGF [120–122]. IL-6 involves STAT-3, MAPK, and PI3K/Akt-mediated signaling resulting in upregulation of numerous anti-apoptotic genes, cyclins, COX-2, PGE2 receptors, hypoxia-inducible factor 1α, and VEGF (reviewed in [123]). It also prevents nuclear export of NFκB by p65 acetylation and inhibits p53 and p21. In terms of angiogenesis, IL-6 has been shown to stimulate tube formation in vitro [124], while its abrogation resulted in compromised post-stroke neovascularization and increased chronic lesion size in vivo [125]. Expression of IL-8 can be drastically upregulated by IL-1, IL-6, and TNFα [126] and seems to be AP-1 rather than NFκB mediated [127]. IL-8 promotes tube formation in vitro [128, 129] and in vivo [130], and its angiogenic effects are, in part, attributed to upregulation of anti-apoptotic Bcl2 and Bcl-xl [129] as well as Glu-Leu-Arg motif that it shares with other angiogenic chemokines [131]. TNFα exerts either anti-proliferative or pro-proliferative effects in a time-concentration-dependent manner [61]. It was shown to upregulate NFκB, VEGF, IL-1, IL-2, and IL-8 [128, 132–134]. In the former study, TNFα-mediated stimulation of angiogenesis partially depended on IL-8, VEGF, Sp1, NFκB, and c-Jun.

Mechanistically, the effects of cytokines including IL-1, IL-6, IL-8, and TNFα are consistently linked to increased activity of COX-2 [135]. COX-2 is intimately implicated in tumor progression [136], and its stimulation was shown to suppress apoptosis and promote proliferation in cancer cells [137, 138]. It also occupies a key position in induction of angiogenesis via directing processing of arachidonic acid toward formation of prostanoids. The dependence of cytokine signaling on COX2 activation was illustrated by Kuwano et al. in endothelial cells and COX-2 deficient mice [135]. In this study, IL-1β enhanced COX2 expression, production of prostanoids, and angiogenesis by HUVECs, and these effects were canceled by COX-2 inhibitors. Implantation of IL-1β into cornea induced robust angiogenic response in WT animals but was largely ineffective in COX2 knockout mice.

An additional mechanism for enhanced proliferation/angiogenesis is suggested by our microarray study (not published). Low concentrations of ox-LDL appear to stimulate receptors for a number of pro-proliferative growth factors, components of extracellular matrix, and cytokines including VEGFR2, KIT, PDGFB, ROBO4, NOTCH1, NOTCH4, CXCR4, EPHB2, ENG, and TNFRSF25. With the exception of NOTCH, upregulation of these receptors, however, was not accompanied with stimulation of their ligands. This suggests that, in addition to direct pro-proliferative and angiogenic action via ROS/NfkB/MAPK, ox-LDL also renders cells hypersensitive to external proliferative stimuli.

19.7 Extravasation

LOX-1 has been shown to act as an adhesive molecule and drastically increase attachment of monocytes to a LOX-1-coated surface [139] and to normally adhesion-negative cells transfected with LOX-1 expression vector [140]. Also ox-LDL/LOX-1 signaling involves stimulation of key molecules such as intercellular adhesion molecule (ICAM), vascular cell adhesion molecule (VCAM), P-selectin, and monocyte chemotactic protein-1 (MCP-1) that are necessary for transendothelial migration [113, 141]. Ox-LDL has been shown to play a key role in increasing adhesiveness of endothelium and recruitment/transendothelial migration of monocytes [113].

Information on mechanisms governing extravasation of detached tumor cells is scarce, but it is likely to be mechanistically similar to the process described for leukocytes. One of the key factors involved in both monocyte and cancer cell migration is MCP-1 [142] which is a transcriptional target of ox-LDL/LOX-1 [143]. In in vitro studies, invasiveness of prostate cancer cells correlated with induced expression of

MCP-1 in bone marrow endothelial cells and was suppressed in presence of MCP-1 antibody [144]. It has been reported that tethering of breast cancer cells to endothelium is mediated by E-selectin binding to the CD44 variant 4 [145]. It has also been shown that upregulation of LOX-1 by TNFα promotes both adhesion and transendothelial migration of MDA-MB-231 breast cancer cells [146]. Among known ligands for LOX-1, phosphatidylserine is a potential counterpart responsible for the adhesive interaction of endothelium and cancer cells. On the other hand, cancer cells display overabundance of phosphatidylserine on their outer membranes [147]. Chronic stimulation of LOX-1 and other scavenger receptors by elevated ox-LDL in circulation of obese individuals can thus make endothelial cells more receptive with regard to cancer cells and promote tumor dissemination.

19.8 Potential Combined Action of LDL Resident miRNAs

miRNAs can form stable ternary complexes with constituents of lipoproteins such as phospholipids [148]. A recent analysis by Vickens et al. identified multiple miRNAs bound to lipoprotein particles in dyslipidemic human subjects and mice [149]. What is interesting is that miRNA profiles substantially differ between HDL and LDL. LDL is a predominant source of lipoproteins for many peripheral tissues, and its import is further enhanced by internalization of its oxidatively modified form via scavenger receptors-mediated mechanisms.

We have analyzed available data in the context of potential oncogenicity of LDL miRNA complement associated with obesity (Fig. 19.2). Several unique or significantly overrepresented miRNAs in LDL particles including miR-19, miR-146b-5p, miR-574, miR-135a, miR-21, miR-20, mi-27a, and miR-221 have been shown to exert oncogenic effects [150, 151] (Fig. 19.2a). Several miRNAs from this group had overlapping targets (Fig. 19.2b) including tumor-suppressor genes.

Of particular interest, combined effects of LDL miRNAs concern several protocadherins including most members of PCDHα gene cluster PCDHA1-8, PCDHA10-13, and PCDHAC1-2 (5q31) as well as non-clustered protocadherins PCDH7 (4p15), PCDH9 (19p14), PCDH10 (4q28), and PCDH17 (13q14) (Fig. 19.2c). These genes are primarily expressed in CNS and have been shown to be engaged in synapse formation [152]. There is also a growing body of evidence that most protocadherins are powerful suppressors of transformation and carcinogenesis. PCDHα cluster is a subject to epigenetic silencing at the early stages of the process the development of various neoplasms [153, 154]. In gliomas, loss of expression for PCDH9 was associated with higher grade and poorer prognosis [155]. Early promoter hypermethylation has been also reported for PCDH10 in multiple cancers [156–160]. On the other hand, overexpression of PCDH10 in cancer cells impaired proliferation and migration and promoted apoptosis [157, 159].

The mechanism by which protocadherins exert their tumor-suppressor activity is not known. They do not bind β-catenin and, therefore, are unlikely to influence Wnt/β-catenin signaling pathway in a manner typical for tumor suppressive action

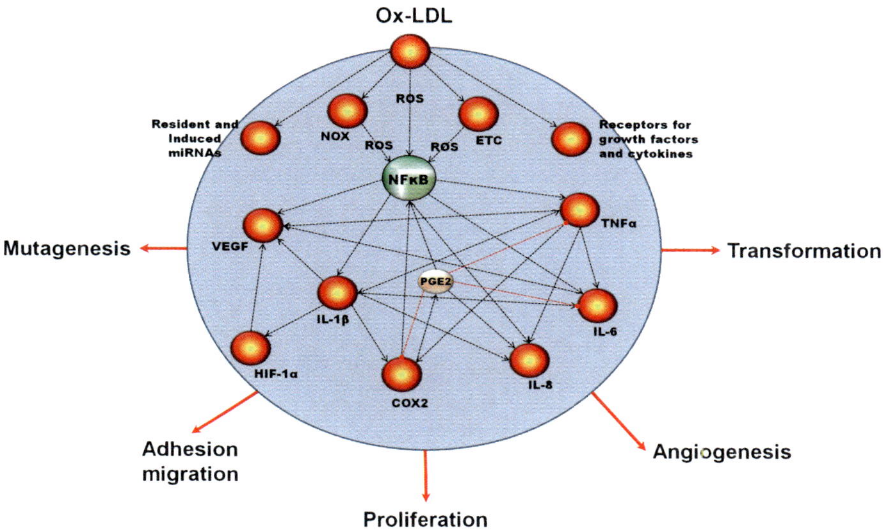

Fig. 19.2 Analysis of target genes for miRNAs overrepresented in LDL particles [149]. (**a**) Radar plot for miRNA composition in HDL (*red squares*) and LDL (*blue squares*). miRNAs known to have oncogenic properties are highlighted by *red circles*. (**b**) Genes targeted by more than six miRNAs; (**c**) protocadherins targeted by multiple LDL resident miRNAs. The lists for miRNA target genes were obtained using TargetScan database

of cadherins. However, two studies provide evidence that antitumor signaling of protocadherins might be linked to modulation of canonical Wnt and mTOR pathways [154, 161]. In these experiments, knockdown of γ-PCDHs in Wit49 cells resulted in increase of β-catenin/Tcf reporter activity, upregulation of β-catenin and its target genes CCND1, CMYC, and PAX8 [154]. In another study, overexpression of PCDHGC3 inhibited both β-catenin/Tcf-driven transcription and mTOR [161].

19.9 miRNAs Involved in Ox-LDL Signaling

miRNA composition in commercially available ox-LDL preparations is not known. It is likely, however, that prolonged periods of oxidation by copper sulfate and dialysis would essentially deplete ox-LDL from all resident RNA molecules. In this context, experimental data on miRNA upregulation might truly indicate intrinsic stimulatory activity of ox-LDL.

In recent years, there have been few studies reporting effects of ox-LDL on miRNA expression in vascular cells (Table 19.1). Most of the studies utilized ox-LDL at concentrations that are normally found toxic in vitro. There is almost no overlap between two available microarray studies [162, 163] with the exception of

Table 19.2 MiRNAs deregulated in cell exposed to ox-LDL

Cell type	Exposure [mg/ml] × [hours]	Deregulated miRNAs Up [red] and Down [blue]	Ref
HUVEC	[50] × [48]	142-3p, 365, 1207-3p, Let-7c 18b, 32, 33a, 155, 222, 374a, 375, 429, 589 590-5p, 1197	[162]
Macrophages,	[30] × [48]	9, 9*, 21, 99b, 125a-5p, 125a-3p, 132, 139-5p,146b-5p, 146a, 155, 181d, 214, 574-3p, 1225-3p, 15a, 28-3p, 28-5p, 32, 128, 132*, 138-2*.143*, 143, 145, 151-3p, 194*, 221*, 222*, 323-5p, 376c, 423-5p, 629*	[163]
Human primary monocyte-derived DCs	[30] × [48]	146a	[164]
THP1 macrophages	[50] × [48]	155	[165]
Primary human aortic SMCs	[25–50] × [24]	let-7g	[166]
Mouse EC lines HSV and b.END.3	[10] × [6–48]	125a-5p 125b-5p	[167]
Primary human aortic SMCs	[40] × [24]	29b	[168]

similarly downregulated miR-32 and miR 222 and showing opposite trends miR-155. This probably reflects cell type-specific tolerances for ox-LDL and the extent of interference from expression shifts caused by execution of apoptotic program. Recently published string of analyses of individual miRNAs (despite similar objections concerning concentration in most cases) suggests that ox-LDL stimulates miR-146a, miR-125a-5p, miR-155, and miR-29b and inhibits let-7g as well as miR-125b-5p (Table 19.2). It is likely that at near-physiological concentrations ox-LDL generates different miRNA expression profiles. In our studies (unpublished), exposure of human coronary artery endothelial cells to low concentration ox-LDL (5 μg/mL) for 12 h resulted in mild upregulation of only three miRNAs including miR-1974, miR-1978, and oncogenic miR-21. Clearly, more studies are required to determine true expression profile generated by nontoxic concentrations of ox-LDL.

19.10 Concluding Remarks

Observing a flood of information concerning major signaling pathways, one can't help but notice that many of them (including NFκB, PI3K/Akt, and MAPK) are almost compulsorily triggered by virtually any stimuli whereas the functional outcomes of such signals are nonetheless specific. There are uncomfortably too many "master switches" for any given process in the literature. The accumulated data, however, force a shift in perception of cellular signaling from the concept of linear chain of reactions to a concept of "grid" where everything regulates everything to a different extent (much like plucking a spider web at different locations) creating a more or less unique response pattern for individual stimuli. In this context, determination of relative seniority of components of ox-LDL mediated angiogenic response is probably inconsequential, if not misleading.

Accumulated data suggests that in many aspects, ox-LDL (and not LDL) is an active entity in pathobiology of increased susceptibility to cancer and tumor progression mediated by dyslipidemia. As we attempted to show, there are multiple potential mechanisms by which ox-LDL can promote carcinogenesis. First, the payload of ox-LDL particle contains multiple species of modified lipids with mutagenic and signaling properties. Second, ox-LDL causes a wave of endogenous production of ROS originating from NADPH oxidase and mitochondria followed by activation of redox-sensitive pathways and leading to upregulation of several pro-oncogenic and pro-angiogenic cytokines. Third, ox-LDL particle, as a carrier of miRNAs, introduces into the cells a distinct panel of miRNAs which display significant overlaps in targeting tumor-suppressor genes. Fourth, constituents of ox-LDL particle themselves appear to activate several miRNAs with oncogenic properties. Finally, from the tumor/host perspective, ox-LDL may act as a molecule that forces cooperation between growing tumor and cardiovascular system in terms of promoting neovascularization and creating permissive environment for adhesion, transendothelial migration, and formation of metastases.

References

1. Renehan AG, Tyson M, Egger M et al (2008) Body-mass index and incidence of cancer: a systematic review and meta-analysis of prospective observational studies. Lancet 371:569–578
2. Renehan AG, Soerjomataram I, Leitzmann MF (2010) Interpreting the epidemiological evidence linking obesity and cancer: a framework for population-attributable risk estimations in Europe. Eur J Cancer 46:2581–2592
3. Calle EE, Rodriguez C, Walker-Thurmond K et al (2003) Overweight, obesity, and mortality from cancer in a prospectively studied cohort of U.S. adults. N Engl J Med 348:1625–1638
4. Grek CL, Tew KD (2010) Redox metabolism and malignancy. Curr Opin Pharmacol 4:362–368
5. Trinchieri G (2012) Cancer and inflammation: an old intuition with rapidly evolving new concepts. Annu Rev Immunol 30:677–706
6. Kelesidis I, Kelesidis T, Mantzoros CS (2006) Adiponectin and cancer: a systematic review. Br J Cancer 94:1221–1225
7. Braun S, Bitton-Worms K, LeRoith D (2011) The link between the metabolic syndrome and cancer. Int J Biol Sci 7:1003–1015
8. Wang D, Dubois RN (2012) The role of the PGE2-aromatase pathway in obesity-associated breast inflammation. Cancer Discov 2:308–310
9. Brown MS, Kovanen PT, Goldstein JL (1981) Regulation of plasma cholesterol by lipoprotein receptors. Science 212:628–635

10. Vykhovanets EV, Shankar E, Vykhovanets OV et al (2011) High-fat diet increases NF-κB signaling in the prostate of reporter mice. Prostate 71:147–156
11. Bełtowski J, Wójcicka G, Górny D et al (2000) The effect of dietary-induced obesity on lipid peroxidation, antioxidant enzymes and total plasma antioxidant capacity. J Physiol Pharmacol 51:883–896
12. Brinkley TE, Kume N, Mitsuoka H et al (2008) Elevated soluble lectin-like oxidized LDL receptor-1 [sLOX-1] levels in obese postmenopausal women. Obesity (Silver Spring) 16:1454–1456
13. Holvoet P (2008) Relations between metabolic syndrome, oxidative stress and inflammation and cardiovascular disease. Verh K Acad Geneeskd Belg 70:193–219
14. Holvoet P, Lee DH, Steffes M et al (2008) Association between circulating oxidized low-density lipoprotein and incidence of the metabolic syndrome. JAMA 299:2287–2293
15. Holvoet P, Kritchevsky SB, Tracy RP et al (2004) The metabolic syndrome, circulating oxidized LDL and risk of myocardial infarction in well-functioning elderly people in the health, aging, and body composition cohort. Diabetes 53:1068–1073
16. Meisinger C, Baumert J, Khuseyinova N et al (2005) Plasma oxidized low-density lipoprotein, a strong predictor for acute coronary heart disease events in apparently healthy, middle-aged men from the general population. Circulation 112:651–657
17. Steinberg D, Parthasarathy S, Carew TE et al (1989) Beyond cholesterol: modifications of low density lipoprotein that increase its atherogenicity. N Engl J Med 320:915–924
18. Haberland ME, Fogelman AM, Edwards PA (1982) Specificity of receptor-mediated recognition of malondialdehyde-modified low density lipoproteins. Proc Natl Acad Sci U S A 79:1712–1716
19. Parthasarathy S, Fong LG, Otero D et al (1987) Recognition of solubilized apoproteins from delipidated, oxidized low density lipoprotein [LDL] by the acetyl-LDL receptor. Proc Natl Acad Sci U S A 84:537–540
20. Li D, Mehta JL (2000) Upregulation of endothelial receptor for oxidized LDL [LOX-1] by oxidized LDL and implications in apoptosis of human coronary artery endothelial cells. Arterioscler Thromb Vasc Biol 20:1116–1122
21. Nicholson AC, Han J, Febbraio M et al (2001) Role of CD36, the macrophage class B scavenger receptor, in atherosclerosis. Ann N Y Acad Sci 947:224–228
22. Marnett LJ (1999) Chemistry and biology of DNA damage by malondialdehyde. IARC Sci Publ 150:17–27
23. Esterbauer H, Schaur RJ, Zollner H (1991) Chemistry and biochemistry of 4-hydroxynonenal, malonaldehyde and related aldehydes. Free Radic Biol Med 11(1):81–128
24. Feng Z, Hu W, Tang MS (2004) Trans-4-hydroxy-2-nonenal inhibits nucleotide excision repair in human cells: a possible mechanism for lipid peroxidation-induced carcinogenesis. Proc Natl Acad Sci U S A 101:8598–8602
25. Chen KH, Srivastava DK, Singhal RK et al (2000) Modulation of base excision repair by low density lipoprotein, oxidized low density lipoprotein and antioxidants in mouse monocytes. Carcinogenesis 21:1017–1022
26. Thum T, Borlak J (2008) LOX-1 receptor blockade abrogates oxLDL-induced oxidative DNA damage and prevents activation of the transcriptional repressor Oct-1 in human coronary arterial endothelium. Biol Chem 283:19456–19464
27. Inoue T, Inoue K, Maeda H et al (2001) Immunological response to oxidized LDL occurs in association with oxidative DNA damage independently of serum LDL concentrations in dyslipidemic patients. Clin Chim Acta 305:115–121
28. Rueckschloss U, Galle J, Holtz J et al (2001) Induction of NAD[P]H oxidase by oxidized low-density lipoprotein in human endothelial cells: antioxidative potential of hydroxymethylglutaryl coenzyme A reductase inhibitor therapy. Circulation 104:1767–1772
29. Bae YS, Lee JH, Choi SH et al (2009) Macrophages generate reactive oxygen species in response to minimally oxidized low-density lipoprotein: toll-like receptor 4- and spleen tyrosine kinase-dependent activation of NADPH oxidase 2. Circ Res 104:210–218

30. Zhao R, Moghadasian MH, Shen GX (2011) Involvement of NADPH oxidase in up-regulation of plasminogen activator inhibitor-1 and heat shock factor-1 in mouse embryo fibroblasts induced by oxidized LDL and in apolipoprotein E-deficient mice. Free Radic Res 45:1013–1023
31. Landar A, Zmijewski JW, Dickinson DA et al (2006) Interaction of electrophilic lipid oxidation products with mitochondria in endothelial cells and formation of reactive oxygen species. Am J Physiol Heart Circ Physiol 290:H1777–H1787
32. Zmijewski JW, Moellering DR, Le Goffe C et al (2005) Oxidized LDL induces mitochondrially associated reactive oxygen/nitrogen species formation in endothelial cells. Am J Physiol Heart Circ Physiol 289:H852–H861
33. Takabe W, Li R, Ai L et al (2010) Oxidized low-density lipoprotein-activated c-Jun NH2-terminal kinase regulates manganese superoxide dismutase ubiquitination: implication for mitochondrial redox status and apoptosis. Arterioscler Thromb Vasc Biol 30:436–441
34. Ceaser EK, Ramachandran A, Levonen AL et al (2003) Oxidized low-density lipoprotein and 15-deoxy-Delta 12,14-PGJ2 increase mitochondrial complex I activity in endothelial cells. Am J Physiol Heart Circ Physiol 285:H2298–H2308
35. Jiang F, Lim HK, Morris MJ et al (2011) Systemic upregulation of NADPH oxidase in diet-induced obesity in rats. Redox Rep 16:223–229
36. Kamata T (2009) Roles of Nox1 and other Nox isoforms in cancer development. Cancer Sci 100:1382–1388
37. Ranjan P, Anathy V, Burch PM et al (2006) Redox-dependent expression of cyclin D1 and cell proliferation by Nox1 in mouse lung epithelial cells. Antioxid Redox Signal 8:1447–1459
38. Mitsushita J, Lambeth JD, Kamata T (2004) The superoxide-generating oxidase Nox1 is functionally required for Ras oncogene transformation. Cancer Res 64:3580–3585
39. Irani K, Xia Y, Zweier JL et al (1997) Mitogenic signaling mediated by oxidants in Ras-transformed fibroblasts. Science 275:1649–1652
40. Suh YA, Arnold RS, Lassegue B et al (1999) Cell transformation by the superoxide-generating oxidase Mox1. Nature 401:79–82
41. Puca R, Nardinocchi L, Starace G et al (2010) Nox1 is involved in p53 deacetylation and suppression of its transcriptional activity and apoptosis. Free Radic Biol Med 48:1338–1346
42. Graham KA, Kulawiec M, Owens KM et al (2010) NADPH oxidase 4 is an oncoprotein loc. lized to mitochondria. Cancer Biol Ther 10:223–231
43. Roy Chowdhury SK, Sangle GV, Xie X et al (2010) Effects of extensively oxidized low-density lipoprotein on mitochondrial function and reactive oxygen species in porcine aortic endothelial cells. Am J Physiol Endocrinol Metab 298:E89–E98
44. Nowsheen S, Aziz K, Kryston TB et al (2012) The interplay between inflammation and oxidative stress in carcinogenesis. Curr Mol Med 12:672–680
45. Reedy J (1975) Galen on cancer and related diseases. Clio Med 10:227–238
46. Balkwill F, Mantovani A (2001) Inflammation and cancer: back to Virchow? Lancet 357:539–545
47. Norris AL, Steinberger J, Steffen LM et al (2011) Circulating oxidized LDL and inflammation in extreme pediatric obesity. Obesity (Silver Spring) 19:1415–1419
48. Hulthe J, Fagerberg B (2002) Circulating oxidized LDL is associated with subclinical atherosclerosis development and inflammatory cytokines [AIR Study] arteriosclerosis. Arterioscler Thromb Vasc Biol 22:1162–1167
49. Ishikawa M, Ito H, Akiyoshi M et al (2012) Lectin-like oxidized low-density lipoprotein receptor 1 signal is a potent biomarker and therapeutic target for human rheumatoid arthritis. Arthritis Rheum 64:1024–1034
50. Cominacini L, Pasini AF, Garbin U et al (2000) Oxidized low density lipoprotein [ox-LDL] binding to ox-LDL receptor-1 in endothelial cells induces the activation of NF-κB through an increased production of intracellular reactive oxygen species. J Biol Chem 275:12633–12638

51. Matsunaga T, Hokari S, Koyama I et al (2003) NF-kappa B activation in endothelial cells treated with oxidized high-density lipoprotein. Biochem Biophys Res Commun 303:313–319
52. http://www.bu.edu/nf-kb/gene-resources
53. Rayet B, Gélinas C (1999) Aberrant rel/nfkb genes and activity in human cancer. Oncogene 18:6938–6947
54. Sun XF, Zhang H (2007) NFKB and NFKBI polymorphisms in relation to susceptibility of tumour and other diseases. Histol Histopathol 22:1387–1398
55. Hirsch HA, Iliopoulos D, Joshi A et al (2010) A transcriptional signature and common gene networks link cancer with lipid metabolism and diverse human diseases. Cancer Cell 17:348–356
56. Kang BY, Hu C, Prayaga S et al (2009) LOX-1 dependent overexpression of immunoglobulins in cardiomyocytes in response to angiotensin II. Biochem Biophys Res Commun 379:395–399
57. Khaidakov M, Mitra S, Kadlubar S et al (2011) Oxidized LDL receptor 1 [LOX-1] as a possible link between obesity and cancer. PLoS One 6(5):e20277
58. Maziere C, Auclair M, Djavaheri-Mergny M et al (1996) Oxidized low density lipoprotein induces activation of the transcription factor NF kappa B in fibroblasts, endothelial and smooth muscle cells. Biochem Mol Biol Int 39:1201–1207
59. Calara F, Dimayuga P, Niemann A et al (1998) An animal model to study local oxidation of LDL and its biological effects in the arterial wall. Arterioscler Thromb Vasc Biol 18:884–893
60. Nishimura S, Akagi M, Yoshida K et al (2004) Oxidized low-density lipoprotein [ox-LDL] binding to lectin-like ox-LDL receptor-1 [LOX-1] in cultured bovine articular chondrocytes increases production of intracellular reactive oxygen species [ROS] resulting in the activation of NF-kappaB. Osteoarthritis Cartilage 12:568–576
61. Fajardo LF, Kwan HH, Kowalski J et al (1992) Dual role of tumor necrosis factor-alpha in angiogenesis. Am J Pathol 140:539–544
62. Beg AA, Finco TS, Nantermet PV et al (1993) Tumor necrosis factor and interleukin-1 lead to phosphorylation and loss of I kappa B alpha: a mechanism for NF-kappa B activation. Mol Cell Biol 13:3301–3310
63. Moore RJ, Owens DM, Stamp G et al (1999) Mice deficient in tumor necrosis factor-alpha are resistant to skin carcinogenesis. Nat Med 5:828–831
64. Schmid MC, Avraamides CJ, Foubert P et al (2011) Combined blockade of integrin-$\alpha4\beta1$ plus cytokines SDF-1α or IL-1β potently inhibits tumor inflammation and growth. Cancer Res 71:6965–6975
65. Tsirakis G, Pappa CA, Kaparou M et al (2011) Assessment of proliferating cell nuclear antigen and its relationship with proinflammatory cytokines and parameters of disease activity in multiple myeloma patients. Eur J Histochem 55:e21
66. Heikkilä K, Ebrahim S, Lawlor DA (2008) Systematic review of the association between circulating interleukin-6 [IL-6] and cancer. Eur J Cancer 44:937–945
67. Kishimoto T (2005) Interleukin-6: from basic science to medicine–40 years in immunology. Annu Rev Immunol 23:1–21
68. Becker C, Fantini MC, Wirtz S et al (2005) IL-6 signaling promotes tumor growth in colorectal cancer. Cell Cycle 4:217–220
69. Yun UJ, Park SE, Jo YS et al (2012) DNA damage induces the IL-6/STAT3 signaling pathway, which has anti-senescence and growth-promoting functions in human tumors. Cancer Lett 323:155–160
70. Waugh DJJ, Wilson C (2008) The interleukin-8 pathway in cancer. Clin Cancer Res 14:6735–6741
71. Kuniyasu A, Hayashi S, Nakayama H (2002) Adipocytes recognize and degrade oxidized low density lipoprotein through CD36. Biochem Biophys Res Commun 295:319–323
72. Terkeltaub R, Banka CL, Solan J et al (1994) Oxidized LDL induces monocytic cell expression of interleukin-8, a chemokine with T-lymphocyte chemotactic activity. Arterioscler Thromb 14:47–53

73. van Tits LJ, Stienstra R, van Lent PL et al (2011) Oxidized LDL enhances pro-inflammatory responses of alternatively activated M2 macrophages: a crucial role for Krüppel-like factor 2. Atherosclerosis 214:345–349
74. Shao Q, Shen LH, Hu LH et al (2010) Nuclear receptor Nur77 suppresses inflammatory response dependent on COX-2 in macrophages induced by oxLDL. J Mol Cell Cardiol 49:304–311
75. Xie C, Ng H, Nagarajan S (2011) OxLDL or TLR2-induced cytokine response is enhanced by oxLDL-independent novel domain on mouse CD36. Immunol Lett 137:15–27
76. Chávez-Sánchez L, Chávez-Rueda K, Legorreta-Haquet MV et al (2010) The activation of CD14, TLR4, and TLR2 by mmLDL induces IL-1β, IL-6, and IL-10 secretion in human monocytes and macrophages. Lipids Health Dis 9:117
77. Li HX, Lei L, Yan FH (2010) Changes in cytokine levels during foam cells formation induced by oxidized low density lipoprotein stimulation. Xi Bao Yu Fen Zi Mian Yi Xue Za Zhi 26:742–745
78. Uyemura K, Demer LL, Castle SC et al (1996) Cross-regulatory roles of interleukin [IL]-12 and IL-10 in atherosclerosis. J Clin Invest 97:2130–2138
79. Jovinge S, Ares MPS, Kallin B et al (1996) Human monocytes/macrophages release TNF-α in response to Ox-LDL. Arterioscler Thromb Vasc Biol 6:1573–1579
80. Jiang YR, Miao Y, Yang L et al (2012) Effect of chinese herbal drug-containing serum for activating-blood and dispelling-toxin on ox-LDL-induced inflammatory factors' expression in endothelial cells. Chin J Integr Med 18:30–33
81. Hoffmann CJ, Hohberg M, Chlench S et al (2011) Suppression of zinc finger protein 580 by high oxLDL/LDL-ratios is followed by enhanced expression of endothelial IL-8. Atherosclerosis 216:103–108
82. Costa S, Zimetti F, Pedrelli M et al (2010) Manidipine reduces pro-inflammatory cytokines secretion in human endothelial cells and macrophages. Pharmacol Res 62:265–270
83. Mattaliano MD, Wooters J, Shih HH et al (2010) ROCK2 associates with lectin-like oxidized LDL receptor-1 and mediates oxidized LDL-induced IL-8 production. Am J Physiol Cell Physiol 298:C1180–C1187
84. Lappalainen J, Lindstedt KA, Oksjoki R et al (2011) OxLDL-IgG immune complexes induce expression and secretion of proatherogenic cytokines by cultured human mast cells. Atherosclerosis 214:357–363
85. Nagahama Y, Obama T, Usui M et al (2011) Oxidized low-density lipoprotein-induced periodontal inflammation is associated with the up-regulation of cyclooxygenase-2 and microsomal prostaglandin synthase 1 in human gingival epithelial cells. Biochem Biophys Res Commun 413:566–571
86. Suzuki K, Sakiyama Y, Usui M et al (2010) Oxidized low-density lipoprotein increases interleukin-8 production in human gingival epithelial cell line Ca9-22. J Periodontal Res 45:488–495
87. Yimin, Furumaki H, Matsuoka S et al (2010) A novel murine model for non-alcoholic steatohepatitis developed by combination of a high-fat diet and oxidized low-density lipoprotein. Lab Invest 92:265–281
88. Park J, Euhus DM, Scherer PE (2011) Paracrine and endocrine effects of adipose tissue on cancer development and progression. Endocr Rev 32:550–570
89. Matsubara M, Maruoka S, Katayose S (2002) Inverse relationship between plasma adiponectin and leptin concentrations in normal-weight and obese women. Eur J Endocrinol 147:173–180
90. Wauman J, Tavernier J (2011) Leptin receptor signaling: pathways to leptin resistance. Front Biosci 17:2771–2793
91. Ukropec J, Seböková E, Klimes I (2001) Nutrient sensing, leptin and insulin action. Arch Physiol Biochem 109:38–51
92. Tajmir P, Ceddia RB, Li RK et al (2004) Leptin increases cardiomyocyte hyperplasia via extracellular signal-regulated kinase and phosphatidylinositol 3-kinase-dependent signaling pathways. Endocrinology 145:1550–1555

93. Barouch LA, Gao D, Chen L et al (2006) Cardiac myocyte apoptosis is associated with increased DNA damage and decreased survival in murine models of obesity. Circ Res 98:119–124
94. Matsuzawa Y, Funahashi T, Kihara S, Shimomura I (2004) Adiponectin and metabolic syndrome. Arterioscler Thromb Vasc Biol 24:29–33
95. Ryo M, Nakamura T, Kihara S et al (2004) Adiponectin as a biomarker of the metabolic syndrome. Circ J 68:975–981
96. Bruun JM, Lihn AS, Verdich C et al (2003) Regulation of adiponectin by adipose tissue-derived cytokines: in vivo and in vitro investigations in humans. Am J Physiol Endocrinol Metab 285(3):E527–E533
97. Motoshima H, Wu X, Mahadev K et al (2004) Adiponectin suppresses proliferation and superoxide generation and enhances eNOS activity in endothelial cells treated with oxidized LDL. Biochem Biophys Res Commun 315:264–271
98. Wang Y, Lam KS, Xu JY et al (2005) Adiponectin inhibits cell proliferation by interacting with several growth factors in an oligomerization-dependent manner. J Biol Chem 280:18341–18347
99. Niu K, Asada M, Okazaki T et al (2012) Adiponectin pathway attenuates malignant mesothelioma cell growth. Am J Respir Cell Mol Biol 46:515–523
100. Wang Y, Lam JB, Lam KS et al (2006) Adiponectin modulates the glycogen synthase kinase-3beta/beta-catenin signaling pathway and attenuates mammary tumorigenesis of MDA-MB-231 cells in nude mice. Cancer Res 66:11462–11470
101. Zhang D, Guo M, Zhang W et al (2011) Adiponectin stimulates proliferation of adult hippocampal neural stem/progenitor cells through activation of p38 mitogen-activated protein kinase [p38MAPK]/glycogen synthase kinase 3β [GSK-3β]/β-catenin signaling cascade. J Biol Chem 286:44913–44920
102. Chen MJ, Yeh YT, Lee KT et al (2012) The promoting effect of adiponectin in hepatocellular carcinoma. J Surg Oncol 106:181–187
103. Lautamäki R, Rönnemaa T, Huupponen R et al (2007) Low serum adiponectin is associated with high circulating oxidized low-density lipoprotein in patients with type 2 diabetes mellitus and coronary artery disease. Metabolism 56:881–886
104. Kaur S, Zilmer K, Leping V et al (2011) The levels of adiponectin and leptin and their relation to other markers of cardiovascular risk in patients with psoriasis. J Eur Acad Dermatol Venereol 25:1328–1333
105. Porreca E, Di Febbo C, Moretta V et al (2004) Circulating leptin is associated with oxidized LDL in postmenopausal women. Atherosclerosis 175:139–143
106. Nakhjavani M, Morteza A, Asgarani F et al (2011) Metformin restores the correlation between serum-oxidized LDL and leptin levels in type 2 diabetic patients. Redox Rep 16:193–200
107. Takanabe-Mori R, Ono K, Sowa N et al (2010) Lectin-like oxidized low-density lipoprotein receptor-1 is required for the adipose tissue expression of proinflammatory cytokines in high-fat diet-induced obese mice. Biochem Biophys Res Commun 398:576–580
108. Masella R, Varì R, D'Archivio M et al (2006) Oxidised LDL modulate adipogenesis in 3T3-L1 preadipocytes by affecting the balance between cell proliferation and differentiation. FEBS Lett 580:2421–2429
109. Chui PC, Guan HP, Lehrke M et al (2005) PPARgamma regulates adipocyte cholesterol metabolism via oxidized LDL receptor 1. J Clin Invest 115:2244–2256
110. Scazzocchio B, Varì R, D'Archivio M et al (2009) Oxidized LDL impair adipocyte response to insulin by activating serine/threonine kinases. J Lipid Res 50(5):832–845
111. Scoles DR, Xu X, Wang H et al (2010) Liver X receptor agonist inhibits proliferation of ovarian carcinoma cells stimulated by oxidized low density lipoprotein. Gynecol Oncol 116:109–116
112. Dandapat A, Hu C, Sun L et al (2007) Small concentrations of ox-LDL induce capillary tube formation from endothelial cells via LOX-1-dependent redox-sensitive pathway. Arterioscler Thromb Vasc Biol 27:2435–2442

113. Mehta JL (2004) The role of LOX-1, a novel lectin-like receptor for oxidized low density lipoprotein in atherosclerosis. Can J Cardiol 20(suppl B):32B–36B
114. Watanabe T, Pakala R, Katagiri T et al (2002) Lysophosphatidylcholine is a major contributor to the synergistic effect of mildly oxidized low-density lipoprotein with endothelin-1 on vascular smooth muscle cell proliferation. J Cardiovasc Pharmacol 39:449–459
115. Stiko A, Regnström J, Shah PK et al (1996) Active oxygen species and lysophosphatidylcholine are involved in oxidized low density lipoprotein activation of smooth muscle cell DNA synthesis. Arterioscler Thromb Vasc Biol 16:194–200
116. Bochkov VN, Philippova M, Oskolkova O et al (2006) Oxidized phospholipids stimulate angiogenesis via autocrine mechanisms, implicating a novel role for lipid oxidation in the evolution of atherosclerotic lesions. Circ Res 99:900–908
117. Davies SS, Pontsler AV, Marathe GK et al (2001) Oxidized alkyl phospholipids are specific, high affinity peroxisome proliferator-activated receptor gamma ligands and agonists. J Biol Chem 276:16015–16023
118. Auron PE (1998) The interleukin 1 receptor: ligand interactions and signal transduction. Cytokine Growth Factor Rev 9:221–237
119. Janssens S, Beyaert R (2003) Functional diversity and regulation of different interleukin-1 receptor-associated kinase [IRAK] family members. Mol Cell 11:293–302
120. Carmi Y, Voronov E, Dotan S et al (2009) The role of macrophage-derived IL-1 in induction and maintenance of angiogenesis. J Immunol 183:4705–4714
121. Voronov E, Carmi Y, Apte RN (2007) Role of IL-1-mediated inflammation in tumor angiogenesis. Adv Exp Med Biol 601:265–270
122. Jagielska J, Kapopara PR, Salguero G et al (2012) Interleukin-1 assembles a proangiogenic signaling module consisting of caveolin-1, tumor necrosis factor receptor-associated factor 6, p38-mitogen-activated protein kinase [MAPK], and MAPK-activated protein kinase 2 in endothelial cells. Arterioscler Thromb Vasc Biol 32:1280–1288
123. Ara T, Declerck YA (2010) Interleukin-6 in bone metastasis and cancer progression. Eur J Cancer 46:1223–1231
124. Fan Y, Ye J, Shen F et al (2008) Interleukin-6 stimulates circulating blood-derived endothelial progenitor cell angiogenesis in vitro. J Cereb Blood Flow Metab 28:90–98
125. Gertz K, Kronenberg G, Kälin RE et al (2012) Essential role of interleukin-6 in post-stroke angiogenesis. Brain 135:1964–1980
126. Baggiolini M, Dewald B, Moser B (1994) Interleukin-8 and related chemotactic cytokines—CXC and CC chemokines. Adv Immunol 55:97–179
127. Desbaillets I, Diserens AC, de Tribolet N et al (1999) Regulation of interleukin-8 expression by reduced oxygen pressure in human glioblastoma. Oncogene 18:1447–1456
128. Yoshida S, OnoM ST et al (1997) Involvement of interleukin-8, vascular endothelial growth factor, and basic fibroblast growth factor in tumor necrosis factor alpha-dependent angiogenesis. Mol Cell Biol 17:4015–4023
129. Li A, Dubey S, Varney ML et al (2003) IL-8 directly enhanced endothelial cell survival, proliferation, and matrix metalloproteinases production and regulated angiogenesis. J Immunol 170:3369–3376
130. Koch AE, Polverini PJ, Kunkel SL et al (1992) Interleukin-8 as a macrophage derived mediator of angiogenesis. Science 258:1798–1801
131. Strieter RM, Polverini PG, Kunkel SL et al (1995) The functional role of the ELR motif in CXC chemokine-mediated angiogenesis. J Biol Chem 270:27348–27357
132. Libby P, Ordovas JM, Auger KR et al (1986) Endotoxin and tumor necrosis factor induce interleukin-1 gene expression in adult human vascular endothelial cells. Am J Pathol 124:179–185
133. Hamanaka R, Kohno K, Seguchi T et al (1992) Induction of low density lipoprotein receptor and a transcription factor SP-1 by tumor necrosis factor in human microvascular endothelial cells. J Biol Chem 267:13160–13165

134. Lowenthal JW, Ballard DW, Bogerd H et al (1989) Tumor necrosis factor-alpha activation of the IL-2 receptor-alpha gene involves the induction of kappa B-specific DNA binding proteins. J Immunol 142:3121–3128
135. Kuwano T, Nakao S, Yamamoto H et al (2004) Cyclooxygenase 2 is a key enzyme for inflammatory cytokine-induced angiogenesis. FASEB J 18:300–310
136. Ghosh N, Chaki R, Mandal V et al (2010) COX-2 as a target for cancer chemotherapy. Pharmacol Rep 62:233–244
137. Sheng H, Shao J, Washington MK et al (2001) Prostaglandin E2 increases growth and motility of colorectal carcinoma cells. J Biol Chem 276:18075–18081
138. Sheng H, Shao J, Morrow JD et al (1998) Modulation of apoptosis and Bcl-2 expression by prostaglandin E2 in human colon cancer cells. Cancer Res 58:362–366
139. Honjo M, Nakamura K, Yamashiro K et al (2003) Lectin-like oxidized LDL receptor-1 is a cell-adhesion molecule involved in endotoxin-induced inflammation. Proc Natl Acad Sci U S A 100:1274–1279
140. Hayashida K, Kume N, Minami M et al (2002) Lectin-like oxidized LDL receptor-1 [LOX-1] supports adhesion of mononuclear leukocytes and a monocyte-like cell line THP-1 cells under static and flow conditions. FEBS Lett 511:133–138
141. Sawamura T, Kume N, Aoyama T et al (1997) An endothelial receptor for oxidized low-density lipoprotein. Nature 386:73–77
142. Deshmane SL, Kremlev S, Amini S et al (2009) Monocyte chemoattractant protein-1 [MCP-1]: an overview. J Interferon Cytokine Res 29:313–326
143. Li D, Mehta JL (2000) Antisense to LOX-1 inhibits oxidized LDL-mediated upregulation of monocyte chemoattractant protein-1 and monocyte adhesion to human coronary artery endothelial cells. Circulation 101:2889–2895
144. Lu Y, Cai Z, Xiao G et al (2007) Monocyte chemotactic protein-1 mediates prostate cancer-induced bone resorption. Cancer Res 67:3646–3653
145. Zen K, Liu DQ, Guo YL et al (2008) CD44v4 is a major E-selectin ligand that mediates breast cancer cell transendothelial migration. PLoS One 3:e1826
146. Liang M, Zhang P, Fu J (2007) Up-regulation of LOX-1 expression by TNF-alpha promotes trans-endothelial migration of MDA-MB-231 breast cancer cells. Cancer Lett 258:31–37
147. Utsugi T, Schroit AJ, Connor J et al (1991) Elevated expression of phosphatidylserine in the outer membrane leaflet of human tumor cells and recognition by activated human blood monocytes. Cancer Res 51:3062–3068
148. Janas T, Janas T, Yarus M (2006) Specific RNA binding to ordered phospholipid bilayers. Nucleic Acids Res 34:2128–2136
149. Vickers KC, Palmisano BT, Shoucri BM et al (2011) MicroRNAs are transported in plasma and delivered to recipient cells by high-density lipoproteins. Nat Cell Biol 13:423–433
150. Lovat F, Valeri N, Croce CM (2011) MicroRNAs in the pathogenesis of cancer. Semin Oncol 38:724–733
151. Tie J, Fan D (2011) Big roles of microRNAs in tumorigenesis and tumor development. Histol Histopathol 26:1353–1361
152. Zipursky SL, Sanes JR (2010) Chemoaffinity revisited: dscams, protocadherins, and neural circuit assembly. Cell 143:343–353
153. Novak P, Jensen T, Oshiro MM et al (2008) Agglomerative epigenetic aberrations are a common event in human breast cancer. Cancer Res 68:8616–8625
154. Dallosso AR, Hancock AL, Szemes M et al (2009) Frequent long-range epigenetic silencing of protocadherin gene clusters on chromosome 5q31 in Wilms' tumor. PLoS Genet 5:e1000745
155. Wang C, Yu G, Liu J et al (2012) Downregulation of PCDH9 predicts prognosis for patients with glioma. J Clin Neurosci 19:541–545
156. Li Z, Chim JC, Yang M et al (2012) Role of PCDH10 and its hypermethylation in human gastric cancer. Biochim Biophys Acta 1823:298–305

157. Ying J, Li H, Seng TJ et al (2006) Functional epigenetics identifies a protocadherin PCDH10 as a candidate tumor suppressor for nasopharyngeal, esophageal and multiple other carcinomas with frequent methylation. Oncogene 25:1070–1080
158. Ying J, Gao Z, Li H et al (2007) Frequent epigenetic silencing of protocadherin 10 by methylation in multiple haematologic malignancies. Br J Haematol 136:829–832
159. Yu JS, Koujak S, Nagase S et al (2008) PCDH8, the human homolog of PAPC, is a candidate tumor suppressor of breast cancer. Oncogene 27:4657–4665
160. Narayan G, Freddy AJ, Xie D et al (2011) Promoter methylation-mediated inactivation of PCDH10 in acute lymphoblastic leukemia contributes to chemotherapy resistance. Genes Chromosomes Cancer 50:1043–1053
161. Dallosso AR, Oster B, Greenhough A et al (2012) Long-range epigenetic silencing of chromosome 5q31 protocadherins is involved in early and late stages of colorectal tumorigenesis through modulation of oncogenic pathways. Oncogene 31:4406–4419
162. Qin B, Xiao B, Liang D et al (2011) MicroRNAs expression in ox-LDL treated HUVECs: MiR-365 modulates apoptosis and Bcl-2 expression. Biochem Biophys Res Commun 410:127–133
163. Chen T, Yan H, Li Z et al (2011) MicroRNA-155 regulates lipid uptake, adhesion/chemokine marker secretion and SCG2 expression in oxLDL-stimulated dendritic cells/macrophages. Int J Cardiol 147:446–447
164. Chen T, Li Z, Jing T et al (2011) MicroRNA-146a regulates the maturation process and pro-inflammatory cytokine secretion by targeting CD40L in oxLDL-stimulated dendritic cells. FEBS Lett 585:567–573
165. Huang RS, Hu GQ, Lin B et al (2010) MicroRNA-155 silencing enhances inflammatory response and lipid uptake in oxidized low-density lipoprotein-stimulated human THP-1 macrophages. J Investig Med 58:961–967
166. Chen KC, Hsieh IC, Hsi E et al (2011) Negative feedback regulation between microRNA let-7g and the oxLDL receptor LOX-1. J Cell Sci 124:4115–4124
167. Li D, Yang P, Xiong Q et al (2010) MicroRNA-125a/b-5p inhibits endothelin-1 expression in vascular endothelial cells. J Hypertens 28:1646–1654
168. Chen KC, Wang YS, Hu CY et al (2011) OxLDL up-regulates microRNA-29b, leading to epigenetic modifications of MMP-2/MMP-9 genes: a novel mechanism for cardiovascular diseases. FASEB J 25:1718–1728

Chapter 20
Role of Genetic Polymorphisms in the Angiogenesis Pathway and Non-small-Cell Lung Cancer Tumor Behavior: Implications in Risk Assessment and Clinical Outcome

Ramon Andrade de Mello, Michael Luis, António Araújo, Rui Manuel Reis, and Venceslau Hespanhol

Abstract Lung cancer is a highly prevalent disease worldwide. Currently, it is the leading cause of cancer-related deaths in Western nations. Non-small-cell lung cancer (NSCLC) corresponds to approximately 85% of all histological types. Risk factors are usually associated with tobacco consumption, occupational exposure, radon, and passive smoking. Lung cancer diagnosis often occurs in the advanced stages IIIB and IV. Thus systemic therapies, such as cytotoxic agents and therapeutic targets, play a major role in NSCLC management. To date, many factors influence NSCLC behavior and, therefore, clinical response to therapeutic targets, such as epidermal growth factor (EGF) and its receptor (EGFR) and vascular endothelial

R.A. de Mello (✉)
Department of Medical Oncology, Portuguese Oncology Institute, Rua Dr. António Bernardino de Almeida, 4200-072 Porto, Portugal

Department of Medicine, University of Porto, Alameda Prof. Hernani Monteiro 4200-319 Porto, Portugal
e-mail: ramondemello@gmail.com

M. Luis • A. Araújo
Department of Medical Oncology, Portuguese Oncology Institute, Rua Dr. António Bernardino de Almeida, 4200-072 Porto, Portugal

R.M. Reis
Life and Health Sciences Research Institute (ICVS), School of Health Sciences, University of Minho, Campus de Gualtar, 4710-057 Braga, Portugal

Molecular Oncology Research Center, Barretos Cancer Hospital, Rua Antenor Duarte Vilella, 1331, 14784-400, Barretos-SP, Brazil

V. Hespanhol
Department of Medicine, University of Porto, Alameda Prof. Hernani Monteiro, 4200-319 Porto, Portugal

Service of Pneumology, Hospital São João, Alameda Prof. Hernani Monteiro, 4200-319 Porto, Portugal

J.L. Mehta and N.S. Dhalla (eds.), *Biochemical Basis and Therapeutic Implications of Angiogenesis*, Advances in Biochemistry in Health and Disease 6,
DOI 10.1007/978-1-4614-5857-9_20,

growth factor (VEGF) and its receptor (VEGFR). Angiogenesis-related genetic polymorphisms are of primary interest in NSCLC research. Recently, a Portuguese study identified the *EGF* +61 A/G polymorphism as a risk factor for NSCLC patients with advanced stages of cancer. Furthermore, other genetic polymorphisms, such as *VEGF* −2578 C/A and *VEGF* −1154 G/A, were correlated with increased tumor VEGF expression, vascular density, and poor survival. This topic will address research involving angiogenesis genetic polymorphisms and NSCLC behavior.

Keywords Lung cancer • Non-small-cell lung cancer • Epidermal growth factor • Epidermal growth factor receptor • Vascular endothelial growth factor • Vascular endothelial growth factor receptor • Erlotinib • Gefitinib • Bevacizumab

Abbreviations

EGF	Epidermal growth factor
EGFR	EGF receptor
GWAS	Genome-wide associate study
HBO	Hyperbaric oxygen
HIF	Hypoxia inducible factor
KDR receptor	Kinase insert domain receptor
mTOR	Mammalian target of rapamycin
NSCLC	Non-small-cell lung cancer
OS	Overall survival
PFS	Progression-free survival
PIK3	Phosphoinositide 3-kinase
PlGF	Placental growth factor
SNP	Single nucleotide polymorphism
SOS	Guanine nucleotide exchange factor sos
TKI	Tyrosine kinase inhibitor
VEGF	Vascular endothelial growth factor
VEGFR	VEGF receptor

20.1 Introduction

Lung cancer management has changed in the last decade [1–4]. Advances in molecular tools and targeted therapies have provided new insight for the field of oncology [5]. Despite these advancements, surgery is still the only effective curative approach, whether used alone or alongside a combination of other treatment modalities, such as radiotherapy and chemotherapy. Targeted therapies play a major role in this field [3, 5]. Further options are also promising for lung cancer treatment, including stereotaxic radiotherapy and its use in combination with induction chemoradiotherapy,

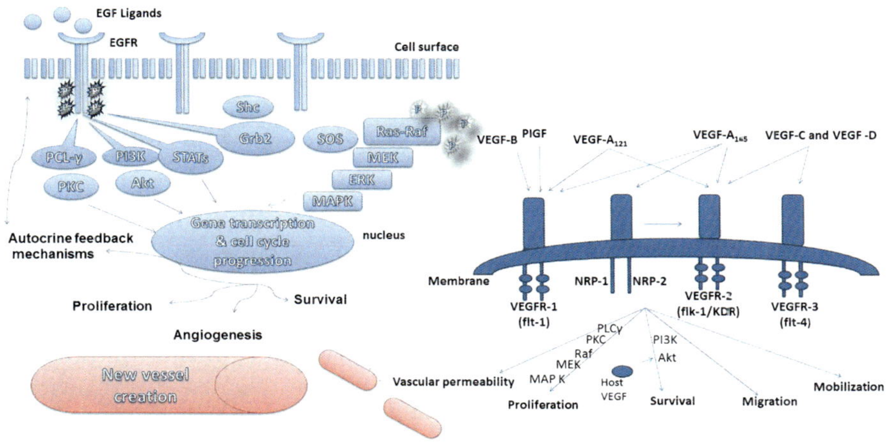

Fig. 20.1 Summary of the main VEGF and EGF pathways. *EGF* epidermal growth factor; *EGFR* EGF receptor; *VEGF* vascular endothelial growth factor; *VEGFR* VEGF receptor; *KDR receptor* kinase insert domain receptor; *PI3K* phosphoinositide 3-kinase; *mTOR* mammalian target of rapamycin; *Shc* Src homologous and collagen protein; *PCL-γ* phospholipase Cγ; *PI3K* phosphatidylinositol 3-kinase; *STATs* signal transducer and activator of transcription; *PKC* protein kinase C; *Grb2* growth factor receptor bound protein 2; *SOS* guanine nucleotide exchange factor sos; *MAPK* mitogen-activated protein kinase

as recently reported in a Japanese trial [6]. In light of the current knowledge, clinicians have begun selecting optimized populations that could receive improved benefits from these innovative therapeutic modalities [5, 7]. Tumor carcinogenesis pathways are now of primary interest in the field of oncology [2, 8–11]. First, angiogenesis pathways seem to be one of the key players in many cancers, including lung cancer [10]. Vascular endothelial growth factor (VEGF) and its receptor (VEGFR) are indeed important in the angiogenesis cascade [11]. Additionally, epidermal growth factor (EGF) and its receptor (EGFR) were demonstrated to be useful in the angiogenic mechanisms and clinical management of non-small-cell lung cancer (NSCLC) [12]. Figure 20.1 summarizes the most important VEGF pathways. Mutations in exon 19 and exon 21 of the *EGFR* gene are a predictive factor for a positive response to EGFR tyrosine kinase inhibitors (TKIs) [13, 14]. Recent trials showed favorable outcomes regarding overall survival (OS) and progression-free survival (PFS) in selected patients treated with gefitinib [15, 16] and erlotinib [17, 18], two EGFR TKIs [19, 20]. In this framework, it is important to focus on angiogenesis research. Until recently, the role of many angiogenesis genetic polymorphisms was under investigation in NSCLC patients [12, 21, 22]. A polymorphism is a variation in a gene that leads to two or more alleles existing at a frequency of at least 1% in the general population [23–25]. A variation at a single nucleotide is referred to as a single nucleotide polymorphism (SNP) [26]. The human genome contains more than 1,000,000 SNPs, and at least 60,000 of these are in exons [25, 27].

A SNP has a functional consequence if it exists within a coding or regulatory region of a gene [27]. Otherwise, an SNP may be silent if it resides in a noncoding region of the DNA or if it results in a synonymous amino acid substitution, except in rare situations [25]. Thus, this chapter will provide a discussion regarding the state of genetic polymorphisms related to angiogenesis and non-small-cell lung cancer tumor behavior, with a particular focus on risk assessment and prognosis.

20.2 Effects of Tumor Hypoxia and Clinical Issues

Interesting insights into the complex relationship of tumors with hypoxia can be gained by addressing the role of hypoxia as a predictor of response to different treatments. One field of active research focuses on the effect of oxygen on the response to radiotherapy (Fig. 20.2).

Tumor hypoxia influences the outcome of treatment not only in the setting of large tumors with extensive necrotic areas, but also in small tumors [26]. Hypoxia generally occurs as a result of insufficient vascularization, indicative of a long distance between the tumor cells and a functional blood vessel. Besides areas of chronic hypoxia, areas with intermittent hypoxia within tumors have also been described, thereby establishing persistent or transient resistance to radiotherapy [26, 28, 29].

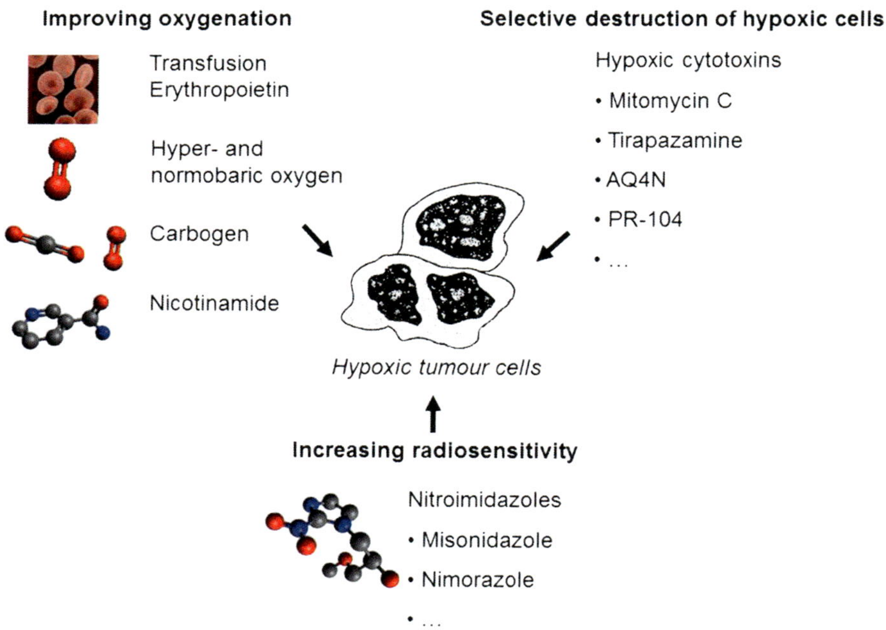

Fig. 20.2 Mechanisms of tumor hypoxia

The role of molecular oxygen (O_2) as a critical determinant to the response of cells to radiation has been described as early as the beginning of the last century [30, 31]. The radio-sensitizing effect of O_2 is explained by the chemical properties of the molecule, whose affinity for electrons predisposes it to participate in chemical reactions that lead to DNA damage if cells are exposed to ionizing radiation. Hypoxia reduces the radiosensitivity of cells, which require approximately three times as much radiation to become sensitized as cells with normal oxygen tension. In addition, hypoxia has been described as an important factor for tumor aggressiveness, a fact that should be taken into account when analyzing the poor response of hypoxic tumors to radiotherapy [26, 32, 33].

To circumvent the radioprotective approaches of hypoxia, three main approaches of hypoxic modification have been studied: improving tumor oxygenation, increasing the radiosensitivity of hypoxic cells, and selectively killing hypoxic tumor cells.

Improvement of tumor oxygenation can be achieved by pursuing hematocrit values above a certain threshold in patients. The hematocrit is an important clinical measure, considering that 40–60% of cancer patients are anemic at the beginning of radiotherapy [34]. Clinical trials examining the effects of transfusions and erythropoietin on the outcome of therapy have produced mixed results; nevertheless, hematocrit monitoring is a standard procedure in patients treated with radiotherapy [34, 35].

Hyperbaric oxygen (HBO) breathing yielded encouraging results beginning in the 1950s [36]. The physiological effects of HBO can be divided into short-term effects, such as enhanced oxygen delivery, phagocytosis activation, and anti-inflammatory effects, and long-term effects, including neovascularization and stimulation of collagen production by fibroblasts [37]. Issues have been raised concerning a possible enhancement of tumor growth or metastasis by HBO, but current evidence precludes such an effect [38]. Procedural difficulties in delivering both HBO therapy and radiotherapy as well as a reduced number of patients limited the usefulness of data obtained in the first clinical trials. Meta-analyses of more recent trials showed evidence that HBO improves local tumor control and reduces patient mortality for cancers of the head and neck, local tumor recurrence in cancers of the head and neck, and cervical cancer [39, 40]. Administration of normobaric oxygen and carbogen has also been studied, with results in animal models and clinical trials showing greater variability in outcome than with HBO [41]. Carbogen breathing is often combined with other modalities that might improve response, particularly nicotinamide (a vitamin B6 analog with vasoactive properties). The accelerated radiotherapy with carbogen and nicotinamide (ARCON) protocol consists of combining accelerated therapy and nicotinamide. Phase I and II trials with this protocol reported promising results in head and neck and bladder cancer [42, 43].

Increasing the radiosensitivity of hypoxic cells has been pursued with oxygen-mimetic radiosensitizers, particularly 2-nitroimidazoles such as misonidazole and nimorazole, as shown in Fig. 20.2. Clinical trials note a possible role for these radiosensitizers as adjuncts to radiotherapy, with promising results in some tumors but little improvement in loc.l tumor control and disease-free survival. Delayed peripheral neuropathy is an important concern with these agents, limiting their routine clinical use [41, 43].

The third approach, selectively killing hypoxic cells, involves hypoxic cytotoxins, which could play an important role in combination with radiotherapy and conventional cytotoxic chemotherapeutic agents, as these are generally more toxic to rapidly proliferating and well-perfused cells. Promising results have been reported with mitomycin C, a quinone that yielded positive results in loc.l control and disease-free survival in head and neck and cervical carcinomas [44, 45]. Another agent, the benzotriazine tirapazamine, has a different mechanism of action on hypoxic cells from quinones and requires less pronounced hypoxia. Preclinical studies demonstrated the efficacy of tirapazamine in combination with radiotherapy and chemotherapy; however, phase III trials revealed issues of toxicity, and conflicting results have been reported [46]. Other agents in this group, such as the alkylaminoanthraquinone N-oxide AQ4N and the 3,5-dinitrobenzamide-2-mustard PR-104, are being actively explored [47, 48].

In a systematic review, Overgaard [41] identified 10,108 patients in 86 randomized trials designed to modify tumor hypoxia in patients treated with curative attempted primary radiation therapy alone. Overall modification of tumor hypoxia significantly improved the effect of radiotherapy on the outcome of local regional control and was associated with a significant OS benefit. No significant influence was found on the incidence of distant metastases or on the risk of radiation-related complications.

Importantly, the implementation of these therapeutic modalities results in a better understanding of the molecular processes involved in the relationship between hypoxia and tumor growth. An appropriate selection of patients is also important: various minimally invasive and noninvasive methods hold promise for monitoring tumor oxygen distribution before and after treatment begins, thereby allowing the clinician to tailor specific regimens for different patients [29, 43].

20.3 Genetic Polymorphisms Related to Angiogenesis Pathways

20.3.1 A Brief Consideration of Angiogenesis in Lung Cancer

Angiogenesis is a crucial step for lung cancer progression [49–51]. The creation of new vessels from preexisting vessels influences tumor growth, invasion, and metastasis [52]. Recently, many studies have shown that NSCLC patients presented with higher serum VEGF than healthy controls [21, 53, 54]. Additionally, microvessel density (MVD) was demonstrated to be linked with tumor aggressiveness in some studies [55, 56]. In light of this knowledge, it was hypothesized that VEGF expression could have clinical implications in disease outcome [57]. In addition, the interaction between EGF and EGFR is associated with tumor angiogenesis, as shown in Fig. 20.2. The EGF/EGFR intracellular signaling stimulates gene transcription in the nucleus and is responsible for the creation of new vessels, cell migration, and metastasis [58]. Tables 20.1 and 20.2 summarize the main genetic polymorphisms studied in the lung cancer field.

Table 20.1 Summary of *VEGF* polymorphisms in lung cancer

Polymorphism	Tumor type	No. cases/ controls	Region	Effects on tumor risk[a]	Functional and clinical effects (VEGF levels, tumor behavior, survival)[a]	References
rs699947 (−2578 C/A)	NSCLC	36/0	N/A	N.Det.	CA genotype increased tumor VEGF expression and vascular density	[115]
	NSCLC	126/0	N/A	N.Det.	AA genotype associated with worse survival	[104]
	NSCLC	566/0	Asian	N.Det.	N.S.	[62]
rs1005230 (−2489 C/T)	NSCLC	566/0	Asian	N.Det.	N.S.	[62]
rs1570360 (−1154 G/A)	NSCLC	36/0	N/A	N.Det.	AA and GA genotypes increased tumor VEGF expression	[115]
	NSCLC	126/0	N/A	N.Det.	AA and GA genotypes associated with worse survival	[104]
rs833061 (−460 C/T)	Lung cancer	432/432	Asian	N.S.	N.Det.	[113]
	NSCLC	1900/1458	Caucasian	N.S.	N.Det.	[93]
	NSCLC	462/0	N/A	N.Det.	N.S.	[111]
	NSCLC	126/0	N/A	N.Det.	CC genotype associated with worse survival	[104]
rs25648 (−7 C/T)	NSCLC	568/0	Asian	N.Det.	N.S.	[62]

(continued)

Table 20.1 (continued)

rs2010963 (405 G/C)	NSCLC	36/0	N/A	N.Det.	GC genotype associated with increased tumor VEGF expression and vascular density	[115]
	Lung cancer	432/432	Asian	GG genotype reduces risk of SCC	N.Det.	[62]
	NSCLC	1900/1458	Caucasian	N.S.	N.Det.	[93]
	NSCLC	462/0	N/A	N.Det.	C allele associated with improved survival	[111]
	NSCLC	462/0	N/A	N.Det.	GC and combined GC+CC genotypes associated with improved survival	[111]
	Lung cancer	88/0	N/A	N.Det.	N.S.	[112]
	NSCLC	126/0	N/A	N.Det.	N.S.	[104]
	NSCLC	566/0	Asian	N.Det.	N.S.	[62]
rs3025039 (936 C/T)	Lung cancer	432/432	Asian	CT or CT+TT genotypes reduce risk of SCC	N.Det.	[62]
	NSCLC	1900/1458	Caucasian	N.S.	N.Det.	[93]
	NSCLC	462/0	N/A	N.Det.	N.S.	[111]
	Lung cancer	88/0	N/A	N.Det.	T allele increased *VEGF* gene expression and VEGF serum levels	[112]
	NSCLC	126/0	N/A	N.Det.	N.S.	[104]
	NSCLC	568/0	Asian	N.Det.	N.S.	[62]
rs10434 (1612 G/A)	NSCLC	560/0	Asian	N.Det.	N.S.	[62]
rs833069 (intron 2 G/A)	NSCLC	560/0	Asian	N.Det.	N.S.	[62]
rs833070 (intron 2 G/A)	NSCLC	560/0	Asian	N.Det.	N.S.	[62]
rs3024994 (intron 2 C/T)	NSCLC	568/0	Asian	N.Det.	N.S.	[62]
rs3025010 (intron 5 C/T)	NSCLC	567/0	Asian	N.Det.	N.S.	[62]

rs3025035 (intron 7 C/T)	NSCLC	568/0	Asian	N.Det.	N.S.	[62]
rs3025040 (3′-UTR T/C)	NSCLC	568/0	Asian	N.Det.	N.S.	[62]
rs3025053 (3′-UTR G/A)	NSCLC	568/0	Asian	N.Det.	N.S.	[62]
rs2736100	Adenocarcinoma	13,300/19,666	USA	Increased risk of adenocarcinoma	N.S.	[79]
rs31489	Lung cancer	13,300/19,666	USA	Increased risk	N.S.	[79]
rs4635969	Lung cancer	13,300/19,666	USA	Increased risk	N.S.	[79]
rs12914385	All lung cancers	13,300/19,666	USA	Increased risk	N.S.	[79]
rs1051730	All lung cancers	13,300/19,666	USA	Increased risk	N.S.	[79]
rs8034191	All lung cancers	13,300/19,666	USA	Increased risk	N.S.	[79]
rs12914385	Lung cancer	716/716	Japan	Increased risk	N.S.	[80]
rs931794	Lung cancer	716/716	Japan	Increased risk	N.S.	[80]
rs931794	Lung cancer	716/716	Japan	Increased risk	N.S.	[80]
rs12914385	Lung cancer	716/716	Japan	Increased risk	N.S.	[80]
rs2736100	Lung cancer	1094/1100	Korean	Increased risk	N.S.	[81]
rs402710	Lung cancer	1094/1100	Korean	Increased risk	N.S.	[81]
rs401681	Lung cancer	1094/1100	Korean	Increased risk	N.S.	[81]
rs31489	Lung cancer	1094/1100	Korean	Increased risk	N.S.	[81]
rs9295740	Lung cancer	1094/1100	Korean	Increased risk	N.S.	[81]

NSCLC non-small-cell lung cancer; *SCC* small cell carcinoma

[a]Significant differences were considered when $P<0.05$

Table 20.2 Summary of *VEGF* haplotypes in lung cancer

Haplotypes	Comparisons	Tumor type	No. cases/controls	Ethnicity	Effects on tumor risk[a]	Functional and clinical effects (VEGF levels, tumor behavior, survival)[a]	References
−460,405,936	TCT vs. all other haplotypes	Lung cancer	432/432	Asian	TCT reduces risk of lung cancer	N.Det.	[113]
	TGT vs. all other haplotypes	Lung cancer	432/432	Asian	TGT increases risk of lung cancer	N.Det.	[113]
	Each haplotype vs. all other haplotypes	AC	432/432	Asian	N.S.	N.Det.	[113]
	CGT vs. all other haplotypes	SCC	432/432	Asian	CGT reduces risk of SCC	N.Det.	[113]
	TCC vs. all other haplotypes	SCC	432/432	Asian	TCC increases risk of SCC	N.Det.	[113]
	TCC vs. CGC (reference)	NSCLC	1900/1458	Caucasian	N.S.	N.Det.	[83]
	TGC vs. CGC (reference)	NSCLC	1900/1458	Caucasian	N.S.	N.Det.	[83]
	CGT vs. CGC (reference)	NSCLC	1900/1458	Caucasian	N.S.	N.Det.	[83]
	TCT vs. CGC (reference)	NSCLC	1900/1458	Caucasian	N.S.	N.Det.	[83]
405,936	All variant alleles vs. GC (reference)	NSCLC	462/0	N/A	N.Det.	Higher number of variant alleles increased overall survival	[111]

NSCLC non-small-cell lung cancer; *AC* adenocarcinoma; *SCC* small cell carcinoma; *N/A* not available; *N.Det.* not determined; *N.S.* not significant
[a]Significant differences were considered when $P<0.05$

20.3.2 VEGF and VEGFR Genetic Polymorphisms in Cancer

The interaction between VEGF and VEGFR is one of the key regulators of angiogenesis [59–62]. Genetic variability of VEGF may play an important role in modifying cancer development and progression [57, 61, 63, 64]. The VEGF gene is located on chromosome 6p21.3 and contains eight exons [65]. It is highly polymorphic, with more than 20 polymorphisms showing up in cancer risk studies. The polymorphisms located on regulatory regions, such as promoter regions or 5′ and 3′ untranslated regions, are of primary interest in cancer susceptibility and outcome because they may modulate VEGF expression [65]. Tables 20.1 and 20.2 summarize the main VEGF polymorphisms and their influence on cancer.

20.3.3 Variants in VEGFR2 and Micro Vessel Density in Lung Cancer

Recent studies assessed the role of VEGFR2 genetic variants and NSCLC [21, 56]. MVD is considered a good experimental tool for evaluation of angiogenesis in NSCLC tissue. An American study [56] had defined VEGFR2 genetic variation in three populations (African American, Caucasian, and Asian) and identified common variants that impact tumor VEGFR2 expression and vascularization. VEGFR2 re-sequencing led to the discovery of 120 genetic variants, of which 25 had not been previously reported. The Q472H point mutation increased VEGFR-2 protein phosphorylation, and also was associated with increased MVD in NSCLC tumor samples. −2854C and −2455A increased luciferase expression and were associated with higher KDR mRNA levels in NSCLC samples. −271A reduced luciferase expression and was associated with lower VEGFR-2 levels in NSCLC samples. −906C and 23408G were associated with higher KDR mRNA levels in NSCLC samples. These findings are important because they elucidate a genetic signature that may influence clinical phenotypes related to VEGFR2 function.

20.3.4 EGF+61 A/G Genetic Polymorphisms and Cancer Risk

The *EGF* gene is located on chromosome 4q25-q27 [66], and *EGF* polymorphisms are associated with cancer susceptibility [66–69]. In 2002, Shahbazi et al. elucidated the role of the *EGF+61* A/G polymorphism in melanoma patients [66]. They suggested that cells from individuals homozygous for the +61 A allele produce significantly less EGF than cells from homozygous +61 G ($p=0.0004$) or heterozygous 61A/G individuals ($p=0.001$). Furthermore, Shahbazi's study showed that elevated EGF production is important for melanoma development [66]. In 2004, Bhowmick et al. demonstrated that the *EGF* +61 A/G polymorphisms were associated with an increased frequency of glioblastomas [70]. In 2007, our group also described the role of the *EGF* +61A/G polymorphism in glioma susceptibility in a Portuguese population and showed that the +61 G allele was associated with an elevated EGF expression level in vitro [69]. Other studies revealed that this genotype, along with the +61 G allele, is likely associated with a risk of pancreatic cancer development [71]. Furthermore, in 2009, Wu et al. also found a statistically significant association between both the *EGF* +61 G/G genotype and the +61 G allele and the risk for colorectal cancer [71].

Initial studies on the role of *EGF* +61A/G polymorphisms in lung cancer risk were controversial and restricted to Asian populations [72, 73]. In 2012, a Portuguese study first showed that the cumulative influence of *EGF* +61 A/G polymorphisms in Caucasian NSCLC patients increased susceptibility [74]. This is in agreement with a South Korean study that found a similar result in a comparative study in a schizophrenic population and healthy controls [73].

20.3.5 EGFR and EGFR Genetic Polymorphisms

EGFR is a 170 kDa transmembrane glycoprotein with an intracellular tyrosine protein kinase domain that plays critical cellular signaling roles in diverse pathways [20]. EGFR activation leads to many biological processes, including cell cycle progression, cell invasion, metastasis, angiogenesis, and cell differentiation, all of which are closely related to tumor progression [5]. In transgenic mice, overexpression of EGFR initiates the formation of oligodendroglioma [75] and breast cancer [76]. In humans, EGFR is overexpressed in 50–81% of NSCLC, and its overexpression correlates with poor prognosis of NSCLC [14]. Currently, EGFR is a noteworthy predictive biomarker in NSCLC management. In particular, in nonsmoking women with adenocarcinoma histology, a mutation in exon 19 or 21 of the *EGFR* gene is predictive of an improved response to EGFR TKIs. A study conducted among three major hospitals of north Portugal retrospectively assessed *EGFR* status in 621 NSCLC patients [77]. This study showed that *EGFR* was mutated in 14.3% of all NSCLC patients (17.5% adenocarcinoma and 9.5% squamous cell carcinoma). Exon 19 was mutated in 43.8% of patients, and exons 20, 21, and 18 were mutated in 37.1%, 21.4%, and 6.7% of patients, respectively [77]. Given these results, *EGFR* genetic polymorphisms are also of key interest. In 2012, a Chinese study enrolled 568 NSCLC patients and examined 54 SNPs of the *EGFR* and *VEGF* genes [78]. The authors found that subjects carrying *EGFR* rs3735061 A/A and rs6958497 A/G and G/G genotypes showed significantly shorter survival times (median survival time [MST]: 22.2 and 19.4 months, respectively) than those carrying rs3735061 A/G and G/G (MST: 25.1 months) and rs6958497 A/A (MST: 25.9 months) genotypes (log-rank $p=0.015$ for rs3735061 and log-rank $p=0.028$ for rs6958497). Nevertheless, subjects carrying *EGFR* rs759165 A/G and A/A genotypes survived significantly longer (MST: 38.7 months) than those carrying the rs759165 G/G genotype (MST: 24.7 months, log-rank $p=0.024$). Multivariate Cox regression analyses showed that the genotypes of rs3735061 A/A and rs6958497 A/G and G/G were associated with a significantly increased risk of death from

NSCLC (hazard ratio [HR] = 2.82, 95% confidence interval [CI] = 1.66–4.78 for rs3735061 A/A and HR = 1.69, 95% CI 1.26–2.28 for rs6958497 A/G and G/G), whereas the rs759165 A/G and A/A genotypes were associated with a significant 44% decreased risk of death from NSCLC (HR = 0.56, 95% CI 5 0.39–0.83). Furthermore, a stepwise Cox regression analysis suggested that *EGFR* rs373506, rs759165, and rs6958497 may be independent candidate biomarkers to predict NSCLC survival in this population [78].

20.4 Genetic Polymorphisms and Risk of Lung Cancer

20.4.1 GWAS Variants and Risk of Lung Cancer

Since 2009, the genome-wide association study (GWAS) group has found significant associations in chromosomes 5p15.33, 15q25, and 6p21 variants and lung cancer risk [79–82]. In the 5p15.33 locus, the most prominent were rs4635969 (OR 0.87, 95% confidence interval [CI], 0.82–0.93, $p = 9.80 \times 10^{-5}$), followed by rs31489 in CLPTM1L (also named CCR9 [MIM 612585], OR 0.90, 95% CI 0.86–0.95, $p = 2.80 \times 10^{-4}$) and rs2736100 in TERT (MIM 187270, OR 1.09, 95% CI, 1.03–1.15, $p = 0.001$) [79]. Additionally, the rs2736100 SNP was associated only with adenocarcinoma histology (OR 1.23, 95% CI, 1.13–1.33, $p = 3.02 \times 10^{-7}$) but not other histologic types (OR 1.01, $p = 0.84$ and OR = 1.00, $p = 0.93$, for squamous cell carcinoma and small cell carcinoma, respectively; $p = 0.001$. The test for heterogeneity across histologies was corrected for multiple comparisons.), as shown in Table 20.1 [79]. Thus, this study associated the rs2736100 variant in chromosome 5p15.33 with lung adenocarcinoma risk. The rs2736100 variant is located in intron 2 of the TERT gene [83]. TERT is a ribonucleoprotein that extends TTAGGG nucleotide repeats at telomeres, which progressively shorten with each cell division. Telomere shortening is associated with increased genomic instability and therefore increased risk of overall cancer development [84]. In cancer cells, reactivated TERT is linked to cellular proliferation and abnormal telomere maintenance [85]. Interestingly, TERT expression is lower in adenocarcinoma than in other histological subtypes [86]. Its re-expression may be related to the progression from bronchiolo-alveolar carcinoma to adenocarcinoma [87]. Moreover, Landi et al. also reported that other nicotinic acetylcholine receptor gene variants in chromosome 15q25 (rs4635969, CLPTM1L gene; rs12914385, CGRNA3 gene; rs1051730, CHRNA5 gene; and rs8034191, LOC 123688) were associated with an elevated overall risk of lung cancer [79]. These SNPs were also strongly associated with all major histology groups of patients who were current and former smokers [79]. In rs4324798 on chromosome 6p21.33, the association with lung cancer was weak and inconsistent throughout the studies [79, 88]. In 2012, Ito et al. reported a study in Japan with 716 lung cancer patients and 716 controls [80]. It was found that the variants of rs12914385 and rs931794 on 15q25 modified the effect of cumulative tobacco smoking on lung cancer risk, but that these two loci showed no statistically significant effects on lung cancer risk. Furthermore, association of the TERT-CLPM1L locus on 5p15 with lung cancer risk in Japanese patients was of a magnitude similar to that in Caucasians. Therefore, Ito's study confirmed the contribution of 15q25 and 5p15 to lung cancer susceptibility [82].

20.4.2 VEGF Polymorphisms and Controversies in Assessing the Risk of Lung Cancer

The VEGF gene is located on chromosome 6p21.3 and consists of eight exons that undergo alternate splicing to form a family of proteins [89]. Several genetic polymorphisms have been described, and they are associated with variation of VEGF production and expression [8, 10, 11, 21, 65, 78, 90–93] (Tables 20.1 and 20.2). Regarding the *VEGF* −1154 G/A polymorphism, the data are inconclusive and a recent meta-analysis conducted by Hong et al. [65] suggested that although *VEGF* −1154 G/A may not be associated with cancer risk in the general population, the −1154 GG homozygote allele may confer an elevated risk of cancer in non-Caucasians compared to −1154 A carriers.

20.4.3 EGF+61 Genetic Polymorphisms and Their Role in Lung Cancer Susceptibility

In 2002, *EGF*+61 A/G polymorphisms were first associated with cancer susceptibility in a melanoma model [66]. One decade later, only two Asian studies have addressed their role in lung cancer risk [72, 73]. However, in 2012, a Portuguese group showed that *EGF*+61 A/G polymorphisms are also associated with the risk of NSCLC in a population from North Portugal [74]. It was the first study in a Caucasian population and confirmed the results reported by Lim and colleagues [73] in a Korean population. Only a small number of studies addressed the role of *EGF*+61 A/G polymorphisms and lung cancer risk because *EGF*+61 A/G and *EGF*+61 G/G genotypes influence high EGF serum levels [66] and thus favor carcinogenesis in the lung [68, 94, 95].

20.5 Genetic Polymorphisms and Non-small-Cell Lung Cancer Prognosis

20.5.1 Current Lung Cancer Prognostic Factors

Many factors influence lung cancer prognosis [2, 4, 14, 58, 94, 96–105]. The tumor node metastasis (TNM) classification system [97], smoking status [106], and performance status [105] give physicians important clues to improve the clinical approach [4]. Currently, many platinum-based chemotherapy protocols with or without bevacizumab, EGFR TKIs, and crizotinib are interesting weapons against disease aggressiveness [5, 21]. Moreover, many efforts have been undertaken toward the genomic and molecular research to find prognostics and predictive biomarkers for lung cancer management [15, 103, 104, 106–110]. As they play a main role in angiogenesis and the overall biology of NSCLC, genetic polymorphisms in the VEGF [21, 90, 93, 111–113], VEGFR [92, 114], EGF [12, 72–74], and EGFR genes [78] will be discussed in the following paragraphs.

20.5.2 VEGF −2578 C/A and VEGF −1154 G/A Polymorphisms and NSCLC Survival

Many VEGF genetic polymorphisms have been studied alongside NSCLC survival and outcome [78, 104, 115]. In 2004, *VEGF* −2578 C/A and *VEGF*-1154 G/A played a key role in lung cancer vasculature [115]. Koukourakis et al. reported that tumors with the *VEGF* −2578 C/A genotype had significantly higher VEGF expression than those expressing the *VEGF* −2578 A/A genotype. Furthermore, the authors reported that the *VEGF*-1154 G/A polymorphism seemed to be related to poor vascularization, but the difference was not significant [115]. In 2009, a Japanese group demonstrated that the *VEGF* −1154 AA and AG genotypes (HR, 1.482; 95% CI, 1.144–1.897; $P=0.0034$) and the *VEGF* −2578AA genotype (HR, 1.797; 95% CI, 1.219–2.495; $P=0.0047$) had a significant prognostic effect on survival based on univariate analysis. Based on multivariate analysis of a current and former smoker (HR, 1.407; 95% CI, 1.095–1.840; $P=0.0070$), poor PS (HR, 2.249; 95% CI, 1.309–3.468; $P=0.0058$) and the *VEGF* −1154AA and AG genotypes (HR, 1.419; 95% CI, 1.033–1.901; $P=0.0316$) were significant independent prognostic factors for survival [104]. However, in 2010, Dong et al. studied 56 SNPs in 568 NSCLC patients [78], but the *VEGF* −2578 C/A polymorphisms did not influence survival and prognosis.

20.5.3 VEGF Variants [−460 T/C (rs833061), 405 G/C (rs2010963), and +936 C/T (rs3025039), −2489 C/T (rs1005230)] and Survival

In a previous study, VEGF serum levels were also associated with the OS of NSCLC patients treated with bevacizumab [53]. The reason for this association was that some genotypes, such as VEGF (−2578 C/A, −1498 C/T, −1154 G/A, −634 G/C, 936 C/T, and VEGF −1498 T/T), were correlated with toxicity, bevacizumab-related grades 3 and 4 hypertension, and disease outcome in patients treated for metastatic cancer [92]. In spite of controversial studies [21, 61, 90], *VEGF* −460 T/C polymorphisms were associated with a better OS in loc.lly advanced NSCLC after chemoradiotherapy [61]. In 2012, Naik et al. reported that the *VEGF* +405 C/G SNP, also called *VEGF −634 G/C* SNP, showed an association with age, pathological grade, and stage [90]. However, no results were consistent with survival [21]. Furthermore, *VEGF* −2489 C/T polymorphisms were not associated with OS in advanced NSCLC [78].

20.5.4 VEGFR2 Copy Number as a Predictive and Prognostic Factor

In 2011, experimental and translational research concerning angiogenesis and NSCLC was published to assess the role of VEGFR2 in tumor behavior [56, 116]. In vitro, VEGFR2 copy number gains were significantly associated with resistance to platinum chemotherapy as well as increased levels of nuclear HIF-1α in both NSCLC tumor specimens and cell lines [116]. VEGFR2 genetic variation (Q472H, −2854 C, and −2455 A) in three populations had an impact on tumor VEGFR2 expression and vascularization [56]. In light of these data, it is understandable that the role of angiogenesis and NSCLC is not completely elucidated. It is true that bevacizumab did not show a satisfactory cost-effectiveness relationship in advanced NSCLC treatment [21, 117]. Clinicians should expect further improved research about enhanced approaches involving anti-angiogenic therapies and NSCLC.

20.5.5 EGF+61 A/G and EGFR Polymorphisms and Outcome

Although the current role of *EGFR* mutations in exons 19 and 21 in advanced NSCLC treatment [5] has been elucidated, the roles of *EGF*+61 A/G polymorphisms, *EGFR* A/A (rs3735061) polymorphisms, and *EGFR* A/G (rs6958497) polymorphisms and outcome still need clarification [12, 78, 118]. In 2010, a Chinese study showed that the *EGFR* genotypes rs3735061AA and rs6958497AG/GG were associated with a significantly increased risk of death for NSCLC (hazard ratio [HR] = 2.82, 95% confidence interval [CI] = 1.66–4.78 for rs3735061AA and HR = 1.69, 95% CI = 1.26–2.28 for rs6958497AG/GG), whereas the rs759165AG/AA genotype was associated with a 44% decreased risk of death of NSCLC (HR = 0.56, 95% CI 5 0.39–0.83) [78]. Furthermore, other genetic polymorphisms in genes involved in the EGFR pathway (MAP3K1, RAF1, NRAS, and GPX7) were associated with NSCLC survival in a cohort study involving 1,076 lung cancer patients [119]. In addition, a Chinese study involving 88 NSCLC patients treated with gefitinib showed that the *EGFR* rs2293347 G/G genotype was associated with the efficacy of gefitinib [120]. The response rate for the rs2293347 GG genotype was significantly higher than that of the GA or AA genotypes (71.4% vs. 36.0%, $p = 0.002$).

The rs2293347 GG genotype was also associated with a longer PFS compared with the GA or AA genotypes (10 months vs. 3 months, $p=0.005$). However, no significant difference was observed regarding OS, $p=0.409$ [120]. Moreover, another Chinese study [121] with 115 NSCLC patients treated with EGFR TKIs showed a synergistic effect of the *CYP1A1*2A* and the *EGFR* intron 1 (CA) repeat polymorphisms, and this effect influenced patient clinical responses to EGFR-TKIs. Nevertheless, the *EGF*+61 A/G polymorphism was only assessed with regard to lung cancer risk [72–74] and not clinical outcome [118]. Thus, changes in the expression of EGF–EGFR pathway components may be very important in NSCLC carcinogenesis and clinical outcome. Further studies are warranted to validate this hypothesis.

20.6 Conclusions

Lung cancer is currently one of the major cancers around the world. Genomic profiling to improve its characterization is of particular interest in clinical practice [21, 118]. In the new era of targeted therapies, a thorough understanding of NSCLC molecular issues is crucial for clinical management [122]. It is true that angiogenic molecular mechanisms throughout the VEGF and EGFR pathways are key players of NSCLC pathogenesis. Therefore, studying genetic polymorphisms that regulate VEGF, VEGFR, EGF, and EGFR expression may be useful for improving screening strategies, therapeutic selection, and patients' prognoses. Additionally, GWAS variants were identified as risk factors for lung cancer in general. Thus, further research in this field is needed to improve methods to bridge genomic approaches and clinical practice.

Competing Interest The authors declare that they do not have any competing interests.

References

1. Alberg A, Brock M, Samet J (2005) Epidemiology of lung cancer: looking to the future. J Clin Oncol 23:3175–3185
2. Lopez A, Collishaw N, Piha T (1994) A descriptive model of the cigarette epidemic in developed countries. Tob Control 3:242–247
3. Asmis TR, Ding K, Seymour L et al (2008) Age and comorbidity as independent prognostic factors in the treatment of non-small-cell lung cancer: a review of National Cancer Institute of Canada Clinical Trials Group trials. J Clin Oncol 26:54–59
4. Brundage M, Davies D, Mackillop W (2002) Prognostic factors in non-small cell lung cancer. Chest 122:1037–1057
5. de Mello RA, Marques DS, Medeiros R, Araújo AM (2011) Epidermal growth factor receptor and K-Ras in non-small cell lung cancer-molecular pathways involved and targeted therapies. World J Clin Oncol 2:367–376
6. Futamura Y, Sawa T, Hasegawa T et al (2011) Stereotactic radiotherapy following chemoradiotherapy for lymph node metastasis of stage III non-small-cell lung cancer. Gan To Kagaku Ryoho 38:2191–2193

7. Hanagiri T, Sugio K, Mizukami M et al (2008) Significance of smoking as a postoperative prognostic factor in patients with non-small cell lung cancer. J Thorac Oncol 3:1127–1132
8. Rotunno M, Yu K, Lubin JH et al (2009) Phase I metabolic genes and risk of lung cancer: multiple polymorphisms and mRNA expression. PLoS One 4:e5652
9. Almog N, Ma L, Raychowdhury R et al (2009) Transcriptional switch of dormant tumors to fast-growing angiogenic phenotype. Cancer Res 69:836
10. Balasubramanian S, Brown N, Reed M (2002) Role of genetic polymorphisms in tumour angiogenesis. Br J Cancer 87:1057–1065
11. Bonnesen B, Pappot H, Holmstav J, Skov BG (2009) Vascular endothelial growth factor A and vascular endothelial growth factor receptor 2 expression in non-small cell lung cancer patients: relation to prognosis. Lung Cancer 66:314–318
12. Araujo A, Ribeiro R, Azevedo I et al (2007) Genetic polymorphisms of the epidermal growth factor and related receptor in non-small cell lung cancer—a review of the literature. Oncologist 12:201–210
13. Zhu C, Ding K, da Cunha Santos G et al (2008) Role of KRAS and EGFR as biomarkers of response to erlotinib in National Cancer Institute of Canada Clinical Trials Group Study BR. 21. J Clin Oncol 26:4268–4275
14. Selvaggi G, Novello S, Torri V et al (2004) Epidermal growth factor receptor overexpression correlates with a poor prognosis in completely resected non-small-cell lung cancer. Ann Oncol 15:28–32
15. Mitsudomi T, Morita S, Yatabe Y et al (2010) Gefitinib versus cisplatin plus docetaxel in patients with non-small-cell lung cancer harbouring mutations of the epidermal growth factor receptor (WJTOG3405): an open label, randomised phase 3 trial. Lancet Oncol 11:121–128
16. Maemondo M, Inoue A, Kobayashi K et al (2010) Gefitinib or chemotherapy for non-small-cell lung cancer with mutated EGFR. N Engl J Med 362:2380–2388
17. Shepherd F, Rodrigues Pereira J, Ciuleanu T et al (2005) Erlotinib in previously treated non-small-cell lung cancer. N Engl J Med 353:123–132
18. Zhou C, Wu Y, Chen G et al (2011) Erlotinib versus chemotherapy as first-line treatment for patients with advanced EGFR mutation-positive non-small-cell lung cancer (OPTIMAL, CTONG-0802): a multicentre, open-label, randomised, phase 3 study. Lancet Oncol 12:735–742
19. Mok T, Wu Y, Thongprasert S et al (2009) Gefitinib or carboplatin-paclitaxel in pulmonary adenocarcinoma. N Engl J Med 361:947–957
20. Herbst RS, Fukuoka M, Baselga J (2004) Gefitinib—a novel targeted approach to treating cancer. Nat Rev Cancer 4:979–987
21. de Mello RA, Costa BM, Reis RM, Hespanhol V (2012) Insights into angiogenesis in non-small cell lung cancer: molecular mechanisms, polymorphic genes, and targeted therapies. Recent Pat Anticancer Drug Discov 7:118–131
22. Schneider B, Wang M, Radovich M et al (2008) Association of vascular endothelial growth factor and vascular endothelial growth factor receptor-2 genetic polymorphisms with outcome in a trial of paclitaxel compared with paclitaxel plus bevacizumab in advanced breast cancer: ECOG 2100. J Clin Oncol 26:4672–4678
23. Sheppard PM (1955) Genetic variability and polymorphism: synthesis. Cold Spring Harb Symp Quant Biol 20:271–275
24. Hsu TC, Klatt O (1958) Mammalian chromosomes in vitro. IX. On genetic polymorphism in cell populations. J Natl Cancer Inst 21:437–473
25. El-Khoueiry A, Lenz H (2011) Pharmacogenomics. In: DeVita VT, Lawrence TS, Rosenberg SA (eds) Cancer: principles and practice of oncology, 9th edn. Lippincott Williams & Wilkins, Philadelphia
26. Marques Santos DS (2012) Papel prognóstico e preditivo do polimorfismo da IL-8-251 T/A no carcinoma epitelial do ovário. Master degree, Instituto de Ciências Biomédicas Abel Salazar, University of Porto
27. Sachidanandam R, Weissman D, Schmidt SC et al (2001) A map of human genome sequence variation containing 1.42 million single nucleotide polymorphisms. Nature 409:928–933

28. Harris AL (2002) Hypoxia—a key regulatory factor in tumour growth. Nat Rev Cancer 2:38–47
29. Rockwell S, Dobrucki IT, Kim EY et al (2009) Hypoxia and radiation therapy: past history, ongoing research, and future promise. Curr Mol Med 9:442–458
30. Crabtree H, Cramer W (1933) The action of radium on cancer cells. II.–some factors determining the susceptibility of cancer cells to radium. Proc R Soc Ser B 113:238–250
31. Schwarz G (1909) Über Desensibilisierung gegen Röntgen- und Radiumstrahlen. Munch Med Wochenschr 24:1–2
32. Chan DA, Giaccia AJ (2007) Hypoxia, gene expression, and metastasis. Cancer Metastasis Rev 26:333–339
33. Kim J, Gao P, Dang CV (2007) Effects of hypoxia on tumor metabolism. Cancer Metastasis Rev 26:291–298
34. Varlotto J, Stevenson MA (2005) Anemia, tumor hypoxemia, and the cancer patient. Int J Radiat Oncol Biol Phys 63:25–36
35. Boogaerts M, Mittelman M, Vaupel P (2005) Beyond anaemia management: evolving role of erythropoietin therapy in neurological disorders, multiple myeloma and tumour hypoxia models. Oncology 69:22–30
36. Gray L, Conger A, Ebert M et al (1953) The concentration of oxygen dissolved in tissues at the time of irradiation as a factor in radiotherapy. Br J Radiol 26:638–648
37. Mayer R, Hamilton-Farrell MR, van der Kleij AJ et al (2005) Hyperbaric oxygen and radiotherapy. Strahlenther Onkol 181:113–123
38. Feldmeier J, Carl U, Hartmann K, Sminia P (2003) Hyperbaric oxygen: does it promote growth or recurrence of malignancy? Undersea Hyperb Med 30:1–18
39. Bennett M, Feldmeier J, Smee R, Milross C (2005) Hyperbaric oxygenation for tumour sensitisation to radiotherapy. Cochrane Database Syst Rev (4):CD005007
40. Overgaard J, Horsman MR (1996) Modification of hypoxia-induced radioresistance in tumors by the use of oxygen and sensitizers. Semin Radiat Oncol 6:10–21
41. Overgaard J (2007) Hypoxic radiosensitization: adored and ignored. J Clin Oncol 25:4066–4074
42. Kaanders JHAM, Bussink J, van der Kogel AJ (2002) ARCON: a novel biology-based approach in radiotherapy. Lancet Oncol 3:728–737
43. Jordan BF, Sonveaux P (2012) Targeting tumor perfusion and oxygenation to improve the outcome of anticancer therapy. Front Pharmacol 3:94
44. Rewari AN, Haffty BG, Wilson LD et al (2006) Postoperative concurrent chemoradiotherapy with mitomycin in advanced squamous cell carcinoma of the head and neck: results from three prospective randomized trials. Cancer J 12:123–129
45. Roberts KB, Urdaneta N, Vera R et al (2000) Interim results of a randomized trial of mitomycin C as an adjunct to radical radiotherapy in the treatment of locally advanced squamous cell carcinoma of the cervix. Int J Cancer 90:206–223
46. Bennewith KL, Dedhar S (2011) Targeting hypoxic tumour cells to overcome metastasis. BMC Cancer 11:504
47. Albertella MR, Loadman PM, Jones PH et al (2008) Hypoxia-selective targeting by the bioreductive prodrug AQ4N in patients with solid tumors: results of a phase I study. Clin Cancer Res 14:1096–1104
48. Jameson MB, Rischin D, Pegram M et al (2010) A phase I trial of PR-104, a nitrogen mustard prodrug activated by both hypoxia and aldo-keto reductase 1C3, in patients with solid tumors. Cancer Chemother Pharmacol 65:791–801
49. Chen CH, Lai JM, Chou TY et al (2009) VEGFA upregulates FLJ10540 and modulates migration and invasion of lung cancer via PI3K/AKT pathway. PLoS One 4:e5052
50. Choong N, Salgia R, Vokes E (2008) Key signaling pathways and targets in lung cancer therapy. Clin Lung Cancer 8:52–60
51. Kerbel R (2008) Tumor angiogenesis. N Engl J Med 358:2039–2049
52. González RP, Leyva A, Melo RAB et al (2000) Método para estudo in vivo da angiogênese: indução de neovascularização na córnea de coelho. Acta Cir Bras 15:168–173

53. An SJ, Huang YS, Chen ZH et al (2012) Posttreatment plasma VEGF levels may be associated with the overall survival of patients with advanced non-small cell lung cancer treated with bevacizumab plus chemotherapy. Med Oncol 29:627–632
54. Bremnes R, Camps C, Sirera R (2006) Angiogenesis in non-small cell lung cancer: the prognostic impact of neoangiogenesis and the cytokines VEGF and bFGF in tumours and blood. Lung Cancer 51:143–158
55. Dvorak H (2002) Vascular permeability factor/vascular endothelial growth factor: a critical cytokine in tumor angiogenesis and a potential target for diagnosis and therapy. J Clin Oncol 20:4368–4380
56. Glubb DM, Cerri E, Giese A et al (2011) Novel functional germline variants in the VEGF receptor 2 gene and their effect on gene expression and microvessel density in lung cancer. Clin Cancer Res 17:5257–5267
57. Dowlati A, Gray R, Sandler A et al (2008) Cell adhesion molecules, vascular endothelial growth factor, and basic fibroblast growth factor in patients with non-small cell lung cancer treated with chemotherapy with or without bevacizumab—an Eastern Cooperative Oncology Group Study. Clin Cancer Res 14:1407–1412
58. Duda D, Jain R, Willett C (2007) Antiangiogenics: the potential role of integrating this novel treatment modality with chemoradiation for solid cancers. J Clin Oncol 25:4033–4042
59. Folkman J (2008) Tumor angiogenesis: from bench to bedside. In: Marmé D, Fusenig N (eds) Tumor angiogenesis: basic concepts and cancer therapy, 8th edn. Springer, Berlin, pp 3–28
60. Folkman J, Watson K, Ingber D, Hanahan D (1989) Induction of angiogenesis during the transition from hyperplasia to neoplasia. Nature 339:58–61
61. Guan X, Yin M, Wei Q et al (2010) Genotypes and haplotypes of the VEGF gene and survival in loc.lly advanced non-small cell lung cancer patients treated with chemoradiotherapy. BMC Cancer 10:431
62. Dong Q, Feng J, Huang J et al (2002) [Vascular endothelial growth factor promotes hematogenous metastasis of cancer cells in patients with non-small cell lung cancer]. Zhonghua Zhong Liu Za Zhi 24:142–146
63. Forgacs E, Zöchbauer-Müller S, Oláh E, Minna J (2001) Molecular genetic abnormalities in the pathogenesis of human lung cancer. Pathol Oncol Res 7:6–13
64. Gessner C, Rechner B, Hammerschmidt S et al (2009) Angiogenic markers in breath condensate identify non-small cell lung cancer. Lung Cancer 68:177–184
65. Hong TT, Zhang RX, Wu XH, Hua D (2011) Polymorphism of vascular endothelial growth factor −1154 G>A (rs1570360) with cancer risk: a meta-analysis of 16 case–control studies. Mol Biol Rep 2012; 39(5):5283–5289
66. Shahbazi M, Pravica V, Nasreen N et al (2002) Association between functional polymorphism in EGF gene and malignant melanoma. Lancet 359:397–401
67. Araújo A, Costa B, Pinto-Correia A et al (2011) Association between EGF +61A/G polymorphism and gastric cancer in caucasians. World J Gastroenterol 17:488–492
68. Zhang Y, Cao C, Liang K (2010) Genetic polymorphism of epidermal growth factor 61A>G and cancer risk: a meta-analysis. Cancer Epidemiol 34:150–156
69. Costa B, Ferreira P, Costa S et al (2007) Association between functional EGF+61 polymorphism and glioma risk. Clin Cancer Res 13:2621–2626
70. Bhowmick D, Zhuang Z, Wait S, Weil R (2004) A functional polymorphism in the EGF gene is found with increased frequency in glioblastoma multiforme patients and is associated with more aggressive disease. Cancer Res 64:1220–1223
71. Wu G, Hasenberg T, Magdeburg R et al (2009) Association between EGF, TGF-beta1. VEGF gene polymorphism and colorectal cancer. World J Surg 33:124–129
72. Kang H, Choi J, Lee W et al (2007) +61A>G polymorphism in the EGF gene does not increase the risk of lung cancer. Respirology 12:902–905
73. Lim Y, Kim J, Song J et al (2005) Epidermal growth factor gene polymorphism is different between schizophrenia and lung cancer patients in Korean population. Neurosci Lett 374:157–160

74. de Mello RA, Ferreira M, Costa S et al (2012) Association between EGF +61 genetic polymorphisms and non-small cell lung cancer increased risk in a Portuguese population: a case-control study. Tumour Biol 33:1341–1348
75. Weiss WA, Burns MJ, Hackett C et al (2003) Genetic determinants of malignancy in a mouse model for oligodendroglioma. Cancer Res 63:1589–1595
76. Brandt R, Eisenbrandt R, Leenders F et al (2000) Mammary gland specific hEGF receptor transgene expression induces neoplasia and inhibits differentiation. Oncogene 19:2129–2137
77. Araújo A, Coelho A, de Mello RA et al (2012) Personalizing medicine—strategies for implementing the evaluation of ALK rearrangement in non-small-cell lung cancer in Portugal. Rev Port Pneumol 18:244–246. doi:10.1016/j.rppneu.2012.04.011
78. Dong J, Dai J, Shu Y et al (2010) Polymorphisms in EGFR and VEGF contribute to non-small cell lung cancer survival in a Chinese Population. Carcinogenesis 31:1080–1086
79. Landi MT, Chatterjee N, Yu K et al (2009) A genome-wide association study of lung cancer identifies a region of chromosome 5p15 associated with risk for adenocarcinoma. Am J Hum Genet 85:679–691
80. Ito H, MacKay JD, Hosono S et al (2012) Association between a genome-wide association study-identified locus and the risk of lung cancer in Japanese population. J Thorac Oncol 7:790–798
81. Bae EY, Lee SY, Kang BK et al (2012) Replication of results of genome wide association studies on lung cancer susceptibility loci in a Korean population. Respirology 17:699–706
82. Li Y, Sheu C, Ye Y et al (2010) Genetic variants and risk of lung cancer in never smokers: a genome-wide association study. Lancet Oncol 11:321–330
83. Taylor J, Tyekucheva S, King DC et al (2006) ESPERR: learning strong and weak signals in genomic sequence alignments to identify functional elements. Genome Res 16:1596–1604
84. Feldser DM, Hackett JA, Greider CW (2003) Telomere dysfunction and the initiation of genome instability. Nat Rev Cancer 3:623–627
85. Fernandez-Garcia I, Ortiz-de-Solorzano C, Montuenga LM (2008) Telomeres and telomerase in lung cancer. J Thorac Oncol 3:1085–1088
86. Lantuejoul S, Soria J, Moro-Sibilot D et al (2004) Differential expression of telomerase reverse transcriptase (hTERT) in lung tumours. Br J Cancer 90:1222–1229
87. Aviel-Ronen S, Coe BP, Lau SK et al (2008) Genomic markers for malignant progression in pulmonary adenocarcinoma with bronchioloalveolar features. Proc Natl Acad Sci USA 105:10155–10160
88. Wang Y, Broderick P, Webb E et al (2008) Common 5p15.33 and 6p21.33 variants influence lung cancer risk. Nat Genet 40:1407–1409
89. Vincenti V, Cassano C, Rocchi M, Persico MG (1996) Assignment of the vascular endothelial growth factor gene to human chromosome 6p21.3. Circulation 93:1493–1495
90. Naik NA, Bhat IA, Afroze D et al (2012) Vascular endothelial growth factor A gene (VEGFA) polymorphisms and expression of VEGFA gene in lung cancer patients of Kashmir Valley (India). Tumour Biol 33:833–839
91. Stevens A, Soden J, Brenchley P et al (2003) Haplotype analysis of the polymorphic human vascular endothelial growth factor gene promoter. Cancer Res 63:812–816
92. Vaziri S, Kim J, Ganapathi M, Ganapathi R (2010) Vascular endothelial growth factor polymorphisms: Role in response and toxicity of tyrosine kinase inhibitors. Curr Oncol Rep 12:102–108
93. Zhai R, Liu G, Zhou W et al (2008) Vascular endothelial growth factor genotypes, haplotypes, gender, and the risk of non-small cell lung cancer. Clin Cancer Res 14:612–617
94. Toschi L, Cappuzzo F (2007) Understanding the new genetics of responsiveness to epidermal growth factor receptor tyrosine kinase inhibitors. Oncologist 12:211–220
95. Harris R, Chung E, Coffey R (2003) EGF receptor ligands. Exp Cell Res 284:2–13
96. Shepherd FA, Rosell R (2007) Weighing tumor biology in treatment decisions for patients with non-small cell lung cancer. J Thorac Oncol 2:S68–S76

97. Sculier J, Chansky K, Crowley J et al (2008) The impact of additional prognostic factors on survival and their relationship with the anatomical extent of disease expressed by the 6th Edition of the TNM classification of malignant tumors and the proposals for the 7th Edition. J Thorac Oncol 3:457–466
98. Singhal S, Vachani A, Antin-Ozerkis D et al (2005) Prognostic implications of cell cycle, apoptosis, and angiogenesis biomarkers in non-small cell lung cancer: a review. Clin Cancer Res 11:3974–3986
99. Shen L, Ji H (2011) More on crizotinib. N Engl J Med 364:777
100. Shaw AT, Yeap BY, Mino-Kenudson M et al (2009) Clinical features and outcome of patients with non-small-cell lung cancer who harbor EML4-ALK. J Clin Oncol 27:4247–4253
101. Schrump D, Giaccone G, Kelsey K, Marks L (2008) Non small cell lung cancer. In: DeVita V, Lawrence T, Rosenberg S et al (eds) DeVita, Hellman, and Rosenberg's cancer: principles & practice of oncology, 8th edn. Lippincott Williams & Wilkins, Philadelphia, pp 896–939
102. Pao W, Miller V, Politi K et al (2005) Acquired resistance of lung adenocarcinomas to gefitinib or erlotinib is associated with a second mutation in the EGFR kinase domain. PLoS Med 2:225–235
103. Olaussen KA, Dunant A, Fouret P et al (2006) DNA repair by ERCC1 in non-small-cell lung cancer and cisplatin-based adjuvant chemotherapy. N Engl J Med 355:983–991
104. Masago K, Fujita S, Kim Y et al (2009) Effect of vascular endothelial growth factor polymorphisms on survival in advanced stage non small cell lung cancer. Cancer Sci 100:1917–1922
105. Dajczman E, Kasymjanova G, Kreisman H et al (2008) Should patient-rated performance status affect treatment decisions in advanced lung cancer? J Thorac Oncol 3:1133–1136
106. Pérez-Soler R, Chachoua A, Hammond L et al (2004) Determinants of tumor response and survival with erlotinib in patients with non-small-cell lung cancer. J Clin Oncol 22:3238–3247
107. Miller V, Riely G, Zakowski M et al (2008) Molecular characteristics of bronchioloalveolar carcinoma and adenocarcinoma, bronchioloalveolar carcinoma subtype, predict response to erlotinib. J Clin Oncol 26:1472–1478
108. Brugger W, Triller N, Blasinska-Morawiec M et al (2011) Prospective molecular marker analyses of EGFR and KRAS from a randomized, placebo-controlled study of erlotinib maintenance therapy in advanced non-small-cell lung cancer. J Clin Oncol 29:4113–4120
109. Jain L, Vargo CA, Danesi R et al (2009) The role of vascular endothelial growth factor SNPs as predictive and prognostic markers for major solid tumors. Mol Cancer Ther 8:2496–2508
110. Zacharatos P, Kotsinas A, Tsantoulis P et al (2001) Relationship of the K-ras/c-mos expression patterns with angiogenesis in non-small cell lung carcinomas. Mol Med 7:590–597
111. Heist RS, Zhai R, Liu G et al (2008) VEGF polymorphisms and survival in early-stage non-small-cell lung cancer. J Clin Oncol 26:856–862
112. Bieniasz M, Oszajca K, Eusebio M et al (2009) The positive correlation between gene expression of the two angiogenic factors: VEGF and BMP-2 in lung cancer patients. Lung Cancer 66:319–326
113. Lee SJ, Lee SY, Jeon HS et al (2005) Vascular endothelial growth factor gene polymorphisms and risk of primary lung cancer. Cancer Epidemiol Biomarkers Prev 14:571–575
114. Rodrigues P, Furriol J, Tormo E et al (2012) The single-nucleotide polymorphisms +936 C/T VEGF and -710 C/T VEGFR1 are associated with breast cancer protection in a Spanish population. Breast Cancer Res Treat 2012 Jun;133(2):769–778
115. Koukourakis M, Papazoglou D, Giatromanolaki A et al (2004) VEGF gene sequence variation defines VEGF gene expression status and angiogenic activity in non-small cell lung cancer. Lung Cancer 46:293–298
116. Yang F, Tang X, Riquelme E et al (2011) Increased VEGFR-2 gene copy is associated with chemoresistance and shorter survival in patients with non-small-cell lung carcinoma who receive adjuvant chemotherapy. Cancer Res 71:5512–5521
117. Reck M, von Pawel J, Zatloukal P et al (2009) Phase III trial of cisplatin plus gemcitabine with either placebo or bevacizumab as first-line therapy for nonsquamous non-small-cell lung cancer: AVAil. J Clin Oncol 27:1227–1234

118. Couraud S, Zalcman G, Milleron B et al (2012) Lung cancer in never smokers—a review. Eur J Cancer 48:1299–1311
119. Li Y, Sun Z, Cunningham JM et al (2011) Genetic variations in multiple drug action pathways and survival in advanced stage non-small cell lung cancer treated with chemotherapy. Clin Cancer Res 17:3830–3840
120. Ma F, Xu B, Lin D et al (2011) [Effect of rs2293347 polymorphism in EGFR on the clinical efficacy of gefitinib in patients with non-small cell lung cancer]. Zhongguo Fei Ai Za Zhi 14:642–645
121. Nie Q, Yang X, An S et al (2011) CYP1A1* 2A polymorphism as a predictor of clinical outcome in advanced lung cancer patients treated with EGFR-TKI and its combined effects with EGFR intron 1 (CA) n polymorphism. Eur J Cancer 47:1962–1970
122. Dy G, Adjei A (2009) Emerging therapeutic targets in non-small cell lung cancer. Proc Am Thorac Soc 6:218–223

Part III
Pharmacology and Therapy of Angiogenesis

Chapter 21
Angiogenic and Arteriogenic Factors: Their Role in the Treatment of Cancer and Arterial Regeneration

Wolfgang Schaper

Abstract Angiogenesis is modulated by hypoxia and is concerned with the development of capillaries, whereas arteriogenesis is regulated by fluid shear stress and is concerned with the formation of arteries from arterioles. It is hoped that treatment of cancer will emerge with the development of agents based on knowledge concerning cancer genome in addition to antiangiogenic therapy. On the other hand, novel approaches based on molecular mechanisms for fluid shear stress as well as the development of new growth factors are needed for the therapy of cardiovascular diseases.

Keywords Arteriogenesis • Angiogenesis • Capillaries • Arterioles • Hypoxia • Fluid shear stress • Endothelium • Smooth muscle cell proliferation

Angiogenesis [1], the development of capillaries by sprouting through endothelial mitosis, or intussusception, and arteriogenesis [2], the development of arteries from arterioles by proliferation of smooth muscle and endothelial cells, occur under physiological conditions mainly in the adult female reproductive system. Angiogenesis is important for the development of the embryo, fetus, and placenta in the uterus, and arteriogenesis is important for the growth of the uterine artery to meet the enormous demands of the growing uterus and the fetus [3, 4]. After birth the uterine artery undergoes marked inward remodeling in parallel with the involution of the uterus. The embryonal development of the circulatory system starts with angiogenesis from which arteries and veins finally mature. In the literature both

W. Schaper (✉)
Division of Arteriogenesis Research, The Max Plank Institute for Heart Research and Lung Research, Bad Nauheim, Germany
e-mail: wolfgang.schaper@mpi-bn.mpg.de

J.L. Mehta and N.S. Dhalla (eds.), *Biochemical Basis and Therapeutic Implications of Angiogenesis*, Advances in Biochemistry in Health and Disease 6,
DOI 10.1007/978-1-4614-5857-9_21,

types of vascular growth are often subsumed under the heading angiogenesis. Both types of vascular growth assume new roles under pathological conditions: wound healing, tumor growth, and arterial regeneration via collateral vessels following arterial occlusions in the wake of atherosclerosis [5]. Angiogenesis and arteriogenesis differ in several aspects: angiogenesis, but not arteriogenesis, is stimulated by tissue hypoxia [6], whereas arteriogenesis is controlled by fluid shear stress [7]. Arterioles undergoing positive remodeling because of increased shear stress are largely located in normoxic tissue, and they carry fully oxygenated blood.

Arteriogenesis depends on the availability of smooth muscle, which undergoes mitosis and proliferation when stimulated by shear stress. Both types of vascular growth are indirectly dependent on the availability of NO [8]. Blockade of eNOS plus blockade of iNOS completely inhibits both types of vascular growth, and some NO-donors are able to stimulate [9]. The discovery of peptide growth factors and the design of growth factor inhibitors kindled the hope that the scourges of mankind, i.e., atherosclerosis and cancer, could be cured by the unifying principle of angiogenesis and antiangiogenesis. Angiogenesis and especially arteriogenesis should cure atherosclerosis and antiangiogenesis should cure cancer [10].

Unfortunately these hopes were dashed, because clinical studies with growth factors in coronary and peripheral artery disease did not produce convincing results [11, 12]. These results were, at least in part, foreseeable because rigorous animal experiments had shown that long-term application of high doses of growth factors in dogs with chronic coronary occlusion produced only a 10 % point increase in collateral flow with a remaining flow deficit of 50 %, i.e., only 50 % of the normal dilatory reserve was restored by growth factor therapy, up from 40 % [13]. None of the presently available growth factors are able to fully restore arterial function. Mode of delivery and the total amount of growth factors delivered in animals starkly contrasts with the "on-shot" application in human patients during catheterization.

Similar disappointing results were reported from the cancer field: antiangiogenic therapy prolongs life of cancer patients only in the order of months instead of years [14], and treatment may lead to the selection of resistant forms of cancer growth not needing capillaries [15]. Antiangiogenic cancer therapy leads to increased number of hypoxic cancer cells, a primary aim of the therapeutic principle. However, these were found to increase metastasis and hence increase the malignancy of tumors [16–20]. Cancer vessels are leaky, and antiangiogenic therapy was believed to normalize vascular permeability thereby increasing the chance of delivery of cytotoxic anticancer drugs [21]. However, the opposite was also reported [22, 23]. VEGF-receptor2 blockade increases the chance of escape pathways using VEGFR3, promoting the formation of new blood vessels [24, 25]. Collateral arteries differ from normal arteries in that they are narrowly connected with the average daily blood flow demand of the subtended tissue, which is usually not more than less than half of the vasodilatory reserve of normal arteries [5]. Experiments with artificially increased blood flow (AV shunts) showed that the growth of collateral arteries was stimulated to such a degree that even the normal maximal vasodilatory reserve was doubled [26]. However, on restoration of normal flow, all vessels in excess of those necessary for the daily demand are pruned away. The reductionist view of growth factors as the leading principle of vascular growth and its inhibition can be said as being surpassed by realizing the enormous complexity of the interacting pathways. A genome-wide screening of the transcriptome isolated from growing collateral vessels showed over 300 differentially expressed genes, and the study of their interdependence is an enormous task. The discovery of the actin-binding protein abra [27] and the shear stress-sensitive calcium channel TRPV4 [28] as players in this interaction is just the beginning. The situation becomes even more complex when considering the fact that transcripts are not necessarily linearly translated because of the presence (and upregulation in collateral artery tissue) of microRNAs that interfere with the stability or the translation of mRNAs.

Future research should continue to search for principles that stimulate vascular growth that fully restore vasodilatory reserve, but this must be followed by factors that inhibit pruning. Angiogenesis proper, i.e., the production of more capillaries, has probably no place in the treatment of arterial occlusions because there is no shortage of capillaries and the addition of new ones should, theoretically, increase resistance rather than diminish it. This may differ when capillaries in the border zone between well- and poorly perfused regions enlarge. The looming problem with growth stimulation of blood vessels is the possibility of awakening dormant cancer cells.

21.1 The Primary Pathway of Arteriogenesis Remains a Mystery

Although fluid shear stress is a very potent modulator of collateral artery development (high FSS stimulates and low FSS inhibits collateral growth), its molecular pathway is only partially known, and the identified pathways have little in common with each other and some even remain contradictory. Nitric oxide inhibitors like L-NAME completely block the effects of increased fluid shear stress, which strongly suggests a leading role of NO in arteriogenesis [2, 8]. However, the targeted deletion of the eNOS gene produces mainly vasoconstriction but interferes little in the growth of collaterals after femoral artery occlusion in mice, and transgenic overexpression does not accelerate or increase the extent of collateral formation [29]. The acute application of NO-donors in eNOS ko mice unveils the slowed recovery as produced by vasoconstriction. Only the combined inhibition of both eNOS and iNOS by treating eNOS ko mice with the iNOS-specific L-NIL resulted in the complete blockade of arteriogenesis [9], thereby confirming the results obtained with L-NAME which is a nonselective NOS inhibitor. L-NAME, by the way, would make an excellent agent against the vascularization of tumors as demonstrated by Rubanyi and Kauser (in a non-indexed journal). Scanning the transcriptome of growing collaterals with microarrays, the NOS genes showed late upregulation which contrasts with the early shutdown of arteriogenesis following femoral occlusion in the double-targeted mice [9].

Genome screening of strains of mice with a well-developed network of brain collaterals under normal conditions (BL 6) versus those with a poorly developed network (Balb C) showed a gene cluster that explains the difference [30]. However, this cluster did not contain any of the known genes whose transcripts are upregulated when collateral growth is stimulated by high FSS, like abra, TRPV4, members of the notch family, and Klf2. It may be that brain collaterals differ from those in the vascular periphery, already suggested by the lack of abra regulation with high FSS in the circle of Willis [31]. Since smooth muscle in the brain circulation originates from the neuroectoderm whereas those from the lower body region originate from the mesoderm, different pathways may be the answer to the apparent contrast.

21.2 The Future of Angiogenesis and Arteriogenesis

In spite of the disappointments in the wake of high hopes that cancer and arterial occlusions can be cured by the unifying paradigm of vascular modulation, research continues because the escape pathways that kick in when antiangiogenic cancer therapy fails are now known and suggest future drug development. It should be kept in mind that antiangiogenic therapy is not an attack at the core of the cancer problem and new approaches will most probably come from the new knowledge of the cancer genome and its pathways. Cancer treatment of the future will most probably consist of attacks from all possible sites, as it already exists, plus new genomic knowledge including antiangiogenesis. It must be ceded that the efforts by clinical cancer and basic scientists had produced more insights into the problem compared to their cardiovascular colleagues who virtually abandoned the field after the first negative clinical trials with growth factors that were limited in scope. Furthermore, technical developments in interventional cardiology, radiology, and surgery offered alternative ways of treatment, which are indeed very successful and lowered the incentive to go molecular. Nevertheless, the technological approach to arterial occlusions and subsequent heart failure will meet its limits, and molecular alternatives will have a place. In view of the ease with which the vasculature responds with growth to increases in fluid shear stress and knowledge of its molecular pathway, new therapies, hopefully more active and lasting, will become available.

References

1. Carmeliet P (2003) Angiogenesis in health and disease. Nat Med 9:653–660
2. Schaper W, Schaper J (2004) Arteriogenesis. Kluwer, Boston
3. Nanaev A, Chwalisz K, Frank HG et al (1995) Physiological dilation of uteroplacental arteries in the guinea pig depends on nitric oxide synthase activity of extravillous trophoblast. Cell Tissue Res 282:407–421
4. Nienartowicz A, Link S, Moll W (1989) Adaptation of the uterine arcade in rats to pregnancy. J Dev Physiol 12:101–108

5. Schaper W (2009) Collateral circulation: past and present. Basic Res Cardiol 104:5–21
6. Ito WD, Arras M, Scholz D et al (1997) Angiogenesis but not collateral growth is associated with ischemia after femoral artery occlusion. Am J Physiol Heart Circ Physiol 273:H1255–H1265
7. Pipp F, Boehm S, Cai WJ et al (2004) Elevated fluid shear stress enhances postocclusive collateral artery growth and gene expression in the pig hind limb. Arterioscler Thromb Vasc Biol 24:1664–1668
8. Prior BM, Lloyd PG, Ren J et al (2003) Arteriogenesis: role of nitric oxide. Endothelium 10:207–216
9. Troidl K, Tribulova S, Cai WJ et al (2010) Effects of endogenous nitric oxide and of DETA NONOate in arteriogenesis. J Cardiovasc Pharmacol 55:153–160
10. Folkman J (2007) Angiogenesis: an organizing principle for drug discovery? Nat Rev Drug Discov 6:273–286
11. Henry TD, Annex BH, McKendall GR et al (2003) The VIVA trial: vascular endothelial growth factor in ischemia for vascular angiogenesis. Circulation 107:1359–1365
12. Henry TD, Rocha-Singh K, Isner JM et al (2001) Intracoronary administration of recombinant human vascular endothelial growth factor to patients with coronary artery disease. Am Heart J 142:872–880
13. Shou M, Thirumurti V, Rajanayagam S et al (1997) Effect of basic fibroblast growth factor on myocardial angiogenesis in dogs with mature collateral vessels. J Am Coll Cardiol 29:1102–1106
14. Carmeliet P, Jain RK (2011) Molecular mechanisms and clinical applications of angiogenesis. Nature 473:298–307
15. Casanovas O (2012) Cancer: limitations of therapies exposed. Nature 484:44–46
16. Facciabene A, Peng X, Hagemann IS et al (2011) Tumour hypoxia promotes tolerance and angiogenesis via CCL28 and T(reg) cells. Nature 475:226–230
17. Ebos JM, Lee CR, Cruz-Munoz W et al (2009) Accelerated metastasis after short-term treatment with a potent inhibitor of tumor angiogenesis. Cancer Cell 15:232–239
18. Paez-Ribes M, Allen E, Hudock J, Takeda T, Okuyama H, Vinals F, Inoue M, Bergers G, Hanahan D, Casanovas O (2009) Antiangiogenic therapy elicits malignant progression of tumors to increased local invasion and distant metastasis. Cancer Cell 15:220–231
19. Rini BI, Atkins MB (2009) Resistance to targeted therapy in renal-cell carcinoma. Lancet Oncol 10:992–1000
20. Conley SJ, Gheordunescu E, Kakarala P et al (2012) Antiangiogenic agents increase breast cancer stem cells via the generation of tumor hypoxia. Proc Natl Acad Sci USA 109:2784–2789
21. Jain RK (2005) Normalization of tumor vasculature: an emerging concept in antiangiogenic therapy. Science 307:58–62
22. Van der Veldt AA, Lubberink M, Bahce I et al (2012) Rapid decrease in delivery of chemotherapy to tumors after anti-VEGF therapy: implications for scheduling of anti-angiogenic drugs. Cancer Cell 21:82–91
23. Padera TP, Kuo AH, Hoshida T et al (2008) Differential response of primary tumor versus lymphatic metastasis to VEGFR-2 and VEGFR-3 kinase inhibitors cediranib and vandetanib. Mol Cancer Ther 7:2272–2279
24. Benedito R, Rocha SF, Woeste M et al (2012) Notch-dependent VEGFR3 upregulation allows angiogenesis without VEGF-VEGFR2 signalling. Nature 484:110–114
25. Bergers G, Hanahan D (2008) Modes of resistance to anti-angiogenic therapy. Nat Rev Cancer 8:592–603
26. Eitenmuller I, Volger O, Kluge A, Troidl K, Barancik M, Cai WJ, Heil M, Pipp F, Fischer S, Horrevoets AJ, Schmitz-Rixen T, Schaper W (2006) The range of adaptation by collateral vessels after femoral artery occlusion. Circ Res 99:656–662
27. Troidl K, Ruding I, Cai WJ et al (2009) Actin-binding rho activating protein (Abra) is essential for fluid shear stress-induced arteriogenesis. Arterioscler Thromb Vasc Biol 29:2093–2101
28. Troidl C, Troidl K, Schierling W et al (2009) Trpv4 induces collateral vessel growth during regeneration of the arterial circulation. J Cell Mol Med 13:2613–2621

29. Mees B, Wagner S, Ninci E et al (2007) Endothelial nitric oxide synthase activity is essential for vasodilation during blood flow recovery but not for arteriogenesis. Arterioscler Thromb Vasc Biol 27:1926–1933
30. Wang S, Zhang H, Wiltshire T et al (2012) Genetic dissection of the Canq1 locus governing variation in extent of the collateral circulation. PLoS One 7:e31910
31. Schierling W, Troidl K, Mueller C, Troidl C, Wustrack H, Bachmann G, Kasprzak PM, Schaper W, Schmitz-Rixen T (2009) Increased intravascular flow rate triggers cerebral arteriogenesis. J Cereb Blood Flow Metab 29:726–737

Chapter 22
Physiological Roles and Therapeutic Implications of Hepatocyte Growth Factor for Angiogenesis

Hiroyuki Ohnishi, Shinya Mizuno, Kiyomasa Oka, and Toshikazu Nakamura

Abstract Angiogenesis is an important event for organ development and regeneration via supply of nutrition and oxygen. Appropriate cellular responses, such as endothelial cell (EC) growth, are involved in the angiogenic actions, especially under hypoxic conditions. Hepatocyte growth factor (HGF) is now an organotrophic factor that is essential for organ development and tissue regeneration. In addition, HGF is the most potent angiogenic growth factor in culture of ECs, with mitogenic, and morphogenic functions via activation of c-Met/HGF-receptor. In hypoxic diseases, such as coronary arterial disease (CAD), up-regulation of HGF production is transient, followed by the HGF-deficient conditions, along with the lowered angiogenesis and enhanced hypoxia. Enhancement of angiogenesis by exogenous HGF leads to the improvement in heart dysfunction. Such angio-therapeutic effects of HGF are also seen in various hypoxic diseases, such as skin ulcer and lung emphysema. In tumor tissues, angiogenesis is required for cancer growth and metastasis. Indeed, an HGF-antagonist, NK4, potently inhibits tumor angiogenesis and metastasis in mouse models of malignant tumors. These provide a rationale why HGF is useful for the improvement in CAD and other ischemic diseases and NK4 for the suppression

H. Ohnishi
Division for Regenerative Drug Discovery, Center for Advanced Science and Innovation, Osaka University, 2-1 Yamadaoka, Suita 565-0871, Japan

College of Pharmacy, Kinjo Gakuin University, 2-1723 Omori, Moriyama-ku, Nagoya 463-8521, Japan

S. Mizuno
Division of Virology, Department of Microbiology and Immunology, Osaka University Graduate School of Medicine, 2-2-B7 Yamadaoka, Suita 565-0871, Japan

K. Oka • T. Nakamura (✉)
Division for Regenerative Drug Discovery, Center for Advanced Science and Innovation, Osaka University, 2-1 Yamadaoka, Suita 565-0871, Japan
e-mail: nakamura@casi.osaka-u.ac.jp

J.L. Mehta and N.S. Dhalla (eds.), *Biochemical Basis and Therapeutic Implications of Angiogenesis*, Advances in Biochemistry in Health and Disease 6,
DOI 10.1007/978-1-4614-5857-9_22,

of tumor malignancy. Importantly, appropriate HGF-c-Met signal is required for sustaining homeostasis in normal tissue. Herein, we provide a concept that amount of HGF may determine the predisposition to ischemic disease, healthy conditions, or neoplastic disease, all of which depend on the degree of c-Met activation, and in part, concomitant angiogenic activity.

Keywords HGF • c-Met • CAD • PAD • Angiogenesis • Hypoxia • Regeneration • Anti-apoptosis

22.1 Introduction

Organ development, tissue repair and regeneration, and pathological conditions depend on, more or less, angiogenesis (i.e., formation of new blood vessels) to supply nutrients and oxygen. The major blood vessels are lined by endothelial cells (ECs) and are surrounded by mural cells such as pericytes and vascular smooth muscle cells (VSMCs). The proper interaction between ECs and mural cells are required for the formation and physiological function of blood vessels [1]. Thus, sequential events, such as growth, migration, and morphogenesis of the vascular cells, should be tightly regulated during angiogenesis. In contrast, abnormal angiogenesis is an increased risk for vascular disorder, most of which are related to preexisting diseases including hypercholesterolemia, diabetes, and hypertension. The end-stage vascular deterioration leads to the onset of a life-threatening disease, such as peripheral arterial disease (PAD) and coronary arterial disease (CAD) [2]. In addition, aberrant angiogenesis also participates in tumor growth, invasion, and metastasis [3]. For the medical control of these diseases, it is important to elucidate the molecular mechanism whereby angiogenesis is induced, maintained, or deteriorated under pathological conditions.

HGF is discovered as a mitogen for rat hepatocytes in primary culture [4, 5]. HGF cDNA was cloned, based on the purified HGF protein [6, 7]. On the other hand, c-Met was identified as an oncogenic protein that induces malignant formation of normal cells [8]. c-Met was identified as a functional receptor for HGF [9, 10]. HGF is a multifunctional growth factor that exerts mitogenic, motogenic, and morphogenic functions in various cells via c-Met signaling (Fig. 22.1) [11, 12], contributing to embryogenesis and organ regeneration. Using animal models, we have demonstrated that organ failures become evident, due to an insufficient production of HGF, while HGF supplemental therapy improves these pathological states [12, 13]. In this process, HGF exerts regenerative and protective effects on parenchymal functional cells, such as epithelial cells, cardiomyocytes, and neurons [13–15]. Thus, our initial emphasis was that such direct actions of HGF to functional cells are required for therapeutic outcomes of HGF.

HGF targets not only parenchymal cells but also ECs to induce angiogenic actions at least in vitro [16, 17]. Morishita and our group accumulated evidence that HGF administration is useful for the attenuation of CAD, PAD, and other ischemic diseases [18], as discussed later. The angiogenic effect of HGF is also involved in tumor development, while NK4, an HGF-antagonist, inhibits tumor growth and

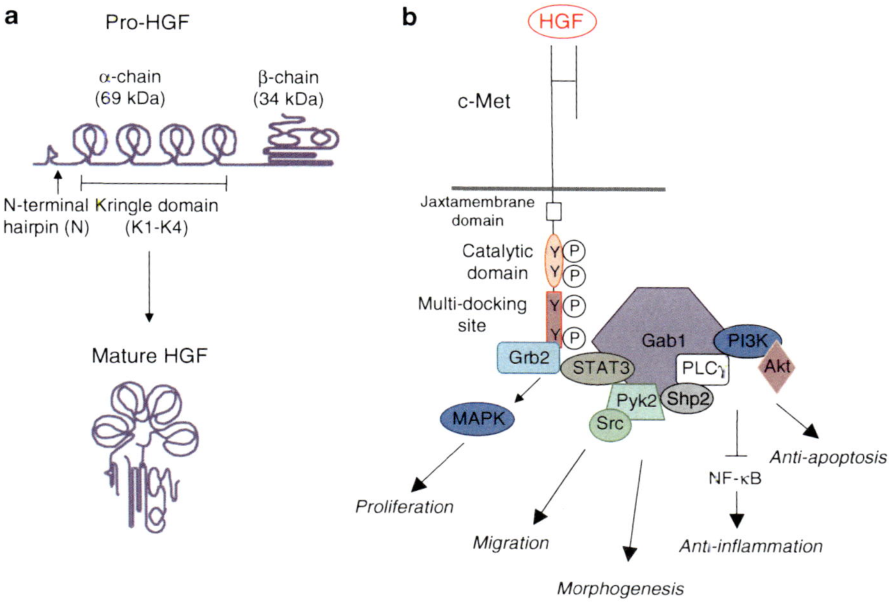

Fig. 22.1 Structure and pleiotropic function of HGF. (**a**) Schematic presentation of the domain structure of HGF. HGF is produced as a single-chain form (pro-HGF). HGF, formed by proteolytic processing of pro-HGF, is a disulfide-linked heterodimer composed of a 69 kDa α-chain and a 34 kDa β-chain. The α-chain contains the N-terminal hairpin domain and the subsequent four kringle domains, and the β-chain contains a serine protease-like domain with no enzymatic activity. (**b**) Pleiotropic functions of HGF mediated by c-Met. The accurate binding of HGF to c-Met triggers signaling transduction. The ATP-dependent phosphorylation at three residues in the c-Met active loop kinase domain; Tyr-1230/34/35 is an initial step for activating c-Met. Phosphorylation at Tyr-1349/56 in the C-terminal docking site is required for various bio-functions via recruiting downstream adaptors. For example, phospho-Tyr1349/56-dependent recruitment of Grb2-SOS activates Ras-ERK cascades, leading to cellular proliferation. Association and tyrosine phosphorylation of Gab-1, a docking protein that couples c-Met with multiple signaling proteins, such as PI-3kinase, PLC-γ, Shp-2, and Crk-2, plays definite roles in HGF-induced morphogenesis and motility [11, 12]

metastasis [19]. This review describes a new concept that loss of HGF-c-Met function causes CAD or PAD associated with decreased angiogenesis, while over-gain of HGF-c-Met function leads to tumor development, in part, via the enhancement of angiogenesis. This is a rationale why HGF is useful for the improvement in CAD or PAD, and NK4/HGF-antagonist for the control of tumor invasive growth.

22.2 Biological Activities of HGF on Vascular Cells

One of the most unique functions of HGF is to induce branching tubulogenesis of parenchymal epithelial cells [20]. Angiogenesis is also one of these morphogenetic events, with a vascular tube formation from preexisting vessel, in response to hypoxic stresses. In this process, sequential events, such as migration, proliferation, and morphogenesis of ECs, along with recruiting VSMCs, are required for functional maturation of vascular beds. Growing evidence indicates that HGF has various actions on vascular cells, required for angiogenesis without any permeability and inflammation (i.e., physiology-based angiogenesis).

22.2.1 HGF/c-Met Expression in Vascular Cells

Our initial study indicates that HGF is produced mainly by ECs, while c-Met is expressed by both ECs and VSMCs, hence indicating an autocrine loop in EC-EC and paracrine loop in EC-VSMC [21]. Activation of c-Met-dependent signaling cascade by HGF can be regulated by cell–cell contact [22]. In normal vessels, ECs express c-Met at basal levels, but do not show migratory actions at tight cell–cell adhesion, associated with the expression of VE-cadherin, a vascular-specific junctional molecule [23].

Angiogenesis is a physiological response to the oxygen demand. In the tissues with low oxygen tension, a transcription factor hypoxia inducible factor-1 (HIF-1) is stabilized and induces c-Met [24]. In contrast, HGF transcription is decreased under such an ischemic condition [25] probably via the decrease in cAMP levels [26]. Overall, hypoxia causes up-regulation of c-Met via the HIF-1-dependent cascade, while HGF is down-regulated, and such a reciprocal regulation accelerates ischemia-associated damages, including apoptosis and inflammation. Thus, hypoxia-induced c-Met induction provides a rationale why HGF administration is logical to attenuate ischemic organ failures (see Sect. 22.3).

22.2.2 Angiogenic Actions of HGF

Mitogenesis: The mitogenic activity of HGF on ECs was demonstrated in various types of ECs [16]. HGF has more potent mitogenic effects on ECs compared to VEGF and b-FGF [17]. HGF-mediated growth of ECs depends on MAPK-associated cascades. Indeed, the rapid activation of ERK1/2 by HGF resulted in EC proliferation, while such mitogenic effects were abolished by MAPK-inhibitors. Importantly, MAPK mediates the HGF-induced phosphorylation of STAT3 and *c-fos* promoter activation [27]. Thus, it is likely that HGF-c-Met-MAPK-STAT3 cascade is required for EC proliferation, associated with the c-Fos-related gene transcription. Another possibility is that HGF induces VEGF production, and such an indirect effect also participates in angiogenesis [28]. In contrast to the effect seen in ECs, HGF does not stimulate the DNA synthesis of VSMCs and pericytes even though these cells express c-Met [17, 29].

Migration: The following processes are required for the successful migration of ECs: cell shape change; promotion of motility; reduction in cell adhesion; and extracellular matrix (ECM) degradation. In the initial step, HGF reduces the expression of adhesion molecule VE-cadherin, leading to the loss in cell–cell contact and stimulation of migration [30]. ECM degradation is also required for ECs to invade in neighboring tissue across the basement membranes. In this process, HGF induces ECM-degrading enzymes such as MT1-MMP and MMP2 in ECs [31, 32]. Additional interest is that HGF-induced nitric oxide (NO) is required for EC migration, because HGF-induced migratory actions are abolished by NO-synthase inhibitors [33].

In addition to ECs, HGF also induces motility of VSMCs or pericytes. VSMCs form lamellipodia and show migratory phenotypes post-HGF stimulation [34, 35]. PI3K-Akt pathway or MAPK-mediated activation of FAK and Pyk2 is involved in this process. Angiopoietin-1 (Angp1) is a key molecule involved in the recruitment of VSMCs to EC wall. Notably, Angp1 up-regulates the expression of HGF. Indeed, Angp1-induced VSMC migration was largely abolished by anti-HGF antibody. Overall, HGF was shown to be an essential downstream effector during Angp1-induced VSMC recruitment [36].

Morphogenesis: HGF has a unique morphogenetic activity, which forms polarized, tubular and branching structure of epithelial cells in collagen gels. HGF also induces in vitro and in vivo capillary tube formation of ECs [31]. In matrigel plug in vitro assay, HGF induces blood vessel formation that contains vasculature with surrounding VSMCs [31]. The adaptor protein Gab1 is required for the morphogenesis of ECs induced by HGF. EC-specific Gab1-knockout mice (Gab1-ecKO mice) showed no abnormality of vascular development, but showed little angiogenic response in a limb ischemia model. Indeed, HGF did not induce the migration, morphogenesis, and angiogenic sprouting in a culture of Gab1-null ECs [37].

Src signaling pathway is required for the HGF-mediated morphogenic actions, because Src gene deletion led to the loss in HGF-mediated vessel formation [38]. HGF-induced vascular lumen and cord formation, sprouting of vessels were also inhibited by the several inhibitors that targets Rho kinase and MMPs, which is a key player for migration or ECM degradation [39]. Overall, HGF-c-Met-Gab1-induced downstream responses (such as Rho activation and MMP induction) are required for vessel formation and functional maturation.

22.2.3 Anti-cell Death

In addition to the morphogenic activities, HGF acts as a survival factor for ECs. For example, exposure of ECs to high glucose results in a decrease in cell number. In this situation, HGF suppresses the glucose-induced cell death, indicating that HGF has an activity to inhibit cellular injury and death [40]. The rapid induction of Bcl-2, an anti-apoptotic molecule, by HGF leads to the inhibition of apoptotic death in culture of ECs [25]. Likewise, HGF prevents hypoxia- or starvation-induced cell death via PI3K/Akt-dependent pathway [26, 27]. HGF also prevents necrotic cell death via up-regulating de-toxication molecules, such as SOD [12]. Such anti-apoptotic and anti-necrotic effects of HGF on ECs (and other functional cells) will contribute to prohibit or reverse numerous ischemic diseases through the maintenance of local blood flow (see Sect. 22.3).

22.2.4 Endothelial Barrier Function

Vascular ECs form inner lining of vessels, and it serves as a barrier to regulate the exchange of the fluid and blood components between blood and tissues. The loss of EC barrier function leads to the exacerbation of hyper-permeability, edema formation, and inflammation during organ diseases. The transendothelial electrical resistance (TER) is used as an indicator of the permeability of EC monolayers in vitro. VEGF causes decrease in TER as its original name indicates, VPF (vascular permeability factor). In contrast, HGF treatment results in the increase of TER, indicating the vascular integrity and the enhanced barrier function [41].

The increase of TER by HGF is dependent on Tiam1, a nucleotide exchange factor of Rac-GTPase. HGF inhibits thrombin-induced disruption of EC barrier by activating Tiam1-mediated Rac1 pathway that counteracts with Rho-GTPase signaling by which forms stress fiber and loosens cell–cell junctions [42]. Such an inhibitory effect of HGF-Rac1 axis on VEGF-induced Rho activation is mediated via an assembly between c-Met and CD44v10 [43].

As a result, HGF blocks VEGF-mediated vascular permeability and impairment of EC barrier function (Fig. 22.2a). Activation of Rac1 signaling by HGF leads to redistribution of focal adhesions to cell periphery. Indeed, VEGF-induced Rho-dominant events (such as actin cytoskeleton reorganization and randomized focal adhesions) were counteracted by HGF-Rac1 axis [44]. Thus, prevention of permeability and enhancement of EC barrier function are advantageous for the therapeutic applications of HGF on angiogenesis.

22.2.5 Anti-inflammation

Several studies reported the unexpected inflammatory role of angiogenic factors. VEGF promotes leukocyte adhesion to ECs by up-regulating the expression of ICAM-1 and VCAM-1 in a NF-κB-dependent manner, which contributes to the tissue inflammation [45]. In addition, b-FGF evokes inflammatory response by activating NF-κB signaling and increasing levels of pro-inflammatory cytokines (such as IL-8 and MCP-1) in VSMCs [46].

In contrast, HGF shows anti-inflammatory actions on ECs. In culture of HUVEC, TNF-α-induced ICAM-1, a key molecule involved in the transendothelial migration of leukocytes, associated with the activation of NF-κB. Of note, HGF counteracted the TNF-α-mediated induction of ICAM-1 through the inactivation of NF-κB. As a result, inflammation-induced ICAM-1 up-regulation was blocked by HGF in culture model of HUVEC [47]. Such an anti-inflammatory effect of HGF on ECs was

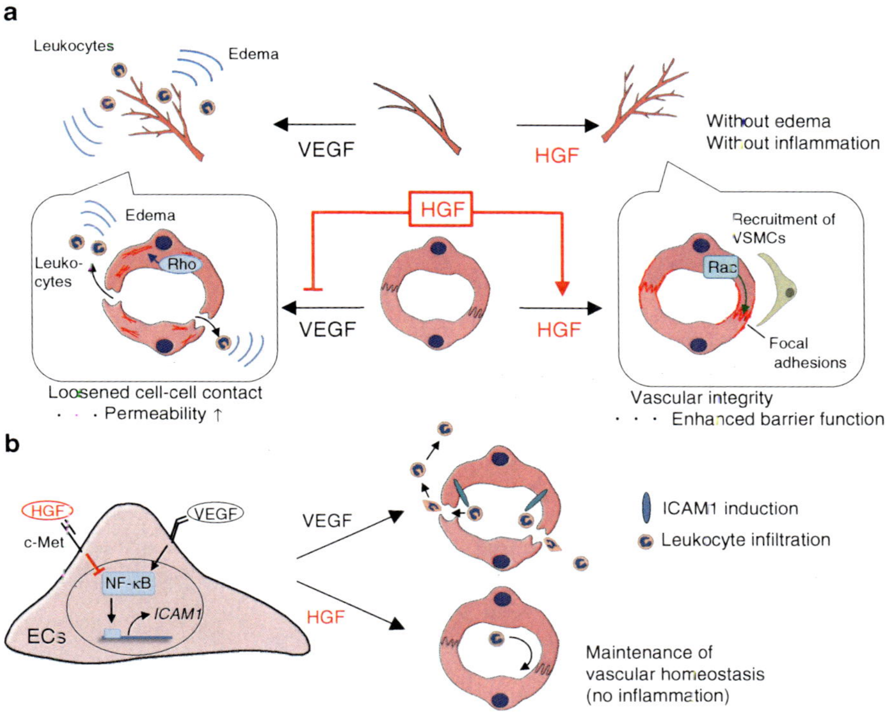

Fig. 22.2 Comparison of angiogenic properties between VEGF and HGF. (**a**) In contrast to edema formation by VEGF, HGF can induce angiogenesis with maintenance of vascular homeostasis. In ECs, VEGF signaling induces Rho activation, formation of actin stress fiber and paracellular gaps, leading to the loss of endothelial cell–cell contacts and the enhancement of permeability. On the other hand, HGF activates Rac1 and accumulates focal adhesions to cell periphery, and additionally it counteracts with VEGF-mediated signaling, which results in tight cellular contacts and enhanced barrier function. In addition, HGF stimulates the migration of VSMCs and their recruitment, eventually inducing vascular maturation and less permeable vasculature. (**b**) VEGF and HGF show reciprocal effects on NF-κB signaling in ECs. HGF inhibited, whereas VEGF activated, NF-κB signaling and HGF induces angiogenesis without up-regulation of ICAM1. ICAM1 induction potentiates leukocyte adhesion and infiltration, thus HGF has advantages in inducing angiogenesis in that it shows no inflammatory actions. Rather, HGF inhibits inflammation by counteracting VEGF actions on ECs. HGF also suppresses inflammatory cytokine productions from macrophages and shows anti-inflammatory activities (see, text)

also seen in a mouse model of acute renal failure [47]. As mentioned, VEGF induces endothelial inflammation, while HGF prohibits the VEGF-induced inflammatory events in ECs [48] (Fig. 22.2b).

Furthermore, HGF directly targets activated macrophages and suppresses the production of pro-inflammatory cytokines, such as IL-1β and IL-6, at least in part, via induction of anti-inflammatory enzyme, heme oxygenase-1 [49]. These anti-inflammatory but not pro-inflammatory actions of HGF have a beneficial effect on the protection of vascular and organ cells (see Sect. 22.3).

HGF is produced mainly by ECs in vessel tissues, and EC-secreted HGF is critical for sustaining EC integrity and inducing branching tubulogenesis during angiogenesis. HGF itself stimulates HGF production in ECs via an autocrine manner [50]. HGF-mediated recruitment of VSMCs contributes to the inhibition of vascular permeability and inflammation, possibly through the paracrine mechanism. The direct effects of HGF on vascular resident cells were summarized in Table 22.1.

22.3 Therapeutic Angiogenesis in Heart Disease

Direct effects of HGF on vascular cells lead to the improvement in cardiovascular diseases. CAD is the most common type of heart disease and is now the leading cause of death worldwide in both men and women. CAD is caused by the accumulation of plaque within the coronary arterial walls. The decrease in coronary artery flow causes hypoxia and subsequent loss in cardiomyocytes, leading to acute myocardial infarction (MI), or chronic MI with cardiac dysfunction and fibrosis. Forced induction of angiogenesis in intact coronary arteries by HGF leads to improved hypoxia, associated with the myocyte protection and improvement of cardiac dysfunction in acute MI and chronic angina, as follows.

22.3.1 Regulation of HGF-c-Met System During CAD

Circulating HGF is increased in the patients of acute MI [53]. Likewise, plasma HGF levels arise after 1 h of cardiac ischemia in a rat model of acute MI [15]. Monocytes isolated from peripheral blood of infarcted patients release more HGF into cultured medium, indicating monocytes as a source of HGF during acute MI [54]. In the rat ischemic heart, HGF expression is transiently up-regulated in ECs, fibroblasts, and macrophages in the early phase of MI [55]. Of note, anti-HGF antibody markedly worsened the MI-associated pathological phenotypes, such as apoptosis, inflammation, and loss in left ventricular (LV) function, in the rats, thus identifying HGF as an intrinsic cardioprotective factor [15]. In the late phase of heart diseases, the expression of local HGF becomes faint [56, 57]. Inversely, the expression of c-Met is augmented in the capillary ECs and myocardium after heart ischemia [15, 55], suggesting the role of exogenous HGF for the promotion of angiogenesis and cardioprotection during CAD.

22.3.2 Therapeutic Activities of HGF on Acute CAD

The angiogenic actions of HGF in the heart were shown in vivo in the model of acute MI [56]. HGF gene transfection into rat hearts by HVJ-liposome method

Table 22.1 Biological activities of HGF on vascular cells in vitro and their mechanisms

Target cells	Effect	Involved mechanisms	References
ECs	Proliferation	Activation of MAPK pathway	Bussolino-et al. [16]
		MAPK-mediated STAT3 and c-fos activation	Nakamura-et al.[17]
			Kuba-et al. [51]
			Nakagami-et al. [27]
	Migration	Gab1-SHP2→ERK1/2	Bussolino-et al. [16]
		Gab1-p85→Akt	Grant-et al. [31]
		Induction of matrix-degrading enzymes	Wang and Keiser [32]
		Activation of JNK, Pyk and Src	Martin-et al. [30]
		Impairment of cell adhesion by down-regulating VE-cadherin transcription	Purdie-et al. [33]
			Rush-et al. [52]
		Induction of iNOS via PI3K pathway	
	Morphogenesis	Activation of Src	Grant-et al. [31]
		Rho kinase and MMP activity	Somlyo-et al. [39]
		Gab1→ERK1/2 and Akt	Kanda-et al. [38]
			Zhao-et al. [37]
	Anti-cell death	Induction of Bcl-2	Morishita-et al. [40]
		ERK1/2 and PI3K-Akt pathways	Morishita-et al. [25]
			Hayashi-et al. [26]
			Nakagami-et al. [27]
	Anti-inflammation	Down-regulation of leukocyte adhesion molecules through inhibition of NF-κB	Mizuno and Nakamura [47]
			Min-et al. [48]
	Enhancement of EC barrier function	Rac/Tiam1-mediated inhibition of Rho signaling	Liu et al. [41]
			Birukova-et al. [42]
		Tightening cell–cell junction by GSK-3β	Singleton-et al. [43]
			Birukova-et al. [44]
VSMCs	Migration	PI3K activation	Aoyagi-et al. [34]
		Activation of FAK and Pyk2 via ERK1/2	Taher-et al. [35]
Pericytes	Migration	PI3K-Akt pathway	Liu-et al. [29]

increased the number of proliferating ECs post-ischemia. As a result, HGF gene transfection increased the number of vessels in the ischemic hearts, especially near the infarcted myocardium. Importantly, administration of HGF gene resulted in the increase in cardiac blood flow and in the improvement of cardiac functions, as measured by the LV ejection fraction [56]. Such an HGF-mediated angiogenic activity was achieved via the activation of a transcription factor, Ets-1, which is one of the key cascades leading to angiogenesis. In the rat model, the up-regulation of HGF production is transient and insufficient. Thus, forced transfer of HGF gene into infarcted myocardium can induce beneficial responses, including angiogenesis, in order to compensate for the loss of endogenous HGF.

22.3.3 Therapeutic Activities of HGF on Chronic CAD

As mentioned above, HGF was shown to be effective in controlling the early phase of acute MI, but its usefulness for post-diagnostic therapy of MI was still unclear. To address this question, Li et al. examined the effect of HGF gene therapy after the onset of MI in mice. An adenoviral vector containing human HGF cDNA was injected into the limb muscles 3 days after heart ischemia, resulting in a persistent increase in plasma human HGF in treated mice [58]. At 4 weeks after MI, exogenous HGF increased the number of coronary vessels, accompanied by a reduction in cardiac fibrosis. This systemic HGF gene therapy inhibited the myocardial atrophy, especially near the infarcted areas, associated with the enhanced angiogenesis [58]. Overall, LV remodeling and dysfunction was improved in the HGF-treated mice compared with controls, as indicated by the smaller LV cavity, greater percentage of fractional shortening and LV+/−d*P*/d*t*.

The usefulness of delayed HGF treatment was also reported in other models of advanced MI. Strikingly, HGF supplement was effective in improving chronic MI, even when the adenoviral vector-containing HGF gene was injected 3 weeks after the induction of MI in rats [59], or when intracardial injection of recombinant HGF was started 4 weeks post-MI in dogs [60]. Forced induction of angiogenesis by exogenous HGF was shown to be contributable for improving dilative cardiomyopathy in hamsters, possibly via the inhibition of cardiac remodeling and fibrosis [57, 61]. Given that persistent hypoxia causes apoptosis and inflammation, release of hypoxia via HGF-induced therapeutic angiogenesis should be considered as a new strategy to attenuate CAD [18].

22.3.4 Therapeutic Activities of HGF on Restenosis

Percutaneous transluminal coronary angioplasty (PTCA), such as metal stent placement or balloon-dilated catheter therapy, greatly contributes to improvement in CAD. However, restenosis of coronary arteries post-PTCA is still serious due to reoccurrence of CAD. Restenosis is characterized by neointimal formation, along with over-proliferation of VSMCs, in response to the defect of EC integrity in the vessel lumens. Thus, rapid induction of EC regeneration may lead to prevention of restenosis post-PTCA.

In 2000, two groups provided evidence that the administration of HGF led to prevention of neointimal formation, using animal models of PTCA. When recombinant

HGF was administered into the balloon-injured iliac artery of rabbits using a drug delivery catheter, EC regeneration was rapidly induced in the injured arteries, followed by the inhibition of neointimal formation [62]. Likewise, HGF gene therapy is also useful for suppressing neointimal formation. HGF cDNA-containing HVJ-liposome was injected into the carotid artery of rabbits soon after the balloon injury [63]. As a result, neointimal hyperplasia was suppressed in the HGF-supplemented rabbits, accompanied by the enhancement of re-endothelialization, indicating that a rapid restoration of EC damage by HGF treatment is a reasonable strategy to suppress restenosis after angioplasty.

The molecular mechanisms of HGF therapy against restenosis should be discussed. While HGF elicits a rapid re-sheet of the defective endothelial continuity, VSMCs in the neointima also express c-Met [34], suggesting a direct effect of HGF on VSMC-based pathogenic events. Platelet-derived growth factor (PDGF) is a potent mitogen for stromal cells including VSMCs, while HGF inhibits the PDGF-mediated overgrowth of stroma cells [64] via the suppression of ERK1/2-phosphorylation. HGF also inhibits angiotensin-II-induced proliferation of VSMCs in vitro [65]. These anti-mitogenic effects of HGF on VSMCs may explain the therapeutic outcome after PTCA. NO is important to attenuate intimal formation via relaxation of VSMCs, while HGF increases endothelial NO levels during arterial stenosis in rats [63]. HGF inhibits production of endothelin-1, a constrictor of VSMCs in vascular tissues. Thus, up-regulation of NO, or down-regulation of endothelins, by HGF may contribute to restenosis post-PTCA, possibly via the relaxation of VMSCs.

22.3.5 Other Mechanisms of HGF to Improve Heart Disease

Inducible angiogenesis is one of the key mechanisms whereby HGF produces beneficial effects in various types of heart diseases. Under heart injuries, c-Met is noted not only in ECs but also in other cells, such as cardiomyocytes, myofibroblasts (MyoFB), and macrophages. Using a rat model of MI, we provided the first evidence that a direct protection of cardiomyocytes by HGF leads to inhibition of apoptosis and improvement in LV function [15]. In this process, HGF induces Bcl-xL, an anti-apoptotic molecule that inhibits cytochrome-c release, followed by the avoidance of apoptosis and cardiac atrophy. The similar effect was also seen in a hamster model of cardiomyopathy [57].

MyoFB play a central role in tissue fibrosis as a major source of ECM, such as collagens. Importantly, HGF induces apoptosis in MyoFB through the MMP-induced degradation of ECM (i.e., anoikis-like death) [66]. TGF-β is a key regulator to elicit organ fibrosis through ECM production and MyoFB induction. Angiotensin-II up-regulates TGF-β production in MyoFB, while HGF inhibits the induction of TGF-β by angiotensin-II [57]. Such an anti-fibrotic effect by HGF facilitates neo-vascular formation via the reduction of scar formation in chronic MI and cardiomyopathy.

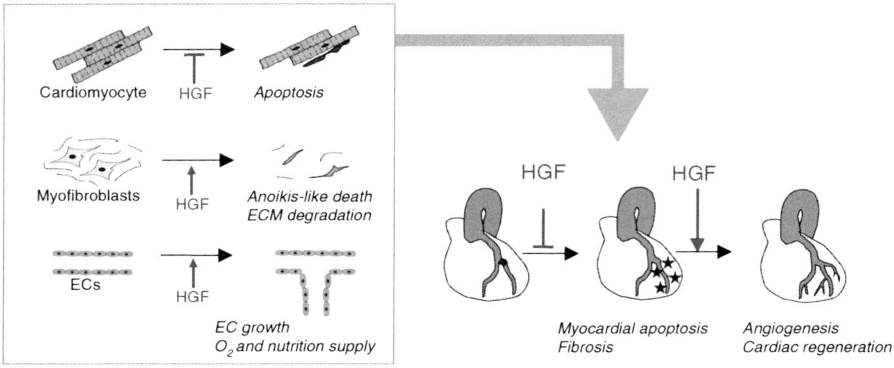

Fig. 22.3 Target cells and therapeutic actions of HGF on CAD. Coronary artery obstruction often caused by plaque accumulation induces myocardial apoptosis and infarction. In acute phase of CAD, c-Met is up-regulated in cardiomyocytes and ECs. HGF inhibits apoptosis of cardiomyocytes and protects the organ from dysfunction. Furthermore, HGF induces angiogenesis and enables the supply of oxygen and nutrition, which prevents cell death. In the late phase, tissue fibrosis progresses for the compensation of parenchymal cells. HGF increased the expression of ECM-degrading enzymes and decreased the expression of TGF-β, which evokes anoikis-like cell death of myofibroblasts and prevents fibrosis, leading to the attenuation of chronic tissue remodeling and further dysfunction. In CAD, HGF expression is transient and insufficient, thus supplementation of HGF becomes therapeutic options for the treatment of this disease via enhancing tissue protection or regeneration

According to each pathological stage, HGF can alter the target cells, along with the change in c-Met localization. In an inflammatory stage, HGF targets macrophages to suppress NF-κB activation [67]. In the middle stage of post-PTCA, VMSCs may be deactivated by HGF to prohibit neointimal formation. In a fibrotic stage, HGF targets MyoFB to induce anoikis-like death. Such multiple effects of HGF would contribute to the recovery from cardiac failures (Fig. 22.3).

22.4 HGF-Mediated Therapeutic Angiogenesis in Other Diseases

The decrease in HGF production by persistent hypoxia, in a reciprocal increase in c-Met, is probably the common pathogenesis in ischemic organ diseases, where blood flow is impaired. Exogenous HGF can restore the impairment in HGF production by ECs [50]. Thus, supplementation of HGF is logical to reverse HGF-deficiency in an autocrine manner, or via the recovery from hypoxia, associated with neovascularization. There is now ample evidence to show that HGF is useful for attenuating not only CAD but also other hypoxic diseases, as described below.

22.4.1 Peripheral Arterial Disease

PAD is a typical disease that develops when the inner wall of arteries, frequently in the legs, are occluded by the plaques made of fat, cholesterol, and calcium. Diabetes, hypertension, and obesity often cause this disease. Drugs with antiplatelet or vasodilator effect and vascular surgery with balloon catheter or stent are used for the treatment of PAD. Neovascularization induced by growth factors is considered as a promising therapy for PAD. Here, we describe the regulation of HGF production under the PAD-related conditions and the therapeutic effects of HGF on PAD.

22.4.1.1 Regulation of Endogenous HGF in PAD

Production of HGF in stroma cells is ordinarily increased in response to the organ injury [12]. In the animal models of PAD, however, HGF production in vascular tissues is decreased, likely due to tissue hypoxia [25, 26]. In patients with arteriosclerosis obliterans (ASO), HGF transcription and protein production in hypoxic legs are also decreased than normal tissues [25]. Thus, it is likely that the local self-defensive system (i.e., paracrine system) is impaired due to the persistent hypoxia.

Oppositely, circulating levels of HGF is elevated in patients of PAD, particularly with collateral formation (i.e., enhancement of endocrine system) [68]. High serum concentration of HGF is positively correlated with collateral vessel formation. Thus, serum HGF could be a marker for PAD and/or collateral formation. The increase in serum HGF levels might be a compensation for the loss of "local" HGF production in the vasculature, but this is not sufficient for the complete inhibition of skin ulcer, as seen in diabetic mice [69].

22.4.1.2 The Therapeutic Effects of HGF in PAD

Several lines of evidence indicate that HGF prevents disease progression in preclinical studies of PAD. Intra-arterial injection of recombinant HGF into ischemic limbs resulted in the increase in capillary density, improvement of regional blood flow, and reduction in muscle atrophy [28]. Moreover, an intravenous administration of recombinant HGF is also effective in inducing collateral formation in this model of PAD [25]. Besides the induction of angiogenesis, HGF has protective effects on ECs and reduces muscle cell necrosis induced by hypoxia [26]. These therapeutic effects of HGF are obtained by the gene delivery method, as described below.

HGF has beneficial effects on PAD associated with diabetes. In a diabetic PAD, HGF production is decreased, because exposure of high concentration of glucose to ECs increases TGF-β expression and it reciprocally decreases HGF production. As a result, capillary vessel formation is impaired during diabetes. In diabetic animals, the expressions of MMP-1 and Ets-1, key molecules involved in angiogenic response, are reduced. Injection of HGF plasmid prohibited the loss of angiogenic

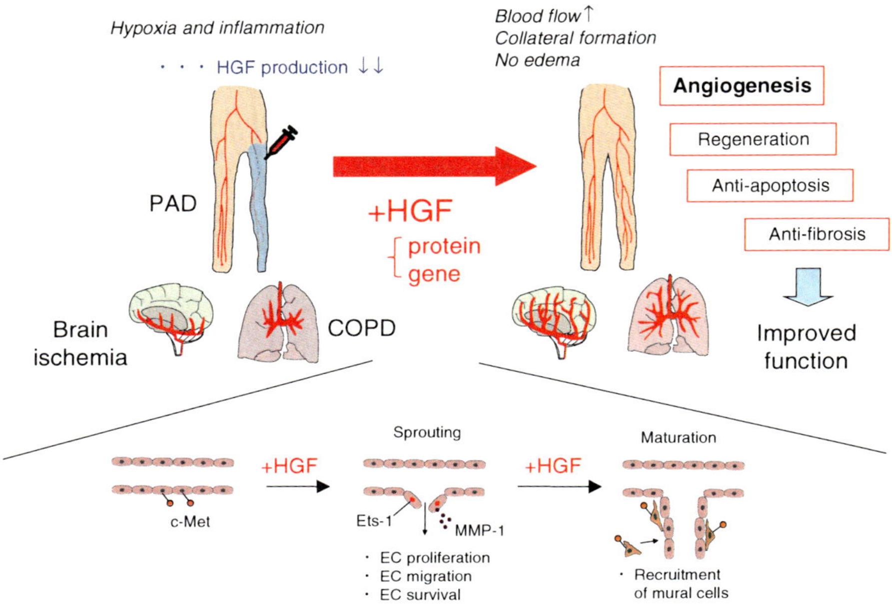

Fig. 22.4 Angio-therapeutic effects of HGF on hypoxic diseases. In the tissues of PAD, occlusion of arteries causes hypoxia and reduction in HGF production. In this condition, further hypoxia and exacerbation of the disease occurs, due to the loss of HGF-mediated protection of the vasculature. Administration of HGF into PAD enhances EC proliferation, migration, and survival by activating Ets-1 and up-regulating MMP-1, which resulted in the induction of angiogenesis and collateral formation. These effects are involved in the improvement of blood flow. Importantly, these therapeutic effects of HGF are obtained without side effects such as edema and pain, because of the recruitment of VSMCs or pericytes. The angiogenic activities of HGF are beneficial in treating other diseases that accompanies tissue hypoxia, including brain ischemia, stomach ulcer, diabetic cutaneous ulcer, and lung emphysema. In these diseases, moreover, parenchymal cells such as keratinocytes, neurons, and alveolar epithelial cells express c-Met, and HGF suppresses their apoptotic death and tissue fibrotic change or stimulates their growth [12]. These multi-target effects of HGF lead to the efficient tissue regeneration and improvement of organ function

capacity in diabetic rats, via the MMP-1 and Ets-1 inductions, suggesting the efficiency of HGF gene transfer even in the diabetic PAD [70] (Fig. 22.4).

22.4.1.3 Little Side Effect of HGF During the Treatment of PAD

Effectiveness and tolerability of HGF gene therapy on PAD might be greater than other angiogenic factors: promotion of angiogenesis (increase in capillary density and collateral formation) is more potent in HGF treatment group than in VEGF group [28]. Importantly, HGF has no obvious side effects onto the vascular system.

The treatment of b-FGF or VEGF augments angiogenesis and enables the improvement of blood flow in the animal models of PAD, but disappointingly side effects of them have been reported. VEGF induced vascular permeability, which resulted in edema [71]. Administration of b-FGF resulted in the significant growth of VSMCs and the vascular inflammation by the production of inflammatory chemokines, such as IL-8 and MCP-1, via activation of NF-κB pathway. In contrast, HGF induced suppression of leukocyte adhesion molecule expression, and vascular maturation via the recruitment of mural cells, which leads to the enhancement of EC barrier function. Consequently, treatment of HGF results in the formation of less-inflamed and less-permeable vasculature [46]. HGF is now a new promising therapeutic candidate for treating PAD patients with no vascular complications (see Sect. 22.6).

22.4.2 Cerebrovascular Diseases

Cerebrovascular disease is a pathological condition that affects the arteries supplying the brain, frequently caused by a clot deposition in the brain (thrombosis or embolism). These vascular disorders reduce local nutrients and oxygen tension, leading to the brain ischemia, stroke with hemorrhage or embolization. Sustained disease leads to massive neuronal cell death, cerebral edema and increase in intracranial pressure, which is a risk for attendant complications such as lethal brain herniation occasionally resulting in brain death. Thus, therapeutic strategies adopted for the treatment of cerebrovascular disease are to prevent clot deposition by using thrombolytic or anti-platelet agents, to inhibit neuronal cell death and to recover the cerebral blood flow while decreasing vascular permeability.

In a cerebral ischemia model, intra-ventricular administration of HGF stimulated angiogenesis and increased vascular lumens [72]. Pre-ischemic transfection of HGF gene using HVJ-liposome method into subarachnoid space also resulted in the increase in vessel number and blood flow after occlusion of carotid artery [73]. HGF gene transfer immediately after occlusion also stimulated angiogenesis on the brain surface and improved cerebral blood flow [73]. Importantly, angiogenic action of HGF on the brain is obtained without edematous formation and disruption of the brain–blood barrier [74]. Rather, HGF protects ECs from apoptotic cell death and inhibits the leakage of brain–blood barrier [75].

HGF has beneficial activities on neurons in addition to ECs. HGF prolonged survival of hippocampal neurons in primary culture [14]. Furthermore, neuronal cell death induced by forebrain ischemia was inhibited by recombinant HGF administration [76]. HGF enhances the expression of pro-survival factor Bcl-2 in neurons [72]. These angiogenic and neuroprotective activities of HGF attribute to the reduction in infarcted area and the improvement of behavioral symptoms [74, 75]. Such an angiogenic effect of HGF was also observed in the mouse model of Alzheimer's disease, followed by the prevention of behavior dysfunction [77].

22.4.3 Lung Emphysema and Other Respiratory Disease

Lung emphysema, mostly caused by a long-term smoking, is a type of chronic pulmonary disease in which alveolar architecture is destructed and overinflated, along with the impairment of gas exchange (hypoventilation). Reconstruction of alveoli is generally believed to be difficult, thus palliative or conservative treatment is chosen in the clinical setting of this disease. The alveoli, made of epithelial layer and capillary ECs, constitute a minimum functional unit in the lung. For supporting the integration of the alveolar structure and ventilation, expansion of capillary ECs would have therapeutic significance on the pathology of emphysema.

Transfection of HGF gene into normal lung increased capillary density and blood flow, suggesting that HGF administration assists the recovery of respiratory diseases through the induction of angiogenesis [78]. In an elastase-induced lung emphysema model, endogenous HGF levels were transiently elevated, but decreased along with the disease progression [79], implying the beneficial effect of HGF treatment on lung emphysema. Indeed, transfection of HGF gene into the rats with lung emphysema increased vascular density [79]. Interestingly, HGF also increased the fraction of circulating EC progenitor cells (Sca-1^+, c-kit^+, Flk^+) and their engraftment into the lung capillary endothelium [80]. Thus, HGF promotes angiogenesis by enhancing the in situ proliferation of resident ECs, and by inducing progenitor cell recruitment.

HGF also has a therapeutic effect on pulmonary hypertension model. HGF promoted the growth of ECs but not of VSMCs. The selective proliferative actions of HGF on ECs are meaningful because the overgrowth of VSMCs in the lungs is involved in the pathogenesis of pulmonary hypertension [81]. The multiple actions of HGF on lung cells are effective in the treatment of lung diseases such as lung emphysema.

In summary, HGF supplement therapy was found to be effective in various types of ischemic diseases, such as CAD, PAD, brain ischemia, lung emphysema, etc., through angiogenesis without edema and inflammation. The therapeutic outcomes of HGF obtained from preclinical studies were summarized in Table 22.2.

22.5 Inhibitory Effect of NK4 on Tumor Angiogenesis and Metastasis

HGF is a key regulator for the induction of angiogenesis, without edema and inflammation, during several types of ischemic organ diseases. On the other hand, tumor angiogenesis is important for cancers to acquire malignant phenotypes, such as invasive growth and metastasis. In this process, stroma-derived HGF promotes tumor angiogenesis that has been induced by VEGF or b-FGF [19, 82]. Based on the past background, we prepared NK4 as a fragment of HGF that antagonizes HGF-c-Met signaling.

Table 22.2 Therapeutic effects of HGF in preclinical studies of ischemic diseases

Model of disease		HGF	Way of administration	Effects	References
PAD	Hindlimb ischemia	Protein	Intra-arterial	Increased capillary density Collateral formation Increased blood flow	Van Belle-et al. [28]
	Hindlimb ischemia with diabetes	HVJ-liposome (gene)	Intra-muscular	Increased capillary density Increased blood flow	Taniyama-et al. [70]
CAD	Coronary I/R (acure MI)	Protein	i.v.	Anti-apoptosis on cardiomyocytes Reduction in infarct area Improved LV function	Nakamura-et al. [15]
	Coronary I/R (acure MI)	HVJ-liposome (gene)	Intra-myocardial	Increased capillary density Increased blood flow	Aoki-et al. [56]
	Coronary I/R (chronic MI)	Adenoviral expression	Intra-muscular	Increased capillary density Anti-fibrosis Improved LV function	Li-et al. [58]
Cardiomyopathy	Mutant model (TO-2 hamster)	HVJ-liposome (gene)	Intra-myocardial	Increased capillary density Anti-fibrosis Increased blood flow	Taniyama-et al. [61]
	Mutant model (TO-2 hamster)	Protein	s.c.	Anti-apoptosis on cardiomyocytes Anti-fibrosis Anti-Inflammation	Nakamura-et al. [57]
Restenosis	Balloon injury (Iliac artery)	Protein	Local injection	Prevention of neointimal formation Re-endothelialization	Yasuda-et al. [62]

(continued)

Table 22.2 (continued)

Model of disease		HGF	Way of administration	Effects	References
Cerebral Ischemia	Carotid artery occlusion	Protein	Intra-striatal	Prevention of neuronal cell death	Miyazawa-et al. [76]
	Middle cerebral artery occlusion	HVJ-liposome (gene)	Sub-arachnoid	Increased capillary density Anti-apoptosis Reduction in infarcted area	Shimamura-et al. [74]
Alzheimer's disease	Aβ-infusion (cognitive dysfunction)	HVJ-liposome (gene)	Ultrasound-mediated gene transfer	Increased capillary density BDNF up-regulation Reduced oxidative stress	Takeuchi-et al. [77]
Lung emphysema	Elastase instillation	Protein	i.p.	Proliferation of resident ECs Engraftment of EPCs Decreased tissue destruction	Ishizawa-et al. [80]
	Elastase instillation	HVJ-liposome (gene)	Intra-arterial	Increased capillary density Anti-apoptosis Improved pulmonary function	Shigemura-et al. [79]
Pulmonary hypertension	Injection of monocrotaline	HVJ-liposome (gene)	Intra-arterial	Increased capillary density Inhibition of VSMC overgrowth Increased pulmonary function	Ono-et al. [81]

HVJ hemagglutinating virus of Japan; *I/R* ischemia-reperfusion; *MI* myocardial infarction; *LV* left ventricular; *i.v.* intravenous; *i.p.* intraperitoneal; *s.c.* subcutaneous; *EPCs* endothelial progenitor cells; see text for other abbreviations

22.5.1 Angiogenic Roles of HGF During Tumor Progression

Needless to say, tumor angiogenesis is a principle event for invasive growth by primary tumors via supply of nutrients, oxygen, etc. Both vascular and lymphatic angiogenesis in growing tumors facilitate cancer cell metastasis via intra-vasation (i.e., in primary tumors) and extra-vasation (i.e., in distant metastatic organs) [82]. Stroma-derived HGF is critical for both vascular and lymphatic angiogenesis via angiogenic actions, such as mitogen and morphogen [16, 83]. Thrombospondin 1 (TSP1) is a negative regulator for tumor angiogenesis, while down-regulation of TSP1 by HGF contributes to tumor angiogenesis [84].

In addition to these angiogenic actions, HGF-c-Met signal in ECs provides an environmental situation in vessels that facilitates cancer cell migration. For instance, HGF stimulates adhesion of cancer cells to ECs via the activation of integrin signaling [85]. HGF increases tumor blood volume [86]. According to a pathological situation, HGF can decrease the expression of endothelial occludin, a cell–cell junction molecule [87]. Moreover, HGF enhances activities of MMPs (including membrane-type1 MMP/MMP14) in ECs [32]. Overall, it is likely that HGF reduces the transendothelial resistance of tumor vessels, and then cancer invasion across the basement membrane barrier (i.e., intra-vasation) could be facilitated for successful metastasis.

Herein, we described the multiple functions of HGF with a focus on vascular ECs. In addition, we will not exclude the direct effects of HGF on cancer cells. HGF targets tumor cells to stimulate the following sequential events: (1) cancer cell scattering; (2) invasive migration via MMP-dependent destruction of mucosal and endothelial basement membrane; and (3) anti-anoikis in vessels during blood flow. For detailed information, see a current review [82].

22.5.2 Preparation of NK4

NK4 was purified as a fragment from elastase-digested HGF samples [88, 89]. The amino acid sequence revealed that NK4 is cleaved between Val^{478} and Asn^{479} by elastase. The N-terminal structure of NK4 is the same as undigested HGF (i.e., 32nd pyroglutamate), indicating that NK4 is composed of the N-terminal hairpin domain and 4-kringle domains (thus designated NK4) (Fig. 22.5a). Precursor HGF is cleaved between Arg^{494} and Val^{495}, while NK4 is cleaved between Val^{478} and Asn^{479}. Thus, NK4 is identical to HGF α-chain that lacks C-terminal 16 amino acids (i.e., Asn^{479} to Arg^{494}). The domains that are responsible for high-affinity binding to c-Met are the N-terminal hairpin and the K1 domains in NK4.

Although NK4 is a c-Met-binder, NK4 alone does not phosphorylate c-Met tyrosine residues, thus meaning that NK4 is not agonistic and acts as a complete HGF-antagonist (Fig. 22.5b). Indeed, NK4 inhibits HGF-mediated c-Met tyrosine phosphorylation and biological actions in vitro: c-Met tyrosine phosphorylation occurs in lung cancer cells within 10 min post-HGF addition, while NK4 inhibits

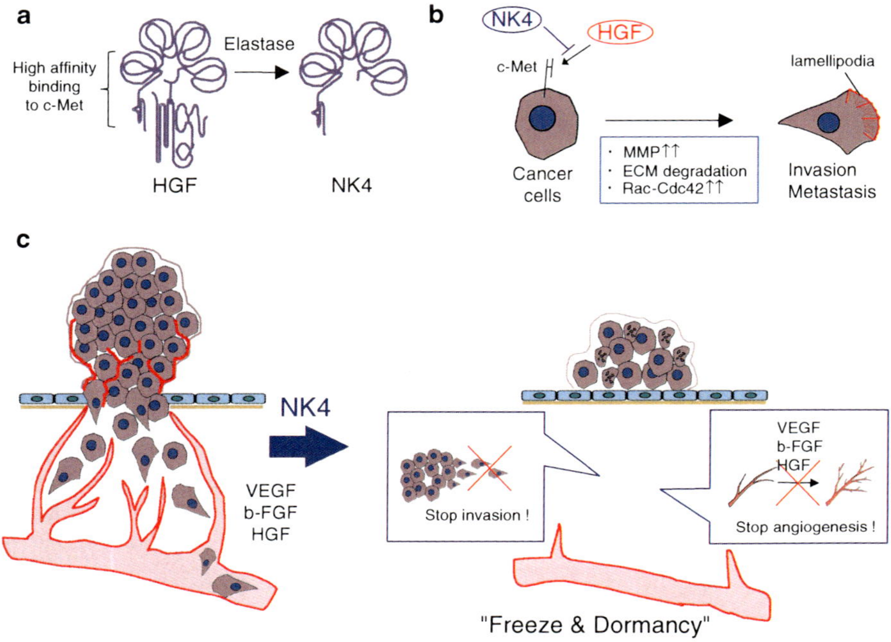

Fig. 22.5 NK4-mediated anti-angiogenesis and anti-tumor actions. (**a**) The schematic structure of NK4. NK4 was originally prepared by the enzymatic digestion of HGF with elastase. It retains N-terminal high affinity c-Met binding sites of HGF, but no agonistic activity. (**b**) Antagonistic activity of NK4 on HGF-c-Met axis and its anti-invasive actions on the tumor cells. NK4 competitively inhibits the HGF activities and thus inhibits HGF-mediated up-regulation of matrix degrading enzymes and activation of Rac-cdc42 signaling, leading to the suppression of cell movement and cell invasion. (**c**) The conceptual model of "freeze and dormancy therapy" with NK4. NK4 acts as an inhibitor of tumor–stromal interactions mediated by HGF. NK4 antagonizes HGF-c-Met signaling and prevents tumor invasion and metastasis ("Freeze"). In addition, NK4 suppresses tumor angiogenesis and their growth induced by various angiogenic molecules such as VEGF and b-FGF ("Dormancy"). These bifunctional effects of NK4 are quite effective and show therapeutic outcomes in a variety of tumor models [98]

the HGF-mediated c-Met activation [88]. Furthermore, HGF induces motility of the gallbladder cancer cells in Matri-gels, while NK4 inhibits HGF-induced invasion. These anti-invasive effects of NK4 are also seen in distinct types of cancer cells, strengthening the common role of NK4 in cancer migration.

HGF is a potent growth factor that induces mitogenic responses in ECs, but not VSMCs. Actually, forced expression of HGF gene enhanced the tumor angiogenesis in a mouse model of glioblastoma [90]. NK4 is an antagonist for the inhibition of HGF-c-Met complex. As expected, NK4 inhibited the HGF-induced migration and wound healing of ECs in vitro [91]. Furthermore, induction of endothelial duct formation by HGF in a collagen gel was impaired by NK4 treatment. These findings clearly indicated that NK4 is a powerful antagonist against HGF-induced angiogenic properties, at least in vitro.

22.5.3 *Anti-tumor Effect of NK4 on Malignant Tumors*

As mentioned above, we accumulated in the mid-1990s evidence to show that stroma-derived HGF is responsible for cancer cell dissociation and migration at least in vitro [92, 93]. Of importance, we found in 1998 that an HGF-antagonist inhibits invasive growth of malignant cancers, using NK4 in a mouse model of gall-bladder carcinoma [94]. In brief, recombinant NK4 inhibited the growth and muscular invasion in mice bearing GB-d1 carcinoma cells. Consistent with growth arrest, apoptotic change was evident in NK4-treated mice. This report provided the first proof-of-concept to show that HGF-antagonist is useful for anti-tumor therapy. In other words, stroma-derived HGF is responsible for cancer growth, invasion and metastasis. Thus, the blockade of HGF-c-Met activation with NK4 (or other HGF-antagonists or c-Met-inhibitors) is reasonable as a pathogenesis-based method for the inhibition of malignant phenotypes, such as tumor invasion and metastasis.

In addition to the anti-invasive effect of NK4, NK4 administration elicits a reduction in vessel density in the primary or secondary tumor masses [51, 95–97], as described below. These findings indicate that NK4 treatment is useful not only for the prevention of tumor invasion but also for the suppression of tumor angiogenesis (i.e., freeze and dormancy therapy) [82, 98]. There are now >30 reports to demonstrate the anti-metastatic effects of NK4 on distinct types of malignant cancers (Table 22.3 and Fig. 22.5c [99–106]).

22.5.4 *Anti-angiogenic Mechanism of NK4 via Perlecan-Dependent Pathways*

As mentioned above, the first evidence to show the anti-angiogenic effects of NK4 was obtained in a mouse model of lung cancer [51]. With regard to this, we found that NK4-mediated anti-angiogenic action is mediated via c-Met-independent pathways. It is well known that VEGF and b-FGF stimulate mitogenic and migrative response in culture of ECs. Notably, NK4 inhibited the EC proliferation and migration, induced by other angiogenic growth factors, such as b-FGF and VEGF [51]. In this process, HGF and VEGF phosphorylate c-Met and KDR/VEGF-receptor, respectively, while NK4 inhibits HGF-induced c-Met tyrosine phosphorylation, but not KDR activation. Nevertheless, NK4 was able to inhibit VEGF-induced proliferation, without alteration of VEGF-mediated ERK1/2 activation. Thus, there may be another mechanism whereby NK4 successfully arrests VEGF-induced (or b-FGF-mediated) angiogenesis. Such an alternative pathway is demonstrated in 2009, as follows.

The fibronectin-integrin signals in ECs are known to be required for angiogenesis. The NK4-mediated growth arrest of ECs is due to a loss of the fibronectin-integrin signal [107]. Actually, the affinity purification with NK4-immobilized beads revealed that perlecan is a counterpart for the binding to NK4. Perlecan is a

Table 22.3 Anti-tumor activities of NK4 in experimental models

Tumor types	Inoculation	NK4	Way of administration	Effects	References
Breast cancer	s.c.	Protein	s.c. (near tumor)	Inhibition of tumor growth Anti-angiogenesis Anti-metastasis	Kuba-et al. [51]
Colon cancer	Spleen	Naked plasmid	i.v.	Inhibition of tumor growth Suppression of tumor invasion Anti-angiogenesis Anti-metastasis Prolonged survival	Wen-et al. [96]
	s.c.	Adenoviral expression	Intra-tumoral administration with dendritic cells	Inhibition of tumor growth Anti-angiogenesis Increased cytotoxic T lymphocyte activity	Kikuchi-et al. [99]
Gallbladder cancer	s.c.	Protein	s.c. (near tumor)	Inhibition of tumor growth Anti-angiogenesis	Date-et al. [94]
Gastric cancer	s.c.	Stable gene expression or adenoviral expression	Expression by tumor cells, or intra-tumoral	Inhibition of tumor growth	Hirao-et al. [100]
	i.p.	Gene expression by cationic lipid	i.p.	Inhibition of tumor growth Anti-angiogenesis Inhibition of ascites accumulation Suppression of gefitinib resistance Prolonged survival	Namiki-et al. [101]
Glioma	Orthotopic	Protein	Intra-tumoral	Inhibition of tumor growth Anti-angiogenesis	Brockmann-et al. [102]

Hepatoma	s.c.	Adenoviral expression	Intra-tumoral	Inhibition of tumor growth Anti-angiogenesis	Heideman-et al. [103]
Lung cancer	s.c.	Protein	s.c. (near tumor)	Inhibition of tumor growth Anti-angiogenesis Anti-metastasis	Kuba-et al. [51]
Lymphoma	s.c.	Adenoviral expression	Intra-tumoral administration with dendritic cells	Inhibition of tumor growth Anti-angiogenesis Increased cytotoxic T lymphocyte activity	Kikuchi-et al. [99]
Melanoma	s.c.	Adenoviral expression	Intra-tumoral administration with dendritic cells	Inhibition of tumor growth T lymphocyte activity	Kikuchi-et al. [99]
Multiple myeloma	s.c.	Adenoviral expression	Intra-muscular	Inhibition of tumor growth Anti-angiogenesis	Du-et al. [104]
Ovarian cancer	i.p.	Stable gene expression	Expression by tumor cells	Anti-metastasis	Saga-et al. [105]
Pancreatic cancer	Orthotopic	Protein	i.p.	Inhibition of tumor growth Anti-angiogenesis Anti-metastasis Suppression of tumor invasion Prolonged survival	Tomioka-et al. [95]
Prostate cancer	s.c.	Protein	s.c. (near tumor)	Inhibition of tumor growth Anti-angiogenesis	Davies-et al. [106]

i.p. intraperitoneal; *i.v.* intravenous; *s.c.* subcutaneous

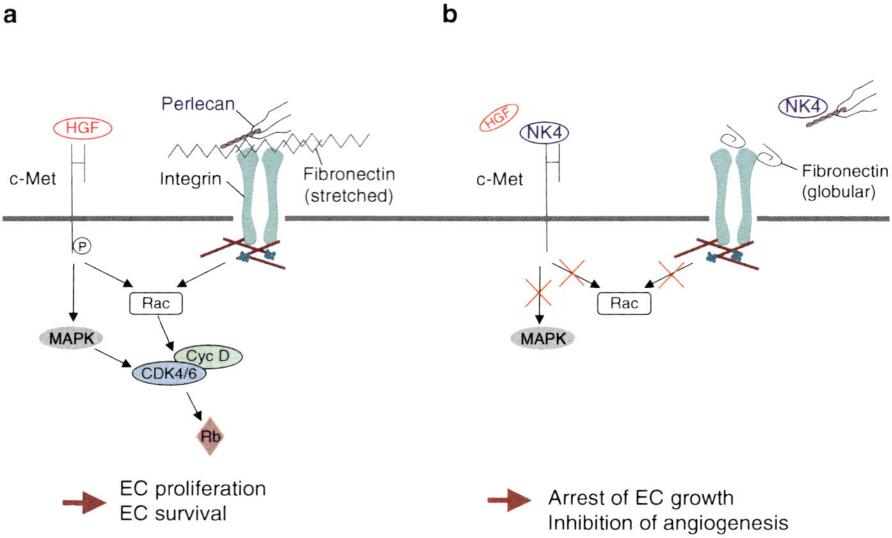

Fig. 22.6 The molecular mechanisms of NK4-mediated angio-inhibitory actions. (**a**) The process of cell proliferation or survival is affected by the signaling from integrin–ECM interaction (i.e., outside-in signal) besides growth factor-receptor signaling. In ECs, perlecan is involved in the interaction between integrin and fibronectin by inducing fibronectin assembly with stretched conformation, which enables their tight adhesion on ECMs. In these condition, integrin-mediated adhesion signaling activates Rac, both of which pathway lead to the cell cycle progression and cell spreading. (**b**) In the presence of NK4, HGF/c-Met signaling is antagonized. Moreover, NK4-perlecan interaction prevents the fibronectin assembly, thus inhibiting integrin-mediated outside-in signals and EC proliferation

heparan sulfate proteoglycan that interacts with basement membrane ECM, such as fibronectin. Suppression of perlecan expression by siRNA diminished the fibronectin assembly and EC spreading, indicating a key role of fibronectin–perlecan interaction during EC migration. Notably, NK4–perlecan interaction reduced the assembly of fibronectin by perlecan. As a result, FAK activation became faint, due to the inhibition of perlecan–fibronectin coupling with NK4. Under the loss of integrin signals, EC growth was impaired, even in the presence of VEGF or b-FGF [107] (Fig. 22.6).

It has been 15 years since NK4 was identified as an HGF-antagonist that blocks the tumor growth in vivo [88, 94]. These preclinical studies prompted pharmaceutical companies to search HGF-antagonists or c-Met-inhibitors, such as c-Met decoy, c-Met tyrosine kinase inhibitors, etc. Indeed, NK4 has a dual role during the control of tumor malignancy: (1) as an HGF-antagonist for suppressing tumor cell motility and migration; and (2) as a perlecan-binder for arresting tumor angiogenesis. Such a dual function of NK4 (i.e., malignostatin) contributes to freeze and dormancy therapy in various types of carcinoma and sarcoma (Fig. 22.5c).

22.6 Conclusion and Perspective

For the future development of angiogenic therapy, it is important to discuss how exogenous HGF is efficiently delivered into hypoxic tissues. Blood HGF levels are increased in patients with ischemic diseases, while HGF is present in blood in the form of precursor (i.e., pro-HGF). Thus, recombinant form of HGF will be recommended as an active HGF, especially in acute diseases, such as acute MI, because apoptotic death occurs within several hours [15]. By contrast, in subacute and chronic phases of ischemic diseases, HGF gene therapy may be suitable, because of stable and long-term expression of HGF at local sites in CAD or PAD [56, 70]. Autologous cell implant therapy is also available for this purpose. For example, ex vivo transfection of HGF cDNA in myoblasts leads to angiogenesis in a rat model of MI via the paracrine pathway [108].

PAD may be one of the most promising models for clinical trials of HGF, because NK4-like fragment in effusive fluid from the skin may impair epidermal repair via antagonizing HGF's actions [109]. The first clinical trial of HGF was tested in Sweden: topical application of recombinant HGF in ulcer areas resulted in the excellent or partial healing in 8 out of 11 patients with chronic leg ulcer [110]. Naked HGF plasmid DNA is also promising. A medium-size clinical trial, tested in Japan, revealed the usefulness of HGF gene therapy in patients with critical limb ischemia. The overall improvement rate of the primary endpoint was 70.4% (19/27) in HGF group and 30.8% (4/13) in placebo group, with a significant difference [111]. The efficacy and safety of HGF gene therapy in the patients with severe PAD were further demonstrated in long-term follow-up analysis [112]. Likewise, intra-myocardial injection of adenovirus with HGF cDNA improved the cardiac blood flow in CAD patients after coronary artery bypass surgery [113].

In contrast to ischemic organ disease, HGF production levels are increased in tumor tissues via tumor-secreted mediators, such as IL-1β [93]. Such an HGF-sufficient condition promotes tumor angiogenesis and growth, thus providing a rationale of HGF-antagonistic strategy for inhibiting tumor malignancy. Angiogenesis inhibition with VEGF-antagonist leads to the reduction of metastasis, but long-term anti-angiogenesis may cause hypoxic adaptation via HIF1-dependent c-Met up-regulation and its signal activation by HGF [82]. In addition to anti-angiogenic effects, NK4 is useful as an HGF-antagonist for the prevention of hypoxia-primed metastasis via its counteraction of HGF actions.

We return to the importance of HGF-c-Met system in living body. Appropriate and constitutive activation of HGF-c-Met system is required for maintenance of tissue homeostasis. The decrease in HGF-c-Met signaling causes pathological conditions in various organs, along with a decrease in vascularity. Inversely, over-activation of HGF-c-Met system might lead to tumorigenesis, in part, via enhanced angiogenesis. In other words, the degree of c-Met activation levels may determine predisposition to "regressive" disease (including CAD) or "progressive" disease (such as tumors). Thus, HGF-agonist is useful for the treatment of CAD and PAD, while HGF-antagonist is hopeful for anti-cancer strategy. Further basic and clinical studies using HGF or NK4 will provide paradoxical insights into angiogenic networks connected to ischemic disease and malignant tumors (like two sides of a coin).

Acknowledgements This work was supported by grants from the Ministry of Education, Science, Technology, Sports and Culture of Japan (No. 23590458 to S.M. and No. 21390079 and Global COE program to T.N.). We also thank Mrs. Yoshiko Yoneda for secretarial assistance.

References

1. Ando J, Kamiya A (1993) Blood flow and vascular endothelial cell function. Front Med Biol Eng 5:245–264
2. Staudacher DL, Preis M, Lewis BS et al (2006) Cellular and molecular therapeutic modalities for arterial obstructive syndromes. Pharmacol Ther 109:263–273
3. Folkman J (1995) Angiogenesis in cancer, vascular, rheumatoid and other disease. Nat Med 1:27–31
4. Nakamura T, Nawa K, Ichihara A (1984) Partial purification and characterization of hepatocyte growth factor from serum of hepatectomized rats. Biochem Biophys Res Commun 122:1450–1459
5. Nakamura T, Nawa K, Ichihara A et al (1987) Purification and subunit structure of hepatocyte growth factor from rat platelets. FEBS Lett 224:311–316
6. Nakamura T, Nishizawa T, Hagiya M et al (1989) Molecular cloning and expression of human hepatocyte growth factor. Nature 342:440–443
7. Tashiro K, Hagiya M, Nishizawa T et al (1990) Deduced primary structure of rat hepatocyte growth factor and expression of the mRNA in rat tissues. Proc Natl Acad Sci USA 87:3200–3204
8. Dean M, Park M, Le Beau MM et al (1985) The human met oncogene is related to the tyrosine kinase oncogenes. Nature 318:385–388
9. Bottaro DP, Rubin JS, Faletto DL et al (1991) Identification of the hepatocyte growth factor receptor as the c-met proto-oncogene product. Science 251:802–804
10. Higuchi O, Mizuno K, Vande Woude GF, Nakamura T (1992) Expression of c-met proto-oncogene in COS cells induces the signal transducing high-affinity receptor for hepatocyte growth factor. FEBS Lett 301:282–286
11. Schaeper U, Gehring NH, Fuchs KP et al (2000) Coupling of Gab1 to c-Met, Grb2, and Shp2 mediates biological responses. J Cell Biol 149:1419–1432
12. Nakamura T, Mizuno S (2010) The discovery of hepatocyte growth factor (HGF) and its significance for cell biology, life sciences and clinical medicine. Proc Jpn Acad Ser B Phys Biol Sci 86:588–610
13. Ishiki Y, Ohnishi H, Muto Y et al (1992) Direct evidence that hepatocyte growth factor is a hepatotrophic factor for liver regeneration and has a potent antihepatitis effect in vivo. Hepatology 16:1227–1235
14. Honda S, Kagoshima M, Wanaka A et al (1995) Localization and functional coupling of HGF and c-Met/HGF receptor in rat brain: implication as neurotrophic factor. Brain Res Mol Brain Res 32:197–210
15. Nakamura T, Mizuno S, Matsumoto K et al (2000) Myocardial protection from ischemia/reperfusion injury by endogenous and exogenous HGF. J Clin Invest 106:1511–1519
16. Bussolino F, Di Renzo MF, Ziche M et al (1992) Hepatocyte growth factor is a potent angiogenic factor which stimulates endothelial cell motility and growth. J Cell Biol 119:629–641
17. Nakamura Y, Morishita R, Nakamura S et al (1996) A vascular modulator, hepatocyte growth factor, is associated with systolic pressure. Hypertension 28:409–413
18. Morishita R, Aoki M, Hashiya N et al (2004) Therapeutic angiogenesis using hepatocyte growth factor (HGF). Curr Gene Ther 4:199–206
19. Matsumoto K, Nakamura T (2003) NK4 (HGF-antagonist/angiogenesis inhibitor) in cancer biology and therapeutics. Cancer Sci 94:321–327

20. Montesano R, Matsumoto K, Nakamura T, Orci L (1991) Identification of a fibroblast-derived epithelial morphogen as hepatocyte growth factor. Cell 67:901–908
21. Nakamura Y, Morishita R, Higaki J et al (1995) Expression of local hepatocyte growth factor system in vascular tissues. Biochem Biophys Res Commun 215:483–488
22. Machide M, Hashigasako A, Matsumoto K, Nakamura T (2006) Contact inhibition of hepatocyte growth regulated by functional association of the c-Met/hepatocyte growth factor receptor and LAR protein-tyrosine phosphatase. J Biol Chem 281:8765–8772
23. Ding S, Merkulova-Rainon T, Han ZC et al (2003) HGF receptor up-regulation contributes to the angiogenic phenotype of human endothelial cells and promotes angiogenesis in vitro. Blood 101:4816–4822
24. Scarpino S, Cancellario d'Alena F, Di Napoli A et al (2004) Increased expression of Met protein is associated with up-regulation of hypoxia inducible factor-1 (HIF-1) in tumour cells in papillary carcinoma of the thyroid. J Pathol 202:352–358
25. Morishita R, Nakamura S, Hayashi S et al (1999) Therapeutic angiogenesis induced by human recombinant hepatocyte growth factor in rabbit hind limb ischemia model as cytokine supplement therapy. Hypertension 33:1379–1384
26. Hayashi S, Morishita R, Nakamura S et al (1999) Potential role of hepatocyte growth factor, a novel angiogenic growth factor, in peripheral arterial disease: downregulation of HGF in response to hypoxia in vascular cells. Circulation 100:II301–II308
27. Nakagami H, Morishita R, Yamamoto K et al (2001) Mitogenic and antiapoptotic actions of hepatocyte growth factor through ERK, STAT3, and AKT in endothelial cells. Hypertension 37:581–586
28. Van Belle E, Witzenbichler B, Chen D et al (1998) Potentiated angiogenic effect of scatter factor/hepatocyte growth factor via induction of vascular endothelial growth factor: the case for paracrine amplification of angiogenesis. Circulation 97:381–390
29. Liu Y, Wilkinson FL, Kirton JP et al (2007) Hepatocyte growth factor and c-Met expression in pericytes: implications for atherosclerotic plaque development. J Pathol 212:12–19
30. Martin TA, Mansel R, Jiang WG (2001) Hepatocyte growth factor modulates vascular endothelial-cadherin expression in human endothelial cells. Clin Cancer Res 7:734–737
31. Grant DS, Kleinman HK, Goldberg ID et al (1993) Scatter factor induces blood vessel formation in vivo. Proc Natl Acad Sci USA 90:1937–1941
32. Wang H, Keiser JA (2000) Hepatocyte growth factor enhances MMP activity in human endothelial cells. Biochem Biophys Res Commun 272:900–905
33. Purdie KJ, Whitley GS, Johnstone AP, Cartwright JE (2002) Hepatocyte growth factor-induced endothelial cell motility is mediated by the upregulation of inducible nitric oxide synthase expression. Cardiovasc Res 54:659–668
34. Aoyagi M, Yamamoto S, Azuma H et al (1999) Localization and effects of hepatocyte growth factor on smooth muscle cells during neointimal formation after balloon denudation. Histochem Cell Biol 111:419–428
35. Taher TE, Derksen PW, de Boer OJ et al (2002) Hepatocyte growth factor triggers signaling cascades mediating vascular smooth muscle cell migration. Biochem Biophys Res Commun 298:80–86
36. Kobayashi H, DeBusk LM, Babichev YO et al (2006) Hepatocyte growth factor mediates angiopoietin-induced smooth muscle cell recruitment. Blood 108:1260–1266
37. Zhao J, Wang W, Ha CH et al (2011) Endothelial Grb2-associated binder 1 is crucial for postnatal angiogenesis. Arterioscler Thromb Vasc Biol 31:1016–1023
38. Kanda S, Kanetake H, Miyata Y (2006) HGF-induced capillary morphogenesis of endothelial cells is regulated by Src. Biochem Biophys Res Commun 344:617–622
39. Somlyo AV, Phelps C, Dipierro C et al (2003) Rho kinase and matrix metalloproteinase inhibitors cooperate to inhibit angiogenesis and growth of human prostate cancer xenotransplants. FASEB J 17:223–234
40. Morishita R, Nakamura S, Nakamura Y et al (1997) Potential role of an endothelium-specific growth factor, hepatocyte growth factor, on endothelial damage in diabetes. Diabetes 46:138–142

41. Liu F, Schaphorst KL, Verin AD et al (2002) Hepatocyte growth factor enhances endothelial cell barrier function and cortical cytoskeletal rearrangement: potential role of glycogen synthase kinase-3beta. FASEB J 16:950–962
42. Birukova AA, Alekseeva E, Mikaelyan A, Birukov KG (2007) HGF attenuates thrombin-induced endothelial permeability by Tiam1-mediated activation of the Rac pathway and by Tiam1/Rac-dependent inhibition of the Rho pathway. FASEB J 21:2776–2786
43. Singleton PA, Salgia R, Moreno-Vinasco L et al (2007) CD44 regulates hepatocyte growth factor-mediated vascular integrity. Role of c-Met, Tiam1/Rac1, dynamin 2, and cortactin. J Biol Chem 282:30643–30657
44. Birukova AA, Cokic I, Moldobaeva N, Birukov KG (2009) Paxillin is involved in the differential regulation of endothelial barrier by HGF and VEGF. Am J Respir Cell Mol Biol 40:99–107
45. Kim I, Moon SO, Kim SH et al (2001) Vascular endothelial growth factor expression of intercellular adhesion molecule 1 (ICAM-1), vascular cell adhesion molecule 1 (VCAM-1), and E-selectin through nuclear factor-kappa B activation in endothelial cells. J Biol Chem 276:7614–7620
46. Kaga T, Kawano H, Sakaguchi M et al (2012) Hepatocyte growth factor stimulated angiogenesis without inflammation: differential actions between hepatocyte growth factor, vascular endothelial growth factor and basic fibroblast growth factor. Vascul Pharmacol 57:3–9
47. Mizuno S, Nakamura T (2005) Prevention of neutrophil extravasation by hepatocyte growth factor leads to attenuations of tubular apoptosis and renal dysfunction in mouse ischemic kidneys. Am J Pathol 166:1895–1905
48. Min JK, Lee YM, Kim JH et al (2005) Hepatocyte growth factor suppresses vascular endothelial growth factor-induced expression of endothelial ICAM-1 and VCAM-1 by inhibiting the nuclear factor-kappaB pathway. Circ Res 96:300–307
49. Kamimoto M, Mizuno S, Matsumoto K, Nakamura T (2009) Hepatocyte growth factor prevents multiple organ injuries in endotoxemic mice through a heme oxygenase-1-dependent mechanism. Biochem Biophys Res Commun 380:333–337
50. Hayashi S, Morishita R, Higaki J et al (1996) Autocrine-paracrine effects of overexpression of hepatocyte growth factor gene on growth of endothelial cells. Biochem Biophys Res Commun 220:539–545
51. Kuba K, Matsumoto K, Date K et al (2000) HGF/NK4, a four-kringle antagonist of hepatocyte growth factor, is an angiogenesis inhibitor that suppresses tumor growth and metastasis in mice. Cancer Res 60:6737–6743
52. Rush S, Khan G, Bamisaiye A et al (2007) c-jun amino-terminal kinase and mitogen activated protein kinase 1/2 mediate hepatocyte growth factor-induced migration of brain endothelial cells. Exp Cell Res 313:121–132
53. Soeki T, Tamura Y, Shinohara H et al (2000) Role of circulating vascular endothelial growth factor and hepatocyte growth factor in patients with coronary artery disease. Heart Vessels 15:105–111
54. Zhu Y, Hojo Y, Ikeda U, Shimada K (2000) Production of hepatocyte growth factor during acute myocardial infarction. Heart 83:450–455
55. Ono K, Matsumori A, Shioi T et al (1997) Enhanced expression of hepatocyte growth factor/c-Met by myocardial ischemia and reperfusion in a rat model. Circulation 95:2552–2558
56. Aoki M, Morishita R, Taniyama Y et al (2000) Angiogenesis induced by hepatocyte growth factor in non-infarcted myocardium and infarcted myocardium: up-regulation of essential transcription factor for angiogenesis, ets. Gene Ther 7:417–427
57. Nakamura T, Matsumoto K, Mizuno S et al (2005) Hepatocyte growth factor prevents tissue fibrosis, remodeling, and dysfunction in cardiomyopathic hamster hearts. Am J Physiol Heart Circ Physiol 288:H2131–H2139
58. Li Y, Takemura G, Kosai K et al (2003) Postinfarction treatment with an adenoviral vector expressing hepatocyte growth factor relieves chronic left ventricular remodeling and dysfunction in mice. Circulation 107:2499–2506

59. Jayasankar V, Woo YJ, Pirolli TJ et al (2005) Induction of angiogenesis and inhibition of apoptosis by hepatocyte growth factor effectively treats postischemic heart failure. J Card Surg 20:93–101
60. Yamaguchi T, Sawa Y, Miyamoto Y et al (2005) Therapeutic angiogenesis induced by injecting hepatocyte growth factor in ischemic canine hearts. Surg Today 35:855–860
61. Taniyama Y, Morishita R, Aoki M et al (2002) Angiogenesis and antifibrotic action by hepatocyte growth factor in cardiomyopathy. Hypertension 40:47–53
62. Yasuda S, Noguchi T, Gohda M et al (2000) Single low-dose administration of human recombinant hepatocyte growth factor attenuates intimal hyperplasia in a balloon-injured rabbit iliac artery model. Circulation 101:2546–2549
63. Hayashi K, Nakamura S, Morishita R et al (2000) In vivo transfer of human hepatocyte growth factor gene accelerates re-endothelialization and inhibits neointimal formation after balloon injury in rat model. Gene Ther 7:1664–1671
64. Bessho K, Mizuno S, Matsumoto K, Nakamura T (2003) Counteractive effects of HGF on PDGF-induced mesangial cell proliferation in a rat model of glomerulonephritis. Am J Physiol Renal Physiol 284:F1171–F1180
65. Sanada F, Taniyama Y, Iekushi K et al (2009) Negative action of hepatocyte growth factor/c-Met system on angiotensin II signaling via ligand-dependent epithelial growth factor receptor degradation mechanism in vascular smooth muscle cells. Circ Res 105:667–675
66. Mizuno S, Matsumoto K, Li MY, Nakamura T (2005) HGF reduces advancing lung fibrosis in mice: a potential role for MMP-dependent myofibroblast apoptosis. FASEB J 19:580–582
67. Mizuno S, Nakamura T (2012) Improvement of sepsis by hepatocyte growth factor, an anti-inflammatory regulator: emerging insights and therapeutic potential. Gastroenterol Res Pract 2012:909350
68. Yoshitomi Y, Kojima S, Umemoto T et al (1999) Serum hepatocyte growth factor in patients with peripheral arterial occlusive disease. J Clin Endocrinol Metab 84:2425–2428
69. Yoshida S, Matsumoto K, Tomioka D et al (2004) Recombinant hepatocyte growth factor accelerates cutaneous wound healing in a diabetic mouse model. Growth Factors 22:111–119
70. Taniyama Y, Morishita R, Hiraoka K et al (2001) Therapeutic angiogenesis induced by human hepatocyte growth factor gene in rat diabetic hind limb ischemia model: molecular mechanisms of delayed angiogenesis in diabetes. Circulation 104:2344–2350
71. Vajanto I, Rissanen TT, Rutanen J et al (2002) Evaluation of angiogenesis and side effects in ischemic rabbit hindlimbs after intramuscular injection of adenoviral vectors encoding VEGF and LacZ. J Gene Med 4:371–380
72. Tsuzuki N, Miyazawa T, Matsumoto K et al (2001) Hepatocyte growth factor reduces the infarct volume after transient focal cerebral ischemia in rats. Neurol Res 23:417–424
73. Yoshimura S, Morishita R, Hayashi K et al (2002) Gene transfer of hepatocyte growth factor to subarachnoid space in cerebral hypoperfusion model. Hypertension 39:1028–1034
74. Shimamura M, Sato N, Oshima K et al (2004) Novel therapeutic strategy to treat brain ischemia: overexpression of hepatocyte growth factor gene reduced ischemic injury without cerebral edema in rat model. Circulation 109:424–431
75. Date I, Takagi N, Takagi K et al (2004) Hepatocyte growth factor attenuates cerebral ischemia-induced learning dysfunction. Biochem Biophys Res Commun 319:1152–1158
76. Miyazawa T, Matsumoto K, Ohmichi H et al (1998) Protection of hippocampal neurons from ischemia-induced delayed neuronal death by hepatocyte growth factor: a novel neurotrophic factor. J Cereb Blood Flow Metab 18:345–348
77. Takeuchi D, Sato N, Shimamura M et al (2008) Alleviation of Abeta-induced cognitive impairment by ultrasound-mediated gene transfer of HGF in a mouse model. Gene Ther 15:561–571
78. Ono M, Sawa Y, Matsumoto K et al (2002) In vivo gene transfection with hepatocyte growth factor via the pulmonary artery induces angiogenesis in the rat lung. Circulation 106:I264–I269

79. Shigemura N, Sawa Y, Mizuno S et al (2005) Amelioration of pulmonary emphysema by in vivo gene transfection with hepatocyte growth factor in rats. Circulation 111:1407–1414
80. Ishizawa K, Kubo H, Yamada M et al (2004) Hepatocyte growth factor induces angiogenesis in injured lungs through mobilizing endothelial progenitor cells. Biochem Biophys Res Commun 324:276–280
81. Ono M, Sawa Y, Mizuno S et al (2004) Hepatocyte growth factor suppresses vascular medial hyperplasia and matrix accumulation in advanced pulmonary hypertension of rats. Circulation 110:2896–2902
82. Mizuno S, Nakamura T (2011) Hydrodynamics, Chap 6. In: Schulz HA, Anfrade AL, Lobosco RJ (eds) Endocrine delivery system of NK4, an HGF antagonistic and anti-angiogenic regulator, for controlling tumor growth and metastasis, 1st edn. InTECH publisher, Rijeka, Croatia, pp 119–142
83. Kajiya K, Hirakawa S, Ma B et al (2005) Hepatocyte growth factor promotes lymphatic vessel formation and function. EMBO J 24:2885–2895
84. Zhang YW, Su Y, Volpert OV, Vande Woude GF (2003) Hepatocyte growth factor/scatter factor mediates angiogenesis through positive VEGF and negative thrombospondin 1 regulation. Proc Natl Acad Sci USA 100:12718–12723
85. Lee WJ, Chen WK, Wang CJ et al (2008) Apigenin inhibits HGF-promoted invasive growth and metastasis involving blocking PI3K/Akt pathway and beta 4 integrin function in MDA-MB-231 breast cancer cells. Toxicol Appl Pharmacol 226:178–191
86. Tsarfaty G, Stein GY, Moshitch-Moshkovitz S et al (2006) HGF/SF increases tumor blood volume: a novel tool for the in vivo functional molecular imaging of Met. Neoplasia 8:344–352
87. Jiang WG, Martin TA, Matsumoto K et al (1999) Hepatocyte growth factor/scatter factor decreases the expression of occludin and transendothelial resistance (TER) and increases paracellular permeability in human vascular endothelial cells. J Cell Physiol 181:319–329
88. Date K, Matsumoto K, Shimura H et al (1997) HGF/NK4 is a specific antagonist for pleiotrophic actions of hepatocyte growth factor. FEBS Lett 420:1–6
89. Matsumoto K, Kataoka H, Date K, Nakamura T (1998) Cooperative interaction between alpha- and beta-chains of hepatocyte growth factor on c-Met receptor confers ligand-induced receptor tyrosine phosphorylation and multiple biological responses. J Biol Chem 273:22913–22920
90. Laterra J, Nam M, Rosen E et al (1997) Scatter factor/hepatocyte growth factor gene transfer enhances glioma growth and angiogenesis in vivo. Lab Invest 76:565–577
91. Jiang WG, Hiscox SE, Parr C et al (1999) Antagonistic effect of NK4, a novel hepatocyte growth factor variant, on in vitro angiogenesis of human vascular endothelial cells. Clin Cancer Res 5:3695–3703
92. Matsumoto K, Matsumoto K, Nakamura T, Kramer RH (1994) Hepatocyte growth factor/scatter factor induces tyrosine phosphorylation of focal adhesion kinase (p125FAK) and promotes migration and invasion by oral squamous cell carcinoma cells. J Biol Chem 269:31807–31813
93. Nakamura T, Matsumoto K, Kiritoshi A et al (1997) Induction of hepatocyte growth factor in fibroblasts by tumor-derived factors affects invasive growth of tumor cells: in vitro analysis of tumor-stromal interactions. Cancer Res 57:3305–3313
94. Date K, Matsumoto K, Kuba K et al (1998) Inhibition of tumor growth and invasion by a four-kringle antagonist (HGF/NK4) for hepatocyte growth factor. Oncogene 17:3045–3054
95. Tomioka D, Maehara N, Kuba K et al (2001) Inhibition of growth, invasion, and metastasis of human pancreatic carcinoma cells by NK4 in an orthotopic mouse model. Cancer Res 61:7518–7524
96. Wen J, Matsumoto K, Taniura N et al (2004) Hepatic gene expression of NK4, an HGF-antagonist/angiogenesis inhibitor, suppresses liver metastasis and invasive growth of colon cancer in mice. Cancer Gene Ther 11:419–430
97. Suzuki Y, Sakai K, Ueki J et al (2010) Inhibition of Met/HGF receptor and angiogenesis by NK4 leads to suppression of tumor growth and migration in malignant pleural mesothelioma. Int J Cancer 127:1948–1957

98. Matsumoto K, Nakamura T (2005) Mechanisms and significance of bifunctional NK4 in cancer treatment. Biochem Biophys Res Commun 333:316–327
99. Kikuchi T, Maemondo M, Narumi K et al (2002) Tumor suppression induced by intratumor administration of adenovirus vector expressing NK4, a 4-kringle antagonist of hepatocyte growth factor, and naive dendritic cells. Blood 100:3950–3959
100. Hirao S, Yamada Y, Koyama F et al (2002) Tumor suppression effect using NK4, a molecule acting as an antagonist of HGF, on human gastric carcinomas. Cancer Gene Ther 9:700–707
101. Namiki Y, Namiki T, Yoshida H et al (2006) Preclinical study of a "tailor-made" combination of NK4-expressing gene therapy and gefitinib (ZD1839, Iressa) for disseminated peritoneal scirrhous gastric cancer. Int J Cancer 118:1545–1555
102. Brockmann MA, Papadimitriou A, Brandt M et al (2003) Inhibition of intracerebral glioblastoma growth by local treatment with the scatter factor/hepatocyte growth factor-antagonist NK4. Clin Cancer Res 9:4578–4585
103. Heideman DA, Overmeer RM, van Beusechem VW et al (2005) Inhibition of angiogenesis and HGF-cMET-elicited malignant processes in human hepatocellular carcinoma cells using adenoviral vector-mediated NK4 gene therapy. Cancer Gene Ther 12:954–962
104. Du W, Hattori Y, Yamada T et al (2007) NK4, an antagonist of hepatocyte growth factor (HGF), inhibits growth of multiple myeloma cells: molecular targeting of angiogenic growth factor. Blood 109:3042–3049
105. Saga Y, Mizukami H, Suzuki M et al (2001) Expression of HGF/NK4 in ovarian cancer cells suppresses intraperitoneal dissemination and extends host survival. Gene Ther 8:1450–1455
106. Davies G, Mason MD, Martin TA et al (2003) The HGF/SF antagonist NK4 reverses fibroblast- and HGF-induced prostate tumor growth and angiogenesis in vivo. Int J Cancer 106:348–354
107. Sakai K, Nakamura T, Matsumoto K, Nakamura T (2009) Angioinhibitory action of NK4 involves impaired extracellular assembly of fibronectin mediated by perlecan-NK4 association. J Biol Chem 284:22491–22499
108. Siltanen A, Kitabayashi K, Lakkisto P et al (2011) hHGF overexpression in myoblast sheets enhances their angiogenic potential in rat chronicheart failure. PLoS One 6:e19161
109. Buchstein N, Hoffmann D, Smola H et al (2009) Alternative proteolytic processing of hepatocyte growth factor during wound repair. Am J Pathol 174:2116–2128
110. Nayeri F, Strömberg T, Larsson M et al (2002) Hepatocyte growth factor may accelerate healing in chronic leg ulcers: a pilot study. J Dermatolog Treat 13:81–86
111. Shigematsu H, Yasuda K, Iwai T et al (2010) Randomized, double-blind, placebo-controlled clinical trial of hepatocyte growth factor plasmid for critical limb ischemia. Gene Ther 17:1152–1161
112. Makino H, Aoki M, Hashiya N et al (2012) Long-term follow-up evaluation of results from clinical trial using hepatocyte growth factor gene to treat severe peripheral arterial disease. Arterioscler Thromb Vasc Biol 32:2503–2509
113. Yuan B, Zhao Z, Zhang YR et al (2008) Short-term safety and curative effect of recombinant adenovirus carrying hepatocyte growth factor gene on ischemic cardiac disease. In Vivo 22:629–632

Chapter 23
Harnessing Endothelial Progenitor Cells for Therapeutic Angiogenesis

Alan W. Stitt and Reinhold J. Medina

Abstract Tissue ischaemia is associated with a range of important conditions such as occlusive retinopathies, stroke, peripheral artery disease and myocardial infarction. These diseases represent major healthcare issues since they affect millions of individuals each year and impart a considerable economic burden to healthcare systems worldwide. Treating ischaemic organs remains difficult and a worthy goal is to address the underlying vascular insufficiency before it leads to irreversible tissue damage and organ failure. Recently, there has been a growing interest in the therapeutic potential of using a cell-based approach to achieve 'therapeutic angiogenesis' and tissue regeneration. So-called endothelial progenitor cells have clinical potential because they can be harnessed as a cellular therapy for promoting vascular repair. This chapter will examine the clinical potential of several stem and progenitor cells that may be utilised to regulate angiogenesis, regenerate degenerative vasculature and restore blood flow to ischaemic tissues.

Keywords Angiogenesis • Ischaemia • Stem cell • Endothelial progenitor • Cell therapy

23.1 Introduction

Ischaemia is characterised by a restriction in blood flow to tissues and organs, depriving them of vital oxygen and nutrients and capacity for removal of metabolites. Ischaemia-related pathology is central to many diseases and the clinical consequences pose a challenge for healthcare systems. Diseases such as myocardial

A.W. Stitt (✉) • R.J. Medina
Centre for Vision and Vascular Science, School of Medicine, Dentistry, and Biomedical Science, Queen's University Belfast, Royal Victoria Hospital, Belfast BT12 6BA, Northern Ireland, UK
e-mail: a.stitt@qub.ac.uk

J.L. Mehta and N.S. Dhalla (eds.), *Biochemical Basis and Therapeutic Implications of Angiogenesis*, Advances in Biochemistry in Health and Disease 6,

infarction, angina, stroke and ischaemic retinopathies are common and represent a major cause of morbidity and/or mortality worldwide. For example, ischaemic heart disease (IHD) continues to affect millions of individuals each year and is the primary cause of death in the Western world [1].

Treating ischaemic diseases remains a major challenge with the main goal being the restoration of underlying vascular insufficiency before it leads to irreversible tissue damage. Therefore, a range of strategies have been used to promote 'therapeutic angiogenesis' and achieve regeneration of the ischaemic tissues. Such an approach is now commonplace and several strategies have been used to stimulate the growth of new blood vessels and promote regeneration in ischaemic tissues [2]. However, recent clinical trials studying exogenous angiogenic agents such as recombinant proteins VEGF [3], FGF-2 [4] and adenoviral constructs for gene transfer of VEGF165 [5] and FGF-4 [6] have, thus far, only demonstrated partial benefit. Nevertheless, the potential is clear and considerable optimism remains around the therapeutic angiogenesis concept. Recently, there has been a growing interest in the potential of using a cell-based approach to achieve regenerative angiogenesis and treat ischaemic disorders.

23.2 Harnessing Donor Cells to Enhance Therapeutic Angiogenesis

Cell therapy to achieve therapeutic angiogenesis and vascular regeneration offers an exciting new prospect in regenerative medicine. There are already a considerable number of ongoing clinical trials using various cells to treat myocardial ischaemia and peripheral limb ischaemia [7, 8].

The underpinning rationale behind cell therapy to achieve reparative angiogenesis involves delivery of vascular progenitor cells which will undergo endothelial differentiation and ultimately integrate with and repair the damaged resident vasculature. However, a wealth of recent evidence suggests that there are many other mechanisms to explain clinical benefits even when delivered cells do not actually incorporate into resident vasculature. Therapeutic benefit can occur in a number of ways:

1. Transplanted cells integrate completely into the vascular complex differentiating into endothelium, pericytes or vascular smooth muscle.
2. Cells do not integrate but act in a paracrine manner producing growth factors that will promote angiogenesis and inhibit vascular cell apoptosis.
3. Cells secrete cytokines that function as immunomodulatory agents that regulate inflammatory processes to promote tissue repair.
4. Cells can themselves mobilise endogenous progenitor cells and control recruitment to injured tissue.
5. Transplanted cells can interact with damaged host cells through nanotubes or microparticles that will deliver functional organelles, such as mitochondria or specific miRNAs, to promote cell survival and tissue repair.

Taking all these observations into account, many therapies that claim to facilitate repair by the use of stem/progenitor cells should be more precisely referred to as pro-angiogenic/accessory/reparative/immunomodulatory cells since they may not actually differentiate into endothelium or vascular mural cells such as pericytes or smooth muscle. Nevertheless, many of these cell therapy approaches hold promise to reperfuse ischaemic tissues.

23.3 Cells with Pro-Angiogenic Potential

Unspecified bone marrow (BM) cells with pro-angiogenic capacity have also been isolated and used to enhance therapeutic angiogenesis in an autologous treatment regime [9]. For example, it has been demonstrated in animal models that BM cell delivery to ischaemic tissues significantly promote reparative angiogenesis [10], and subsequent clinical trials in patients with heart disease [10, 11] and peripheral artery disease [12, 13] have suggested this therapy to be feasible and safe. However, recent meta-analysis of these clinical trials suggests only modest improvements for patients with ischaemic limbs or myocardial infarction [8]. This is most likely related to the nondescript, heterogenous nature of the cells in these BM fractions.

A diverse range of cell types have been tested in preclinical models for therapeutic angiogenesis. These include mesenchymal stem cells (MSCs) [14, 15], endothelial progenitor cells (EPCs) [16], embryonic stem cells (ESCs) [17] and induced pluripotent stem cells (iPSCs) [18].

23.4 Endothelial Progenitor Cells

EPCs are circulating cells with considerable diagnostic and therapeutic value. These cells represent a minor population (~0.05 cells/mL of blood) relative to the cell populations in whole blood, but they have been shown to play a major role in therapeutic angiogenesis and vascular repair [19]. EPCs have potential as circulating biomarkers for disease and to be harnessed for their ability to home to areas of vascular insufficiency and to participate in endothelial repair and therapeutic angiogenesis. Interest in EPCs started with the suggestion that angioblasts and haematopoietic stem cells (HSCs) originate from a common precursor [20]. These cells were isolated using $CD34^+$ and in subsequent culture, they expressed endothelial markers such as CD31, E-selectin and Tie-2, showed uptake of acetylated LDL (AcLDL) and were active at sites of angiogenesis in vivo [20]. Subsequent reports also demonstrated circulating bone marrow-derived EPCs from adult peripheral blood, and this suggested that subgroups of cells circulate in adult blood and participate in vascular repair and neovascularisation.

It is important to underline that there has been considerable controversy in the EPC field concerning the precise nature of these cells. Preclinical and clinical investigations evaluating the cell therapy potential of EPCs have yielded inconsistent and contradictory outcomes [16, 21]. To a large extent, this is due to the fact that EPCs lack a uniform definition and there is no definitive markers used to isolate these cells [22].

23.4.1 Isolation and Characterisation of EPCs

EPCs have been obtained using either cell sorting technology or in vitro isolation of blood mononuclear cells using specific substrates and media. The sorting approach relies on the type and number of surface markers used. Since there is still debate about the most appropriate markers that define an EPC, ambiguity is inevitable [23]. For example, some researchers have used CD34, VEGFR2 and CD133 to separate cells from peripheral blood, but it remains uncertain if this approach isolates EPCs or if it simply enriches for haematopoietic progenitors [24]. Some groups have attempted to enrich for EPCs by sorting lineage-negative haematopoietic cells or negative selection based on CD45 to exclude differentiated blood cells although such populations are mainly composed of myeloid progenitor cells. Nevertheless, umbilical cord blood-derived $CD34^{+}/CD45^{-}$ cells have been reported to be enriched for endothelial cells that display high proliferative potential akin to EPCs [25]. Irrespective of the cell surface markers used to isolate putative EPCs, sorting remains a challenging approach because of the paucity of EPCs circulating in blood and because various markers overlap between different populations.

EPC isolation from adult peripheral and umbilical cord blood using initial propagation of heterogeneous cells in vitro has provided more homogeneity. This approach has isolated several distinct EPC populations with different angiogenic properties [19, 26]. For example, one cell type appears after a few days in culture and has been referred to as early EPCs (eEPCs). Some subsequent modifications in the EPC isolation protocol by Hill gave rise to the descriptor 'colony-forming unit-Hill' cells (CFU-Hill). Recently, new names for the eEPC subtype have been suggested which reflect their phenotype; these include pro-angiogenic monocytes, vascular accessory cells, pro-angiogenic haematopoietic cells and circulating angiogenic cells (CACs) myeloid angiogenic cells (MACs). The other EPC type, appearing much later, is called outgrowth endothelial cells (OECs) or endothelial colony-forming cells (ECFCs) [19, 27]. There are major differences between these cell groups (although they are still referred to as 'EPCs'), and they may influence angiogenesis by the release of paracrine factors or differentiate into mature endothelium and integrate into the vessel lumen.

23.5 Nonintegrating Progenitors with Pro-angiogenic Properties

The eEPC population is not endothelial in origin [28]. The confusion has arisen from the fact that this subset of circulating cells express typical endothelial markers such as CD31, VEGFR2 and Tie2, although these markers are not necessarily endothelial specific and mononuclear cells in culture may 'acquire' these markers

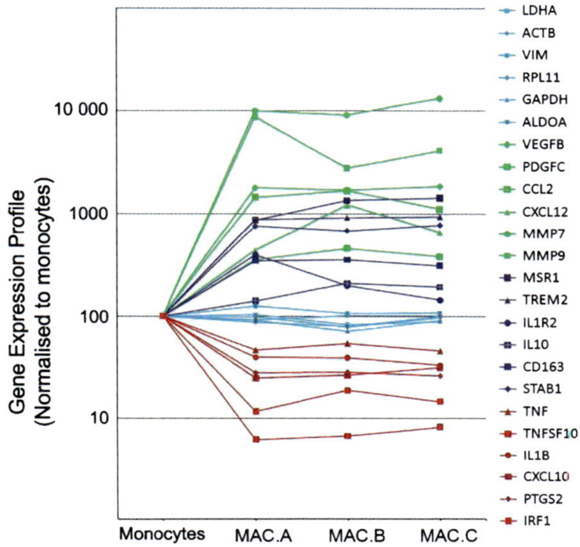

Fig. 23.1 Transcriptomics demonstrates that MACs are similar to M2-activated macrophages. The MAC gene signature is highly enriched for M2 markers, while expression of M1 macrophage markers is low. Pro-angiogenic genes were also highly expressed. This figure shows a subset of expressed transcripts from human MACS (from three different donors) and normalised to monocytes. There is upregulation of pro-angiogenic genes (*green*) and M2 markers (*purple*) and downregulation of M1 markers (*red*) when compared to housekeeping genes (*blue*)

through platelet microparticle uptake [29]. The true nature of eEPCs is revealed by the fact that they express haematopoietic markers CD14 and CD45 and have low proliferative potential. Moreover, eEPCs are distinct from endothelial cells, in terms of their gene expression, proteomic profile and ultrastructure [28]. Their molecular fingerprint is that of haematopoietic cells with a profile akin to pro-angiogenic M2 macrophages [30] (Fig. 23.1). Indeed, it has been suggested that eEPCs should be termed myeloid angiogenic cells (MACs) which befits their true identity.

MACs have a major influence on angiogenesis by the release of paracrine cytokines and growth factors [30, 31]. These cells have been shown to home to sites of ischaemia and stimulate regeneration of existing vasculature through the secretion of pro-angiogenic cytokines and growth factors such as IL-8, HGF, IGF and GCSF [30, 32]. While these cells do not incorporate directly in the vasculature, they remain proximal or loosely attached to the damaged tissue in a perivascular position (Fig. 23.2). In this manner, MACs have been shown to facilitate therapeutic angiogenesis and promote reperfusion in critical limb ischaemia, ischaemic retinopathy and after myocardial infarction [30, 33, 34]. However, it must be noted that as these cells are M2 macrophage-like cells [30], they are likely to be highly plastic, and therefore, in the presence of an inflammatory or hypoxic tissue environment, they could serve to enhance inflammation as M1 macrophages [35]. This remains a valid concern when using these cells in a clinical context.

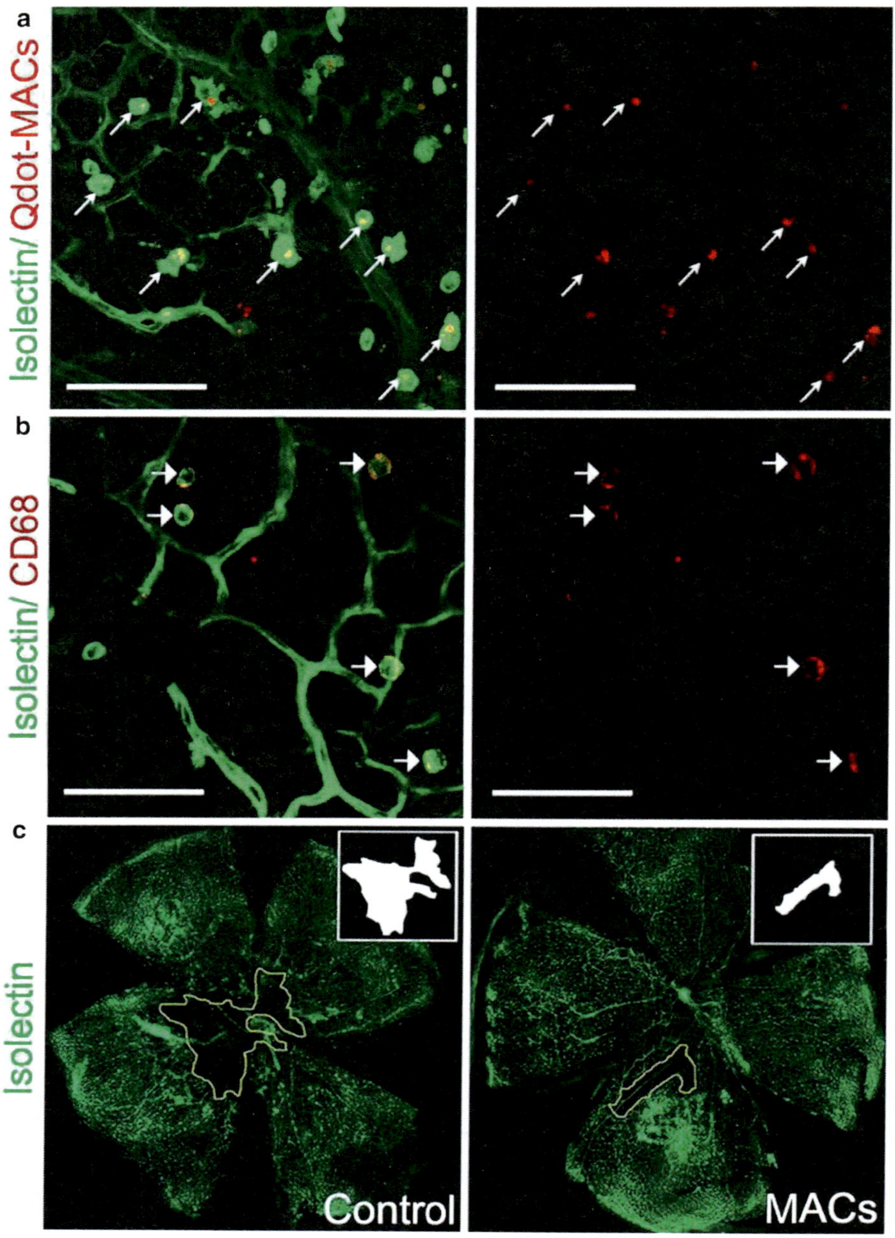

Fig. 23.2 MACs do not integrate into the resident vasculature but can promote vascular repair in the ischaemic retina. C57BL/6 mice were subjected to oxygen-induced retinopathy. MACs (or vehicle control) were intra-vitreally delivered to the ischaemic retina. (**a**) Injected MACs labelled with red Qdots did not incorporate into the resident vascular network labelled in *green* with isolectin. Scale bars: 100 μm. (**b**) Injected human MACs retained a myeloid phenotype as demonstrated

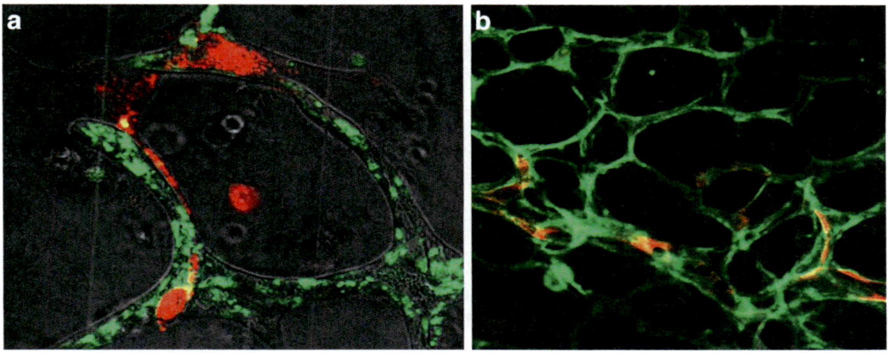

Fig. 23.3 OECs integrate into the vasculature. (**a**) In an in vitro assay of tubule formation, OECs and mature capillary endothelium were co-cultured. The red-labelled OECs and green-labelled endothelial cells integrate with each other in the vascular tube forming process. (**b**) Using a human cell-specific marker (CD105), there is clear integration of human blood-derived OECs into resident murine vasculature (stained *green* using isolectin)

23.6 Integrating Progenitor Cells

OECs/ECFCs probably represent *bona fide* endothelial progenitor cells [36]. From a molecular viewpoint, this is demonstrated by genome-wide transcriptomics, proteomics and ultrastructural evaluation [28]. Functionally, OEC/ECFCs display endothelial characteristics with inherent angiogenic potential based on integration into pre-existing vessels or de novo tube formation (Fig. 23.3) [37, 38]. OEC/ECFCs directly participate in angiogenesis in ischaemic tissue where they promote vascular repair (Fig. 23.4) [26, 37, 39]. While OEC/ECFCs may also influence resident vessels by releasing paracrine factors [40], they unequivocally belong to the endothelial lineage and consistently associate with endothelial monolayers and tubes in vitro and homing to sites of ischaemia and incorporating into mature blood vessels in vivo [39]. This strongly indicates that OECs possess intrinsic vasoreparative and proangiogenic properties, and transplanted cells can produce a significant decrease in ischaemia and a concomitant increase in normal vasculature (Fig. 23.4).

Persistence with the resident vasculature has been demonstrated when OEC/ECFCs were injected into the systemic circulation of non-obese diabetic (NOD)/severe combined immunodeficient (SCID) mice. The donor cells incorporated and survived in nine different vascular beds for up to 7 months after intravenous tail vein

Fig. 23.2 (continued) by expression of human-specific CD68 and typical amoeboid morphology. Scale bars: 75 μm. (**c**) Representative flat-mounted retinas injected with vehicle (Control) or MACs, respectively. Isolectin staining in *green* identifies retinal vasculature, and avascular regions outlined in yellow. Insets show ischaemic areas in *white*. (**d**) Quantification of avascular areas and ratio of avascular/total retinal area shows MACs significantly induced vascular repair when compared to vehicle injected eyes

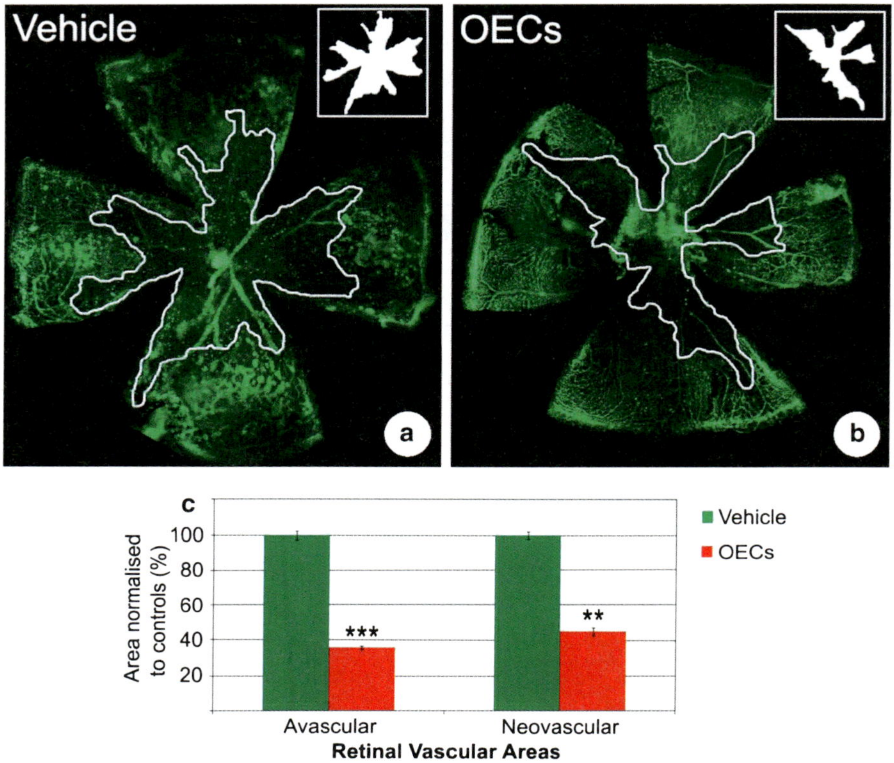

Fig. 23.4 OECs can promote revascularisation of the ischaemic retina and reduce pre-retinal neovascularisation. (**a, b**) Representative flat-mounted retinas of C57BL/6 mice subjected to oxygen-induced retinopathy. OECs or vehicle was intra-vitreally delivered to the ischaemic retina. Postinjection, lectin staining in *green* identifies retinal vasculature, and the avascular regions are surrounded with a *white line* (*insets* show the areas of ischaemia). (**c**) Quantification of avascular areas and extent of pre-retinal neovascularisation is shown. $^{***}p<0.001$, $^{**}p<0.01$ when comparing vehicle-treated against OEC-treated eyes

injection without inducing thrombosis or infarcts [41]. This highlights the potential benefits of a novel cytotherapy using a well-defined population of EPCs for patients with ischaemia-related pathology.

It should be noted that OEC/ECFCs are relatively difficult to isolate using culture methodologies, and they can take up to a month to appear in vitro. However, once isolated, OEC/ECFCs colonies can be expanded to yield a pure population of cells with a high proliferation rate. For example, some umbilical cord blood-derived colonies have been expanded up to 80 population doublings (PDLs) in just 90 days. Although many reports highlight the remarkable proliferative capacity of OEC/ECFCs [19, 37], the implications for their long-term expansion remain largely unknown.

23.7 EPCs as Disease Biomarkers

EPCs in blood vary with disease, and these cells have been suggested to be useful as diagnostic/prognostic biomarkers for cardiovascular dysfunction [10] and cancer progression [42]. Numbers of circulating EPCs are reduced and show depressed function in patients who smoke tobacco or have diabetes, hypercholesterolemia or hypertension [43–45]. Functionally, EPCs isolated from diabetic donors show significantly impaired vasoreparative potential [46] and are susceptible to premature senescence [47]. Type 1 diabetic patients may demonstrate almost 50% less circulating EPCs when compared to age-matched nondiabetic controls [48], and these levels are negatively associated with severity of vascular complications [49]. While they may vary with disease, the concept is further complicated by the fact that normal physiological function influence blood EPCs. For example, chronological age [50], gender [51], exercise [52] and diurnal rhythm [53] may all influence EPC function.

23.8 Cell Therapy for Vascular Mural Cell Regeneration

The differentiation potential of mesenchymal stem cells (MSCs) into mural cells such as smooth muscle cells (SMCs) and pericytes has been reported [54, 55]. Mural cells play a key role in the context of vascular regeneration by providing structural support to the vasculature and regulating new blood vessel formation, maturation and stabilisation. It has been suggested that building a new blood vessel requires the interaction of both endothelial cells and mural cells.

For example, an exciting new approach is to promote repair by using human pluripotent stem cells (hPSCs) which differentiate to mesodermal precursors and can then be expanded to produce cells with a pericyte phenotype expressing CD146/NG2/PDGFR-β [56]. Implantation of these cells into a murine model of ischaemic limb injury induces vascular regeneration independent of EPCs [56]. Furthermore, it has also been shown that human arterial adventitia-derived $CD44^+$ multipotent stem cells can differentiate into pericytes [57], while others have suggested that human saphenous vein-derived pericyte precursors have prolonged therapeutic benefit in a model of myocardial infarction [58]. However, methods for isolating these cells are highly variable and lead to different populations. At present, there is no combination of markers that permits identification and isolation of equivalent MSCs from mouse, rat, rabbit and human, much less a population of cells with pericyte differentiation potential, and this remains an important area for continued research.

It is possible that a mixed transplantation of EPCs and mesenchymal progenitors could offer a beneficial synergistic effect on angiogenesis and vessel stabilisation in [59, 60]. Different EPC types demonstrate distinct functions regarding revascularisation of ischaemic tissues. It has been demonstrated that a mixed transplantation of eEPCs and OECs/ECFCs results in superior neovascularisation in vivo than any single cell type transplantation [61].

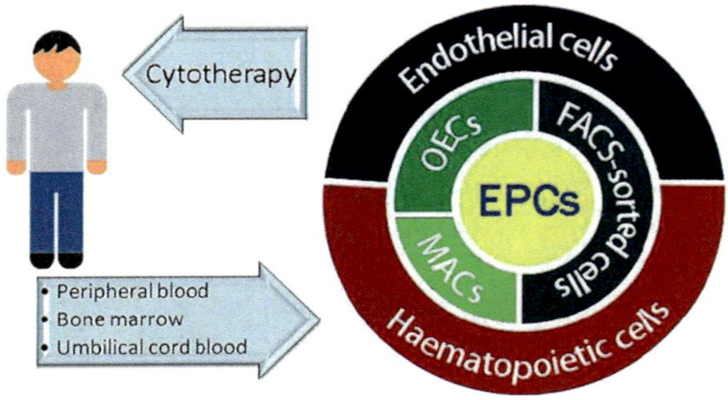

Fig. 23.5 Different EPC subsets with cell therapy potential. Endothelial progenitors can be isolated by sorting techniques using specific antibodies such as CD34 or by in vitro cell culture. EPC cellular sources are human peripheral blood, bone marrow and umbilical cord blood. Two distinct populations have been clearly defined when using cell culture methodology: MACs of haematopoietic nature and OECs with intrinsic endothelial phenotype

23.9 Conclusion

There have been many important advances in the field of therapeutic angiogenesis. However, a strategy for using delivery of defined cytokines and growth factors to diseased tissue has many limitations, and there is increasing recognition that tissue regeneration could be significantly enhanced if accompanied with a cell-based approach [62]. Cell therapy to revascularise ischaemic tissues is an exciting new prospect in regenerative medicine [63], and considerable effort has been put into trying to identify 'the best cell' for reparative angiogenesis and tissue regeneration (Fig. 23.5). Among the many candidates are BM mononucleated cells, monocytes, EPCs and a great variety of other stem/progenitor cells. However, the delivery of the correct cell type to the precise area of injury or vascular insufficiency is difficult and many factors need to be considered:

1. *Cell choice*: This is a critical aspect of any cell therapy. The correct cell must be chosen for its phenotype, cell characteristics and biological functions.
2. *Efficacy*: Cells for vascular therapy must be able to home to ischaemic or damaged tissue and engage in vessel formation alone or in unison with resident vasculature to achieve a controlled and functional reperfusion event, without causing pathological angiogenesis (e.g. proliferative retinopathy in the eye).
3. *Timing of delivery and cell number*: Delivering cells to promote angiogenesis and revascularise ischaemic tissue may have a defined therapeutic window in order to avoid extensive tissue damage, fibrosis and necrosis. The evaluation of the most suitable timing of cell delivery as well as the number of cells needed to integrate into resident vasculature and promote revascularisation of specific tissues needs careful optimisation and evaluation.

4. *Administration*: Studies using vascular progenitor cells have shown that a local delivery results in increased homing as the cells are directly delivered to the ischaemic area or tissue environment that is experiencing the disease [64]. A systemic delivery strategy is based on the capacity of the cells to be mobilised and directed via chemokines to the ischaemic area; however, the drawback of this approach is that this may result in cells localising to nontarget organs such as the liver, kidneys, spleen and lung.

With a focus on EPCs, in particular, clinical trials are underway although the precise definition of these cells and their biological properties remains an important hurdle. We foresee even more developments in the near future, as basic and clinical science continues to reveal new information about EPCs and their therapeutic potential for enhancing reparative angiogenesis and tissue regeneration in ischaemic disease.

References

1. Joggerst SJ, Hatzopoulos AK (2009) Stem cell therapy for cardiac repair: benefits and barriers. Expert Rev Mol Med 11:e20
2. Carmeliet P (2004) Manipulating angiogenesis in medicine. J Intern Med 255:538–561
3. Henry TD, Annex BH, McKendall GR et al (2003) The VIVA trial: vascular endothelial growth factor in ischemia for vascular angiogenesis. Circulation 107:1359–1365
4. Simons M, Annex BH, Laham RJ et al (2002) Pharmacological treatment of coronary artery disease with recombinant fibroblast growth factor-2: double-blind, randomized, controlled clinical trial. Circulation 105:788–793
5. Makinen K, Manninen H, Hedman M et al (2002) Increased vascularity detected by digital subtraction angiography after VEGF gene transfer to human lower limb artery: a randomized, placebo-controlled, double-blinded phase II study. Mol Ther 6:127–133
6. Grines CL, Watkins MW, Helmer G et al (2002) Angiogenic Gene Therapy (AGENT) trial in patients with stable angina pectoris. Circulation 105:1291–1297
7. Tongers J, Losordo DW, Landmesser U (2011) Stem and progenitor cell-based therapy in ischaemic heart disease: promise, uncertainties, and challenges. Eur Heart J 32:1197–1206
8. Fadini GP, Agostini C, Avogaro A (2010) Autologous stem cell therapy for peripheral arterial disease meta-analysis and systematic review of the literature. Atherosclerosis 209:10–17
9. Beeres SL, Atsma DE, van Ramshorst J et al (2008) Cell therapy for ischaemic heart disease. Heart 94:1214–1226
10. Sekiguchi H, Ii M, Losordo DW (2009) The relative potency and safety of endothelial progenitor cells and unselected mononuclear cells for recovery from myocardial infarction and ischemia. J Cell Physiol 219:235–242
11. Hansson EM, Lindsay ME, Chien KR (2009) Regeneration next: toward heart stem cell therapeutics. Cell Stem Cell 5:364–377
12. Lasala GP, Silva JA, Gardner PA et al (2010) Combination stem cell therapy for the treatment of severe limb ischemia: safety and efficacy analysis. Angiology 61:551–556
13. Tateishi-Yuyama E, Matsubara H, Murohara T et al (2002) Therapeutic angiogenesis for patients with limb ischaemia by autologous transplantation of bone-marrow cells: a pilot study and a randomised controlled trial. Lancet 360:427–435
14. Toma C, Pittenger MF, Cahill KS et al (2002) Human mesenchymal stem cells differentiate to a cardiomyocyte phenotype in the adult murine heart. Circulation 105:93–98

15. Aranguren XL, McCue JD, Hendrickx B et al (2008) Multipotent adult progenitor cells sustain function of ischemic limbs in mice. J Clin Invest 118:505–514
16. Kalka C, Masuda H, Takahashi T et al (2000) Transplantation of ex vivo expanded endothelial progenitor cells for therapeutic neovascularization. Proc Natl Acad Sci USA 97:3422–3427
17. Caspi O, Huber I, Kehat I et al (2007) Transplantation of human embryonic stem cell-derived cardiomyocytes improves myocardial performance in infarcted rat hearts. J Am Coll Cardiol 50:1884–1893
18. Rufaihah AJ, Huang NF, Jame S et al (2011) Endothelial cells derived from human iPSCS increase capillary density and improve perfusion in a mouse model of peripheral arterial disease. Arterioscler Thromb Vasc Biol 31:e72–e79
19. Yoder MC, Mead LE, Prater D et al (2007) Redefining endothelial progenitor cells via clonal analysis and hematopoietic stem/progenitor cell principals. Blood 109:1801–1809
20. Asahara T, Murohara T, Sullivan A et al (1997) Isolation of putative progenitor endothelial cells for angiogenesis. Science 275:964–967
21. Ziegelhoeffer T, Fernandez B, Kostin S et al (2004) Bone marrow-derived cells do not incorporate into the adult growing vasculature. Circ Res 94:230–238
22. Ingram DA, Caplice NM, Yoder MC (2005) Unresolved questions, changing definitions, and novel paradigms for defining endothelial progenitor cells. Blood 106:1525–1531
23. Van Craenenbroeck EM, Conraads VM, Van Bockstaele DR et al (2008) Quantification of circulating endothelial progenitor cells: a methodological comparison of six flow cytometric approaches. J Immunol Methods 332:31–40
24. Yoder MC (2009) Defining human endothelial progenitor cells. J Thromb Haemost 7(Suppl 1): 49–52
25. Hirschi KK, Ingram DA, Yoder MC (2008) Assessing identity, phenotype, and fate of endothelial progenitor cells. Arterioscler Thromb Vasc Biol 28:1584–1595
26. Hur J, Yoon CH, Kim HS et al (2004) Characterization of two types of endothelial progenitor cells and their different contributions to neovasculogenesis. Arterioscler Thromb Vasc Biol 24:288–293
27. Stitt AW, O'Neill CL, O'Doherty MT et al (2011) Vascular stem cells and ischaemic retinopathies. Prog Retin Eye Res 30:149–166
28. Medina RJ, O'Neill CL, Sweeney M et al (2010) Molecular analysis of endothelial progenitor cell (EPC) subtypes reveals two distinct cell populations with different identities. BMC Med Genomics 3:18
29. Prokopi M, Pula G, Mayr U et al (2009) Proteomic analysis reveals presence of platelet microparticles in endothelial progenitor cell cultures. Blood 114:723–732
30. Medina RJ, O'Neill CL, O'Doherty TM et al (2011) Myeloid angiogenic cells act as alternative M2 macrophages and modulate angiogenesis through interleukin-8. Mol Med 17:1045–1455
31. Sieveking DP, Buckle A, Celermajer DS et al (2008) Strikingly different angiogenic properties of endothelial progenitor cell subpopulations: insights from a novel human angiogenesis assay. J Am Coll Cardiol 51:660–668
32. Urbich C, Aicher A, Heeschen C et al (2005) Soluble factors released by endothelial progenitor cells promote migration of endothelial cells and cardiac resident progenitor cells. J Mol Cell Cardiol 39:733–742
33. Urbich C, Heeschen C, Aicher A et al (2003) Relevance of monocytic features for neovascularization capacity of circulating endothelial progenitor cells. Circulation 108:2511–2516
34. Assmus B, Schachinger V, Teupe C et al (2002) Transplantation of progenitor cells and regeneration enhancement in acute myocardial infarction (TOPCARE-AMI). Circulation 106:3009–3017
35. Awad O, Jiao C, Ma N et al (2005) Obese diabetic mouse environment differentially affects primitive and monocytic endothelial cell progenitors. Stem Cells 23:575–583
36. Lin Y, Weisdorf DJ, Solovey A et al (2000) Origins of circulating endothelial cells and endothelial outgrowth from blood. J Clin Invest 105:71–77
37. Reinisch A, Hofmann NA, Obenauf AC et al (2009) Humanized large-scale expanded endothelial colony-forming cells function in vitro and in vivo. Blood 113:6716–6725

38. Ingram DA, Mead LE, Tanaka H et al (2004) Identification of a novel hierarchy of endothelial progenitor cells using human peripheral and umbilical cord blood. Blood 104:2752–2760
39. Medina R, O'Neill CL, Humphreys MW et al (2010) Outgrowth endothelial cells: characterisation and their potential for reversing ischaemic retinopathy. Invest Ophthalmol Vis Sci 51:5906–5913
40. He T, Lu T (2008) d'Uscio LV, et al., Angiogenic function of prostacyclin biosynthesis in human endothelial progenitor cells. Circ Res 103:80–88
41. Milbauer LC, Enenstein JA, Roney M et al (2009) Blood outgrowth endothelial cell migration and trapping in vivo: a window into gene therapy. Transl Res 153:179–189
42. Strijbos MH, Gratama JW, Kraan J et al (2008) Circulating endothelial cells in oncology: pitfalls and promises. Br J Cancer 98:1731–1735
43. Besler C, Doerries C, Giannotti G et al (2008) Pharmacological approaches to improve endothelial repair mechanisms. Expert Rev Cardiovasc Ther 6:1071–1082
44. Mohler ER 3rd, Shi Y, Moore J et al (2009) Diabetes reduces bone marrow and circulating porcine endothelial progenitor cells, an effect ameliorated by atorvastatin and independent of cholesterol. Cytometry A 75:75–82
45. Vasa M, Fichtlscherer S, Aicher A et al (2001) Number and migratory activity of circulating endothelial progenitor cells inversely correlate with risk factors for coronary artery disease. Circ Res 89:E1–E7
46. Tepper OM, Galiano RD, Capla JM et al (2002) Human endothelial progenitor cells from type II diabetics exhibit impaired proliferation, adhesion, and incorporation into vascular structures. Circulation 106(22):2781–2786
47. Ingram DA, Lien IZ, Mead LE et al (2008) In vitro hyperglycemia or a diabetic intrauterine environment reduces neonatal endothelial colony-forming cell numbers and function. Diabetes 57:724–731
48. Loomans CJ, de Koning EJ, Staal FJ et al (2004) Endothelial progenitor cell dysfunction: a novel concept in the pathogenesis of vascular complications of type 1 diabetes. Diabetes 53:195–199
49. Egan CG, Lavery R, Caporali F et al (2008) Generalised reduction of putative endothelial progenitors and CXCR4-positive peripheral blood cells in type 2 diabetes. Diabetologia 51:1296–1305
50. Jie KE, Goossens MH, van Oostrom O et al (2009) Circulating endothelial progenitor cell levels are higher during childhood than in adult life. Atherosclerosis 202:345–347
51. Masuda H, Kalka C, Takahashi T et al (2007) Estrogen-mediated endothelial progenitor cell biology and kinetics for physiological postnatal vasculogenesis. Circ Res 101:598–606
52. Hoetzer GL, Van Guilder GP, Irmiger HM et al (2007) Aging, exercise, and endothelial progenitor cell clonogenic and migratory capacity in men. J Appl Physiol 102:847–852
53. Busik JV, Tikhonenko M, Bhatwadekar A et al (2009) Diabetic retinopathy is associated with bone marrow neuropathy and a depressed peripheral clock. J Exp Med 206:2897–2906
54. Quevedo HC, Hatzistergos KE, Oskouei BN et al (2009) Allogeneic mesenchymal stem cells restore cardiac function in chronic ischemic cardiomyopathy via trilineage differentiating capacity. Proc Natl Acad Sci USA 106:14022–14027
55. Crisan M, Chen CW, Corselli M et al (2009) Perivascular multipotent progenitor cells in human organs. Ann N Y Acad Sci 1176:118–123
56. Dar A, Domev H, Ben-Yosef O et al (2012) Multipotent Vasculogenic Pericytes from Human Pluripotent Stem Cells Promote Recovery of Murine Ischemic Limb. Circulation 125:87–99
57. Klein D, Weisshardt P, Kleff V et al (2011) Vascular wall-resident $CD44^{+}$ multipotent stem cells give rise to pericytes and smooth muscle cells and contribute to new vessel maturation. PLoS One 6:e20540
58. Katare R, Riu F, Mitchell K et al (2011) Transplantation of human pericyte progenitor cells improves the repair of infarcted heart through activation of an angiogenic program involving micro-RNA-132. Circ Res 109:894–906
59. Campagnolo P, Cesselli D (2010) Al Haj Zen A, et al., Human adult vena saphena contains perivascular progenitor cells endowed with clonogenic and proangiogenic potential. Circulation 121:1735–1745

60. Melero-Martin JM, De Obaldia ME, Kang SY et al (2008) Engineering robust and functional vascular networks in vivo with human adult and cord blood-derived progenitor cells. Circ Res 103:194–202
61. Yoon CH, Hur J, Park KW et al (2005) Synergistic neovascularization by mixed transplantation of early endothelial progenitor cells and late outgrowth endothelial cells: the role of angiogenic cytokines and matrix metalloproteinases. Circulation 112:1618–1627
62. Madeddu P (2005) Therapeutic angiogenesis and vasculogenesis for tissue regeneration. Exp Physiol 90:315–326
63. Gulati R, Simari RD (2007) Cell therapy for acute myocardial infarction. Med Clin North Am 91(xiii):769–785
64. Aicher A, Brenner W, Zuhayra M et al (2003) Assessment of the tissue distribution of transplanted human endothelial progenitor cells by radioactive labeling. Circulation 107:2134–2139

Chapter 24
Microarray, MicroRNA, and Angiogenesis

Zufeng Ding, Xianwei Wang, and Jawahar L. Mehta

Abstract Angiogenesis is defined as the formation and the growth of new blood vessels that sprout from existing vascular network. Angiogenesis plays a very important role in the physiological and pathological situations such as ischemia, atherosclerosis, wound healing, and cancer growth and metastasis. Microarray is a new molecular biological technique that can analyze expression of hundreds to thousands of genes or proteins simultaneously in a single experiment. Microarray includes DNA microarray, protein microarray, tissue microarray, and cell microarray. Many different types of microarray chips are now commercially available, some designed specifically for angiogenesis. MicroRNAs (miRNAs) are endogenous, short, noncoding RNAs found in eukaryotic cells. MiRNAs are posttranscriptional regulators that negatively regulate gene expression by binding to their target messenger RNAs for degradation and/or translational repression. The main function of miRNAs is gene regulation. MiRNAs have been found to modulate many pathophysiological process including cell differentiation, contraction, migration, proliferation, apoptosis, and tissue inflammation. There are more than 1,000 miRNAs in human genome, some of them are involved in angiogenesis. In this review, we will summarize the application of microarray assay and function of miRNAs in angiogenesis.

Keywords Microarray • MicroRNA • Angiogenesis • Vascular endothelial growth factor

---

Zufeng Ding and Xianwei Wang contributed equally to this work

Z. Ding • J.L. Mehta • X. Wang (✉)
Division of Cardiovascular Medicine, University of Arkansas for Medical Sciences
and the Central Arkansas Veterans Healthcare System, 4301 West Markham,
Little Rock, AR 72205, USA
e-mail: XWang2@uams.edu

J.L. Mehta and N.S. Dhalla (eds.), *Biochemical Basis and Therapeutic Implications of Angiogenesis*, Advances in Biochemistry in Health and Disease 6,
DOI 10.1007/978-1-4614-5857-9_24,

Abbreviations

Ang	Angiotensin
bFGF	Basic fibroblast growth factor
CTGF	Connective tissue growth factor
ECs	Endothelial cells
EGF	Endothelial growth factor
EGFL7	Epidermal growth factor-like domain 7
eNOS	Endothelial nitric oxide synthase
FLT1	FMS-related tyrosine kinase 1
HGS	Hepatocyte growth factor-regulated tyrosine kinase substrate
HIF-1	Hypoxia-inducible factor-1
IFN-γ	Interferon γ
IGF-1	Insulin-like growth factor 1
MCP-1	Monocyte chemoattractant protein 1
MMPs	Matrix metalloproteinases
MRE	MicroRNA responsive element
PDGF	Platelet-derived growth factor
PIK3R2	Phosphoinositol-3 kinase regulatory subunit 2
PGF	Placental growth factor
PE	Preeclamptic
RCC	Renal cell carcinoma
RL	Renilla luciferase
ROS	Reactive oxygen species
SCF	Stem cell factor
SGA	Small-for-gestational-age
TGF-ß1	Transforming growth factor ß1
TIMP	Tissue inhibitor of metalloproteinase
VEGF	Vascular endothelial growth factor

24.1 Introduction

Angiogenesis is defined as the formation of new blood vessels that sprout from existing vascular network. It is a complex process that involves differentiation, proliferation, migration, and maturation of endothelial cells [1]. Angiogenesis is an important phenomenon not only in physiological situations but also in pathological conditions like myocardial, cerebrovascular, renal, and limb ischemia [2]. As described elsewhere in this book, angiogenesis plays a central role in the development of certain cancers and cancer metastasis. The basic fibroblast growth factor (bFGF) was the first growth factor identified to play an important role in angiogenesis, followed by vascular endothelial growth factor (VEGF), platelet-derived growth factor (PDGF), tyrosine kinase receptors Tie-1 and Tie-2, and the Tie-2 angiopoietin ligands (Ang-1 and Ang-2) [3, 4].

Endothelial migration and proliferation enhance the generation of primary capillaries that undergo remodeling by sprouting, branching, or intussusception [5]. There is increasing evidence to support the concept that angiogenesis participates in the progression of atherosclerotic plaque [6]. Enhanced angiogenesis is a major cause of tumor growth and progression defining an angiogenic switch [7]. The development of plaque angiogenesis in atherosclerosis is regulated by multiple signals such as hypoxia, reactive oxygen species (ROS), and inflammation, which are also closely associated with the development of certain cancers [2, 4].

Microarray is a high-throughput technique that provides information on hundreds to thousands of genes or proteins simultaneously in a single experiment. This technique has been widely used in disease diagnosis, drug evaluation, and toxicological research. Microarray can also provide information on gene (or protein) profile and novel pathways in the development of a host of diseases.

MicroRNAs (miRNAs or miRs) are endogenous, small, noncoding RNAs that negatively regulate gene expression by binding to their target messenger RNAs for degradation and/or translational repression [8]. MiRNAs are critical modulators for vascular functions such as cell differentiation, contraction, migration, proliferation, and apoptosis [9]. MiRNAs have been associated with inflammation, oxidative stress, and angiogenesis. Several miRNAs are involved in vascular function, and some of them, such as miR-16, -17, -92, -296, -130a, -378, -210, -214, -221, -222, -424 Let7f, and -126, have been identified to participate in angiogenesis [10–14].

24.2 Microarray and Angiogenesis

24.2.1 Status of Microarray Assay in Angiogenesis

Microarray assay is a high-throughput technique that facilitates screening of genomic genes and proteins and to identify specific genes and proteins. Currently, this assay is widely used as a common molecular biological technique to systematically analyze gene or protein expression in a variety of human diseases [15]. Microarray includes DNA microarray, protein microarray, tissue microarray, or cell microarray. The most commonly used microarrays in the study of angiogenesis are DNA microarray, protein microarray, and tissue microarray.

24.2.2 DNA Microarray and Angiogenesis

DNA microarray is the quickest method to analyze expression of multiple genes (hundreds to thousands genes in a single experiment). It has widely been used to identify genetic basis of a variety of complex polygenic disease states, including

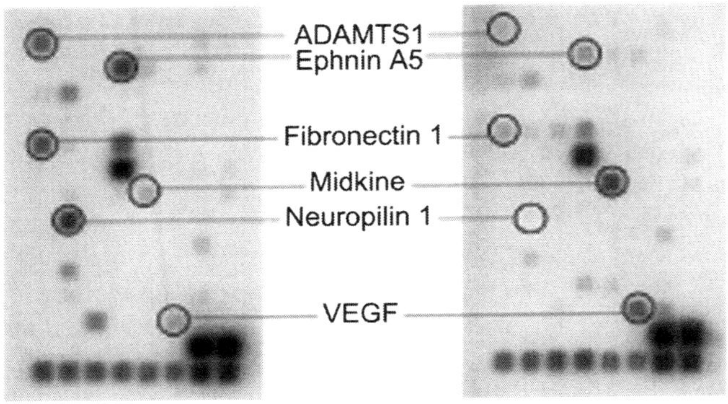

Fig. 24.1 A gene microarray shows angiogenesis-related genes in LNCaP and LNCaP-19 cancer cell lines (Gustavsson H, Jennbacken K, Welén K, et al. Altered expression of genes regulating angiogenesis in experimental androgen-independent prostate cancer. Prostate. 2008, 68:161–170)

cardiovascular diseases and cancers. There are many DNA microarray chips available commercially; some of them are specific for study of angiogenesis (Fig. 24.1). Currently, angiogenesis DNA microarray analysis is also used to evaluate the progression of many disease states as a reference.

24.2.2.1 DNA Microarray, Angiogenesis, and Cardiovascular Diseases

DNA microarray is a novel research tool for cardiovascular scientists and physicians [16]. In 1995, Schena et al. firstly used DNA microarray to quantity gene expression [17]. Subsequently, it was used to study several cardiovascular disease states, including myocardial infarction, cardiac hypertrophy, heart failure, hypertension, atherosclerotic vascular dysfunction, and cardiovascular inflammation [16]. It is now evident that angiogenesis plays an important role in cardiac remodeling as well as in the progression of atherosclerosis. Angiogenic therapy has been widely viewed as an attractive approach to treat many cardiovascular diseases [18]. DNA microarray analysis can identify the expression of candidate genes to regulate angiogenesis in the causation of cardiovascular diseases also in angiogenic therapy.

Sustained hypertension and ischemic heart disease lead to a group of complex alterations, collectively known as cardiac remodeling which if left untreated leads to heart failure. Since mortality from hypertensive stroke and myocardial infarction has declined dramatically over the last four decades, heart failure resulting from cardiac remodeling has now become the leading cause of mortality worldwide. Myocardial ischemia is associated with enhanced angiogenesis as nature's way to deliver more oxygen to the ischemic tissues; however, pathologic cardiac hypertrophy, especially in hypertension, may be associated with impaired angiogenesis resulting in myocyte death [19]. Both cardiac size and function in cardiac remodeling are dependent on angiogenesis [19].

As mentioned earlier, DNA microarray assay has been used to evaluate angiogenic factors in cardiac remodeling. Wang et al. evaluated angiogenic gene expression in patients with chronic ischemia [20]. VEGF is an essential growth and survival factor for endothelial cells, and has been viewed as one of the most important growth factors for angiogenesis, but Wang's study did not show any change in VEGF gene expression in chronic ischemia [20]. However, Shiojima et al. showed that VEGF blockade can inhibit angiogenesis in the acutely ischemic heart and lead to an impaired cardiac hypertrophy. Another group of investigators [21] demonstrated that VEGF-R deletion also causes impaired angiogenesis during ischemia-reperfusion. Other DNA microarray studies showed that VEGF plays an important function in angiogenesis, but it does not imply that that VEGF or VEGF-R gene abnormalities can be used as diagnostic indicator/s for cardiac remodeling.

In addition to myocardial ischemia, DNA microarray has been used to profile angiogenesis-related genes or factors in brain ischemia. It is well known that angiogenesis commonly takes place in the ischemic brain tissues and the boundary area surrounding the ischemic zone. The formation of new blood vessels in the ischemic brain, especially in the ischemic boundary area, helps increase oxygen and nutrient supply to the infarct tissues and facilitates functional recovery, and thus plays a pivotal role against additional ischemic injury [22]. Hayashi et al. used cDNA array assay to study the expression of angiogenesis-related genes in acute (1 h) and chronic (21 days) cerebral ischemia [23]. In acutely ischemic brain tissues, 42 genes (including angiopoietin-1, angiopoietin-2, VEGF, PDGF-β, IGF-1, EGF, CTGF, FGF-1, FGF-3, FGF-4, bFGF, FGFR-1, FGFR-3, TGF-β-1, TGF-β-2, HIF-1-α, integrin-β3, IFN-γ) among the assayed 96 angiogenesis-related genes were found to be increased, and none were decreased. In the chronically ischemic tissues, 29 genes (angiopoietin-2, IGF, PGF, TGF-βR, integrin-β3, IFN-α, neuropilin, and others) of the 96 genes were increased and two genes (PDGFR-α and ephrin B2) were decreased [23]. Recently, Ozduman et al. utilized the neonatal rodent retina as a model to study cerebral angiogenesis and also evaluated angiogenesis-related genes at different time points (day 0, 2, 4, 6, 8,10, 12, 14, 16, 18, 20, and 32) of retinal vascularization [24]. They found that 22 genes among the assayed 96 angiogenesis-related genes, including 1 antiangiogenic gene (vasostatin), 12 growth factors or their receptors (VEGF, VEGFR,TGFB2, TGFB3, TGFBR2, FGF7, FGF16, FGF-R, EGF, IGF-1, EFNA5, and KDR), 4 cytokines/adhesion molecule (FNA1, IFNR, PARC, and Cadherin 5), 2 transcription factors (HIF-1A and ID3), and 3 other angiogenic genes (SR-A, NOS-3, and Restin), were changed in this process [24]. These altered genes exhibited a different change pattern: VEGFR, EGF, FGFR1, IFNA1, IFNR, Cadherin, and SR-A at day 8; VEGFC, EFNA5, FGF7, FGF16, IGF-1, ID, and vasostatin at day 10; TGFB2 at day 12; KDR, TGFB3, TGFBR2, Restin, and SPARC at day 16; and HIF-1A at day 32 reached their highest level [24].

Atherosclerosis is the leading cause of mortality in the developed countries. It causes about 75% of cardiovascular-related deaths. Angiogenesis in the atherosclerotic plaques contributes to the development of the plaques and makes them unstable and rupture-prone [18, 25]. Previous studies have suggested that angiogenic factors such as VEGF play a critical role in the progression of atherosclerotic plaque as well as in the evolution of intraplaque hemorrhage [26]. Plaque angiogenesis is also viewed as a marker of the late stages of carotid atherosclerotic process [27]. Although there is a strong association between angiogenesis and atherosclerotic disease progression and vulnerability to rupture, not many angiogenic DNA microarray assays have been carried out in atherosclerotic plaque. Türeyen et al. [27] showed that 31 angiogenesis-related genes, including angiopoietin 2, angiogenic inducer 61, BDNF, FGF-20, EGF receptor, FGF receptor, and TGF-binding protein 4, were significantly increased in the carotid atherosclerotic plaques from symptomatic patients. The 31 angiogenic genes that were upregulated on microarray study were also further confirmed by qPCR assay.

24.2.2.2 DNA Microarray, Angiogenesis, and Tumors

Angiogenesis DNA microarray is mostly used in cancer diagnosis. It is now widely accepted that angiogenesis is an important factor in the development of cancers, especially solid tumors. Capillaries in solid tumors supply oxygen and nutrients necessary for survival of tumor cells. The growth and metastasis of malignant tumors are also angiogenesis dependent, and blocking angiogenesis has been used as a therapeutic strategy to inhibit tumor growth and spread [28]. New blood vessel formation is also viewed as a hallmark for tumor progression. Currently, DNA microarray on angiogenic factors is utilized as a routine tool to evaluate the progress of many tumors, such as lung cancer [29, 30], ovarian cancer [31], thyroid cancer [32], colorectal cancer [33–36], prostate cancer [37], endometrial cancer [38], oral cancer [39], gastric cancer [40], laryngopharyngeal carcinoma [41], neuroblastoma [42], and tumor cell lines [43].

24.2.2.3 DNA Microarray, Angiogenesis, and Other Disease States

DNA microarray assay has also been utilized to evaluate the expression of angiogenesis-related genes in other disease states. For example, Goessler et al. used DNA array to analyze the angiogenic genes in radiation-induced dermal wounds and found that some angiogenic factors, such as TGF, FGF, and VEGF, were decreased significantly in irradiated skin as the expression of MMPs (MMP-2, MMP-7, MMP-12, and MMP-13) was enhanced [44]. Their study suggested that irradiated wound is associated with a low level of pro-angiogenic factors and thus heals poorly. Toft et al. utilized genome-wide microarray to evaluate angiogenesis-regulating genes in placental tissues taken from preeclamptic (PE) and small-for-gestational-age (SGA) pregnancies, but the microarray data failed to show any significant differences in angiogenic gene profile (including VEGF, endoglin, FLT1, PlGF) between different study groups [45]. But, qPCR showed that antiangiogenic genes endoglin and FLT1 were significantly increased in pregnancies with

PE+SGA than those with PE or SGA alone [45]. This study also suggested that statistical analysis of DNA microarray data at times presents a false discovery [46]. The expression of target genes assayed by DNA microarray needs to be confirmed by PCR assay and molecular biological methods.

24.2.3 Protein Microarray and Angiogenesis

Protein microarray is a high-throughput tool for proteomics. A protein microarray comprises many special reagents, such as antibodies or peptides spotted on a high-affinity solid support [47]. Protein microarray for angiogenesis has been mainly used to identify potential makers for different cancers. Protein microarray is more sensitive than the measurement of individual proteins in the serum. Many malignant tumors (solid tumors) have high density of capillaries and secrete many angiogenic factors locally as well as in the circulation. Measurement of angiogenic factors in blood may indicate cancers progress. A protein array study by Crocker et al. showed that the serum levels of PDGF, PDGFR, Ang1, TIMP-1, and TIMP2 were increased in patients with glioblastoma; however, the most important angiogenic factors VEGF and bFGF and their soluble receptors were not [48]. Vihinen et al. utilized multicytokine antibody to screen 79 cytokines including some pro-angiogenic factors (interleukins, angiogenins, VEGF, PDGF) in patients with stage IV melanoma [49]. They did not show any changes in angiogenic factors except angiogenins that were increased in patients with progressive cancers [49]. Lukesová et al. also observed no significant changes in serum angiogenic factors including VEGF, TGF-ß1, EGF, PDGF, bFGF, IFN-γ, IL-6, IL-8, insulin-like growth factor 1 (IGF-1), leptin, and monocyte chemoattractant protein 1 (MCP-1) except angiogenin in patients with renal cell carcinoma (RCC) [50]. Another group of investigators [51] observed that angiogenic factors such as Ang-2, IL-8, and MMP-1 were increased, but not VEGF and VEGF-R in the medium of the cultured colorectal cancer cell lines. These studies based on protein array collectively suggest that angiogenins, pro-inflammatory cytokines, and TIMP, but not VEGF and bFGF, are indicators of malignant tumors.

24.2.4 Tissue Microarray, Angiogenesis, and Cancers

Tissue microarray offers another high-throughput tool to evaluate protein and gene expression in a histological analysis fashion. Tissue microarray consists of conventional histological paraffin blocks in which a large number of histological samples from different patients are spotted on the same slide (Fig. 24.2). This technique has been used widely in cancer diagnosis, including evaluation of drug therapy, analysis of clinical relevance between potential factors (such as angiogenic factors) and

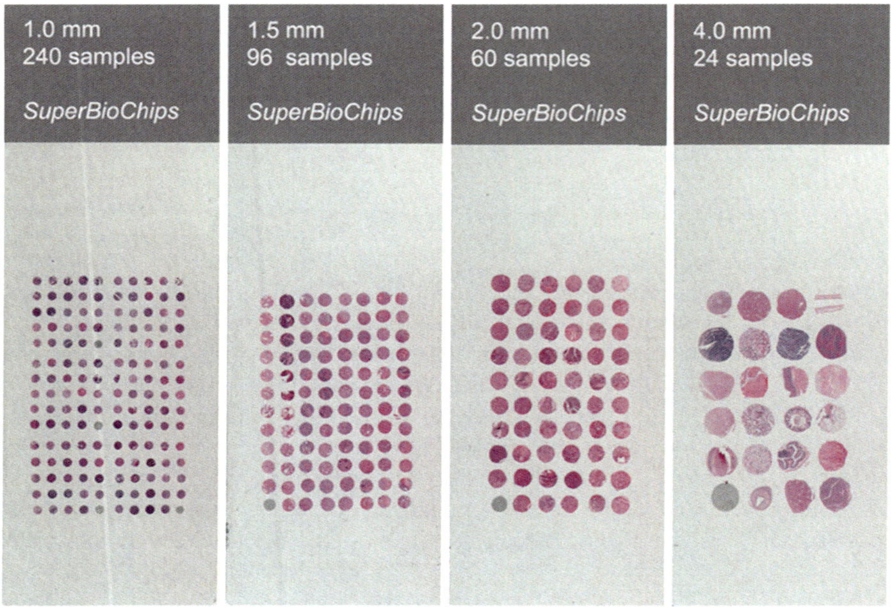

Fig. 24.2 Tissue microarray slices (2004–2005. SuperBioChips Laboratories, Yongdu-Dong, Seoul, Korea)

specific cancers, or looking for new protein or gene markers for different tumors. Here are some examples of use of this technique to study certain tumors.

Hypoxia-inducible factor-1 (HIF-1) is an important transcription factor that stimulates angiogenesis and tumor growth. Vleugel et al. used a tissue microarray to analyze HIF-1 expression in 94 patients with invasive breast cancer [52]. Their study showed that the expression of HIF-1alpha protein was upregulated in 62% of patients, but there was no amplification of HIF-1alpha mRNA, which suggests that there may not be an association between HIF protein and gene in human breast cancer [52]. However, tissue array studies from other groups have shown that HIF-1alpha mRNA was significantly increased in other tumors [53]. For example, Luo et al. found that HIF-1alpha mRNA was positively expressed in the border region and inside the epithelial ovarian tumor, and the expression of HIF-1alpha positively related with the presence of angiogenesis and tumor histological grade [53]. In addition to the identification of HIF, tissue microarray has also been used to check other angiogenic markers such as VEGF-1, EGFR and STAT3 [54], VEGFR1 and VEGFR2 [55], Ets-1 [56], caveolin-1 and VEGF-C [57], angiopoietins-1 and -2 and Tie-2 receptor [58], thymosin β4 [59], and breast cancer metastasis suppressor 1 (BRMS1) [60] for different cancers diagnosis.

24.3 MicroRNA and Angiogenesis

24.3.1 MiR-15, 16, 107, and Angiogenesis

MiR-15 and miR-16 were firstly reported to inhibit VEGF expression in human carcinoma cell line [61, 62]. Several studies profiling miRNA expression in endothelial cells demonstrated that these miRNAs are expressed at high levels, indicating their potential importance in angiogenesis [63, 64]. Overexpression of miRNA-15/16 has been demonstrated to decrease hypoxia-induced VEGF expression, while their knockdown results in an increase of VEGF level [65]. Multiple computational programs have predicted that potential regulation of VEGF by miRNA-16 through an miRNA responsive element (MRE) located ~260 bases downstream of the translation stop in the VEGF 3′UTR. Karaa et al. verified this regulation by using VEGF expression together with miR-16 in HeLa cells, which validated the targeting of miRNA-16 to a predicted binding site in the VEGF 3′UTR [62].

The miR-15/107 group of miRs regulates gene expression involved in cell division, proliferation, metabolism, stress response, and angiogenesis [66]. The miR-15/107 group has also been implicated in human cancers, cardiovascular diseases, and neurodegenerative diseases [67]. An experimental study by Yamakuchi et al. [68] suggested that miR-107 can mediate p53 regulation of hypoxic signaling and tumor angiogenesis. A study by Finnerty et al. [67] provides insights into upstream regulation of miR-107 expression through specific p53-responsive promoter regions. Karaa et al. [62] reported that miR-16 strongly regulates VEGF expression in angiogenesis. In addition to miR-15/107, miR-17-92 cluster, miR-378, miR-126, miR-221, and miR-222 have also been shown to participate in blood vessel development in normal and/or tumor tissues [63, 69] (Fig. 24.3).

24.3.2 MiR-17-92 Cluster and Angiogenesis

MiRs are frequently transcribed together as polycistronic primary transcripts that are processed into multiple individual mature miRs in animals. The genomic organization of these miR clusters is often highly conserved, indicating an important role for coordinated regulation and function. As a polycistronic miR gene in human genome, the miR-17-92 cluster encodes six miRs, including miR-17, miR-18a, miR-19a, miR-20a, miR-19b-1, and miR-92-1, which are tightly grouped within an 800 base-pair region in human chromosome 13 [70, 71]. Both sequences of these mature miRNAs and their organization are highly conserved in all vertebrates [70, 71]. MiRs encoded by the miR-17-92 cluster and its paralogs are known to act as oncogenes. Expression of these miRs promotes cell proliferation, suppresses apoptosis of cancer cells, and induces tumor angiogenesis [72]. C13orf25 gene expresses in association with genomic amplification and may play an important role in

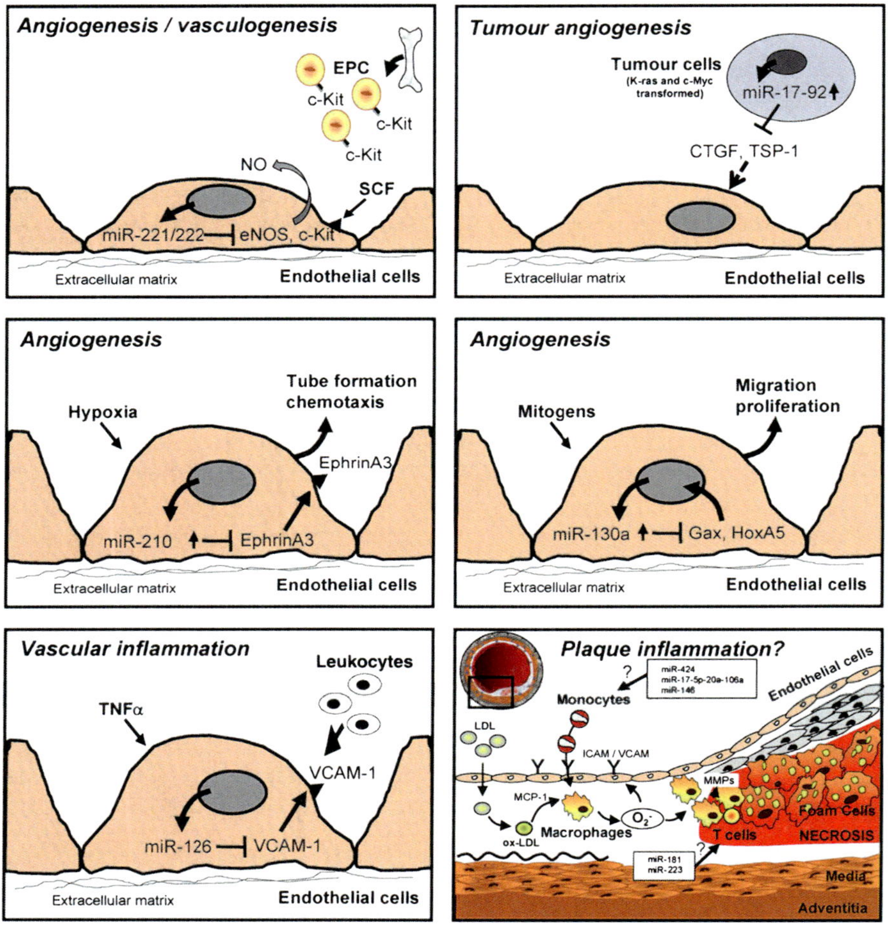

Fig. 24.3 Role of miRNAs for vascular biology. Schematic illustration of the specific functions of miR-221/miR-222, miR-17-92, miR-210, miR-130a, and miR-126 for vascular biology (Urbich C, Kuehbacher A, Dimmeler S. Role of microRNAs in vascular diseases, inflammation, and angiogenesis. Cardiovasc Res. 2008 Sep 1;79:581–8)

tumorigenesis and resulting poor prognosis [73]. The highly conserved human miR-17-92 cluster is located in the third intron of an approximately 7 kb primary transcript known as C13orf25 [74]. It evolved two miR-17-92 cluster paralogs in mammals due to ancient gene duplications: (1) the miR-106b-25 cluster that is located within the thirteenth intron of the protein-coding gene MCM7 (chromosome 7) and (2) the miR-106a-363 cluster that is located on the X chromosome. MiR-17-92 and miR-106b-25 clusters are both abundantly expressed across many tissues and cell types. Of note, the miR-106a-363 cluster is undetectable or expressed at trace levels in all settings that have been examined [75].

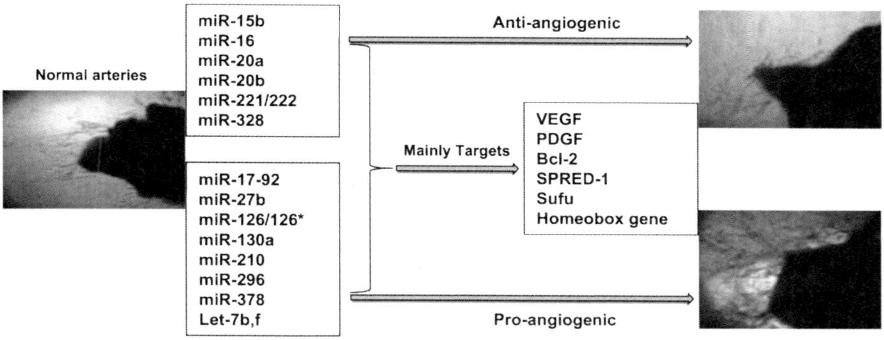

Fig. 24.4 Regulation of angiogenesis by antiangiogenic miRNAs (miR-15b, miR-16, miR-20a, miR-20b, miR-221/222, miR-328) and pro-angiogenic miRNAs (miR-17-92, miR-27b, miR-126/126*, miR-130a, miR-210, miR-296, miR-378, let-7b,f). MiRNAs regulate the angiogenic responses to growth factors by targeting angiogenic factors, receptors, and signaling molecules

The miR-17-92 cluster first attracted attention following a series of observations linking these miRs to cancer pathogenesis [76]. Overexpression of miR-17-92 in Ras expressing murine carcinoma cells resulted in enhanced tumor angiogenesis in vivo in a non-cell autonomous manner. The downregulation of the potent endogenous inhibitor of angiogenesis thrombospondin-1 together with several proteins containing thrombospondin type 1 repeats has been shown to be involved in this pathway [77]. Transfection of endothelial cells with components of the miR-17-92 cluster, induced by VEGF treatment, rescued the induced expression of thrombospondin-1 and the defect in endothelial cell proliferation and morphogenesis initiated by the loss of Dicer. Inhibition of miR-17 and miR-20a increased the number of blood vessels in Matrigel plugs, but antagomiRs that target miR-18a and miR-19a were less effective [78]. The aforementioned study support that miR-17–92 promotes tumor angiogenesis by targeting antiangiogenic proteins thrombospondin-1 and connective tissue growth factor (CTGF), therefore regulating angiogenesis in a non-cell-autonomous manner [79] (Fig. 24.4).

24.3.3 MiR-126/miR-126 and Angiogenesis*

Recently, a protein named epidermal growth factor-like domain 7 (EGFL7, also known as VE-statin, MEGF7, Notch4-like protein, or Zneu1) was described as a novel endothelial cell-derived factor that is involved in the regulation of the spatial arrangement of cells during vascular tube assembly or blood vessel formation [80]. This protein is conserved among vertebrates but an orthologue is also found in *Drosophila melanogaster* [81]. In humans, there are three alternative isoforms containing the same open reading frame but are transcribed from separate promoters [82]. Gene analysis revealed that EGFL7 is expressed within the neurons of adult mice, indicating that EGFL7 serves diverse biological functions in various tissues and not only in the vascular system [83]. Soncin et al. [84] showed that EGFL7 inhibits human aortic smooth muscle cells' migration, but not proliferation, indicating that EGFL7 might have a role in vessel maturation. Parker et al. [85] established the role of egfl7 as an important tubulogenic factor in the process of vasculogenesis. Recently, Schmidt et al. provided another compelling clue for resolving the function of EGFL7 protein in angiogenesis [86].

MiR-126 (also referred to as miR-126-3p) and its complement miR-126* (miR-126-5p or miR-123) are derived from EGFL7 gene and harbor both miRNAs within intron 7 in all vertebrates [87]. Both miR-126 and miR-126* are relevant for the development of the cardiovascular system, cardiovascular diseases, and the formation of certain cancers [88]. MiR-126 performs an abundant expression level in highly vascularized tissues, and it is the only miRNA known to be expressed specifically in the endothelial lineage and hematopoietic progenitor cells [89].

The function of miR-126/miR-126* in angiogenesis, though it remained an enigma for quite some time, has now been clarified. MiR-126 seems to regulate endothelial cell angiogenic activity in response to angiogenic growth factors such as VEGF and bFGF, through targeting multiple proteins that modulate angiogenesis and vascular integrity [88]. Fish et al. [90] found that miR-126 regulates the response of endothelial cells to VEGF. In addition, knockdown of miR-126 in zebrafish resulted in loss of vascular integrity and hemorrhage during embryonic development. MiR-126 functioned in part by directly repressing negative regulators of the VEGF pathway, including the Sprouty-related protein, SPRED1, and phosphoinositol-3 kinase regulatory subunit 2 (PIK3R2). SPRED1 contains a predicted target sequence for miR-126, and it plays a key role in miR-126 mediated pro-angiogenic actions [91]. Increased expression of SPRED1 or inhibition of VEGF signaling in zebrafish resulted in defects similar to miR-126 knockdown [87].

Studies by Parker et al. [85] also showed that loss of egfl17 function in zebrafish embryos specifically blocks vascular tubulogenesis. EGFL7 is downregulated in quiescent endothelial cells, but is upregulated in the endothelium of proliferating tissue, such as some tumors. Angiogenesis and vascular integrity can be disrupted through modulation of miR-126 expression. Since miR-126 is embedded within EGFL7, it is possible that miR-126 increases the sensitivity of these activated endothelial cells to VEGF or other growth factors through repression of SPRED1 and/or PIK3R2 expression. Work by Nicoli et al. [92] demonstrated that mediated by the mechanosensitive zinc finger-containing transcription factor klf2a, miR-126 can be induced by blood flow, leading to the activation of VEGF signaling in the endothelium. Due to damage of vascular integrity and defects in endothelial cell proliferation, migration, and angiogenesis, endothelial-specific deletion of miR-126 in mice causes leaky vessels, hemorrhage, and partial embryonic lethality. Subsequent miR-126 knockdown studies in zebrafish showed that miR-126 induces hemorrhage and collapse of lumen-containing vascular structures [93]. These observations are consistent with the concept of miR-126 binding to the 3′ UTR of the VEGF-A mRNA [94]. Considering the key role of miR-126 in the regulation of angiogenesis and vascular integrity, it has been proposed that miR-126 may be an important target for pro- or antiangiogenic therapies.

24.3.4 MiR-221/222 and Angiogenesis

MiR-221 and miR-222 belong to the same family and control common targets, which are located in close proximity on Xp11.3 chromosome and might be regulated in a coordinated manner [95]. MiR-221/222 are highly expressed in human umbilical vein endothelial cells and are known to regulate the angiogenesis. These miRs inhibit endothelial cell migration, proliferation, and angiogenesis in vitro by targeting stem cell factor (SCF) receptor, c-kit (a receptor tyrosine kinase that binds stem cell factor and mediates VEGF expression) [96]. Antisense miR-221 oligonucleotide was shown to reduce the expression of miR-221, to restore c-kit expression in HUVECs, and to abolish the inhibitory effect of high glucose on HUVECs transmigration [97]. Interestingly, Li et al. observed that miR-221 expression was induced by high glucose while c-kit expression was reduced, indicating that miR-221-c-kit pathway may play an important role in diabetes-associated vascular dysfunction [98].

24.3.5 MiR-378 and Angiogenesis

miR-378 is highly expressed in CD34$^+$ hematopoietic progenitor cells, which function as an oncogene by enhancing angiogenesis, tumor cell survival, and tumor growth [99]. By inhibiting the expression of two tumor suppressors, Sufu (suppressor of fused) and Fus-1 (tumor suppressor candidate 2), miR-378 reduced caspase-3 activity and enhanced cell survival, tumor growth, and angiogenesis [100]. Cell survival assays showed that transfection with a construct expressing an antisense sequence against miR-378, cell survival decreased significantly [101]. Sonic hedgehog (Shh) is a prototypical morphogen known to regulate epithelial/mesenchymal interactions during embryonic development. Sufu functions as a negative regulator of Shh signaling. Shh promotes large-diameter vessel formation by inducing expression of angiogenic cytokines, including VEGF and angiopoietin-1 (Ang-1) and-2 (Ang-2) [102]. MiR-378 can also repress the effect of Sufu in the presence of the 3′-UTR, which promotes cell survival, confirming that the Sufu 3′-UTR is a target of miR-378 [103]. These findings suggest that miRNA-378 promotes cell survival and regulates tumor angiogenesis through regulating Sufu and Fus-1.

24.3.6 Other miRs Related to Angiogenesis

MiR-210, a hypoxia-induced miRNA, is a crucial element of endothelial cell response to hypoxia, affecting cell survival, migration, and differentiation [104]. In normoxic endothelial cells, overexpression of miR-210 was shown to stimulate VEGF-induced cell migration and formation of capillary-like structures (angiogenesis). Conversely, blockade of miR-210 by transfection of anti-miRNA inhibited tube formation stimulated by hypoxia and cell migration in response to VEGF [105].

Endothelial nitric oxide synthase (eNOS) is one of the key factors for angiogenesis. Inhibition of eNOS production in vivo inhibits angiogenesis [106]. MiRNA-214 expressed highly in vascular system [107] has been demonstrated to inhibit angiogenesis via inhibition of eNOS production [108]. Recently, van Mil et al. showed that miRNA-214 can negatively regulate in vivo and in vitro angiogenesis by inhibiting the expression of pro-angiogenic growth factors such as VEGF and PDGF [108].

MiR-296 is also an important regulator for angiogenesis [109]. Hepatocyte growth factor-regulated tyrosine kinase substrate (HGS), which mediates the degradative sorting of PDGFR as well as VEGFR and EGFR, has been identified as a target for miR-296 that mediates angiogenic function [110]. MiR-296 is upregulated in human gliomas tumor ECs, and it seems that this miR downregulates HGS expression and upregulates VEGFR2 and PDGFRβ in glioma blood vessels. Other studies have confirmed that EGF could induce miR-296, suggesting a role for miR-296 in promoting angiogenesis in tumors [110, 111].

As potential pro-angiomiRs, miR-21 and miR-31 are upregulated in various cancers, and miR-21 can stimulate invasion and metastasis in cancer [112, 113]. Let-7f and miR-27b are also involved in angiogenesis. Inhibition of let-7f significantly reduces vascular sprouting while let-7b controls EC proliferation and motility and affects tube formation by regulating antiangiogenic factor TIMP-1 [114, 115]. MiR-130a, an miRNA strongly upregulated after exposure to fetal bovine serum in ECs, is able to antagonize the antiangiogenic activity by regulating the expression of its target gene homeobox gene GAX (growth arrest-specific homeobox) and homeobox A5 (HOXA5) [116, 117]. MiR-20a is a member of the miR-17-92 cluster, and miR-20b is a member of the miR-106a cluster located on X chromosome, both of them are potential antiangiomiRs by targeting VEGF for repression [118, 119].

24.4 Conclusion

A number of studies summarized here have provided much needed information on pathways involved in angiogenesis. Discovery of miRs has led to discoveries of new therapeutic targets. Obviously, we need to await human trials to see if these targets are successful in ameliorating human disease states associated with angiogenesis.

References

1. Liu Y, Petreaca M, Martins-Green M (2009) Cell and molecular mechanisms of insulin-induced angiogenesis. J Cell Mol Med 13:4492–4504
2. Dejana E (2010) The role of wnt signaling in physiological and pathological angiogenesis. Circ Res 107:943–952
3. Hoeben A, Landuyt B, Highley MS, Wildiers H, Van Oosterom AT, De Bruijn EA (2004) Vascular endothelial growth factor and angiogenesis. Pharmacol Rev 56:549–580
4. Shim WS, Ho IA, Wong PE (2007) Angiopoietin: a TIE(d) balance in tumor angiogenesis. Mol Cancer Res 5:655–665

5. Yokomori H, Oda M, Yoshimura K et al (2012) Enhanced expressions of apelin on proliferative hepatic arterial capillaries in human cirrhotic liver. Hepatol Res 42:508–514
6. Jiang J, Yan M, Mehta JL et al (2011) Angiogenesis is a link between atherosclerosis and tumorigenesis: Role of LOX-1. Cardiovasc Drugs Ther 25:461–468
7. Chau CH, Fig. WD (2012) Angiogenesis inhibitors increase tumor stem cells. Cancer Biol Ther 13(8):586–587
8. Nazari-Jahantigh M, Wei Y, Schober A (2012) The role of microRNAs in arterial remodelling. Thromb Haemost 107:611–618
9. Abdellatif M (2012) Differential expression of microRNAs in different disease states. Circ Res 110:638–650
10. Le Bot N (2012) MicroRNAs in angiogenesis. Nat Cell Biol 14:342
11. Ptak-Belowska A, Targosz A, Brzozowski T (2012) MicroRNAs and chronic inflammation contribution to gastrointestinal integrity. Curr Med Chem 9:133–136
12. Biyashev D, Qin G (2011) E2F and microRNA regulation of angiogenesis. Am J Cardiovasc Dis 1:110–118
13. Weis SM, Cheresh DA (2011) Tumor angiogenesis: molecular pathways and therapeutic targets. Nat Med 17:1359–1370
14. Hartmann D, Thum T (2011) MicroRNAs and vascular (dys)function. Vascul Pharmacol 5:92–105
15. Ye SQ, Lavoie T, Usher DC et al (2002) Microarray. SAGE and their applications to cardiovascular diseases. Cell Res 12:105–115
16. Napoli C, Lermman LO, Sica V et al (2003) Microarray analysis: a novel research tool for cardiovascular scientists and physicians. Microarray analysis: a novel research tool for cardiovascular scientists and physicians. Heart 89:597–604
17. Schena M, Shalon D, Davis RW et al (1995) Quantitative monitoring of gene expression patterns with a complementary DNA microarray. Science 270:467–470
18. Khurana R, Simons M, Martin JF et al (2005) Role of angiogenesis in cardiovascular disease: a critical appraisal. Circulation 112:1813–1824
19. Shiojima I, Sato K, Izumiya Y et al (2005) Disruption of coordinated cardiac hypertrophy and angiogenesis contributes to the transition to heart failure. J Clin Invest 115:2108–2118
20. Wang Y, Gabrielsen A, Lawler RR et al (2006) Myocardial gene expression of angiogenic factors in human chronic ischemic myocardium: influence of acute ischemia/cardioplegia and reperfusion. Microcirculation 13:187–197
21. Addya S, Shiroto K, Turoczi T et al (2005) Ischemic preconditioning-mediated cardioprotection is disrupted in heterozygous Flt-1 (VEGFR-1) knockout mice. J Mol Cell Cardiol 38:345–351
22. Beck H, Plate KH (2009) Angiogenesis after cerebral ischemia. Acta Neuropathol 117:481–496
23. Hayashi T, Noshita N, Sugawara T et al (2003) Temporal profile of angiogenesis and expression of related genes in the brain after ischemia. J Cereb Blood Flow Metab 23:166–180
24. Ozduman K, Ozkan A, Yildirim O et al (2010) Temporal expression of angiogenesis-related genes in developing neonatal rodent retina: a novel in vivo model to study cerebral vascular development. Neurosurgery 66:538–543
25. Slevin M, Krupinski J, Badimon L (2009) Controlling the angiogenic switch in developing atherosclerotic plaques: possible targets for therapeutic intervention. J Angiogenes Res 1:4
26. Hiyama T, Tanaka T, Endo S et al (2010) Angiogenesis in atherosclerosis plaque obtained from carotid endarterectomy: association between symptomatology and plaque morphology. Neurol Med Chir (Tokyo) 50:1056–1061
27. Türeyen K, Vemuganti R, Salamat MS et al (2006) Increased angiogenesis and angiogenic gene expression in carotid artery plaques from symptomatic stroke patients. Neurosurgery 58:971–977
28. Carmeliet P, Jain RK (2000) Angiogenesis in cancer and other diseases. Nature 407:249–257
29. Campioni M, Ambrogi V, Pompeo E et al (2008) Identification of genes down-regulated during lung cancer progression: a cDNA array study. J Exp Clin Cancer Res 27:38

30. Lioglou T, Bediaga NG, Brown B et al (2012) Epigenetic biomarkers in lung cancer. Cancer Lett; Epub ahead of print
31. Martoglio AM, Tom BD, Starkey M et al (2000) Changes in tumorigenesis- and angiogenesis-related gene transcript abundance profiles in ovarian cancer detected by tailored high density cDNA arrays. Mol Med 6:750–765
32. Kim DS, Franklyn JA, Stratford AL et al (2006) Pituitary tumor-transforming gene regulates multiple downstream angiogenic genes in thyroid cancer. J Clin Endocrinol Metab 91:1119–1128
33. Tsunoda T, Nakamura T, Ishimoto K et al (2001) Upregulation expression of angiogenesis genes and down regulation of cell cycle genes in human colorectal cancer tissue determined by cDNA macroarray. Anticancer Res 21:137–143
34. Strauss LG, Koczan D, Klippel S et al (2008) Impact of angiogenesis-related gene expression on the tracer kinetics of 18F-FDG in colorectal tumors. J Nucl Med 49:1238–1244
35. GrÐne J, Doebler O, Loddenkemper C et al (2006) Robo1/Robo4: differential expression of angiogenic markers in colorectal cancer. Oncol Rep 15:1437–1443
36. Yamamoto H, Imsumran A, Fukushima H et al (2002) Analysis of gene expression in human colorectal cancer tissues by cDNA array. J Gastroenterol 37(Suppl 14):83–86
37. Morrissey C, True LD, Roudier MP et al (2008) Differential expression of angiogenesis associated genes in prostate cancer bone, liver and lymph node metastases. Clin Exp Metastasis 25:377–388
38. Mannelgvist M, Stefansson IM, Bredholt G et al (2011) Gene expression patterns related to vascular invasion and aggressive features in endometrial cancer. Am J Pathol 178:861–871
39. Tsai WC, Tsai ST, Ko JY et al (2004) The mRNA profile of genes in betel quid chewing oral cancer patients. Oral Oncol 40:418–426
40. Chen CN, Chang CC, Su TE et al (2009) Identification of calreticulin as a prognosis marker and angiogenic regulator in human gastric cancer. Ann Surg Oncol 16:524–533
41. Ganly I, Talbot S, Carlson D et al (2007) Identification of angiogenesis/metastases genes predicting chemoradiotherapy response in patients with laryngopharyngeal carcinoma. J Clin Oncol 25:1369–1376
42. Takahashi F, Akutagawa S, Fukumoto H et al (2002) Osteopontin induces angiogenesis of murine neuroblastoma cells in mice. Int J Cancer 98:707–712
43. Anfosso L, Efferth T, Albini A et al (2006) Microarray expression profiles of angiogenesis-related genes predict tumor cell response to artemisinins. Pharmacogenomics J 6:268–278
44. Goessler UR, Bugert P, Kassner S et al (2010) In vitro analysis of radiation-induced dermal wounds. Otolaryngol Head Neck Surg 142:845–850
45. Toft JH, Lian IA, Tarca AL et al (2008) Whole-gene microarray and targeted analysis of angiogenesis-regulating gene expression (ENG, FLT1, VEGF, PIGF) in placentas from pre-eclamptic and small-for-gestational-age pregnancies. J Matern Fetal Neonatal Med 21:267–273
46. Leha A, Beissbarth T, Jung K (2011) Sequential interim analysis of survival of data in DNA microarray experiments. BMC Bioinformatics 29:127
47. MacBeath G (2002) Protein microarrays and Proteomics. Nat Genet 32(Suppl):526–532
48. Crocker M, Ashley S, Giddings I et al (2011) Serum angiogenic profile of patients with glioblastoma identifies distinct tumor subtypes and shows that TIMP-1 is a prognostic factor. Neuro Oncol 13:99–108
49. Vihinen P, Kallioinen M, Vuoristo MS et al (2007) Serum angiogenin levels predict treatment response in patients with stage IV melanoma. Clin Exp Metastasis 24:567–574
50. Lukesová S, Kopecký O, Vroblová V et al (2008) Determination of angiogenic factors in serum by protein array in patients with renal cell carcinoma. Folia Biol (Praha) 54:134–140
51. Abajo A, Bitarte N, Zarate R et al (2012) Identification of colorectal cancer metastasis markers by angiogenesis-related cytokine-antibody array. World J Gastroenterol 28:637–645
52. Vleugel MM, Bos R, Buerger H et al (2004) No amplifications of hypoxia-inducible factor-1alpha gene in invasive breast cancer: a tissue microarray study. Cell Oncol 26:347–351

53. Luo J, Peng ZL, Yang KX et al (2005) Relation between the expression of hypoxia inducible factor-1alpha and angiogenesis in ovarian cancer using tissue microarray. Zhonghua Fu Chan Ke Za Zhi 40:38–41
54. Kim HS, Park YH, Lee J et al (2010) Clinical impact of phosphorylated signal transducer and activator of transcription 3, epidermal grow23th factor receptor, P53, and vascular endothelial growth factor receptor 1 expression in resected adenocarcinoma of lung by using tissue microarray. Cancer 116:676–685
55. Zhang Z, Neiva KG, Lingen MW et al (2010) VEGF-dependent tumor angiogenesis requires inverse and reciprocal regulation of VEGFR1 and VEGFR2. Cell Death Differ 17:499–512
56. Ren TW, Zhou YN, Wu J et al (2009) Relationship between Est-1 expression and angiogenesis, clinicopathological features and survival of patients with gastric carcinoma. Zhonghua Zhong Liu Za Zhi 31:674–678
57. Li M, Chen H, Diao L et al (2010) Caveolin-1 and VEGF-C promote lymph node metastasis in the absence of intratumoral lymphangiogenesis in non-small cell lung cancer. Tumori 96:734–743
58. Staton CA, Hoh L, Baldwin A et al (2011) Angiopoietins 1 and 2 and Tie-2 receptor expression in human ductal breast disease. Histopathology 59:256–263
59. Jo JO, Kang YJ, Ock MS et al (2011) Thymosin β4 expression in human tissues and tumors using tissue microarrays. Appl Immunohistochem Mol Morphol 19:160–167
60. Li J, Cheng Y, Tai D et al (2011) Prognostic significance of BRMS1 expression in human melanoma and its role in tumor angiogenesis. Oncogene 30:896–906
61. Ofir M, Hacohen D, Ginsberg D (2011) MiR-15 and miR-16 are direct transcriptional targets of E2F1 that limit E2F-induced proliferation by targeting cyclin E. Mol Cancer Res 9:440–447
62. Karaa ZS, Iacovoni JS, Bastide A et al (2009) The VEGF IRESes are differentially susceptible to translation inhibition by miR-16. RNA 15:249–254
63. Chang SH, Hla T (2011) Gene regulation by RNA binding proteins and microRNAs in angiogenesis. Trends Mol Med 17:650–658
64. Vincenti S, Brillante N, Lanza V et al (2011) HUVEC respond to radiation by inducing the expression of pro-angiogenic microRNAs. Radiat Res 175:535–546
65. Aqeilan RI, Calin GA, Croce CM (2010) miR-15a and miR-16-1 in cancer: discovery, function and future perspectives. Cell Death Differ 17:215–220
66. Nelson PT, Wang WX, Mao G, Wilfred BR, Xie K, Jennings MH, Gao Z, Wang X (2011) Specific sequence determinants of miR-15/107 microRNA gene group targets. Nucleic Acids Res 39:8163–8172
67. Finnerty JR, Wang WX, Hébert SS et al (2010) The miR-15/107 group of microRNA genes: evolutionary biology, cellular functions, and roles in human diseases. J Mol Biol 402:491–509
68. Yamakuchi M, Lotterman CD, Bao C et al (2010) P53-induced microRNA-107 inhibits HIF-1 and tumor angiogenesis. Proc Natl Acad Sci USA 107:6334–6339
69. Wu F, Yang Z, Li G (2009) Role of specific microRNAs for endothelial function and angiogenesis. Biochem Biophys Res Commun 386:549–553
70. Xiang J, Wu J (2010) Feud or friend? The role of the miR-17-92 cluster in tumorigenesis. Curr Genomics 1 :129–135
71. Osada H, Takahashi T (2011) let-7 and miR-17-92: small-sized major players in lung cancer development. Cancer Sci 102:9–17
72. Mendell JT (2008) miRiad roles for the miR-17-92 cluster in development and disease. Cell 133:217–222
73. Rinaldi A, Poretti G, Kwee I et al (2007) Concomitant MYC and microRNA cluster miR-17-92 (C13orf25) amplification in human mantle cell lymphoma. Leuk Lymphoma 48:410–412
74. Ota A, Tagawa H, Karnan S et al (2004) Identification and characterization of a novel gene, C13orf25, as a target for 13q31-q32 amplification in malignant lymphoma. Cancer Res 64:3087–3095
75. Mu P, Han YC, Betel D et al (2009) Genetic dissection of the miR-17-92 cluster of microRNAs in Myc-induced B-cell lymphomas. Genes Dev 23:2806–2811

76. Mendell JT (2008) miRiad roles for the miR-17-92 cluster in development and disease. Cell 133:217–222
77. Dews M, Homayouni A, Yu D et al (2006) Augmentation of tumor angiogenesis by a Myc-activated microRNA cluster. Nat Genet 38:1060–1065
78. Suarez Y, Fernandez-Hernando C, Yu J et al (2008) Dicer-dependent endothelial microRNAs are necessary for postnatal angiogenesis. Proc Natl Acad Sci USA 105:14082–14087
79. Bonauer A, Carmona G, Iwasaki M et al (2009) MicroRNA-92a controls angiogenesis and functional recovery of ischemic tissues in mice. Science 324:1710–1713
80. Nikolic I, Plate KH, Schmidt MH (2010) EGFL7 meets miRNA-126: an angiogenesis alliance. J Angiogenes Res 2:9
81. Nichol D, Stuhlmann H (2012) EGFL7: a unique angiogenic signaling factor in vascular development and disease. Blood 119:1345–1352
82. Davis GE (2010) Vascular balancing act: EGFL7 and Notch. Blood 116:5791–5793
83. Nichol D, Shawber C, Fitch MJ et al (2010) Impaired angiogenesis and altered Notch signaling in mice overexpressing endothelial Egfl7. Blood 116:6133–6143
84. Soncin F, Mattot V, Lionneton F et al (2003) VE-statin, an endothelial repressor of smooth muscle cell migration. EMBO J 22:5700–5711
85. Parker LH, Schmidt M, Jin SW et al (2004) The endothelial-cell-derived secreted factor Egfl7 regulates vascular tube formation. Nature 428:754–758
86. Schmidt M, Paes K, De Mazière A et al (2007) EGFL7 regulates the collective migration of endothelial cells by restricting their spatial distribution. Development 134:2913–2923
87. Wang S, Aurora AB, Johnson BA et al (2008) The endothelial-specific microRNA miR-126 governs vascular integrity and angiogenesis. Dev Cell 15:261–271
88. Meister J, Schmidt MH (2010) miR-126 and miR-126*: new players in cancer. Scientific World Journal 10:2090–2100
89. Musiyenko A, Bitko V, Barik S (2008) Ectopic expression of miR-126*, an intronic product of the vascular endothelial EGF-like 7 gene, regulates protein translation and invasiveness of prostate cancer LNCaP cells. J Mol Med 86:313–322
90. Fish JE, Santoro MM, Morton SU et al (2008) miR-126 regulates angiogenic signaling and vascular integrity. Dev Cell 15:272–284
91. Kuhnert F, Mancuso MR, Hampton J et al (2008) Attribution of vascular phenotypes of the murine Egfl7 locus to the microRNA miR-126. Development 135:3989–3993
92. Nicoli S, Standley C, Walker P et al (2010) MicroRNA-mediated integration of haemodynamics and Vegf signalling during angiogenesis. Nature 464:1196–1200
93. Grabher C, Payne EM, Johnston AB et al (2011) Zebrafish microRNA-126 determines hematopoietic cell fate through c-Myb. Leukemia 25:506–514
94. Donnem T, Lonvik K, Eklo K et al (2011) Independent and tissue-specific prognostic impact of miR-126 in nonsmall cell lung cancer: coexpression with vascular endothelial growth factor-A predicts poor survival. Cancer 117:3193–3200
95. Altuvia Y, Landgraf P, Lithwick G et al (2005) Clustering and conservation patterns of human microRNAs. Nucleic Acids Res 33:2697–2706
96. Felicetti F, Errico MC, Bottero L et al (2008) The promyelocytic leukemia zinc finger-microRNA-221/-222 pathway controls melanoma progression through multiple oncogenic mechanisms. Cancer Res 68:2745–2754
97. Zhu N, Zhang D, Chen S et al (2011) Endothelial enriched microRNAs regulate angiotensin II-induced endothelial inflammation and migration. Atherosclerosis 215:286–293
98. Li Y, Song YH, Li F et al (2009) MicroRNA-221 regulates high glucose-induced endothelial dysfunction. Biochem Biophys Res Commun 381:81–83
99. Merkerova M, Krejcik Z, Votavova H et al (2011) Distinctive microRNA expression profiles in CD34+ bone marrow cells from patients with myelodysplastic syndrome. Eur J Hum Genet 19:313–319
100. Fang J, Song XW, Tian J et al (2012) Overexpression of microRNA-378 attenuates ischemia-induced apoptosis by inhibiting caspase-3 expression in cardiac myocytes. Apoptosis 17:410–423

101. Lee DY, Deng Z, Wang CH, Yang BB (2007) MicroRNA-378 promotes cell survival, tumor growth, and angiogenesis by targeting SuFu and Fus-1 expression. Proc Natl Acad Sci USA 104:20350–20355
102. Feng M, Li Z, Aau M et al (2011) Myc/miR-378/TOB2/cyclin D1 functional module regulates oncogenic transformation. Oncogene 30:2242–2251
103. Xu S, Linher-Melville K, Yang BB et al (2011) Micro-RNA378 (miR-378) regulates ovarian estradiol production by targeting aromatase. Endocrinology 152:3941–3951
104. Yamasaki K, Nakasa T, Miyaki S et al (2012) Angiogenic microRNA-210 is present in cells surrounding osteonecrosis. J Orthop Res 30(8):1263–1270
105. Liu F, Lou YL, Wu J et al (2012) Upregulation of MicroRNA-210 regulates renal angiogenesis mediated by activation of VEGF signaling pathway under ischemia/perfusion injury in vivo and in vitro. Kidney Blood Press Res 35:182–191
106. Fukumura D, Gohong T, Kadambi A et al (2001) Predominant role of endothelial nitric oxide synthase in vascular endothelial growth factor-induced angiogenesis and vascular permeability. Proc Natl Acad Sci USA 98:2604–2609
107. Chan LS, Yue PY, Mak NK et al (2009) Role of microRNA-214 in ainsenoside-Rg1-induced angiogenesis. Eur J Pharm Sci 38:370–377
108. van Mil A, Grundmann S, Goumans MJ et al (2012) MicroRNA-214 inhibits angiogenesis by targeting quaking and reducing angiogenic growth factor release. Cardiovasc Res 93:655–665
109. Langenkamp E, Zwiers PJ, Moorlag HE et al (2012) Vascular endothelial growth factor receptor 2 inhibition in-vivo affects tumor vasculature in a tumor type-dependent way and downregulates vascular endothelial growth factor receptor 2 protein without a prominent role for miR-296. Anticancer Drugs 23:161–172
110. Würdinger T, Tannous BA, Saydam O et al (2008) miR-296 regulates growth factor receptor overexpression in angiogenic endothelial cells. Cancer Cell 14:382–393
111. Langenkamp E, Zwiers PJ, Moorlag HE et al (2012) Vascular endothelial growth factor receptor 2 inhibition in-vivo affects tumor vasculature in a tumor type-dependent way and downregulates vascular endothelial growth factor receptor 2 protein without a prominent role for miR-296. Anticancer Drugs 23:161–172
112. Cottonham CL, Kaneko S, Xu L (2010) miR-21 and miR-31 converge on TIAM1 to regulate migration and invasion of colon carcinoma cells. J Biol Chem 285:35293–35302
113. Slaby O, Svoboda M, Fabian P et al (2007) Altered expression of miR-21, miR-31, miR-143 and miR-145 is related to clinicopathologic features of colorectal cancer. Oncology 72:397–402
114. Kuehbacher A, Urbich C, Zeiher AM et al (2007) Role of Dicer and Drosha for endothelial microRNA expression and angiogenesis. Circ Res 101:59–68
115. Melo SA, Kalluri R (2012) Angiogenesis is controlled by miR-27b associated with endothelial tip cells. Blood 119:2439–2440
116. Urbich C, Kuehbacher A, Dimmeler S (2008) Role of microRNAs in vascular diseases, inflammation, and angiogenesis. Cardiovasc Res 79:581–588
117. Chen Y, Gorski DH (2008) Regulation of angiogenesis through a microRNA (miR-130a) that down-regulates antiangiogenic homeobox genes GAX and HOXA5. Blood 111:1217–1226
118. Doebele C, Bonauer A, Fischer A et al (2010) Members of the microRNA-17-92 cluster exhibit a cell-intrinsic antiangiogenic function in endothelial cells. Blood 115:4944–4950
119. Kang SG, Lee WH, Lee YH et al (2012) Hypoxia-inducible factor-1α inhibition by a pyrrolopyrazine metabolite of oltipraz as a consequence of microRNAs 199a-5p and 20a induction. Carcinogenesis 33:661–669

Chapter 25
MicroRNA and Cardiovascular Disorders with a Focus on Angiogenesis

Elena Cavarretta, Giacomo Frati, and Gianluigi Condorelli

Abstract MicroRNAs are endogenous short regulatory RNA molecules that govern gene expression at the posttranscriptional level. They play central roles in a variety of biological processes, and their dysregulation is a feature of many pathologies. This chapter briefly outlines microRNA biogenesis and highlights some microRNAs involved in the pathophysiology of the cardiovascular system.

Keywords microRNA • Cardiovascular diseases • Angiogenesis • Nucleic acid • Messenger RNA • Cardiomyocyte • Endothelial cells

Abbreviations

Ago	Argonaute
AMO	Anti-miRNA oligonucleotide
DGCR8	DiGeorge syndrome critical region gene 8

E. Cavarretta (✉)
Department of Medical-Surgical Sciences and Biotechnologies, University of Rome "Sapienza", Corso della Repubblica 79, 04100 Latina, Italy
e-mail: elena.cavarretta@gmail.com

G. Frati
Department of Medical-Surgical Sciences and Biotechnologies, University of Rome "Sapienza", Corso della Repubblica 79, 04100 Latina, Italy

Department of Angiocardioneurology, IRCCS Neuromed, Pozzilli, Isernia, Italy

G. Condorelli
Humanitas Clinical and Research Center and Institute of Genetics and Biomedical Research, National Research Council of Italy, Via Manzoni 56, 20089 Rozzano (Mi), Italy

Department of Medicine, University of California San Diego, La Jolla, CA 92093-0613C, USA

J.L. Mehta and N.S. Dhalla (eds.), *Biochemical Basis and Therapeutic Implications of Angiogenesis*, Advances in Biochemistry in Health and Disease 6,
DOI 10.1007/978-1-4614-5857-9_25,

ds	Double-stranded
Dyrk1a	Tyrosine-(Y)-phosphorylation regulated kinase 1a
Egfl7	Epidermal growth factor-like domain 7
endo-shRNA	Endogenous short hairpin RNA
endo-siRNA	Endogenous-small-interfering (*endo-si*) RNA
esi-RNA	Endoribonuclease-prepared siRNAs
ERK/MAPK	Extracellular signal-regulated kinase/mitogen-activated protein kinase
ETS	E26 transformation-specific sequence
Exp-5	Exportin-5
GTPase	Guanine triphosphatase
H_2O_2	Hydrogen peroxide
HPC	Hypoxia preconditioning
HUVEC	Human umbilical vein endothelial cells
LNA	Locked nucleic acid
Loqs	Loquacious
miRISC	miRNA-programmed RISC
miRNA	microRNA
miRNome	miRNA transcriptome
MTg-AMO	Multiple-target AMO
ncRNA	Nonprotein-coding RNA
NCX1	Sodium–calcium exchanger 1
NFAT	Nuclear factor of activated T-cells
nt	Nucleotide
PDCD4	Programmed cell death 4
PI3KR2	PI3K regulatory subunit 2
pre-miRNA	Precursor miRNA
pri-miRNA	Primary miRNA
RBD	RNA-binding domain
RISC	RNA-induced silencing complex
RLC	RISC loading complex
RNAi	RNA interference
ROS	Reactive oxygen species
semiRNA	Sibling of esiRNA
SERCA2a	Sarcoplasmic reticulum calcium ATPase 2a
simtrons	Splicing-independent mirtron-like miRNAs
Sirt1	Silent information regulator 1
snoRNA	Small nucleolar RNA
TAR	Transactivation response
TGF-β	Transforming growth factor-beta
TNF-α	Tumor necrosis factor alpha
TRBP	TAR RNA-binding protein
UTR	Untranslated region
VCAM-1	Vascular cell adhesion protein 1
VEGF	Vascular endothelial growth factor

25.1 Introduction

25.1.1 A Brief Historical Perspective

For much of the second half of the twentieth century, RNA was thought either as an intermediate in the flow of information from DNA to the amino acid sequence of proteins (i.e., messenger RNA) or as an infrastructural element (i.e., transfer RNA and ribosomal RNA). Even if there had been suggestions that regulatory genes encoding nonprotein-coding RNA (ncRNA) might exist [1] and that RNA might participate in gene expression [2], the RNA sequences not encoding for a protein (the introns) were considered to have no function. The discovery in the late 1980s that RNA could behave as an enzyme marked a first RNA revolution [3]; later, the discovery of RNA interference (RNAi), the realization that ncRNA comprises the majority of functional transcripts of a cell and the finding that these RNAs perform regulatory tasks, changed the way RNA would be regarded for ever [4].

The proportion of transcribed nonprotein-coding genes within the genome increases with developmental complexity. In mammals, nucleotides (nts) of known protein-coding genes represent only ~2 % of all transcribed bases. A majority is constituted by ncRNA, which accounts for ~70 % of all genes [5]. Long and small ncRNA species have been identified in mammals; one species of ~22 nts-long ncRNA is particularly abundant and represents a pivotal class of gene expression regulators—the microRNAs (miRNAs).

The first miRNA—*lin-4*—was discovered in the nematode *Caenorhabditis elegans* in 1993 [6]. *lin-4* acts by negatively regulating the level of lin-14 protein. The absence of a proper open reading frame in the fragment and the presence of sequences complementary to a repeated sequence element in the 3′ untranslated region (UTR) of *lin-14* RNA suggested an antisense RNA–RNA interaction. The second miRNA—*let-7*—was discovered only in 2000 [7], but since then, miRNA discovery has increased dramatically. miRNAs have now been documented in a variety of animals, plants, and viruses, and roles in developmental biology, in the regulation of physiological processes and in pathogenesis have been elucidated. Several online databases (e.g., http://www.mirbase.org [8], http://www.microrna.org and http://www.ebi.ac.uk/enright-srv/microcosm/htdocs/targets/v5/) contain lists of miRNAs found in over 160 species. Over 1,100 entries have been made hitherto for human miRNAs.

25.2 microRNA Biogenesis

The organization of the mammalian genome is extremely complex: transcripts overlap each other on the same and/or opposite DNA strand and introns of one gene might contain the exons of another [5]. miRNA genes can be found within the introns or exons of host genes, or in intergenic regions [9]. Half of all miRNA genes are found

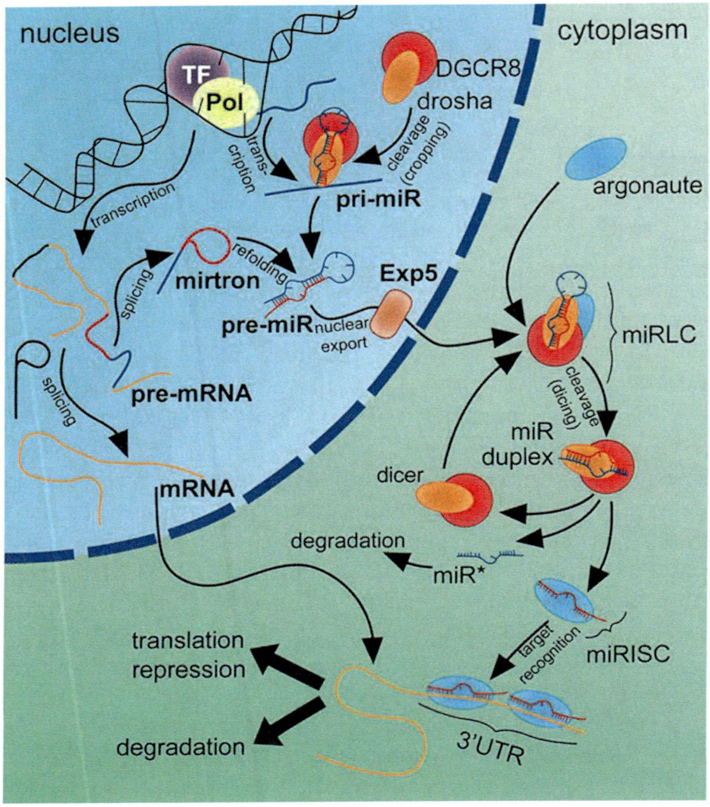

Fig. 25.1 Schematic of microRNA biogenesis and action. See text for explanation. The mature microRNA sequence is given in *red*. *TF* transcription factor; *Pol* RNA polymerase III; *Exp5* exportin 5. (Courtesy of Condorelli G, Latronico MV, Dorn GW 2nd. microRNAs in heart disease: putative novel therapeutic targets? Eur Heart J. (2010) 31(6):649–58)

clustered in polycistronic units, and some miRNA genes are present in multiple copies. Once transcribed, the production of mature miRNAs from small RNAs is a rather complicated process, and a number of pathways have been described (Fig. 25.1).

25.2.1 The Canonical Pathway

RNA polymerase II is mainly responsible for the transcription of miRNA genes; in some cases, the processes are RNA polymerase III-dependent [10, 11]. The primary transcript, a several-kilobases-long primary miRNA (pri-miRNA), is typically spliced, capped, and polyadenylated, and contains one or more long hairpin structures [12]. Pri-miRNAs undergo a two-step processing, first within the nucleus and then in the cytoplasm. The pri-miRNA hairpins have a unique structural feature that distinguishes them from other RNA stem-loop-like structures existing within the nucleus: they harbor a single-strand RNA flanking sequence attached to an extensive stem-loop structure that is recognized by a multiprotein complex called the Microprocessor. The Microprocessor is composed of the ribonuclease Drosha (an RNAse III enzyme), the double-stranded RNA-binding domain (dsRBD) protein Pasha/DGCR8 (DiGeorge syndrome critical region gene 8) and other proteins [13]. The cleavage site recognized by DGCR8 is located approximately 11 bp from the single-strand–double-strand RNA junction [14]. Drosha cleaves the pri-miRNA hairpin co-transcriptionally, before splicing of the host RNA, to generate a 60–100 nts-long hairpin termed a precursor miRNA (pre-miRNA) [15]. RNase III cleavage products contain 5′-phosphate and 3′-hydroxyl ends with a few nts of overhang. This overhang is usually 2 nts long on regular RNA helices, but it can vary from 1 to 4 nts when the helix is distorted [16]. Exportin-5 (Exp-5) recognizes as essential structural elements the 3′ overhang and the double-stranded stem of the pre-miRNA, and in association with the guanine triphosphatase (GTPase) Ran, transports the pre-miRNA to the cytoplasm [17]. RNA recognition by Exp-5:Ran-GTP is independent of the sequence, implying that Exp-5:Ran-GTP can recognize a variety of pre-miRNAs [18]. In the cytoplasm, the final processing needed to obtain a mature miRNA is conducted by another RNase III enzyme that is evolutionarily conserved in worms, flies, plants, fungi, and mammals—the ~200 kDa ribonuclease Dicer. Dicer pairs with the dsRBD protein TRBP/PACT/Loquacious (Loqs) [19]. Recognizing the 3′ ends generated by Drosha, Dicer cleaves the pre-miRNA near the terminal loop to generate a miRNA–miRNA* duplex having 2-nt-long 3′ overhangs at both ends [20, 21]. Dicer acts in cooperation with different proteins to form an RNA-induced silencing complex (RISC) loading complex (RLC): members of the Argonaute (Ago) family, HIV-1 transactivation response (TAR) RNA-binding protein (TRBP), and others [22]. miRNA* (or the passenger strand) is then stripped away and degraded, leaving only one miRNA (or guide) strand loaded in Ago to constitute a miRNA-programmed RISC (or miRISC). miRNA strand selection is probably dependent upon the thermodynamic stability of the miRNA at the 5′ end and other structural features of the miRNA–miRNA* duplex, such as base position mismatch, the 5′ nucleotide identity and sequence composition [23, 24]. The miRNA maturation process is controlled at several levels: in particular, recent studies highlighted that each step is regulated independently, determining a further level of complexity in miRNA-dependent gene expression regulation.

25.2.2 The Noncanonical Pathway

In addition to the canonical pathway, other, "non-canonical," biogenesis pathways exist [25]. Deep-sequencing technologies, followed by extensive bioinformatics-based characterization, have identified several miRNA classes that bypass key steps of the canonical biogenesis pathway—the Drosha- and the Dicer-independent pathways.

In the first category, a class of short introns, called "mirtrons," are spliced out of their host gene in the form of a lariat by a spliceosome and then are debranched and refolded into pre-miRNA hairpins suitable for Dicer. In this way, the Microprocessor step is bypassed. The mirtron pathway converges with the canonical pathway during hairpin export by Exp-5 [26]. Mirtrons have been found in plants, flies, and mammals; in humans, 13 mirtrons have been predicted to exist based on structure, location, and other evidence. Among these are miR-877, miR-1224, and miR-1226 [27].

A Dicer-independent pathway has also been recently identified. Hitherto, only two predicted miRNAs—miR-1225 and miR-1228—are known to be generated by this pathway. These miRNAs have been termed "simtrons" (splicing-independent mirtron-like miRNAs) because they do not require splicing for their biogenesis; in addition, simtron processing does not require DGCR8, Exp-5, or Ago2 [28].

Some small nucleolar RNAs (snoRNAs) can also generate pre-miRNAs that are independent of the Microprocessor step. snoRNAs are ~70–200 nts long and act in ribonucleoprotein complexes to guide enzymatic modulation of ribosomal RNA. Some have a secondary structure characterized by two pre-miRNA-like hairpins linked by a hinge [29], which can be processed by Ago1 and Ago2 into miRNAs.

Another category of Dicer-dependent, DGCR-independent miRNAs derive from endogenous short hairpin RNAs (endo-shRNAs) and tRNA precursors, that, like snoRNAs, are a dual-function molecule, behaving as both miRNAs and tRNA precursors. The transcript forms a secondary structure with two potential pre-miRNA-like hairpins, but only the second one produces miRNAs (miR-1980, miR-320) [30].

Last but not least, miRNAs can derive from a terminal hairpin of an endogenous small interfering RNA precursor (endo-siRNAs or esiRNAs), which are a product of long endogenous dsRNAs. Endo-siRNAs are produced by Dicer2 and loaded onto Ago2. Moreover, they depend on Loqs, the canonical partner of Dicer1 in the miRNA pathway, which is associated with the Loqs isoforms Loqs-PA and -PB. Dicer2 interacts with another isoform, Loqs-PD. The production of endo-siRNAs strongly depends on Loqs-PD, but this complex cannot process the terminal hairpins, which are instead identified as miRNA precursors and processed into miRNAs by the Dicer1-Loqs-PA/PB complex [25]. This unique subclass is named "siblings of esiRNAs" or "semiRNAs." One example of independency from Dicer cleavage is the processing of miR-451, where Drosha catalyzes the excision of pre-miR-451 from its primary transcript, which is then cleaved by Ago2 instead.

25.3 Mechanisms of microRNA Action

Once integrated into the miRISC (or mirgonaute), the miRNA directs the ribonucleoprotein complex to binding sites that are usually, but not always, found on the 3′ UTR of the target mRNAs. Typically, there are more than one complementary binding sites present on the target 3′UTR, so that several miRNAs can bind synergically to provide stronger repression. miRNAs act as down-regulators of gene expression, through a number of mechanisms: translational arrest, mRNA deadenylation, destabilization, and degradation. Based on ribosome profiling, it seems that the predominant mechanism is the destabilization of the target mRNA, leading to impaired protein production. A few miRNAs are known to act reversely and the miRISC may even activate target expression in some circumstances [31]. Because of the 5′ end-restricted complementarity to mRNA targets, it has been proposed that miRNAs can regulate a large number of genes, but the targets with biological relevance are fewer in animals compared with plants. Algorithms based on different criteria have been developed in order to find putative targets bioinformatically: these include Diana-MicroT (http://www.diana.pcbi.upenn.edu/cgi-bin/micro_t.cgi), PicTar (http://pictar.bio.nyu.edu), miRanda (http://www.microrna.org/miranda_new.html), and TargetScan (http://www.targetscan.org) [32]. However, not all complementary sequences identified by these algorithms are necessarily bona fide binding sites, so targets must be validated experimentally.

MiRNAs can act as on-off switches to degrade specific mRNAs ("switch targets") that should not be expressed in that particular cell type or at that moment. miRNAs can also operate to fine-tune mRNA abundance ("tuning targets"), adjusting levels depending on environmental setting, within physiological ranges. Other mRNAs ("neutral targets") can be simple bystanders, in which the down-regulation induced by miRNAs is inconsistent. mRNAs that are pivotal for the activity of a specific cell type ("anti-targets") must not contain binding sites for the miRNAs expressed by that cell [33]. Thus, each cell type has a specific miRNA profile—the miRNA transcriptome or "miRNome."

25.4 MicroRNA and Cardiac Pathophysiology

MiRNAs are crucial for the correct development of animals and humans, regulating the structural growth and function of tissue and organs. Progressively, data in the literature are corroborating, in adult animals, the regulation of a multitude of pathophysiological processes, such as metabolism, survival, and the immune response, by miRNA. The ~1,000 miRNAs estimated to be encoded in the human genome are predicted to regulate at least a third of all genes [34]. Reports have recently started to be published on the role of miRNAs in the cardiovascular system.

25.4.1 MicroRNAs in Cardiac Development

MiRNA-mediated fine modulation of gene expression starts from an early phase of embryogenesis and has been shown to play a pivotal role in cell lineage commitment and embryonic tissue development by temporal activation or inactivation of specific targets. A number of miRNAs have been shown to be involved in cardiogenesis, such as miR-1, let-7, miR-133, miR-126, miR-30c, miR-26a, and miR-208 [35]. MiR-1 and miR-133, which are encoded by the same duplicated bicistronic locus, play essential roles in muscle proliferation and differentiation [36].

25.4.2 microRNAs in Cardiac Hypertrophy and Failure

Under physiological stimuli—such as physical training—or pathological states—such as hypertension and heart failure—the myocardium reacts with gene expression profile changes. This micro-adaptation results macroscopically in cardiac remodeling, which is characterized by structural alterations of myocardial tissue, modification of the extracellular matrix, and reshaping of left ventricular geometry and performance. Several reports have highlighted fundamental roles for miRNA in cardiac hypertrophic growth and heart failure. Surprisingly, a strong similarity has been found between the miRNA expression pattern in human failing heart and that of 12–14-week-old fetuses [37]. In particular, miR-1, miR-30, miR-133, and miR-150 are often down-regulated, while miR-21, miR-23a, miR-125, miR-195, miR-199, and miR-214 are often up-regulated. Some of these miRNAs have been tested in model systems to see whether or not their misexpression is sufficient to produce an effect on the phenotype. For example, inhibition of miR-133, which has been found down-regulated in human hypertrophic cardiomyopathy, was sufficient to provoke hypertrophy in wild-type mice [38]. Databases are being developed to provide comprehensive lists of the dysregulated miRNAs found in various diseases and models, including those of the heart (for example, http://www.mir2Disease.org) [39]. Furthermore, miR-1 expression has been inversely correlated with cardiac hypertrophy and associated with regulation of hypertrophy-associated genes, such as calmodulin, Mef2a, and Gata4, key transcription factors that mediate calcium-dependent changes in gene expression [40]. Although some authors [36, 41] reported up-regulation of miR-1, other authors found a reduction in miR-1 in ischemic and non-ischemic dilated cardiomyopathies, which are associated with a significant reduction in ejection fraction [42–44]. Impaired myocardial sarcoplasmic reticulum calcium ATPase 2a (SERCA2a) activity is a hallmark of failing hearts, and SERCA2a gene therapy improves cardiac function in animals and patients with heart failure. SERCA2a gene therapy of failing hearts restored miR-1 expression by an Akt/FoxO3A-dependent pathway, which is associated with normalized enhanced sodium–calcium exchanger 1 (NCX1) expression and improved cardiac function [45].

The first report of a role for miR-21 in cardiac pathology was conducted on a murine model of pressure overload-induced heart failure: Thum et al. showed that miR-21 is a critical regulator of extracellular signal-regulated kinase/mitogen-activated protein kinase (ERK/MAPK) activity in cardiac fibroblasts, enhancing fibrosis via inhibition of Sprouty 1 [46]. Inhibition of miR-21, which is usually up-regulated in cardiac hypertrophy, was sufficient to inhibit hypertrophic growth in vitro. miR-21, which has been implicated also in human tumorigenesis and progression, is up-regulated by transforming growth factor-beta (TGF-β). In the endothelium, this growth factor increases expression of both primary and mature miR-21: after exposure to TGF-β, mature miR-21 expression is much higher in comparison with the primary transcript [47].

In vivo, over-expression of miR-195, which is usually up-regulated in cardiac hypertrophy, was also found to be sufficient to lead to hypertrophy, whereas a gene mutation or "decoy" approach has demonstrated roles for miR-208 and miR-133 too [48]. Although miR-1 and miR-133 are enriched in both skeletal and cardiac muscle, miR-208 is cardiac restricted. It has two isoforms, miR-208a and miR-208b, which are contained within the introns of cardiac muscle myosin heavy chain genes *Myh6* (*αMyh*) and *Myh7* (*βMyh*), respectively [49]. Together with miR-499, they have been termed "myomiRs."

Whereas miR-133a regulates cardiac hypertrophy, miR-29b is involved in cardiac fibrosis. A relationship between miR-133a and collagen 1A1 has been demonstrated, suggesting that myocardial fibrosis occurring in angiotensin II-dependent hypertension is regulated by the down-regulation of miR-133a and miR-29b through the modulation of collagen 1A1 expression [50]. miR-23, which is clustered with miR-27a and miR-24, is also up-regulated in pressure overload- and isoproterenol-induced hypertrophy [51]. The nuclear factor of activated T-cells (NFAT) c3 could directly activate miR-23a expression through the transcriptional machinery. The muscle-specific ring finger protein 1, an anti-hypertrophic protein, was identified as a target of miR-23a, and if suppressed can lead to cardiac hypertrophy. Moreover, Da Costa Martins et al. [52] demonstrated that miR-199b, which is up-regulated during cardiac hypertrophy, is also a direct transcriptional target of NFAT. In particular, miR-199b targets and suppresses the NFAT kinase dual specificity tyrosine-(Y)-phosphorylation regulated kinase 1a (Dyrk1a), which is a negative regulator of NFAT. Conversely, antisense "antagomir" treatment (see below) normalized miR-199a levels and not only prevented hypertrophy and failure, but also almost completely reversed them [53].

25.4.3 miRNAs in Ischemic Heart Disease

MiRNAs are rapidly altered upon exposure of cells or organs to hypoxia or ischemia, which provoke apoptotic, autophagic, and necrotic cell death. There is a bimodal response to hypoxia and ischemia: initially there is an adaptive, conditioning reaction that transitions to cell death upon persistence of the ischemic insult. Based on this, brief exposure to hypoxia or ischemia leads to an anti-apoptotic response, usually with AKT pathway activation, which protects the cell against protracted periods of hypoxia or ischemia that may follow [54, 55]. This phenomenon is termed early hypoxia preconditioning (HPC) or ischemia preconditioning, and miRNAs are involved in the regulation of this process. In particular, miR-199a-5p is acutely down-regulated through a posttranscriptional mechanism, and its effects are mediated by AKT [56]. Conversely, AKT induces up-regulation of miR-21 [57], which inhibits apoptosis by targeting Fas ligand and programmed cell death 4 (PDCD4). In addition, miR-21 is up-regulated after treatment with hydrogen peroxide (H_2O_2) and participates in H_2O_2-mediated gene regulation and functional modulation in the protection of cardiac myocytes against heart diseases related to reactive oxygen species (ROS) [58].

Similarly to miR-21, miR-494 becomes down-regulated in the infarct area. Although miR-494 targets both pro-apoptotic (PTEN, ROCK1, and CaMKIIδ) and anti-apoptotic proteins (FGFR2 and LIF), the ultimate consequence is activation of the AKT pathway, leading to cardio-protective effects against ischemia/reperfusion-induced injury [59]. Normalizing its levels by transgenic over-expression reduced infarct size and improved contractility.

Moreover, miR-214 becomes up-regulated during ischemic injury and heart failure. The cardio-protective roles of miR-214 during ischemia/reperfusion injury were attributed to suppression of the mRNA encoding NCX1, a key regulator of Ca^{2+} influx, and to the repression of several downstream effectors of Ca^{2+} signaling that mediates cell death [60].

The miR-15 family, which includes six closely related miRNAs, is also regulated in infarcted zones. Systemic delivery of miR-15 anti-miRNAs dose-dependently repressed miR-15 in cardiac tissue of mice and pigs, whereas therapeutic targeting of miR-15 in mice reduces infarct size and cardiac remodeling and enhances cardiac function [61]. miRNAs also regulate angiogenesis and, thus, influence the rate of tissue healing after ischemia.

25.5 microRNA and the Vascular System

25.5.1 microRNAs in Vascular Development

Several miRNAs have been implicated in the development of the vascular system, which requires the formation and remodeling of a contiguous series of lumenized tubes constituted by endothelial cells [62]. At the same time, vascular smooth muscle cells (VSMC) are recruited to the endothelial plexus to give the necessary tone and contractility to the vessel. The first evidence that linked miRNA to endothelial cells was the observation that silencing Dicer impairs angiogenesis in vitro [63]. A pivotal role is attributed to the endothelial cell-specific miRNA miR-126, which is highly expressed in endothelial cells in murine lung and heart, and to a lesser extent in the brain, liver, and kidney, and regulates angiogenesis and vascular inflammation. The miR-126 gene is found in an intron of *epidermal growth factor-like domain 7* (*Egfl7*), which encodes for a protein involved in the spatial control of cell migration [64]. miR-126 is induced by blood flow and is required in a flow-induced genetic pathway to develop the angiogenic sprouting of aortic arch vessels [65] by stimulation of vascular endothelial growth factor (VEGF) signaling. VEGF-mediated down-regulation of miR-101 causes pro-angiogenic effects and provides a functional link between growth factor-mediated signaling, posttranscriptional silencing, and histone methylation [66].

Moreover, miR-143 and miR-145—encoded by a bicistronic pre-miRNA—are expressed specifically in VSMCs and target several regulators of actin signaling [67].

They act as integral components of the regulatory network whereby serum response factor and its coactivator, myocardin, control cytoskeletal remodeling, and phenotypic switching of VSMCs. In addition, miR-24 and miR-23a are selectively expressed in microvascular endothelial cells in vivo, whereas miR-145 is expressed in pericytes [68]. miR-133 is robustly expressed in VSMCs in vitro and in vivo and plays a mechanistic role in VSMC growth, whereas miR-1 vascular levels are negligible. miR-133 has a potent inhibitory role on VSMC phenotypic switching in vitro and in vivo. miR-133 decreases when VSMCs are primed to proliferate in vitro and following vascular injury in vivo; it increases when VSMCs are coaxed back to quiescence in vitro and in vivo [69].

25.5.2 microRNAs in Angiogenesis

Endothelial miRNAs can be classified on the basis of the bidirectional relationship between miRNAs and angiogenesis into two groups: (1) miRNAs that target genes implicated in angiogenesis (i.e., "angiomiRs"); and (2) miRNAs whose expression can be modulated by pro-angiogenic or anti-angiogenic stimuli [70].

25.5.2.1 AngioMirs

"AngiomiRs" regulate angiogenesis posttranscriptionally, either cell-autonomously or non cell-autonomously. In particular, pro-angiomiRs suppress negative regulators in angiogenic signaling pathways, whereas anti-angiomiRs inhibit angiogenesis by targeting positive regulators of angiogenesis [71]. Representative angiomiRs include miR-126, miR-221/222, miR-23/27, and the miR-17-92 cluster. miR-126 is required for angiogenesis and maintenance of vascular integrity. miR-126 pro-angiogenic action is in part mediated by promoting MAP kinase and PI3K signaling in response to VEGF and FGF, targeting the Sprouty-related EVH domain-containing protein Spred-1 and PI3K regulatory subunit 2 (PI3KR2/p85-β), which are negative regulators of the signaling pathways [72]. The transcription factors E26 transformation-specific sequence (ETS)-1 and ETS-2 play a key role in controlling the expression of miR-126 [73] as it is induced by inflammatory mediators, such as tumor necrosis factor alpha (TNF-α). miR-126 targets also vascular cell adhesion protein 1 (VCAM-1) to regulate leukocyte homing in the vessel wall, suggesting a role in vascular inflammation, by reducing TNF-α-induced expression of VCAM-1 [74].

The multifunctional miRNA miR-126 has also been implicated in tumourigenesis: it is down-regulated in many cancer lines, such as those from stomach, lung, cervix, pancreas, and prostate [75–77]. Various molecular signal cascades seem involved, and miR-126 has opposite biological effects on VEGF regulation in different cancer cell lines [78].

The polycistronic miR-17–92 cluster, also named "oncomiR-1" [79], is composed of seven mature miRNAs: miR-17-5p and -3p, miR-18, miR-19a and b, miR20a, and miR-92a. Some of the first observations to support the oncogenic action of miR-17–92 came from mouse B cell lymphoma [79], in which it collaborated with the c-myc oncogene to accelerate B lymphoma genesis. Over-expression of miR-17–92 in Ras-expressing cells promotes tumor angiogenesis in vivo [80]. c-Myc up-regulates the miR-17–92 cluster and down-regulates the angiogenesis inhibitor thrombospondin-1, along with other members of the thrombospondin type 1 repeat superfamily. Recently, a pathway has been described in which c-Myc activation of miR-17–92 attenuates the TGF-β signaling pathway to shut down clusterin expression, thereby stimulating angiogenesis and tumor cell growth [81]. The exact biological activities of each miR-17–92 component have not been functionally dissected. Recently, a role for miR-17–92 has been suggested also in cardiac development, in developmental angiogenesis, and endothelial cell proliferation.

The miR-23-27-24 cluster has also been reported to have an important role in angiogenesis in that miR-23 and miR-27 enhance angiogenesis by targeting Sprouty2 and Sema6A proteins, which promote anti-angiogenic activity [82]. Differently, over-expression of the anti-angiogenic miR-221/222 suppressed the angiogenic responses to stem cell factor [83].

miR-181b has been implicated in chronic lymphocytic leukemia, hepatic stellate cell proliferation and has been found to regulate NF-κB-mediated endothelial cell activation and vascular inflammation in response to pro-inflammatory stimuli. In this setting, miR-181b expression could provide a new target for anti-inflammatory therapy and critical illness [84].

25.5.2.2 microRNAs Modulated by Pro-Angiogenic Stimuli and by Hypoxia

VEGF-A induces time-dependent expression of miR-191, miR-155, miR-31, miR-17-5p, miR-18a, and miR-20a, with little change in miR-126 and miR-222 [70, 85]. In the presence of VEGF-A and epidermal growth factor, the expression of the pro-angiogenic miRNA miR-296 was significantly increased in brain endothelial cells [86]. VEGF-mediated down-regulation of miR-101 caused pro-angiogenic effects; in particular, the suppression operated by miR-101 of the histone methyltransferase EZH2—a member of the Polycomb group family—increased methylation of histone H3 at lysine 27 and transcriptome alterations [66].

miR-210, which is induced by hypoxia, is sufficient for inducing tubulogenesis of endothelial cells [87]. Moreover, recent findings show that supplementing the myocardium with miR-210 increases angiogenesis in the peri-infarct zone, reduces apoptosis and infarct size, and results in improved cardiac function following left anterior descending coronary artery ligation [88].

25.5.2.3 microRNAs Modulated by Anti-Angiogenic Stimuli

ROS increased miR-200c and the co-transcribed miR-141 more than eightfold, with miR-200c over-expression inducing growth arrest, apoptosis, and senescence in human umbilical vein endothelial cells (HUVEC). The mechanism of H_2O_2-mediated miR-200c up-regulation involves p53 and retinoblastoma proteins [89]. Very recently, miR-200c was recognized as a novel negative regulator of the pro-apoptotic Bcl-2 family member Noxa [90].

Diabetes is a pathology characterized by damaged endothelial function. Conditions mimicking diabetes mellitus—high D-glucose—and ischemia-associated starvation—low growth factors—were found to up-regulate miR-503, which impairs endothelial cell proliferation, migration, adhesion, and network formation capacities. miR-503 also reduced VSMC proliferation and migration, which are instrumental for arteriogenesis [91]. miR-34a, recently reported as a tumor suppressor, has been found to target silent information regulator 1 (Sirt1), leading to cell cycle arrest or apoptosis and impairment of endothelial progenitor cells-mediated angiogenesis by induction of senescence [92].

25.6 Therapeutic Prospective

The discovery of miRNAs has opened up avenues for diagnostic tools and targeted molecular therapies. On the one hand, miRNA profile expression has been shown to serve as an impressive phenotypic signature; on the other, specific miRNAs might have prognostic significance, defining the response to therapy [93]. Another exciting way of translating the discovery of miRNAs to the clinic lies in the development of miRNA-based therapies and therapeutics aimed at modifying miRNA expression [93]. There are essentially two potential strategies to regulate these regulators, based on whether the specific miRNAs are up- or down-regulated with disease (Fig. 25.2). In the case of up-regulation, the effects of excessively expressed miRNAs have to be antagonized at the binding site of the mRNA targets [94]. Anti-miRNA oligonucleotides (AMOs or anti-miRs) are antisense nucleotides that are perfectly complementary to a specific mature miRNA. They act as competitive inhibitors, scavenging miRNAs and hampering the hybridization with the 3′UTR binding site. Unfortunately, AMOs are susceptible to nuclease degradation, so a chemical modification must be introduced to improve their pharmacokinetic characteristics.

AntagomiRs are AMOs synthesized with a 2′-*O*-methyl-group—which improves resistance to nucleases—partial phosphorothioate linkage—to ameliorate tissue distribution—and conjugated to cholesterol at the 3′ end of the single strand—to augment cellular uptake and to reduce renal clearance. However, accurate prediction of a comprehensive set of putative antagomiRs against miRNAs is still an open problem. A list of putative antagomiR–miRNA heterodimers is available at http://bioinfopresidencycollegekolkata.edu.in/antagomirs.html [95].

So-called "miRNA sponges" have been synthesized to address the need for multiple miRNA-binding site inhibition. These have mismatches at sites normally cleaved by Ago2 to interfere with the degradation. Similarly, multiple antisense unit constructs have been engineered, the multiple-target AMOs (MTg-AMOs), for the muscle-specific miRNAs miR-1 and miR-133 [96].

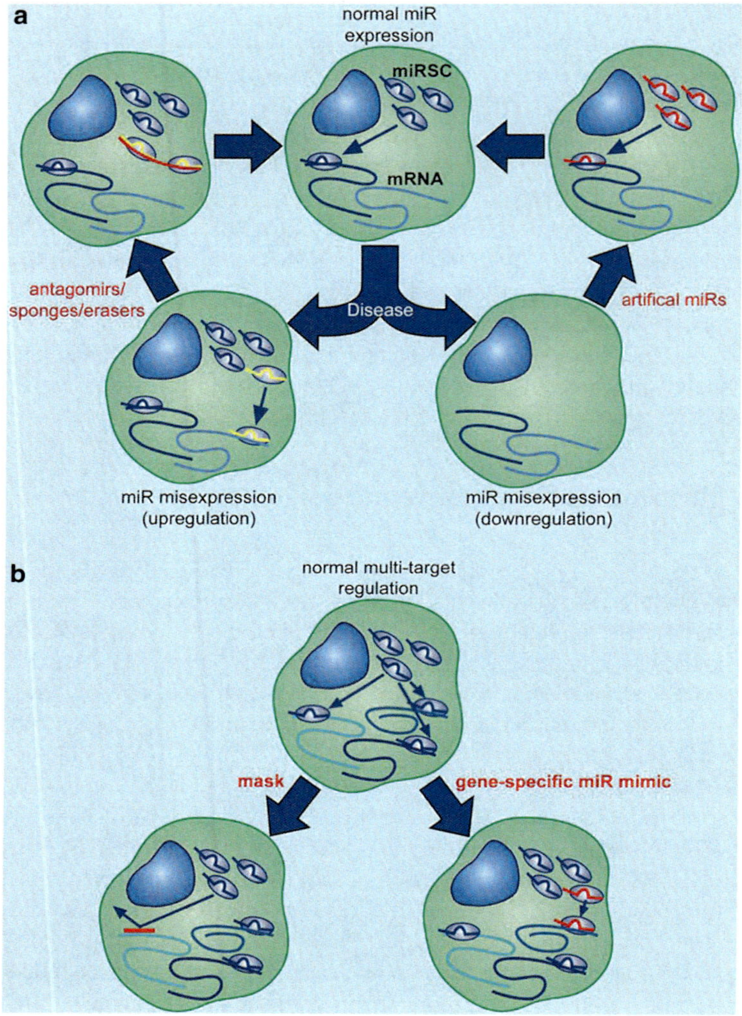

Fig. 25.2 Schematic overview of strategies used to alter microRNA expression. (**a**) Cells express a microRNA profile that can become altered with disease. Antisense oligonucleotides, such as antagomirs, sponges, and erasers (in *red*) can capture microRNAs for knockdown or sequestrate inappropriately overexpressed microRNAs, whereas artificially introduced microRNAs (in *red*) can be used to overexpress microRNAs or, potentially, to replace expression of down-regulated ones. These strategies have the potential to affect large numbers of different targets (for simplicity, only one target mRNA per microRNA is represented). (**b**) Masks and gene-specific microRNA mimics (in *red*) can be used to affect single targets specifically (mRNAs in different shades of blue represent a set affected by a given microRNA). (Courtesy of Condorelli G, Latronico MV, Dorn GW 2nd. microRNAs in heart disease: putative novel therapeutic targets? Eur Heart J. (2010) 31(6):649–58)

Two other main chemical oligonucleotide alterations used to synthesize AMOs are 2′-*O*-methoxyethyl and locked nucleic acid (LNA) modifications. The therapeutic power of the LNA oligonucleotides, so-called "third generation" modified nucleotide analogs, have recently been demonstrated [97].

Conversely, if the miRNAs are down-regulated during disease, exogenous miRNA-mimicking molecules must be supplied to counterbalance the missing miRNAs. In order to be incorporated into the RISC, miRNA mimics must be introduced into the cell as short double-stranded oligonucleotides that resemble miRNA precursors. Moreover, it is necessary to employ viral vectors to focus on tissue- or cell-specific delivery, and thereby limit off-target effects. The most promising vectors are those bioengineered from adeno-associated viruses [98]. While the methods to down-regulate miRNA expression are reliable, mimic technology requires further development. Much hope has been laid on miRNA-based therapeutics, but the clinical potential still needs to be fulfilled.

Acknowledgments This work was supported by Foundation LeDucq and Fondazione CARIPLO (GC). We thank Michael V.G. Latronico for critical reading and suggestions.

References

1. Jacob F, Monod J (1961) Genetic regulatory mechanisms in the synthesis of proteins. J Mol Biol 3:318–356
2. Britten RJ, Davidson EH (1969) Gene regulation for higher cells: a theory. Science 165:349–357
3. Zaug AJ, Cech TR (1986) The intervening sequence RNA of Tetrahymena is an enzyme. Science 231:470–475
4. Kong Y, Han JH (2005) MicroRNA: biological and computational perspective. Genomics Proteomics Bioinformatics 3:62–72
5. Frith MC, Pheasant M, Mattick JS (2005) The amazing complexity of the human transcriptome. Eur J Hum Genet 13:894–897
6. Lee RC, Feinbaum RL, Ambros V (1993) The C. elegans heterochronic gene lin-4 encodes small RNAs with antisense complementarity to lin-14. Cell 75(5):843–854
7. Reinhart BJ, Slack FJ, Basson M et al (2000) The 21-nucleotide let-7 RNA regulates developmental timing in Caenorhabditis elegans. Nature 403:901–906
8. Griffiths-Jones S, Saini HK, van Dongen S, Enright AJ (2008) miRBase: tools for microRNA genomics. Nucleic Acids Res 36:D154–D158
9. Kim VN, Nam JW (2006) Genomics of microRNA. Trends Genet 22:165–173
10. Lee Y, Kim M, Han J et al (2004) MicroRNA genes are transcribed by RNA polymerase II. EMBO J 23:4051–4060
11. Borchert GM, Lanier W, Davidson BL (2006) RNA polymerase III transcribes human microRNAs. Nat Struct Mol Biol 13:1097–1101
12. Cai X, Hagedorn CH, Cullen BR (2004) Human microRNAs are processed from capped, polyadenylated transcripts that can also function as mRNAs. RNA 10:1957–1966
13. Zeng Y, Cullen BR (2005) Efficient processing of primary microRNA hairpins by Drosha requires flanking nonstructured RNA sequences. J Biol Chem 280:27595–27603
14. Han J, Lee Y, Yeom KH et al (2006) Molecular basis for the recognition of primary microRNAs by the Drosha-DGCR8 complex. Cell 125:887–901

15. Lee Y, Ahn C, Han J, Choi H et al (2003) The nuclear RNase III Drosha initiates microRNA processing. Nature 425:415–419
16. Basyuk E, Suavet F, Doglio A et al (2003) Human let-7 stem-loop precursors harbor features of RNase III cleavage products. Nucleic Acids Res 31:6593–6597
17. Zeng Y, Cullen BR (2004) Structural requirements for pre-microRNA binding and nuclear export by Exportin 5. Nucleic Acids Res 32:4776–4785
18. Okada C, Yamashita E, Lee SJ et al (2009) A high-resolution structure of the pre-microRNA nuclear export machinery. Science 326:1275–1279
19. Murphy D, Dancis B, Brown JR (2008) The evolution of core proteins involved in microRNA biogenesis. BMC Evol Biol 8:92
20. Bernstein E, Caudy AA, Hammond SM, Hannon GJ (2001) Role for a bidentate ribonuclease in the initiation step of RNA interference. Nature 409:363–366
21. Hutvágner G, McLachlan J, Pasquinelli AE, Bálint É et al (2001) A cellular function for the RNA-interference enzyme Dicer in the maturation of the let-7 small temporal RNA. Science 293:834–838
22. Starega-Roslan J, Koscianska E, Kozlowski P, Krzyzosiak WJ (2011) The role of the precursor structure in the biogenesis of microRNA. Cell Mol Life Sci 68:2859–2871
23. Schwarz DS, Hutvágner G, Du T, Xu Z et al (2003) Asymmetry in the assembly of the RNAi enzyme complex. Cell 115:199–208
24. Okamura K, Liu N, Lai EC (2009) Distinct mechanisms for microRNA strand selection by Drosophila Argonautes. Mol Cell 36:431–444
25. Miyoshi K, Miyoshi T, Siomi H (2010) Many ways to generate microRNA-like small RNAs: non-canonical pathways for microRNA production. Mol Genet Genomics 284:95–103
26. Westholm JO, Lai EC (2011) Mirtrons: microRNA biogenesis via splicing. Biochimie 93:1897–1904
27. Sibley CR, Seow Y, Saayman S et al (2012) The biogenesis and characterization of mammalian microRNAs of mirtron origin. Nucleic Acids Res 4:438–448
28. Havens MA, Reich AA, Duelli DM, Hastings ML (2012) Biogenesis of mammalian microRNAs by a non-canonical processing pathway. Nucleic Acids Res 40:4626–4640
29. Ender C, Krek A, Friedländer MR, Beitzinger M et al (2008) A human snoRNA with microRNA-like functions. Mol Cell 32:519–528
30. Babiarz JE, Ruby JG, Wang Y et al (2008) Mouse ES cells express endogenous shRNAs, siRNAs, and other Microprocessor-independent, Dicer-dependent small RNAs. Genes Dev 22:2773–2785
31. Vasudevan S, Tong Y, Steitz JA (2007) Switching from repression to activation: microRNAs can up-regulate translation. Science 318:1931–1934
32. Latronico MV, Catalucci D, Condorelli G (2007) Emerging role of microRNAs in cardiovascular biology. Circ Res 101:1225–1236
33. Bartel DP, Chen CZ (2004) Micromanagers of gene expression: the potentially widespread inf. uence of metazoan microrNAs. Nat Rev Genet 5:396–400
34. Catalucci D, Gallo P, Condorelli G (2009) MicroRNAs in cardiovascular biology and heart disease. Circ Cardiovasc Genet 2:402–408
35. Thum T, Catalucci D, Bauersachs J (2008) MicroRNAs: novel regulators in cardiac development and disease. Cardiovasc Res 79:562–570
36. Chen JF, Mandel EM, Thomson JM et al (2006) The role of microRNA-1 and microRNA-133 in skeletal muscle proliferation and differentiation. Nat Genet 38:228–233
37. Thum T, Galuppo P, Wolf C et al (2007) MicroRNAs in the human heart: a clue to fetal gene reprogramming in heart failure. Circulation 116:258–267
38. Carè A, Catalucci D, Felicetti F et al (2007) MicroRNA-133 controls cardiac hypertrophy. Nat Med 13:613–618
39. Jiang Q, Wang Y, Hao Y et al (2009) miR2Disease: a manually curated database for microRNA deregulation in human disease. Nucleic Acids Res 37:D98–D104
40. Ikeda S, He A, Kong SW et al (2009) MicroRNA-1 negatively regulates expression of the hypertrophy-associated calmodulin and Mef2a genes. Mol Cell Biol 29:2193–2204

41. Matkovich SJ, Van Booven DJ, Youker KA et al (2009) Reciprocal regulation of myocardial microRNAs and messenger RNA in human cardiomyopathy and reversal of the microRNA signature by biomechanical support. Circulation 119:1263–1271
42. Ikeda S, Kong SW, Lu J et al (2007) Altered microRNA expression in human heart disease. Physiol Genomics 31:367–373
43. Sucharov C, Bristow MR, Port JD (2008) miRNA expression in the failing human heart: functional correlates. J Mol Cell Cardiol 45:185–192
44. Naga Prasad SV, Duan ZH, Gupta MK et al (2009) Unique microRNA profile in end-stage heart failure indicates alterations in specific cardiovascular signaling networks. J Biol Chem 284:27487–27499
45. Kumarswamy R, Lyon AR, Volkmann I et al (2012) SERCA2a gene therapy restores microRNA-1 expression in heart failure via an Akt/FoxO3A-dependent pathway. Eur Heart J 33:1067–1075
46. Thum T, Gross C, Fiedler J et al (2008) MicroRNA-21 contributes to myocardial disease by stimulating MAP kinase signalling in fibroblasts. Nature 456:980–984
47. Cavarretta E, Latronico MV, Condorelli G (2012) Endothelial-to-mesenchymal transition and microRNA-21: the game is on again. Arterioscler Thromb Vasc Biol 32:165–166
48. Zhang C (2010) MicroRNAs in vascular biology and vascular disease. J Cardiovasc Transl Res 3:235–240
49. Callis TE, Pandya K, Seok HY et al (2009) MicroRNA-208a is a regulator of cardiac hypertrophy and conduction in mice. J Clin Invest 119:2772–2786
50. Castoldi G, Di Gioia CR, Bombardi C et al (2012) MiR-133a regulates collagen 1A1: potential role of miR-133a in myocardial fibrosis in angiotensin II-dependent hypertension. J Cell Physiol 227:850–856
51. Lin Z, Murtaza I, Wang K, Jiao J et al (2009) miR-23a functions downstream of NFATc3 to regulate cardiac hypertrophy. Proc Natl Acad Sci USA 106:12103–12108
52. da Costa Martins PA, Salic K, Gladka MM et al (2010) MicroRNA-199b targets the nuclear kinase Dyrk1a in an auto-amplification loop promoting calcineurin/NFAT signalling. Nat Cell Biol 12:1220–1227
53. Han M, Toli J, Abdellatif M (2011) MicroRNAs in the cardiovascular system. Curr Opin Cardiol 26:181–189
54. Ye Y, Perez-Polo JR, Qian J, Birnbaum Y (2011) The role of microRNA in modulating myocardial ischemia-reperfusion injury. Physiol Genomics 43:534–542
55. Hausenloy DJ, Tsang A, Mocanu MM, Yellon DM (2005) Ischemic preconditioning protects by activating prosurvival kinases at reperfusion. Am J Physiol Heart Circ Physiol 288:H971–H976
56. Rane S, He M, Sayed D et al (2010) An antagonism between the AKT and beta-adrenergic signaling pathways mediated through their reciprocal effects on miR-199a-5p. Cell Signal 22:1054–1062
57. Sayed D, He M, Hong C, Gao S et al (2010) MicroRNA-21 is a downstream effector of AKT that mediates its antiapoptotic effects via suppression of Fas ligand. J Biol Chem 285:20281–20290
58. Cheng Y, Liu X, Zhang S et al (2009) MicroRNA-21 protects against the H(2)O(2)-induced injury on cardiac myocytes via its target gene PDCD4. J Mol Cell Cardiol 47:5–14
59. Wang X, Zhang X, Ren XP et al (2010) MicroRNA-494 targeting both proapoptotic and antiapoptotic proteins protect against ischemia/reperfusion-induced cardiac injury. Circulation 122:1308–1318
60. Aurora AB, Mahmoud AI, Luo X et al (2012) MicroRNA-214 protects the mouse heart from ischemic injury by controlling Ca^{2+} overload and cell death. J Clin Invest 122:1222–1232
61. Hullinger TG, Montgomery RL, Seto AG et al (2012) Inhibition of miR-15 protects against cardiac ischemic injury. Circ Res 110:71–81
62. Small EM, Olson EN (2011) Pervasive roles of microRNAs in cardiovascular biology. Nature 469:336–342
63. Kuehbacher A, Urbich C, Zeiher AM, Dimmeler S (2007) Role of Dicer and Drosha for endothelial microRNA expression and angiogenesis. Circ Res 101:59–68

64. Schmidt M, Paes K, De Mazière A et al (2007) EGFL7 regulates the collective migration of endothelial cells by restricting their spatial distribution. Development 134:2913–2923
65. Nicoli S, Standley C, Walker P et al (2010) MicroRNA-mediated integration of haemodynamics and Vegf signalling during angiogenesis. Nature 464(7292):1196–1200
66. Smits M, Mir SE, Nilsson RJ et al (2011) Down-regulation of miR-101 in endothelial cells promotes blood vessel formation through reduced repression of EZH2. PLoS One 6:e16282
67. Xin M, Small EM, Sutherland LB et al (2009) MicroRNAs miR-143 and miR-145 modulate cytoskeletal dynamics and responsiveness of smooth muscle cells to injury. Genes Dev 23:2166–2178
68. Larsson E, Fredlund Fuchs P et al (2009) Discovery of microvascular miRNAs using public gene expression data: miR-145 is expressed in pericytes and is a regulator of Fli1. Genome Med 1:108
69. Torella D, Iaconetti C, Catalucci D et al (2011) MicroRNA-133 controls vascular smooth muscle cell phenotypic switch in vitro and vascular remodeling in vivo. Circ Res 109(8):880–893
70. Caporali A, Emanueli C (2011) MicroRNA regulation in angiogenesis. Vascul Pharmacol 55:79–86
71. Wang S, Olson EN (2009) AngiomiRs—key regulators of angiogenesis. Curr Opin Genet Dev 19:205–211
72. Fish JE, Santoro MM, Morton SU et al (2008) miR-126 regulates angiogenic signaling and vascular integrity. Dev Cell 15:272–284
73. Harris TA, Yamakuchi M, Kondo M et al (2010) Ets-1 and Ets-2 regulate the expression of microRNA-126 in endothelial cells. Arterioscler Thromb Vasc Biol 30(10):1990–1997
74. Harris TA, Yamakuchi M, Ferlito M et al (2008) MicroRNA-126 regulates endothelial expression of vascular cell adhesion molecule 1. Proc Natl Acad Sci USA 105:1516–1521
75. Watahiki A, Wang Y, Morris J, Dennis K et al (2011) MicroRNAs associated with metastatic prostate cancer. PLoS One 6:e24950
76. Hamada S, Satoh K, Fujibuchi W et al (2012) MiR-126 acts as a tumor suppressor in pancreatic cancer cells via the regulation of ADAM9. Mol Cancer Res 10:3–10
77. Meister J, Schmidt MH (2010) miR-126 and miR-126*: new players in cancer. Scientific World Journal 10:2090–2100
78. Quintavalle C, Garofalo M, Croce CM, Condorelli G (2011) "ApoptomiRs" in vascular cells: their role in physiological and pathological angiogenesis. Vascul Pharmacol 55:87–91
79. He L, Thomson JM, Hemann MT et al (2005) A microRNA polycistron as a potential human oncogene. Nature 435:828–833
80. Dews M, Homayouni A, Yu D, Murphy D et al (2006) Augmentation of tumor angiogenesis by a Myc-activated microRNA cluster. Nat Genet 38:1060–1065
81. Dews M, Fox JL, Hultine S et al (2010) The myc-miR-17 92 axis blunts TGF{beta} signaling and production of multiple TGF{beta}-dependent antiangiogenic factors. Cancer Res 70:8233–8246
82. Zhou Q, Gallagher R, Ufret-Vincenty R et al (2011) Regulation of angiogenesis and choroidal neovascularization by members of microRNA-23 27 24 clusters. Proc Natl Acad Sci USA 108:8287–8592
83. Poliseno L, Tuccoli A, Mariani L et al (2006) MicroRNAs modulate the angiogenic properties of HUVECs. Blood 108:3068–3071
84. Sun X, Icli B, Wara AK et al (2012) MicroRNA-181b regulates NF-κB-mediated vascular inflammation. J Clin Invest 122:1973–1990
85. Suárez Y, Fernández-Hernando C, Yu J et al (2008) Dicer-dependent endothelial microRNAs are necessary for postnatal angiogenesis. Proc Natl Acad Sci USA 105:14082–14087
86. Würdinger T, Tannous BA, Saydam O et al (2008) miR-296 regulates growth factor receptor overexpression in angiogenic endothelial cells. Cancer Cell 14:382–393
87. Fasanaro P, D'Alessandra Y, Di Stefano V et al (2008) MicroRNA-210 modulates endothelial cell response to hypoxia and inhibits the receptor tyrosine kinase ligand Ephrin-A3. J Biol Chem 283:15878–15883
88. Hu S, Huang M, Li Z, Jia F et al (2010) MicroRNA-210 as a novel therapy for treatment of ischemic heart disease. Circulation 122(11 Suppl):S124–S131

89. Magenta A, Cencioni C, Fasanaro P et al (2011) miR-200c is upregulated by oxidative stress and induces endothelial cell apoptosis and senescence via ZEB1 inhibition. Cell Death Differ 18(10):1628–1639
90. Lerner M, Haneklaus M, Harada M, Grandér D (2012) MiR-200c regulates Noxa expression and sensitivity to proteasomal inhibitors. PLoS One 7:e36490
91. Caporali A, Meloni M, Völlenkle C et al (2011) Deregulation of microRNA-503 contributes to diabetes mellitus-induced impairment of endothelial function and reparative angiogenesis after limb ischemia. Circulation 123:282–291
92. Zhao T, Li J, Chen AF (2010) MicroRNA-34a induces endothelial progenitor cell senescence and impedes its angiogenesis via suppressing silent information regulator 1. Am J Physiol Endocrinol Metab 299:E110–E116
93. Latronico MV, Condorelli G (2009) MicroRNAs and cardiac pathology. Nat Rev Cardiol 6(6):419–429
94. Latronico MV, Condorelli G (2011) Therapeutic use of microRNAs in myocardial diseases. Curr Heart Fail Rep 8:193–197
95. Ganguli S, Mitra S, Datta A (2011) Antagomirbase- a putative antagomir database. Bioinformation 7:41–43
96. Lu Y, Xiao J, Lin H et al (2009) A single anti-microRNA antisense oligodeoxyribonucleotide (AMO) targeting multiple microRNAs offers an improved approach for microRNA interference. Nucleic Acids Res 37:e24
97. Lanford RE, Hildebrandt-Eriksen ES, Petri A et al (2010) Therapeutic silencing of microRNA-122 in primates with chronic hepatitis C virus infection. Science 327:198–201
98. Hacker U, Büning H (2011) Optimizing the adeno-associated viral vector system: a brief summary. Ther Deliv 2:967–970

Chapter 26
Overview of Angiogenesis Inhibitors from Natural Sources

Dipti Gatne and Veeranjaneyulu Addepalli

Abstract Angiogenesis, the biological process of formation of new blood vessels from pre-existing ones, is essential for many diseases and its modulation is considered to be an important therapeutic strategy. It is controlled by various pro and anti-angiogenic factors. Plants, animals, microbes, and marine are the natural sources of various chemical constituents possessing anti-angiogenic activity which may act as leads in future angiogenesis inhibitors therapy. This review highlights the anti-angiogenic activity of many constituents from natural sources along with their biochemical mechanisms.

Keywords Angiogenesis • Natural molecules • Phytoconstituents • Terpenes • Polyphenolics

26.1 Introduction

The formation of new blood vessels from pre-existing ones, known as angiogenesis, is a biological process essential for wound healing, organ growth and repair, embryonic development, and reproduction. The process of angiogenesis is tightly regulated by a balance between pro-angiogenic and anti-angiogenic factors [1]. Uncontrolled angiogenesis, by virtue of its ability to supply oxygen and nutrients to ischemic

D. Gatne
Department of Pharmaceutical Chemistry, SPP School of Pharmacy & Technology Management, SVKM's NMIMS, V. L. Mehta Road, Mumbai, Maharashtra, India

V. Addepalli (✉)
Department of Pharmacology, SPP School of Pharmacy & Technology Management, SVKM's NMIMS, V. L. Mehta Road, Mumbai 400056, Maharashtra, India
e-mail: v.addepalli@nmims.edu

J.L. Mehta and N.S. Dhalla (eds.), *Biochemical Basis and Therapeutic Implications of Angiogenesis*, Advances in Biochemistry in Health and Disease 6,

areas, is a major contributor towards a number of major pathological diseases such as tumor progression, inflammatory disorders, obesity, asthma, diabetes, and proliferative retinopathies [2]. Therefore, modulation of angiogenesis is considered to be a therapeutic strategy of great importance.

Numerous inducers of angiogenesis have been reported in literature including vascular endothelial growth factor (VEGF) family, angiopoietins, transforming growth factors (TGF), platelet-derived growth factor (PDGF), tumor necrosis factor-α (TNF-α), interleukins, and the members of the fibroblast growth factor (FGF) family [3–5]. Angiogenesis inhibitors are able to inhibit the key events of the angiogenic processes such as proliferation, migration, and tube formation of endothelial cells; expression of the growth factors such as VEGF, FGF, EGF, PDGF; activation of matrix metalloproteinases (MMPs), PI3K/Akt signaling pathway, farnesyltransferase, etc. [6–9].

Extensive research in the field of angiogenesis has led to discovery of several mechanisms which confer resistance to the anti-angiogenic agents. These mechanisms, both intrinsic and acquired include intratumor hypoxia, upregulation of anti-apoptotic as well as compensatory pro-angiogenic factors, increased pericyte recruitment in tumor vasculature, and altered energy metabolism. Role of intratumor hypoxia, hypoxia inducible factor-1 (HIF-1), redundancy of growth factors like FGF, PDGF, placental growth factor (PIGF), increased pericyte coverage, tumor associated fibroblast (TAF), survival kinase, Pim-1 and vascular mimicry (VM) has also been emphasized in resistance to anti-angiogenic therapy [10–12].

In the given scenario, it is highly unlikely to conceive that a single strategy would be sufficient to effectively tackle uncontrolled angiogenesis. Hence, for an effective anti-angiogenic therapy, a multi pronged approach is desirable. This can be achieved by use of various agents obtained from synthetic or natural origin in combination for direct and indirect anti-angiogenic effects.

Natural molecules represent an important source of leads and provide diverse structures in library design suitable for further modifications in new drug development. Phytochemicals, by virtue of their ability to act on multiple pathways tend to reduce development of resistance. Hence, traditional therapy of natural products should be scientifically transformed into therapeutic advantage by appropriate selection of chemical compounds acting on multiple targets.

The natural sources like plants, animals, microbes, and marine contain various constituents such as alkaloids, glycosides, polyphenols, terpenes and terpenoids, pigments, and polysaccharides which have been reported to possess the anti-angiogenic property.

This review highlights various naturally occurring molecules exhibiting anti-angiogenic activity, some of which may develop into lead molecules in future. There are many reports of preparation of semisynthetic and synthetic derivatives of these chemical constituents which will definitely aid in designing future anti-angiogenic strategy. The structure-activity relationship of these molecules remains a challenge for which molecular modeling techniques have been reported in literature.

26.2 Terpenes and Terpenoids

Terpenes are the structurally diverse compounds derived from five-carbon isoprene units. Many types of terpenes like sesquiterpenes, diterpene, triterpene, and terpenoids have shown anti-angiogenic potential by affecting various pathways involved in the mechanism.

Sesquiterpenes like torilin, purified from fruits of *Torilis japonica* and fumagillin, isolated from the microbial organism *Aspergillus fumigatus* demonstrated anti-angiogenic activity in several in vivo and in vitro assays. Torilin has shown downregulation of the expression of hypoxia-inducible VEGF and insulin-like growth factor-II [13, 14]. Also, synthetic bicyclic analogues of fumagiline have been reported to show some anti-angiogenic potential [15].

Artemisinin and dihydroartemisinin, the natural sesquiterpenes isolated from *Artemisia annua* have inhibited angiogenesis in a dose-dependent manner via inhibition of human umbilical vein endothelial cells (HUVECs) growth [16]. Various semi-synthetic thio-acetal, non-acetal, dimers, and fullerene derivatives of artemisinin have also exhibited anti-angiogenic activity [17, 18].

Vernolide-A is a sesquiterpene lactone present in the plant *Vernonia cinerea*. It showed inhibition of angiogenesis by downregulation of the production of pro-inflammatory cytokines, VEGF and MMPs and upregulation of the endogenous anti-angiogenic factors like interleukin-2 (IL-2) and tissue inhibitors of metalloproteinases-1 (TIMP-1) [19].

Diterpenoids from Chinese herbs like, dihydrotanshinone I extracted from *Salvia miltiorrhiza* and triptolide from *Tripterygium wilfordii* also exhibited anti-angiogenic activity. Dihydrotanshinone I suppressed migration, invasion, and tube formation of HUVECs, whereas triptolide inhibited VEGFR-2 signaling pathway [20–22]. Furthermore, triptolide suppressed nuclear factor kappa B (NF-κB) pathway along with the target genes associated with endothelial cell mobilization in HUVECs and impaired VEGF expression in thyroid carcinoma TA-K cells [23].

Cheiradone, a natural diterpene isolated from euphorbia species was found to inhibit all stages of VEGF-induced angiogenesis and VEGF binding to VEGF receptor-1 (VEGFR-1) and VEGF receptor-2 (VEGFR-2) [24].

The diterpenoids, coronarin A and epi-coronarin A, isolated from rhizomes of the Brazilian anti-rheumatic medicinal plant, *Hedychium coronarium* and andrographolide, from *Andrographis paniculata* exhibited anti-angiogenic activity in many in vitro and in vivo models [25, 26] (Fig. 26.1).

Two sesterterpenes, leucosesterterpenone and leucosterlactone isolated from the hexane extract of the medicinal plant *Leucosceptrum canum* showed inhibition of FGF-2/fibroblast growth factor receptor 1 (FGFR-1) interaction with leucosesterterpenone having higher affinity for FGFR-1 than leucosterlactone [27].

Norcantharidin, a demethylated analogue of naturally occurring terpenoids and toxic compound cantharidin, isolated from Chinese blister beetel *Mylabris phalerata* is reported to block angiogenesis and metastasis. The mouse experiments have demonstrated that norcantharidin prolonged the survival and reduced the plasma VEGF levels in mice bearing pulmonary metastasis [28].

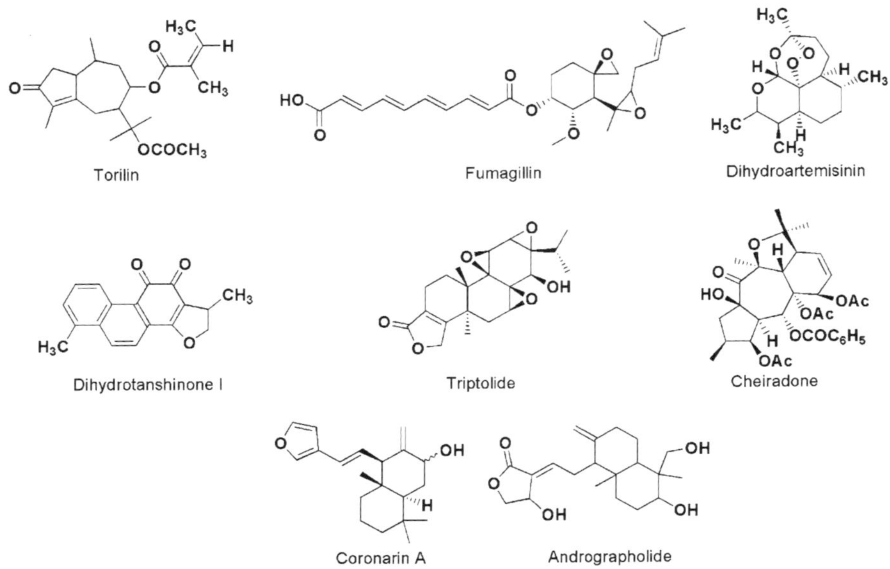

Fig. 26.1 Sesquiterpenes and diterpenoids from natural sources as anti-angiogenic agents

Acetyl-11-keto-β-boswellic acid, a triterpene and the active component of *Boswellia serrata* inhibited tumor angiogenesis through suppression of tumor growth in the human prostate tumor xenograft mice. It was found to inhibit VEGFR-2 activation and mTOR signaling pathways. The downstream protein kinases of VEGFR-2 like Src family kinase, focal adhesion kinase, extracellular signal-related kinase, AKT kinase, mammalian target of rapamycin, and ribosomal protein S6 kinase were suppressed by this derivative of boswellic acid [29].

Betulinic acid, a pentacyclic triterpene has been reported to inhibit the enzyme, aminopeptidase N, which is involved in the regulation of angiogenesis and overexpressed in several cancers. It was also reported to exert anti-angiogenic effects by inhibiting growth factor-induced in vitro angiogenesis in endothelial cells [30]. Some semisynthetic imidazole carboxylic esters and *N*-acylimidazole derivatives have been reported for cytotoxic activity whereas 3-hydrazono-20,29-dihydro, 17-carboxyacryloyl ester, 3-*O*-acyl, 3-benzylidene, 3-hydrazine, derivatives of betulinic acid are reported for anti-angiogenic activity [31–34].

The anti-angiogenic properties of julibroside J 8, a triterpenoid saponin, isolated from *Albizia julibrissin* were established by dose-dependent inhibition of growth, migration, and tube formation in human microvascular endothelial cells [35].

Sootepin B, a naturally occurring triterpene isolated from apical buds of *Gardenia sootepensis* and exudate of *Gardenia obtusifolia* displayed anti-angiogenic activity by inhibition of microvessel sprouting from rat aortic rings in a dose-dependent manner [36] (Fig. 26.2). Study of anti-angiogenic activity of nomilin, a triterpenoid present in common edible citrus fruits revealed inhibition of tumor-directed capillary

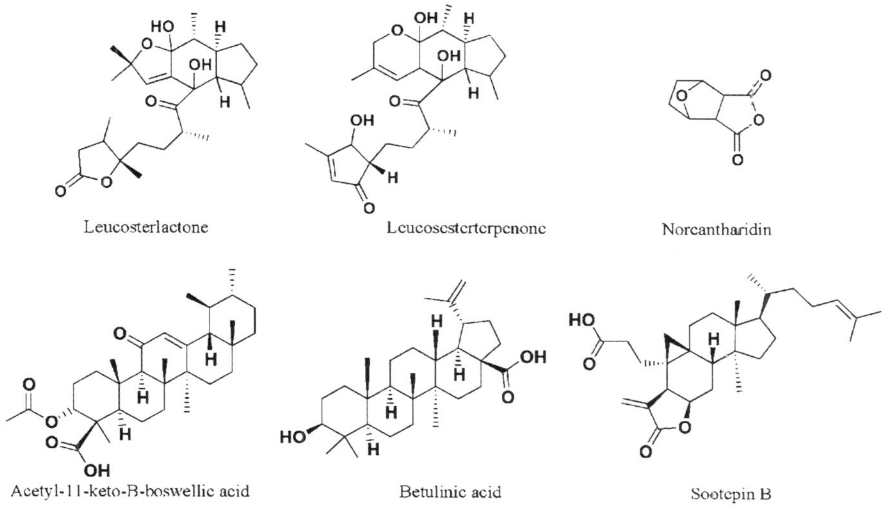

Fig. 26.2 Sesterterpenes and triterpenoids from natural sources as anti-angiogenic molecules

formation, reduction of serum proinflammatory cytokines such as IL-1β, IL-6, TNF-α, and pro-angiogenic factor VEGF, and increased anti-angiogenic factors IL-2 and TIMP-1 [37].

Presence of isoprene units and chemical modifications in terpenes has shown inhibitory effect on angiogenesis. The terpenes have shown the effect mainly on VEGFR-2 signaling pathway inhibiting various intracellular protein kinases.

26.3 Phenols and Polyphenolics

Phenolic compounds are ubiquitously found in various natural sources. The polyphenolic compounds are classified according to structure of aglycon into three main types as phenolic acid, flavonoids, and stilbenoids [38]. As the polyphenolic compounds are well-known antioxidants, many of these molecules have been reported for anti-angiogenic activity (Fig. 26.3).

Paeonol, a major phenolic component isolated from the roots of *Paeonia suffruticosa*, showed inhibition of proliferation, migration, and tube formation of bFGF-stimulated HUVECs in vitro. It demonstrated downregulation of Akt phosphorylation in bFGF-stimulated HUVECs and reduced expression of MMP-2 and MMP-9 in HT1080 human fibrosarcoma cells [39].

Eugenol, a phenolic constituent of clove oil, cinnamon, basil, and nutmeg exhibited anti-angiogenic potential by inhibiting cell proliferation via suppression of NF-κB signaling in a rat model of gastric carcinogenesis and inducing mitochondrial apoptosis by modulating the expression of Bcl-2 family proteins, Apaf-1,

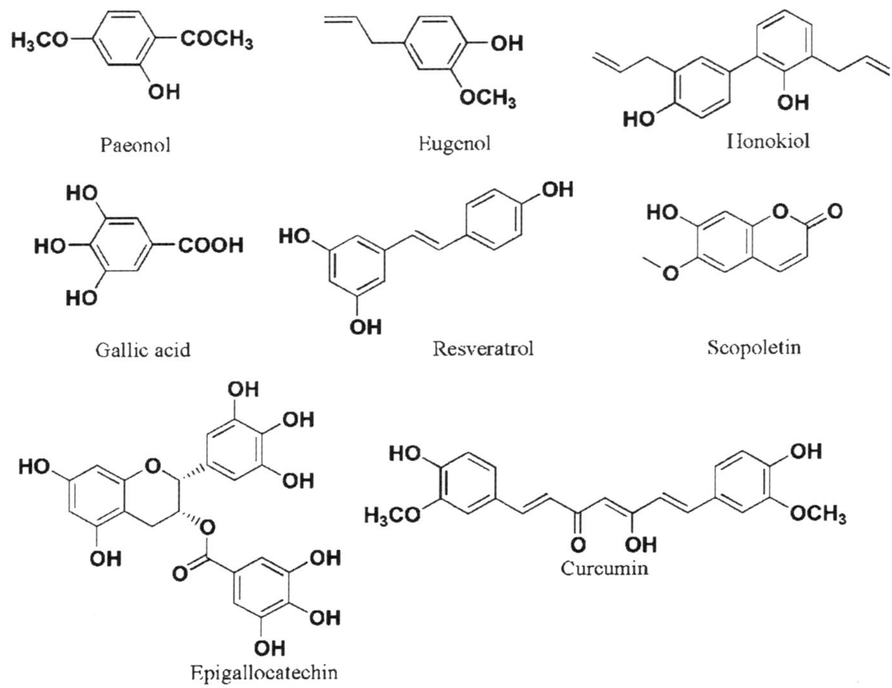

Fig. 26.3 Phenolic compounds with anti-angiogenic activity

cytochrome C, and caspases. It was also found to inhibit invasion and angiogenesis by altering the expression of MMP-2 and MMP-9, VEGF, VEGFR-1, TIMP-2, and reversion-inducing cysteine-rich protein with Kazal motifs (RECK) [40]. Honokiol, a biphenyl neolignan and the active principle of aqueous extract of *Magnolia grandiflora* exhibited anti-angiogenic activity in endothelial cell proliferation assay and angiosarcoma model in nude mice [41].

Sugar moiety of the flavonoids may have a role in blocking VEGF secretion which was indicated in a study of two molecules, naringin (rhamnoglucoside of naringenin flavone) and rutin (rhamnoglucoside of quercetin). The glycosylated forms were found to be potent inhibitors of VEGF release and aglycone, naringenin is reported to show no inhibitory activity. Of course, the systemic availability of these glycosides is limited when given by oral route. Overall, solubility, polarity, and hydrophilicity of the molecules related to number of hydroxyl groups and structural parameters may have role to play in the anti-angiogenic activity [42].

Scopoletin is a coumarin constituent extracted from stems of *Erycibe obtusifolia* reported to possess many biological activities. It was observed to block VEGF-induced autophosphorylation of VEGFR-2 and downregulate the activation of ERK1/2, p38 MAPK, and endothelial nitric oxide synthase along with production of nitric oxide in HUVECs [43]. Resveratrol, a naturally occurring polyphenolic phytoalexin found in grapes and wine inhibited growth of HUVECs and decreased

Fig. 26.4 Various flavones and isoflavones with angiogenesis inhibition activity

MMP-2 activity. In another study, it was reported to modulate angiogenesis in HUVECs through GSK3b/b-catenin/TCF pathways, which is in turn regulated by activation of PI3K/Akt and MEK/ERK signaling pathways [44, 45]. Curcumin, the active ingredient of *Curcuma longa* which is traditionally used for treatment of inflammation and wound healing has demonstrated the anti-angiogenic property. Curcumin inhibited expression of VEGF in HCT 116 human colon cancer cells exposed to angiogenic arsenic in the ELISA assay. There have been few reports of proposing mechanisms of anti-angiogenic activity of curcumin including effect on gene expression [46, 47]. Several synthetic curcumin analogues have also demonstrated anticancer and anti-angiogenesis activity [48–51]. Epigallocatechin, the most abundant and active component of green tea suppressed tumor growth by inhibiting proliferation and inducing apoptosis in a mouse model. It was found to inhibit VEGF-induced VEGFR-1 and VEGFR-2 phosphorylation on endothelial cells. It also suppressed growth of gastric cancer by targeting multiple steps of angiogenesis [52].

Flavonoids are naturally occurring polyphenolic components found in fruits and vegetables. Depending on hydroxylation and variation in chromane ring, they are subdivided into flavones, flavanols, anthocyanidins, isoflavonoids, and neoflavonoids [53]. Several reports in literature mention about the anti-angiogenic activity of flavonoids (Figs. 26.4 and 26.5).

Chalcones, the derivatives of phenylpropanoids are precursors of the flavonoids in higher plants and display a wide variety of pharmacological effects. One of the mechanisms of the anticancer activity of chalcone is suppression of angiogenesis.

2′,4′-dihydroxy-6′-methoxy-3′,5′-dimethylchalcone (ON III) extracted from dry flowers of Chinese plant *Cleistocalyx operculatus* reversibly inhibited tyrosine kinase phosphorylation [54]. 2,5-Dihydroxychalcones are reported to posses anti-angiogenic activity [55]. In a study, chalcone inhibited the proliferation of human breast cancer cell lines by inducing apoptosis and blocking cell cycle progression in the G2/M phase [56]. The anti-angiogenic effects of xanthohumol and isoxanthohumol, the chalcones isolated from hop plant (*Humulus lupulus*) were established by inhibition of invasion and migration of endothelial cell and formation and growth of a network of tubular-like structures [57, 58].

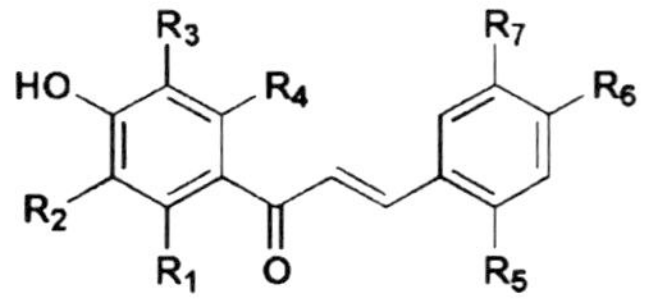

Isobavachalcone: R_1, R_2, R_5, R_7 - H, R_3 - CH_2-CH=C$(CH_3)_2$, R_4, R_6 - OH
Kanzonol C: R_1, R_2, R_5 - H R_4, R_6 - OH, R_3, R_7 - CH_2-CH=C$(CH_3)_2$
Stipulin: R_1, R_3, R_5 - H R_4, R_6 - OH, R_2, R_7 - CH_2-CH=C$(CH_3)_2$
Licochalcone A: R_1, R_2, R_3, R_4 - H R_5 - OCH_3, R_6 - OH, R_7 - C$(CH_3)_2$-CH=CH_2
Xanthohumol: R_1, R_6 - OH, R_2 - CH_2-CH=C$(CH_3)_2$, R_3, R_5, R_7 - H, R_4 - OCH_3
ON III: R_1 - OH, R_2, R_3 - CH_3, R_4 - OCH_3, R_5, R_6, R_7 - H

Fig. 26.5 Chalcone analogues showing anti-angiogenic activity

A licochalcone E, a new retrochalcone isolated from the roots of *Glycyrrhiza inflata*, was found to be an inducer of apoptosis in endothelial cells. It showed anti-angiogenic activity by inhibiting proliferation, migration, and tube formation of HUVECs, microvessel growth of rat aortic rings, and growth of colon cancer implant [59]. Anti-angiogenic effect of 2-hydroxy-4-methoxychalcone was observed under in vitro as well as in vivo conditions using chick embryo in the CAM assay and bFGF-induced vessel formation in the mouse matrigel plug assay [60]. Hydroxychalcones were tested for melanoma cell line (B16-F10) where they induced apoptosis in B16-F10 melanoma cells via glutathione and ATP depletion [61].

Chalcones and their analogues extracted from the twigs of *Dorstenia barteri* were investigated for their capacity to inhibit MMP-2 secretion in brain tumor-derived glioblastoma cells. The structure–activity relationships established role of hydroxyl, 2,3-double bond, prenyl group, and its positioning in the activity. The combined extract of dichloromethane and methanol contained isobavachalcone, 4-hydroxylonchocarpin, kanzonol C, paratocarpin C, stipulin, and dorsmannin A [62]. Methoxychalcones inhibited certain steps of angiogenesis including cell migration, micro capillary tube formation, and MMP expression [63]. 1-Furan-2-yl-3-pyridin-2-yl-propenone was found to inhibit the invasion and migration of HT1080 human fibrosarcoma cells by inhibiting MMP-2 and MMP-9 modulating the process of angiogenesis [64]. Few quinolyl–thienyl chalcones derivatives designed using molecular modeling techniques have been reported for anti-angiogenic activity. Utilizing molecular modeling techniques in the design of molecules especially from the natural sources seems to be advantageous [65].

We carried out a molecular modeling study to understand the interactions between chalcone analogues and VEGFR-2. The synthetic chalcone analogues were designed and the study included effect of side chains and various substituents on the aromatic rings present in chalcone analogues. The polycyclic analogues of chalcone showed better binding interactions with VEGFR-2 as compared to phenyl or substituted phenyl rings. The designed molecules were synthesized and tested for effect on blood vessel capillary formation using CAM assay model. All the synthesized analogues were found to be more active than the backbone and comparable to known angiogenesis inhibitors. Further biochemical investigation for these molecules is planned to understand the details of anti-angiogenic activity [66].

Genistein, daidzein, 6,7,40-trihydroxyisoflavone, 7,8,40-trihydroxyisoflavone, and orobol, the isoflavones from a tempeh (fermented soyabean) extract exhibited anti-angiogenic activity in CAM assay and endothelial cell proliferation assay [67]. At non-cytotoxic concentrations, genistein induced dose dependent reduction of motility in B16 melanoma and F3II mammary carcinoma mouse models, where as biochanin A, an isoflavone, has been reported to reduce cell proliferation in vitro assays acting as anti-angiogenic molecule [68].

Morelloflavone, a bioactive biflavonoid from *Garcinia dulcis* inhibited VEGF-induced cell proliferation, migration, invasion, and capillary-like tube formation of primary-cultured HUVECs in a dose-dependent manner. It is reported to inhibit angiogenesis of prostate cancer cells in xenograft mouse tumor model in vivo [69]. Baicalein, the major flavonoid of *Scutellaria baicalensis*, exhibited anti-angiogenic, antiproliferative, and apoptogenic effects. In biochemical analysis, they showed a dose-dependent decrease in the MMP-2 activity [70].

Hispidulin, an active component and flavone from *Artemisia vestita*, a tibetan medicinal plant, inhibited human pancreatic tumor growth in xenograft mice with inhibition of angiogenesis. It showed inhibition of VEGF-induced cell migration, invasion, and capillary-like structure formation of HUVECs, suppressed VEGF-induced microvessel sprouting of rat aortic rings and corneal neovascularization in C57/BL6 mice [71]. Chrysin, a naturally widely distributed flavonoid has been reported to possess anticancer, antibactericidal, anti-inflammatory, antiallergic, and antioxidant effects. It exhibited suppression of HIF-1α expression of tumor cells in vitro and inhibition of tumor cell-induced angiogenesis in vivo [72]. Wogonin, a naturally occurring monoflavonoid extracted from *Scutellariae radix*, inhibited the growth and tumor angiogenesis of human gastric carcinoma in nude mice. Wogonin suppressed the VEGF-stimulated migration, tube formation of HUVECs, and VEGF-induced tyrosine phosphorylation of VEGFR-2 [73]. Wogonoside, a glycosylated flavonoid derived from the roots of *S. baicalensis* Georgi, was found to suppress the lipopolysaccharide-stimulated migration and tube formation of HUVECs and microvessel sprouting from rat aortic rings in vitro [74].

A nutritional supplement obtained from fruits juice extracts containing phenolic compounds such as myricetin, quercetin, kaempferol, isorhamnetin, (+) catechin, (−) epicatechin, malvidin-3-glucoside, caffeic acid, transreveratrol, *cis*-resveratrol, chrysin, galangin, apigenin, fisetin, and luteolin suppressed basic steps in angiogenesis. The mixture showed anti-apoptotic activity along with inhibition of endothelial cell migration, capillary tube formation, and MMP-9 activity [75].

A study of receptor-oriented pharmacophore based in silico screening for identification of hVEGFR-2 inhibitors from flavonols database is reported. The designed pharmacophore map included four features, a hydrogen bond acceptor on Cys917, two hydrogen bond donors on Glu917 and Glu883, and one hydrophobic interaction with Val846, Ala864, Val897, Val914, and Phe1045 of hVEGFR-2. Five flavonols, kaempferol, quercetin, fisetin, morin, and rhamnetin were proposed to be the hVEGFR-2 inhibitors using this pharmacophore model. Also, 3-OH of B-ring and 4′-OH of C-ring as shown in Fig. 26.4 of flavones were found to be the essential features for hVEGFR-2 inhibition [76].

Another molecular modeling study reports 3-hydroxy substitution of flavones to be responsible for anti-angiogenic activity in CAM model. The docking experiments demonstrated that flavones appear to bind tightly with VEGFR-2 especially with 3-hydroxy substitution as compared to VEGF and HIF-1 [77].

26.4 Alkaloids

Alkaloids are group of natural compounds with basic nitrogen atoms. Some of the alkaloids have been reported for angiogenesis inhibitory activity (Fig. 26.6).

Sanguinarine, the benzophenanthridine alkaloid is reported to inhibit angiogenesis in a mouse model and useful in the therapy of melanoma cancer [78]. Derivatives of graveoline and graveolinine, the alkaloids isolated from *Ruta graveolens* with reported antitumor activities were tested using cell adhesion, migration, and CAM assay for anti-angiogenic activity [79].

Wondonins A and B, two imidazole alkaloids isolated from the sponges *Poecillastra wondoensis* and *Poecillastra jaspis* exhibited anti-angiogenic activity against HUVECs [80]. It also decreased HIF-1α protein and VEGF expression [81].

Naturally occurring guanidine alkaloid pterogynidine, isolated from the Brazilian plant *Alchornea glandulosa* showed anti-angiogenic activity in HUVEC cell viability, proliferation, apoptosis, invasion, and capillary-like structures formation assay [82]. Vinflunine, a new alkaloid from vinca family has shown anti-angiogenic and anti-metastatic properties which can be used in combination therapy for cancer treatment [83].

An isoindoline derivative and alkaloid in the extract of a Chinese shrub, *Fissistigma cavaleriei* roots isolated by bioassay-guided fractionation showed anti-angiogenic and antitumor activity [84].

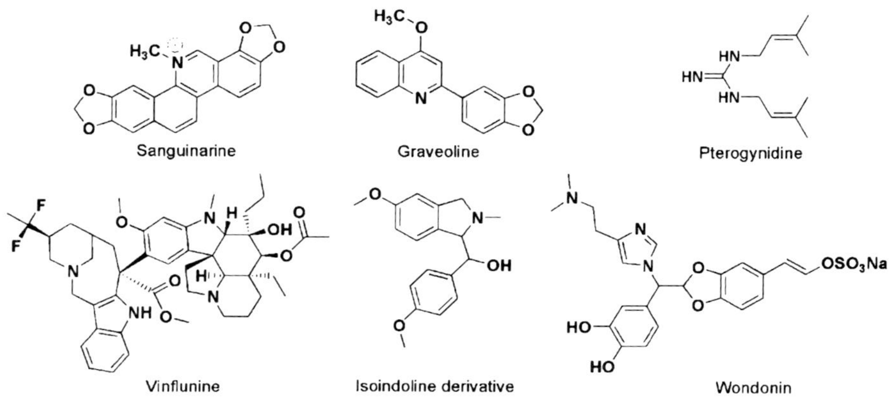

Fig. 26.6 Alkaloids showing anti-angiogenic activity

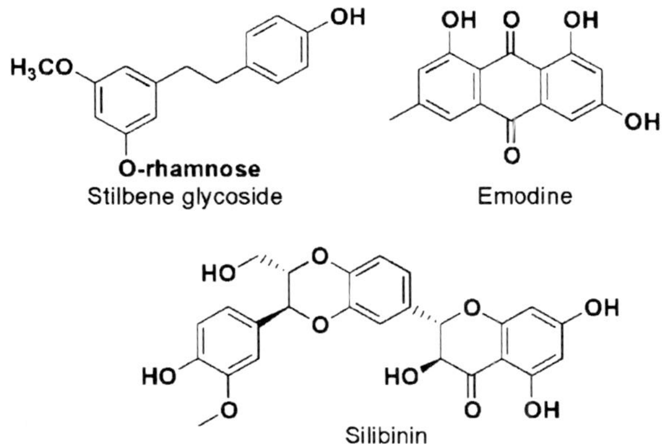

Fig. 26.7 Few reported glycosides as anti-angiogenic agents

26.5 Glycosides

Though, few of the glycosides have shown anti-angiogenic activity, an elaborate study of their mechanism is required to develop glycosides as therapeutic molecules in future.

The anti-angiogenic activity of a naturally occurring stilbene glycoside isolated from the medicinal plant *Boswellia papyrifera* was explored by inhibition of FGF-2-induced proliferation in various in vitro models [85]. Emodin, an anthraquinone derived from the rhizome of *Rheum palmatum*, inhibited endothelial cell proliferation, migration, and tube formation in a dose-dependent manner and expression of MMP-9 in polymerase chain reaction with reverse transcription (RT-PCR) and gelatin zymography [86, 87]. Silibinin, the major active constituent of silymarin and flavonolignan, showed anti-angiogenic activity by concentration-dependent inhibition of HUVECs migration and differentiation assay [88] (Fig. 26.7).

26.6 Miscellaneous Compounds

26.6.1 Pigments

Marine resources are reported for several beneficial effects as being the source of bioactive compounds such as carotenoids, dietary fibers, amino acids, essential fatty acids, pigments, vitamins, and minerals.

Fascaplysin, a red pigment isolated from Fijian marine sponge fascaplysinopsis species with a pentacyclic ring system showed anti-angiogenic activity in CAM assay, HUVECs proliferation assay, and sarcoma mice model [89–91]. Cochlioquinone A1, a pigment isolated from the culture extract of *Bipolaris zeicola* inhibited in vitro angiogenesis of bovine aortic endothelial cells (BAECs), bFGF-induced tube formation, and invasion without cytotoxicity [92].

26.6.2 Peptides

Aplidine, dehydrodidemnin B, a cyclic depsipeptide isolated from marine Mediterranean tunicate *Aplidium albicans* exhibited anti-angiogenic effects by inhibition of the secretion of VEGF by human leukemia cells. It inhibited endothelial cell proliferation and migration and blocked the production of MMP-2 and MMP-9 by endothelial cells [93–96]. The peptides, azumamides A–E isolated from the marine sponge *Mycale izuensis* exhibited anti-angiogenic activity in vascular organization model [97].

26.6.3 Polysaccharide

Anti-angiogenic activity of the polysaccharopeptide isolated from the edible mushroom *Coriolus versicolor* was revealed by suppression of VEGF expression in tumors [98].

The anti-angiogenic and antitumoral properties of SargA, a polysaccharide extracted from the brown marine alga *Sargassum stenophyllum*, studied in CAM assay and mice experiment are reported [99]. The polysaccharides from *Antrodia cinnamomea*, a medicinal mushroom showed anti-angiogenic activity in tube formation and CAM assay. The polysaccharides decreased the secretion of VEGF in human leukemia cells and inhibited the matrigel tube formation in HUVECs [100].

26.6.4 Other Agents

Gambogic acid, the main active compound of *Gamboge hanburyi*, a polyprenylated xanthone, inhibited HUVEC proliferation, migration, invasion, tube formation, and microvessel growth. It inhibited the activations of VEGFR-2 and its downstream protein kinases, c-Src, FAK, and AKT [101]. The polyketide macrolide, borrelidin-1 exhibited anti-angiogenic activity by inhibition of capillary tube formation and disruption of newly formed capillary tubes, in a dose-dependent manner using in vitro assay models [102].

Octacosanol, a long-chain aliphatic alcohol isolated from the plant *Tinospora cordifolia* was evaluated for anti-angiogenic effects. It inhibited proliferation of endothelial cells and Ehrlich ascites tumor cells, neovascularization using in vivo models and secretion of ascites fluid in the growing tumor cells. At the molecular level, octacosanol was found to inhibit activity of MMPs, translocation of transcription factor nuclear factor-kappa B (NF-κB) to nucleus and secretion of VEGF into ascites fluid by the tumor cells [103]. Acharan sulfate, a glycosaminoglycan from African snail, *Achatina fulica* showed anti-angiogenic activity in CAM assay, calf pulmonary artery endothelial cells proliferation assay, and matrigel plug assay [104]. Lupulone, a beta acid induced concentration-dependent inhibition of HUVEC proliferation, chemotaxis, and reduction of closed capillary-like structures in matrigel indicating inhibitory effect on neovascularization [105].

Chrysanthones the secondary metabolites, isolated from *Ascochyta chrysanthemi*, with a methyl-benzoisoquinoline were tested for their cytotoxicity properties on endothelial cells (EC) along with two different tumor cell lines and for their ability to inhibit EC migration. SAR study suggested the importance of methylisoquinolinic moiety for the cytotoxic activity and the methylisochromene moiety for endothelial selectivity [106]. Aeroplysinin-1, a secondary metabolite present in marine sponges has shown anti-angiogenic activity. Its epoxyketone and azalactone analogues are reported for angiogenesis inhibition [107]. The natural product withaferin A, a steroidal lactone exhibited antitumor and anti-angiogenic activity in vivo [108].

Tocotrienol, a natural analogue of tocopherol inhibited the proliferation of BAECs and formation of networks of elongated endothelial cells within 3D collagen gels [109]. Ascorbic acid, one of the well-known antioxidant, is found to have angiostatic effect by inhibiting the formation of vessel-like tubular structures of endothelial cells cultured on matrigel but did not show any effect on proliferation or migration of endothelial cell cultures [110].

15-Hydroxy-tetracosa-6,9,12,16,18-pentaenoic acid, an oxylipin isolated from the soft coral *Sinularia numerosa* inhibited EC tube formation exhibiting anti-angiogenic activity [111]. A metabolite, 4-senecioyloxymethyl-6,7-dimethoxycoumarin from the plant *Crinum latifolium* showed cytotoxicity against HUVECs. 6- and 7-substituted analogues of 4-senecioyloxymethylcoumarin have been studied and SAR is drawn for anti-angiogenic activity [112]. Taspine, a tetracyclic natural product isolated from *Rhizoma leonticis* and its ring-opened derivatives showed anti-angiogenic activity [113]. Rhizoxin isolated from *Rhizopus chinensis*, suppressed angiogenesis in CAM assay. It has been reported to affect functions of endothelial cells [114]. The polyacetylenes, 1,2-dihydroxytrideca-5,7,9,11-tetrayne and 1,3-dihydroxy-6(*E*)-tetradecene-8,10,12-triyne from ethyl acetate fraction of plant *Bidens pilosa* showed anti-angiogenic activity by inhibition of cell proliferation and tube formation [115].

A series of synthesized 7-diethylaminocoumarin compounds showed inhibition of HUVECs proliferation [116]. Riccardin D, a macrocyclic bisbibenzyl compound extracted from Chinese liverwort plant *Dumortiera hirsuta* exhibited anticancer and anti-angiogenic activity. It showed inhibitory effect on growth of human lung carcinoma cells, decreased the motility and migration of HUVEC cells, and reduced expressions of VEGF, phospho-VEGFR-2, EGF receptor, and MMP-2 [117]. Azaspirene, a fungal metabolite, isolated from *Neosartorya* species showed anti-angiogenic effect by inhibition of HUVECs migration and reduction of tumor-induced blood vessel formation. The molecular mechanism of azaspirene showed suppression of Raf-1 activation induced by VEGF without affecting the activation of VEGFR-2 [118, 119]. Siphonaxanthin, a ketocarotenoid compound isolated from siphonaceous green algae *Codium fragile* exhibited anti-angiogenic effect by suppressing HUVEC proliferation and tube formation possibly by reducing angiogenic mediators [120] (Fig. 26.8).

Apart from these constituents derived from natural sources, extracts of several natural sources have been reported for anti-angiogenic activity (Table 26.1).

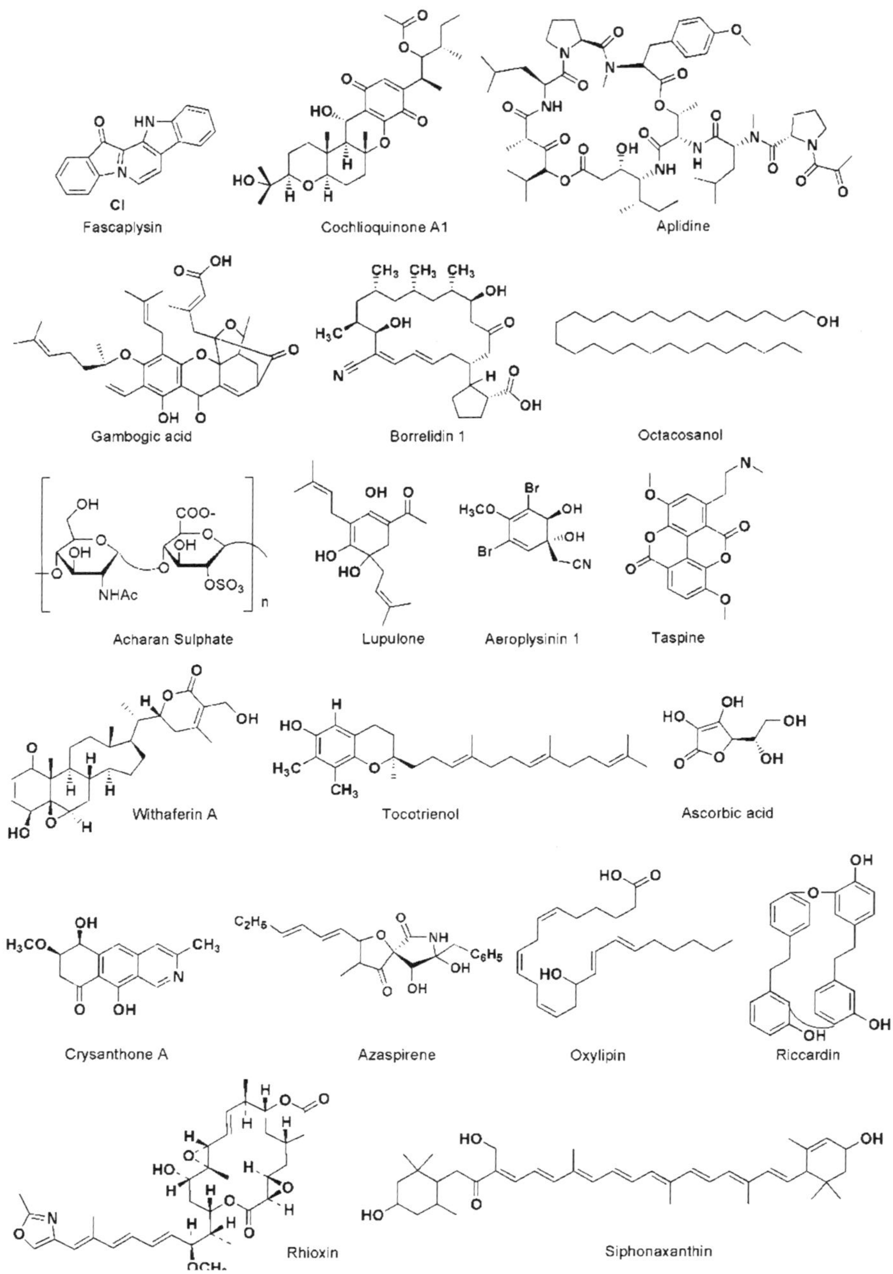

Fig. 26.8 Miscellaneous compounds reported to possess anti-angiogenic activity

Table 26.1 Various extracts prepared from natural sources reported for anti-angiogenic activity

Sr. no.	Source and extract	Constituents reported for anti-angiogenic activity	Anti-angiogenic effect
1.	Ethanolic and aqueous extracts of leaves of *Justicia gendarussa*	Lignans, phenolic dimmers, triterpenoids, friedelin, β-sitosterol, lupeol, and *O*-substituted aromatic amines	CAM model [122]
2.	Juice of the fruit, *Morinda citrifolia* (Noni)	–	Three-dimensional fibrin clot matrix model using human placental vein and human breast tumor explants [123]
3.	Ethanolic extract of fruiting bodies of *Phellinus linteus*	–	CAM assay, AKT signaling inhibition [124, 125]
4.	Crude extract of marine sponge	New bioactive alkaloids *N*-methyl-epi-manzamine D1 and epi-manzamine D2	HeLa and B16F10 mammalian cell line assay [126]
5.	Methanol, methylene chloride, ethyl acetate, *n*-butanol extracts of mushroom *P. linteus*	–	Proliferation, migration, tube formation, and phosphorylation of VEGFR-2 in HUVECs, matrigel plug model [127]
6.	Ethanol extract of *Salvia plebeian* and *Salvia officinalis*	–	CAM assay and HUVECs migration inhibition [128, 129]
7.	Extract of *Trifolium pratense* (Red clove)	Non-methylated isoflavones daidzein and genistein, mononentin and biochanin A	CAM assay [130]
8.	Fresh and fermented fractions from pomegranate *Punica granatum* fruit	Flavonoids	human HL-60 promyelocytic leukemia cells [131]
9.	*n*-Butanol fraction of rhizome of *Gastrodia elata*	Phenolic compounds such as, 4-hydroxybenzyl alcohol, 4-hydroxybenzaldehyde, 4-hydroxy-3-methoxybenzaldehyde, and gastrodin [4-(d-glucopyranosyl) benzyl alcohol]	CAM assay [132]
10.	Water extract of *P. linteus* (fungus)	–	Cell migration and secretion of MMPs [133]
11.	Ethanol extract of *Siegesbeckiae herba*	–	Matrigel plug assay [134]
12.	Methanolic extract of *Pfaffia paniculata* (Brazilian ginseng) roots	Stigmasterol, sitosterol, and their glycosides allantoin, pfaffic acid, and saponins—pfaffosides A, B, C, D, E, and F	Corneal angiogenesis in mice [135]

(continued)

Table 26.1 (continued)

Sr. no.	Source and extract	Constituents reported for anti-angiogenic activity	Anti-angiogenic effect
13.	Methanolic extract of *Piper longum*	Polyphenolic alkaloid	Tube formation, endothelial cell migration assay, B16F-10 melanoma cell-induced capillary formation in C57BL/6 mice [136]
14.	Oil extract of *Myristica fragrans*	–	Rat aorta assay [137]

CAM chorioallantoic membrane assay; *VEGFR-2* vascular endothelial growth factor-2

The extracts of *Epimedium sagittatum*, *Trichosanthes kirilowii*, and *Dalbergia odorifera* showed strong anti-angiogenic activity both in CAM and BAECs models; and the aqueous extracts of *Berberis paraspecta*, *Catharanthus roseus*, *Coptis chinensis*, *Taxus chinensis*, *S. baicalensis*, *Polygonum cuspidatum*, *Scrophularia ningpoensis* and many more elicited significant inhibition of angiogenesis [121].

26.7 Conclusion

Angiogenesis inhibitors have proven to be a new and improved line of treatment for various diseases. New anti-angiogenic molecules with novel structures from natural sources can be a valuable tool in future to study angiogenesis as well as for the development of new anti-angiogenic therapeutic molecules. Further, structure–activity relationship studies for these molecules obtained from natural sources will help to focus on better understanding of activity profiles of the molecules and designing of new anti-angiogenesis inhibitors.

Acknowledgments We express our gratitude to Department of Biotechnology (DBT), Govt. of India for providing financial support and SPP School of Pharmacy & Technology Management, Mumbai, India for providing facilities to carry out the research work.

References

1. Ferrara N, Kerbel RS (2005) Angiogenesis as a therapeutic target. Nature 438:967–974
2. Griffioen AW, Molema G (2000) Angiogenesis: potentials for pharmacologic intervention in the treatment of cancer, cardiovascular diseases, and chronic inflammation. Pharmacol Rev 52:237–268
3. Otrock ZK, Mahfouz RA, Makarem JA, Shamseddine AI (2007) Understanding the biology of angiogenesis: review of the most important molecular mechanisms. Blood Cells Mol Dis 39:212–220

4. Liekens S, Clercq ED, Neyts J (2001) Angiogenesis: regulators and clinical applications. Biochem Pharmacol 61:253–270
5. Cristofaro B, Emanueli C (2009) Possible novel targets for therapeutic angiogenesis. Curr Opin Pharmacol 9:102–108
6. Raffetto JD, Khalil RA (2008) Matrix metalloproteinases and their inhibitors in vascular remodeling and vascular disease. Biochem Pharmacol 75:346–356
7. Bikfalvi A (2004) Recent developments in the inhibition of angiogenesis: examples from studies on platelet factor-4 and the VEGF/VEGFR system. Biochem Pharmacol 68:1017–1021
8. Larsen AK, Ouaret D, El Ouadrani K, Petitprez A (2011) Targeting EGFR and VEGF(R) pathway cross-talk in tumor survival and angiogenesis. Pharmacol Ther 131:80–90
9. Peng G, Ren Y, Sun X, Zhou J, Li D (2012) Inhibition of farnesyltransferase reduces angiogenesis by interrupting endothelial cell migration. Biochem Pharmacol 83:1374–1382
10. Berger G, Hanahan D (2008) Modes of resistance to antiangiogenic therapy. Nat Rev Cancer 8:592–603
11. Abdollahia A, Folkman J (2010) Evading tumor evasion: current concepts and perspectives of anti-angiogenic cancer therapy. Drug Resist Updat 13:16–28
12. Qu B, Guo L, Ma J, Lv Y (2010) Antiangiogenesis therapy might have the unintended effect of promoting tumor metastasis by increasing an alternative circulatory system. Med Hypotheses 74:360–361
13. Myoung SK, You ML, Eun-Joung M et al (2000) Anti-angiogenic activity of Torilin, a sesquiterpene compound isolated from *Torilis japonica*. Int J Cancer 87:269–275
14. Bedel O, Haudrechy A, Pouilhes A, Langlois Y (2005) Syntheses of antiangiogenic or cytotoxic natural products: fumagillin and bengacarboline. Pure Appl Chem 77:1139–1152
15. Furness MS, Robinson TP, Ehlers T et al (2005) Anti-angiogenic agents: studies on fumagillin and curcumin analogs. Curr Pharm Des 11:357–373
16. Huan-Huan C, Hui-Jun Z, Xin F (2003) Inhibition of human cancer cell line growth and human umbilical vein endothelial cell angiogenesis by artemisinin derivatives in vitro. Pharmacol Res 48:231–236
17. Sangtae O, In HJ, Chan MA, Woon-Seob S, Seokjoon L (2004) Synthesis and anti-angiogenic activity of thioacetal artemisinin derivatives. Bioorg Med Chem 12:3783–3790
18. Mankil J, Jungae T, Won-Yoon C, Kwang-Kyun P (2006) Antiangiogenic activity of deoxoartemisinin derivatives on chorioallantoic membrane. Bioorg Med Chem Lett 16:1227–1230
19. Pratheeshkumar P, Kuttan G (2011) Vernolide-A inhibits tumour specific angiogenesis by regulating pro-inflammatory cytokines, VEGF, MMPs and TIMP. Eur J Pharmacol 656:10–18
20. Bian W, Chen F, Bai L, Zhang P, Qin W (2008) Dihydrotanshinone I inhibits angiogenesis both *in vitro* and *in vivo*. Acta Biochim Biophys Sin 40:1–6
21. Ming-Fang H, Yi-Hsien H, Li-Wha W et al (2010) Triptolide functions as a potent angiogenesis inhibitor. Int J Cancer 126:266–278
22. Ming-Fang H, Lin L, Wei G et al (2009) Antiangiogenic activity of *Tripterygium wilfordii* and its terpenoids. Ethnopharmacology 121:61–68
23. Wenbo Z, Songmin H, Yan L et al (2010) Anti-angiogenic activity of triptolide in anaplastic thyroid carcinoma is mediated by targeting vascular endothelial and tumor cells. Vascul Pharmacol 52:46–54
24. Hussain S, Slevin M, Mesaik MA et al (2008) Cheiradone: a vascular endothelial cell growth factor receptor antagonist. BMC 9:7. http://www.biomedcentral.com/1471-2121/9/7
25. Sangtae O, In HJ, Woon-Seob S, Seokjoon L (2003) A study on the synthesis of antiangiogenic (+)-coronarin A and congeners from (+)-sclareolide. Bioorg Med Chem Lett 13:2009–2012
26. Rong P, XingHua G, Dan L et al (2007) Prevention of FGF-2-induced angiogenesis by scopoletin, a coumarin compound isolated from Erycibe obtusifolia Benth, and its mechanism of action. Immunopharmacology 7:211–221
27. Hussain S, Slevin M, Matou S et al (2008) Anti-angiogenic activity of sesterterpenes; natural product inhibitors of FGF-2-induced angiogenesis. Angiogenesis 11:245–256

28. Yu-Jen C, Yin-Meng T, Cheng-Deng K et al (2009) Norcantharidin is a small-molecule synthetic compound with anti-angiogenesis effect. Life Sci 85:642–651
29. Xiufeng P, Zhengfang Y, Xiaoli Z et al (2009) Acetyl-11-keto-B-boswellic acid inhibits prostate tumor growth by suppressing vascular endothelial growth factor receptor 2 mediated angiogenesis. Cancer Res 69:5893–5900
30. Santos RC, Salvador JA, Marín S, Cascante M (2009) Novel semisynthetic derivatives of betulin and betulinic acid with cytotoxic activity. Bioorg Med Chem 17:6241–6250
31. Santos RC, Salvador JA, Marin S, Cascante M (2010) Synthesis and structure–activity relationship study of novel cytotoxic carbamate and N-acylheterocyclic bearing derivatives of betulin and betulinic acid. Bioorg Med Chem 18:4385–4396
32. Mukherjee R, Jaggi M, Rajendran P et al (2004) Betulinic acid and its derivatives as anti-angiogenic agents. Bioorg Med Chem Lett 14:2181–2184
33. Mukherjee R, Jaggi M, Rajendran P et al (2004) Synthesis of 3-O-acyl/3-benzylidene/3-hydrazone/3-hydrazine/17-carboxyacryloyl ester derivatives of betulinic acid as anti-angiogenic agents. Bioorg Med Chem Lett 14:3169–3172
34. Fulda S (2008) Betulinic acid for cancer treatment and prevention. Int J Mol Sci 9: 1096–1107
35. Hui H, Lei F, Xiao-ping Z, Lian-fen Z, Jian J (2009) Anti-angiogenic activity of julibrosideJ8, a natural product isolated from *Albizia julibrissin*. Phytomedicine 16:703–711
36. Pudhom K, Nuanyai T, Matsubara K, Vilaivan T (2012) Antiangiogenic activity of 3,4-seco-cycloartane triterpenes from Thai Gardenia spp. and their semi-synthetic analogs. Bioorg Med Chem Lett 22:512–517
37. Pratheeshkumar P, Kuttan G (2011) Nomilin inhibits tumor-specific angiogenesis by downregulating VEGF, NO and proinflammatory cytokine profile and also by inhibiting the activation of MMP-2 and MMP-9. Eur J Pharmacol 668:450–458
38. Sahib HB, Harchan NAH, Atraqchi SAM, Abbas AA (2010) The role of medicinal herbs in angiogenesis related diseases. Int J Pharmacol 6:616–623
39. Kim SA, Lee HJ, Ahn KS et al (2009) Paeonol exerts anti-angiogenic and anti-metastatic activities through down modulation of Akt activation and inactivation of matrix metalloproteinases. Biol Pharm Bull 32:1142–1147
40. Manikandan P, Murugan RS, Priyadarsini RV, Vinothini G, Nagini S (2010) Eugenol induces apoptosis and inhibits invasion and angiogenesis in a rat model of gastric carcinogenesis induced by MNNG. Life Sci 86:936–941
41. Bai X, Cerimele F, Ushio-Fukai M et al (2003) Honokiol, a small molecular weight natural product, inhibits angiogenesis *in vitro* and tumor growth *in vivo*. J Biol Chem 278:35501–35507
42. Schindler R, Mentlein R (2006) Flavonoids and vitamin E reduce the release of the angiogenic peptide vascular endothelial growth factor from human tumor cells. Nutrition 136:1477–1482
43. Rong P, Yue D, Xing-Hua G, Dan L, Yu-Feng X (2011) Inhibition of vascular endothelial growth factor-induced angiogenesis by scopoletin through interrupting the autophosphorylation of VEGF receptor 2 and its downstream signaling pathways. Vascul Pharmacol 54:18–28
44. Yu C, Zhao-Di F, Fang W, Hong-Yan L, Rui H (2005) Anti-angiogenic activity of resveratrol, a natural compound from medicinal plants. J Asian Nat Prod Res 7:205–213
45. Wang H, Zhou H, Zou Y et al (2010) Resveratrol modulates angiogenesis through the GSK3b/b-catenin/TCF dependent pathway in human endothelial cells. Biochem Pharmacol 80:1386–1395
46. Pantazis P, Varman A, Simpson-Durand C et al (2010) Curcumin and turmeric attenuate arsenic induced angiogenesis *in ovo*. Altern Ther Health Med 16:12–14
47. Gururaj AE, Belakavadi M, Venkatesh DA, Marm D, Salimath BP (2002) Molecular mechanisms of anti-angiogenic effect of curcumin. Biochem Biophys Res Commun 297:934–942
48. Adams BK, Ferst EM, Davis MC et al (2004) Synthesis and biological evaluation of novel curcumin analogs as anti-cancer and anti-angiogenesis agents. Bioorg Med Chem 12:3871–3883

49. Chandru H, Sharada AC, Bettadaiah BK (2007) In vivo growth inhibitory and anti-angiogenic effects of synthetic novel dienone cyclopropoxy curcumin analogs on mouse Ehrlich ascites tumor. Bioorg Med Chem 15:7696–7703
50. Chandru H, Sharada AC (2007) Antiangiogenic effects of synthetic analogs of curcumin *in vivo*. Afr J Biomed Res 10:241–248
51. Furness MS, Robinson TP, Ehlers T (2005) Antiangiogenic agents: studies on fumagillin and curcumin analogs. Curr Pharm Des 11:357–373
52. Bao-He Z, Wen-Hua Z, Zheng-Rong L (2007) (−)-Epigallocatechin-3-gallate inhibits growth of gastric cancer by reducing VEGF production and angiogenesis. World J Gastroenterol 13:1162–1169
53. Rong T (2010) Chemistry and biochemistry of dietary polyphenols. Nutrients 2:1231–1246
54. Ye CL, Liu JW, Wei DZ, Lu YH, Qian F (2004) In vitro anti-tumor activity of 2′,4′-dihydroxy-6′-methoxy-3′,5′-dimethylchalcone against six established human cancer cell lines. Pharmacol Res 50:505–510
55. Nam NH, Kim Y, You YJ et al (2003) Cytotoxic 2′,5′-dihydroxychalcones with unexpected anti-angiogenic activity. Eur J Med Chem 38:179–187
56. Hsu YL, Kuo PL, Tzeng WS, Lin CC (2006) Chalcone inhibits the proliferation of human breast cancer cell by blocking cell cycle progression and inducing apoptosis. Food Chem Toxicol 44:704–713
57. Albini A, Eva RD, Vene R et al (2005) Mechanisms of the antiangiogenic activity by the hop flavonoid xanthohumol: NF-kB and Akt as targets. FASEB J 20:527–529
58. Albini A, Dell ER, Vene R (2006) Mechanisms of the antiangiogenic activity by the hop flavonoid xanthohumol: NF-kB and Akt as targets. FASEB J 20(3):527–529. doi:10.1096/fj.05-5128fje
59. Kim YH, Shin EK, Kim DH et al (2010) Anti-angiogenic effect of licochalcone A. Biochem Pharmacol 80:1152–1159
60. Lee YS, Lim SS, Shin KH et al (2006) Anti-angiogenic and anti-tumor activities of 2-hydroxy-4-methoxychalcone. Biol Pharm Bull 29:1028–1031
61. Lilian A, Navarini F, Chiaradia LD et al (2009) Hydroxychalcones induces apoptosis in B16-F10 melanoma cells via GSH and ATP depletion. Eur J Med Chem 44:1630–1637
62. Ngameni B, Touaibia M, Belkaid A et al (2007) Inhibition of matrix metalloproteinase-2 secretion by chalcones from the twigs of *Dorstenia barteri*. ARKIVOC ix:91–103
63. Mojzisa J, Varinskaa L, Mojzisova G, Kostovac I, Mirossaya L (2008) Anti-angiogenic effects of flavonoids and chalcones. Pharmacol Res 57:259–265
64. Park BC, Park SY, Lee JS et al (2009) The anti-angiogenic effects of 1-furan-2-yl-3-pyridin-2-yl-propenone are mediated through the suppression of both VEGF production and VEGF-induced signaling. Vascul Pharmacol 50:123–131
65. Rizvi SUF, Siddiqui HL, Nisar M, Khan N, Khan I (2012) Discovery and molecular docking of quinolyl-thienyl chalcones as anti-angiogenic agents targeting VEGFR-2 tyrosine kinase. Bioorg Med Chem Lett 22:942–944
66. Gatne D, Ghone S, Mohanraj K, Addepalli V (2010) Design and synthesis of VEGFR-2 ligands as tumor angiogenesis inhibitors. In: Conference on frontiers in medicinal chemistry, organized by German Pharma Society, Germany
67. Kiriakidis S, Hogemeier O, Starcke S et al (2005) Novel tempeh (fermented soyabean) isoflavones inhibit in vivo angiogenesis in the chicken chorioallantoic membrane assay. Br J Nutr 93(3):317–323
68. Farina HG, Pomies M, Alonso DF, Gomez DE (2006) Antitumor and antiangiogenic activity of soy isoflavones genistein in mouse models of melanoma and breast cancer. Oncol Rep 16:885–891
69. Pang X, Yi T, Yi Z et al (2009) Morelloflavone, a biflavonoid, inhibits tumor angiogenesis by targeting Rho GTPases and ERK signaling pathways. Cancer Res 69:518–525
70. Jun-Jen L, Tien-Shang H, Wen-Fang C, Fung-Jou L (2003) Baicalein and baicalin are potent inhibitors of angiogenesis: inhibition of endothelial cell proliferation, migration and differentiation. Int J Cancer 106:559–565

71. He L, Wu Y, Lin L et al (2011) Hispidulin, a small flavonoid molecule, suppresses the angiogenesis and growth of human pancreatic cancer by targeting vascular endothelial growth factor receptor 2-mediated PI3K/Akt/mTOR signaling pathway. Cancer Sci 102:219–225
72. Fu B, Xue J, Li Z et al (2007) Chrysin inhibits expression of hypoxia-inducible factor-1α through reducing hypoxia-inducible factor-1α stability and inhibiting its protein synthesis. Mol Cancer Ther 6:220–226
73. Na L, Ying G, Yun L et al (2008) Wogonin suppresses tumor growth in vivo and VEGF-induced angiogenesis through inhibiting tyrosine phosphorylation of VEGFR-2. Life Sci 82:956–963
74. Yan C, Na L, Yun L et al (2009) Wogonoside inhibits lipopolysaccharide-induced angiogenesis *in vitro* and *in vivo* via toll-like receptor 4 signal transduction. Toxicology 259:10–17
75. Mojzis J, Sarissky M, Pilatova M et al (2008) *In vitro* antiproliferative and antiangiogenic effects of flavin7®. Physiol Res 57:413–420
76. Jee-Young L, Ki-Woong J, Woonghee K, Yong SH, Yangmee K (2009) Binding models of flavonols to human vascular endothelial growth factor receptor-2. Bull Korean Chem Soc 30:2083–2086
77. Gacche R, Shegokar H, Gond D, Archana J, Ghole V (2010) Effect of hydroxyl substitution of flavone on angiogenesis and free radical scavenging activities: a structure–activity relationship studies using computational tools. Eur J Pharm Sci 39:37–44
78. Stefano ID, Raspaglio G, Zannoni GF et al (2009) Antiproliferative and antiangiogenic effects of the benzophenanthridine alkaloid sanguinarine in melanoma. Biochem Pharmacol 78:1374–1381
79. Zeng-Yun A, Yi-Yong Y, Dan P et al (2010) Synthesis and evaluation of graveoline and graveolinine derivatives with potent anti-angiogenesis activities. Eur J Med Chem 45: 3895–3903
80. Jongheon S, Jung-Rae R, Youngwan S et al (2001) Wondonins A and B, new bis(dihydroxystyryl)-imidazoles from a two-sponge association. Tetrahedron Lett 42: 1965–1968
81. Hyoung-Oh J, Younghwa K, Yoo-Wook K et al (2007) Wondonin, a novel compound, inhibits hypoxia-induced angiogenesis through hypoxia-inducible factor 1 alpha. FEBS Lett 581:4977–4982
82. Lopes F, Rocha A, Pirraco A et al (2009) Anti-angiogenic effects of pterogynidine alkaloid isolated from *Alchornea glandulosa*. BMC Complement Altern Med 9:15. doi:10.1186/1472-6882-9-15
83. Ng JS (2010) Vinflunine: review of a new vinca alkaloid and its potential role in oncology. J Oncol Pharm Pract 17:209–224
84. Yang Z, Lu W, Ma X, Song D (2012) Bioassay-guided isolation of an alkaloid with antiangiogenic and antitumor activities from the extract of *Fissistigma cavaleriei* root. Phytomedicine 19:301–305
85. Hussain S, Slevin M, Ahmed N et al (2009) Stilbene glycosides are natural product inhibitors of FGF-2-induced angiogenesis. BMC Cell Biol 10:30. doi:10.1186/1471-2121-10-30
86. Kaneshiro T, Morioka T, Inamine M et al (2006) Anthraquinone derivative emodin inhibits tumor-associated angiogenesis through inhibition of extracellular signal-regulated kinase 1/2 phosphorylation. Eur J Pharmacol 553:46–53
87. Suboj P, Babykutty S, Roshan D et al (2012) Aloe emodin inhibits colon cancer cell migration/angiogenesis by downregulating MMP-2/9, RhoB and VEGF via reduced DNA binding activity of NF-jB. Eur J Pharm Sci 45:581–591
88. Shung-Haur Y, Jen-Kou L, Wei-Shone C, Jen-Hwey C (2003) Anti-angiogenic effect of silymarin on colon cancer LoVo cell line. J Surg Res 113:133–138
89. Jing L, Xiao-Jun Y, Hai-Min C (2007) Fascaplysin, a selective CDK4 inhibitor, exhibit anti-angiogenic activity in vitro and in vivo. Cancer Chemother Pharmacol 59:439–445
90. Xiaojun Y, Haimin C, Xiaoling L et al (2011) Fascaplysin exert anti-tumor effects through apoptotic and anti-angiogenesis pathways in sarcoma mice model. Eur J Pharm Sci 43:251–259

91. Zheng YL, Lu XL, Lin J et al (2010) Direct effects of fascaplysin on human umbilical vein endothelial cells attributing the anti-angiogenesis activity. Biomed Pharmacother 64:527–533
92. Hye JJ, Hyang BL, Chi-Hwan L, Chang-Jin K, Ho JK (2003) Cochlioquinone A1, a new anti-angiogenic agent from *Bipolaris zeicola*. Bioorg Med Chem 11:4743–4747
93. Taraboletti G, Poli M, Dossi R et al (2004) Antiangiogenic activity of aplidine, a new agent of marine origin. Br J Cancer 90:2418–2424
94. Caers J, Menu E, Raeve DH et al (2008) Antitumour and antiangiogenic effects of Aplidin® in the 5TMM syngeneic models of multiple myeloma. Br J Cancer 98:1966–1974
95. Soto-Matos A, Szyldergemajn S, Extremera S et al (2011) Plitidepsin has a safe cardiac profile: a comprehensive analysis. Mar Drugs 9:1007–1023
96. Biscardi M, Caporale R, Balestri F (2005) VEGF inhibition and cytotoxic effect of aplidin in leukemia cell lines and cells from acute myeloid leukemia. Ann Oncol 16:1667–1674
97. Nakao Y, Narazaki G, Hoshino T et al (2008) Evaluation of antiangiogenic activity of azumamides by the in vitro vascular organization model using mouse induced pluripotent stem (iPS) cells. Bioorg Med Chem Lett 18:2982–2984
98. Ho JCK, Konerding MA, Gaumann A, Groth M, Liu WK (2004) Fungal polysaccharopeptide inhibits tumor angiogenesis and tumor growth in mice. Life Sci 75:1343–1356
99. Dias PF, Siqueira JM Jr, Vendruscolo LF et al (2005) Antiangiogenic and antitumoral properties of a polysaccharide isolated from the seaweed Sargassum stenophyllum. Cancer Chemother Pharmacol 56:436–446
100. Chih-Min Y, Yu-Jie Z, Ran-Juh W, Miao-Lin H (2009) Anti-angiogenic effects and mechanisms of polysaccharides from *Antrodia cinnamomea* with different molecular weights. J Ethnopharmacol 123:407–412
101. Yi T, Yi Z, Sung-Gook C et al (2008) Gambogic acid inhibits angiogenesis and prostate tumor growth by suppressing VEGFR2 signaling. Cancer Res 68:1843–1850
102. Moss SJ, Carletti I, Olano C et al (2006) Biosynthesis of the angiogenesis inhibitor borrelidin: directed biosynthesis of novel analogues. Chem Commun 22:2341–2343
103. Thippeswamy G, Sheela ML, Salimath BP (2008) Octacosanol isolated from *Tinospora cordifolia* down regulates VEGF gene expression by inhibiting nuclear translocation of NF-κB and its DNA binding activity. Eur J Pharmacol 588:141–150
104. Lee YS, Yang HO, Shin KH et al (2003) Suppression of tumor growth by a new glycosaminoglycan isolated from the African giant snail Achatina fulica. Eur J Pharmacol 465:191–198
105. Siegel L, Miternique-Grosse A, Griffon C et al (2008) Antiangiogenic properties of Lupulone, a bitter acid of hop cones. Anticancer Res 28:289–294
106. Gianninia G, Pencoa S, Pisano C (2003) Chrysanthones, a new source of fungal metabolites with potential antitumor and antiangiogenesis properties. Fitoterapia 74:323–327
107. Cordoba R, Tormo NS, Medarde AF, Plumet J (2007) Antiangiogenic versus cytotoxic activity in analogues of aeroplysinin-1. Bioorg Med Chem 15:5300–5315
108. Bargagna-Mohan P, Hamza A, Yang-eon K et al (2007) The tumor inhibitor and antiangiogenic agent Withaferin A targets the intermediate filament protein vimentin. Chem Biol 14:623–634
109. Miyazawa T, Inokuchi H, Hirokane H et al (2004) Anti-angiogenic potential of pocotrienol *in vitro*. Biochemistry 69:67–69
110. Ashino H, Shimamura M, Nakajima H et al (2003) Novel function of ascorbic acid as an angiostatic factor. Angiogenesis 6:259–269
111. Yamashita T, Nakao Y, Matsunaga S et al (2009) A new antiangiogenic C24 oxylipin from the soft coral Sinularia numerosa. Bioorg Med Chem 17:2181–2184
112. Nguyen-Hai N, Yong K, Young-Jae Y et al (2002) Preliminary structure–antiangiogenic activity relationships of 4-senecioyloxymethyl-6,7-dimethoxycoumarin. Bioorg Med Chem Lett 12:2345–2348
113. Zhang J, Zhang Y, Zhang S, Wang S, He L (2010) Discovery of novel taspine derivatives as antiangiogenic agents. Bioorg Med Chem Lett 20:718–721
114. Aokia K, Watanabe K, Sato M et al (2003) Effects of rhizoxin, a microbial angiogenesis inhibitor, on angiogenic endothelial cell functions. Eur J Pharmacol 459:131–138

115. Li-Wha W, Yi-Ming C, Hsiao-Ching C et al (2004) Polyacetylenes function as anti-angiogenic agents. Pharm Res 21:2112–2119
116. Seokjoon L, Krishnamoorthy S, Woon-Seob S, Fang X, Qian W (2006) Synthesis and anti-angiogenesis activity of coumarin derivatives. Bioorg Med Chem Lett 16:4596–4599
117. Cui-cui S, Yu-sheng Z, Xia X et al (2011) Inhibition of angiogenesis involves in anticancer activity of riccardin D, a macrocyclic bisbibenzyl, in human lung carcinoma. Eur J Pharmacol 667:136–143
118. Hayashi Y, Shoji M, Yamaguchi J et al (2002) Asymmetric total synthesis of (−)-azaspirene, a novel angiogenesis inhibitor. J Am Chem Soc 124:12078–12079
119. Asami Y, Kakeya H, Komi Y et al (2008) Azaspirene, a fungal product, inhibits angiogenesis by blocking Raf-1 activation. Cancer Sci 99:1853–1858
120. Ganesana P, Matsubarab K, Ohkubo T et al (2010) Anti-angiogenic effect of siphonaxanthin from green alga, Codium fragile. Phytomedicine 17:1140–1144
121. Wang S, Zheng Z, Weng Y et al (2004) Angiogenesis and anti-angiogenesis activity of Chinese medicinal herbal extracts. Life Sci 74:2467–2478
122. Periyanayagam K, Umamaheswari B, Suseela L, Padmini M, Ismail M (2009) Evaluation of anti-angiogenic effect of the leaves of *Justicia gendarussa* (*Burm. f*) (Acanthaceae) by chorioallantoic membrane method. Am J Infect Dis 5:180–182
123. Hornick CA, Myers A, Sadowska-Krowicka H, Anthony CT, Woltering EA (2003) Inhibition of angiogenic initiation and disruption of newly established human vascular networks by juice from morinda citrifolia (noni). Angiogenesis 6:143–149
124. Yun SS, Sun-Hyoung K, Jae-Hoon S et al (2003) Anti-angiogenic, antioxidant and xanthine oxidase inhibition activities of the mushroom *Phellinus linteus*. J Ethnopharmacol 88:113–116
125. Sliva D, Jedinak A, Kawasaki J, Harvey K, Slivova V (2008) Phellinus linteus suppresses growth, angiogenesis and invasive behaviour of breast cancer cells through the inhibition of AKT signalling. Br J Cancer 98:1348–1356
126. Bing-Nan Z, Carla S, Randall KJ, Mattern MR, Kingston DGI (2000) New cytotoxic manzamine alkaloids from a palaun sponge. Tetrahedron 56:5781–5784
127. Yeon SL, Yoon HK, Eun KS et al (2010) Anti-angiogenic activity of methanol extract of *Phellinus linteus* and its fractions. J Ethnopharmacol 131:56–62
128. Hyun-Joo J, Yun SS, Chang-Jin L, Eun-Hee P (2009) Anti-inflammatory, anti-angiogenic and anti-nociceptive activities of an ethanol extract of *Salvia plebeian*. J Ethnopharmacol 126:355–360
129. Keshavarz M, Bidmeshkipour A, Mostafaie A et al (2011) Anti tumor activity of *salvia officinalis* is due to its anti-angiogenic, anti-migratory and anti-proliferative effects. Cell 12:477–482
130. Krenn L, Paper DH (2009) Inhibition of angiogenesis and inflammation by an extract of red clover (Trifolium pratense L.). Phytomedicine 16:1083–1088
131. Kawaii S, Lansky EP (2004) Differentiation-promoting activity of pomegranate (*Punica granatum*) fruit extracts in HL-60 human promyelocytic leukemia cells. J Med Food 7:13–18
132. Eun-Kyoung A, Hye-Jin J, Eun-Ju L et al (2007) Anti-inflammatory and anti-angiogenic activities of *Gastrodia elata* blume. J Ethnopharmacol 110:476–482
133. Suk HL, In SO, Young IK et al (2007) Phellinus extracts inhibit migration and matrix metalloproteinase secretion in porcine coronary artery endothelial cells. Biotechnol Bioprocess Eng 12:100–105
134. Jae-Ho H, Jong-Deog K (2011) Inhibitory effects of *Siegesbeckiae herba* extract on angiogenesis and adipogenesis. Biotechnol Bioprocess Eng 16:144–152
135. Carneiroa CS, Costa-Pintoa FA, Silva AP et al (2007) *Pfaffia paniculata* (Brazilian ginseng) methanolic extract reduces angiogenesis in mice. Exp Toxicol Pathol 58:427–431
136. Sunila ES, Kuttan G (2006) Piper longum inhibits VEGF and proinflammatory cytokines and tumor-induced angiogenesis in C57BL/6 mice. Int Immunopharmacol 6:733–741
137. Al-Rawi SS, Ibrahim AH, Nik Ab Rahman NN et al (2011) The effect of supercritical fluid extraction parameters on the nutmeg oil extraction and its cytotoxic and antiangiogenic properties. Proc Food Sci 1:1946–1952

Index

J.L. Mehta and N.S. Dhalla (eds.), *Biochemical Basis and Therapeutic Implications of Angiogenesis*, Advances in Biochemistry in Health and Disease 6,
DOI 10.1007/978-1-4614-5857-9, © Springer Science+Business Media New York 2013

F

G

Printed by Publishers' Graphics LLC
MO20130412.15.36.33